NITROGEN FIXATION RESEARCH PROGRESS

CURRENT PLANT SCIENCE AND BIOTECHNOLOGY IN AGRICULTURE

Evans, H.J., Bottomley, P.J. and Newton, W.E. (eds): Nitrogen fixation research progress, 1985. ISBN 90-247-3255-7

Nitrogen fixation research progress

Proceedings of the 6th international symposium on Nitrogen Fixation, Corvallis, OR 97331, August 4–10, 1985

edited by

H.J. EVANS

Laboratory for Nitrogen Fixation Research
Oregon State University
Corvallis, OR 97331, USA

P.J. BOTTOMLEY

Departments of Microbiology and Soil Science
Oregon State University
Corvallis, OR 97331, USA

W.E. NEWTON

Western Regional Research Center
USDA, ARS
Albany, CA 94710, USA

1985 **MARTINUS NIJHOFF PUBLISHERS**
a member of the KLUWER ACADEMIC PUBLISHERS GROUP
DORDRECHT / BOSTON / LANCASTER

Distributors

for the United States and Canada: Kluwer Academic Publishers, 190 Old Derby Street, Hingham, MA 02043, USA
for the UK and Ireland: Kluwer Academic Publishers, MTP Press Limited, Falcon House, Queen Square, Lancaster LA1 1RN, UK
for all other countries: Kluwer Academic Publishers Group, Distribution Center, P.O. Box 322, 3300 AH Dordrecht, The Netherlands

Library of Congress Catalogue Card Number 85-21794

ISBN-13: 978-94-010-8790-2   e-ISBN-13: 978-94-009-5175-4
DOI: 10.1007/978-94-009-5175-4

PREFACE

This Symposium, held August 4-10, 1985 on the campus of Oregon
State University in Corvallis, is the sixth of a series of international
symposia concerned with broad aspects of the fixation of nitrogen gas by
biological and chemical means. The first symposium of this series was
held in Pullman, Washington (1974), the second in Salamanca, Spain
(1976), the third in Madison, Wisconsin (1978), the fourth in Canberra,
Australia (1980) and the fifth in Noordwijkerhout, The Netherlands
(1983). Prior to the organization of these symposia, small groups of
usually no more than 10 or 12 of the now "old guard" in the field met in
some obscure places, including Butternut Lake, Wisconsin, Sanabel Island,
Florida and Camp Sagehen in California, to discuss developments in the
field. Concern about an energy crisis in the nineteen seventies served
as an impetus for the organization of workshops and preparatioi. of
publications urging government agencies to provide funds for the support
of several neglected areas in the field, including the genetics of
nitrogen-fixing organisms and the biology of <u>Frankia</u>. In looking back,
it becomes apparent that there have been drastic changes in the extent of
research support in the field and in the contents of the programs of the
continuing series of symposia.

In planning for the program for the Sixth Symposium, we solicited
input from the Advisory Committe, considered the activity and real
progress in the various areas of the field, then concluded that it was
essential to emphasize genetic research in the program. In the first
symposium of this series, only ten papers were devoted to the genetics of
nitrogen-fixing organisms. In contrast, 48 percent of the 50 invited
papers and 40 percent of the 420 poster abstracts presented in this sixth
symposium concerned the genetics of host legumes, N_2-fixing endophytes,
and free-living N_2-fixing bacteria. This shift in emphasis does not
imply a declining importance of the classical areas of the field, but
represents a normal and probably temporary surge in the relative
advancement of one area of the field.

Without the generous financial support provided by our sponsors
listed on page xiv, this Symposium would not have been possible. We are
especially grateful to the Oregon Agricultural Experiment Station and to
Associate Director Ralph Berry and Director Robert Witters for provision
of the administrative skills necessary for a functional and fiscally
responsible organization. Also, we appreciate the financial support
provided by the Oregon State University Research Office, The College of
Science and the Office of International Agriculture. Special thanks are
due to Dr. Timothy Righetti and Miss Patti Schuttler for arrangements for
registration, housing and meals, and entertainment; to Dr. David Hannaway
for the excursions; and to Dr. David Mok who made all the travel
arrangements. An enormous clerical and organization load was admirably
borne by the Symposium Secretariat consisting of Kathy Marshall, Chemae
Prime and Sheri Woods-Haas. As usual, Mrs. Vicky Newton played her
essential and important role in the assembly of the Proceedings
manuscripts. Numerous duties ranging from computer programming to a
variety of office and managerial arrangements were abley carried out by
Mr. Joe Hanus and Mr. Sterling Russell. We also thank Dr. Julie Searcy
and her staff for their expert operation and management of the facilities
in the LaSells Stewart Center. Finally, we thank the contributors,
convenors and chairpersons of the program sessions for their

participation and cooperation in the completion of the scientific
program.

There seems to be general agreement that the value of a symposium
proceedings is inversely related to the length of the period required for
publication. As a consequence, the editors have decided to do the best
job possible without taking the time to approach perfection.

Harold J. Evans   Corvallis, Oregon, USA

Peter J. Bottomley   Corvallis, Oregon, USA

William E. Newton   Albany, California, USA

                                         09/04/85

TABLE OF CONTENTS

GENETICS OF NITROGEN FIXATION IN FREE-LIVING BACTERIA

ENZYMOLOGY AND CHEMISTRY OF NITROGEN FIXATION

A. NITROGENASE

SPONSORS OF THE SYMPOSIUM

    Oregon State University - Laboratory for Nitrogen Fixation Research
                                  Agricultural Experiment Station
                                  Research Office
                                  College of Science
                                  Office of International Agriculture
    U.S. Department of Agriculture - Competitive Grants Office and CSRS
    Tennessee Valley Authority
    C.F. Kettering Foundation
    Allied Corporation
    Agrigenetics Corporation
    Weyerhaeuser Company Foundation
    Rockefeller Foundation
    Monsanto Agricultural Products Company
    E.I. du Pont de Nemours & Company
    Pioneer Hybrid Seed Company
    Exxon Research and Engineering
    Battelle-Kettering Laboratory
    Beckman Instruments Inc.
    New England Nuclear

ORGANIZING COMMITTEE

Administrative Chairman	- Robert E. Witters
Associate Administrative Chairman	- Ralph E. Berry
Scientific Program	- Harold J. Evans (Co-chairman)
	- Peter J. Bottomley (Co-Chairman)
	- Lyle R. Brown
	- Sterling A. Russell
Excursions and Entertainment	- David B. Hannaway (Chairman)
	- Kermit Cromack
	- James M. Trappe
Registration, Housing and Meals	- Timothy L. Righetti (Chairman
	- Robert Tarrant
Communications	- David D. Myrold (Chairman)
	- Adolph Ferro
Transportation	- David Mok (Chairman)
	- Larry Boersma
Symposium Secretariat	- Kathy Marshall
	- Chemae Prime
	- Sheri Woods-Haas

INTERNATIONAL ADVISORY BOARD

W.E. Newton	- Berkeley, California USA (Chairman)
R.H. Burris	- Madison, Wisconsin USA
F.C. Cannon	- Cambridge, Massachusetts USA
A.H. Gibson	- Canberra, ACT, Australia
A. van Kammen	- Wageningen, The Netherlands
W.H. Orme-Johnson	- Cambridge, Massachusetts USA
R. Palacios	- Cuernavaca, Morelos, Mexico
J.R. Postgate	- Brighton, United Kingdom
C. Rodriquez-Barrueco	- Salamanca, Spain
J. Torrey	- Harvard Forest, Massachusetts USA
C. Veeger	- Wageningen, The Netherlands

PLANT GENETICS RELEVANT TO
SYMBIOTIC NITROGEN FIXATION

A. BREEDING APPROACHES

INVITED PAPERS

POSTER SUMMARIES:

(i)   Genetic improvement of legume-<u>Rhizobium</u> symbiosis.
(ii) Legume nodulation mutants.

B. MOLECULAR APPROACHES

INVITED PAPERS

POSTER SUMMARIES.

DISCUSSION GROUP SUMMARY

HOST PLANT CHARACTERISTICS OF COMMON BEAN LINES SELECTED USING INDIRECT
MEASURES OF N_2 FIXATION

J. Attewell and F.A. Bliss
Department of Horticulture, University of Wisconsin, Madison.

1. INTRODUCTION

The genetic variability in common bean (Phaseolus vulgaris L.) germplasm
for the amount of atmospheric N fixed by the bean - Rhizobium phaseoli
symbiosis has been estimated using direct methods such as growth on
N-free medium (Rennie and Kemp, 1981) and ^{15}N isotope dilution
(Rennie and Kemp, 1983; Westermann et al., 1981), and indirect methods
to analyze plant traits that affect fixation potential. Although the
correlations between amounts of N_2 fixed and indirect measures such
as plant mass, nodule mass and acetylene reduction activity (ARA) vary
depending on a great many factors that affect symbiotic expression,
indirect methods have been used to select plants with enhanced fixation
potential (McFerson, 1983). Although the indirect methods offer
relative simplicity, low cost and utility for large numbers of samples,
the effectiveness of selection when they are used as selection criteria
must be determined by measuring the superiority of selected progenies
relative to their parents and other standard genotypes.

In the grain legumes, apart from the demonstration of genetic
variability for amount of N fixed (LaRue and Patterson, 1981) and that
traits related to fixation are heritable (Ronis et al., 1985), no
results of selection for increased fixation have been reported
previously. To demonstrate effective selection for increased symbiotic
response in variable breeding populations, improvement must be
demonstrated for three criteria; 1) total plant N from fixation, 2) %
of total N derived from fixation and 3) seed yield due to fixed N.

2. BREEDING POPULATIONS

The effective use of an appropriate breeding method to increase
N_2-fixation requires the ability to discriminate among genetically
different phenotypes and the consideration of the type of breeding
population and experimental units upon which selection is practiced.
The use of single plant selection is difficult because of the large
effects of environmental factors on symbiotic expression and the need
to evaluate root systems using destructive sampling procedures.
Selection based on family performance is expected to be more effective
than single plant selection since accuracy can be increased through
replication, subsamples using destructive methods can be taken from row
plots (families), and each family (rows) can be evaluated for multiple
traits (e.g. fixation, yield etc.) (McFerson et al., 1982).

Several methods, including the modified pedigree method (Brim, 1966) and the inbred backcross line method (IBLM) (Bliss, 1981), have been used in self-pollinated crops to produce populations of genetically variable families. The IBLM was used to develop populations in which to select for increased N_2-fixation combined with desirable plant traits. 'Porrillo Sintetico' was used as the recurrent parent in population 21 and 'Sanilac' as the recurrent parent in population 24. 'Puebla 152' was used as the high fixing donor parent in both crosses (Bliss, 1985).

3. SELECTION FOR INCREASED FIXATION POTENTIAL

Each family in the two inbred backcross populations was evaluated for N_2-fixation potential using several criteria (McFerson, 1983). Selection of the best families was based on replicated ARA values (μ M plant^{-1}hr^{-1}) determined from 5-plant root samples taken from each row plot. The plants remaining in each row (family) were scored for other traits such as days to maturity, plant type, root and shoot wt., and seed yield. The progeny families in each population with high ARA values and high seed yields were selected to be evaluated in later field trials for increased symbiotic response. These families and other lines with contrasting traits were grown in separate experiments to study the host traits that are either conducive to or restrictive of N_2-fixation.

4. EVALUATION OF SELECTED PROGENY LINES

4.1 Symbiotic Response Measured by ^{15}N Isotope Method.

The application of ^{15}N-depleted $(NH_4)_2SO_4$ to bean plants and to a non-nodulating soybean line growing in the field at the University of Wisconsin, Hancock Research Station allowed the estimation of total plant N from fixation, percent of total N from fixation (% NDFA) and seed yield (Table 1). The selected progenies in each population were superior to the respective recurrent parents, "Porrillo Sintetico" for population 21 and "Sanilac" for population 24 for all three criteria of increased symbiotic response. In population 21, the four selected progenies each fixed from 25 to 40% more than "Porrillo Sintetico" but less than the high fixing donor parent. In population 24, the selected lines fixed from 3- to 6-fold more total N than "Sanilac", but also substantially less than the donor, "Puebla 152". Seed yields and the % NDFA of the selected lines were also superior to the recurrent parents. These results demonstrate clearly that substantial improvement in the amount of N_2-fixation can be made through breeding and selection.

Since the lines had been chosen originally using ARA as a main selection criterion, the correlations among direct and indirect measures on the plants grown in this experiment were calculated (Table 2). All correlations were positive and highly significant. Although there was a positive correlation between ARA values and total N fixed and % NDFA, this must be interpreted carefully. Inspection of the values shows that the large positive correlation results from two distinct groups of values, one quite high and the other low. It becomes more difficult to distinguish between line means that are more similar in ARA values. The large C.V. for ARA demonstrates the need for adequate replication to reduce experimental error as much as possible. Selection based on family means allows extensive replication which is necessary to improve selection precision.

Table 1. Total nitrogen fixed, % of total N derived from fixation and seed yield of selected inbred backcross lines from 2 populations and their respective parents. Field, Wisconsin, 1984.

Parent or line	50% Bloom (R3 Stage)		Maturity (R9 Stage)		
	Total N fixed	%N DFA	Total N fixed	%N DFA	Seed Yield
	(mg Plant^{-1})		(mg Plant^{-1})		(g Plant^{-1})
Puebla 152	300	42	852	57	38
Porr. Sint.	175	33	558	48	28
21-16	187	33	779	50	35
21-38	178	33	779	56	32
21-43	213	36	741	54	33
21-58	243	38	712	54	30
Sanilac	4	2	76	12	18
24-17	46	13	583	48	31
24-21	98	25	216	25	19
24-48	54	14	211	22	23
24-55	91	24	192	22	23
24-65	44	12	279	31	25
Non-nod. Soybean					13
Duncan Critical Value$_{.05}$	96	18	180	9	8
C.V. (%)	42	42	21	14	17

[1] $\mu M C_2 H_2 \cdot Plant^{-1} \cdot HR^{-1}$

Source: DuBois et al., personal communication.

Table 2. Correlations between direct measures of nitrogen fixation and indirect measures of fixation potential, ARA and total plant N, R3 and R9 growth stages. Selected lines and parents of common bean, Wisc., 1984.

	ARA	Total Plant N	Total N fixed	% NDFA
ARA (R3)		.85**	.73**	.61**
Total Plant N (R9)			.97**	.92**
Total N Fixed (R9)				.98**

Source: DuBois et al., personal communication.

4.2. Traits Associated with N_2-Fixation in Selected Lines.

Three lines studied in the ^{15}N isotope experiment, 24-17, 24-21 and 24-55, were chosen in addition to five other lines from population 24 for further study of N accumulation in a separate field experiment conducted at the Hancock Research Station in 1984. Although they have similar genetic backgrounds because of backcrosses to the recurrent parent Sanilac, they differed for maturity and vining habit as well as other traits (Table 3).

Table 3. Developmental stages of progeny lines and parents of population 24. Field, Wisc., 1984.

| Line | Days to [3] | | | | Vining[1] Habit |
	R1[2]	R3	R7	R9	
24-40	47	50	72	92	−
24-55	48	51	73	94	−
24-21	49	52	77	91	−
24-17	45	49	87	110	+
24-5	48	52	75	97	−
24-12	44	47	49	100	+
24-4	45	49	77	102	+
24-19	50	53	89	109	+
Sanilac	45	47	67	85	−
Puebla	54	59	91	120	+

[1] + = Vining; − = Determinate.
[2] R1 = Beginning of flowering; R9 = Seed maturity.
[3] Represents days from planting to each developmental stage.

Since total plant N was correlated positively with N_2 fixed based on isotope estimates, total plant N was used as an indrect measure of fixation. Line 24-40, selected as a poor fixing line based on ARA value, accumulated the least total N. All other progeny lines had been selected for high ARA values and they accumulated more than Sanilac, suggesting increased fixation (Table 4). Not only were there differences in total plant N at the R7, but also there were differences in accumulation relative to different developmental stages. By the R3 stage Puebla 152, 24-4, 24-21, 24-40, and 24-55 had accumulated a larger proportion of the total N at R7 (R3/R7) than the other lines.

Four progeny lines were chosen for comparisons with the parent lines and with each other. Puebla 152, 24-17, and 24-12 show vining growth habits while 24-21, 24-5 and Sanilac show determinate growth. The length of time (days) from the R3 to the R7 stages varied from 20 to 38 days among these 6 lines (Table 5). The lines that were vining accumulated more total N and more total plant dry wt. than the determinate lines during the R3-R7 period. The exception was that 24-21 showed a higher rate of accumulation than 24-17, these lines being compared because the total plant N of each were similar at the R7 stage (Table 4).

Table 4. Total nitrogen accumulation in plants of progeny lines and parents of population 24. Field, Wisc., 1984.

| Line | Total Plant N at | | | R3/R7 |
| | R1 | R3 | R7 | |
	(mg·Plant^{-1})			
24-40	36	289	552	.52
24-55	32	322	668	.48
24-21	50	393	1045	.38
24-17	49	172	1068	.16
24-5	54	466	1149	.41
24-12	57	198	1388	.14
24-4	57	282	1396	.20
24-19	37	265	1471	.18
Sanilac	45	207	591	.35
Puebla 152	68	646	1429	.45

Table 5. Accumulation of toal plant nitrogen and dry wt. from R3-R7 stages in progeny lines with contrasting plant growth and development. 1984.

		Accumulation (R3-R7)			
		Plant N		Plant Dry wt.	
		Total	Per Day	Total	Per Day
	Days	(mg Plant^{-1})		(mg Plant^{-1})	
24-21	25	652	26.1	18768	751
24-17	38	896	23.6	28413	748
24-5	23	683	39.7	19194	834
24-12	32	1190	37.2	34745	1086
Sanilac	20	384	19.2	1120	560
Puebla 152	32	783	24.5	26816	838

The accumulation of seed N and seed dry wt. of these 6 lines during the R7 to R9 (maturity) stages was examined (Table 6). Here the differences between vining and determinate lines were less pronounced. Line 24-5 showed more total accumulation of seed N and dry wt. than 24-12. The seed yields of all the progeny lines were greater than Sanilac the recurrent parent, but all were later maturing (Table 7). It remains to be seen whether lines with higher yield and N_2-fixation but as early as Sanilac can be developed through crossing and selection.

The amount of N resulting from fixation in each of 5 lines was estimated using Sanilac as the poor fixing check line (Table 8). Comparison of the figures to the total plant N of Sanilac (in parentheses) shows that fixation in these lines was substantially greater than in the recurrent parent, again demonstrating the level of improvement in N_2-fixation from selection.

Table 6. Accumulation of total seed nitrogen and dry wt. from R7-R9 stages in progeny lines with contrasting plant growth and development. 1984.

		Accumulation (R7-R9)			
		Seed N		Seed Dry wt.	
	Days	Total (mg Plant^{-1})	Per Day	Total (mg Plant^{-1})	Per Day
24-21	14	66	4.7	3453	247
24-17	23	153	6.7	5755	250
24-5	22	204	9.3	7956	362
24-12	21	188	9.0	6619	315
Sanilac	18	102	5.7	4159	231
Puebla 152	29	473	16.5	13410	462

Table 7. Seed wt. (yield) at R7 and R9 (maturity) of progeny lines and parents. 1984

	Seed Wt.			Vining Habit	Days to Maturity
	R7	R9	R7/R9		
	(g Plant^{-1}))				
24-21	12.7	16.1	.79	−	91
24-17	15.1	20.8	.73	+	110
24-5	13.3	21.3	.62	−	97
24-12	14.7	21.4	.69	+	100
Sanilac	6.5	10.7	.61	−	85
Puebla 152	15.8	29.2	.54	+	120

+ = Vining habit; − = Determinate habit.

Table 8. Estimated nitrogen from fixation based on differences between amounts accumulated in each line and in Sanilac.

	N from fixation	
Line	Plant (R7)	Seed (R9)
24-21	454	253
24-17	477	325
25-5	558	403
24-12	797	436
Sanilac	0 (591)[1]	0 (316)[1]
Puebla 152	838	729

[1] Total plant N at R7 and R9 stages.

5. CONCLUSIONS

Breeding lines of common bean resulting from the cross of poor fixing commercial cultivars with a high fixing donor parent were selected using ARA and seed yield as indirect measures of increased fixation potential. The symbiotic response of the selected lines compared to the parents using both the ^{15}N isotope method and measurement of total plant N showed that the selected lines fixed more total N, had a higher % NDFA and produced more seed than the recurrent parent cultivar.

Patterns of fixation and N accumulation differed considerably among the lines. Although it has been stated that ARA usually declines rapidly after the R3 stage, the patterns of N accumulation suggest that some lines fix a large proportion of the seasonal after R3 (early pod development). Measurement of nodule mass, primarily in the crown and primary root areas, showed that in some high fixing lines there was considerable nodule mass up through the R7 stage. However in other high fixing lines nodule mass in the crown and primary root area was rather small. Nodulation of secondary lateral roots may account for some of the large amount of fixation during the later stages of development. Analyses of the data are continuing to determine plant genotype related factors that have the greatest influence on fixation.

6. REFERENCES

Bliss FA (1981) HortScience 16, 129-132.
Bliss FA (1985) In Ludden PW and Burris JE, eds, Nitrogen Fixation and CO_2 Metabolism, pp. 303-310, Elsevier, New York.
Brim CA (1966) Crop Sci. 6, 220.
LaRue TA and Patterson TG (1981) Adv. Agron. 34, 15-38.
McFerson et al (1982) In Graham P and Harris S, eds, Biological Nitrogen Fixation Technology for Tropical Agriculture, pp. 39-44, CIAT.
McFerson JR (1983) Ph.D. Thesis, University of Wisconsin, Madison.
Rennie RJ and Kemp GA (1981) Euphytica 30, 87-95.
Rennie RJ and Kemp GA (1983) Agron. J. 75, 645-649.
Ronis et al (1985) Crop Sci 25, 1-4.
Westermann et al (1981) Agron. J. 73, 660-663.

7. ACKNOWLEDGEMENTS

The authors wish to acknowledge the contributions of J. DuBois and R.H. Burris, Dept. of Biochemistry and Dina St. Clair, J.C. Rosas and K.A. Kmiecik, Dept. of Horticulture, University of Wisconsin. Support for this research was received from the McKnight Foundation, the USAID/CSRS Special Grant No. 59-2551-1-5-006-0, and from the College of Agricultural and Life Sciences.

GENETIC IMPROVEMENT OF SYMBIOTIC NITROGEN FIXATION IN LEGUMES

DONALD A. PHILLIPS AND LARRY R. TEUBER, DEPARTMENT OF AGRONOMY, UNIVERSITY OF CALIFORNIA, DAVIS, USA

N_2 fixation by the Rhizobium-legume symbiosis is influenced by the genotype of both the bacterium and the plant. This review assesses genetic variation in N_2 fixation associated with R. meliloti and alfalfa (Medicago sativa), indicates the limitations to improving Rhizobium, and suggests that genetic traits in the host legume can be used to control Rhizobium symbionts. The conclusion is that our present inability to manage indigenous soil rhizobia temporarily argues against producing Rhizobium strains with superior N_2 fixation and supports the concept of breeding legumes. One particularly effective method may be to develop legumes that increase N_2 fixation in diverse Rhizobium strains.

Genetic Variation in N_2 Fixation

Many workers have considered the question of whether R. meliloti or alfalfa has the greater genetic variation for traits influencing N_2 fixation (Burton, Wilson 1939; Erdman, Means 1953; Tan 1981; Gibson 1962; Mytton et al 1984). Those investigations found that both plant cultivars and bacterial strains had significant effects on N_2 fixation. The genetic effects on variation in N_2 fixation can be quantified by fitting the data to a fixed-effects analysis-of-variance model. Calculations from data reported in three of those studies show interesting similarities and contrasts (Table 1). In this type of analysis, the Rhizobium or the legume is concluded to be a more suitable source of genetic diversity if it contributes a greater portion of the total variation. Thus the Burton, Wilson (1939) and Erdman, Means (1953) data suggest that genetic improvement of Rhizobium should have priority over plant breeding efforts, while the opposite conclusion is supported by the Tan (1981) data (Table 1). Those divergent results probably reflect the fact that different groups of strains and cultivars were examined by the various investigators. All three studies, however, show that Rhizobium-strain X alfalfa-cultivar interactions were responsible for a large part of the variation. Mytton (1978) described a method for breeding to enhance such interactions, but the procedures require essentially twice as much work as improving only the plant or the bacterium. We conclude from this rationale that it is more reasonable to improve N_2 fixation by genetically altering either Rhizobium or the plant than by selecting for favorable interactions.

Genetic Improvement of Rhizobium

Numerous attempts have been made to increase N_2 fixation by selecting or developing superior Rhizobium strains. Significant progress has been achieved in many cases by screening naturally occurring Rhizobium isolates for N_2-fixation capacity. Such efforts form the basis of the commercial inoculant industry, but the factors responsible for superior N_2 fixation in those strains are not known.

Table 1. Percentage of total phenotypic variance effects contributed by alfalfa cultivar, Rhizobium strain, and their interaction calculated from published studies using a fixed-effects model.

Source of Variation	Trait		
	1 Total N	2 Forage Dry Wt.	3 Total N
	------------------- % -------------------		
Alfalfa cultivar	8.7***	3.1	51.9***
Rhizobium strain	19.7**	21.8***	6.4***
Interaction	66.2**	53.0***	41.1***
Error	5.5	2.1	0.6

[1]Burton, Wilson 1939; [2]Erdman, Means 1953; [3]Tan 1981 **, *** Mean squares for the source of variation was significant at P <0.01 or 0.001.

A modern program designed to increase N_2 fixation by genetically altering Rhizobium will have several parts. First, specific biochemical traits associated with increased N_2 fixation must be identified. Second, suitable genetic techniques must be used to introduce those traits into bacterial strains already known to have other outstanding characteristics. Third, increases in N_2 fixation must be produced by the modified organism under field conditions where the legume crop normally is grown. Presently there is no example in which all three steps have been accomplished.

Recent work with the Hup[+] phenotype in R. japonicum (Evans et al 1985) is the only case in which a specific biochemical trait has been associated with increased N_2 fixation. However, the trait has not been stably integrated into superior Hup[-] rhizobia, and the 11% Hup[+] benefit has not been demonstrated under normal field conditions.

The largest increases in N_2 fixation associated with a genetically "defined" change in Rhizobium are the 31-128% enhancements of N_2 fixation produced by the recombinant plasmid pIJ1008 in R. leguminosarum[2] under controlled conditions (DeJong et al 1982). Because pIJ1008 conferred the Hup[+] phenotype on the Nod Fix Hup[-] recipients in that study, it is tempting to attribute the increase in N_2 fixation to that trait. However, subsequent experiments using isogenic Hup[-] strains produced by Tn5-mob mutagenesis suggest that the Hup[+] phenotype associated with pIJ1008 is not responsible for the increase in N_2 fixation (Cunningham et al 1985), and the biochemical basis of the enhancement phenomenon is not known. Benefits of pIJ1008 have not been tested under field conditions.

The only genetically-altered Rhizobium that has increased agronomic yield of a legume grown under field conditions is the mutant R. japonicum C33 (Williams, Phillips 1983). The biochemical basis of the increased N_2 fixation in that organism is not known (Scott et al 1979), and the soils used for the California field tests have far fewer indigenous rhizobia than soils in normal soybean production areas. It seems unlikely that the significant 12% increase in seed yield measured on plants in which more than 80% of the nodules contained C33 would be observed in regions with indigenous rhizobia.

In fact, the problem most likely to prevent the successful introduction and use of improved Rhizobium strains is the presence of competitive indigenous rhizobia in legume production areas. The effectiveness of indigenous rhizobia has been known for many years (Thornton 1929), and their competitiveness has been quantified (Johnson et al 1965). Detailed studies of the problem have been initiated (Ellis et al 1984), but until our understanding of microbial ecology in the soil increases greatly, it will be difficult to derive agronomic benefits from genetically improved rhizobia. We conclude that although it is important from a basic viewpoint to understand how N$_2$ fixation can be increased by changing Rhizobium, there presently is little justification for claiming practical advantages will result from such efforts.

Genetic Improvement of the Host Legume

Many investigators have tried to increase symbiotic N$_2$ fixation by genetically modifying the host legume, but as yet agronomic benefits have not been derived from those efforts. However, several concepts that may be important for future work have been produced. Those ideas include, first, the suggestion that the legume may provide a mechanism for managing indigenous rhizobia and, second, an understanding that increased N$_2$ fixation capability is only one component of an improved plant.

Managing Rhizobium with the Legume. Many options exist for breeding host legumes to control indigenous soil rhizobia. Possibilities range from one extreme where plant genes prevent root nodule development by all but the desired strain (Devine, Weber 1977) to the opposite extreme in which selected plants increase N$_2$ fixation by any effective rhizobia that form nodules (Phillips et al 1985a,b). Within that spectrum lies the possibility for breeding legumes that form a higher proportion of nodules with the most effective rhizobia available (Hardarson et al 1982). The opportunity for using nonnodulating plant genotypes to control indigenous rhizobia has been described for soybean (Glycine max) (Devine, Weber 1977) and will not be discussed further. The other options may also apply to grain legumes, but they have been explored more completely in outcrossing forage legumes, such as red clover (Trifolium pratense) and alfalfa, where the genetic variation within each population can be exploited by accumulating alleles that affect quantitatively controlled traits such as root nodule development. In every case, however, the concept of using the legume to manage Rhizobium has been only one part of a larger effort to improve total plant growth. Thus it is impossible to analyze any one component separately, and several aspects of the three relevant programs will be summarized.

Red Clover: Breeding for Increased N$_2$ fixation. P. S. Nutman (1984) recently reviewed attempts to increase N$_2$ fixation in red clover. Those efforts were initiated in test tubes where a single strain of R. trifolii could be manipulated, but later work in open pot systems was more productive. Tests of populations produced by crosses among high-yielding lines indicated some progress was made toward increasing leaf area and dry weight of plants grown on N_2 with strains of R. trifolii that had not been present during the selection process. The same clover populations also tended to show increased growth on NO_3^-. Those very promising results, which unfortunately did not include Kjeldahl N data or rigorous statistical analyses, are consistent with the idea that the host plant can increase N_2 fixation by indigenous rhizobia.

Alfalfa: Breeding for Increased N_2 fixation. Heritable differences in C_2H_2 reduction by alfalfa (Seetin, Barnes 1977) led to a program of separate phenotypic recurrent selections for C_2H_2 reduction, nodule mass, fibrous roots, and top dry weight without making direct measurements of N content. After two cycles of selection using a mixture of Rhizobium genotypes, glasshouse tests showed the desired changes in the selected traits and parallel changes in Kjeldahl N content of the plants (Viands et al 1981). However, field tests of the same materials showed no increase in N_2 fixation or yield (Heichel et al 1981).

Those disappointing field results were counterbalanced in part by the finding that the most effective rhizobia in a mixture of strains formed a greater fraction of root nodules on the selected alfalfa population than on the starting material (Hardarson et al 1982). That observation suggests, but does not prove, that the breeding procedure selected for favorable Rhizobium-strain X plant-genotype interactions. It also raises the interesting possibility that appropriate breeding techniques might produce plants that associate only with desirable rhizobia. The mechanism responsible for such a phenomenon is not known, but an interaction involving the amount of N_2 fixed, rather than the surface antigens of the Rhizobium, probably would have a broader and more practical application. Additional work with the materials already available obviously is the first step toward understanding this potentially important process.

Alfalfa: Breeding for Increased C and N Assimilation. Our group has developed an integrated, direct selection method (Phillips et al 1982) that produces alfalfa with increased forage yield and N content under controlled conditions (Teuber et al 1984; Phillips et al 1985a,b) and in a preliminary field test (Phillips et al 1985a). Much of the improvement results from greater N_2 fixation, but the selection method incorporates the idea that N_2 fixation is only one aspect of plant growth and that C assimilation and the availability of other forms of N also must be considered.

Three concepts, quite different from other attempts to increase N_2 fixation, underlie our selection method. First, because some soluble N is almost always available under field conditions (Munns 1977) and individual alfalfa genotypes often perform differently relative to each other on N_2 and NH_4NO_3 (Sheehy et al 1980; Phillips et al 1982), plants are evaluated for growth and N assimilation separately on both N_2 and 8 mM NH_4NO_3. Second, because a mixture of both Rhizobium and plant genotypes might confound the effects of bacterial and plant genes rather than identify only desirable plant genes, a single Rhizobium genotype is used. Finally, because both dry matter and tissue N concentration are important in a forage crop but are negatively correlated, only those plants with both high dry weight and high N concentration are selected. All measurements are made at flowering, a uniform growth stage, and testing of every genotype on both N_2 and NH_4NO_3 is possible because the perennial alfalfa plant regenerates new shoots after cutting. Thus a treatment of 8 mM NH_4NO_3 given as shoots are harvested at flowering causes nodule senescence within 3 days and allows one to assess forage production potential under NH_4NO_3-dependent growth conditions following the N_2-dependent growth test.

Using the complex protocol, two cycles of phenotypic recurrent selection were completed under glasshouse conditions in African (A) and Hairy Peruvian (HP) alfalfa. The resulting populations termed African 32 (A32) and Hairy Peruvian 32 (HP32) were tested together on both N_2 and 8 mM NH_4NO_3 to determine genetic gains. Orthogonal comparisons across populations in both germplasm sources showed significant (P < 0.01) linear increases in shoot dry matter, shoot N concentration, and total shoot N (Teuber et al 1984). Averaged across both N treatments, HP32 had over 50% more total forage N than HP.

Additional tests with HP and HP32 have assessed how the selection method influenced growth, N assimilation and Rhizobium X legume inter-actions (Phillips et al 1985b). Leonard jar trials showed that, although only R. meliloti strain 102F28 had been present during selection, HP32 increased the amount of N_2 fixed in three other strains of R. mel-iloti by 22 to 53% relative to the same strains in HP (Figure 1). Those increases were not as great as the 62% increase in N_2 fixation by strain 102F28 in HP32, but in every case the increase in N_2 fixation associated with HP32 was statistically significant. The estimates, which were based on Kjeldahl N content, provide a direct measure of N_2 fixation. Results analyzed across all bacterial treatments showed no significant difference in length of the growth period for HP and HP32. Two of the bacterial genotypes tested, strains 414 and 445, were indigenous strains isolated from California alfalfa fields. Thus the data from this ex-periment strongly suggest that HP32 has traits which allow it to improve the symbiotic performance of diverse effective Rhizobium genotypes.

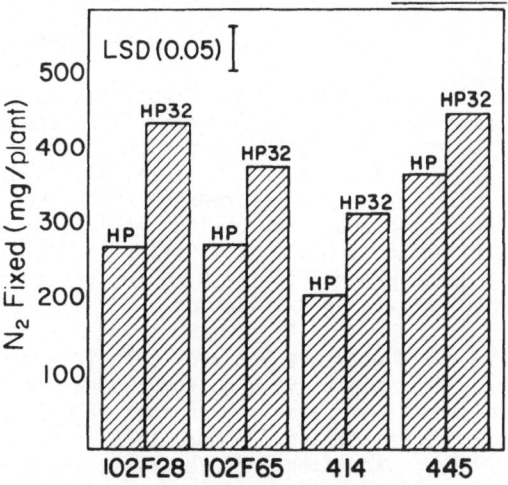

Figure 1.   Total N_2 fixation by the selected alfalfa population Hairy Peruvian 32 (HP32) and the original Hairy Peruvian (HP) grown under N_2-dependent, microbiologically-controlled conditions with various strains of Rhizobium meliloti.

Using a fixed-effects model to analyze data collected from the experiment reported in Figure 1, it was found that more than 70% of the phenotypic variance effects for shoot dry weight, total dry weight, and total Kjeldahl N were associated with the plant cultivars (Table 2). Only 10-15% of the variation in those parameters was associated with the Rhizobium strains, and no Rhizobium-strain X alfalfa-cultivar interactions were detected. Those responses were quite different from the results calculated for length of the growth period, a parameter for which no selection was made. These results provide another indication that the selection protocol used to develop HP32 improved plant growth.

Table 2. Percentage of total phenotypic variance effects contributed by alfalfa cultivar, Rhizobium strain, and their interaction using a fixed-effects model to examine the experiment reported in Figure 1.

Source of Variation	Trait			
	Shoot Dry Wt.	Total Dry Wt.	Total N	Growth Period
	-------------------		%	-------------------
Alfalfa cultivar	74.7**	71.2**	76.8**	26.1**
Rhizobium strain	11.1**	15.9**	13.7**	32.8**
Interaction	1.9	0.6	0.0	24.6**
Error	12.3	12.2	9.5	16.5

** Mean squares for source of variation was significant at $P < 0.01$.

Other Leonard jar tests compared HP and HP32 when both N_2 and NO_3^- were being assimilated (Phillips et al 1985b). For that experiment bacterial strain 102F28 was used in all treatments, and $^{15}NO_3^-$ was supplied to separate N_2 and NO_3^- utilization. The results clearly established that R. meliloti 102F28 fixed significantly more N_2 in HP32 than in HP with 0, 1, 2, and 8 mM NO_3^-. Total N_2 fixed, however, declined in both HP32 and HP with increasing NO_3^-. Thus the genetic changes in HP32 did not produce any obvious modification of the normal inhibition of N_2 fixation by NO_3^-. Utilization of NO_3^- was significantly greater in HP32 than in HP only in the 8 mM NO_3^- treatment, so most of the genetic improvement was in N_2 fixation.

No conclusive field data are yet available from materials selected with our protocol. Preliminary values from two harvests during the year of seedling establishment at Davis showed that HP32 produced an average of 10% more forage dry matter and 22% more total forage N than HP (Phillips et al 1985a). Additional field trials are planned with HP32 and a promising population, Moapa 69-32, produced by two cycles of phenotypic recurrent selection from the agronomically important cultivar Moapa 69. Glasshouse tests showed that, when results were averaged across both N_2- and NH_4NO_3-dependent conditions, Moapa 69-32 produced 42% more dry matter and 40% more total crude protein than the original Moapa 69 material.

One cannot yet conclude that the selection method used to produce HP32 (Figure 1) is an optimum procedure. It may be possible to simplify the technique, but we suggest that direct measurements of parameters that are critical to the integrated functioning of the plant are most likely to produce agronomically useful materials. In addition, as a possible result of exposing the plants to only one Rhizobium genotype during the selection process, an alfalfa population that increases N_2 fixation by diverse Rhizobium strains was produced.

REFERENCES

Burton JC and Wilson PW (1939) Soil Scí. 47, 293-303.
Cunningham SD et al (1985) Appl. Env. Microbiol. In press.
DeJong TM et al (1982) J. Gen. Microbiol. 128, 1829-1838.
Devine TE and Weber DF (1977) Euphytica 26, 527-535.
Ellis WR et al (1984) Agron. J. 76, 573-576.
Erdman LW and Means UM (1953) Agron. J. 45, 625-629.
Evans HJ et al (1985) In Ludden PW and Burris JE, eds, Nitrogen Fixation and CO_2 Metabolism, pp 3-11, Elsevier, New York.
Gibson AH (1962) Aust. J. Agric. Res. 13, 388-399.
Hardarson GH et al (1982) Crop Sci. 22, 55-58.
Heichel GH et al (1981) In Lyons JM et al., eds, Genetic Engineering of Symbiotic Nitrogen Fixation and Conservation of Fixed Nitrogen, pp 217-232, Plenum Press, New York.
Johnson HW et al (1965) Agron. J. 57, 179-185.
Munns DN (1977) In Hardy RWF and Gibson AH, eds, A Treatise on Dinitrogen Fixation, vol IV, pp 353-391, John Wiley, New York.
Mytton LR (1978) Ann. Appl. Biol. 88, 445-447.
Mytton LR et al (1984) Euphytica 33, 401-410.
Nutman PS (1984) Plant Soil 82, 285-301.
Phillips DA et al (1982) Crop Sci. 22, 606-610.
Phillips DA et al (1985a) In Ludden PW and Burris JE, eds, Nitrogen Fixation and CO_2 Metabolism, pp 203-212, Elsevier, New York.
Phillips DA et al (1985b) Crop Sci. In press.
Scott DB et al (1979) Biochim. Biophys, Acta 565, 365-378.
Seetin MW and Barnes DK (1977) Crop Sci. 17, 783-787.
Sheehy JE et al (1980) Crop Sci. 20, 491-495.
Tan GY (1981) Crop Sci. 21, 485-488.
Teuber LR et al (1984) Crop Sci. 24, 553-558.
Thornton HG (1929) J. Agric. Sci. 19, 48-70.
Viands DR et al (1981) USDA Tech. Bull. 1643.
Williams LE and Phillips DA (1983) Crop Sci. 23, 246-250.

ACKNOWLEDGEMENT

This work was supported in part by National Science Foundation grants PFR 77-07301 and PCM 82-17187.

PLANT HOST GENETICS OF NODULATION AND SYMBIOTIC NITROGEN FIXATION IN PEA AND SOYBEAN

Peter M. Gresshoff, David A. Day, Angela C. Delves,
Anne P. Mathews, Jane E. Olsson, G. Dean Price,
Katheryn A. Schuller and Bernard J. Carroll.
Botany Department, Australian National University, Canberra,
Australia

The need to study the plant functions controlling nodulation and symbiotic nitrogen fixation has received progressive recognition and attention. Comprehensive reviews of recent advances can be found in LaRue et al (1985) as well as Miflin and Cullimore (1984) and Verma and Nadler (1984).

The genetic analysis of the plant has centred on two organisms, i.e. pea (_Pisum sativum_) and soybean (_Glycine max_). Both are of economic importance , are self-fertile and have an extensive background of botanical, physiological and biochemical investigation. The pea system has the advantage of more convenient genetic analysis.

THE PEA SYSTEM

Studies on induced nodulation mutants in _P. sativum_ date back to work by LaRue and Holl in the early 1970s (Holl 1975) . Substantial progress has been made in the last year. Kneen and LaRue(1984) used ethylmethane sulfonate (EMS) and gamma irradiation to mutagenise pea variety "Sparkle" and isolated nodulation resistant mutants using visual examination. They observed that EMS treatment (1.0%, 60 min) yielded more mutants than gamma radiation (15 kilorads).

A stable nodulation mutant (sym5) was nonallelic to the sym2 gene found previously in pea variety Afghanistan. Sym5 is a single, mendelian recessive gene and its activity is associated with the roots and not the shoots of mutant plants. Additionally Kneen and LaRue found another gene (sym7) giving nonnodulation in pea.

Similar nonnodulation mutants of pea variety "Frisson" were isolated after EMS mutagenesis and visual selection (0.1%, 360 min) by Messager and collaborators (Messager,1985). They found 7 nodulation deficient mutants, which define seven separate complementation groups. All were recessive mendelian, except for mutant F.4.218, which was dominant. Grafting of reciprocal shoots and roots showed that the nod minus phenotype was root controlled in all 7 mutants.

Feenstra and Jacobsen (1985) isolated two nonnodulation mutants in pea variety "Rondo". Again EMS mutagenesis and M2 seed selection were used as the selection strategy. Mutant K5 was nonnodulated in general, but produced occasional nodules on lateral roots. This is a relatively common occurrence seen with nonnodulating legume material and will be discussed later in relation to soybean mutants. Mutant K5 showed root hair curling and infection thread formation similar to wildtype plants. Another mutant (K24) showed a complete nonnodulation phenotype under all studied conditions. Both nod minus mutants are monogenic and recessive. Similar, but less characterised, material was reported by Engvild (1985).

The mutant screening by Messager (1985) also yielded 8 nodulated, but nitrogen fixation deficient mutants, which were all monogenic, recessive and in 8 separate complementation groups. Their phenotype was always root controlled (ascertained through grafting) and mutant F261 was characterised by a suppression of its mutant phenotype, if inoculated with <u>Rhizobium leguminosarum</u> strains 79 or 94. Both strains were selected from soils of divergent geograghic origin by the mutant plant.

Several super- or hypernodulation mutants were isolated by different groups. Jacobsen and Feenstra (1984) described the first supernodulation mutant of pea (using variety "Rondo"). This mutant developed substantially more nodule mass and increased nodule number in the presence or absence of exogenous nitrate (15 mM KNO_3). The increased symbiotic parameters in the absence of nitrate warrants the term "supernodulating", rather than just "nitrate tolerant". Nitrogen fixation as measured by acetylene reduction was also nitrate tolerant giving increases of up to 20 fold in the presence of nitrate. Of interest with mutant nod3 is the observation that in the absence of nitrate, acetylene reduction was similar between mutant and wildtype plants. The nod3 mutant is monogenic and recessive. The analysis of <u>in vitro</u> translation products from total RNA isolated from 5 day old roots grown on nitrate in the absence of <u>Rhizobium</u> indicated the absence of one peptide in the mutant.

The mutant screen by Messager (1985) yielded two supernodulation mutants in cultivar "Frisson". These were nitrate tolerant, developing 3-5 times the nodule number compared to control plants in the presence or absence of nitrate (12mM). The type of nodulation was termed "hypernodulation". Both mutants (HN191F and HN190F) were closely associated with flower fasciation, were allelic (if not identical) and behaved as mendelian recessive alleles. Grafting showed that supernodulation was entirely shoot controlled.

Figure 1. Supernodulated root systems of soybean mutants. a) mutant nts1007 grown in soil at full plant maturity. Similar plants but at 89 days age were used for Table 1. b) nts1007 grown in high nitrate, black soil illustrating heavy nodulation into the soil profile.

THE SOYBEAN SYSTEM

Genetic variability affecting nodulation and nitrogen fixation in soybean has been available for the last 30 years. Mutant alleles rj_1, Rj_2, Rj_3 and Rj_4, all found in varietal selections or their crosses, have been described previously (Verma, Nadler 1984, LaRue et al 1985). The latter three, all controlled by dominant alleles, affect the fixation ability of the nodule and show strong strain specificity. Mutant rj_1 in contrast is recessive, fails to nodulate with most strains, but can be nodulated with selected strains , if the inoculant density is sufficiently high. All four mutations are root controlled and have been bred into various soybean cultivars. The nonnodulation isolines thus produced are of agronomic value as they permit the evaluation of the contribution of nitrogen fixation to the total nitrogen of the plant.

In 1981 we started an induced mutagenesis programme to isolate mutants of soybean cultivar "Bragg" altered in (a) constitutive nitrate reductase activity, (b) nodulation initiation and development , (c) nodule function and (d) nitrate sensitivity of nodule initiation and function. We tested three mutagens (gamma-rays, EMS and sodium azide). An EMS treatment of 0.5% for 360 min. at 28C in a well-aerated phosphate buffer using 6 hour pregerminated seeds yielded M2 families with a frequency of chlorophyll deficiency of 2.8%. We chose to collect separate M2 families as this permitted the subsequent genetic analysis of M2 wildtype segregants in the M3 and it furthermore gave us assurance that each mutant stemmed from a separate mutagenic event.

TABLE 1
Comparison of "Bragg" and supernodulation mutant nts1007 in soil

parameter tested	Bragg(n=11)	nts1007(n=8)
shoot dry weight (g)	8.71(2.70)**	6.06(1.37)
root dry weight (g)	0.89(0.28)**	0.71(0.17)
shoot/root (dw)	9.7.(0.64)**	8.6 (0.78)
shoot/root+nodule (dw)	7.0 (0.62)	3.6 (0.3)
nodule number	108 (41)	694 (201)
nodule fresh weight (g)	2.00(0.68)	4.98(0.94)
nodule dry weight (mg)	350 (140)	980 (150)
acetylene reduction*	26.6(3.9)**	29.8(6.1)
plant height (cm)	78.2(5.0)	67.6(6.5)
node number	11.5(0.5)**	10.5(0.4)
days to flowering	75 (1.0)	71 (2.0)

* nanomoles ethylene produced per minute per plant. All plants were harvested at 89 days after planting.This was equivalent to the R2 stage of development.** indicates that relevant means are not significantly different at the 0.05% level using a small sample T-test. (+ S.D.).

Using a direct leaf disc assay for nitrate reductase deficiency we isolated two mutants (nr328 and nr345), which have reduced constitutive nitrate reductase activity (Carroll, Gresshoff 1986a). Mutant nr345 has broad similarities to mutant lnr2, isolated by Ryan et al (1983), but mutant nr328 appears to differ as it shows nitrate toxicity in leaves and affects the inducible nitrate reductase activity (D. Whitmore-Smith, unpublished).

The mutagenised M2 families of "Bragg" (including a set of families, which showed 0.9% chlorophyll deficiency) were screened for nonnodulation and nitrogen fixation inabilities. We isolated four nod minus mutants (nod45, nod49, nod139 and nod773) by visual screening (Carroll, Gresshoff 1986b). Preliminary segregation data from the M2 suggested that all mutants were controlled by recessive alleles. Mutant nod45 has general fitness problems, when grown on nitrate or urea . Mutant nod49 was studied in the greatest detail. It lacked curled roothairs, infection threads and pseudoinfection sites. Grafting , as with the pea mutants and soybean mutant rj_1 , showed root control in mutant nod49 and nod139, which was not affected by shoots of supernodulating soybean mutants nts382 and nts1007. When a wide range of Rhizobium strains was tested for nodulation on mutant nod49 in Leonard jars, most failed to produce nodules. Occasionally large nodules developed at low numbers (never greater than 3, mostly one), but reisolates from such nodules always gave the original phenotype upon reinoculation. In this respect mutant nod49 was similar to mutant rj_1 . However, strains , which specifically suppressed the nonnodulation phenotype of mutant rj_1 failed to do so on mutant nod49 (i.e. strain USDA76). At moderate bacterial cell densities (about 1×10^9 cells per seedling) some Rhizobium strains like USDA110 and USDA138 gave moderate nodulation (at times up to 10 nodules per 4 week old plant). Nodules formed by strain USDA110 on mutant nod49 were effective in nitrogen fixation and had normal morphology.

Nitrate has long been known to affect nodulation and symbiotic nitrogen fixation. We attempted to break the nexus between the plant's ability to use either soil nitrogen or atmospheric nitrogen. Recent research implied the plant as the major intermediary in this process. We felt that specific plant mutants, which fail to react to the otherwise inhibitory effects of nitrate, would be helpful in the further analysis. We thus screened our M2 families (2500 in total involving approx. 25000 plants) for the ability to nodulate well in the presence of about 5 mM nitrate, which was supplied regularly. A preliminary summary of our strategy and procedures as well as some characterisation of one mutant are published (Carroll et al 1984, Gresshoff et al 1985, Carroll et al 1985a,b)

Fifteen independent mutants showing the nitrate tolerant phenotype (termed nts for nitrate tolerant symbiosis) were isolated. Genetic analysis indicated that mutants nts382, nts1116, nts1007, nts501 and nts2264 were inherited as monogenic recessives. Mutant nts733 appeared to be semidominant, giving an intermediate phenotype in the heterozygote. The mutants fell into at least 3 complementation groups with nts382, nts1007 and nts501 being in group A, nts1116 group B and nts733 group C.

Mutant nts1007 exhibited its nodulation vigour in soil, vermiculite, sand and growth pouches. Fig.1a shows a root system of mutant nts1007 grown to maturity in soil. Fig.1 b shows a supernodulation response in nitrate rich, volcanic black soil

inoculated with Rhizobium strain USDA110. Both figures illustrate that supernodulation occurs in soil conditions. Nitrogen
content of such root systems is increased in comparison to wild
type roots demonstrating that the additionally fixed nitrogen
may have beneficial carry-over effects to the subsequent crop.

Supernodulation mutants are characterised by extensive nodulation in the presence of 5mM nitrate as well as in its
absence. Table 1 compares mutant nts1007 plants with "Bragg" at
89 days after planting in low nitrogen soil, surface irrigation
and strain USDA110 inoculation. The soil was not previously
used for soybean culture and was considered Rhizobium -free.
These data indicate as well that under low N conditions in the
soil, plant performance lags only slightly behind "Bragg" . All
nodule parameters other than acetylene reduction are increased
in the mutant. The equivalent levels of nitrogen fixation
reflect other data described by Carroll et al (1985b) on 0 mM
nitrate.

Table 2 shows a nodule parameter comparison between "Bragg"
and mutant nts1007 at 30 days after planting. Accelerated nodulation on both taproot and lateral roots is clearly seen.

The supernodulation response does not represent an inability
to reject bacterial invasion as plants can grow well in natural
conditions without pathological symptoms. A range of Rhizobium
strains such as USDA110 , USDA138 , USDA191 , CB1809 , and
61A76 have nodulated nts mutants. Strain USDA191 is only partially effective on "Bragg" and this phenotype is maintained in
the nts mutant.

TABLE 2
Early nodule development in soil-grown Bragg and nts1007

parameter tested	Bragg	nts1007
nodule number-taproot	2	19
lateral root	21	193
root fresh weight (mg)	914	750
nodule fresh weight (mg)	171	390
max. nodule diam. (mm)	3.0	2.5

plant age was 30 days after planting in low nitrogen soil inoculated with strain USDA110.

Grafting of the shoots of mutants nts382 and nts1116 onto
wildtype roots resulted in supernodulation (Fig. 2). The
cotyledons did not affect the result. Grafting can occur as
early as two days after germination and can be done with plants
raised in growth pouches (W.D.Bauer, unpublished). Reciprocal
grafts failed to give the supernodulation phenotype. Even the
intermediate phenotype of mutant nts1116 was transfered to the
root via a shoot signal. Thus at least 2 genes (or rather complementation groups) acting in the shoot of soybean are controlling nodulation proliferation and indirectly nitrate sensitivity of the symbiosis. The supernodulation phenotype is also
expressed in roots of cultivar Williams, showing that this is
not a phenomenon specific to cultivar "Bragg".

In general it appears that excessive supernodulation is counterproductive in terms of plant growth and especially root development. However, several parameters seem to interact. For example, growth in the presence or absence of nitrate alters the differences seen between mutant and wild type. Likewise growth in soil as presented in Table 1 lessens plant differences mainly by restricting vigorous root growth, which is normally observed in sand cultures or hydroponics. Some mutants like nts1116 and the heterozygote of nts733 show more intermediate levels of supernodulation (200-300 nodules instead of 800-1000 per plant after 4-6 weeks of sand culture). This material grows relatively strong roots and shoot mass is equivalent to wild type controls in preliminary tests under field conditions. We also found that in the absence of <u>Rhizobium</u> but in the presence of nitrate ,mutant nts382 and "Bragg" grew equally well, suggesting that growth reductions are a result of pleiotropic effects and not a residue of the original mutagenesis.

Anatomical studies of mutant nts382 showed that infection and nodule initiation processes were qualitatively similar to wild type. The mutant plants exhibited a pronounced delay of the autoregulation response normally seen in legumes and most precisely described by Bauer (1981) , Pierce and Bauer (1983) and Kosslak and Bohlool (1984). The extent of delay may depend on the culture system and the degree of irrigation. Studies in our laboratory point to an increased nodulation interval, if irrigation is by flooding from the top.

Figure 2. Summary of soybean grafting studies.

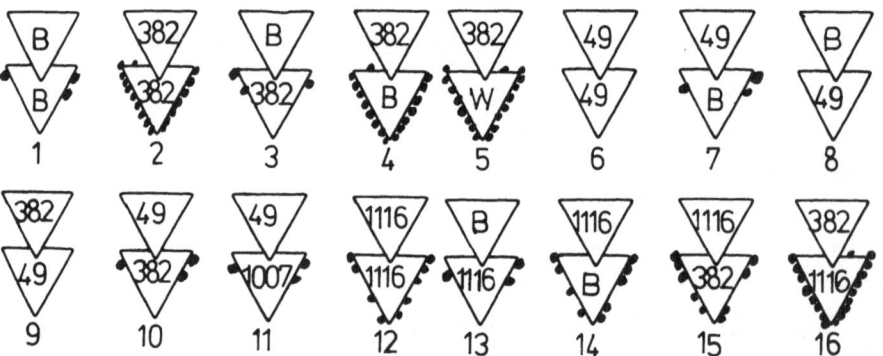

B= Bragg; W= Williams; 382= nts382; 1007= nts1007, 1116= nts1116; 49= nod49;. nts382 and nts1116 are non-allelic and give supernodulation and hypernodulation respectively. Results for nod49 also hold true for nod139. Allelic mutants nts1007 and nts 382 gave identical results if grafted with nod49 or nod139.

Nitrogenase activities of nodules induced on mutant nts382 under 0 mM nitrate supplementation were about one third that of control nodules. This specific nitrogenase activity was not altered if the mutant was raised continuously on 5 mM nitrate . In contrast wild type nodules had 25 - 35% specific nitrogenase activity of 0 mM nitrate controls, when grown on 5 mM nitrate .

What is of interest is the fact that in terms of specific
nitrogenase activity the mutant nodules are nitrate tolerant .
Similar tolerance is observed in other mutants. The general
observation was confirmed by Schuller (unpublished), who grew
soybean symbiotically and then applied nitrate. Several nodule
parameters, including enzymatic functions related to ureide
metabolism , carbohydrate supply , ammonia assimilation and
nitrogenase activity, were followed relative to untreated controls. "Bragg" nodules were inhibited by nitrate only in terms
of acetylene reduction activity after 2 days nitrate exposure.
Other parameters were similar to controls except that a significant cessation of nodule fresh weight increase was detected.
Nitrogenase activity was assayed by 15 N incorporation (done in
collaboration with Dr. Alan Gibson, CSIRO) and acetylene reduction . Similar inhibition values were obtained (Schuller et al
1986). Yet when mutant nts382 was studied , no significant
differences to untreated plants were observed . Eventually,
after prolonged nitrate exposure (about 14 days), total nodule
senescence occurred. Similar irreversible changes occurred in
wild type soybean . The major question as to why a single mutation , which affects early nodule initiation and autoregulation
, also has effects at late, nitrogen fixation related processes
remains unanswered.

REFERENCES
Bauer WD (1981) Ann. Rev. Pl. Physiol. 32, 407-449.
Carroll BJ et al (1984) In Ghai BS, ed ,Symbiotic nitrogen
fixation I, pp 43-50, USG Publishers and
distributors, Ludhiana, India .
Carroll BJ et al (1985a) Plant Physiol. 78, 34-40.
Carroll BJ et al (1985b) Proc. Nat. Acad. Sci., 82, 4162-4166
Carroll BJ and Gresshoff PM (1986a) Plant Physiol. submitted.
Carroll BJ and Gresshoff PM (1986b) Plant Sci.Lett. submitted.
Engvild KC (1985) In Analysis of the plant genes involved in
the legume-Rhizobium symbiosis.pg.49,OECD Publ.,Paris,France.
Feenstra WJ and Jacobsen E (1985) ibid.,pp.50-51.
Gresshoff PM et al (1985) In Szalay AA and Legocki RP, eds.,
Advances in molecular genetics of plant-bacterial interactions
pp.86-88,Media Services, Cornell Univ. Publ. Ithaca ,NY., USA.
Holl FB (1975) Euphytica 24,767-770.
Jacobsen E and Feenstra WJ (1984)Plant Sci.Lett.33,337-344.
Kneen BE and LaRue TA (1984) J. Heredity,75,238-240.
Kosslak RM and Bohlool BB (1984) Plant Physiol. 75,125-130.
LaRue TA et al (1985) In Analysis of the plant genes
involved in the legume-Rhizobium symbiosis. pp.39-48. OECD
Publication , Paris , France .
Messager A (1985) ibid.,pp. 52-59.
Miflin BJ and Cullimore JV(1984) In Verma DPS and Hohn T,eds.
Genes involved in plant-microbe interactions. Plant Gene
Research Vol.I, pp. 129-178, Springer Verlag. Berlin,Wien.
Pierce M and Bauer WD (1983) Plant Physiol. 73, 286-290.
Ryan SA et al (1983) Plant Physiol. 72, 510-514.
Schuller KA et al (1986) Plant Physiol. submitted.
Verma DPS and Nadler K (1984) In Verma DPS and Hohn T,eds.,
Genes involved in plant-microbe interactions. Plant Gene
Research Vol. I.,pp. 57-93, Springer Verlag.Berlin,Wien.

SELECTION FOR MULTIPLE TRAITS INCREASES ALFALFA N_2-FIXATION AND YIELD

D.K. BARNES, D.L. JESSEN, G.H. HEICHEL, and C.P. VANCE, USDA-ARS and University of Minnesota, Borlaug Hall, 1991 Buford Circle, St. Paul, MN 55108 U.S.A.

A breeding program was developed for improving the coordinated responses among physiological and biochemical traits affecting N_2-fixation (Barnes et al, 1984). Four alfalfa germplasm sources were used to evaluate the effectiveness of the program. Three of them were the cultivars: 'Blazer', 'Citation', and 'Saranac AR'. The fourth (MNPL10 x MNNC-7) was a cross between two subpopulations, each selected for three cycles for improved N_2-fixation (Viands et al, 1981). The generations of selection were: GS (unselected base population); CO (selected in a stepwise program in a nil-nitrate growth medium in a glasshouse for: superior seedling vigor, preference for highly effective Rhizobium strains, large shoot yield, many fibrous roots, increased nodule mass, and high nitrogenase activity (measured by acetylene reduction); C1L and C1H (one cycle of selection in the CO generation for low and high nodule glutamate synthase (GOGAT) and phosphoenolpyruvate carboxylase (PEPC) activity, respectively). Four population crosses (GS x GS, CO x CO, C1L x C1L, and C1H x C1H) were made between each of the three cultivars and the MNPL10 x MNNC-7 population.

The 28 subpopulations were grown in a randomized complete-block design in a split-plot with five replicates. Two soil N treatments (0 and 100 kg supplemental N ha^{-1}) were used as whole plots. Whole plots were 1-m^2 plots and subplots were 3-plant hills spaced on 11 cm centers. Six-week-old seedlings were transplanted to the field on May 5 and the shoots harvested on July 5, July 29, September 16. Whole plants were harvested on October 21. N_2-fixation was measured by the difference method using an ineffectively nodulating alfalfa as the control (Henson and Heichel, 1984).

The mean proportion of N derived from N_2-fixation for all entries was 96 and 86% at 0 and 100 N, respectively. The GS and GS x GS generations fixed more N_2 at 100 N than at 0 N. In contrast, the CO, C1L, and C1H generations each fixed about 15% more N_2 at 0 N than at 100 N. At 0 N the CO, C1L and C1H generations each fixed about 35% more N_2 than GS, illustrating the effectiveness of the multiple trait selection for improving N_2 fixation. The GS x GS crosses fixed 17% more N_2 than the GS generation, illustrating the heterosis effects of population crosses. The COxCO crosses fixed 17% more N_2 than the GS x GS crosses. The C1H x C1H crosses fixed 7% more N_2 than the CO x CO crosses, whereas the C1L x C1L crosses had a 20% decrease. The responses to selection for GOGAT and PEPC activity illustrated the importance of selecting for enzymes of both carbon and nitrogen assimilation in nodules.

REFERENCES

Barnes, D.K. et al, (1984) Plant and Soil 82, 303-314.

Henson, R.A. and Heichel, G.H., (1984) Field Crops Res. 9, 333-346.

Viands, D.R. et al, (1981) USDA Tech. Bull. 1643. 18p.

GENETIC VARIABILITY FOR NITROGEN FIXATION IN PEA.

R. COUSIN, INSTITUT NATIONAL DE LA RECHERCHE AGRONOMIQUE,A. MESSAGER, ELF. STATION DE GENETIQUE ET D'AMELIORATION DES PLANTES, ROUTE DE SAINT-CYR-78000 VERSAILLES, AND J. PICARD, INSTITUT NATIONAL DE LA RECHERCHE AGRONOMIQUE, STATION D'AMELIORATION DES PLANTES, BP 7540 21034 DIJON (FRANCE).

In order to assess whether improved nitrogen fixation could contribute to an increase in yield and protein content or both, 50 varieties were grown in glasshouses in hydroponic solutions with or without nitrogen fertilizer. Large variation in nitrogen assimilation and atmospheric nitrogen fixation was recorded. Symbiotic nitrogen fixation does not seem to be correlated with nitrogen assimilation. However, nitrogen fixation was observed to be negatively correlated with maturity earliness. Nevertheless, with Rhizobium alone, the weight of seeds varied from 40 to 100% of the weight obtained when nitrogen fertilizer was also added, depending on the variety. Several mutants : non nodulating (nod⁻) and inactive nodule (fix⁻) of the winter pea variety "Frisson" were obtained. Line F261-fix⁻, grown in the glasshouse in hydroponic solution with nitrogen fertilizer gives the same amount of yield as "Frisson". However, under field conditions with 200 Nitrogen units, the yield obtained with F261 is 66% of that obtained in the case of "Frisson". Nitrogen fixation activity under field conditions, was also studied in about twenty varieties. Large variation in the fixation activity of different varieties is recorded. The seed yield seems to be positively correlated with the accumulated activities of nitrogen fixation. With Afila type, on the other hand, no reduction of seed yeild was observed. However, nitrogen fixation and leaf area were reduced in the same proportion.

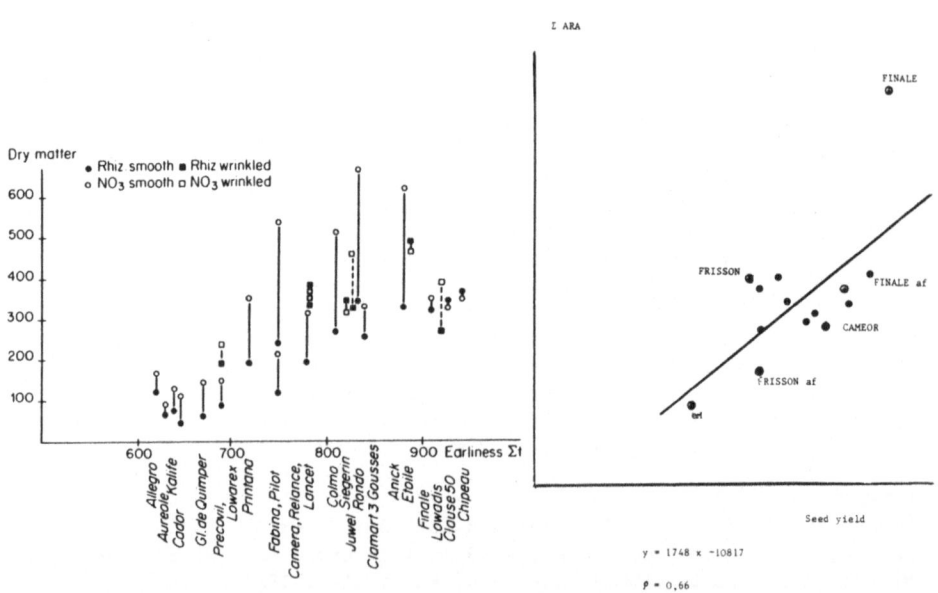

SEASONAL N_2 FIXATION AND N PARTITIONING IN RECOMBINANT BEAN LINES USING ^{15}N-DEPLETED $(NH_4)_2SO_4$

JOHN D. DUBOIS[1]/DINA A. ST.CLAIR[2]/JUAN C. ROSAS[2]/FRED A. BLISS[2]/ROBERT H. BURRIS[1].
[1]DEPARTMENT OF BIOCHEMISTRY AND [2]DEPARTMENT OF HORTICULTURE, COLLEGE OF AGRICULTURAL AND LIFE SCIENCES, UNIVERSITY OF WISCONSIN-MADISON, MADISON, WI 53706, USA

This report describes a portion of an ongoing project aimed at evaluating parental and progeny lines of common beans as to their N_2 fixation and yield potentials and overall field performance. Progeny lines (Pop. 24) were developed using the Inbred Backcross Line Method. The objectives of this study were to assess the seasonal rates of N_2 fixation, the seasonal pattern of N_2 fixation activity and the partitioning of fixed N at the whole plant level in 7 lines of common bean. Plants (bean lines and a nod⁻ soybean control) were grown in loamy sand field plots and were fertilized weekly with ^{15}N-depleted $(NH_4)_2SO_4$ solution. Plants were sampled at the R3 (50% bloom) and R9 (maturity) stages, separated into roots, stems, leaves, and reproductive tissues and analyzed for dry wt., N content, N fixed and % N from fixation.

There were substantial improvements in seed yields, total N and N fixed for Pop. 24 over the recurrent parent, Sanilac. Sanilac was similar to the nod⁻ soybean for dry wt., yield and total N and considering its low N_2 fixation (9.0% N from N_2 fixation), it could be used as a control under the conditions of this experiment. The lines also showed differences in the time of maximal N_2 fixation. Some lines fixed over 1/3 of their total fixed N by the R3 stage, whereas other lines fixed less than 10% of their total fixed N by the R3 stage. By maturity, the Pop. 24 lines as a group showed significant improvement in mg N fixed/plant over the recurrent parent. Individually, only 2 of the 5 Pop. 24 lines tested (UW24-17, UW 24-65) showed significant improvement in mg N fixed/plant over the recurrent parent.

Major differences occurred among the lines in the partitioning of fixed N at the whole plant level. The donor parent, Puebla 152 Black, had the greatest amount of N_2 fixation, yet it partitioned almost 20% of its fixed N into leaflets and less than 70% into seeds. The recurrent parent, Sanilac, had the lowest amount of N_2 fixation, yet it partitioned approximately 85% of its fixed N into seeds and less than 5% into leaflets. The progeny lines showed various patterns of partitioning with the UW24-17 line displaying the optimal pattern of approximately 5% of total fixed N into leaflets and 80% into seeds.

In conclusion, the inbred backcross line method can be used to improve N_2 fixation and yield abilities while retaining favorable traits of the recurrent parent. Along with improving the N_2-fixation potential, both the time of maximal N_2 fixation and the partitioning of fixed N can also be changed during the breeding program. The R3 (50% bloom) stage is not necessarily the peak in N_2 fixation in common bean, and routine sampling of plants at this stage may not be satisfactory for comparing different lines of common beans.

This research was supported by a grant from the McKnight Foundation.

CHARACTERIZATION OF RHIZOBIUM FREDII SYMBIOSIS WITH WILD AND
CULTIVATED SOYBEAN

Nancy M. DuTeau, Reid G. Palmer and Alan G. Atherly, Department of Genetics,
Iowa State University, Ames, IA 50011

Soybean improvement programs are searching for soybean-Rhizobium genotype combinations that optimize nitrogen fixation. This includes a search for
resistance (exclusion) and co-evolved, efficient symbioses. A new species of fast-
growing Rhizobium that nodulates soybeans Glycine max (L.) Merr. was isolated
from nodules collected in the People's Republic of China. The rapid growth rate
and potential of these strains for genetic manipulation may lead to improved
inoculum for soybeans. However, several of the fast–growing strains form
ineffective nodules on North American cultivars. But, one strain, USDA 191,
forms effective nodules on commercial cultivars. The symbiotic performance of
these strains on a variety of host genotypes needs to be tested to assess their
nitrogen-fixing ability. We have examined the nitrogen fixation of 13 soybean
cultivars and wild soybean (G. soja Sieb. and Zucc.) plant introductions inoculated
with Bradyrhizobium japonicum strain 61A76 and 4 fast–growing strains of
Rhizobium fredii to identify pairs of genotypes with high nitrogen-fixing ability.
We examined several G. max cultivars including 2 with host–ineffective alleles
('Hardee', Rj2 Rj2, Rj3 Rj3, and 'Hill', Rj4 Rj4) and 1 non-nodulating isoline
('Harosoy', rj1 rj1). We also tested several plant introductions of G. soja that
showed low, medium, or high levels of nodulation in Iowa soils. Five-week–old
plants were assessed for nodule morphology, acetylene reduction rate, nodule
number, nodule fresh weight, and plant-top dry weight.

The combinations of soybean-Rhizobium genotypes that were clearly superior for nitrogen fixation were Evans, Hill, and Hardee with the slow-growing
control, B. japonicum strain 61A76. Fast-growing R. fredii strain USDA191 fixed
nitrogen as well as strain 61A76 in combination with 'Peking', 'Virigina', 'Harosoy
63', 'Rampage', and PI's 342622A and 101404B. These two plant introductions
showed low levels of nodulation in Iowa soils. All fast–growing strains formed
effective nodules on Peking, Virginia, and Hardee. The cultivar Virginia is a
genetically unimproved genotype from the same geographic location as Peking.
The cultivar Hardee contains the host-ineffective alleles Rj2 and Rj3. The
formation of effective nodules on Hardee was unexpected, and an explanation was
not evident. None of the soybean genotypes showed superior nitrogen fixation
with the fast-growing strains. No nodules were formed on Harosoy rj1 rj1 by the
fast-growing strains. Fast-growing strain USDA191 formed effective nodules on
all cultivars, but the nitrogen fixation was not efficient. This strain fixed less
nitrogen than the slow-growing control with cultivars 'Evans', 'Williams', 'Hardee',
'Hill', and G. soja PI 407217 and PI 81762. The fast–growing strains of R. fredii
are poor nitrogen-fixing symbionts of soybeans.

PHOTOASSIMILATE AND PHYTOMASS ALLOCATION IN ALFALFA
POPULATIONS WITH CONTRASTING NODULE MASSES

G.H. HEICHEL[+]/HARRY T. CRALLE[++]/D.K. BARNES[+]
[+]USDA-ARS, 411 BORLAUG HALL, 1991 BUFORD CIRCLE, ST. PAUL, MN 55108, USA
[++]DEPT OF SOIL AND CROP SCIENCES, TEXAS A&M UNIV., COLLEGE STN, TX,
77843, USA

INTRODUCTION. Alfalfa (_Medicago sativa_ L.) N_2 fixation can be improved by recurrent selection for associated characters, especially nodule mass. Because of the interdependence of nodule metabolism and shoot CO_2 assimilation, selection for nodule mass might have correlated effects on photoassimilate partitioning to dry matter. This possibility was examined by investigating leaf CO_2 exchange, dry matter distribution among organs, and photoassimilate partitioning in experimental alfalfa populations developed for high and low nodule mass.

METHODS. Experimental populations with low (LNM), intermediate (INM), and high (HNM) nodule mass/plant were developed by crossing two subpopulations produced from two different germplasm sources. The subpopulations had been developed for high and low nodule mass by two cycles of divergent recurrent selection. Glasshouse-grown plants at two stages of ontogeny were radiolabeled and analyzed for allocation of radiolabel as previously described (Cralle, Heichel; 1985, 1986).

RESULTS. Selection for HNM yielded plants with larger organs than those of the INM control, and selection for LNM yielded plants with smaller organs. Compared to the LNM population, the HNM population had 140% greater nodule mass, 129% larger leaf mass, 134% greater mass of crown shoots, 135% larger crown mass, 141% larger root mass, and 135% larger plants.
    Despite the direct response of nodule mass and the correlated responses of other organs to selection for nodule mass, the proportional distribution of whole-plant dry matter among organs was unaffected by selection. Furthermore, selection for nodule mass did not materially change the partitioning of photoassimilate to dry matter among organs of the experimental populations, leaf CO_2 assimilation, or export of photoassimilate by labeled leaves.
    These results suggest homeostatic control(s) in alfalfa to maintain source-sink proportionality between nodules and other organs. This idea is supported by the results that photoassimilate partitioning was unaltered by selection of experimental populations differing by 140% in nodule mass. The results also suggest that selection of alfalfa for greater N_2 fixation capability by selection for nodule mass should maintain or increase yields without changes in C allocation.

REFERENCES: Cralle HT and Heichel GH (1985) Plant Physiol [In Press].
              Cralle HT and Heichel GH (1986) Crop Sci. [In Press].
              Partly supported by USDA-S&E, CRGO Grant 59-2177-0-1-417-0.

NITRATE TOLERANCE IN SOYBEAN: VARIATION BETWEEN GENOTYPES

DAVID F. HERRIDGE/JOHN H. BETTS
N.S.W. DEPT. AGRICULTURE, RMB 944, TAMWORTH, NSW, AUSTRALIA, 2340.

The proportion of soybean crop N derived from N fixation rarely exceeds 60% (Herridge, 1982). The remainder is taken up as nitrate from the soil and the net effect of soybean cropping is depletion of the soil N pool. In order to achieve higher rates of N fixation, symbioses which are capable of maintaining N fixation activity in the presence of soil nitrate must be developed. We report on a three year study involving 489 genotypes of soybean of diverse origin, all of which were taken from the collection of parent lines held by the N.S.W. Department of Agriculture.

MATERIALS AND METHODS

In the first year, the 489 genotypes were divided into three maturity groups and sown into sand in 14 L free-draining pots. Sowing dates were October, 1982 (early), December (mid) and February, 1983 (late). Pots were inoculated at sowing with R.japonicum CB1809 and watered with a complete nutrient solution containing either zero or 2.5 mM NO_3^--N. Plants were sampled at R2 or equivalent. Symbiotic activity was determined for each treatment by combining data on nodulation and N fixation (relative ureides for xylem exudate, shoot axes and nodulated roots)(Betts, Herridge, 1984). In the second year, 87 of the original 489 genotypes, comprising 66 'tolerants', 9 'susceptibles' and 12 others were similarly screened. In year three, 32 of the most tolerant lines were inoculated and sown into a Rhizobium-free, high nitrate vertisol (260 kg NO_3^--N/ha; 0-120 cm depth) in the field. Symbiotic activity was assessed at R5-6 as before. Plant growth, grain yield and soil nitrate levels were measured at appropriate times.

RESULTS AND DISCUSSION

Large ranges of genotypic responses to nitrate supply were measured in the first two years of screening. The second year of screening provided 32 elite lines for further testing in the field. The superior lines from the field study were of Korean origin and showed up to 17 fold increases in nodulation and 20 fold increases in N fixation, equivalent growth and reduced grain yield relative to 'Bragg' (Table 1). Soil nitrate levels, measured subsequent to grain harvest, were up to 33 kg N/ha higher in the plots containing the superior lines relative to Bragg.

TABLE 1. Measurements of growth and symbiotic activities of the nitrate 'tolerant' genotypes, Bragg and Hardee in a high nitrate field soil.

Genotype	Nodule mass (mg/plant)	Nodule no. (/plant)	Ureide index (%)	Shoot mass (g/plant)
Korean Line 1	376	34.5	36	45.9
2	254	16.8	27	43.3
3	176	19.5	30	41.6
Bragg	24	2.0	11	39.7
Hardee	0	0	12	67.1

REFERENCES

Betts J H and Herridge D F (1984) A.I.A.S. Occ.Publ. 12, 65-66.
Herridge D F (1982) Plant Physiol. 70, 7-11.

NODULATION SPECIFICITY IN PISUM SATIVUM L.

SHAUN L.A. HOBBS
PLANT BIOTECHNOLOGY INSTITUTE, NATIONAL RESEARCH COUNCIL OF CANADA,
SASKATOON, SASK., CANADA, S7N 0W9

Legume cultivars that nodulate with rare or engineered rhizobial strains but not with indigenous soil strains could reduce competition between improved inocula and indigenous bacteria (Devine and Kuykendall 1982). In pea, cv. Afghanistan is nodulated by the Tom strain of Rhizobium leguminosarum but not by most other strains. However, some non-nodulating strains will still compete with Tom and reduce its ability to infect Afghanistan (Winaro and Lie, 1979). Such competition may preclude the use of the Afghanistan-Tom specificity as an inoculum delivery system.

Materials and Methods: Afghanistan in the field were inoculated with: Tom; a local nodulating isolate; Nitragin commercial inoculum; all possible combinations of these; or nothing (control). Acetylene reduction and nodulation were measured during the season. In controlled environment conditions, Afghanistan in Leonard jars containing -N nutrient medium were inoculated with: Tom; the local isolate; strains that do not nodulate Afghanistan but which have relatively high (128C52) or low (10004) acetylene reduction ability with normal peas; or combinations of these nodulating and non-nodulating strains. Acetylene reduction was assayed at 24 days.

Results and Discussion: In the field, little nodulation of Afghanistan occurred in non-inoculated or Nitragin treatments despite abundant nodulation of normal pea varieties in the guard rows. Plants inoculated with Tom or the local isolate nodulated abundantly and reduced acetylene effectively (23 and 14 μmol C_2H_4 g root^{-1} h^{-1}, respectively) and there was no significant decrease when Nitragin was also added. In controlled conditions effective nodulation occurred with Tom and the local isolate but not with 128C52 or 10004. Nodulation and acetylene reduction were not significantly reduced when nodulating and non-nodulating strains were added together. All isolates taken from single nodules formed in the 128C52 + Tom treatment formed effective nodules on Afghanistan and were shown by intrinsic antibiotic resistance to contain the Tom type. These results indicate that the Afghanistan-Tom specificity could be used to produce a delivery system for a specific inoculum for peas in some field conditions.

References
Devine TE and Kuykendall LD (1982). In Clark KW and Stephens JHC (eds) Proc. 8th Am. Rhiz. Conf., 7-13.
Winaro R and Lie TA (1979). Plant and Soil 51,135-142.

PLANT GENOTYPE AND THE CONTROL OF NITROGEN FIXATION IN SOYBEAN

John Imsande, Department of Genetics, Iowa State University, Ames, Iowa 50011, U.S.A.

Nitrate (i.e., soil nitrogen) represses nodulation and nitrogen fixation in soybean. Consequently, the purpose of the experiments described was to isolate soybean genotypes that, when grown in the presence of high nitrate, fix nitrogen at a relatively high rate and yet maintain a vigorous growth rate. The genetic stocks used in these screenings were not mutagenized because, 1) genetic heterogeneity exists within all cultivars, 2) selective mutagenesis for high nitrogen fixation in soybean is currently not possible, and 3) spontaneous mutations do occur.

Various cultivars, lines, and genetic crosses were grown hydroponically in the presence of 2 to 3 mM nitrate and screened with the nondestructive acetylene reduction assay for high nitrogen fixation (Imsande, Ralston 1981; Ralston, Imsande 1983). Seeds used in this study were either germinated aseptically in paper rolls or in the field (Ralston, Imsande 1983). All aseptically grown plants were inoculated at the V-5 stage with Rhizobium japonicum USDA strain 138, whereas all field-germinated plants were inoculated only by soil-borne R. japonicum, the majority of which are USDA strain 123 (Moawad et al., 1984). At beginning bloom (R1) soil-grown plants were carefully removed from the soil (Imsande, Ralston 1982) and, without further inoculation, transferred to hydroponic growth. Nitrogenase activity of all hydroponically grown plants was assayed twice weekly (Imsande, Ralston 1982; Ralston, Imsande 1983).

Several different genetic lines of soybean (Glycine max L. Merr.) were grown either hydroponically in growth chambers or hydroponically in the field, and the temporal profile of nitrogenase activity for each plant was determined. The rate of acetylene reduction of most plants fluctuated from day to day. Also, the nitrogenase activities changed continuously with plant development. Thus, determination of a plant's potential for nitrogen fixation requires periodic assay of nitrogenase activity throughout plant development. High nitrogen-fixation capacity in a soybean plant correlated weakly with early onset of nodulation, whereas for each genetic line a strong positive correlation (r = approximately 0.8) existed between nitrogen-fixation capacity and plant fresh weight. Hence, a temporal partitioning profile (i.e., μmoles of acetylene reduced/h•g plant fresh weight vs. time) and a partitioning index (i.e., total measured μmoles of acetylene reduced/g maximum plant fresh weight) were defined to aid identification of plants with a genetic enhancement for high nitrogen fixation. Using temporal profiles of nitrogenase activity, temporal partitioning profiles, and partitioning indices, we show that several newly selected lines fix nitrogen at a higher rate (P<0.05) than the parental lines when grown hydroponically in the presence of high nitrate (Imsande, In preparation).

REFERENCES

Imsande J, EJ Ralston (1981) Plant Physiol. 68, 1380-1384.
Imsande J, EJ Ralston (1982) Plant Physiol. 69, 745-746.
Moawad HA et al (1984) Appl. Environ. Microbiol. 47, 607-612.
Ralston EJ, J Imsande (1982) J. Exp. Botany 33, 208-214.
Ralston EJ, J Imsande (1983) J. Exp. Botany 34, 1371-1378.

DANISH RHIZOBIUM LEGUMINOSARUM STRAINS NODULATING 'AFGHANISTAN' PEA (PISUM SATIVUM L.)

E. S. JENSEN AND K. C. ENGVILD
AGRICULTURAL RESEARCH DEPARTMENT, RISØ NATIONAL LABORATORY
DK-4000 ROSKILDE, DENMARK.

1. INTRODUCTION

A primitive pea genotype native to Afghanistan usually does not nodulate with Rhizobium leguminosarum strains from Europe, Africa or USA (Lie, 1978), but are nodulated by strains from Middle East locations, e.g. the 'Tom' strain from Turkey (Winarno, Lie, 1979). We found that 'Afghanistan' nodulated early and abundantly in one particular field at Risø, where peas had been grown once 7 years previous to 'Afghanistan'. In another field, which has not been cropped with R. leguminosarum hosts for at least 27 years, 'Afghanistan' was nodulated late and sporadically.

2. MATERIALS AND METHODS

Rhizobium leguminosarum was isolated from nodules of field-grown 'Afghanistan' and tested for nodulation ability in small Leonard jar assemblies on 'Bodil' and 'Afghanistan' pea under aseptic conditions. The R. leguminosarum isoenzyme types were determined as described by Engvild, Nielsen (1985).

3. RESULTS AND CONCLUSION

Isolates from nodules of field-grown 'Afghanistan' peas formed effective nodules abundantly on 'Afghanstian' and 'Bodil' peas on reinfection under aseptic conditions. Five types differing in isoenzyme composition pattern were found among 15 isolates. None were identical with the 'Tom' strain from Turkey. The five types were also different with respect to isoenzyme pattern from twenty R. leguminosarum strains isolated from the modern pea cultivar 'Bodil' in the same field.

The resistance of 'Afghanistan' to nodulation with European R. leguminosarum is not absolute. 'Afghanistan' infecting rhizobia occur sporadically, and their number may be increased by cultivation of R. leguminosarum hosts.

4. REFERENCES

Engvild KC and Nielsen G (1985) Plant and Soil, in press.
Lie TA (1978) Ann. Appl. Biol. 88, 462-465.
Winarno R and Lie TA (1979) Plant and Soil 51, 135-142.

GENETIC IMPROVEMENTS IN PLANT GROWTH AND NITROGEN ASSIMILATION

Yoram Kapulnik, Larry R. Teuber and Donald A. Phillips
Dept. of Agronomy, Univ. of California, Davis, CA 95616

Hairy Peruvian 32 (HP32) alfalfa (Medicago sativa L.) was produced from the original Hairy Peruvian (HP) cultivar by two generations of phenotypic recurrent selection for high forage dry weight and increased forage N concentration under both N_2- and NH_4NO_3-dependent growth conditions (1,2). In 3 years of experiments under glasshouse conditions, HP32 showed consistently higher plant productivity and greater N assimilation than HP under all N regimes (3). The purpose of the current study was to describe the physiological basis of that superior growth potential by comparing rates of C and N assimilation in HP and HP32.

Absolute rates of forage dry matter production and N assimilation were compared for HP32 and HP during the seedling growth period and each of two subsequent regrowth cycles for plants grown with Rhizobium meliloti and 0, 1, 2 or 8 mM 15N-nitrate. N_2 fixation was estimated by the 15N-dilution method. All rates estimated for HP32 were consistently, and generally significantly, greater than for HP in tests over 2 years. Although only above-ground plant material was analyzed during each test, measurements at the end of the second regrowth cycle showed that HP32 had a larger root system with a higher N concentration than HP, so the superior forage production by HP32 was not achieved at the expense of the root or crown.

More detailed studies of the seedling growth period were conducted to separate factors such as plant size, time growth is initiated, and underlying physiological differences which can affect absolute growth rate measurements. Those experiments with totally N_2-dependent plants showed that HP32 did not have a consistently larger relative growth rate or relative N assimilation rate than HP. A $14CO_2$-labeling study during the log phase of seedling growth did not show any difference between HP and HP32 in partitioning of recent photosynthate or in sink strength of the root nodules. Although HP32 had a larger mass of root nodules on each plant than HP, specific C_2H_2-reduction activity in the two populations was similar. The superiority of HP32 was evident from the youngest stage with an earlier development of the first trifoliate leaf and the root nodules. Those early differences surely contributed to the subsequent performance of HP32, but whether they are the primary causal factors cannot be stated.

REFERENCES

1. Phillips DA et al (1982) Crop Sci. 22, 606-610.
2. Teuber LR et al (1984) Crop Sci. 24, 553-558.
3. Phillips DA et al (1985) Crop Sci. 25, In press.

AN EVALUATION OF TROPICAL PHASEOLUS VULGARIS GENOTYPES FROM A BREEDING PROGRAM FOR IMPROVED NITROGEN FIXATION

J. A. Kipe-Nolt and K. E. Giller

CIAT, AA 6713, Cali, Colombia and Rothamsted Expt. St. Harpenden, U.K.

P. vulgaris genotypes (RIZ lines) from a program of crossing and recurrent selection for N_2-fixation (CIAT 1983) were evaluated for their N_2-fixation ability together with early ancestral genotypes. Comparisons were limited to genotypes with similar growth habit and duration. To improve the selection criteria for N_2 fixation breeding the following were determined: nodulation patterns, plant growth and N-uptake curves, and efficiency of carbon use for fixation.

METHODS

In two experiments in adjacent plots, four RIZ lines, five standard check parent genotypes, and a dwarf sorghum variety (to estimate available soil nitrogen) were included as treatments. A mixed strain inoculant was used. In one experiment, plots were randomized within 6 replicate blocks. ^{15}N isotope-labeled fertilizer, (10 kgNha^{-1}) was applied with sucrose 7 days prior to sowing. The central 0.8m^2/plot was harvested at 56 days. In the second trial, plots (7m x 2.5m) were randomized within 4 replicate blocks. Unlabeled fertilizer was added. Harvests were taken at 13 days and subsequently at weekly intervals until 56 days. A grain harvest (7m^2) was taken at 93 days. Using the method of Witty et al. (1983), the efficiency of fixation in 5 genotypes was compared. Plants were grown in sterilized purlite, inoculated with R. phaseoli strain 899, and watered with N-free nutrient solution. Four replicate harvests were taken during flowering.

RESULTS AND DISCUSSION

Field measurements using isotope dilution techniques indicated that P. vulgaris fixed 18-36 kgNha^{-1} (32-47% of plant N) in 56 days. The RIZ lines, had generally greater fixation both in proportion and amount of N than the parental genotypes. The one parental genotype A268, which did fix a large amount of N, was selected under low fertility conditions. There were large differences between genotypes in total shoot N, with the ranking distinct from the ranking for fixation. In most genotypes 30-40% of plant N was in the pods at 56 days. However, there were differences in N partitioning with BAT1297, a poor fixing genotype, having 52% of plant N in the pods. Fertilizer recovery ranged from 7.5% to 10.6% and was correlated with root growth. Nodules formed early (10-15/plant at 13 days) and in most genotypes the number and weight reached maximal values at the 41 day harvest. Low amounts of fixation were found in genotypes slow to form nodules, although correlations between fixation and the nodule parameters measured were not significant (5% level). Substantial variation between genotypes in the efficiency of carbon utilization was observed (2.3 to 3.2 moles CO_2/mole C_2H_4). RIZ30, a good fixer, had more efficient fixation and higher nodule specific activity than BAT76 and BAT1297, both poor fixers. All had similar nodule weights. Two other genotypes had more efficient fixation but nodule weights were very low. The breeding program has resulted in genotypes with improved N_2-fixation potential. However, refinements in the selection criteria are indicated, and should lead to a more effective program.

REFERENCES

CIAT (1983) Annual Report of the bean program, pp 86-90.

Witty J et al (1983) J. Expt. Bot. 34, 951-963.

A HOST GENE IN PISUM SATIVUM L. CONTROLLING NITROGEN FIXATION

T.A. LIE and P.C.J.M. TIMMERMANS
DEPT. MICROBIOLOGY, AGRIC. UN., WAGENINGEN, THE NETHERLANDS

1. INTRODUCTION

Modern pea cultivars show little variation in symbiotic performance with Rhizobium. Recently we observed that the seemingly symbiotic uniformity in cultivated pea lines only pertains to symbiosis with European Rhizobium strains(Lie et. al. 1984). In contrast, with Rhizobium strains from the Middle East symbiotic variability can be observed in the cultivated pea lines (Lie,Göktan 1984).We make use of this variability to study the inheritance of a host gene in a Swedish pea line, controlling nitrogen fixation.

2.MATERIAL AND METHODS

Two modern pea lines from the Netherlands (Rondo) and Sweden (L110) and the primitive Afghan pea were tested with Rhizobium strains from the Netherlands (PF_2), Sweden (Waitor) and Turkey (Tom). The plants were hand crossed to obtain a F_1 and F_2 populations . The symbiotic performance was assayed in a N-free nutrient solution under aseptic conditions at 20°C using Rhizobium strain Tom.

3. RESULTS

Table 1. N_2-fixation of pea cvs. Rondo,L110 and Afghan. in association with Rhizobium strains PF_2, Waitor and Tom,respectively.

Rhizobium	Pea cultivar		
	Rondo	L110	Afghan
PF_2	E	E	-
Waitor	E	E	-
Tom	E	I	E

-,no nodules,I, ineffective and E, effective nodules

Table 2.Segregation of the ability to fix nitrogen in a F_2-population of a cross between pea cvs.L110 and Afghan.,inoculated with strain Tom.

Plants	Nodule Total	Number Red nod. (%)	Non-nodula-ted plants (%)
L110	7	0	38
Afghan.	19	74	0
F_1	29	62	0

F_2-Population Class	Number of individuals	
	Found	Expected
"Afghan"type	81	82
"L110" type	28	27

4. CONCLUSION

The results suggest that a single recessive gene in pea L110 controls nitrogen fixation when in association with Rhizobium strain Tom.

5. REFERENCES

Lie TA et al. (1984) Anton. v. Leeuw. J. Microbiol. 50,489-503
Lie TA, Göktan D (1984) Plant Soil 82,359-367

HOST GENETICS AND PHYSIOLOGICAL STUDIES ON NITRATE INHIBITION OF NODULATION AND NITROGEN FIXATION IN SOYBEAN

B.J. CARROLL/D.L. McNEIL/D. WHITMORE SMITH/P.M. GRESSHOFF
BOTANY DEPARTMENT, AUSTRALIAN NATIONAL UNIVERSITY, P.O. BOX 4,
CANBERRA, AUSTRALIA.

In this paper, the nodulation and nitrate metabolism characteristic of nitrate-tolerant symbiosis (nts) and constitutive nitrate reductase (cNR) mutants are described. In addition, data are reported indicating that oxygen supply to the nodule is a major factor limiting specific nitrogenase (acetylene reduction) activity in symbiotically-stressed wild-type soybeans.

All twelve of the nts mutants (2, 3) thus far characterized nodulated more than the parent cultivar Bragg in the presence and absence of nitrate. Therefore, we also use the term supernodulator to describe these mutant lines. Nitrate did, however, accentuate the difference between the wild type and the nts mutants. Nitrate stimulated growth in the nts lines and the mutants have nitrate reductase (NR) activity. Thus, the nts mutants described here are mutants in the regulation of nodulation, rather than in nitrate metabolism.

Two independent cNR-deficient mutants were isolated using an in vivo NR assay (1). These mutants were designated NR328 and NR345. NR345 is similar to the previously isolated nr_1 (isolated by Nelson et al. (5) and formerly known as LNR-2), whereas NR328 is different. NR328 is leaky and, furthermore, develops necrosis on the leaf margins during culture on nitrate. Lack of cNR activity did not confer nitrate-tolerant nodulation (also see ref. 7).

Physiological studies on the wild type indicated that O_2 supply is a major factor limiting specific nitrogenase activity in nitrate- and dark-inhibited plants (4). It has been reported that O_2 supply also limits activity in water-stressed soybeans (6). Perhaps oxygen limitation is a common regulatory phenomenon controlling nitrogenase activity in stressed soybean plants.

nitrate level[a]	nmol C_2H_4·g nod.FW^{-1}·min^{-1}[b]		
(mM)	20.5% O_2	32% O_2	$LSD_{0.05}$
0	127	138	NS[c]
6	25	71	30

[a] wild-type plants were cultured in the presence or absence of 6mM KNO_3.
[b] acetylene reduction was measured at 20.5 or 32% O_2.   [c] not significant.

1. Carroll B, Gresshoff P (1985) Submitted to Plant Physiol.
2. Carroll B, et al (1985 a) PNAS 82: 4162-4166.
3. Carroll B, et al (1985 b) Plant Physiol. 78: 34-40.
4. Carroll B, et al (1985 c) Submitted to Aust J. Plant Physiol.
5. Nelson R, et al (1983) Plant Physiol. 72: 503-509.
6. Pankhurst C, Sprent J (1975) J. Exp. Bot. 26: 287-304.
7. Ryan S, et al (1983) Plant Physiol. 72: 510-514.

HOST GENES AFFECTING ROOT NODULE FORMATION AND FUNCTION
IN CHICKPEA (Cicer arietinum L.).

THOMAS M. DAVIS
PLANT SCIENCE DEPARTMENT, UNIVERSITY OF NEW HAMPSHIRE
DURHAM, NH   03824-3597

For the purpose of studying host genetic control of
nodule formation and function in chickpea, mutations were
induced in the highly inbred accession line P502 (ICC 640)
by gamma-irradiation of seed, and putative nodulation
mutants were identified as described by Davis et al (1985).
Experimental plants were grown in a greenhouse in N-free
sand or vermiculite, supplied with N-free but otherwise
complete nutrient solution, and inoculated with chickpea
Rhizobium strain CC1192.

Non-nodulating mutants PM233, PM665, and PM679, and
ineffectively nodulating mutants PM405 and PM796 were
easily distinguishable from wild type P502 by their severe
shoot N-deficiency symptoms. PM665 was also distinguished
by a co-segregating complex of aberrant shoot characters
including deformed leaf epidermal trichomes, lighter leaf
color, and slightly crinkled pods.

When root temperature was controlled, effective nodules
were formed by mutants PM665 and PM679 at 24°C, but nodulation was strongly suppressed (PM665) or eliminated (PM679)
at 29°C. Wild type parent P502 was effectively nodulated
at both temperatures but lacked nodules at 34°C. PM233
lacked nodules under all conditions tested. At 24°C root
temperature, ineffective nodulation was manifested by
severe (87%) reduction (PM796) or complete absence (PM405)
of nodule specific acetylene reduction activity. PM405
nodules lacked apparent leghemoglobin pigmentation.

Results of mutant x P502 crosses indicated recessive,
single gene control of aberrant phenotypes in all five
mutants. Mutant x mutant crosses (which constituted genetic
complementation tests) produced normal F_1 plants, and two-
gene segregation ratios further demonstrated absence of
allelism in crosses among PM233, PM665, and PM679, and
between PM405 and PM796. The gene symbols rn1, rn2, rn3,
rn4, and rn5 were provisionally assigned to the operative
loci in these five mutants, respectively.

REFERENCE

Davis, T.M., K.W. Foster, and D.A. Phillips. 1985.
Nodulation mutants in chickpea. Crop Sci. 25:345-348.

REGULATION OF NODULATION AND NITROGEN FIXATION IN NITRATE TOLERANT, SUPERNODULATING SOYBEANS.

Angela C. Delves, David, A. Day, G. Dean Price, Bernard J. Carroll and Peter M. Gresshoff.
Botany Department, Australian National University,
Canberra, A.C.T., 2601, Australia

A number of mutants which supernodulate and are tolerant to nitrate (nitrate tolerant symbiosis - nts) have been isolated from soybean cultivar Bragg (1). One of these (nts382) has been used to investigate the regulation of nodulation.

Experiments in which shoots from nts382 were grafted onto Bragg root stocks, and vice versa, have shown that both supernodulation and nitrate tolerance are determined by shoot factor(s). Bragg roots grafted with nts shoots had many more nodules than wildtype controls and these covered the entire length of the tap root. Increased nodulation occurred even in the presence of applied nitrate. Nts roots grafted with Bragg shoots had fewer nodules, which were restricted to the upper portion of the root, than nts controls.

Acetylene reduction is greater in nts382, on a whole plant basis, but specific activity (acetylene reduced per g nodules) is less than that in wild type plants. The total N content (both per plant and per g plant part) is significantly higher in nts382 than in Bragg.

Uninoculated nts382 plants (grown on nitrate) have significantly greater numbers of lateral roots and a higher shoot/root ratio than Bragg plants grown under identical conditions.

The lower specific nitrogenase activity on nts382 nodules was shown to be due to a smaller bacteroid content (mg bacteroid protein per g nodule); in situ bacteroid activity is the same in nts382 as in Bragg. The haem content of nts382 nodules is also lower than that of Bragg, infected cell size is smaller, and there are fewer bacteroids within each peribacteroid envelope. The nodules of nts382 resemble developing wildtype nodules.

The results show that shoot factor(s) control the number and extent of nodulation and that the supernodulating mutants are developmentally altered in some root parameters.

(1). Carroll, B.J., McNeil, D.L, and Gresshoff, P.M. (1985).
Plant Physiol. 78: 34-40.

NODULATION MUTANTS OF PEA

KJELD C. ENGVILD
RISØ NATIONAL LABORATORY, DK-4000 ROSKILDE, DENMARK.

Nodulation mutants, legumes with changed nitrogen fixation patterns, are now studied in several laboratories. Peas (Jacobsen 1984, Kneen, LaRue 1984, Messager 1985) and chickpeas (Davis et al. 1985) may not nodulate, or nodulated peas may not fix nitrogen (Messager 1985), or peas (Jacobsen 1984) and soybeans (Carroll et al. 1985) may nodulate abundantly at high levels of nitrate.

Pisum sativum L. 'Finale' (Cebeco, Rotterdam) was chosen for mutation studies because of its stable yields and wide adaptation in many European countries. It is a short, determinate, white flowered cultivar with round, green seeds.

The seeds were treated with the mutagens Ethyl methanesulphonate (EMS), diethyl sulphate (DES), ethyl nitroso urea (ENU) or sodium azide at pH 3 (NaN_3) according to the table

Mutagen treatments

dry seeds			presoaked seeds		
EMS	0.1%	16 hours	EMS	0.5%	1 hour
EMS	0.2%	16 hours	EMS	0.5%	2 hours
DES	0.2%	16 hours	EMS	1.0%	1 hour
ENU	0.05%	16 hours	EMS	1.0%	2 hours
ENU	0.025%	16 hours			
NaN_3	2 mM	3 hours	NaN_3	2 mM	2 hours

After washing and treatment with Captan fungicide the seeds were sown wet in the field. One pod was harvested per plant and M_2 seeds sown in sand in the greenhouse after treatment with fungicide. Chlorophyll mutants were counted and removed. Plants showing nitrogen deficiency symptoms, yellowing from the bottom up, were pulled out at the 7-8 leaf stage and examined for nodulation. Prospective mutants were planted in gardeners soil. Mutants were verified in the M_3 by acetylene reduction on whole plants. Among 28000 plants following EMS treatment was found:

> 3 non-nodulating mutants
> 2 non-fixing mutants
> 1 mutant with high acetylene reduction

More than 10 mutants after other treatments await final verification. A mutant which nodulates heavily in the presence of nitrate has also been isolated. Many non-confirmed mutants have been "slow" chlorophyll mutants or plants with nodules destroyed by fungi.

Carroll BJ et al (1985) Plant Physiol. 78, 34-40.
Davis TM et al (1985) Crop Sci. 25, 345-348.
Jacobsen E (1984) Plant Soil 82, 427-438.
Kneen BE, LaRue TA (1984) Heredity 52, 383-389.
Messager A (1985) In "Analysis of the Plant Genes Involved in the Legume-Rhizobium Symbiosis", pp 52-60, OECD, Paris.

GENETICAL AND GRAFTING EXPERIMENTS WITH PEA MUTANTS IN STUDIES ON SYMBIOSIS.

E. JACOBSEN, J.G. POSTMA, H. NIJDAM and W.J. FEENSTRA
Department of Genetics, University of Groningen,
The Netherlands.

For the detection of genetical and physiological factors involved in the symbiosis between <u>Pisum sativum</u> and <u>Rhizobium
leguminosarum</u>, mutants with an altered interaction between
both partners are a good help. In the host only a few of such
mutants (Jacobsen, 1984; Kneen and LaRue, 1984) are available.
To obtain more genetical variation for nodulation characters
in pea, mutants have been selected in M_2-families of EMS-
treated cv Rondo after inoculation with R. leguminosarum
strain PF_2.
a. <u>A mutant stimulated in nodulation</u>. In mutant nod_3 (Jacobsen, 1984), the nodulation is stimulated both in nitrogen-free
and nitrate containing medium; as a pleiotropic effect, the
root system is more intensively branched and the shoot is more
compact. These mutant root characters are the result of a
factor synthesized in the root as grafts with shoots of cv
Rondo showed. This factor also moves into the shoot as the
altered shoot morphology of cv Rondo grafted on nod_3-root-
stocks indicated. All these changes are probably the result
of an altered hormonal balance in the mutant plant.
b. <u>Mutants disturbed in nodulation</u>. A relatively high frequency of potentially nodulation resistant (nod^-) mutants (8
per 1000 M_2-families) has been found. The monogenic and re-
cessively inherited mutants K_5 and K_{24} are two of them. In
mutant K_5, the interaction with <u>Rhizobium</u> bacteria is up to
the infection thread formation normal. Sometimes, this mutant
nodulated delayed, which seems to be dependent on seed age.
Mutant K_{24} was in all experiments nod^-, however, when it was
grown in the presence of nodulating plants swellings on the
root surface appeared. From grafting experiments it appeared
that in the shoot of mutant K_{24} a transposable factor is
formed inhibiting nodulation.
What is the phenotype of a plant in which a nod^- gene has
been combined with the nod_3 gene? The question could be
answered by crossing K_5 x nod_3: the root morphology of the
recombinant resembled, even after inoculation, that of a
nodule-free nod_3 plant. It means that in the recombinant the
nod_3 mutation is expressed but that the expected stimulated
nodulation is suppressed by the mutated gene of K_5. In the F_2-
generation, both genes segregated independently, whereas four
phenotypes appeared: Rondo-, nod_3^-, K_5-types and the recombinant.

REFERENCES

Jacobsen E (1984) Plant and Soil 82, 427-438.
Kneen BE and LaRue TA (1984) J.of Heredity 75, 238-240.

INDUCED MUTANTS OF <u>PISUM</u> <u>SATIVUM</u> DEFECTIVE IN NODULATION

BARBARA E. KNEEN and THOMAS A. LARUE
BOYCE THOMPSON INSTITUTE, ITHACA, NY 14853, U.S.A.

The availability of both bacterial and plant mutants defective in nodulation and fixation is essential to studying the molecular basis of nitrogen fixing symbioses. To obtain host mutants, seeds of <u>Pisum</u> <u>sativum</u> cv. 'Sparkle' were treated with radiation (γ or n) or a chemical mutagen (EMS or nitroso urea). M_2 populations were screened with <u>Rhizobium</u> <u>leguminosarum</u> strain 128C53 for nodulation defective plants, and selections which were stable and fertile to M_3 are being characterized. We found well nodulated ineffective mutants which have little or no nitrogenase (C_2H_2) activity. There are 29 selections with few or no nodules; 4 of these have shortened roots and/or shoots. F_1 and F_2 progeny from reciprocal crosses among mutants are being examined to determine allelism.

Three non-nodulating lines have single gene recessive mutations at unique loci. All the F_1 plants from crosses between the mutants and 'Sparkle' or 'Rondo' nodulate, and segregation of the F_2 progenies approximates 3:1 nod:non-nod. Crosses among selections E69, R25, R72 and <u>sym</u>-2 and <u>sym</u>-5 lines yield F_1 nodulating plants. The new loci are designated <u>sym</u>-7, <u>sym</u>-8, and <u>sym</u>-9. None of these mutants exhibit strain specificity in infectivity tests with nine strains of <u>R.</u> <u>leguminosarum</u> including four Middle Eastern strains. The non-nodulating phenotypes were not temperature dependent in an experiment comparing plants grown at a 25C/20C day/night regime or plants whose roots were kept at 9C, with plants grown at the normal 20C/15C regime. Grafting experiments on E69, and R25 demonstrated that non-nodulation is not affected by substances transported from the shoot.

Six mutants, five obtained by EMS and one by γ radiation, are allelic with the <u>sym</u>-5 gene of our first reported non-nod mutant. The reason for the high frequency of mutation at this locus is unknown. Two mutants, one from EMS and one from γ radiation, are allelic at <u>sym</u>-8.

Two mutants with few nodules and abnormal roots are monogenic recessives, based on crosses with 'Sparkle' or 'Trapper'. R50 has shortened internodes and shortened lower lateral roots. The leaves are pale with raised veins. Very few (0-5) white nodules are found near the crown of the roots. E151 appears normal in shoot growth, but the lower lateral roots are variable in length, averaging approx. 1/2 that of Sparkle. The number of pale pink or white nodules varies (0-30). These low nodulation phenotypes are always associated with shortened roots in segregating F_2 populations.

References
Kneen BE and TA LaRue (1984) Heredity 52,383-389.
Kneen BE and TA LaRue (1984) J. Heredity 75,238-240.

Acknowledgements: Supported in part by USDA-CRGO 83-1-1279.

ISOLATION AND CHARACTERIZATION OF NON-NODULATING MUTANTS OF SOYBEANS

ANNE MATHEWS, BERNARD J. CARROLL, DAVID McNEIL AND PETER M. GRESSHOFF
BOTANY DEPARTMENT, THE AUSTRALIAN NATIONAL UNIVERSITY,
CANBERRA, A.C.T, 2601, AUSTRALIA

Three stable non-nodulating mutants (nod49, nod139 and nod772) were isolated when approximately 25,000 M_2 soybean plants were screened for non-nodulation in the presence of Rhizobium japonicum strain CB1809(=USDA136) (Carroll et al., 1985). This strain effectively nodulated the parent cultivar Bragg. R. japonicum strains USDA110 and USDA123 also failed to nodulate the mutants to any extent.

Plant Genotype	Nodule number per plant[a] Rhizobium japonicum strain		
	CB1809 (=USDA136)	USDA110	USDA123
Bragg	28[b]	55[b]	45[b]
nts382	404[b]	380[b]	339[b]
nod49	1.0[c]	0.5[d]	0
nod772	1.0[e]	2.0[f]	4.0[g]
nod139	4.9[h]	0	0

[a] mean of 3 to 7 plants, unless noted.
[b] 5 out of 5 plants were nodulated
[c] 2 out of 4 plants were nodulated
[d] 1 out of 2 plants was nodulated
[e] single plant tested had 1 small nodule
[f] 3 out of 3 plants were nodulated
[g] 2 out of 3 plants were nodulated
[h] 2 out of 7 plants were nodulated

Of these non-nodulating mutants, nod49 has been characterized in more detail. At standard inoculant doses 23 R. japonicum strains that normally nodulate Bragg failed to nodulate nod49. However, high dose inoculation with some strains resulted in a few nodules being formed on the mutant. Nodules on the non-nodulating mutant nod49 were always located on the root tissue which develops much later during plant growth and never on the upper portion of the tap root. Rhizobium isolated from sparse nodules on nod49 failed to nodulate nod49 to any extent in repeat experiments.

Grafting experiments using Bragg and nod49 indicated that the non-nodulation trait is controlled by the genotype of the root tissue.

Scion genotype (shoot)	Root genotype (root)	Nodule number per plant	Nodule dry weight (mg) per plant
Bragg	Bragg	26	49
nod49	nod49	0	0
nod49	Bragg	23	66
Bragg	nod49	0	0

Reference: Carroll et al., (1985) submitted to Plant Science.

STRUCTURE AND TRANSCRIPTION OF THE SOYBEAN LEGHEMOGLOBIN AND NODULIN GENES

BOJSEN, K., SANDAL, N.N. AND MARCKER, K.A.
DEPARTMENT OF MOLECULAR BIOLOGY AND PLANT PHYSIOLOGY, UNIVERSITY OF AARHUS, C. F. MØLLERS ALLÉ 130, DK-8000 ÅRHUS C, DENMARK

The soybean leghemoglobin (Lb) genes are encoded in the plant genome as a small family of genes which are specifically activated in response to an infection of the root with Rhizobium japonicum. The soybean Lb gene family consists of four functional genes (Lba, Lbc_1, Lbc_2 and Lbc_3), one pseudo gene and at least three truncated genes. About 7-8 days after infection the functional Lb genes are activated in the opposite order to which they are arranged in the soybean genome. About 12 days post infection there is a dramatic increase in the transcription of the Lb genes. In addition to the Lb genes other nodule specific plant genes (nodulins) are activated during development of the nodule. Some of the nodulin genes are also encoded in the soybean genome as small families of genes which are activated at the same time as the amplification of Lb gene transcription.

The association between legumes and Rhizobium results in the development of a nitrogen fixing root nodule. During nodule development several plant and bacterial genes are activated. The most abundant plant gene product within the nodule is leghemoglobin (Lb), a monomeric hemoprotein which is synthesized in root nodules only. Soybean nodules contain four major components of Lbs called Lba, Lbc_1, Lbc_2 and Lbc_3, respectively (Fuchsman, Appleby, 1979). The Lbs are therefore encoded in the soybean genome as a small family of genes (Marcker et al., 1981; Sullivan et al., 1981). The Lbs have a defined function in O_2 transport in nodules whereas the functions of other nodule specific proteins are unknown except for the nodule specific forms of the enzymes uricase (Jochimsen, Rasmussen, 1982) and glutamine synthetase (Cullimore et al., 1983). Several of the nodulin genes are encoded in the plant genome as small families of genes. This paper presents a detailed description of one such family having four members. Some of these genes share a 100% homology over large stretches of their coding sequences, indicating that the genes separated very recently. The activation of the genes present in this family as well as the activation of several other nodule specific genes occur at the same time as the amplification of Lb gene transcription. This may imply that the same regulatory molecule(s) is responsible for both events.

THE CHROMOSOMAL ARRANGEMENT OF THE SOYBEAN Lb GENES

Six Lb genes (four functional, one pseudo and one truncated gene) are arranged in two independent clusters (Fig. 1). Four genes are very closely linked in the order 5' Lba - Lbc_1 - $\psi_1 Lb$ - Lbc_3 - 3' (cluster 1) while two genes are linked in the order 5' - $\psi_2 Lb$ - Lbc_2 - 3' (cluster 2) (Bojsen et al., 1983). The $\psi_2 Lb$ gene is a truncated gene consisting of exon 1, intron 1, exon 2, intron 2 and a part of exon 3. The distances between the genes in both clusters are about 2-3 kb.

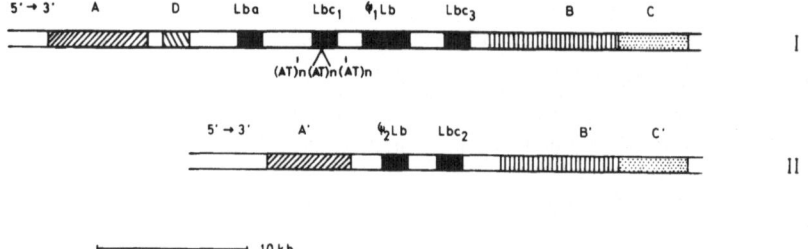

<u>Figure 1</u>. Chromosomal arrangement of six soybean Lb genes. Solid boxes indicate the position of the Lb genes (the three introns present in all Lb genes are not shown). Identical shadings (A and A', B and B', C and C') represent cross hybridizing regions containing non Lb-genes. Region D contains a non-Lb gene. The transcription polarity of the Lb genes is indicated by the two arrows. $(AT)_n$ indicate the positions of poly $(AT)_n$ sequences (about 25 AT pairs each) in cluster I. The poly $(AT)_n$ sequence in Lbc_1 is located in the first intron.

Homologous non-Lb genes which are not nodule specific are closely linked to each Lb gene cluster in corresponding positions. These genes are regulated in a way which differs from that of the Lb genes. This implies that the sequences responsible for Lb gene activation most likely are located close to the structural Lb sequences. Curiously, poly$(AT)_n$ sequences (about 25 AT pairs each) are located in three positions in cluster 1 (Fig. 1). Other such sequences may be present within the two Lb gene loci in addition to other locations within the soybean genome. In this connection it is interesting that such poly$(AT)_n$ sequences where n is 17 or more have a tendency to form cruciform structures (Haniford, Pulleyblank, 1985). Such a transition may serve as a regulatory switch, but at present we have no evidence that this is the case for the AT repeats present in one of the soybean Lb loci.

TRANSCRIPTION OF SOYBEAN Lb GENES DURING NODULE DEVELOPMENT

The rate of appearance of the various Lb proteins during nodule development suggests that the Lb genes are activated at different times during the early stages of nodule development (Fuchsman, Appleby, 1979; Verma et al., 1979). In order to investigate whether this is so, specific probes for each Lb gene was constructed. Northern blotting analysis using RNA extracted from nodules in various stages of development and Lb gene specific probes allowed the detection of the various Lb gene transcriptional products (Marcker et al., 1984). In this way it was demonstrated that Lbc_3 transcriptional products are detected before the corresponding Lba RNAs. Similarly the Lbc_1 and Lbc_2 genes are transcribed before the Lba gene, but the Lbc_2 gene is less well transcribed than the other Lb genes. No transcriptional products could be detected from the $\psi_1 Lb$ gene, indicating that this gene is a pseudogene.

A more detailed analysis of the activation of the Lb genes during nodule development revealed that these genes exhibit an exponential activation curve. Thus, about 7-8 days after infection the Lb genes are activated sequentially during a period of a few days in the order Lbc_3 (Lbc_2) – Lbc_1 – Lba. However, about twelve days post infection there is a dramatic increase in the transcription of the Lbc_1, Lbc_3 and Lba genes, while the transcription of the Lbc_2 gene is not amplified to a similar extent.

Finally, all the Lb genes are active for a long period during the life-time of the nodule. Consequently, the soybean Lb genes are not regulated by a developmental gene switching mechanism as is the case for vertebrate globin gene regulation.

NODULIN GENES

Several other nodule specific plant genes are activated during development of the nodule. Recently we have isolated fourteen different nodulin cDNA clones including a cDNA clone which contain nodule specific uricase sequences. The nodule specific mRNAs corresponding to the isolated cDNA clones vary in size from about 400 bases to about 1700 bases. A number of cDNA clones corresponds to more than one nodule specific mRNA suggesting that some nodulins are encoded in the soybean genome as small families of genes. One such gene family was chosen for further study. One nodulin

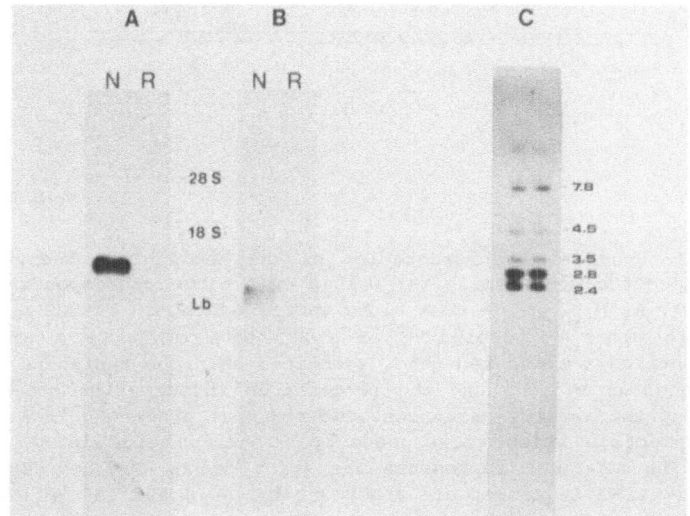

Figure 2. Northern blotting analysis of polyA$^+$ RNA extracted from nodules (N) harvested 17 days post infection and polyA$^+$ RNA extracted from uninfected roots (R). **A:** Hybridization with the N_{12} cDNA clone. **B:** Hybridization with the $N_{(10,8)}$ cDNA clone. **C:** Southern blotting analysis of soybean DNA digested with the restriction endonuclease EcoRI. Hybridization was with a M13 probe which contains a part of the homologous sequence. 28S, 18S and Lb in A and B refer to the positions of the two ribosomal RNAs and Lb mRNA respectively. C numbers refer to the sizes of the hybridizing fragments.

cDNA clone ($N_{10,8}$) corresponds to two mRNAs with sizes of approximately 1000 and 800 bases, respectively, while another cDNA clone (N_{12}) corresponds to a mRNA with a size of about 1200 bases (Fig. 2). Subsequent DNA sequence analysis revealed a substantial homology between clones N_{12} and $N_{(10,8)}$ which indicates that the two clones correspond to transcripts belonging to the same family of genes. Subsequent Southern blotting analysis with a probe containing a homologous sequence revealed the presence of five hybridizing EcoRI DNA fragments in soybean DNA (Fig. 2). Several genomic clones were isolated and subjected to DNA sequence analysis. The results so far obtained are shown in Fig. 3. There are no introns in the genomic clones sequenced so far. One genomic sequence corresponds to the mRNA which is 800 bases long. Closely linked to the 800 bp gene is a truncated gene with a sequence which corresponds precisely to the corresponding sequence present in the former gene. This homology also extends 160 bases upstream from the initiating ATG codon, after which the homology is about 85% in the sequences determined so far.

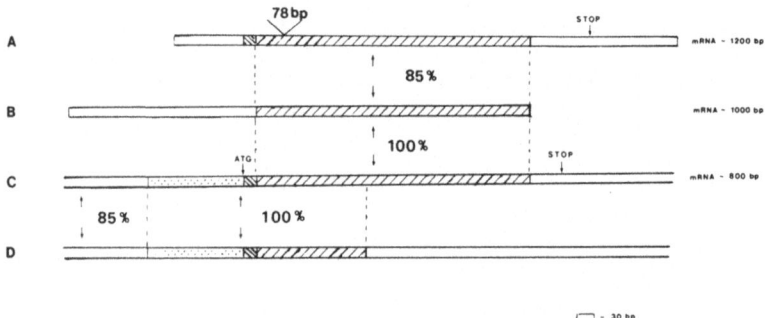

Figure 3. Schematic representation of the homologies observed within a nodule specific gene family. A: N_{12}, a cDNA clone corresponding to a 1200 base mRNA. B: $N_{(10,8)}$, a cDNA clone corresponding to a 1000 base mRNA. C: A genomic clone corresponding to a 800 base mRNA. D: A genomic clone which contains a truncated gene. Identical shadings represent corresponding homologous regions and the percentages indicate the degree of homology among the various sequences. The sequence presented in A contains an 78 bp insertion which corresponds to 26 amino acids in the coding sequence. The ATG in C represents the start codon. The corresponding ATGs in the two cDNA sequences are about 200 bp (N_{12}) and 100 bp (N_{10}) further upstream, from the known sequences.

Recent evidence now suggests that the truncated gene is in fact part of a functional gene, and the reason for its isolation as only part of a gene may be due to a recombinational event during phage propagation. It is therefore quite possible that soybean DNA contains two identical copies of this gene which are about 2 kb apart. This implies that the two genes separated rather recently. The homology between the 800 bp gene and the cDNA corresponding to the 1000 bp cDNA is also 100% in a continous coding sequence of about 500 bp, while no homology is observed close to the initiating ATG in the smaller gene when compared to the corresponding region in the larger gene. The sequence 3' to the observed homology in the 1000 bp gene has not been determined yet, so it is not clear how far further downstream this remarkable homology continues. Nevertheless, the close similarity between these two types of genes also suggests that their separation occurred quite recently. Finally the cDNA corresponding to the 1200 nucleotide long mRNA is about 85% homologous to the other related sequences apart from a 78 bp insertion in the 5' region of this gene. However, about 100 bases 5' to the stop codon, homology with the other genes is no longer apparent.

All the nodulin specific mRNAs investigated are not detected in the early stages of nodulin development but accumulate rapidly from about 12 days post infection. This also includes the appearance of uricase mRNA within the nodule. Thus these nodulin genes are activated at the same time as the transcription of the Lb genes is increased implying that the same mechanism is responsible for both events. For this reason the Lb genes and some of the nodulin genes may share common regulatory sequences in the 5' flanking regions.

In a comparison of the 5' flanking region of the Lbc$_3$ gene with the corresponding regions of two nodulin genes several consensus sequences which may serve such a function were proposed (Mauro et al., 1985). However, extension of this analysis with the 5' flanking regions of the other functional Lb genes and the corresponding regions in the nodulin genes sequenced here indicates that the proposed consensus sequences are not valid. The types of nodulin genes investigated might be regulated differently, but the Lb genes are most likely regulated in a similar way. However, the Lb genes are very homologous in this region, so it is impossible to single out any sequence which may be important for the regulation of these genes. We have noted a sequence which is present in the 5' flanking regions of all the Lb genes as well as in all the nodulin genes examined. However, this sequence consists of only seven nucleotides and is located at varying positions in front of the initiating ATG codons. The statistical significance of the occurrence of this sequence in these genes is therefore uncertain, until its presence has been established in an appropriate position in other nodulin genes.

ACKNOWLEDGEMENTS

We thank Drs.R. Goldberg and R. Fischer, UCLA, for providing the limited EcoRI and AluI/HaeIII soybean libraries. This research was supported financially by the EEC contract GBI-4-025-DK and the Danish State Biomolecular Engineering Programme.

REFERENCES

Bojsen K, Abildsten D, Jensen EØ, Paludan K and Marcker KA (1983) EMBO J. 2, 1165-1168.
Cullimore JV, Lara M, Lea PJ and Miflin BJ (1983) Planta 157, 245-253.
Fuchsman WH and Appleby CA (1979) Biochim. Biophys. Acta 579, 314-324.
Haniford DB and Pulleyblank DE (1985) Nucleic Acids Res. 13, 4343-4363.
Jochimsen B and Rasmussen O (1982) In Abstracts from the 1st Internatinal symposium in Molecular Genetics of the Bacteria-Plant Interaction, p. 83.

Marcker A, Lund M, Jensen EØ and Marcker KA (1984) EMBO J. 3, 1691-1695.

Marcker KA, Gausing K, Jochimsen B, Jørgensen P, Paludan K and Truelsen E (1981) In Panopoulos NJ, ed., Genetic Engineering in the Plant Sciences, pp. 63-71. Praeger Publishers, New York.

Mauro VP, Nguyen T, Katinakis P and Verma DPS (1985) Nucleic Acids Res. 13, 239-249.

Sullivan D, Brisson N, Goodchild B, Verma DPS and Thomas DY (1981) Nature 289, 516-518.

Verma DPS, Ball S, Guérin C and Wanamaker L (1979) Biochemistry 18, 476-483.

NODULIN GENE EXPRESSION IN PISUM SATIVUM.

Ton Bisseling, Henk Franssen, Francine Govers, Ton Gloudemans, Jeanine
Louwerse, Marja Moerman, Jan-Peter Nap and Albert van Kammen.
Department of Molecular Biology, Agricultural University, Wageningen,
The Netherlands.

INTRODUCTION

The symbiotic association of rhizobia and leguminous plants leads to the
formation of specialized root nodules in which both the plant and bacterial cells are highly differentiated. Compared with other plant differentiation processes root nodule formation does not appear a very
complex process. Goldberg et al. (Kamalay, Goldberg, 1980, 1984;
Goldberg et al., 1981) using hybridization techniques for analyzing
messenger RNA populations arrived at the conclusion that 15-25 thousand
genes are involved in the formation of a leaf, a stem or a root. 15-40%
of which are organ-specific. Verma et al. (Auger, Verma, 1981; Verma,
1982) at the other hand showed by similar hybridization experiments that
less than 100 different genes are specifically involved in root nodule
formation and maintenance.
Since the formation of nodules on roots of leguminous plants is the
result of a joint effort of two organisms and is closely regulated in
time and space by Rhizobium as well as the host plant, the involvement
of the bacterial genome would seem to increase the complexity of the
system. In fact the regulatory role of Rhizobium in root nodule formation offers unique possibilities for analyzing the plant differentiation process. Using the attainments of Rhizobium genetics it proves
possible to arrest root nodule formation at different stages of development which facilitates the analysis of the successive steps in the differentiation process of the plant.
In addition to the interesting aspects of symbiotic nitrogen fixation
the formation of a root nodule also represents a very attractive system
for studying plant differentiation by virtue of its relative simplicity
and the possibilities for experimental approach.

In this paper we will focuss on the identification and time-course of
expression of nodulin genes during nodule development, the specificity
of induction of nodulin gene expression and the involvement of Rhizobium
signals in eliciting nodulin gene expression.

Identification of nodulin mRNAs.

Since only small quantities of RNA (< 10 µg/100 plants) could be
obtained from some nodules with a defect in their development, a sensitive method for analyzing small quantities of RNA is essential. In
vitro translation followed by two-dimensional (2-D) gel electrophoresis

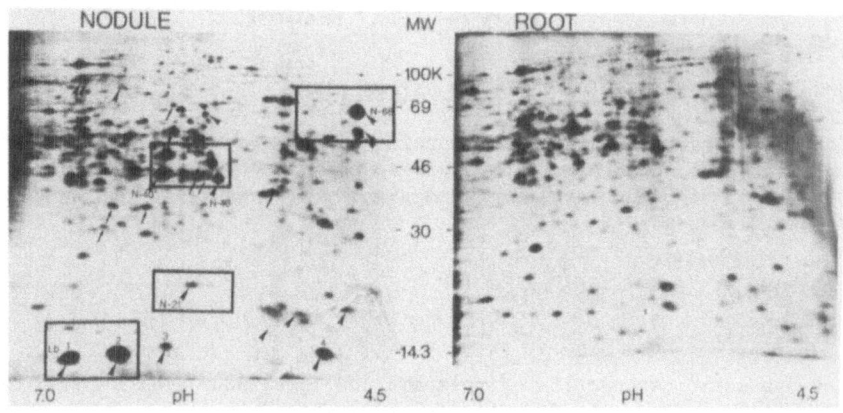

Fig. 1 Identification of nodulin mRNAs. Fluorographs of 2-D gels of in
vitro translation products from total RNA isolated from (A) wild
type effective pea nodules (15 d) and (B) 3-day-old uninfected
pea roots. Nodulin spots are indicated by   nodule stimulated
spots by →.

of the translation products appeared a good method to examine nodulin
gene expression, as it only required minute amounts of RNA (1-2 µg) and
still revealed the large diversity of messenger RNAs.

For identification of nodulin mRNAs (Van Kammen, 1984) total RNA was
isolated from pea root nodules harvested 15 days after sowing and inoculation and then translated in reticulocyte lysate. The translation products were separated by 2-D gel electrophoresis (Fig. 1). The RNA
concentration used for these in vitro translations was chosen in the
range of a linear relationship between the amount of RNA added and the
incorporated radioactivity (^{35}S-methionine). The intensity of a particular spot will therefore be directly proportional to the relative
amount of mRNA present for that particular polypeptide. Using this
method ~ 500 polypeptides could be identified in a reproducible manner.
No detectable polypeptides were synthesized when bacteroid RNA was
translated in the same eukaryotic translation system, indicating that
all in vitro translation products derived from total nodule RNA and
visualized by this method are plant encoded. This was confirmed by analyzing the translation products of nodule poly(A)+ RNA, which produced
an identical pattern of translation products. Comparison of the 2-D pattern of nodule polypeptides with those obtained after translation of
mRNA from 3- and 8-day-old uninfected pea roots showed the majority of
proteins to be present in both roots and nodules (Fig. 1). However, 21
spots are only observed in the pattern of polypeptides obtained after
translation of nodule mRNA and therefore represent the products of nodulin mRNAs. Four of the nodulin spots are leghemoglobins, as shown by
immunoprecipitation of the in vitro translation products with anti-
leghemoglobin serum, and are indicated as Lb-1, Lb-2, Lb-3 and Lb-4

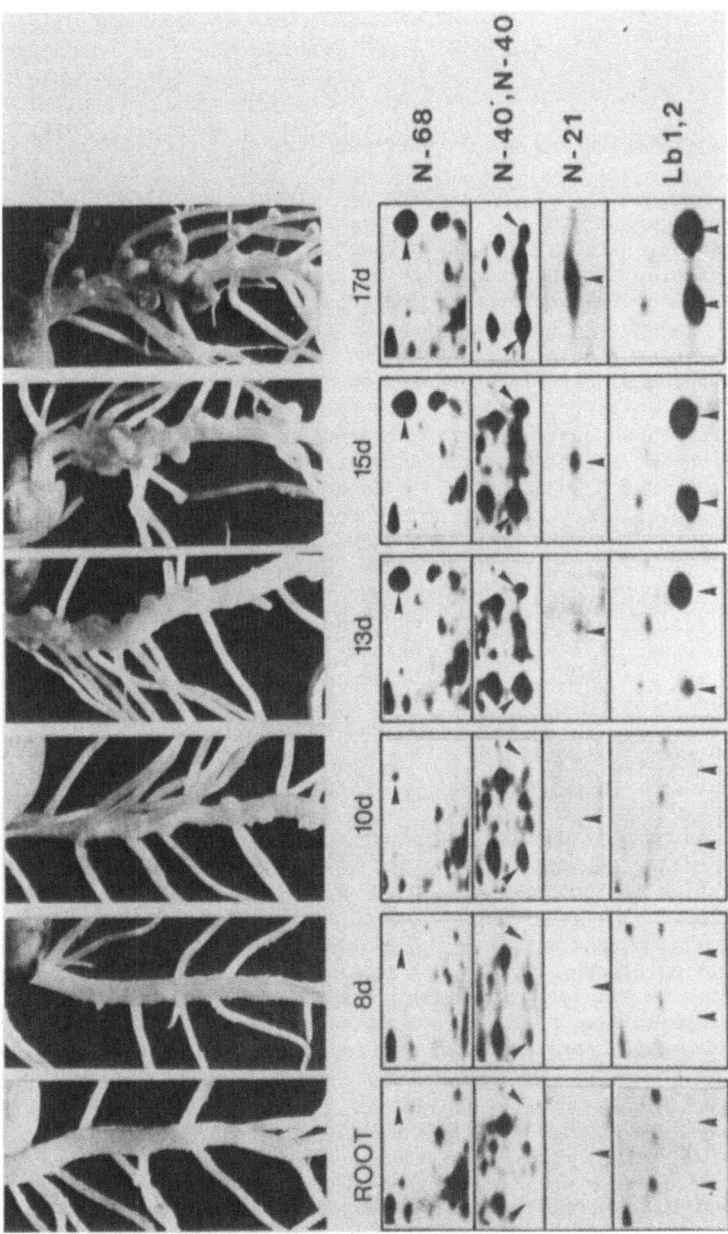

Fig. 2 Expression of nodulin genes during development of nitrogen fixing
pea nodules formed by <u>Rhizobium</u> <u>leguminosarum</u>. Fluorographs of <u>in
vitro</u> translation products from total RNA isolated from uninfected pea roots (8 d) and from root nodules, 8, 10, 13, 15 and
17 days after sowing and inoculation. Only the parts of the gels
within the squares indicated in Fig. 1 are shown. The parts of
the pea main root at different stages of development (as indicated) are shown at the top of the figure.

(Govers et al., 1985). Other major nodulins have apparent mol. wts. of 68,000, 40,000 and 21,000 and are indicated as N-68, N-40, N-40' and N-21 respectively.

Nodulin mRNAs during nodule development

Under our conditions for growing pea plants the nodules produced by R.leguminosarum are confined to a restricted part of the main root. The first nodule-like structures become macroscopically visible 10 days after sowing and inoculation (Fig. 2). At day 12 leghemoglobin is detectable and one day later nitrogenase activity starts to appear (Bisseling et al., 1980). To study the time-course of expression of the nodulin genes during nodule development, we collected the 2.5 cm pieces of root where nodules normally emerge at day 7, 8 and 10 after sowing and inoculation, whereas root nodules were harvested at day 13, 15 and 17. Total RNA from these samples was translated and the products analyzed on 2-D gels, and we directed our attention especially to the major nodulin spots. At day 7 no nodulins are found, but in 8 day old infected tissue N-40' appears as the first nodule specific translation product (Fig. 2). N-40' mRNA increases in amount and produces a rather intense spot at day 10; it increases further until day 15, after which it remains at constant level. N-68 is visible as a minor spot at day 10 and becomes a major spot at day 13, so its mRNA concentration increases drastically during these 3 days. During the next 2 days N-68 further increases only slightly. N-40 and the four Lbs are not yet translated from RNA of 10 day old infected roots, but appear among the translation products of RNA nodules at day 13 (Fig. 2, Lb-3 and Lb-4 not shown). Finally, N-21 is hardly visible in the 2-D pattern from 13-day-old nodules; it is a minor spot at day 15 and becomes a major spot at day 17 (Fig. 2). Therefore the rapid increase in concentration of N-21 mRNA occurs at least 2 days later in comparison with the other nodulin mRNAs. More detailed analyses of 2-D-patterns showed that most nodulin genes follow the time-course of expression of the Lb genes. So far N-40 is the only so-called "early nodulin".
Studies with cDNA clones of nodulin mRNA confirm the results described in this paragraph. On Northern blots the majority of the nodulin-cDNA clones give similar hybridization patterns, if hybridized with nodule RNA isolated from different stages of development indicating that the nodulin mRNAs, including the Lb messengers first appear in nodules 13 days after sowing. Only with a few cDNA clones nodulin mRNA has been detected several days preceding the appearance of Lb mRNA (Franssen et al. and Govers et al., in preparation).

Is nodulin gene expression only induced by Rhizobium ?

The mechanisms by which the expression of nodulin genes are regulated are still unknown. Although there is no doubt that Rhizobium infection induces nodule formation and nodulin gene expression, the nature and the specificity of the signals involved is fully unknown.

It has been suggested that the physiological conditions inside a root nodule may be responsible for the induction of the synthesis of (at least some) nodulins. De Vries et al. (1980) proposed for example that

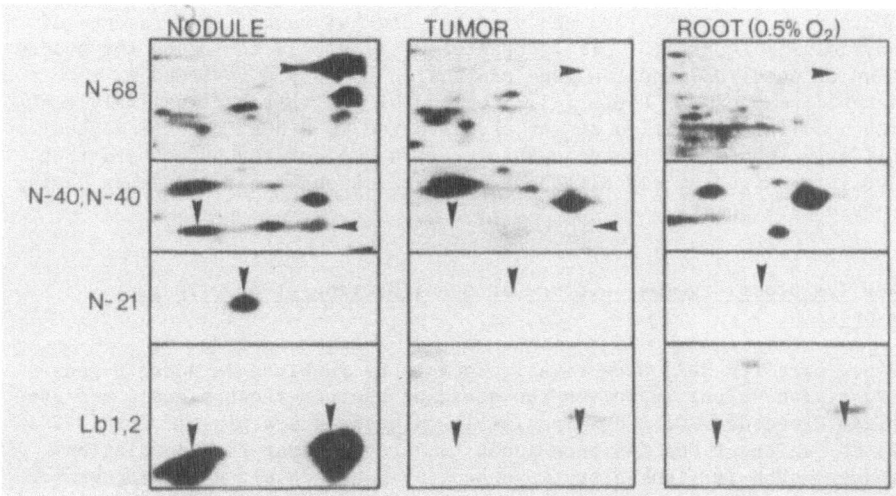

Fig. 3 Expression of nodulin genes in pea tumors and pea roots, exposed
to microaerobic conditions. Fluorographs of in vitro translation
products from total RNA isolated from pea nodules (15 d), pea
tumors formed by Agrobacterium tumefaciens and uninfected roots
exposed to 0.5% O_2.

the low O_2 tension in root nodules caused the expression of some nodule
specific genes. To study the possible role of the O_2 concentration in
nodulin gene expression we analysed RNA obtained from uninoculated pea
roots that were exposed for 40 hours to varying O_2 concentrations
(0.5%, 20% and 1.6%) for the presence of nodulin mRNAs. No nodulin mRNAs
could be detected in any of these microaerobic grown roots (Fig. 3).
Therefore it can be excluded that a low oxygen tension is, if at all,
the only cause of nodulin gene expression.

As a control we examined the expression of the alcoholdehydrogenase
(ADH) gene which is known to be very sensitive to the O_2 concentration.
Using a maize ADH cDNA clone (Gerlach et al., 1982) as a probe we could
not observe any increase of this gene in root nodules whereas there was
a markedly higher expression of the ADH gene in microaerobically grown
roots. Apparently the free O_2 level in developing root nodules is not
low enough to enhance ADH gene expression. Probably due to the presence
of Lb the situation in root nodules is not comparable to microaerobic
conditions.

Other authors are inclined to point out the similarities between the
Rhizobium/legume-interaction and other plant/bacterium- or
plant/pathogen-interactions. They view upon nodulation as a "beneficial
plant disease" (Vance, 1983). From this point of view it might be argued
that at least some nodulins could be involved in the host defence
reaction. If nodulins are really associated with existing plant defence
mechanisms, this must implicate that other bacteria would just as well
be able to induce the expression of nodulin genes. To test this hypothe-

sis, we have looked for nodulin mRNAs in tumors formed on pea by
Agrobacterium tumefaciens. Tumors were induced by wounding the stem of
14 day old pea plants with Agrobacterium. The in vitro translation products from tumor RNA were analyzed but did not reveal the presence of
any nodulin transcript. If Agrobacterium is unable to induce the production of nodulins, nodulin gene expression cannot be regarded as a
general response to bacterial infection. It is therefore unlikely that
the nodulins identified so far are involved in a host defence mechanism.
All experiments described in this section support the hypothesis that
specific Rhizobium signals cause the expression of nodulin genes during
nodule development.

Are sym-plasmid genes involved in the induction of nodulin gene expression ?

Since specific Rhizobium signals seem to be involved in nodulin gene
expression we put ourselves the question whether these signals are sym-
plasmid encoded. On the Rhizobium leguminosarum sym-plasmid (pRL1JI) a
45 Kb region of DNA has been shown to carry a cluster of nodulation
genes, which is flanked by two groups of genes involved in nitrogen
fixation (Downie et al., 1983). Downie et al. have cloned the entire
sym-plasmid pRL1JI into a cosmid library. Two overlapping clones,
pJI1089 and pJI1085 with a 10 Kb region in common, are upon transfer
into a cured R.phaseoli strain able to form on pea roots ineffective
nodules in which the rhizobia are released from the infection thread and
surrounded by a peribacteroid membrane. In the absence of the 10 Kb
region no nodules can be formed and by deduction Downie et al. concluded
that the 10 Kb region is the only part of the sym-plasmid that is essential for nodule formation. We used both R.phaseoli/pJI1089 and
R.phaseoli/pJI1085 to determine whether these strains can induce
expression of nodulin genes. For that purpose total RNA isolated from
these nodules was translated in a reticulocyte lysate and the translation products were separated by 2-D gel electrophoresis. The pattern of
nodulin spots from pea root nodules formed by R.phaseoli/pJI1089 and
R.phaseoli/pJI1085 was similar to that produced by wild type
R.leguminosarum, although N-21 is only detectable at strongly reduced
levels. In previously published experiments using Rhizobium mutants
which produce ineffective (fix⁻) root nodules N-21 mRNA was always below
the detection level and it was suggested that the expression of the N-21
gene was controlled by the nitrogen fixation process (Govers et al.,
1985). However, the present results show that this conclusion is not
valid.
Since the expression of all nodulin genes is induced in the nodules
formed by R.phaseoli/pIJ1089 and R.phaseoli/pIJ1085 it can be concluded
that the 10 Kb region of the sym-plasmid with the nod-region in the
R.phaseoli chromosomal background is sufficient to induce nodulin gene
expression. The nif- and fix-genes as well as other sym-plasmid genes
are apparently not involved in the induction of expression of these host
genes.

Several groups have described that empty root nodules are formed on
alfalfa roots by Agrobacterium tumefaciens carrying clones with the nod
region of Rhizobium meliloti (Truchet et al., 1984; Hirsch et al.,
1985). We have found that this can occur in pea roots as well.
Inoculation of pea roots with an Agrobacterium tumefaciens transcon

jugant (LBA2712) (Hooykaas et al., 1982) carrying a Rhizobium leguminosarum sym-plasmid (pSym1) resulted in the development of small ineffective nodules. Histological examinations of nodules formed by the
Agrobacterium tumefaciens transconjugants showed that these nodules have
several vascular bundles at the periphery of the nodule which is characteristically found in normal nodules. The plant cells are however devoid
of bacteria, although infection threads can be observed inside the
nodule.

We investigated whether nodulin genes are expressed in these empty pea
nodules formed by the Agrobacterium transconjugant, and especially examined the expression of the early nodulin (N-40') gene, since the mRNA
of this gene can be detected before nodules, formed by Rhizobium, are
macroscopically visible. In root nodules formed by the Agrobacterium
transconjugant neither the N-40' mRNA nor the other nodulin mRNAs were
detected. Therefore the formation of a nodule structure appears to be
independent of the presence of the early nodulin as well as other nodulins, identified with the method used. Furthermore, it can be concluded
that the sym-plasmid in the A.tumefaciens background is not sufficient
to induce nodulin gene expression in the host; so chromosomal genes of
Rhizobium appear to be involved in the induction of expression of the
nodulin genes.

REFERENCES

1.   Auger, S. and Verma, D.P.S. (1981). Biochem. 20, 1300-1306.
2.   Bisseling, T., Moen, A.A., Van den Bos, R.C. and Van Kammen, A.
     (1980). J.Gen.Microbiol. 118, 377-381.
3.   De Vries, G.E., In 't Veld, P. and Kijne, J.W. (1980). Plant Sci.
     Lett. 20, 115-123.
4.   Downie, J.A., Ma, Q-S., Knight, C.D., Hombrecher, G. and Johnston,
     A.W.B. (1983). EMBO J. 2, 947-952.
5.   Gerlach, W.L., Pryor, A.J., Dennis, E.S., Ferl, R.J., Sachs, M.M.
     and Peacock, W.J. (1982). Proc.Natl.Acad.Sci.USA 79, 2981-2985.
6.   Goldberg, R.B., Hoschek, G., Tam, S.H., Ditta, G.S., and
     Breidenbach, R.W. (1981). Develop. Biol. 83, 201-217.
7.   Govers, F., Gloudemans, T., Moerman, M., Van Kammen, A. and
     Bisseling, T. (1985). EMBO J. 4, 861-867.
8.   Hirsch, A.M., Drake, D., Jacobs, T.W., and Long, S.R. (1985). J.
     Bacteriol. 161, 223-230.
9.   Hooykaas, P.J.J., Snijdewint, F.G.M. and Schilperoort, R.A. (1982).
     Plasmid 8, 73-82.
10. Kamalay, J.C. and Goldberg, R.B. (1980). Cell 19, 935-946.
11. Kamalay, J.C. and Goldberg, R.B. (1984). Proc.Natl.Acad.Sci.USA,
     81, 2801-2805.
12. Truchet, G., Rosenberg, C., Vasse, J., Julliot, J.S., Camut, S. and
     Denarie, J. (1984). J. Bacteriol. 157, 134-142.
13. Vance, C.P. (1983). Ann.Rev.Microbiol. 37, 399-424.
14. Van Kammen, A. (1984). In: Veeger, C. and Newton W.E. (eds.).
     Advances in Nitrogen Fixation Research, Nijhoff, The Hague, pp.
     587-588.
15. Verma, D.P.S. (1982). In: Smith, H. and Grierson, D. (eds.). The
     Molec. Biol. of Plant Devel., Bot. Mon., Vol. 18, Blackwell Sc.
     Pub. Oxford pp. 437-466.

RHIZOBIUM AND AGROBACTERIUM STRAINS INDUCE THE SAME NODULINS WITH DIFFERENT KINETICS IN PHASEOLUS VULGARIS

VICTOR CONDE, JAIME PADILLA, ESPERANZA MARTINEZ, MIGUEL LARA AND FEDERICO SANCHEZ.
CENTRO DE INVESTIGACION SOBRE FIJACION DE NITROGENO, UNIVERSIDAD NACIONAL AUTONOMA DE MEXICO. APARTADO POSTAL 565-A, CUERNAVACA, MORELOS. MEXICO.

Phaseolus vulgaris can form nodules and fix nitrogen in association with Rhizobium phaseoli as well as with various other tropical Rhizobium spp. Some of these promiscuous strains were compared to two different R. phaseoli wild type strains with respect to the nodulin pattern they induce on the host plant, P. vulgaris. One symbiotic mutant of R. phaseoli and two Agrobacterium tumefaciens transconjungants, harbouring different R. phaseoli Sym plasmids, were included in the same analysis, carried out by SDS-PAGE. Nitrogenase, glutamine synthetase and uricase activities were determined and the levels of the last two enzymes were immunodetected by specific antisera in the same days. All effective strains induce the same nodulin pattern on a qualitative point of view, although the kinetic of expression is delayed in the case of the Agrobacteria and in one Rhizobium spp. induced nodules. The analysis of comparative patterns of soluble and bacteroidal protein, reveals the existence of common polypeptides. This fact demonstrates their bacteroidal origin and the variable expression depending on the strain. Nevertheless, the very conserved nodulin pattern obtained in this study suggests that the induction signals for the establishment and maintenance of the symbiotic association are conserved, whatever the eliciting strain.

EXPRESSION OF NODULE SPECIFIC SEQUENCES DURING PEA ROOT NODULE DEVELOPMENT

FRANCINE GOVERS/MARJA MOERMAN/JAN-PETER NAP/ALBERT VAN KAMMEN/TON BISSELING
DEPARTMENT OF MOLECULAR BIOLOGY, AGRICULTURAL UNIVERSITY, DE DREYEN 11
6703 BC WAGENINGEN, THE NETHERLANDS

During the development of root nodules on leguminous plants several plant genes, the so-called nodulin genes are specifically expressed. Among these are the leghemoglobin genes. Previously we have shown that analysis on two dimensional gels of in vitro translation products from RNA isolated from pea root nodules leads to the identification of 21 nodule-specific mRNAs (Govers et al. 1985). Our aim is to determine the regulation of expression of the nodulin genes. We assume that specific signals from Rhizobium are involved in this process. This hypothesis is supported by our studies on gene expression in roots that have been exposed to microaerobic conditions and tumors formed by Agrobacterium tumefaciens.

The RNA population from pea root nodules was compared with that from tumors on pea induced by Agrobacterium tumefaciens. In vitro translation products from tumor RNA were analysed on 2-D gels and no nodulin transcripts were found. This suggests that nodulin gene expression is specifically elicited by Rhizobium and not as a general response upon a bacterial infection.

To study whether the microaerobic environment in the root nodule leads to expression of specific genes we analysed the RNA from uninfected roots grown under microaerobic conditions (provided by J. Hooymans, University of Leiden). No nodulin transcripts were detected so the microaerobic environment in root nodules is not inducing nodulin gene expression. Furthermore we did not observe an increase of expression of the alcohol-dehydrogenase gene(s) in pea root nodules, whereas this gene is expressed at a higher level in microaerobically grown roots. This indicates that the supply of oxygen is not limited in the root nodule and apparently leghemoglobin is an efficient oxygen carrier for the plant cell as well as for the bacteroid.

Useful tools to study the expression of nodulin genes are nodule specific cDNA clones. We selected clones from a cDNA library in pBR322. This library was constructed from poly A$^+$ RNA isolated from 15 day old pea root nodules. Five different cDNA clones hybridize to mRNA ranging in size from 440 to 800 bases. One clone pPsNod10 hybridizes to 3 different mRNAs. During nodule development the genes corresponding to the selected clones show the same pattern of expression as the leghemoglobin genes except one, pPsNod13, which mRNA appears 2 days later. In vitro translation studies have shown that during nodule development at least one nodulin gene is expressed before the leghemoglobin genes (Govers et al. 1985). Therefore search for cDNA clones from genes that are expressed in earlier developmental stages is in progress.

Govers F et al (1985) EMBO J. 4, 861-867.

ANALYSIS OF LEGHEMOGLOBIN AND NODULIN C-DNA CLONES OF VICIA FABA

J. Kuhse, S. Mohapatra and A. Pühler
Universität Bielefeld, D-4800 Bielefeld 1, FRG

Introduction

The development of the nitrogen fixing root nodules by legumes involves specific synthesis of a group of plant proteins, called nodulins. To investigate Vicia faba nodulins in our laboratory a c-DNA library was constructed. This report focuses on molecular analysis of c-DNA clones coding for leghemoglobinsand two other recently detected nodulin clones.

Results and Discussion

A c-DNA library comprising of about 2600 clones was constructed from nodule polyA+ RNA. By screening 200 clones with pea Lb clone pPSL101 (kindly provided by T. Bisseling) a clone pCK277 was identified. Northern blot analysis indicated this clone to be nodule specific. Hybrid released translation experiments showed that the family of the Vicia faba leghemoglobins includes at least two different forms of different molecular weights. Using pCK277 as a probe five more leghemoglobin clones were isolated. Four of these clones namely pCK277,1957,1947 and pCK1349 were sequenced using Maxam-Gilbert procedure. The former three clones possess identical sequences thus represent one gene, while pCK1349 represents a different leghemoglobin gene.

The DNA sequence comparison of pCK277 and pCK1349 with the coding sequence of the Lbc3 gene of soybean shows a varying homology in respect to the different exons of Lbc3. The sequence of exon 4 of Lbc3 shows a homology of about 84 % while exon three and two are conserved to about 57 % and 59 % respectively.

The comparison of the 3' untranslated regions of pCK277 and 1349 with that from Lbc3 reveals three conserved sequence motives. There are two conserved sequences of 9 bp which involve the putative polyadenylation signal, located 12 and 11 bp upstream from a poly A-tail found in the c-DNA clones.

Additionally, this library was screened for other nodulins by differential hybridisation and using heterologous nodulin probes from soybean (kindly provided by K. A. Marcker). On a dot-blot analysis three out of twelve soybean nodulin c-DNA clones hybridized to nodule but not to root m-RNA of broadbean. Six clones from the library hybridized to these soybean c-DNA clones. The hybridisation signals from the differential screening and northern blot analysis showed two of these, pCK1168 and pCK1169 to be nodule specific. These clones are being further studied by hybrid released translation and sequencing.

ARE SYM-PLASMID GENES INVOLVED IN THE INDUCTION OF NODULIN GENE EXPRESSION?

Marja Moerman, Francine Govers, Albert van Kammen, Ton Bisseling.
Department of Molecular Biology, Agricultural University, Wageningen,
The Netherlands.

We studied whether Rhizobium signals involved in nodulin gene expression
are sym-plasmid encoded. Downie et al. have cloned the entire sym-
plasmid pRL1JI into a cosmid library. Two overlapping clones, pJI1089
and pJI1085 with a 10 Kb region in common are upon transfer into a cured
R.phaseoli strain able to form on pea roots ineffective nodules in which
the rhizobia are released from the infection thread and surrounded by a
peribacteroid membrane. We used both R.phaseoli/pJI1089 and
R.phaseoli/pJI1085 to determine whether these strains can induce
expression of nodulin genes. For that purpose total RNA isolated from
these nodules was translated in a reticulocyte lysate and the translation products were separated by two-dimensional (2-D) gel electrophoresis. The pattern of nodulin spots from pea root nodules formed by
R.phaseoli/pJI1089 and R.phaseoli/pJI1085 was similar to that produced
by wild type R.leguminosarum (Govers et al., 1985). Since all nodulin
genes are induced in the nodules formed by R.phaseoli/pIJ1089 and
R.phaseoli/pIJ1085 it can be concluded that the 10 Kb region of the sym-
plasmid with the nod-region in the R.phaseoli chromosomal background is
sufficient to induce nodulin gene expression. The nif- and fix-genes as
well as other sym-plasmid genes are apparently not involved in the
induction of expression of these host genes.

Inoculation of pea roots with an Agrobacterium tumefaciens transconjugant (LBA2712) carrying a Rhizobium leguminosarum sym-plasmid (pSym1)
resulted in the development of small ineffective nodules. Histological
examinations of nodules formed by the Agrobacterium tumefaciens
transconjugants showed that thes nodules have several vascular bundles
at the periphery of the nodule which is characteristically found in normal nodules. The plant cells are however devoid of bacteria, although
infection threads can be observed inside the nodule.

We investigated whether nodulin genes are expressed in the empty pea
nodules formed by the Agrobacterium transconjugant, and especially examined the expression of the early nodulin (N-40') gene, since the mRNA
of this gene can be detected before nodules, formed by Rhizobium, are
macroscopically visible. In root nodules formed by the Agrobacterium
transconjugant neither the N-40' mRNA nor the other nodulin mRNAs were
detected. Therefore the formation of a nodule structure appears to be
independent of the presence of the early nodulin as well as other nodulins, detectable with the used method. Furthermore, it can be concluded
that the sym-plasmid in the A.tumefaciens background is not sufficient
to induce nodulin gene expression in the host; so chromosomal genes of
Rhizobium appear to be involved in the induction of expression of the
nodulin genes.

Downie, J.A., Ma, Q-S., Knight, C.D., Hombrecher, G., and Johnston,
A.W.B. (1983). EMBO J. 2, 947-952.
Govers, F., Gloudemans, T., Moerman, M., Van Kammen, A. and Bisseling,
T., (1985). EMBO J. 4, 861-867.

NODULIN-35, A NODULE-SPECIFIC URICASE OF SOYBEAN: STRUCTURE, BIOSYNTHESIS AND LOCATION

Truyen Nguyen, Victoria Foster, Hanna Bergmann, Maria Zelechowska and Desh Pal S. Verma, Centre for Plant Molecular Biology, Department of Biology, McGill University, 1205 Docteur Penfield Avenue, Montreal, CANADA H3A 1B1

Nodulin-35 represents a 35 kd protein which is induced about ten days after infection of soybean by *Rhizobium japonicum* (1). It was shown to be a subunit of the nodule-specific uricase (2). Subcellular localization studies done using nodulin-35 specific antibody and protein-A gold revealed the presence of this protein in the peroxisomes of the uninfected cells of nodules. This protein is found to be synthesized on free polysomes. A cDNA clone (pNod35) was isolated using mRNA from immuno-precipitated polysomes. That this clone represents nodulin-35 was confirmed by comparing the deduced amino acid sequence with the partial sequence of a CNBr-cleaved peptide of purified nodulin-35. Southern-blot hybridizations with genomic DNA suggest that there are several *EcoRI* fragments containing nodulin-35 sequences. Three of these sequences were isolated from a genomic library of soybean. Nucleotide sequence analysis of one region showed that the complete gene extends almost 5,000 bp, and the coding region (309 codons) is interrupted by seven introns ranging in size from 154 to 1341 bp. Lack of a signal sequence and its translation on free polysomes suggest that nodulin-35 is post-translationally transported to the peroxisomes. Induction of nodulin-35 mRNA occurs between seven and nine days after infection by *Rhizobium*, and it appears to be independent of the effectiveness of nodules in fixing nitrogen. Furthermore, there is no cross-hybridization of nodulin-35 cDNA with RNA from young (three- to four-day) roots and leaves indicating that the observed "uricase" activity in these tissues is due to the product of a different gene (3).

(1) Legocki RP and Verma DPS (1979) Science 205, 190-193.

(2) Bergmann H et al (1983) The EMBO Journal 2, 2333-2339.

(3) Nguyen T et al (1985) Proc. Natl. Acad. Sci. USA (In press).

DIVERSITY IN THE PATTERNS OF GLUTAMINE SYNTHETASE ACTIVITY IN
PHASEOLUS VULGARIS AND PHASEOLUS LUNATUS ROOT NODULES

FRANCOISE M. ROBERT AND PETER P. WONG
DIVISION OF BIOLOGY, KANSAS STATE UNIVERSITY, MANHATTAN, KS 66506, U.S.A.

The glutamine synthetase (GS) activity in root nodule crude extracts
prepared from 64 cultivars of Phaseolus vulgaris and one cultivar of
P. lunatus was analyzed by polyacrylamide gel electrophoresis for the
presence of the nodule (GS_{n1}) and root (GS_{n2})-specific GS observed by
Cullimore et al. (1983) in the nodules of the bean cultivar Bush Blue
Lake 274. The nodules (1 g) were extracted in 0.05 M Tris-HCl buffer
(2 ml) at pH 7.5 containing 1.0 mM dithiothreitol. The GS activity
bands produced in native polyacrylamide gels were detected by the
transferase reaction based on the production of γ-glutamyl hydroxamate.
Reaction with $FeCl_3$ resulted in the formation of brown bands of ferric
chelate of γ-glutamyl hydroxamate (Barratt, 1980). Nodule extracts of
all 64 cultivars of P. vulgaris displayed two distinct GS bands. One
band corresponded to the nodule-specific (GS_{n1}) and another to the
root-specific GS (GS_{n2}). In contrast, P. lunatus nodule extract
exhibited only the nodule-specific (GS_{n1}) band. The electrophoretic
mobility (EM) of the GS_{n1} of P. vulgaris varied depending upon the
cultivars of the plants as well as the stage of nodule development.
In some cultivars, like "Provider", the EM was slow and changed little
when nodules from 2-, 3-, and 4-week-old plants were analyzed. In
other cultivars, e.g. "Kentucky Wonder", the EM of GS_{n1} was slow in
2-week-old plants but the EM increased as nodules developed resulting
in the fusion of GS_{n1} and GS_{n2} bands in 4-week-old plants. GS_{n1}
activity band of nodule crude extracts from P. vulgaris and P. lunatus
coincided with six to nine protein bands when a gel stained for GS
activity was subsequently treated with Coomassie blue. All bands,
however, showed an identical molecular weight of approximately 47,000
daltons as determined by sodium dodecyl sulfate polyacrylamide gel
electrophoresis.

In conclusion, the fact that the electrophoretic mobility of the GS_{n1}
activity bands of P. vulgaris is affected by the age of the plant
suggests that the alteration in the charge of the bands may be due to
a post-translational modification of GS_{n1}.

REFERENCES

Barratt DHP (1980) Plant Sci. Lett. 18, 249-255.
Cullimore JV et al (1983) Planta 157, 245-253.

NUCLEOTIDE SEQUENCE ANALYSIS OF TWO SOYBEAN NODULIN cDNA CLONES

*C. SENGUPTA GOPALAN/D. THOMAS/J. PITAS AND L. HOFFMAN
Agrigenetics Advanced Research Div., 5649 E. Buckeye Rd., Madison, WI 53716. USA
*Present Address: Crop and Soil Sciences/Plant Genetics Engineering Labs, New Mexico State University, Box 3GL/ Las Cruces, New Mexico 88001, U.S.A.

Two clones that are most abundantly represented in our soybean nodule specific, non-leghemoglobin cDNA library, C51 and E27, have been characterized with respect to the RNA and protein that they encode. Furthermore, clone C51 and E27, that contain the entire coding region, were sequenced in order to determine if analysis of the predicted amino acid sequence would provide some information on the structure of the proteins.

RNA blot analysis of poly(A)-RNA from 21d old nodules and uninfected roots of soybean with clones C51 and E27 as probes, showed under low stringency conditions, hybridization to nodule RNA species of mol. wt. 1.6Kb and 1.1Kb. However, under more stringent hybridization conditions, the two clones could be distinguished in that C51 hybridized to the 1.1Kb RNA molecule while E27 hybridized to the 1.6Kb RNA molecule. In hybrid-select translation experiments, clones C51 and E27, selected mRNA from nodule poly(A)-RNA which translated to produce polypeptides of molecular weight 29KD and 42KD, however, clone C51 preferentially selected RNA for the 29KD protein and E27 for the 42KD protein. In spite of the cross-hybridization between the two clones, the proteins encoded by them have different antigenic properties.

Sequence analysis of the two clones showed the absence of poly(A) stretches at the 3' end but did contain the entire protein coding regions. Clone E27 showed the presence of two direct repeats of 269 bases separated by 46 bases and there are two sequences of 35 and 50 bases repeated directly three times. Hydropathic analysis of the deduced amino acid sequence corresponding to the two clones revealed the presence of a very hydrophobic region in the NH_2-terminal end of the protein, typical of a signal peptide. However, the rest of the protein did not show any other stretch of hydrophobic regions long enough to act as 'stop transfer' signal during the passage of the protein across a membrane, thus ruling out the possibility of these proteins being integral membrane proteins. Comparison of clones C51 and E27 showed 50% homology at the nucleotide level which is concentrated mostly at the 5' and 3' ends of the insert and at the amino acid level, the 40% homology between the two was concentrated at the NH_2- and $COOH^-$ ends of the proteins. Since the two proteins are antigenically different, it would imply that the antigenic determinants are probably located in the middle of the protein, where the two share no homology.

Based on the structural predictions from the deduced amino acid sequence, the proteins probably represent organellar proteins that have to be transported in. Since many of the steps leading to ureide biosynthesis in soybean root nodules, take place in peroxisomes and plastids it is possible that these proteins represent enzymes that are involved in ureide biosynthesis. The 50% homology between clones E27 and C51 would suggest that these clones represent heterologous members of a multigene family.

PLANT GENETICS RELEVANT TO SYMBIOTIC NITROGEN FIXATION

Convenor: N.J. BREWIN,
JOHN INNES INSTITUTE, NORWICH NR47UH, U.K.

The discussion theme was plant genes that affect nodule function. If the ultimate objective of this research is to improve symbiotic N_2 fixation under field conditions, then the plant breeders have the most immediate prospects for success because plant molecular biology, although exciting, is still at the analytical stage.

1. PLANT BREEDING

It was recognized by all contributors that legume crops, eg., alfalfa, Phaseolus, Pisum, soybean and green-gram, carried sufficient genetic variability in symbiotic N_2 fixation to form the basis for a breeding program. However, different contributions emphasized different breeding goals - improved N_2 fixation in the absence or presence of soil nitrate, or in response to a specific Rhizobium inoculant. They also differed somewhat in their strategies for selection, although the convenor's view was that direct measurements of enhanced fixation using ^{15}N isotopes were less convenient parameters than such indirect measurements as nodule mass, acetylene reduction activity, plant fresh weight, seed yield, nitrogen content or nodule assimilatory activity.

2. PLANT MOLECULAR BIOLOGY

The chief concern of plant molecular biologists was to identify and characterize those genes (nodulins) specifically expressed at high level within nodules. These were identified from soybean, pea, Vicia or Phaseolus either by the use of a nodule-specific antiserum or by the construction and screening of a c-DNA library from nodule poly A mRNA. The research on nodulins followed three interlocking themes: the patterns and timing of gene expression; the cytological localization of gene products; and the extent of nodulin gene homology revealed by RNA sequence information. A nodule-specific uricase has been identified and localized to the peroxisomes of uninfected nodule cells. The lack of a hydrophobic N-terminal signal peptide sequence suggests that uricase is post-translationally transported to these organelles. With the exception of uricase, glutamine synthetase and the well-characterized leghemeglobin genes, no function has yet been ascribed to any other nodulin. However, DNA sequence information suggests that many nodulins should be grouped into gene families which perhaps have related functions.

3. PLANT MUTAGENESIS

Soybean and pea populations have been screened for symbiotically defective lines after mutagenesis and a number of interesting classes of mutation have emerged. Perhaps the most dramatic are the "supernodulating" mutants, characterized by a 30-fold increase in nodulation, even for plants grown with nitrate at concentrations normally inhibiting nodulation. In both pea and soybean, these mutations may be recessive or co-dominant. In soybean, grafting experiments indicate that phenotype may be controlled by a diffusible factor present in the shoots. Nitrate metabolism was unaffected in these mutants, but a reduction in root vigor was noted. Other isolated mutants include non-nodulating and non-fixing lines and also lines that lack the constitutive nitrate-reductase system.

MICROBIAL GENETICS RELEVANT TO
SYMBIOTIC NITROGEN FIXATION

A. NODULATION (nod) GENES

INVITED PAPERS

POSTER SUMMARIES:

(i)   nod genes in fast-growing rhizobia.
(ii)   nod genes in slow-growing rhizobia.
(iii)   Genes involved in nodule development.
(iv)   Genes involved in competitiveness.
(v)   Miscellaneous

DISCUSSION GROUP SUMMARY

B. ORGANIZATION AND REGULATION OF nif GENES

INVITED PAPERS

POSTER SUMMARIES.

C. GENES INVOLVED IN NODULE FUNCTIONING

INVITED PAPERS

POSTER SUMMARIES:

(i)   Carbon metabolism and ammonia assimilation.
(ii)   Hydrogen metabolism.

DISCUSSION GROUP SUMMARY

IDENTIFICATION AND ORGANIZATION OF RHIZOBIUM MELILOTI GENES RELEVANT TO THE INITIATION AND DEVELOPMENT OF NODULES

ADAM KONDOROSI[1], BEATRIX HORVATH[1], MICHAEL GÖTTFERT[1], PETER PUTNOKY[1], KATALIN ROSTAS[1], ZOLTAN GYÖRGYPAL[1], EVA KONDOROSI[2], ISTVAN TÖRÖK[2], CHRISTIAN BACHEM[3], MICHAEL JOHN[3], JÜRGEN SCHMIDT[3], JEFF SCHELL[3]
INSTITUTES OF GENETICS[1] AND BIOCHEMISTRY[2], BIOLOGICAL RESEARCH CENTER, HUNGARIAN ACADEMY OF SCIENCES, H-6701 SZEGED, P.Q.BOX 521, HUNGARY, MAX-PLANCK INSTITUT FÜR ZÜCHTUNGSFORSCHUNG[3] D-5000 KÖLN 30, GERMANY.

1. INTRODUCTION

The bacterium species Rhizobium meliloti induces nitrogen fixing nodules on the roots of its host plant Medicago sativa (alfalfa). The majority of symbiotic genes of R. meliloti are located on a megaplasmid, including genes coding for early functions in nodulation (nod), the nitrogenase genes (nif) and other genes required for nitrogen fixation (fix) (Banfalvi et al. 1981; Rosenberg et al. 1981). When this megaplasmid was transferred into other Rhizobium species or into Agrobacterium tumefaciens, these latter bacteria became able to induce ineffective nodules on alfalfa, indicating that the essential genes coding for nodule initiation and development are carried by this megaplasmid (Kondorosi et al. 1982). The development of these nodules, however, halted at an early stage: infection threads did not form and neither bacteria nor bacteroids were found in the inner nodule tissue (Wong et al. 1983).
In R. meliloti strain 41 all essential nod genes were located in the vicinity of nif genes (Kondorosi et al. 1984a; Fig.1). In this paper we discuss the identification, fine structure and organization of these nod genes and our approaches to identify their products are also presented.

2. THE ESSENTIAL NODULATION GENES ARE ARRANGED INTO TWO CLUSTERS

Previously, we have isolated R-prime plasmids carrying sections of the megaplasmid and by introducing them into A. tumefaciens those containing the essential nodulation genes were identified (Banfalvi et al. 1983) and the physical map of the inserted DNA segment was established (Kondorosi et al. 1984a). Nod⁻ mutations were localized on this region in two clusters. One cluster is located on an 8.5 kb EcoRI fragment and it was cloned into pRK290 (pKSK5). The other cluster was identified on a 6.8 kb EcoRI fragment (its pRK290 derivative is pEK10; Fig.1).
We constructed a recombinant plasmid (pPP346) in pLAFRI which carries a 24 kb region of the megaplasmid, comprising both nod clusters (Kondorosi et al. 1984b). Upon introduction of pPP346 into Nod⁻ mutants with large deletions, into different Rhizobium species or into A. tumefaciens the derivatives were able to nodulate alfalfa, albeit with varying efficiency.

R. meliloti 41 symbiotic megaplasmid region

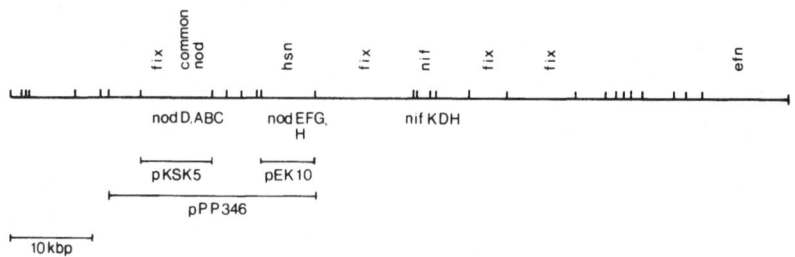

Figure 1. The physical-genetic map of a section of the symbiotic megaplasmid in R. meliloti 41. Vertical bars indicate
EcoRI restriction sites. Regions indicated by nod, hsn, fix,
nif and efn contain more than one gene.

3. NODULATION GENES ON THE 8.5 KB FRAGMENT: COMMON NOD GENES

The nod gene region on the 8.5 kb fragment was delimited by
the method of directed Tn5 mutagenesis (Ruvkun, Ausubel,
1981). More than 80 Tn5 insertions on this fragment were
mapped, the mutated fragments were transferred into the wild
type R. meliloti and the wild type region was exchanged for
the mutated one via homologous recombination. The symbiotic
phenotype of the R. meliloti mutants were determined in plant
assays. Fig. 2 shows that Nod⁻ mutants were localized on a
3.0 kb region, while mutants nodulating with a week delay
were mapped on a 1.0 kb segment. Nod⁻ mutants belonging to
the first class were unable to nodulate alfalfa even after
two months and were defective in root hair curling (Hac⁻)
which is the first observable step of the infection process.
The protein coding regions were determined on the 8.5 kb
fragment by using either E. coli minicells or E. coli in
vitro transcription-translation system (Schmidt et al. 1984).
In these experiments subcloned fragments of this region were
replaced after a strong E. coli promoter. Moreover, the nuc-
leotid sequence of the delimited nod region was also determined (Török et al. 1984; Göttfert et al. 1985). From both
lines of experiments the existence of four nod genes was
deduced (Fig. 2), which were designated as nodA,B,C and D.
These four genes code for proteins of 196, 217, 402 (or 426)
and 311 amino acid residues, respectively. Similar results
were obtained recently for R. meliloti strain 2011 (Egelhoff
et al. 1985; Jacobs et al. 1985).
We had shown previously that nodulation functions coded by
the 8.5 kb fragment are conserved in different Rhizobium
species ("common" nod genes; Banfalvi et al. 1981; Kondorosi
et al. 1984a). It was found that in Nod⁻ deletion mutants
lacking the 8.5 kb fragment the ability to nodulate alfalfa

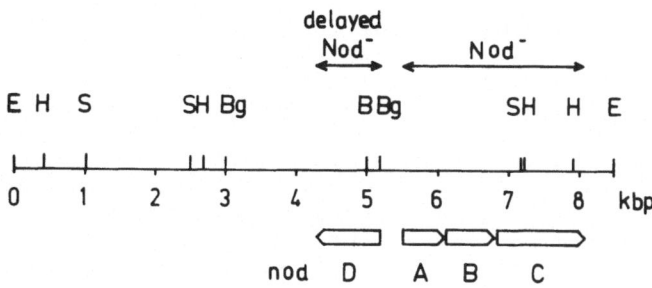

Figure 2. The physical-genetic map of the 8.5 kb EcoRI
fragment carrying the common nod genes. Nod sequences were
delimited by directed Tn5 mutagenesis (Török et al. 1984;
Göttfert et al. 1985). The location and size of nod genes
are from the nucleotide sequence data. Restriction sites:
E: EcoRI, H: HindIII, S: SalI, B: BamHI, Bg: BglII.

was restored upon the introduction of symbiotic plasmids or
cloned nodulation genes from other Rhizobium species. Comparison of the R. meliloti nodA,B,C nucleotide and amino acid
sequences with those of R. leguminosarum (Rossen et al. 1984)
showed 69-72% homology, demostrating that the structure of
the common nod genes are also conserved.
The functional and structural conservation of common nodulation genes provides suitable approaches to identify common
nod genes of other Rhizobium species. Using the cloned R.
meliloti nodC gene as hybridization probe we have identified
the nodC gene of the wide host range Rhizobium strain
MPIK3030. Nucleotide sequencing of the nodC gene of MPIK3030
revealed a 74% homology with the R. meliloti nodC gene. Tn5
insertion in this region resulted in Nod⁻ phenotype. Interspecies complementation also showed that the nod regions
from both species were able to restore nodulation ability
of Tn5-induced nodC mutants from either strain (Bachem et al.
1985). The biochemical function of the nod gene products are
still not known. From the amino acid sequence data, however,
it is possible to gain certain informations, such as the
possible location of these proteins. For instance, the
predicted amino acid sequence of the nodC protein shows that
the carboxy-terminal part is highly hydrophobic, suggesting
that this protein may be associated with the membrane (Török
et al. 1984).
To support this assumption and for the biochemical analysis
of the nodC gene product, the nodC protein has been purified.
First a gene fusion of the λcI repressor sequences to a large
fragment of the nodC gene was constructed. Then it was placed
under the control of the tac promoter which can be induced
by the addition of IPTG in E. coli. The induced E. coli cells

harboring this gene fusion accumulated the λcI-nodC hybrid
protein at large quantities (19% of the total cellular protein). It was observed that the growth of the induced E. coli
cells was inhibited, which was due to the overproduction of
the nodC protein. The fusion protein was purified by gel and
hydroxyapatite chromatography in the presence of SDS.
Antibodies were raised against the purified fusion protein
which were shown to precipitate the nodC proteins either of
R. meliloti or Rhizobium sp. MPIK3030 origin, expressed in
E. coli minicells.
Further biochemical analysis demonstrated that λcI-nodC
fusion protein was associated with the outher membrane of
E. coli cells. For instance, solubilization of the hybrid
protein required an ionic detergent, by cell fractionation
the protein was localized in the outher membrane fraction
of E. coli and in Triton X-114 solution the protein was
recovered from the detergent phase (John et al. 1985).
Detection of the natural nodC protein in free-living
Rhizobium or in nodules was unsuccessful. This is in line
with the unsuccessful attempts to demonstrate nod gene
transcripts under these conditions.
The addition of antibodies against the hybrid protein to
plant growth medium and the inoculum resulted in 50% inhibition of nodulation of alfalfa by R. meliloti or clover by
R. trifolii. These data suggest that the nodC protein is
located in the cell envelope of Rhizobium and a plant substance may be involved in the induction of this protein. One
can speculate that the nodC product plays a role in transmembrane signalling. If it is so, the nodulation inhibition
by the antibodies against nodC could be explained by blocking
a possible receptor site or masking a pore protein.

4. NODULATION GENES ON THE 6.8 KB FRAGMENT: HSN GENES

Directed Tn5 mutagenesis of the 6.8 kb fragment revealed three
nod regions (Fig. 3). Mutations in two regions resulted in
delayed nodulation, while the third mutant class were unable
to nodulate alfalfa even after 6 weeks. Bacteria reisolated
from the nodules infected with delayed Nod⁻ mutants retained
the mutant phenotype.
Mapping of the protein coding regions and nucleotide sequencing revealed that four nod genes are present on the 6.8
kb EcoRI fragment (Fig. 3). These were designated as nodEFG
and H, respectively. The corresponding gene products consist
of 90, 414, 244 and 247 amino acid residues.
Nod⁻ mutants mapped on the 6.8 kb fragment were not comple-
mentable with symbiotic plasmids of other Rhizobium species
tested so far. Moreover, lack of interspecies homology between
nodH and other rhizobia and only week homology to the DNA
fragment carrying nodEFG were observed. Transfer of pEK10
into Rhizobium sp. MPIK3030 resulted in transconjugants with
low level of alfalfa nodulation ability. Preliminary results
indicate that mutations in the nod genes carried by the 6.8
kb fragment influences the nodulation ability of other natural plant hosts of R. meliloti, such as Melilotus or

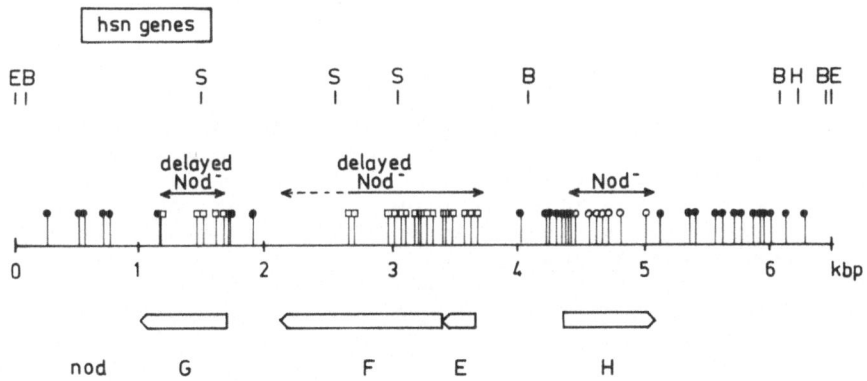

Figure 3. The physical-genetic map of the 6.8 kb EcoRI fragment carrying genes coding for host specificity of nodulation. Vertical bars indicate Tn5 insertions: ●Nod⁺; oNod⁻; □delayed Nod. For further explanations see legend to Fig. 2.

Trigonella, but to different extent (I. Barabas, unpublished), supporting our previous suggestions that the nod genes of the 6.8 kb fragment code for host specificity of nodulation.
Recently we were able to clone the hsn region of Rhizobium sp. MPIK3030. Transfer of this DNA region to other Rhizobium species, rendered the transconjugants to be able to nodulate Macroptilium atropurpureum. The common and hsn genes form separate clusters also in this species.

5. CONSERVED SEQUENCES IN THE 5'-FLANKING REGIONS OF NOD GENES

Nucleotide sequence data suggest that the 8 nod genes described above are organized into 4 or 5 transcription units and regulatory sequences must be present at least in front of the nodA, nodD, nodE and nodH genes. Several lines of indication suggest that most of these genes are specifically turned on during the symbiotic process. Since mutations in these nod genes stop or delay nodule initiation, it is likely that at least some of the nod genes are coordinately regulated. The common regulation may be reflected in the conservation of specific regulatory sequences in the promoter regions of the nod operons. It has been shown for the nif gene cluster that several regions in the promoter sequences of various nif genes which are subject to a common positive regulatory mechanism are conserved (Schofield, Watson, 1985).
By comparing the 5'-flanking regions of the different nod genes, a 25 kb long highly conserved sequence was found in front of the nodA, nodE, and nodH genes, at around 250 basepairs upstream of the proposed translational start codons. To determine the occurrence of this sequence in the

R. meliloti genome the conserved 25 bp oligonucleotide was
synthetized chemically and used as hybridization probe against
total DNA of R. meliloti 41. It was found that this sequence
is present in 6 copies and further analysis demonstrated that
all are located within the nif-nod regions presented on Fig.1.
The nucleotide sequences of the 3 newly detected homologous
regions were determined which indicated that a more extended 50 bp region was conserved.
The synthetic oligonucleotide probe hybridized to DNA from
all Rhizobium species tested so far and using available nod
clones from other Rhizobium species the linkage of the
homologous region to nod genes was also demonstrated.
If we suppose that these conserved sequences in front of the
nod genes are necessary for nod gene expression, insertion
of foreign DNA into the homologous region of R. meliloti
should result in the loss of nodulation ability. We have
inserted the kanamycin resistance gene of Tn5 into the
middle of the conserved sequences of the nodE and nodH 5'-
-flanking regions and the mutated fragments were homo-
genotized back into the wild type R. meliloti which became
Nod⁻. These results show that the conserved promoter sequences are essential for nod gene expression. These conserved sequences were not found in front of the nodD gene,
suggesting that nodD is regulated differently. Experiments
are in progress to show that the conserved sequences are
involved in the coordinated regulation of nod genes by plant
substances.

6. REFERENCES

Bachem C et al (1985) Molec. Gen. Genet. 199, 271-278.
Banfalvi Z et al (1981) Molec. Gen. Genet. 184, 318-325.
Banfalvi Z et al (1983) Molec. Gen. Genet. 189, 129-135.
Egelhoff TT et al (1985) DNA, in press
Göttfert M et al (1985) submitted
Jacobs TW et al (1985) J. Bacteriol., in press
John M et al (1985) EMBO J. in press
Kondorosi A et al (1982) Molec. Gen. Genet. 188, 433-439.
Kondorosi E et al (1984a) Molec. Gen. Genet. 193, 445-452.
Kondorosi A et al (1984b) In Zalewski RI and Skolik JJ, eds,
Natural Products Chemistry, pp 643-654, Elsevier S.P.,
Amsterdam.
Rosenberg C et al (1981) Molec. Gen. Genet. 184, 326-333.
Rossen L et al (1984) Nucleic Acids Res. 12, 9497-9508.
Ruvkun GB and Ausubel FM (1981) Nature (London) 489, 75-78.
Schmidt J et al (1984) EMBO J. 3, 1705-1711.
Schofield P and Watson JM (1985) Nucleic Acids Res. 13,
3407-3418.
Török I et al (1984) Nucleic Acids Res. 12, 9509-9524.
Wong C-H et al (1983) J. Cell Biol. 97, 787-794.

PLANT-SECRETED FACTORS INDUCE THE EXPRESSION OF R.TRIFOLII NODULATION AND HOST-RANGE GENES

B.G. ROLFE, R.W. INNES, P.R. SCHOFIELD, J.W. WATSON, C.L. SARGENT, P.L. KUEMPEL, J. PLAZINSKI, H. CANTER-CREMERS and M.A. DJORDJEVIC
GENETICS DEPARTMENT, RESEARCH SCHOOL OF BIOLOGICAL SCIENCES, AUSTRALIAN NATIONAL UNIVERSITY, CANBERRA, 2601, AUSTRALIA.

SUMMARY:

A 14kb DNA fragment from the Sym plasmid of R.trifolii strain ANU843 was shown to carry several genes which were functionally conserved between different Rhizobium species, as well as, a complex array of genes which determine host specific nodulation ability and strain competitivenes. Extensive mutagenesis of the 14kb Nod region using Tn5 and the mini Mu-lac transposon MudI 1734, revealed a correlation between the site of insertion and the nodulation defect induced. Four distinct regions (designated I, II, III and IV) were identified. The phenotypic analysis of specific subcloned DNA fragments from the Nod region, resulted in the identification of a fifth region (region V) involved in determining host range ability. DNA sequence and mutant analysis showed that region I contained 4 genes (R.trifolii nodA,B,C and D) which affected root hair curling and nodulation ability. Region II mutants were severely debilitated in nodulation ability and predominantly induced short, truncated infection threads. Region III mutants displayed host range properties which differed from the parent strain. In contrast to the parent strain these mutants were able to induce nodules on Pisum sativum including the recalitrant Afghanistan variety. Region IV mutants were consistently delayed in their ability to nodulate clovers. Transcriptional fusion of the E.coli β-galactosidase (lacZ) gene to genes in regions I, II, III and IV were analysed. Appropriately oriented nodD insertions of MudI 1734 were found to be expressed under normal culture conditions while activity from other genes was low. Expression of nodA and B and region II, III and IV genes could only be detected by exposure of bacteria to plant-secreted factors.

RESULTS

Analysis of subcloned fragments from R.trifolii Nod region: In R.trifolii strain ANU843, a 14kb DNA fragment from the Sym plasmid has been shown to carry the genetic determinants which permit successful root hair infection and nodule initiation to occur on clover plants (Schofield et al 1984; Rolfe et al 1985). Transfer of this 14kb fragment to the Nod⁻ Sym plasmid-cured derivative of ANU843 (strain ANU845) restored clover nodulation capacity to this strain. Component fragments isolated from the 14kb Nod region, were subcloned onto broad host range vectors and introduced to strain ANU845. Transconjugants were analysed to determine if a detectable symbiotic response was conferred to strain ANU845 (Fig. 1). Significantly, the introduction of

80

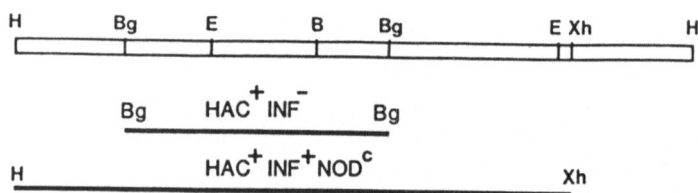

Figure 1: Subcloned DNA fragments. The two subcloned DNA fragments which confer recognisable symbiotic activity in the Sym plasmid cured strain ANU845 are indicated. The 5.3kb BglII fragment confers clover root hair curling ability only while the 11.4kb HindIII-XhoI fragment confers nodulation ability on T.subterraneum but not T.repens.

Figure 2: Genetic analysis of the R.trifolii Nod region. Figure 2(A) indicates the positions of R.trifolii nodA,B,C and D genes (region I) and regions II, III, IV and V. Region I and II mutants are functionally complemented by R.meliloti common nod genes. The subcloned fragments used for reconstruction experiments are indicated. Fig. 2(B) shows the positions and orientations of MudI 1734 transcription fusions. Insertion into nodD are expressed in culture (closed arrow) when oriented left to right. Insertions into nodA,B, region II, III and IV are expressed in the appropriate orientation (open arrows) only in the presence of plant-secreted factors.

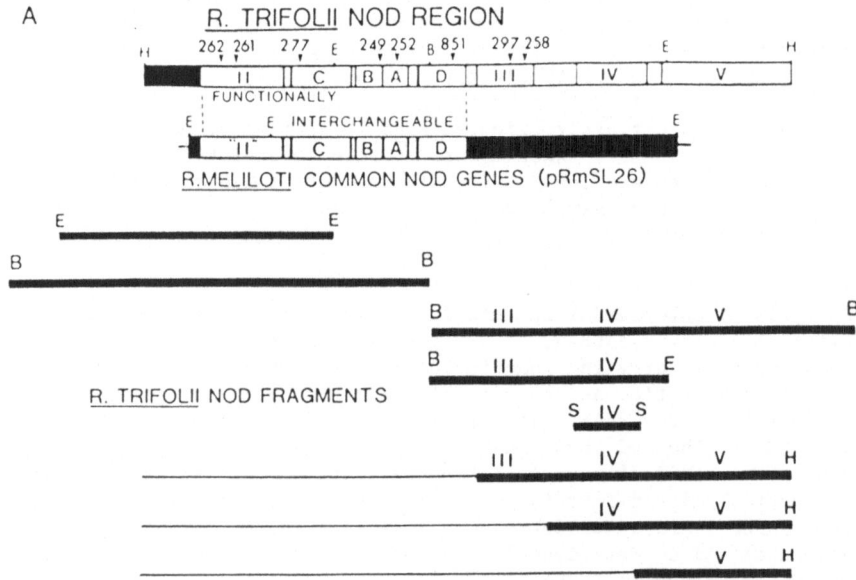

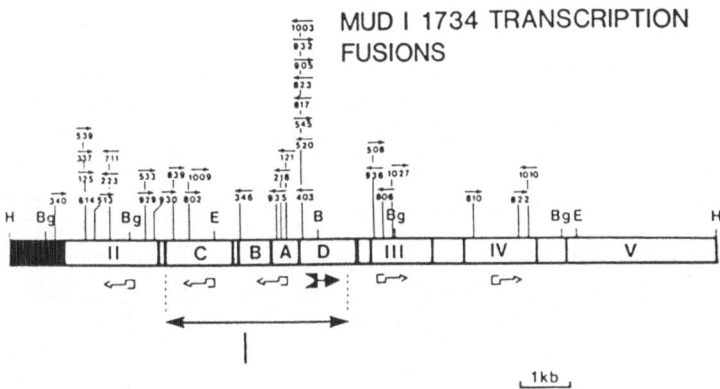

Figure 3: DNA sequence analysis. Fig. 3A shows the size and direction of the nodulation genes and the location (open boxes) of the three reiterated sequences. DNA sequence (Fig. 3B) was determined by the chain termination method. Detail of the sequenced area (3.6kb EcoRI-BglII fragment) is shown. Insertion points of Tn5-induced mutants are indicated as are two of the 75 base pair reiterated sequences (underlined). Translations (single letter) of nodulation genes are shown.

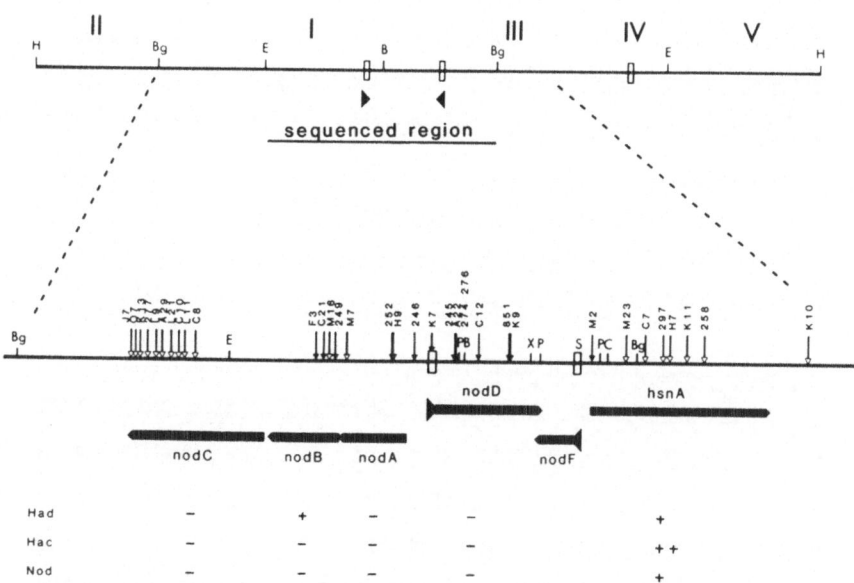

82

a 5.3kb BglII fragment to ANU845 restored root hair curling ability only
to this strain, however, no infection threads or nodules were induced.
This fragment was shown by complementation and DNA sequence analysis to
carry R.trifolii nodA,B,C and D genes. The introduction of an 11.4kb
HindIII-XhoI fragment to ANU845 restored root hair curling and infection
thread formation ability on clovers, however only a small number of
nodules were induced on T.subterraneum plants and none appeared on
T.repens. This indicated that genes located on the remaining 2.6kb
(region V, see below) were important for conferring nodulation
proficiency on all clover species. The introduction to ANU845 of other
subcloned fragments used in this study (see Fig. 2) including an 8.8kb
fragment carrying nodA,B,C, did not induce root hair curling or any
other detectable symbiotic response.

Mutagenesis of the R.trifolii Nod region: An analysis of 48 Tn5-induced
mutants and 33 mutants induced by the bacteriophage transposon MudI 1734
resulted in the identification of four specific regions (designated I,
II, III and IV). A correlation between the mutation sites and the
phenotype generated was established (Djordjevic et al 1985; Innes et al
submitted). Insertions into region I (which contains R.trifolii nodA,B,C
and D genes) abolished normal root hair curling, infection thread
formation and nodulation ability on clovers, thus confirming the
importance of these genes for root hair curling ability. Region II
mutants were severely debilitated in nodulation ability on clovers. An
exaggerated root hair curling response (Hac^{++}) was induced by these
mutants and the poor nodulation ability was correlated with an inability
to induce normal infection threads. Infection threads induced by these
mutants were typically short and truncated. Region I and II mutants
were complemented by the introduction of the recombinant R.meliloti
plasmid pRmSL26 which has been shown to carry R.meliloti nodA,B,C and D
genes (Fig. 2), thus confirming that these genes were functionally
conserved between R.trifolii and R.meliloti.

Region III mutants displayed altered and extended host range
properties when compared to the parent strain. These mutants were
debilitated in nodulation ability on T.repens but not on T.subterraneum.
Unlike region II mutants, region III mutants were able to induce
abundant, normal-looking infection threads on both Trifolium species and
therefore we conclude that the poor nodulation response on T.repens is
due to an inability to induce nodule initiation. Moreover, in contrast
to the parent strain, region III mutants were able to induce pronounced
root hair distortion, infection threads and abundant nodules on
P.sativum including the recalcitrant Afghanistan variety (Djordjevic et
al 1985).

Region IV mutants were consistently delayed in their ability to
induce nodules but only when artificial light intensities were above
$400\mu E/m^2$. At lower light intensities the root hair infection and
nodulation response of these mutants was similar to the parent strain
(Innes et al submitted). Region III and IV mutants were not
complemented by the R.meliloti recombinant plasmid pRmSL26.

DNA sequence analysis: As predicted by complementation data, DNA
sequence analysis of R.trifolii nodA,B,C and D genes shows that they
share extensive sequence homology to their functional analogs in
R.leguminosarum (Rossen et al 1984) and R.meliloti (Torok et al 1984;
Egelhoff et al 1985). A 75 base pair sequence is repeated three times
in the R.trifolii Nod region (see Fig. 3). One copy of the repeat
sequence occurs at the 5' end of nodD and appears to include the

promoter of this gene. A second (inverted repeat) copy appears at the
5' end of an open reading frame which we designate nodF, while the third
copy lies in region IV (Fig. 3). It is possible that these 75 base pair
reiterated sequences are involved in regulating the expression of
adjacent nod genes. An additional gene, hsnA has been identified (Fig.
3). This gene includes the mutants located in region III, a locus which
has been shown to be involved in determining the host range properties
of R.trifolii (Djordjevic et al 1985b).

Region III and IV genes affect strain competiveness: Since the
nodulation phenotype of region III and region IV mutants was more akin
to the parent strain on T.subterraneum plants (as opposed to T.repens),
several tests were conducted on these plants to assay strain competiveness compared to the parent strain ANU843. When parent and mutant
strains were grown under identical conditions, mixed in equal numbers,
and inoculated onto T.subterraneum, nodule occupancy of the mutant
strains (kanamycin-resistant cells) was always lower than that of the
parent strain (C.L. Sargent, unpublished). This result was more
striking when assays were done using plants with "split roots". These
plants were generated by destroying the root meristem of two day old
seedlings and allowing two equally vigorous lateral roots to develop.
Plants possessing split roots provide two equally infectable roots that
occur on the same plant. This was shown by inoculating either the
parent strain or the mutant strain at the same time on both roots of the
one plant. Nodules appeared in equal numbers on both inoculated
roots. However, when either a region III or IV mutant was inoculated on
one side and "competed" with the parent strain on the other side, over
90% of nodules generated always occurred on the root inoculated with the
parent strain. Presumably, the differences in the rates of the
initiation of the infection process by the parent and mutant strains
enabled the wild-type strain to induce nodules first. This is supported
by the finding that infection events occurred on both roots but that
little or no observable nodule initiation was induced by the mutant
strains. The inability of the mutants to initiate nodules may be due to
the production of a systemic plant factor(s) which regulates nodule
initiation.

Reconstruction experiments: Despite the presence on plasmid pRmSL26 of
R.meliloti nodA,B,C and D genes (Egelhoff et al 1985) and genes which
functionally complement region II mutations in R.trifolii, this plasmid
does not confer root hair curling or (normal) nodulation ability to Sym
plasmid cured Rhizobium strains (Hirsch et al 1985; Djordjevic et al
1985a). However, when DNA fragments containing R.trifolii regions III,
IV and V were introduced to the Sym plasmid-cured strain ANU845
(carrying plasmid pRmSL26), the resulting transconjugants were able to
proficiently nodulate clover but not alfalfa plants (see Fig. 2). The
introduction instead, of R.trifolii DNA fragments carrying either
regions III and IV or IV and V conferred poor, but significant,
nodulation on T.subterraneum but not on T.repens. This indicated that
two distinct loci, region III and region V, were required to confer host
range ability on T.repens. The introduction of a 1.4kb fragment
carrying region IV to strain ANU845 (pRmSL26) results in the phenotypic
activation of root hair curling ability on both clover species but not
alfalfa plants. These results clearly indicate that several loci are
involved in determining host range in R.trifolii strain ANU843 and that
genes in regions III, IV and V can interact with the common nod genes
located on plasmid pRmSL26 (Djordjevic et al submitted).

The expression of R.trifolii nod genes is controlled by the plant host:
Transcriptional fusions to R.trifolii nod genes (in the 14kb Nod region)
were generated in vivo in E.coli using the mini Mud-lac bacteriophage
transposon, MudI 1734 (Castihlo et al 1984). Expression of these
fusions was assayed in the Sym plasmid-cured strain ANU845. An analysis
of 33 MudI 1734-induced mutants (Fig. 2, Innes et al submitted) showed
that insertions in the 5' end of nodD in the appropriate orientation
were expressed (500 units of activity) under normal culture conditions.
Mud-lac insertions elsewhere (regardless of orientation) gave little or
no expression of β-galactosidase activity. In the presence of exudate
derived from axenically-germinated clover seedlings, however,
transcriptional fusions in nodA, nodB and regions II, III and IV, showed
at least a 10-fold increase in β-galactosidase levels when inserted in
the appropriate orientation. Insertions into nodC were in the opposite
orientation (as predicted by DNA sequence analysis) to accept expression
from the nodC promoter. Accordingly, these were not expressed under any
conditions. Induction of all Nod operons occurred within 3 hours
exposure to the plant exudate.

A transcription analysis of the R.trifolii nod genes permitted the
determination of the direction of transcription of several Nod operons
and showed that their expression was determined by plant-secreted
factors. Recent evidence shows that R.trifolii Nod promotors are turned
on in the presence of exudate from other legumes including peas, alfalfa
and soybeans. Since expression from region III and IV occurs in the
presence of several different legumes, host specificity is therefore
unlikely to be determined by the selective "turn-on" of bacterial host
range loci by plant factors. Hence we conclude that the bacterial
products from the host range genes directly determine infectability on
specific legumes.

References:

Castihlo BA et al (1984) J Bacteriol 158, 488-495
Djordjevic MA et al (1985a) Plant Mol Biol 4, 147-160
Djordjevic MA et al (1985b) Mol gen Gen (in press)
Egelhoff TT et al (1985) DNA 4, 241-248
Hirsch AM et al (1985) J Bacteriol 161, 223-230
Rolfe BT et al (1985) In: Szalay AA and Legocki RP, eds, Advances in
      the molecular genetics of the bacteria-plant interaction, pp 43-48,
      Media Services, Cornell University, New York
Rossen L et al (1984) Nucleic Acids Res 12, 9497-9508
Schofield PR et al (1984) Plant Mol Biol 3, 3-11
Torok I et al (1984) Nucleic Acids Res 12, 9509-9524

Acknowledgements: This work is supported by an Agrigenetics Sponsored
Research Grant. We thank Jan McIver, Marie Oakes and Elena Gärtner for
technical assistance.

FINE STRUCTURE STUDIES OF R. MELILOTI nodDABC GENES

SHARON R. LONG, THOMAS T. EGELHOFF, ROBERT F. FISHER, THOMAS W. JACOBS,
AND JOHN T. MULLIGAN
DEPARTMENT OF BIOLOGICAL SCIENCES, STANFORD UNIVERSITY, STANFORD CA 94305,
U.S.A.

INTRODUCTION

The ability of Rhizobium bacteria to invade plants and stimulate the
host to develop nodules depends upon a series of genes, including the
nodulation (nod) and host range (host specific nod) genes (Kondorosi et
al, 1984; Jacobs et al, 1985), as well as loci which influence the state
of the bacterial cell surface and probably other genes not yet specified.
The bacterium Rhizobium meliloti invades alfalfa (lucerne, genus
Medicago), and a few other genera of plants. We have analyzed the genes
which R. meliloti requires for symbiosis with alfalfa, and report here our
studies on the map and sequence of four of these genes, together with
identification of their protein products, and the analysis of conditions
for their expression.

Identification and map of nodulation genes in R. meliloti strain 1021.

By complementation of R. meliloti mutants unable to nodulate alfalfa,
a cloned DNA segment was obtained on recombinant plasmid pRmSL26. This
DNA region was localized to within 25 kb of the R. meliloti nif gene
cluster (Long et al, 1982). The pRmSL26 DNA insert segments were subcloned
and physically mapped by restriction enzyme analysis. One EcoRI fragment
from the pRmSL26 segment, 8.7 kb in size, was obtained as recombinant
plasmid pRmJ1 and was used for further study (Jacobs et al, 1985).
In the insert of pRmJ1, 87 mutational insertions were obtained by
treatment of the plasmid with a Tn5 delivery vehicle. These Tn5 mutants
were recombined into the genome of the strain of origin, R. meliloti 1021
(a derivative of wild type SU47). The DNA configurations of the resulting
Tn5-bearing strains were confirmed by restriction analysis and
hybridization, and their phenotypes were tested on plants. We found that
in a 3.5 segment at the nif-proximal portion of the 8.7 kb fragment,
insertion of Tn5 resulted in a non-nodulating phenotype, indicating the
presence of nodulation genes (Jacobs et al, 1985).
We have continued our studies of these genes by complementation
analysis using merodiploids of mutant pairs, one carried in the genome and
the other borne on a cloned nod gene segment cloned in plasmid pRK290.
Our first study of this was carried out as have been the complementation
analyses of nif, that is, in an otherwise wild-type host bacterial
background. However, unlike the case with nif complementations, we found
that pairs of mutations complemented at random, indicating to us that
crossing-over between the mutant segments was yielding wild-type
recombinants; because even a small percentage (1 in 10^5) Nod$^+$ bacteria in
a Nod$^-$ population can result in an apparent Nod$^+$ phenotype, a low amount
of recombination compromises the use of standard complementation tests by
merodiploids.

We have recently used a Rec⁻ background to repeat these tests. By means of a recA::Tn5(Gm^{R}Sp^R) allele constructed by deVos and colleagues at M.I.T., we introduced into our merodiploids a recombination deficient phenotype. The merodiploids were constructed such that each Tn5 mutation in the genome was paired with the Tn5 insertions that were its immediate neighbors on both sides; only where these two Tn5's cross the junction between genetic units will complementation be observed. We found that complementation occurred at two points: between mutants 9B8 and 4C4, and between mutants 2A8 and 7B5. As shown later by sequence analysis these represent borders between gene pair D· and A, and gene pair B and C.

Complementations were also carried out between R. meliloti and R. trifolii mutant strains, using cloned wild-type and transposon mutated nod gene fragments. The R. trifolii cloned nod gene segment from pRt032 (Schofield et al, 1983) was recloned into vector pWB5a, based on pRK290 (Ditta et al, 1980), and was introduced into mutants of R. meliloti from all portions of the nod region. All mutants were restored to wild-type Nod phenotype on alfalfa, indicating that the function was conserved across species. The reciprocal experiment was also carried out, transferring the wild type R. meliloti nod gene segment, borne in pRmJ30, into two different R. trifolii mutants, ANU851 and ANU453. While these mutants both were restored to normal nodulation on clover by pRmJ30, we found that if Tn5 insertions were placed in the nodD region of pRmJ30, then it would no longer complement ANU851. Mutations of the nodA region of pRmJ30 prevented it from complementing ANU453. Thus we were able to establish the correspondance in function between R. meliloti gene D and ANU851, and between R. meliloti nodA and ANU453 (Fisher et al, 1985). It is intriguing that ANU851 is a completely tight Nod⁻ mutant, although nodD mutations in R. meliloti result in a leaky mutant type.

Sequence analysis.

The sequence of all four nodulation genes was determined by a combination of the Maxam-Gilbert chemical cleavage and Sanger dideoxy chain termination techniques. The results are summarized in Figure 1. We found three open reading frames proceeding from nif-distal towards nif-proximal, and these putative genes were designated A, B, and C. In addition, an open reading frame was detected which proceeded in the opposite direction, and which therefore would read off of the opposite strand. This was designated nodD. The predicted sizes of proteins arising from these potential open reading frames were determined by computer analysis to be A: 196 aa (amino acids) = 21.8 kd; B: 217 aa = 23.8 kd; C: 426 aa = 46.7 kd; and D: 308 aa = 34.8 kd. Our sequence data for the nodABC segment are in agreement with those of Torok et al, (1984) and Rossen et al, (1984). The relationship of these open reading frames to the Tn5 mutants was determined by sequencing the junction of 25 of the Tn5 insertions. This permits us to assign these mutations, and other Tn5 insertions which were mapped by comparison to the sequenced set, to potential genes as established by sequencing (Egelhoff et al, 1985).

Among the interesting features of the sequence is a 250 bp segment which lies between the divergently read genes nodD and nodA, and which contains an inverted repeat capable of forming a hairpin loop with an 11 bp stem. This segment would presumably include the promoters of both the D and the A genes. Sequences resembling those for the consensus E. coli promoter and the consensus nif promoter are found in this DNA segment.

Also notable is the hydrophobic nature of the predicted nodC protein

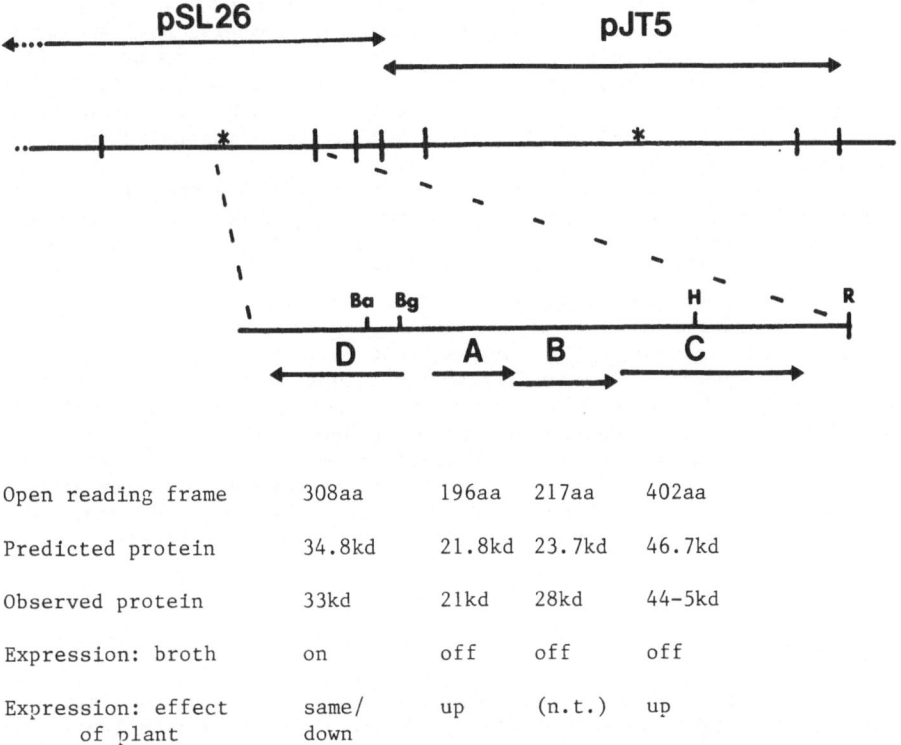

Open reading frame	308aa	196aa	217aa	402aa
Predicted protein	34.8kd	21.8kd	23.7kd	46.7kd
Observed protein	33kd	21kd	28kd	44-5kd
Expression: broth	on	off	off	off
Expression: effect of plant	same/ down	up	(n.t.)	up

Fig.1. Genetic map, gene products and regulation of the Rhizobium
meliloti nodulation genes DABC. Top: linkage of the common nod genes
to the nif loci. * marks position of apparent multiple copies of nodD;
segment between nod and nif is represented by clone pRmJT5, which
probably encodes nodulation genes similar to those reported by
Kondorosi et al (1984). Second line: map of the nodDABC cluster. The
results of sequencing analysis are indicated by predicted length of
open reading frames in aa (amino acids); the predicted protein
molecular weight and the actual proteins observed in expression systems
are compared below. Regulation of the nod genes (see also Table 1) is
lined up with the map in the bottom two lines.

product, especially its C-terminal end. This would be consistent with
membrane localization of this protein. The sequences for the potential A
and B open reading frames show a 4 base pair overlap of the stop codon of
A and the translational start site for B. Such overlap of open reading
frames is often associated with translational coupling of the two protein
products. It is interesting that the mutant pair spanning nodA and nodB
sequences do not show complementation while other gene pairs do, in the
merodiploid constructions mentioned above. This would be consistent with
a much tighter polarity of A on B such as might occur due to translational
coupling. However, coupling of A and B has yet to be demonstrated at the
biochemical level.

The sequence analysis of this multigene segment has provided detailed
information about its organization, and has also yielded as a by-product a
set of clones which can be used a specific probes for individual genes.
These are available from our laboratory for use in detecting the presence
of nodulation genes in Rhizobium, and have been used with success both for
R. meliloti and for other species. In R. meliloti itself, we used an M13
clone specific for the nodD coding sequence to demonstrate the presence of
multiple sequences which hybridize to the coding region of nodD. At high
stringency, two bands are seen, one of approximately 6 kb and the other
the 8.7 kb fragment in which nodD itself is found. At lower stringency,
two more EcoRI fragments appear, of approximate sizes 15 and 10 kb. Honma
and Ausubel (this volume) have identified these cross-hybridizing segments
in R. meliloti, and have cloned and mutagenized the 6 kb EcoRI fragment,
which will permit a test of the symbiotic role of these potential extra
genes.

Protein products.

As described below, current evidence suggests that the nodulation
genes are not highly expressed in R. meliloti during growth in laboratory
media. Thus it is not surprising that it has proved difficult to
associate specific protein products with the individual nodulation genes
in vivo in Rhizobium. Taking advantage of recombinant DNA, protein and
antibody technology, we have used a variety of approaches to identify the
nod gene products.

The DNA segments thought to bear nodulation genes were recloned into
a high copy number vector which utilizes the Salmonella typhimurium trp
promoter to drive expression of inserted genes. Using these recloned
segments, we synthesized protein in maxicell (UV irradiated recombination
deficient) strains of E. coli, and also in a 30,000 g supernatant (S-30)
in vitro transcription-translation system. This permitted the assignment
of several protein products to the nodulation gene region, and these could
be correlated with the gene products predicted from the DNA sequence.

Our approach differs from that often used in the identification of
products, in which protein products of a wild type fragment are compared
with products obtained when a variety of transposons are inserted into the
fragment. Instead, we have created a series of cloned fragments which
were then trimmed or extended at the 5' and 3' ends. This avoids the
problems which may arise due to transposons, such as complex polarity
effect, fused genes, and extra gene products internal to the transposon
itself.

Our data indicated that nodA is a polypeptide migrating at 21 kd,
nodB at 28 kd, nodC at 45-46 kd, and nodD at 33 kd. These are in
agreement with the sequence data with the exception of nodB, which would

be expected to migrate at 24 kd. Isoelectric point, degree of hydrophobicity, and secondary structure may be factors which cause the aberrant migration of this polypeptide in electrophoresis. We found nodD to be a single protein species migrating at 33 kd; this differs from the published results of Downie et al, who assigned a double band (34, 36 kd) to nodD, and of Schmidt et al., who assigned a protein of 17.5 kd to the nodD region. It is likely that differences in the construction of the clones probably account for these varied results, and that nodD is indeed a single species migrating at 33 kd.

Before proceeding very far with studies on such putative gene products, it is essential to confirm that they correspond to genuine Rhizobium proteins. We have taken two approaches to this. We have constructed a broad range expression vector, pTE3, which utilizes the same trp promoter system as described above (Egelhoff and Long, in press). Although this promoter is of enteric bacterial origin, it functions very well in Rhizobium, and we have used it to express the protein products of the chloramphenicol acetyltransferase gene and the nodA gene. The migration of the overexpressed nodA product in Rhizobium was identical to that seen in E. coli systems.

To characterize nodA further, and to obtain definitive evidence of its presence in Rhizobium, we purified the putative nodA product from E. coli cells which expressed it at high level, and raised antiserum to the purified polypeptide. With this antiserum, we examined the presence of cross-reactive material in R. meliloti cells. We found, first, that although no nodA-like protein was found in wild-type cells grown in laboratory media, a 21 kd protein reactive with nodA antiserum was present if the bacteria were exposed for a few hours to exudates from plants, or if the bacteria contained extra copies of the nod genes on a cloned fragment; the level of expression from such clones was weak unless exogenous promoters such as trp were added.

Furthermore, we were able to demonstrate that if the physical DNA sequences identified as the nodA open reading frame were interrupted with a transposable element, then no nodA cross-reactive protein was observed in the cells, even in the presence of root exudate. This combination of genetics and biochemistry establishes the identity of the 21 kd protein as a genuine R. meliloti nodulation gene product. This type of verification will be necessary for all cryptic (non-expressed) gene products such as those for the nodulation genes.

Expression of genes in the nodDABC cluster.

We had previously examined the production of nodulation gene RNA by cells grown in free-living conditions, and had found expression to be so low that it was not detectable by hybridization or sequence protection. Therefore, we turned to gene fusion technology to assay expression in Rhizobium cells of the genes in the nodDABC segment. We constructed two fusions, one to nodC and one to nodD. In these fusions, the reading frame of E. coli lacZ was fused from its 8th codon on to a start site and N-terminal end of a R. meliloti gene. Because the lacZ gene has no start site or ribosome binding site of its own, its expression is dependent upon transcription and translation of the gene into which it is fused. The fusions we constructed were assayed in two conditions: first, while carried on cloned fragments in broad host-range vectors; and second, after recombination into the homologous location in the genome. Cells were typically grown in minimal nutrient media, alone or with added plant exudate prepared by soaking plant material in sterile water overnight. Results of tests are shown in Table 1.

We confirmed by this technique that the nodC gene is not expressed in R. meliloti cells growing in culture. In the presence of exudates from roots or seeds of alfalfa or other legumes, however, the expression of lacZ increases greatly, indicating that nodC is turned on in these conditions. If nodC and its surrounding genes are carried on a broad host-range vector, the basal level of nodC expression is higher, and the amount of induction is 30-fold, which is greater than that seen if all genes are present only in their genomic position and copy number.

NodD, by contrast, is expressed at a substantial basal level by cells growing in laboratory conditions; its expression appears to be, if anything, decreased slightly by exposure to plant exudate. The amount of nodD expression from copies borne on vectors was very high.

We examined the difference between expression on vector-borne and genomic fusions further. We found that if a nodC fusion were carried on the genome, the amount of induction it showed in response to plant exudate was greatly increased when wild-type copies of the nod genes were added in trans on a vector. If nodA, B, or C were mutated on this extra nod gene segment, the increase in inducibility of the genes in the pSym was still great. However, if the extra nod gene segment had its nodD copy deleted or mutated, then the fusion on the genome did not show an exaggerated response to plant exudate anymore. We have concluded from this that nodD is required for the response of nodC to plant exudate, at least in laboratory conditions such as those used here.

Table 1.   Activity in R. meliloti of nod-lac gene fusions.

Background strain	Plasmid	Plant Exudate	lacZ activity
w.t.	none	-	2
		+	2
nodC-lacZ	none	-	2
		+	8
nodD-lacZ	none	-	50
		+	50
w.t.	Nod$^+$(no lac)	-	2
		+	2
w.t.	nodC-lacZ	-	15
		+	450
w.t.	nodD-lacZ	-	800
		+	750
nodC-lacZ	Nod$^+$	-	3
		+	45
nodD-lacZ	Nod$^+$	-	50
		+	47
nodC-lacZ	NodABC$^-$	-	2
		+	43
nodC-lacZ	NodD$^-$	-	2
		+	4

RNA polymerase.

Gene expression is controlled at the molecular level by protein-nucleic acid interactions which affect initiation of transcription, termination, and efficiency of translation, among other steps. We wish to focus on the mechanism of transcription of the nodulation gene segment, and to begin this we have used RNA polymerase from Rhizobium meliloti to express various genes in vitro.

We purified RNA polymerase using the rapid procedure of Gross et al. This yields small amounts of enzyme which are suitable for carrying out several in vitro transcription assays. In Rhizobium this procedure results in an active enzyme preparation as described by Fisher and Long (this volume). The general appearance of the protein preparation resembles those found in more extensive biochemical studies of Rhizobium RNA polymerases (Nielsen and Brown, 1985).

By means of the micropurified enzyme, we were able to show that R. meliloti RNA polymerase produces an in vitro transcript from the E. coli trp leader DNA segment which is identical to that produced in vitro by E. coli RNA polymerase. This indicates precise recognition of the E. coli promoter and rho-independent termination sequences. We further used R. meliloti RNA polymerase to transcribe the nodD-nodA intercistronic segment of R. meliloti, and obtained RNA species indicating in vitro transcription of nodD, which is expected since this gene is expressed well in free-living cells. This in vitro transcription approach should serve as a powerful tool for understanding control of nod gene expression at the molecular level.

REFERENCES

Egelhoff TT et al (1985) DNA 4, 241-248.
Egelhoff TT, Long SR (1985) J. Bacteriol. In press.
Ditta G et al (1980) Proc. Natl. Acad. Sci. 77, 7347-7351.
Downie JA et al (1985) Mol. Gen. Genet. 198, 255-262.
Fisher RF et al (1985) Appl. Env. Microbiol. 49, 1432-1435.
Gross C et al (1976) J. Bacteriol. 128, 382-389.
Jacobs TW et al (1985) J. Bacteriol. 162, 469-476.
Kondorosi E et al (1984) Mol. Gen. Genet. 193, 445-452.
Long SR et al (1982) Nature 298, 485-488.
Mulligan JT, Long SR (1985) P.N.A.S. In press.
Nielsen BL, Brown LR (1985) J. Bacteriol. 162, 645-650.
Rossen L et al (1984) Nucleic Acids Res. 12, 9497-9508.
Schmidt J et al (1984) EMBO J. 3, 1705-1711.
Schofield P et al (1983) Mol. Gen. Genet. 192, 459-465.
Torok I et al (1984) Nucleic Acids Res. 12, 9509-9524.

ACKNOWLEDGMENTS

This work was supported by N.I.H. Grant 1-R01-GM30962 and D.O.E. contract DE-AT03-82ER12084.

THE STRUCTURE AND REGULATION OF THE NODULATION GENES OF RHIZOBIUM LEGUMINOSARUM.

J.A. DOWNIE, L. ROSSEN, C.D. KNIGHT, C. SHEARMAN, I.J. EVANS &
A.W.B. JOHNSTON
JOHN INNES INSTITUTE, COLNEY LANE, NORWICH NR4 7UH, UK.

1. INTRODUCTION

This paper describes our current understanding of the organization of the nodulation genes in a strain of R. leguminosarum, the species that nodulates peas, Vicia, Lens and Lanthyrus. In many of the fast-growing species of Rhizobium it is well established that the genes for nodulation and nitrogen fixation are clustered on large "symbiotic" plasmids one of which is the R. leguminosarum plasmid pRL1JI whose transfer to other species of Rhizobium allows the transconjugants to nodulate peas (Johnston et al., 1978). Further, a region, no more than 10kb in size, was sufficient to confer the ability to nodulate peas when transferred to a strain of R. phaseoli cured of its resident symbiotic plasmid (Downie et al., 1983). The fact that the nodulation and host-range genes of pRL1JI are clustered within such a small region means that they can be analysed in considerable detail using both genetical and physical techniques.

2. RESULTS

2.1 Identification of seven nod genes in pRL1JI

By a combination of cloning and of mapping Tn5-induced nod mutations, the pRL1JI nod genes were localized within a 8kb region of this plasmid (Downie et al., 1983 and 1985). The sequence of this region was established by sonicating the DNA cloning into an M13 vector and sequencing the inserts (Sanger et al., 1983; Rossen et al., 1984; unpublished observations).

Analysis of the DNA sequence showed that this nod DNA contained seven long open reading frames (ORFs), corresponding to genes termed nodABCDEF and G (see Fig. 1). Previously, mutations in these genes had been isolated and their effects on nodulation had been described (Downie et al., 1985). Access to the sequence of this region made it possible to determine exactly the locations of these mutations relative to the positions of the different ORFs. The following describes what is known of the structure, properties and regulation of the individual nod genes in this region of pRL1JI.

2.1.1. nodAB and C. These genes are considered together because they appear to be in a single transcriptional unit since the open reading frames are separated by only 20-30 bps (Rossen et al., 1984). The deduced molecular weights of the polypeptide products of nodAB and C genes were respectively 18K, 23K and 46K . It has been shown that in an in vitro transcription/translation system, the nodC gene directed the synthesis of a polypeptide with a molecular weight of 46,000 (Downie et al., 1985), in good agreement with the value obtained from the DNA sequence. Analysis of the predicted polypeptide sequences of nodAB and C revealed few clues as to their function. None appeared to have a leader sequence but the nodC polypeptide had a hydrophobic carboxy terminus (see also Torok et al., 1984) indicating that it may be a protein associated with the inner membrane of Rhizobium.

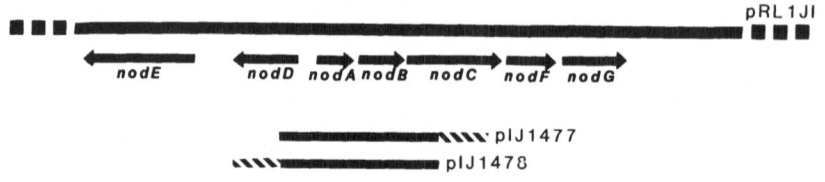

FIGURE 1. Representation of the location and dimensions of the nodABCDEFG
genes of R. leguminosarum and of the nod DNA cloned in the lac fusion
plasmids pIJ1477 and pIJ1478. The dimensions and the orientations of
transcription of the seven nod genes are indicated by arrows. A 2.2kb
BamHI fragment, indicated by the solid line was cloned in pIJ1363 in both
orientations; in pIJ1477, the lacZ genes (indicated by the hatched line)
was fused to nodC and in pIJ1478, lacZ was fused to nodD.

The nodABC genes are required and are sufficient for root hair
deformation and curling since (a) mutations in them abolish root hair
curling and nodulation (Downie et al., 1985) and (b) the nodABC genes,
cloned in the vector pKT230, can induce root hair curling in a strain of
Rhizobium cured of its symbiotic plasmid (Rossen et al., 1984). The nodABC
genes of R. meliloti have also been sequenced (Torok et al., 1984;
Egelhoff et al., 1985) and shown to be very similar to those in R.
leguminosarum (Rossen et al., 1984), the major difference being that in
R. meliloti the nodA gene has an additional 30 amino acids at its amino-
terminal end. This similarity in the sequences of the nodABC genes in
species with different host ranges is consistent with earlier genetic
evidence; when the cloned nod genes of R. leguminosarum were introduced
into a nodC mutant strain of R. meliloti, the transconjugants nodulate
alfafa showing that the nodC genes in these two species are functionally
equivalent (Kondorosi et al., 1984).

It might have been thought that the construction of a Rhizobium strain
containing the cloned nodAB and C root hair curling genes expressed at high
level would be a facile way of 'breeding' increased competitiveness into
Rhizobium strains. Sadly this appears not to be the case. When the
recombinant plasmid pIJ1389 (Rossen et al., 1984), containing the nodABC
genes cloned such that they are transcribed from a promoter in the vector
pKT230, was introduced into a wild-type strain of R. leguminosarum, the
number of nodules/pea was actually reduced by at least 90%. This
inhibition was not seen with strains carrying the cloned nodC gene alone.
Therefore over-expression of nodABC is deleterious to nodulation and this
implies that, for nodulation to occur normally, the expression of these
genes must be regulated (see below).

2.1.2. nodD. Upstream from nodA was found another ORF that corresponded
to a previously identified (Downie et al., 1985) gene, nodD. This ORF
spanned a region in which several nod:Tn5 alleles which abolished
nodulation and caused a delay in root hair curling had been located. The
nodD gene is transcribed in the opposite orientation from nodABC (Fig. 1)
so the space between nodD and nodA must contain the promoters for both
these genes. The deduced molecular weight of the nodD gene product is 33K,
in good agreement with the results of in vitro transcription/translation
experiments (Downie et al., 1985).

As with the nodABC genes, nodD also appears to be a 'common' nod gene; nodD mutants of R. leguminosarum can be corrected by cloned R. phaseoli nod DNA (unpublished) and interspecific complementation of R. trifolli and R. meliloti nodD mutations has also been reported (Fisher et al., 1985). Furthermore, the sequence of nodD in pRL1JI and its location relative to nodABC is highly conserved in R. meliloti (Egelhoff et al., 1985). It had been suggested earlier (Downie et al., 1985) that nodD may be a regulatory gene which acts to induce the expression of the root hair curling genes, nodABC. The reasoning was that the cloned nodABC genes are sufficient for root hair curling (Rossen et al., 1984 and see above) yet nodD:Tn5 mutations severely inhibited root hair deformation. Direct evidence that nodD does indeed regulate nodABC was subsequently obtained (see below).

2.1.3. nodE. To the left of nodD (as drawn in Fig. 1) had been located a nod::Tn5 allele which caused a delay and a reduction in nodulation but which did not appear to inhibit root hair curling (Downie et al., 1985). The corresponding gene was termed nodE and it was shown that it specified a polypeptide of molecular weight 48,000 (Downie et al., 1985). An ORF of the appropriate size corresponding to nodE spanned the site of the nodE::Tn5 mutation. As with the other nod genes, the biochemical function of nodE is unknown. However, it is of interest that the nodE68::Tn5 mutation, which interrupts the coding sequence of nodE (and hence should inactivate the gene completely) led to an apparently 'leaky' phenotype, i.e. nodulation was not abolished, only delayed.

The proposed translational start of nodE is immediately preceded by a stop codon that represents the translational stop of another ORF, which is upstream from nodE and would be transcribed in the same orientation as nodE. No mutations in this ORF have been isolated and its function (if any) in nodulation is unknown. If it does represent a gene, it would specify a polypeptide of molecular weight 14,000 and its 5' end would be so far (0.65kb) from the end of nodD, that it is likely that it (and thus nodE) would be transcribed separately from nodD.

2.1.4. nodF and G. Mutations in these two genes (class V mutations in Downie et al. (1985)) cause only a slight delay (one or two days) in the onset of nodulation and they do not appear to affect root hair curling. These two genes lie downstream from nodC and are both transcribed in the same orientation as nodABC (Fig. 1). It has not been established if nodF and G are in the same transcriptional unit as each other and/or the nodABC genes.

Although the determination of the sequences of the nod genes of pRL1JI has so far yielded few hints on their biochemical function, this information has facilitated the study of their expression.

2.2. nod gene regulation.

In pRL1JI, four genes, nodABC and the divergently transcribed nodD, are required for root hair curling (see above). To study their regulation, the coding sequences of nodC and nodD were fused to the E.coli lacZ gene present in a novel wide host-range translational lac fusion vector. In these constructions, levels of galactosidase activities were under the control of the nodC or nodD promoters.

2.2.1. <u>Construction of a new lac fusion vector and of lacZ fusions to nodC and nodD</u>. The construction of the wide host-range <u>lac</u> fusion vector pIJ1363 is described elsewhere (Rossen <u>et al</u>., 1985). Briefly, it was derived from the P1 group plasmid pRK290 and contains the E.coli <u>lacZ</u> gene minus its promoter and ribosomal binding site and, in addition, it has a transcriptional terminator upstream from <u>lacZ</u>. Translational fusions, resulting in <u>lacZ</u>-hybrid proteins can be made by cloning coding sequences at the unique <u>BamH</u>1 site of pIJ1363, located at the eighth codon of the <u>lacZ</u> gene. Since P1 group plasmids have a wide host-range in Gram negative bacteria, pIJ1363 could be of value in the study of gene expression in many different genera.

To construct the <u>nodC-lacZ</u> and the <u>nodD-lacZ</u> fusion plasmids, a 2.2kb <u>BamH</u>1 fragment was cloned in the <u>BamH</u>1 site of pIJ1363. In one orientation (pIJ1477) <u>lacZ</u> was fused in frame to <u>nodC</u> and in the other (pIJ1478) a <u>nodD-lacZ</u> fusion was formed (Fig. 1). The <u>nod</u> DNA cloned in these two plasmids contained the region between <u>nodD</u> and <u>nodABC</u> and therefore would include the promoters for both sets of genes.

These two plasmids were mobilized into different <u>Rhizboium</u> strains and -galactosidase activities were assayed; to reduce background levels of -galactosidase, the <u>Rhizobium</u> recipient contained a newly isolated <u>lac</u>::Tn<u>5</u> mutation.

2.2.2. <u>nodD is autoregulatory</u>. In <u>Rhizobium</u> strain 8401 <u>lac</u>::Tn<u>5</u> (a derivative of <u>R. phaseoli</u> cured of its symbiotic plasmid) <u>nodD</u> was expressed at substantial levels but this activity was reduced by 60% when the symbiotic plasmid pRL1JI was also present in the strain (Table 1). The gene on pRL1JI responsible for this inhibition was shown to be <u>nodD</u> itself because the presence of pIJ1518 (<u>nodD</u> cloned in the vector pKT230) severely inhibited the expression of <u>nodD</u> (Table 1). Thus <u>nodD</u> is autoregulatory, being capable of inhibiting its own expression.

2.2.3. <u>Induction of nodC requires nodD plus a factor from pea root exudate</u>. Unlike <u>nodD</u>, the <u>nodC</u> gene was expressed at low level when cells were grown in minimal medium and this was the case whether pRL1JI or the cloned <u>nodD</u> gene was present or not (Table 1). However, when strains containing p1J1477 (<u>nodC-lacZ</u> fusion) were grown in sterile pea root exudate (supplemented with a carbon and a nitrogen source) and were assayed for galactosidase, there was a 30-fold enhancement of activity in strains that also carried pRL1JI, or, more specifically, the cloned <u>nodD</u> gene. In contrast, the expression of the <u>nodD-lacZ</u> fusion was unaffected by growth of the cells in root exudate; i.e. <u>nodD-lacZ</u> was still expressed at high level in the strain lacking pRL1JI and <u>nodD</u> was still able to mediate the repression of its own expression (Table 1).

These results show that <u>nodD</u> is a regulatory gene whose product acts to inhibit its own expression constitutively but which, in the presence of a factor in pea root exudate, can activate the expression of the <u>nodABC</u> genes. In <u>R. meliloti</u> it has also been shown that <u>nodD</u> activates the expression of <u>nodABC</u> but, in contrast to R.leguminosarum, it does not inhibit its own transcription (Mulligan and Long, 1985); given the similarity of the <u>nodD</u> genes in the two species (see above) this difference is, initially, surprising and its significance remains to be determined.

TABLE 1.   -galactosidase activities with pIJ1478 (nodD-lacZ) and pIJ1477
(nodC-lacZ)

Strain	Introduced plasmid (genes carried)	-galactosidase activity	
		Minimal medium	Root exudate
8401 lac⁻ pIJ1478	none	600	600
	pRL1JI	200	200
	pIJ1518 (nodD)	40	40
8401 lac⁻ pIJ1477	none	50	70
	pRL1JI	40	3000
	pIJ1518 (nodD)	55	2500

The various strains were grown in minimal liquid medium or in root exudate
obtained from sterile pea seedlings (the root exudate was supplemented with
mannitol (0.2%) and sodium glutamate (0.1%)). The cells were lysed and
assayed for -galactosidase by standard methods (Miller, 1972).

Recent studies, using different nodC-lacZ fusion plasmids have
indicated that the promotor for the nodC gene of pRL1JI lies between nodD
and nodA, confirming that the nodABC genes are in a single transcriptional
unit. Analysis of the sequence betwen nodD and nodABC revealed the
presence of several palindromes and inverted repeats but, in the absence of
S1 mapping data, their role in the function of the nodD and nodABC
promoters remains to be established.

The organization and the mode of regulation of the nodD and nodABC
genes is very similar to that of the araCBAD genes of E.coli (Lee et al.
1981). In this system the regulatory araC gene is transcribed divergently
from araBAD; it represses its own expression constitutively but, in the
presence of arabinose, it can act to induce the expression of the araBAD
operon.

3. CONCLUSIONS

The anatomy of the pRL1JI nodulation genes is fairly clear but the same
cannot be said for the physiology and biochemistry of their gene products.
Insights into nod gene function will come as a result of keen eyes and good
luck, IBM (or similar) and X-gal. Some chance or inspired observation of a
correlation between the presence of nod genes and a particular biochemical
pathway and/or the induction of some aspect of the infection proces (e.g.
thick short roots or root hair curling) by a cell-free component of
Rhizobium may be demonstrated. As computer programs get more
sophisticated, so it may be possible to identify with increasing accuracy
and range, the motifs of amino acid sequences (deduced from DNA sequence)
which point to the role of part or the whole polypeptide. Lastly the fact
that a plant-derived compound is required for the induction of at least
some nod genes is of interest over and above the inherent attraction of
studies of gene regulation. If the components can be identified, it is not
unreasonable (by analogy with a range of gene systems) that the agent

required for the activation of a given set of genes is also biochemically modified by the same genes. The extension of this argument is that the inducer 'X', made by the plant, is not only involved in the regulation of <u>nod</u> genes but may also be converted by one or more of them to 'Y', a molecule responsible for at least one of the early stages of the infection of legumes by <u>Rhizboium</u>.

4. REFERENCES

Downie JA et al (1983) Mol. Gen. Genet. 190, 359-365.
Downie JA et al (1985) Mol. Gen. Genet. 198, 255-262.
Egelhoff TT et al (1985) DNA 4, 241-248.
Fisher RF et al (1985) Appl. Env. Microbiol. In press.
Johnston AWB et al (1978) Nature 276, 634-636.
Kondorosi E et al (1984) Mol. Gen. Genet. 193, 445-452.
Lee NL et al (1981) Proc. Natl. Acad. Sci. USA 78, 752-756.
Miller JH (1972) Experimental in Molecular Genetics, Cold Spring Harbor Laboratory, New York.
Mulligan JT and Long SR (1985) Proc. Natl. Acad. Sci. USA. In press.
Rossen L et al (1984) Nucl. Acid Res. 12, 9497-9508.
Rossen L et al (1985) EMBO J. In press.
Sanger F et al (1980) J. Mol. Biol. 162, 729-773.
Torok I et al. (1984) Nucl Acid Res. 12, 9509-9524.

GENES OF RHIZOBIUM JAPONICUM INVOLVED IN DEVELOPMENT OF NODULES

E. APPELBAUM/N. CHARTRAIN/D. THOMPSON/K. JOHANSEN/M. O'CONNELL[*]/
T. MCLOUGHLIN.
AGRIGENETICS CORPORATION, ADVANCED RESEARCH DIVISION, MADISON, WI 53716,
USA/[*]NATIONAL INSTITUTE FOR HIGHER EDUCATION, GLASNEVIN, DUBLIN 9, IRELAND

The use of bacterial genetic techniques to study the Rhizobium-soybean symbiosis has been hampered by the slow growth rate of Bradyrhizobium japonicum, the classical endosymbiont of soybeans, and by the apparent absence of symbiotic plasmids in this organism. The discovery of fast-growing R. japonicum strains from China (Keyser et al., 1982) raised the possibility that these strains may be better candidates for such genetic studies. In this paper, we summarize the results of our mutagenesis, plasmid mobilization, cloning, and DNA sequencing studies of sym plasmid genes that are required for nodulation and nitrogen fixation in fast-growing R. japonicum.

Pathway of Nodulation. USDA 191 is a fast-growing strain that forms nitrogen-fixing nodules on some commerical soybean cultivars as well as on the undeveloped Chinese cultivar Peking (Yelton et al., 1983). If genetic studies on nodulation by USDA 191 are to be correlated with the extensive existing literature on the structure and function of Bradyrhizobium-induced nodules, it is important to determine whether similar pathways of nodule development are used by both species. Several lines of evidence now indicate that similar pathways are used. First, light microscopic examination of sectioned mature nodules induced by USDA 191 reveals a determinate (spherical) nodule morphology that is indistinguishable from nodules induced by B. japonicum on cultivar Williams (K. Vandenbosch and E. Appelbaum, unpublished data), and infection threads are present during the early stages of infection (Heron, Pueppke, 1984). Second, polyacrylamide gel electrophoresis of soluble proteins in nodule tissues reveals a pattern of polypeptides that is indistinguishable from the pattern in B. japonicum strain USDA 110-induced nodules on cultivars Williams and Peking (C. Sengupta-Gopalan, personal communication). Moreover, western blot analysis of soluble nodule proteins using a mixture of three monoclonal antibodies (specific for leghemoglobin, nodulin-35, and nodulin-29) as probes shows that these proteins appear in the same ratios in both USDA 191-induced and USDA 110-induced nodules (E. Johansen, unpublished). Third, we observed that USDA 191 resembles B. japonicum in that it produces no nodules on a non-nodulating soybean line that is homozygous for the rj_1 gene, and it produces a greatly increased number of nodules on a super-nodulating soybean line containing the nts382 mutation (Carroll et al., 1985).

Symbiotic Plasmid in USDA 191. Analysis of USDA 191 plasmids by gel electrophoresis reveals the presence of several plasmids, including a 200 megadalton plasmid which hybridizes to nif and nod probes from

Klebsiella pneumoniae and R. meliloti (Masterson et al., 1982;
Appelbaum et al., 1984) and has been designated pSym191 (Appelbaum
et al., 1985). Further evidence for the role of pSym191 in nodule
development was obtained by the isolation of symbiotic mutants that
contain alterations in this plasmid (Table 1). Analysis of one Fix⁻
mutant (EA213) containing an insertion of Tn5 in a pSym191 nif gene
and of two Nod⁻ mutants (EJ422 and EA213C1) containing plasmid deletions indicated that genes required for both nodulation and nitrogen
fixation are located within a 75 kilobase region of pSym191.

Table 1.   Symbiotic Mutations[1] in pSym191

Strain	Symbiotic phenotype	Alterations in EcoRI fragments of pSym191 which hybridize to nif and nod probes
EA213	Fix⁻	Insertion of Tn5 in nifD gene on 4.2 kb fragment.
EJ422	Nod⁻	Deletion (about 75 kb) that removes 4.2 kb nif fragment and 3.0 kb and 9.6 kb nod fragments.
EA213C1	Nod⁻	Deletion (about 75 kb) that removes 4.2 kb nif fragment and 3.0 kb, 6.0 kb, and 9.6 kb nod fragments.
EA213C3	Nod⁻	Cured of pSym191. Missing 4.2 kb and 4.7 kb nif fragments and 3.0 kb, 6.0 kb, and 9.6 kb nod fragments.

(1)The isolation and characterization of the mutants is described
   elsewhere (Appelbaum et al., 1984; Appelbaum et al., manuscript
   submitted). The nif probe was pEA105, which contains K. pneumoniae
   genes nifMVSUXNEYKDHJ and hybridizes strongly to two USDA 191
   genomic EcoRI fragments (4.2 and 4.7 kb). The nod probe was
   pRmSL42, which contains R. meliloti nodDABC sequences (Egelhoff
   et al., 1985) and hybridizes strongly to three USDA 191 genomic
   EcoRI fragments (3.0, 6.0, and 9.6 kb). Other weakly hybridizing
   nif and nod fragments are not included in this table.

Expression of pSym191 in R. meliloti and R. leguminosarum. The
Tn5-mob system (Simon, 1984) was used to transfer pSym191 to other
Rhizobium species (Appelbaum et al., 1985). Mobilization of pSym191
to wild-type R. meliloti strain 2011 and to an R. leguminosarum strain
that contains a deletion of symbiotic genes, enabled both transconjugants to form nodule-like structures on soybeans. This shows that
pSym191 controls soybean specificity for nodulation. These structures
were quite different from USDA 191 nodules in appearance. Moreover,
they did not exhibit nitrogenase activity, they did not contain leghemoglobin, nodulin-35, or nodulin-29, and it was not possible to
recover viable bacteria from such nodules. Thus, nodule development
is clearly aberrant in these strains. This is not due to instability
of pSym191, since the plasmid was unaltered in size in these hosts and

could be transferred back to a pSym191-cured mutant of USDA 191, where
it restored the wild-type phenotype. The formation of the aberrant
nodules did not occur on a non-nodulating soybean line and the nodule
numbers were greatly increased on the nts382 line, and the formation
of these structures on wild-type cultivars was completely inhibited by
10 mM KNO_3. It therefore appears that these structures arise by the
same pathway that leads to formation of normal nodules but that nodule
development is arrested or diverted at an early stage. In contrast,
mobilization of pSym191 to ANU265, a pSym-cured derivative of the
broad host range, fast-growing strain NGR234, enabled the transconjugants to form normal nitrogen-fixing nodules on soybeans. This
implies that USDA 191 and NGR234 may be more closely related to each
other than to R. meliloti and R. leguminosarum. This also indicates
that genes which are not located on pSym191 can affect nodule development.

Cultivar Specificity for Nodule Development and Nitrogen Fixation.
USDA 257 is one of several fast-growing R. japonicum strains which
form Fix[+] nodules on cultivar Peking and aberrant Fix[-] nodules
(similar to Rhizobium meliloti (pSym191) nodules) on commerical cultivars. Mobilization of pSym191 into USDA 257 resulted in strains
that had recombinant plasmids that contained both pSym191 and pSym257
sequences (Appelbaum et al., 1985). Some of these transconjugant
strains had the cultivar specificity of USDA 191, which is Fix[+] on
many cultivars, while the other transconjugants had the cultivar
specificity of USDA 257. These results show that plasmid-associated
genes can influence cultivar specificity of nodule development. The
broadened host range could have resulted either from acquisition of
new functions encoded by pSym191, or from loss or inactivation of
USDA 257 genes which limit host range.

Organization of Common nod Genes. Genetic investigation of
R. meliloti, R. leguminosarum, and R. trifolii, has revealed a cluster
of genes, nodD nodA nodB nodC, which controls early steps in nodule
formation (Fisher et al., 1985; Rolfe et al., 1985; Downie et al.,
1985; Torok et al., 1984). These genes are organized in the same way
in all three species, the corresponding genes have similar DNA
sequences, and mutations in one species can, in several cases, be
complemented by introduction of the corresponding gene from another
species. These genes have therefore been described as "common" nod
genes.
    To study the organization of nod genes in USDA 191, total genomic
DNA from USDA 191 was digested with EcoRI, fractionated on an agarose
gel, and hybridized to pRmSL42, a probe that contains all of nodA nodB
and portions of nodD and nodC of R. meliloti strain Rm2011 (Egelhoff
et al., 1985). Three strongly hybridizing EcoRI fragments (9.6 kb,
6.0 kb, 3.0 kb) were present in USDA 191 but absent from mutants containing deletions in pSym191 (Table 1), implying that these fragments
are linked within a 75 kb region of pSym191. Further experiments
showed that the 9.6 kb fragment hybridizes to an R. meliloti fragment
that contains only nodA and nodB sequences and to a fragment that
contains only nodC sequences, but does not hybridize to the nodD
region. The 3.0 kb and 6.0 kb fragments both hybridized to the
R. meliloti nodD region but not to nodA, nodB, or nodC (unpublished

data).

A cosmid clone bank was prepared by packaging partially EcoRI digested genomic USDA 191 DNA with the vector pSUP205. Screening of this bank with the nod-specific fragment of pRmSL42 revealed clones containing USDA 191 common nod genes. A cosmid containing the 3.0 kb fragment and flanking fragments did not contain the 6.0 kb or 9.6 kb fragments. A cosmid containing the 6.0 kb fragment and flanking fragments did not contain the 3.0 kb or 9.6 kb regions. We conclude that USDA 191 contains nodD sequences in two different locations which are separated from each other and from the nodA nodB nodC locus by at least several kilobases (Appelbaum et al., unpublished data).

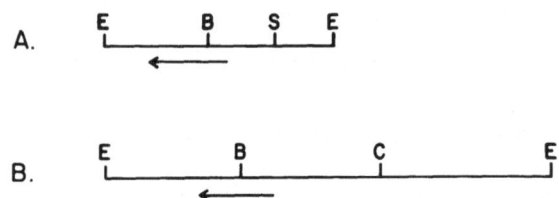

Fig. 1. Restriction maps of EcoRI fragments containing nodD-like genes of USDA 191. A, 3.0 kb fragment; B, 6.0 kb fragment. Arrows designate position and direction of translation of nodD-like open reading frames. Abbreviations: E, EcoRI; B, BamHI; S, SalI; C, ClaI.

```
         .        .        .        .        .        .
61  VPTPRAEALAPAVREALLHIHLSIISWDPFNPAQSDRSFRIILSDFMTLMFFERVVVRVA 120
        I     I        D     QF       M    V Q R   R    VIM V      K L
```

Fig. 2. Predicted partial amino acid sequence of nodD-like gene products of USDA191. Upper line: from 3.0 kb fragment. Lower line: amino acids in 6.0 kb fragment which differ from those in 3.0 kb fragments. Residues number 61 to 120 are shown.

DNA Sequence Analysis of Two nodD-like Genes in USDA 191. The 3.0 kb and 6.0 kb fragments were subcloned and mapped with several restriction enzymes (Fig. 1). The DNA sequences surrounding the central BamHI site in each clone were analyzed by the Maxam and Gilbert technique (Appelbaum et al., unpublished data). Open reading frames were identified which contain predicted amino acid sequences that are homologous to the recently reported (Egelhoff, 1985) sequence of the

R. meliloti nodD gene product. In the partial sequence shown in Fig. 2, for example, the predicted polypeptides on the 3.0 kb and 6.0 kb fragment are 80% and 73% homologous, respectively, to the corresponding region of R. meliloti. The two USDA 191 sequences are 73% homologous to each other in this region. The lengths of the complete nodD-like open reading frames are 321 residues on the 3.0 kb fragment and 312 residues on the 6.0 kb fragment, as compared to 308 residues in the nodD gene of R. meliloti. From this analysis, it appears that USDA 191 contains two complete copies of nodD-like genes.

Functions of nodD-like Genes. Several preliminary observations (Appelbaum et al., unpublished data) suggest that both nodD-like genes are functional, but in different ways.

The 3.0 kb and 6.0 kb fragments were cloned in the broad host range vector pRK290 and transferred to ANU851, a nodD⁻ mutant of R. trifolii (Fisher et al., 1985; Rolfe et al., 1985). The transconjugant containing the 3.0 kb fragment was restored in the ability to form nodules on clover. This suggests that the 3.0 kb EcoRI fragment of pSym191 contains a functional nodD gene. The transconjugant containing the 6.0 kb fragment did not show any complementation. This could reflect either lack of synthesis or lack of activity of the nodD gene product in R. trifolii.

USDA 191 derivatives containing the pRK290 clone of the 6.0 kb fragment had much smaller and much less gummy colonies than wild-type USDA 191 on several different laboratory media. This effect was not seen with pRK290 alone or with clones containing a kanamycin resistance fragment of Tn5 inserted into the BamHI site in the 6.0 kb fragment, or with a pRK290 clone of the 3.0 kb fragment. This indicates that either the nodD-like gene on the 6.0 kb fragment or another downstream gene in the same operon is responsible for this phenotype. The effect was seen regardless of the orientation of the 6.0 kb fragment in pRK290. This strongly suggests that the nodD-like gene in this fragment is expressed from a Rhizobium promoter in free-living cells. The altered appearance of the colonies suggests that the defect is in exopolysaccharide synthesis.

A USDA 191 derivative was constructed which contained a 1.9 kb deletion mutation in the 6.0 kb fragment. A BamHI-ClaI fragment that extends to the right of the BamHI site of the 6.0 kb fragment was deleted and replaced with the kanamycin resistance gene of Tn5. This mutation was crossed back into the USDA 191 genome by double homologous recombination. The structure of the deletion mutation was confirmed by blot hybridization analysis. The symbiotic phenotype of the mutant was then tested on plants. The mutation caused a leaky Fix⁻ phenotype on cultivars Williams and Peking. A time course study showed that nodules induced by the mutant were smaller than wild type nodules at each time point. Thus, nodule development appears to be delayed in these mutants and the nodules do not fully develop, even after 4-6 weeks. These results provide evidence that the nodD-like region of this fragment is functional in nodules.

A computer-assisted comparison (Lipman, Pearson, 1985) of the predicted USDA 191 nodD-like amino acid sequences with a protein sequence data base revealed a weak but apparently significant amount of homology between nodD and the lysR gene of E. coli. The lysR gene product regulates transcription of the lysA and lysR genes (Stragier et al., 1983). Thus, the structural relationship between nodD and lysR suggests that the nodD-like gene products may function as transcriptional regulators in USDA 191. This is consistant with recent results in the R. meliloti-alfalfa system, where it was shown that induction of transcription of the nodC gene by plant exudates can only be detected when a cloned nodD gene is present in high copy number (Mulligan and Long, in press).

Nif Genes of USDA 191. The EcoRI fragment containing the Tn5 insertion in EA213 (Table 1) was cloned, and the DNA sequence of a 2 kb region surrounding Tn5 was determined (Appelbaum et al., unpublished data) and compared with the known sequences of the R. trifolii (Scott et al., 1983) and B. japonicum (Furhmann, Hennecke, 1984) nif genes. The sequence revealed a complete nifH gene, a 99 bp intergenic region downstream of nifH and the first 269 codons of nifD. Tn5 was inserted in nifD. The nifH gene is 89% homologous to the nifH gene of R. trifolii, and 79% homologous to the nifH gene of B. japonicum strain USDA 110. The sequence TTGCACGNNNNTTG, which is characteristic of nif promoters (Schofield, Watson, 1985) was found upstream of the initiation codon of nifH. No similar sequence was seen between nifH and nifD. Thus, nifH and nifD appear to be part of a single operon in USDA 191, just as they are in other fast-growing rhizobia, and in contrast to B. japonicum, where they are present in separate operons (Furhmann, Hennecke, 1984). A second EcoRI fragment (4.7 kb) also contains nif-hybridizable sequences (Table 1; Prakash, Atherly, 1984), but there is no evidence that they are expressed as nitrogenase components in soybean nodules.

Conclusions

1. USDA 191 nodulates soybeans by a pathway similar to that used by slow growing rhizobia.
2. The sym plasmid of USDA 191 contains nif genes, common nod genes, and genes controlling soybean specificity for nodulation. The narrow cultivar specificity for nitrogen fixation of USDA 257 can be broadened by introduction of pSym191 into USDA 257.
3. The organization of common nod genes is different in USDA 191 than in R. meliloti, R. leguminosarum, and R. trifolii. USDA 191 contains two genes that are homologous to nodD, based on DNA sequence analysis. These two nodD-like genes are separated from each other and from nodA nodB nodC sequences on pSym191.
4. A functional nif operon containing nifH and nifD is located on pSym191.

References

Appelbaum, E. et al. (1984) In: Veeger, C. and W. E. Newton (eds.) Advances in Nitrogen Fixation Research, p. 670.
Appelbaum, E. et al. (1985) J. Bacteriol. 163:385-388.
Carroll, B. J. et al. (1985) Plant Physiol. 78:34-40.
Downie, J. A. et al. (1985) Molec. Gen. Genet. 198:255-262.
Egelhoff, T. et al. (1985) DNA 4:241-248.
Fisher, R. F. et al. (1985) Appl. Environ. Microbiol. 49:1432-1435.
Fuhrmann, M. and H. Hennecke (1984) J. Bacteriol. 158:1005-1011.

Heron, D. S. and S. G. Pueppke (1984) J. Bacteriol. 160:1061-1066.
Keyser, H. H. et al. (1982) Science 215:1631-1632.
Lipman, D. and W. Pearson (1985) Science 227:1435-1441.
Masterson et al. (1982) J. Bacteriol. 152:928-931.
Prakash, R. and A. Atherly (1984) J. Bacteriol. 160:785-787.
Rolfe, B. G. et al. (1985) In: Szalay, A. and R. Legocki (eds.)
    Advances in Molecular Genetics of the Bacteria-Plant Interaction,
    pp. 43-48.
Schofield, P. and J. Watson (1985) Nucl. Acids Res. 13:3407-3418.
Scott, K. et al. (1983) DNA 2:149-155.
Simon, R. (1984) Mol. Gen. Genet. 196:413-420.
Stragier, P. and J-C. Patte (1983) J. Mol. Biol. 168:333-350.
Torok, I. et al. (1984) Nucl. Acids Res. 12:9509-9522.
Yelton, M. M. et al. (1983) J. Gen. Microbiol. 129:1537-1547.

Acknowledgements
    We thank J. Lotzer, C. Brown, D. Barker, J. Pitas, S. Alt, and
J. Pertzborn for technical assistance, J. Adang for preparation of
figures, and C. Sengupta-Gopalan and our other colleagues at the
Agrigenetics Advanced Research Division for helpful discussions. We
thank Sharon Long, Barry Rolfe, John Watson, John Shine, Peter
Gresshoff, and Alf Puhler for strains and seeds and for communicating
results prior to publication.

DEVELOPMENTAL BIOLOGY OF THE RHIZOBIUM MELILOTI-ALFALFA SYMBIOSIS :
A JOINT GENETIC AND CYTOLOGICAL APPROACH.

J. BATUT, P. BOISTARD, F. DEBELLE, J. DENARIE, J. GHAI, T. HUGUET,
D. INFANTE, E. MARTINEZ, C. ROSENBERG* Laboratoire de Biologie
Moléculaire CNRS-INRA, BP 27 F-31326 Castanet-Tolosan Cedex, France
J. VASSE, G. TRUCHET Laboratoire de la Différenciation Cellulaire,
Faculté des Sciences Marseille-Luminy, F-13288 Marseille Cedex, France.

1. INTRODUCTION

Alfalfa is the cultivated temperate legume which fixes the largest
amount of nitrogen and produces the largest amount of protein per
hectare. Nitrogen fixation takes place in specialized organs, the root
nodules, the formation of which involves the co-differentiation of both
symbiotic partners. Dissection of this developmental process requires a
joint genetic and cytological approach.

In Rhizobium meliloti 2011 (=SU47) common nodulation (nodABCD)
genes required for early infection steps as well as genes controlling
symbiotic nitrogen fixation (nifHDK, nifA, fixABCE)were found to be
located on a pSym megaplasmid (Rosenberg et al. 1981, Long et al. 1982,
Szeto et al. 1984, Jacobs et al. 1985, Weber et al. 1985). This prompted
us to look for other symbiotic genes on this replicon.

By a series of in vitro-in vivo recombinations we have cloned large
fragments of pSym into the broad host-range plasmid RP4 (Julliot et al.
1984). One of these episomes, pGMI42, carries a 290 kb pSym fragment
containing the nifHDK operon, common nod genes and the genes controlling
the host range (Batut et al. 1985, Truchet et al. 1984). This episome
has been subcloned into pBR322 and into the broad host-range plasmids
pLAFR1 and RP4 for physical mapping of this pSym "42" region, for
providing material for site-directed transposon mutagenesis and for
functional complementation (Batut et al. 1985). Site-directed deletions
of various sizes, from 2 to 300 kb, were also constructed to facilitate
genetic analysis (Huguet et al. submitted ; Renalier et al. in
preparation).

In this paper we describe (1) a nod hsn region, close to nifHDK,
coding for the specific nodulation of alfalfa, the specificity of root
hair curling and infection thread formation and (2) a fix region,
distant of about 200 kb from nifHDK, which contains at least 3 operons
and seems to control differentiation, functioning and persistence of the
nitrogen-fixing central tissue of the nodule.

2. GENETIC CONTROL OF EARLY SYMBIOTIC STEPS

2.1. Cloning of nodulation genes
pGMI71 is a RP4-prime plasmid which contains a 70 kb insert of pSym
(Truchet et al. submitted). Agrobacterium tumefaciens GMI9050 (pGMI71)
transconjugants induced the formation of genuine nodules (cortical
origin, polar meristem, peripheral endodermis and vascular bundles) on

* Authors appear in alphabetical order.

alfalfa, and not on clover. Thus pGMI71 carries the genetic information for triggering nodule formation on a specific host. It restored the Nod[+] phenotype when introduced into R. meliloti GMI766, a Nod[−] derivative carrying a large deletion of pSym (ΔnodABCD nifHDK nifA). pGMI71 was used as a source of DNA for cloning R. meliloti host-specific nodulation genes into pRK290 by the plant selection procedure of Long et al. (1982), using RmGMI766 as recipient. Transconjugants isolated from nodules carried pRK290-primes containing in common the EcoRI fragments of 8.7, 1.8, 1.2, 2.1 and 15.3 kb (see Fig. 1). These fragments lie adjacent to each other on the EcoRI map of the corresponding pSym region (Long et al. 1982, Batut et al. 1985).

One of the smallest episomes, pGMI149 (see Fig. 1) was used in further experiments. It was transferred into A. tumefaciens 9050 (a strain cured of the Ti plasmid), R. leguminosarum 3688, R. phaseoli 8401 (pRL1JI) and R. trifolii ANU843. Transconjugants induced nodule formation on alfalfa. pGMI149 thus carries the host-range nodulation genes required for alfalfa nodulation.

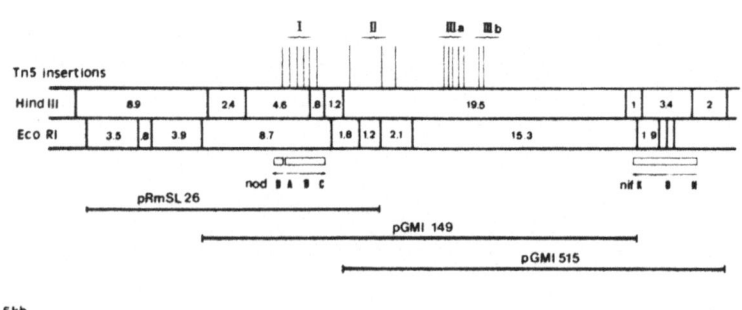

FIGURE 1. Physical and genetic map of the nodulation region.

2.2. Tn5 mutagenesis of the nodulation region

To define nodulation regions present in the 29.1 kb pSym fragment cloned in pGMI149, we mutagenized this plasmid with transposon Tn5 in E. coli and the pGMI149::Tn5 plasmids were then mobilized into R. meliloti GMI766. Among the thousand GMI766 (pGMI149::Tn5) transconjugants tested, 80 were detected, being altered in the nodulation process i.e. lack of nodule formation (Nod[−]) or delay in nodule formation (Nod[d]). To check whether the Tn5 insertion was responsible for the symbiotic defects, Tn5 insertions of the mutants were introduced into the wild-type strain.

Tn5 insertions associated with a nodulation defect are distributed along approximately 16 kb of pSym and clustered in three regions : the common nod region (I) of 3.5 kb (Jacobs et al. 1985) distal to nifHDK, a central region (II) with insertions dispersed over about 3.5 kb, and region III of about 3-3.5 kb, proximal to nifHDK (Fig. 1).

2.3. nod and hsn mutations

All the mutants having a Tn5 insertions in the common nod region were Nod⁻. Mutations in region II resulted in a Nod^d phenotype. Mutations in region III had Nod^d or Nod⁻ phenotype.

The plasmid pIJ1089 is a pRK290 derivative carrying nod genes of R. leguminosarum required for pea and Vicia hirsuta nodulation. It was used in interspecies complementation studies to discriminate amongst Tn5 insertions those altering common (nod) or specific (hsn) nodulation genes as proposed by Kondorosi et al. (1984). A normal alfalfa nodulation was restored for mutations of region I (the "common" region) but not for region III. Hence in strain 2011, as in strain 41 (Kondorosi et al. 1984) a cluster of hsn genes lies between nodABC genes and the nifHDK operon.

To define transcription units genetically, complementation tests were performed with pairs of various mutants. Mutants carrying Tn5 insertions in pSym were made Rec⁻ by introduction of a rec::Tn5-233 mutation (G. de Vos and E.R. Signer, submitted). Scoring for complementation was performed early to avoid the background of nodule formation of Nod^d mutants and of rare recombinants selected by the plant. hsn cluster contains at least two transcription units corresponding to regions IIIa and IIIb.

2.4. Infection phenotypes on alfalfa = Hac⁻and Inf⁻ mutants

Twenty mutants were studied cytologically for their infection properties. The following abbreviations are used to describe the infection phenotypes (Vincent 1980, Truchet et al. submitted) : Had, various root hair deformations such as branching, bulging, moderate curling, swelling, waviness, but not marked hair curling ; Hac, marked root hair curling (=shepherd's crook) ; Inf, infection thread formation. These results are summarized in Table 1 : mutants in region I and IIIb were Hac⁻, mutants in region IIIa were Inf⁻ and mutants in region II showed only a delay in infection.

TABLE 1. Symbiotic properties of nodulation mutants (for legend see text)

Tn5 insertions		M. sativa			T. repens	N. oleracea
		Hac	Inf	Nod	Hac	Nod
I	2310, 2208, 154, 2304, 2217, 2303	−	−	−	−	−
II	2412, 2402	+	+	d	−	+
IIIa	2311, 2309, 135 2205, 2407	+	−	d	+	+
IIIb	2212, 2121	−	−	−	−	+
wild-type control		+	+	+	−	+

2.5. Positive control of hair curling specificity

Long and co-workers have shown that the common nodABCD genes are required for triggering marked root hair curling on alfalfa (Long et al. 1982, Jacobs et al. 1985). pRmSL26 is a derivative of the broad host-range cosmid pLAFR1 containing nodABCD genes (see Fig. 1). Introduction of pRmSL26 into A. tumefaciens GMI9050 and R. meliloti GMI766

enables these strains to produce shepherd's crooks on white clover
(Truchet et al. submitted). Thus the genetic information sufficient to
provoke this plant reaction is contained in pRMSL26 and is likely to be
nodABCD.

Tn5 mutagenesis showed that a second pSym region is involved in the
control of marked root hair curling of alfalfa : the hsn region IIIb
(Fig. 1). pIJ1089 plasmid, containing R. leguminosarum nodulation genes,
restored the Hac+ phenotype when introduced into the Hac⁻ mutants
carrying Tn5 insertions in the common nod genes whereas Hac⁻ mutants of
the hsn region (Tn5 insertions 2121 and 2212) were not complemented.
This suggested that this hsn transcription unit is involved in the
specificity of root hair curling on alfalfa. This was demonstrated by
introducing pGMI515, a RP4 derivative carrying the hsn region, into
R. trifolii ANU843 : ANU843 (pGMI515) transconjugants were Hac+ on
alfalfa. Tn5 insertions of the hsn regions were introduced into
R. trifolii ANU843 (pGMI515). Insertions 2212 and 2121 (see Fig. 1) in
the hsn transcription unit proximal to nifHDK resulted in Hac⁻ phenotype
: this transcription unit is required, in addition to nodABCD, to
provoke shepherd's crook formation on the homologous host.

2.6. Negative control of hair curling specificity

We have shown (Truchet et al. submitted), by transfer of pGMI149
(nodulation regions I, II and III), pRmSL26 (region I) and pGMI515
(regions II and III) into R. meliloti GMI766 and A. tumefaciens GMI9050,
that sequences controlling the formation of shepherd's crook negatively
on an heterologous host as white clover were located in region II and/or
III. We have thus checked the infection phenotype of various mutants on
white clover (Table 1).

Mutants carrying Tn5 insertions in regions I, II and IIIb were
Hac⁻ on clover as the wild-type control. In contrast the Hac+ Inf⁻
mutants carrying Tn5 insertions from 2311 to 2407 in region IIIa induced
genuine shepherd's crooks on clover. Thus gene(s) controlling marked
root hair curling negatively on a heterologous host are located in the
hsn region IIIa.

2.7. Control of infection thread formation

A bright refractile spot can be observed in the middle of shepherd's
crooks and is likely to be the point where rhizobia enter the root hair
and initiate infection thread formation (see Dart 1977, Truchet et al.
submitted). A. tumefaciens (pRmSL26) and R. meliloti GMI766 (pRmSL26)
induce on white clover the formation of shepherd's crooks with a bright
spot but not of infection threads. Insertions in hsn region IIIa, distal
to nifHDK, result in a Hac+ Inf⁻ phenotype on alfalfa : presence of
shepherd's crooks with bright spots and lack of infection threads. Thus
bright spot formation, that is entry of bacteria into the root hair and
initiation of infection thread, seems to be controlled by nodABCD genes
whereas infection thread growth, starting from the spot, is controlled
by an hsn transcription unit located about 10 kb apart. Tn5-induced
mutations in region IIIa are partly complemented by pIJ1089 i.e. a
limited number of threads, most of them unstained by methylene blue,
develop slowly. Thus genes controlling similar functions could be
present in the cluster of R. leguminosarum genes cloned into pIJ1089.
The nodulation phenotype remains unchanged.

2.8. Nodulation of a "crak-in entry" legume

We have shown that R. meliloti strains can induce nodule formation on an aquatic tropical legume Neptunia oleracea (Dreyfus et al. in preparation). The infection process does not seem to require the formation of shepherd's crook and of infection threads through root hairs ; rather, early infection steps seem to involve intercellular penetration of bacteria (Vasse et al. in preparation).

A set of RmGMI766 (pGMI149::Tn5) mutants was tested for nodulation of this "crack-in entry" host. Mutants altered in the common nod region were Nod$^-$ on Neptunia. On the other hand the mutants from regions II and III, which were Nod- and Nod^d on alfalfa, were Nod$^+$ on Neptunia.

Thus the hsn cluster (regions IIIa and b) could be required for specific plant penetration through the root hairs whereas nodABCD genes may control nodule initiation. This is in agreement with the report of Hirsch et al. (1985) that A. tumefaciens carrying small pSym inserts containing nodABCD genes are able to induce, although at a very low frequency, nodules on alfalfa after an intercellular infection.

3. CHARACTERIZATION OF A NEW CLUSTER OF LATE SYMBIOTIC GENES

Recently we demonstrated, using transposon mutagenesis, that a pSym region located more than 200 kb from the nifHDK operon was essential for symbiotic nitrogen fixation. Mutants in this region showed no defect in intermediary metabolism suggesting that the mutated genes directly control symbiotic functions (Batut et al. 1985).

3.1. Genetic organization

In order to define the genetic organization of this fix region we constructed site-directed deletions (Fig. 2, C) and performed Tn5 site-directed mutagenesis (Fig. 2, B). The fix region spans at least 13 kb and is surrounded by regions non essential for symbiotic nitrogen fixation of at least 22.3 kb on its right side (Huguet et al. submitted) and 7.5 kb on its left side. Two fix regions are separated by a 5.5 kb fragment in which neither Tn5 insertions nor a deletion affected the symbiotic efficiency.

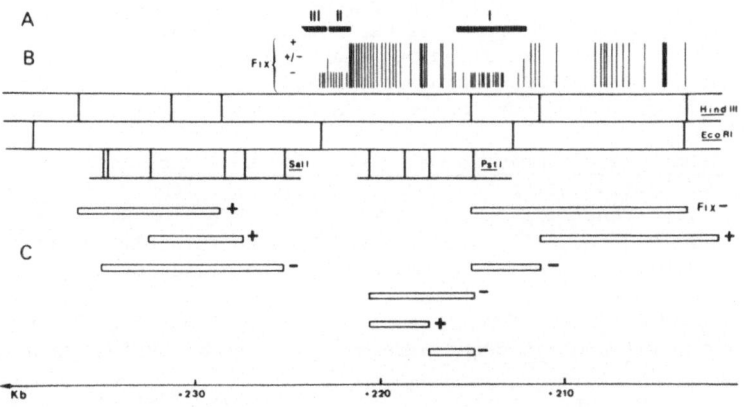

FIGURE 2. Physical and genetic map of the fix region.

The number of transcription units in each fix region was investigated by checking complementation, in a Rec⁻ genetic background, between pSym-located and pLAFR1-located fix::Tn5 mutations. Preliminary results indicate that the right fix region (4 kb) constitutes one transcription unit (I) whereas at least two transcriptions units (II and III) are present in the left fix region (about 3.5 kb), the left border of which is not yet precisely mapped (Fig. 2, A). Therefore the newly identified fix region is a cluster of symbiotic genes like the nif-fix region already described (Szeto et al. 1984, Weber et al. 1985).

3.2. Cytological characterization

In order to correlate the symbiotic defects with alterations in the differentiation process of the nitrogen-fixing nodules, cytology of Fix⁻ nodules induced by mutants affected in the three identified transcription units was examined by photonic and electron microscopy.

Fix⁻ nodules showed the same overall organization as wild-type and the four zones which are classically recognized in the central tissue of a Fix⁺ nodule were present. However zone III which contains the nitrogen-fixing bacteroids in a Fix⁺ nodule was reduced in size and a large senescence zone (IV) could be observed. This early degeneration was particularly important for one mutant located at the right end of transcription unit I.

In all cases bacteria were released from infection threads and were surrounded by a peribacteroid membrane within the host plant cytoplasm. However bacterial evolution towards fully differentiated bacteroids was altered. Two nearby mutations in the left part of transcription unit I prevented the bacteroid cells from reaching the last stage of differentiation and appear to be blocked in a state where the bacteroid cytoplasm is still heterogeneous. With the other mutants, bacteroids similar to those ultimately formed by the wild-type strain could be observed but for all the mutants examined a premature degeneration prevented a great proportion of bacterial cells from reaching the last cytological stage of differentiation and/or from maintaining in this differentiated state.

The new fix cluster controls bacteroid development and/or nodule persistence (see Vincent 1980) : therefore it could contain developmental fix genes. Surprisingly similar phenotypes were observed for R. meliloti mutations in the nifHDK operon as well (Hirsch et al. 1983). Therefore the possibility cannot be ruled out that this fix cluster carries nif genes which, like nifHDK, would control not only nitrogen fixation itself but also nodule development in a way which remains to be elucidated.

3.3. Are these late symbiotic genes specific to R. meliloti ?

By DNA hybridization of a probe consisting in the 3.7 kb HindIII fragment from transcription unit I with total DNA from R. leguminosarum strain 6015 or 3688 (a 6015 derivative carrying the self-transmissible pRL1JI pSym plasmid from R. leguminosarum strain 248, Hirsch et al. 1980) we demonstrated homology with pRL1JI. This homology is present in a pRL1JI region different from that cloned in pIJ1085 and pIJ1089 plasmids carrying the nif and fix genes already described on pRL1JI (Downie et al. 1983).

In addition introduction of a pRL1JI::Tn5 (Nod⁺ Fix⁺) plasmid (Johnston et al. 1978) into R. meliloti Fix⁻ mutants, deleted in transcription unit I of either the 3.7 kb HindIII fragment or of the adjacent 2.4 kb PstI fragment, restored the Fix⁺ phenotype on alfalfa.

This suggests the presence on pRL1JI plasmid of genes which are functionally equivalent to those contained in R. meliloti transcription unit I. This also suggests that the expression of these pRL1JI genes does not require their activation by the specific host plant since these genes which originated from R. leguminosarum, the pea symbiont, could function in alfalfa. Another explanation would be that pRL1JI controls an alternative route for nodule development which does not need the expression of genes equivalent to those present in R. meliloti 2011 transcription unit I.

On the contrary pRL1JI plasmid could not complement a R. meliloti Tn7 mutant affected in transcription unit II and no hybridization could be found between pRL1JI plasmid DNA and a probe encompassing the transcription unit II. However, since homology between this probe and total DNA from strain 6015 was found, it is possible that in R. leguminosarum genes homologous to transcription unit II exist but are not located on this particular pSym plasmid.

4. REFERENCES

Batut J et al. (1985) Mol. Gen. Genet. 199, 232-239.
Dart PJ (1977) In Hardy RWF and Silver WS, eds, A Treatise on Dinitrogen Fixation, section III, pp 367-472, John Wiley, New York.
Downie JA et al. (1983) EMBO J. 2, 947-952.
Hirsch AM et al. (1983) J. Bacteriol. 155, 367-380.
Hirsch AM et al. (1985) J. Bacteriol. 161, 223-230.
Hirsch PR et al. (1980) J. Gen. Microbiol. 120, 403-412.
Jacobs TW et al. (1985) J. Bacteriol. 162, 469-476.
Johnston AWB et al. (1978) Nature 276, 634-636.
Julliot JS et al. (1984) Mol. Gen. Genet. 193, 17-26.
Kondorosi E et al. (1984) Mol. Gen. Genet. 193, 445-452.
Long SR et al. (1982) Nature 298, 485-488.
Rosenberg C et al. (1981) Mol. Gen. Genet. 184, 326-333.
Ruvkun GB et al. (1982) Cell 29, 551-559.
Szeto NW et al. (1984) Cell 36, 1035-1043.
Truchet G et al. (1984) J. Bacteriol. 157, 134-142.
Vincent JM (1980) In Newton WE and Orme-Johnson WH, eds, Nitrogen Fixation, vol II, pp 103-129, University Park Press, Baltimore.
Weber et al. (1985) In Szalay AA and Legocki RP, eds, Advances in the Molecular Genetics of the Bacteria-Plant Interaction, pp 13-15, Cornell University Publishers, Ithaca.

5. ACKNOWLEDGEMENTS

This work was supported by the grant "Fixation Biologique de l'Azote" (Société Nationale Elf Aquitaine, Entreprise Minière et Chimique, Rhône-Poulenc, CdF-Chimie) and by the "European Communities Biomolecular Program". We are grateful to Monique Dénarié and Danièle Moréra for preparing the manuscript and to Sylvie Camut, Michèle Gherardi and Fabienne Maillet for skilful technical assistance. We thank Sharon Long, J.A. Downie and A.W.B. Johnston for providing bacterial strains and bacteriophages. We thank in particular M. David, D. Kahn, G. de Vos and E.R. Signer for providing results and bacterial strains prior to publication.

HOST-RANGE GENES ALSO AFFECT STRAIN COMPETITIVENESS IN <u>RHIZOBIUM</u> <u>TRIFOLII</u>

M.A. DJORDJEVIC, C.L. SARGENT, R.W. INNES, P.L. KUEMPEL AND B.G. ROLFE
GENETICS DEPARTMENT, RESEARCH SCHOOL BIOLOGICAL SCIENCES, AUSTRALIAN
NATIONAL UNIVERSITY, CANBERRA 2601, AUSTRALIA

A 14kb <u>Hind</u>III fragment from the Sym plasmid of the <u>R. trifolii</u> strain
ANU843 codes the genetic determinants required to restore clover nodulation
ability to the pSym⁻ strain ANU845 (Schofield et al, 1984). Surprisingly,
both ANU843 and ANU845 carrying the 14kb Nod fragment were also found to
induce many nodules on <u>Phaseolus</u> beans. This indicated that the 14kb
fragment coded for bean and clover host-range determinants.

Common Nod genes: Four phenotypically distinct regions (designated I, II,
III and IV), were recognized after extensive Tn5 mutagenesis of the 14kb
Nod region, Djordjevic, et al, 1985. Region I mutants, were unable to
curl clover or bean root hairs or induce infection thread formation and
were located in <u>R. trifolii</u> <u>nod</u>A, <u>B</u>, <u>C</u> and <u>D</u> genes (see Rolfe et al, this
volume). The nodulation ability of region II mutants was both poor and
delayed and this correlated with an inability to induce normal infection
threads. The <u>R. meliloti</u> plasmid pRmSL26 complement region I and II mutants indicating that these genes were functionally conserved.

Host-range determinants: Region III mutants displayed altered host range
capabilities. They induced a poor nodulation response on <u>T. repens</u> but induced apparently normal responses on <u>T. subterraneum</u> and beans. In contrast to the parent strain, the region III mutants were able to induce
marked root hair curling, infection threads and nodule formation on <u>Pisum</u>
<u>sativum</u> plants including the recalcitrant Afghanistan variety (Djordjevic,
et al, 1985). Region IV mutants were consistently delayed in forming nodules on clovers but only at light intensities above 400 μE/m². Region IV
mutants were unable to nodulate peas, and bean nodulation ability was unaffected. Region III and IV mutants were not complemented by the <u>R.
meliloti</u> plasmid pRmSL26.

Host range determinants also affect strain competiveness: Since the nodulation response of region III and IV mutants were most similar to the
parent strain on <u>T. subterraneum</u> competition ability of these mutants were
tested on these plants. <u>T. subterraneum</u> plants processing two dominant
lateral roots ("split roots") were generated because they provided two
equally infectable areas on the same plant. This was confirmed by inoculating either the parent strain or the mutant strain onto both roots of
the one plant. Nodules appeared in equal numbers on both roots. However,
when either the region III or IV mutants were inoculated on one side and
the parent on the other, over 90% of the nodules induced were formed by
strain ANU843. Since infection thread formation occurred on both roots
we propose that the plant is preferentially nodulated by the more competitive strain.

References:
Djordjevic et al (1985) Mol. Gen. Genet. (in press).
Schofield et al (1984) Plant Mol Biol. 3; 3-11

PRODUCTS OF EARLY NODULATION GENES OF RHIZOBIUM MELILOTI

THOMAS EGELHOFF AND SHARON LONG
DEPARTMENT OF BIOLOGICAL SCIENCES, STANFORD UNIVERSITY, STANFORD CA
94305, U.S.A.

A set of conserved bacterial nodulation (nod) genes is required for host plant infection by Rhizobium meliloti and other Rhizobium species. Four such genes, nodDABC, have been identified in R. meliloti strain 1021 by genetic analysis and DNA sequencing. An essential step towards understanding nod gene functions is to characterize their protein products. We have used in vitro and maxicell E. coli expression systems to identify protein products corresponding to each of these four genes. E. coli expression analysis was conducted by inserting nod region DNA fragments into colEl based vectors, downstream of the Salmonella typhimurium trp promoter. We found that nodA encodes a protein of 21 kd, and nodB one of 28 kd; the nodC product appears as two polypeptide bands at 44 and 45 kd. Expression of the divergently read nodD yields a single polypeptide of 33 kd. These sizes, determined from SDS-polyacrylamide gel electrophoresis, correspond fairly well to protein sizes predicted from DNA sequence analysis : 21.8 kd for nodA, 23.8 kd for nodB, 46.7 kd for nodC, and 34.8 kd for nodD. The reason for the lower than predicted mobility of nodB is not clear.

We have purified the 21 kd putative nodA protein product by gel electrophoresis, selective precipitation, and ion-exchange chromatography, and have generated antiserum to the purified gene product. This permitted the immunological demonstration that the 21 kd protein is present in wild type cells and in nodB or nodC defective strains, but is absent from nodA::Tn5 mutants, which confirms that the product expressed in E. coli is identical with that produced by R. meliloti nodA. Using antisera detection, we found that the level of nodA protein is increased by exposure of R. meliloti cells to plant exudate, indicating regulation of the bacterial nod genes by the plant host. Induction of nodA expression by plant exudate is dependent upon expression of nodD; overexpression of nodD by the vector pTE3 (see below) allows exaggerated induction of nodA by plant exudate. This finding corresponds to that of Mulligan and Long (P.N.A.S., in press), that induction of nodC-lacZ fusions in Rhizobium by plant exudate is dependent upon nodD expression.

Use of the inc-P trp promoter expression vector pTE3 has allowed overexpression of a nodABC gene fragment in R. meliloti. Western blots with the nodA antisera confirm that this plasmid expresses nodA at high levels. Initial experiments suggest that wild type R. meliloti containing this plasmid is not altered in nodulation behavior.

Work is also in progress to determine the cellular location of the nodA gene product in R. meliloti. Experiments separating membranes from soluble proteins indicate that nodA is membrane bound in strains which express the gene product at low levels. Analysis of overproducing strains seems to indicate nodA product in both membrane and soluble fractions. Further work is in progress to determine more precisely the location of this gene product.

T.E. was supported by an NIH Training Grant (GM072776-09) and by the McKnight Foundation.

TRANSCRIPTION ANALYSIS OF RHIZOBIUM MELILOTI 1021 NODULATION GENES

ROBERT F. FISHER, JOHN T. MULLIGAN, HEIDI BRIERLEY AND SHARON R. LONG
DEPARTMENT OF BIOLOGICAL SCIENCES, STANFORD UNIVERSITY, STANFORD, CA
94305, U.S.A.

Following attachment to alfalfa roots, R. meliloti induces the marked
curling of root hairs and invades host cells via an infection thread.
The genes involved in the initiation of nodulation (nod genes) have been
cloned and localized to a 4-kb segment of the R. meliloti 1021
megaplasmid. Saturation Tn5 mutagenesis coupled with DNA sequencing of
nod genes and of nod::Tn5 insertion sites, has delineated four genes,
designated nodDABC, required for nodulation (Jacobs et al., 1985;
Egelhoff et al., 1985). Interspecific complementation, as well as DNA
sequence analysis, demonstrates that these genes are conserved and
represent allelic equivalents of nod loci in other fast-growing
Rhizobium species (Fisher et al., 1985). The organization of the nod
genes suggests that nodABC are transcribed coordinately, and divergently
from nodD, whose translation initiation site is separated by only 265
intervening nucleotides from that of nodA. Recent studies suggest that
induction of nodABC is dependent on expression of nodD and the addition
of a plant cell exudate (Mulligan, Long, this volume).

We have developed an in vitro transcription system using purified
R. meliloti RNA polymerase and purified restriction fragments which
presumably contain the initiation sites of mRNA synthesis for the nodD
and nodA genes. In this way we can actually visualize on polyacrylamide
gels the run-off transcripts which arise from this region and determine
their size and origin. We have chosen this approach in an effort to
achieve regulated expression of the nod genes in vitro.

RNA polymerase was purified by the method of Gross et al. (1976)
from R. meliloti grown in minimal and rich media, in the presence and
absence of inducing plant exudate. Restriction fragments containing
either the E. coli trp leader region or the R. meliloti nod genes were
used to characterize polymerase activity. The subunit structure is
identical to that reported by Nielsen, Brown (1985) except for the lack
of appearance of the 29-kd tau polypeptide. Since our preparations lack
this subunit yet retain promoter selectivity, polymerization activity,
and rho-independent termination specificity, we think it unlikely that
tau is an integral polymerase subunit.

When RNA polymerase purified from R. meliloti is used to transcribe
BamHI-SstI and BamHI-PvuI restriction fragments which contain the
complete nodD-nodA intercistronic region, we observe an approximately
365 nucleotide long transcript which apparently initiates transcription
about 100 bases upstream from the nodD translation start site. This is
consistent with primer extension analysis of isolated in vivo RNA, which
delineates the same transcription start site.

REFERENCES

Egelhoff TT et al (1985) DNA 4, 241-248.
Fisher RF et al (1985) Appl. Environ. Microbiol. 49, 1432-1435.
Gross C et al (1976) J. Bacteriol. 128, 382-389.
Jacobs TW et al (1985) J. Bacteriol. 162, 469-476.
Nielsen BL and Brown LR (1985) J. Bacteriol. 162, 645-650.

IDENTIFICATION OF RHIZOBIUM MELILOTI IN NODULATION GENES

Mary A. Honma[1], Carol A. Smith[2], Ann M. Hirsch[2], Naomi Lang-Unnasch[1] and Frederick M. Ausubel[1]

[1]Department of Genetics, Harvard Medical School and Department of Molecular Biology, Massachusetts General Hospital, Boston, MA 02114 USA
[2]Department of Biological Sciences, Wellesley College, Wellesley, MA 02181 USA

I. Evidence for Bacterial Release and/or Bacteroid Development Genes on R. meliloti pSym

A R. meliloti pLAFR1 library was conjugated into two R. meliloti nodulation mutants (Rm36a and Rm44d; Nod⁻Fix⁻) which contain large deletions covering the nod-nif region. Cosmids which complemented the Nod⁻ phenotype were isolated by inoculation of alfalfa and retrieval of bacteria from nodules. A single cosmid, pMH36, was isolated from several different ineffective nodules. pMH36 contains the previously identified common nodulation (nodABCD) and host specificity (hsn) genes but does not cover the entire deleted region.

Western blot analysis using nodule specific antiserum to probe Rm36a/pHM36 alfalfa nodule proteins showed that two nodulins were missing and that leghemoglobin was present in reduced amounts. In addition, a 66 kD protein, previously shown to be of bacterial origin, was present in very low amounts. These results indicated that few, if any, bacteria were present in the nodules. Preliminary ultrastructural analysis of nodules elicited by Rm44d/pMH36 confirmed that development was arrested at an early stage. In Melilotus alba nodules elicited by Rm44d/pMH36, no evidence of bacterial penetration was observed. In Medicago sativa, nodules were generally more developed and contained infection threads; however, in no nodules were elongate bacteroids found. These results suggest that nodulation genes for bacterial release and bacteroid development reside on the R. meliloti megaplasmid in the regions deleted in strains Rm36a and Rm44d but outside of the region cloned in pMH36.

II. Evidence for Additional nodD-like Genes in R. meliloti

Hybridization experiments showed that at least two sequences homologous to R. meliloti nodD and to the region between nodD and nodA, reside on R. meliloti pSym. One of these regions appears to contain the host specifity genes. The other hybridizing sequence is located in a 6.7 kb EcoRI fragment that is located outside of the deleted regions in Rm36a and Rm44d. The 6.7 kb EcoRI fragment was cloned from strain Rm36a and then mutagenized with Tn5. One Tn5 insertion in the nodD homologous region of the 6.7 kb EcoRI fragment was obtained. This strain shows no defective phenotype on either Medicago or Melilotus.

One explanation for the leaky phenotype of nodD mutants is that there are several copies of nodD-like genes. If a locus on the 6.7 kb EcoRI fragment is functionally similar to nodD, then a strain containing mutations in both nodD and the 6.7 kb EcoRI region may have a more severe nodulation phenotype than mutations in either alone. One possibility is that nodD belongs to a small gene family (which also contains the hsn genes) which is responsible for determining host range. According to this model, a particular subset of these genes may be required for nodulation of a specific Medicago or Melilotus host.

PLANT FACTORS INDUCE EXPRESSION OF NODULATION AND HOST-RANGE GENES IN RHIZOBIUM TRIFOLII

ROGER W. INNES*, PETER L. KUEMPEL*, HAYO CANTER-CREMERS†, MICHAEL A. DJORDJEVIC† and BARRY G. ROLFE†
*University of Colorado, Boulder, CO 80309 U.S.A. and †Australian National University, Canberra City, ACT 2601, AUSTRALIA

In Rhizobium trifolii ANU843, host specific nodulation capability is encoded within a 14kb HindIII fragment of the symbiosis plasmid (Schofield et al. 1984). To gain a better understanding of the regulation of the nodulation (Nod) genes, we have isolated lac transcriptional fusions to several Nod genes, using the mini-Mu-lac bacteriophage transposon Mu dI1734. Using a broad-host-range vector, fragments containing Mu dI1734 insertions were introduced into R. trifolii ANU845, a derivative of ANU843 which lacks the symbiosis plasmid. Four distinct regions were identified within the Nod fragment, insertions in which resulted in nodulation phenotypes similar to that found previously for Tn5 (Djordjevic et al. 1985). Region I mutants were Nod⁻ and defective in root hair curling (Hac⁻) and corresponded to the nodABC and D genes identified by sequence analysis. Region II mutants showed an exaggerated root hair curling (Hac⁺⁺) response on clover plants and a greatly reduced nodulation ability. Region III mutants were affected in host-range properties, as they gained the ability to nodulate Pisum sativum (peas), but showed only poor nodulation ability on the normal host plant, Trifolium repens (white clover). Region IV mutants showed a delay in the nodulation of Trifolium repens, but only when plants were grown under high light regimes. When ANU845 strains carrying the above Mu dI1734 insertions were grown in standard laboratory media, only insertions in nodD expressed β-galactosidase at high levels. However, when cells were placed in medium in which Trifolium repens was growing, insertions in nodA, nodB, region II, region III, and region IV were all induced from 5-10 times above basal levels. These data indicate that several operons of R. trifolii, which encode genes involved in the nodulation process, are induced directly or indirectly by a plant signal.

REFERENCES

Djordjevic MA et al (1985) Mol. Gen. Genet. In press.
Schofield PR et al (1984) Plant Mol. Biol. 3, 3-11.

NODD ENHANCES INDUCTION OF NODC BY PLANT EXUDATE

JOHN T. MULLIGAN AND SHARON R. LONG
DEPARTMENT OF BIOLOGICAL SCIENCES, STANFORD UNIVERSITY, STANFORD CA 94305, U.S.A.

The soil bacterium Rhizobium meliloti invades and establishes a symbiosis with host plants such as alfalfa. Bacterial nodulation (nod) genes are required for this invasion, but the timing of their expression is not known. We have used translational lacZ fusions to monitor the expression of nodD and nodC, which are located in the cluster of four nod genes on the R. meliloti megaplasmid (pSym).

Activity of the nodC-lacZ fusions is low in broth-grown bacteria and is induced up to thirty-fold when the bacteria are exposed to exudates from aseptically grown plants. A rapid thirty-fold induction of activity from the nodC-lacZ fusion occurs when nodD expression is high. When nodD expression is low the induction of the nodC-lacZ is slow and reduced (2-8 fold over broth-grown cells). NodD is expressed at comparable levels by broth-grown bacteria and by those exposed to plant exudates; however, its expression is 15-fold higher if it is carried on an inc-P vector relative to when it is located on the pSym megaplasmid. NodC expression is similarly affected by which replicon it is located on. NodD expression from vector-borne copies of the nod segment and response of nodC to plant exudate appear to require additional loci on the megaplasmid.

Although the product of the nodD gene appears to be involved in the induction of nodC, an intact nodD gene is not absolutely required for nodulation in R. meliloti. We have found that four EcoRI fragments from 1021 are homologous to the nodD coding region (6.5 Kb, 8.7 Kb, 10 Kb, and 15 Kb). It is possible that a gene on another fragment could partially complement nodD mutations.

We are currently mapping the nodD and nodA promoters, investigating the additional nodD-like genes and screening for mutations altered in the expression of the lacZ fusions. We have used primer extension to map the promoters and have preliminary estimates of their locations which are consistent with in vitro start sites recognized by R. meliloti RNA polymerase. We have shown that the 15 Kb EcoRI fragment between nif and nod hybridizes to nodD at low stringency and are screening a lambda clone bank for the other two nodD homologous fragments. We have developed plate assays for the induction of the nod genes that should make it possible to screen for regulatory mutations.

Our results suggest that regulation of bacterial nod gene expression is an important control mechanism early in symbiosis, and that the biochemical nature of some nod gene products may be cryptic except in cells grown in the presence of plant exudate.

ACKNOWLEDGEMENTS

This work was funded by research and training grants from N.I.H., and the McKnight Foundation.

A 34 KB COSMID INDUCES NODULE FORMATION IN A
RHIZOBIUM PHASEOLI SYM CURED STRAIN

CARMEN QUINTO, MIGUEL ANGEL CEVALLOS, YOLANDA PERALTA, GUADALUPE ESPIN
AND ARACELI DAVALOS.
CENTRO DE INVESTIGACION SOBRE FIJACION DE NITROGENO. UNIVERSIDAD NACIONAL
AUTONOMA DE MEXICO.

In fast-growing Rhizobia it is known that the "common nod genes" are
localized close to the structural nitrogenase genes (nif) (1-4, 6).

We have constructed a genomic library of the strain CE-3, a CFN-42 streptomycin resistant derivative, in the pSUP205 cosmid vehicle (7). Cosmids
containing the nif genes were selected by hybridization with a R. phaseoli
nifH probe. These cosmids were introduced into a strain cured of the Sym
plasmid (CFN-2001 Rif) derived from CFN-42 strain. Two cosmid clones
(927 and 991) were capable of inducing the formation of white nodules in
Phaseolus vulgaris. These two cosmids containing the nodulation functions
were digested with EcoRl and hybridized with the "common nod genes" probe
from R. meliloti (pSKS 5) (2).

Both of them show the same two hybridization fragments of 9.5 kb and 6.5
kb. EcoRl genome blots of strain CFN-42 hybridized with pSKS 5 show
the same hybridizing fragments than the cosmid clones and a smaller band
of approximately 200 bp.

In order to demonstrate the introduction of the cosmids 927 and 991 in
the Sym cured strain, plasmid profiles of the wild type strain, the cured
strain and the transconjugants were obtained.

Thus we have found two cosmid clones that induce nodulate formation in a
Sym cured strain. These cosmids contain one copy of the reiterated nitrogenase structural genes (5). In addition this clones have homology
with the "common nod genes" from R. meliloti.

REFERENCES

1. Downie JA et al (1983b) EMBO J. 2, 927-952.
2. Kondorosi E et al (1984) Mol. Gen. Genet. 193, 445-452.
3. Lamb JW et al (1985) Gene 34, 235-241.
4. Long SR et al (1982) Nature 298, 485-488.
5. Quinto C et al (1982) Nature 299, 724-726.
6. Schofield PR et al (1983) Mol. Gen. Genet. 192, 459-465.
7. Simon R (1984) Mol. Gen. Genet. 196, 413-420.

STRUCTURE AND REGULATION OF R.LEGUMINOSARUM NOD GENES

L.ROSSEN/C.A. SHEARMAN/C.D. KNIGHT/A.W.B. JOHNSTON/J.A. DOWNIE
JOHN INNES INSTITUTE, COLNEY LANE, NORWICH, NH46UH, UK.

In R.leguminosarum, the genes required for the nodulation of peas and the determination of host-range specificity are within a 10 kb region of the symbiotic plasmid pRL1JI. Within this region of DNA, mutations that led to the loss or severe delay of nodulation of peas were within a 6.6 kb fragment which was sequenced in order to define precisely the location of the individual nod genes.

Five coding open reading frames were identified with the corresponding genes nodA,B and C being transcribed divergently from nodD (which is upstream from nodA) and nodE (downstream from nodD).

It had been shown that nodA,B,C are sufficient for root hair curling and it was apparent that these genes are in the same operon. Mutations in nodD abolished nodulation and delayed the onset of nodulation; nodD and nodE are in separate transcriptional units.

To investigate the regulation of nod gene expression a new wide host-range translational fusion vector, pIJ1363, was constructed. Using nodD-lacZ and nodC-lacZ fusions, it was shown that nodD is autoregulatory, being able to repress the expression of its own synthesis. Further, the nodA,B,C genes were induced by a factor in pea root exudate and this induction required the nodD gene.

RHIZOBIUM TRIFOLII NODULATION GENE SEQUENCE OF nodABC AND D AND THEIR TRANSCRIPTIONAL CONTROL

PETER SCHOFIELD, MICHAEL DJORDJEVIC, BARRY ROLFE AND JOHN WATSON
CSIRO DIVISION OF PLANT INDUSTRY AND GENETICS DEPT., RSBS,
AUSTRALIAN NATIONAL UNIVERSITY, CANBERRA, AUSTRALIA.

Clover-specific nodulation genes are encoded on a 14 kb HindIII restriction fragment of the Rhizobium trifolii ANU843 Sym plasmid.

Recombinant plasmids containing Sym plasmid restriction fragments were characterized by: A) Transfer to a Sym plasmid-cured R. trifolii. This allowed the detection of root hair curling determinants (Region I). B) Transfer to an R. trifolii mutant which displays a delayed nodulation onset. This allowed the detection of gene(s) involved in nodule development (Region II). C) Transfer to other Rhizobium species. This allowed the identification of clover host specificity determinants (Regions III, IV and V). The results of this study are consistent with those obtained by Tn5 mutagenesis of the 14 kb nodulation region.

The DNA sequence of a 3.6 kb EcoRI-BglII restriction fragment, which contains portions of Regions I and II, was determined. Most probable open reading frames that are preceded by ribosome binding sites were identified and analysed by Staden's unequal codon usage programme. A synthetic oligonucleotide primer, complementary to the terminal portion of Tn5, was used to determine the exact locations of various Tn5 insertions and their flanking DNA sequences. Using the Tn5 primer, it is possible to immediately access the DNA sequence of previously generated Tn5 mutants. Two copies of a 75 bp reiterated sequence are located in Region I and these contain the promoters and initiation codons of two genes, nodD and nodF. A third copy of this sequence is located in Region IV which contains genes involved in host specificitiy. Four genes involved in root hair curling have been identified: nodA (21.6 kD); nodB (23.5 kD); nodC (ca 45 kD) and nodD (36.1 kD). The nodA and nodD genes are separated by two Nod$^+$::Tn5 insertions, suggesting that these genes belong to separate transcriptional units. Since nodB mutants have a Had$^+$ (hair distortion) phenotype, whereas nodC mutants are Had$^-$, the nodC gene may be expressed from its own promoter. The postulated nodF gene (12.6 kD) has not yet been defined by mutagenesis. A sixth gene, hsnA is involved in clover-specific nodulation and is defined by seven Tn5 insertions in Region III. Genes nodABC and D are all highly conserved with those of R. meliloti, R. leguminosarum and Parasponia Rhizobium.

RNA isolated from ANU845 (Sym$^-$) cells containing plasmid pRt150 (Region I) or pRt032 (14 kb) was hybridized to Southern blots of the nodulation region. Yields of nod gene-specific RNA were very low but differential hybridization is apparent.

The root hair curling genes (Region I), and Region II genes are structurally and functionally conserved between Rhizobium species. Region I genes, can be expressed independently of the bacterial host, on a number of legumes, indicating that when cloned their expression is uncoupled from normal regulation. This is supported by the transcriptional analysis. The presence of specific host range genes can confer on other Rhizobium species the ability to nodulate that specific host. This suggests that host range genes are involved in the activation of expression of the conserved (common) nodulation genes (Regions I and II). Transcriptional analysis suggests that the expression of the root hair curling genes, encoded in this region, is repressed by the product(s) of a gene(s) located elsewhere in the 14 kb nodulation region. To explain the genetics of nod gene expression it is necessary to propose both positive and negative regulatory controls.

TRANSPOSITION OF ISRml INTO THE NODULATION GENES
OF *RHIZOBIUM MELILOTI* JJ1c10.

Roger Wheatcroft and Robert J. Watson
Chemistry and Biology Research Institute
Agriculture Canada
Ottawa, Ontario, K1A 0C6. Canada

Introduction. Plants of *Medicago sativa* (alfalfa) form nitrogen fixing root nodules when infected with the bacterium *Rhizobium meliloti*. Some of the bacterial genes required for nodulation (*nod* genes) have been cloned from *R. meliloti* 1021 in pRmSL26 (Long et al. 1982). This cosmid, used as a probe, enabled us to isolate the common *nod* gene region from wild-type *R. meliloti* JJ1c10 and two Nod⁻ mutants. We have shown that the mutations, which are suppressed by pRmSL26 complementation, are both caused by ISRml. ISRml is an insertion sequence originally found in *R. meliloti* 1021, in which strain it transposes at high frequency and preferentially into the *nif* gene region (Ruvkun et al. 1982). We have found that there are 4 copies of ISRml in the genome of wild-type *R. meliloti* JJ1c10 and that it is transposes preferentially into the *nod* gene region in this strain.

Results and Discussion. pRmSL26 was used to probe the total DNA cut with *Eco*RI from 250 symbiotically defective mutants of *R. meliloti* JJ1c10 isolated after Tn5-mutagenesis. There were no differences from wild-type hybridization, except that in two Nod mutants a 1.3 kb insertion had occurred in the DNA homologous to the 8.7 kb *Eco*RI fragment of the probe. The fragments containing the inserted DNA were cloned from these mutants, together with the corresponding wild-type fragment, and mapped using restriction and heteroduplex analysis. The insertion sequence proved to be structurally identical in the two mutants but located in different, closely adjacent sites. Unique *Eco*RV and *Bgl*1 cleavage sites identified within the insertion sequence enabled us to extract a 0.9 kb internal fragment, which was made blunt-ended and cloned into pUC9. This was used as an insertion-specific probe of pRmR3::ISRml (Ruvkun et al. 1982). Since hybridization only occurred with a 0.9 kb *Eco*RV/*Bgl*1 fragment not found in pRmR3, we concluded the insertion sequence of *R. meliloti* JJ1c10 was identical to ISRml from strain 1021. Further use of the insertion-specific probe showed the wild-type JJ1c10 had 4 copies of the insertion sequence deployed in the genome, whereas the Nod mutants both had 5. Only two other mutants in the collection tested showed evidence of transposition, having 5 copies of the insertion sequence. Thus 2 out of the 4 transposition events resulted in mutational insertions in the *nod* gene region in JJ1c10. There was none in the *nif* gene region, a situation quite different from that reported for ISRml in *R. meliloti* 1021 (Ruvkun et al. 1982) where 32 out of 33 transposition events resulted in insertion in the *nif* gene region, the remaining one occurring in the *nod* gene region.

We conclude that ISRml insertion is target site specific and host dependent, possibly in a way that is determined by the location and deportment of ISRml copies already existing in the host.

References
Long, S.R., Buikema, W.J., Ausubel, F.M. (1982). Nature 298, 485-488.
Ruvkun, G.B., Long, S.R., Meade, H.M., Van den Bos, R.C. and
     Ausubel, F.M. (1982). J. Mol. Appl. Genet. 1, 405-418.

ANALYSIS OF THE NODULATION REGION OF THE RHIZOBIUM LEGUMINOSARUM SYM PLASMID PRL1JI

Carel.A.Wijffelman,Elly Pees,Anton.A.N. van Brussel,
Mieke Priem,Rob Okker and Ben.J.J.Lugtenberg.
Department of Plant Molecular Biology
State University of Leiden, The Netherlands

Seventeen Tn5- or Tn1831 induced nodulation mutants of the Rhizobium leguminosarum Sym plasmid pRL1JI were isolated and examined for their genetic and symbiotic properties. All mutations were located in a 6.6 kb EcoRI fragment of the Sym plasmid. Nine mutants in a 3.5 kb part on the right side of this fragment and located in the nodA, nodB,nodC or nodD genes prevent nodulation on Vicia sativa completely and are unable to induce marked root hair curling. Eight mutations in a 1.5 kb area on the left hand side of the 6.6 kb nod fragment generate other symbiotic defects. Their nodulation behaviour, designated as 'delayed' nodulation can be summarized as follows: 1. The root hair curling is much stronger than that caused by the wild-type. 2. They induce infection threads with a delay of seven days. 3. The first nodules are visible with a delay of one week and are only formed on 10 to 30% of the inoculated plants. All mutants were transferred to two strains, one strain was LPR5039, harbouring the R.trifolii Sym plasmid pRtr5a, the other strain was LPR5039 in which pRmSL26 was introduced, which contains nodulation genes of R.meliloti.The resulting exconjugants were tested for nodulation on V.sativa. All mutants in the 3.5 kb part nodulated V.sativa normally, in contrast to the mutants in the 1.5 kb part, which were not complemented by the nodulation genes of R.trifolii or R.meliloti. Therefore this 1.5 kb area is supposed to play a role in the host specific steps of the nodulation process, whereas the nodA,nodB,nodC and nodD genes are common between R.leguminosarum, R.trifolii and R.meliloti. The Sym plasmid pRL1JI can only be maintained in Agrobacterium as a cointegrate with another plasmid. This allowed us to introduce two nodulation mutants of pRL1JI in one Agrobacterium by making cointegrates with two different Rhizobium plasmids, which are compatible. In this way we constructed a number of pRL1JI diploid strains in the recombination deficient Agrobacterium tumefaciens LBA4301. These strains were tested for nodulation on V.sativa. On the basis of the nodulation pattern the nod area can be divided in three complementation groups: one group in which all host specific mutants are located, a second one in which only nodD, and a third group in which nodA,nodB and nodC are located.

MOLECULAR CHARACTERIZATION OF THE NODULATION REGION FROM <u>BRADYRHIZOBIUM</u> <u>JAPONICUM</u>.

NIRUPAMA DESHMANE, RICHARD HAUGLAND[*], ADRIAN HODGSON, MAUREEN LEAVITT, ANTHONY J. NIEUWKOOP, RUSSEL PASTIAN, PAUL RUSSELL[*], MARIA SCHELL, KARL SIROTKIN, JAE-SEONG SO, AND GARY STACEY, DEPARTMENT OF MICROBIOLOGY, UNIVERSITY OF TENNESSEE, KNOXVILLE, TN AND *ALLIED CORP., SYRACUSE RESEARCH LABORATORIES, SOLVAY, NY.

1. INTRODUCTION

The eventual ability to construct a superior nodulating <u>Bradyrhizobium japonicum</u> strain will depend upon a better understanding, at the molecular level, of the genes involved in the nodulation process. We have begun this analysis by examining 100 Kb of cloned <u>B. japonicum</u> nodulation DNA.

2. RESULTS

The DNA region encoding early nodulation functions of <u>B. japonicum</u> strain I110 was isolated by homology to the functionally similar region from <u>Rhizobium meliloti</u>. The isolation of a number of overlapping clones has allowed the construction a restriction map of the region. Southern hybridization and restriction map analysis have localized the <u>nod</u> homologous region to about 12 Kb of <u>B. japonicum</u> DNA. A mutant of <u>Rhizobium fredii</u> (USDA 201), mutant strain A05B-2, was isolated which is nodulation deficient, being unable to curl soybean root hairs (Nod⁻, Hac⁻). Some of the recombinant DNA clones isolated from <u>B. japonicum</u> were found to restore wild type function to this mutant. Functional complementation was dependent on the presence of a 1.8 Kb <u>EcoRI-HindIII</u> fragment (Russell et al., submitted).

Three clones pRjUT10, pRjUT2, and pRjUT14 are being further characterized for functionally active symbiotic regions by using site directed Tn<u>5</u> mutagenesis. The Tn<u>5</u> insertions are constructed in <u>Escherichia coli</u> and then mobilized into <u>B. japonicum</u> by triparental matings. Thus far 19 site directed Tn<u>5</u> insertion mutants have been tested on soybean plants. An essential nodulation region has been localized in the rightmost region of the 9.5 Kb <u>EcoRI</u> fragment (pRjUT2 and pRjUT10). In addition a region essential for nitrogen fixation has been identified 13 Kb away within the 3.8 Kb <u>EcoRI</u> fragment (pRjUT2 and pRjUT10).

MOLECULAR CLONING OF A CONSERVED NODULATION GENE FROM RHIZOBIUM LOTI AND
BRADYRHIZOBIUM SP (LOTUS)

C.E. PANKHURST, K.Y. CHUA, P.E. MACDONALD AND D.B. SCOTT
APPLIED BIOCHEMISTRY DIVISION, DSIR, PALMERSTON NORTH, NEW ZEALAND.

The pasture legume Lotus pedunculatus forms an effective symbiosis with
both fast-growing (Rhizobium loti) and slow-growing (Bradyrhizobium sp.
[Lotus]) bacteria. Although these two groups of bacteria share less than
5% total DNA homology, their ability to nodulate the same host suggests
that they share common genetic determinants for nodulation.

Using Tn5 mutagenesis we have isolated a Nod⁻ mutant, strain PN233, from
R. loti strain NZP2037.   Examination of the root hairs of L. pedunculatus
seedlings inoculated with PN233 showed that this mutant was unable to
induce root-hair curling.   Tn5 was shown by hybridization to be inserted
into a 7.1kb EcoRl fragment in the PN233 genome.   The 12.8kb nod :: Tn5
EcoRl fragment from PN233 was sub-cloned into pBR328 and used as a
hybridization probe to identify cosmids containing the corresponding nod
region from a pLAFRI gene library of NZP2037 DNA.   A physical map of the
nod region contained in 2 cosmids (pPN305 and pPN306) was thus established
and the site of insertion of Tn5 in PN233 determined.   Site-specific
exchange of the nod :: Tn5 fragment further demonstrated that Tn5 and not
an indigenous IS element was responsible for the nod mutation in PN233.

The nod cosmids pPN305 and pPN306 complemented the Nod⁻ phenotype of PN233
on L. pedunculatus but restoration of the Fix phenotype was variable.
Initially (up to 5 weeks after inoculation) the nodules formed were either
tumour-like (bacteria-free) or Nod⁺Fix⁻ (contained bacteroids), but after 7
weeks some plants developed Nod⁺Fix⁺ nodules.   Bacteria isolated from the
Nod⁺Fix⁻ nodules were Tet^R Neo^R while those isolated from the Nod⁺Fix⁺
nodules were Tet^S Neo^S.   As reversion of mutant PN233 has never been
observed it appears that recombination (i.e. marker rescue) is a
prerequisite for restoration of PN233 to the Nod⁺Fix⁺ phenotype.

Employing the 'in planta' complementation technique, a cosmid able to
complement PN233 was isolated from a pLAFRI gene library of Bradyrhizobium
sp. (Lotus) strain CC814s.   Nodules formed by PN233 containing the CC814s
nod cosmid (pPN330) were predominantly Nod⁺Fix⁻. Hybridization experiments
using the 7.1kb EcoRI nod fragment of NZP2037 as a probe showed that this
nod region was highly conserved in CC814s. However, this same probe showed
only weak homology with total DNA from R. trifolii strains PN100 and
NZP561.   Conversely, a 7.2kb EcoRI fragment containing the nod ABC region
of R. trifolii strain ANU843 showed only weak homology with pPN305, pPN330
and total DNA from NZP2037 and CC814s.   This data, together with that
obtained from complementation experiments where it was shown that pPN305
could not complement the R. trifolli nod D :: Tn5 mutant ANU851, suggests
that the cloned R. loti NZP2037 nod region is functionally (as well as
structurally) different to the nod ABCD region of R. trifolii strain
ANU851.

NODULATION GENES FROM <u>PARASPONIA</u> <u>RHIZOBIUM</u> STRAIN ANU289

K.F. Scott,
Centre for Recombinant DNA Research, Research School of Biological
Sciences, Australian National University, Canberra, ACT 2601, Australia.

Several strains of <u>Rhizobium</u> can effectively nodulate the plant
<u>Parasponia</u> (Trinick, Galbraith, 1980), however, the mechanism by which
non-legume nodulation occurs is unknown. These strains carry nodulation
genes functionally and structurally conserved in both fast and slow-
growing <u>Rhizobium</u> isolates (Marvel <u>et al.</u>, 1984; Rolfe <u>et al.</u>, 1985).
Using cloned nodulation genes from <u>R.trifolii</u> (Djordjevic <u>et al.</u>, this
volume) as a hybridisation probe, we have identified and completely
characterised the conserved nodulation region in the slow-growing isolate
<u>Parasponia</u> <u>Rhizobium</u> strain ANU289. DNA sequence homology to the <u>nod</u>
genes of <u>R.meliloti</u> (Torok <u>et al.</u>, 1984; Egelhoff <u>et al.</u>, 1985), <u>R.legu-</u>
<u>minosarum</u> (Rossen <u>et al.</u>, 1984) and <u>R.trifolii</u> (Schofield <u>et al.</u>, this
volume) has identified four genes, <u>nod</u>A, <u>nod</u>B, <u>nod</u>C and <u>nod</u>D in ANU289.
<u>Nod</u>A,B and C are closely linked and appear to be organized in a single
operon <u>nod</u>ABC, while <u>nod</u>D is located 700bp 5' to <u>nod</u>ABC and is oriented
in the opposite direction. In addition, the ANU289 <u>nod</u> region contains a
novel open reading frame (ORF1) preceding <u>nod</u>A and oriented in the same
direction.

Complementation studies using <u>R.trifolii</u> mutants carrying Tn5 in
<u>nod</u>A, <u>nod</u>C and <u>nod</u>D (Schofield <u>et al.</u>, this volume) resulted in functional
complementation of only <u>nod</u>D mutants by the cloned ANU289 <u>nod</u> region.
Nodulation of white clover by these transconjugants, although delayed and
reduced compared to <u>R.trifolii</u> wild-type strains, demonstrates the
ability of the transcriptional recognition sequences preceding the ANU289
<u>nod</u>D gene to function in <u>R.trifolii</u>. Thus it is interesting to note the
substantial DNA sequence conservation preceding the <u>nod</u>D genes in ANU289,
<u>R.trifolii</u> (Schofield <u>et al.</u>, this volume), <u>R.meliloti</u> (Egelhoff <u>et al.</u>,
1985) and the putative <u>nod</u>F gene in <u>R.trifolii</u> (Schofield <u>et al.</u>, this
volume). As shown below, comparison of these sequences identifies a
consensus sequence.

The high degree of homology, together with the observation that the
sequence is reiterated in <u>R.trifolii</u> indicates that it may play some role
in the transcription of nodulation genes.

References
Egelhoff TT <u>et al.</u> (1985) <u>DNA</u> 4:241-248.
Marvell DJ <u>et al.</u> (1984) Proc. 5th Int. Symp. Nitrogen Fixation (Veeger
    C and Newton WE, eds.), Martinus Nijhoff/W. Junk, The Hague, p.691.
Rolfe BG <u>et al.</u> (1985) In: Advances in the Molecular Genetics of the
    Bacteria Plant Interaction (Szalay AA and Leckocki RP, eds.),
    Cornell University, NY, pp.44-47.
Rossen L <u>et al.</u> (1984) <u>Nucl. Acids Res.</u> 12:9497-9508.
Torok I <u>et al.</u> (1984) <u>Nucl. Acids Res.</u> 12:9509-9524.
Trinick MJ and Galbraith J (1980) <u>New Phytol.</u> 86:17-26.

```
                                                                   Met Arg Phe ..
P.R.  :- ...TCCGCAATCTGGTAAAATCGATTGTTTCGATAGAATACATCCACACGATGGATAGACTCACAAC ATG CGG TTC ..
             :   :   :::::::::: :::::::: ::: : : :::: ::: :::::: ::
R.t.  :- ...CCGAAAGATTGGTAAAATTGATTGTTTGGATCGCAATCATCTACAGCGTGGATCGAGA------ ATG CGT TTT ..
             : ::::::::::::::::: : ::::::: :   ::::: :
R.me. :- ...CTGCAAGATTGGTAAAATTGATTGTTTGGATAACGATCATCTGCGATATGGATGCCGCAC---- ATG CGT TTT ..
             : : : :::::::::::::::::::: :::::::   : :::: :: ::::::::: :
R.t.2 :- ...CCGAAGGATTGGTAAAATTGATTGATTGGATCGAAAGCATCCGCAGTATGGATGAAGGA----- ATG ..........

                    C            C      T  C    CGAT      TA A  A
CONSENSUS :- (5')...-TGGTAAAAT-TGATTG-TT-GAT----NCATC---G-NN-TGGAT---- ------ ATG .....(3')
                    T            T      A  G    AACA      CG G  G
```

TRANSPOSON MUTAGENESIS OF SLOW-GROWING RHIZOBIUM STRAIN NC92; ISOLATION AND CHARACTERIZATION OF HOST SPECIFIC NODULATION (NOD⁻) AND NITROGEN FIXATION (FIX⁻) DEFICIENT MUTANTS.

KATE J. WILSON[1], V. ANJAIAH[2], P.T.C. NAMBIAR[2] AND F.M. AUSUBEL[1].

[1]Department of Molecular Biology, Massachusetts General Hospital, Boston, MA 02114 U.S.A. and [2]ICRISAT, Patancheru P.O., Andhra Pradesh 502 324 INDIA

NC92 is a slow-growing cowpea group Rhizobium strain which nodulates crops of importance in India such as peanut (Arachis hypogaea) and pigeonpea (Cajanus cajan). In an effort to identify genes required for nodulation and nitrogen fixation, transposon mutagenesis was carried out using the Tn5 containing "suicide" plasmid pGS9 (Selveraj G, Iyer VN, 1983). 1000 kanamycin resistant NC92 derivatives were screened for their symbiotic phenotype on the test plant siratro (Macroptilium atropurpureum) and on peanut. Putative mutants were re-tested on peanut, pigeonpea and on siratro to look for host specific mutants.

Ten Fix⁻ mutants were isolated, four of which appear to show host specific defects. One mutant was isolated which is Nod⁻ on all three host plants. It can be complemented on siratro by cloned "common nodulation" genes from slow-growing Parasponia Rhizobium and from fast-growing R. meliloti. The mutant was generated by transposon mutagenesis, and therefore the Nod⁻ phenotype on all three host plants is almost certainly due to the same, single mutation. Thus it seems that common nodulation genes are required for infection both via root hairs (pigeonpea and siratro) and by direct "crack entry" (peanut). This mutant fails to nodulate alfalfa, the normal host plant for R. meliloti, even when carrying a clone which contains both common nod genes and the R. meliloti host specificity genes. This suggests either that some additional R. meliloti recognition factor is lacking, or that NC92 carries some "negative" recognition factor which prohibits alfalfa nodulation.

The most interesting mutant is one which shows an altered host range for nodulation. NC92 # 748 fails to induce any nodules at all on pigeonpea. However, it does induce nodules on siratro, but the nodules are ineffective. Light microscopic examination shows the siratro nodules to be empty of bacteria (A. Hirsch, personal communication), but the arrangement of the vascular bundles indicates that they are true nodules. The phenotype of NC92 # 748 on peanut requires further examination, but appears to be Fix⁺.

REFERENCES

Selveraj G and Iyer VN (1983) J. Bacteriol. 153, 1292-1300.

THE STRUCTURE OF EXOPOLYSACCHARIDES FROM BROAD HOST-RANGE RHIZOBIUM sp. STRAIN ANU280 (NGR234) AND MUTANT ANU2858

M. Batley, H.C. Chen, S.J. Djordjevic, J.W. Redmond and B.G. Rolfe, Genetics Department, Research School of Biological Sciences, Australian National University, ACT 2601; and School of Chemistry, Macquarie University, North Ryde, NSW 2113, Australia.

We have recently reported the use of transposon mutagenesis to isolate over ninety exopolysaccharide-defective strains of the broad host-range, fast-growing cowpea strain NGR234. As part of a program to determine the role of polysaccharide in the altered symbiotic properties of these mutants, we have determined the structure of the exopolysaccharide from Rhizobium sp. strain ANU280, an antibiotic-resistant isolate of NGR234, and mutant ANU2858, which has profoundly altered symbiotic properties.

Characterization of polysaccharides containing uronic acids, such as those produced by other Rhizobium strains, has been hampered by the acid resistance of uronosyl linkages and the instability of the released sugars under the hydrolysis conditions. The present work exploits this acid resistance by carrying out selective partial cleavage of other sugar linkages in native exopolysaccharide. After hydrolysis under optimized conditions, the mixture of neutral and acidic oligosaccharides is separated by a combination of ion-exchange and gel chromatography.

The ^{1}H spectra of key fragments were assigned by proton homonuclear decoupling. Carbon-proton correlation spectroscopy then permitted complete assignment of the carbon spectra. These, in turn, enabled assignment of the ^{13}C spectra. The availability of a series of fragments of increasing size was particularly useful as the addition of a new residue at the reducing terminus of a characterized oligosaccharide immediately gives the anomeric configuration of the penultimate sugar and the site of its attachment.

The extensive overlap between oligosaccharides, together with products of Smith degradation, permitted mapping onto unique repeat units. The repeat unit structure (I) of strain ANU280 (NGR234) resembles those of other fast-growing Rhizobium strains but is unusual in that it has a neutral main backbone chain with a high concentration of acidic groups on a branch. The repeat unit (II) from mutant ANU2858 resembles that of R. japonicum 61A76.

Our analytical procedure lacks the sensitivity of the best protocols based on methylation analysis, but can be carried out on less than 1 gram of polysaccharide. The method is experimentally simple, yields direct information concerning ring types and substitution and is the method of choice when sufficient acidic polysaccharide is available.

PSI, A PLASMID-LINKED RHIZOBIUM PHASEOLI GENE THAT INHIBETS EXOPOLYSACCHARIDE PRODUCTION AND WHICH IS REQUIRED FOR SYMBIOTIC NITROGEN FIXATION

D. BORTHAKUR/J.A. DOWNIE/A.W.B. JOHNSTON/J.W. LAMB
JOHN INNES INSTITUTE, COLNEY LANE, NORWICH, NR47UH, UK.

A strain of R.phaseoli cured of its symbiotic plasmid, pRP2JI, retained the ability to make exopolysaccharide (EPS). However, a region of pRP2JI, when cloned at an increased copy number in wide host-range vectors and transferred to this and other strains of Rhizobium, inhibited EPS synthesis. The gene responsible was termed psi (polysaccharide inhibition) and was located in a region of the symbiotic plasmid close to nodulation and nitrogen fixation genes. psi is important in the symbiosis since a wild type strain containing psi cloned on a multicopy plasmid failed to form Phaseolus nodules, and mutant strains containing psi::Tn5 mutations failed to fix nitrogen in Phaseolus nodules. It is proposed that the function of psi may be to repress in the bacteroid the expression of genes such as those for EPS synthesis which are normally expressed in free-living culture.

AN AGROBACTERIUM TUMEFACIENS VIRULENCE GENE HOMOLOGUE EXISTS IN RHIZOBIUM MELILOTI AND IS ESSENTIAL FOR SYMBIOSIS

TYLER DYLAN, GARY DITTA, LALLAN KASHYAP, CARL DOUGLAS*, MARTY YANOFSKY*, EUGENE NESTER* and DONALD HELINSKI, Univ. of California, San Diego, La Jolla, CA 92093 and *Univ. of Washington, Seattle, WA 98195 U.S.A.

Two closely linked regions of chromosomal DNA in Agrobacterium tumefaciens have been shown to be essential for the oncogenic transformation of dicotyledenous plants (Garfinkel et al., 1980; Douglas et al., 1985). These chromosomal virulence loci, designated chvA and chvB, are particularly interesting because mutations at either locus lead to a dramatic reduction in the ability of A. tumefaciens to attach to plant cells (Douglas et al., 1982). We report here that (a) structural homologues of both chvA and chvB can be identified in various Rhizobium species; (b) that chvB-homologous sequences are essential for symbiotic nitrogen fixation by R. meliloti; and (c) that genes functionally equivalent to those of R. meliloti also exist in R. leguminosarum.

DNA homologous to both chvA and chvB was identified in three fast growing Rhizobium species by Southern blot hybridization. Cloned fragments of A. tumefaciens DNA representing chvA and chvB were used as hybridization probes against restriction enzyme digests of total genomic DNA isolated from R. meliloti, R. leguminosarum, R. trifolii, R. japonicum, and E. coli. Discreet fragments of DNA carrying homology to either chvA or chvB were detected for R. meliloti, R. leguminosarum, and R. trifolii.

Mutations within the chvB-homologous locus of R. meliloti produce a fix⁻ symbiotic phenotype on alfalfa. Four Tn5 insertions, spanning a region of 1.6 kb, were introduced into cloned R. meliloti DNA carrying chvB homology, and then marker exchanged into the R. meliloti genome. When used for infection, each insertion mutant caused the formation of small white root nodules that were devoid of bacteroids and incapable of symbiotic nitrogen fixation.

Aspects of functional equivalence have been demonstrated for chvB and its homologues in A. tumefaciens, R. meliloti, and R. leguminosarum. Using cloned DNA fragments on broad host range plasmids, we have shown that chvB-homologous DNA from R. meliloti can complement avirulent chvB mutants of A. tumefaciens for virulence, while chvB DNA from A. tumefaciens can correspondingly complement the aforementioned fix⁻ mutants of R. meliloti for symbiosis with alfalfa. We have also isolated a cosmid clone from an R. leguminosarum gene bank that will complement the R. meliloti mutants for symbiosis with alfalfa.

Garfinkel and Nester (1980) J. Bacteriol. 144, 732-743.
Douglas et al (1985) J. Bacteriol. 161, 850-860.
Douglas et al (1982) J. Bacteriol. 152, 1265-1275.

FIX LOCI MAP TO A SECOND MEGAPLASMID IN RHIZOBIUM MELILOTI STRAIN SU47.

TURLOUGH M. FINAN[1,2], GUIDO F. DE VOS[1], and ETHAN R. SIGNER[1]
[1]Department of Biology, MIT, Cambridge, MA 02139 and [2]Department of Biology, McMaster University, Hamilton, Ontario L8S 4K1, Canada

In *Rhizobium meliloti* strain SU47 the "common" *nod* and *nif* genes are located on a megaplasmid (pRmeSU47a) (Rosenberg et al., 1982; Long et al. 1982). We recently identified a symbiotic locus in this strain, mutations at which result in the formation of Inf⁻ nodules apparently because of a defect in exopolysaccharide synthesis (ExoB) (Finan et al., 1985). We now report that ExoA, ExoB and ExoF (Leigh et al., 1985), and at least two loci involved in thiamine biosynthesis (Thi) map to a second megaplasmid pRmeSU47b. (ExoC and ExoD map to the chromosome). A map showing the order of transposon inserts (Finan et al., 1985) linked by transduction to the relevant markers is shown below.

Tn5-132	Tn5					Tn5-233
Ω5007	Ω5004	exoB	exoF	exoA	thi-504	Ω5011

Horizontal "Eckhardt" gels of *R. meliloti* strains often revealed a doublet at the megaplasmid position, in which case DNA hybridization with Tn5 as probe showed that insert Ω5004 Tn5 was located on the higher molecular weight plasmid, while a *nif*H::Tn5 insert was on the lower one.

To mobilize pRmeSU47b we used the transposon Tn5-*oriT* which is a derivative of Tn5 containing the origin of transfer of the plasmid RK2 (Yakobson and Guiney, 1984). Insert Ω5007::Tn5-132 (Tc^R) was replaced by Tn5-*oriT* (Nm^R) via homologous recombination between Tn5-*oriT* and Tn5-132 in both inverted repeats (Berg et al., 1980). Replacements Ω5209 and Ω5210 transferred Ω5011::Tn -233 (Gm^R) at frequencies of 10^{-4} and 10^{-6} respectively, suggesting that the two inserts transfer in opposite directions. In matings between Ω5209 and Ω5210::Tn5-*oriT* containing strains and auxotrophs we found linkage to Thi mutations. These mapped to two regions on pRmeSU47b one of which (e.g. *thi*-501) was not transducible with Ω5011 while the other (e.g. *thi*-504) was 80% linked. Cloned fragments which complemented *thi*-504 also complemented ExoA mutations, while *thi*-501 complementing clones also complemented *E. coli*, *thiA*, *thiC* and *thi*-1 mutations.

Plasmid pRmeSU47b was transferred to *Agrobacterium tumefaciens* by mating with *R. meliloti* containing Ω5209::Tn5-*oriT* and selecting for Nm^R transfer. Such transconjugants were unable to form nodules on alfalfa while Fix⁻ nodules were formed by transconjugants containing pRmeSU47a. We are currently examining the symbiotic phenotype of *A. tumefaciens* strains containing both plasmids.

REFERENCES

Berg D et al (1980) J. Bacteriol. 142, 439-446.
Finan T et al (1985) Cell. 40, 869-877.
Leigh J et al (1985) Proc. Natl. Acad. Sci. U.S.A. in press.
Long S et al (1982) Nature. 298, 485-488.
Rosenberg et al (1982) J. Bacteriol. 150, 402-406.
Yakobson EA and Guiney D (1984) J. Bacteriol. 160, 451-453.

IDENTIFICATION OF NODULATION AND NITROGEN FIXATION FUNCTIONS OF RHIZOBIUM SESBANIAE STRAIN ORS571

M. HOLSTERS[1], G. VAN DEN EEDE[1], K. GOETHALS[1], and B. DREYFUS[2]

[1] Laboratorium voor Genetica, Rijksuniversiteit Gent, B-9000 Gent (Belgium); [2] Laboratoire ORSTOM, Dakar (Senegal)

Rhizobium strain ORS571 induces nitrogen-fixing nodules on both stems and roots of the tropical legume, Sesbania rostrata. Under a 97% N_2 −3% O_2 atmosphere, ORS571 can grow in media without nitrogen source, thanks to the fixation and assimilation of atmospheric dinitrogen. After at random mutagenesis with the Tn5 transposon, derivatives of ORS571 affected in nodulation or nitrogen fixation capacities were isolated. Two distinct Nod regions were thus identified. Two types of mutants were found that affect the nitrogen fixation : (i) "Nif⁻" mutants that cannot fix dinitrogen, neither in the free-living nor in the symbiotic state; (ii) "Fix⁻" mutants that can still fix and assimilate nitrogen as free-living organisms, but not anymore in the differentiated state inside the nodules. No overlap has yet been found between the Nif region (which contains among others the kdh genes coding for the nitrogenase subunits) and the two Nod regions. At least some Fix⁻ mutations can be correlated with one or the other Nod region. The symbiotic regions of the ORS571 genome are now extensively studied by reversed genetics, using both the transposon Tn5 and a Tn3 lacz transposon.

THE TWO MEGAPLASMIDS OF Rhizobium meliloti ARE   INVOLVED
IN THE EFFECTIVE NODULATION OF ALFALFA.

M.F.Hynes,P.Müller,K.Niehaus, and A.Pühler. Lehrstuhl
für   Genetik,   Fakultät   Biologie,   Universität
Bielefeld. 4800 Bielefeld 1, FRG.

The general presence of two megaplasmids in various
R.meliloti strains of different geographical origin was
demostrated by physical and genetic means. We were able
to resolve two megaplasmid bands in agarose gels for
almost all strains tested by electrophoresing for 12-14
h at 40 V, using the method described by Hynes et
al.(1985). We were also able to transfer the   two
megaplasmids of the R.meliloti strains 2011, ZB121, and
MVII-1 to plasmid -free Agrobacterium strains (Hynes et
al. 1985) and to other Rhizobium strains using the Tn5-
Mob system (Simon 1984). One megaplasmid in each strain
hybridized to nif gene probes, and conferred the ability
to nodulate alfalfa on Agrobacterium transconjugants.
This corresponds to the pSym described by Rosenberg et
al.(1981) and Banfalvi et   al.(1981).   The   other
megaplasmid was shown to carry genes involved in
extracellular polysaccharide (EPS) production. Transfer
of   this   second megaplasmid into EPS$^-$ strains of
R.meliloti restored them to an EPS$^+$ phenotype. In
addition DNA from cosmid clones which could complement
R.meliloti 2011 mutants producing abnormal ineffective
nodules on alfalfa, and also altered in their EPS
production (eg. mutant 0540-see abstract by P.Müller et
al. this volume) hybridized to this second megaplasmid,
but not to the pSym. This shows that both megaplasmids
are necessary for the effective nodulation of alfalfa by
R.meliloti.

References

Banfalvi Z, Sakanyan V, Koncz C, Kiss A, Dusha I,
Kondorosi A (1981) Mol Gen Genet 184:318-325

Hynes M, Simon R, Pühler A (1985) Plasmid 13:99-105

Rosenberg C, Boistard P, Denarie J, Casse-Delbart F
(1981) Mol Gen Genet 184:326-333

Simon R (1984) Mol Gen Genet 196:413-420

ISOLATION AND CHARACTERIZATION OF MUTANTS OF RHIZOBIUM MELILOTI 2011 OBTAINED BY Tn5-MUTAGENESIS

PETER MÜLLER, KARSTEN NIEHAUS, ALFRED PÜHLER
UNIVERSITÄT BIELEFELD, D-4800 BIELEFELD 1, FRG

Five mutants of Rhizobium meliloti 2011, obtained by random Tn5-mutagenesis and subsequent screening by plant tests, induce the formation of ineffective nodules on alfalfa roots. The morphologically modified nodule structures were studied on a microscopical and an electron microscopical level. Three mutants (0540, 2204 and 2505) are defective in the synthesis of extracellular polysaccharides (EPS) and therefore do not interact with the dye "Cellofluor white". Mutant 101.45, like the previous ones, induces nodules which are devoid of bacteria, though it produces a threefold amount of EPS. Inoculation with mutant 3046 results in elongated nodules,but N_2-fixation does not occur. The mutants 0540, 101.45 and 3046 could be complemented by a wildtype cosmid gene bank to a fully effective symbiosis. These cosmids were also transferred into the other mutants, but they do not crossreact: Each cosmid complements only one mutant. A subclone of cosmid 2-10, carrying a 7.8 kb EcoRI-fragment within the vector pSUP205, still could complement the mutation in strain 0540. Using the same cosmid as a radioactive probe it could be demonstrated that this fragment is homologous to megaplasmid 2 of R.meliloti ZB121 and to the smaller plasmid of this strain, but not to its deleted derivative Sym plasmid (megaplasmid 1). We therefore draw the conclusion that the megaplasmid 2 carries genes which are essential for a regular nodulation and code for the synthesis of EPS.

Cosmid 2-10 has been characterized in more detail: The relative position of the 14 EcoRI-fragments within the entire fragment (32.1 kb) were determined, the 7.8 kb EcoRI-fragment was mapped, and the Tn5-insertion was ascertained: It is located on a 0.8 kb HindIII-fragment. The Tn5 carrying fragment of the complemented mutant 0540 was obtained by transfer of the excising cosmid vector together with the mutated fragment, which can be mobilized by RP4-4-7 into recipient E.coli HB101 Rif[R] at a low frequency. Using the same technique it became evident that in mutant 101.45, Tn5 is inserted in a 6.4 kb EcoRI-fragment. The defective phenotype of mutant 0540 could be complemented both by a 4.6 kb ClaI-fragment and a 3.8 kb XhoI-fragment cloned into vector R56I. This plasmid is a Tc[R] derivative of pSUP401 and does not replicate autonomously in R.meliloti. Fragment specific Tn5-mutagenesis revealed, that besides the HindIII-fragment there is a second region of interest on the 7.8 kb EcoRI-fragment at a distance of about 2.5 kb. Tn5 insertion in that second region which extends about 1.3 kb results in a similar phenotype as mutant 0540: No EPS are detectable or the amounts of EPS are reduced, bacteria do not interact with "Cellofluor white" or give a reduced fluorescence, and the nodules which are induced are devoid of bacteria. They are small and ineffective.

INFECTION MUTANTS OF RHIZOBIUM PHASEOLI

K.D. NOEL, K.A. VANDENBOSCH[*], B.C. KULPACA, J.R. CAVA, R.J. DIEBOLD,
AND E.H. NEWCOMB[*]
Marquette University, Milwaukee, Wisconsin, and University of Wisconsin,
Madison, Wisconsin[*], U.S.A.

Mutations in Rhizobium phaseoli strain CFN42 affecting nodule development have been isolated following Tn5 mutagenesis (Noel, et al,
1984). As described recently (VandenBosch, et al, 1985), three of the
mutants induce initiation of root cortical cell division without eliciting infection thread formation. Colonies of these mutants (class I)
on minimal agar medium are not stained by Calcofluor, a fluorescent
probe for β-linked polysaccharide (Finan, et al, 1985). However, two
other mutants, which induce nodulation that outwardly appears identical,
are stained normally by Calcofluor.

RESULTS

Light microscopic examination of emerging nodules induced by the
latter mutants (class II) reveal infection threads. However, the
threads are abnormally wide, globular structures which abort, usually
within the root hair. Otherwise, the development of nodule tissue induced by these mutants resembles that of class I mutants; normal differentiation is absent.

The class II mutants lack a molecular species that is stained by
periodic acid-Schiff following SDS gel electrophoresis. This molecule
is extracted by the hot phenol-water procedure for preparing lipopolysaccharide (LPS). It co-migrates with ketodeoxyoctonate (KDO) - containing material of this extract on Sepharose 4B. Mild acid hydrolysis
of this extract and fractionation on Sephadex G-50 yields a polysaccharide not found from the class II mutant extracts.

Each mutant of either class contains a single Tn5 insertion,
chromosomally located. Extensive crosses have shown that in four class
I mutants and one class II mutant the Tn5 insertion is the cause of
both the nodule infection defect and the altered physiology in free
culture. Each Tn5 lies on a different EcoRI genomic fragment, each of
which has been cloned.

CONCLUSIONS

One polysaccharide of R. phaseoli, which binds to Calcofluor, appears to be important for infection thread initiation. Another, perhaps the major LPS, is necessary for infection thread development to
persist. When either is altered, bean root cortical cell division appears to initiate normally, but subsequent nodule development is
severely altered.

REFERENCES

Finan TM et al (1985) Cell 40, 869-877.
Noel KD et al (1984) J. Bacteriol. 158, 148-155.
VandenBosch et al (1985) J. Bacteriol 162, 950-959.

GENETIC ANALYSIS OF NITROGEN FIXATION IN SESBANIA RHIZOBIUM ORS571.

FRANCOISE NOREL, ANIL KUSH, PATRICE DENEFLE, GIOVANNI SALZANO AND CLAUDINE ELMERICH. UNITE DE PHYSIOLOGIE CELLULAIRE, INSTITUT PASTEUR, 75724 PARIS CEDEX 15, FRANCE.

The fast growing Rhizobium strain ORS571, isolated from stem nodules of the tropical legume Sesbania rostrata, can grow in the free-living state at the expense of molecular nitrogen. The organization of the nif genes was analyzed by hybridization using Klebsiella pneumoniae nif DNA probes. A cluster containing nifHDKE was identified on a 6.3 kb SalI-BamHI fragment (Figure 1) (Norel et al. 1985). Tn5 insertions in the DNA sequences homologous to nifD, nifK and nifE abolished nitrogen fixation of Rhizobium ORS571 in planta and ex planta. Moreover a deletion of the 4 kb BamHI fragment adjacent to nifE led also to a Nif$^-$ phenotype suggesting the existence of a nif region adjacent to nifE. A second copy of nifH referred to as copy 2, unlinked to the nifHDKE cluster was identified and cloned. A deletion of nifH copy 1 was obtained and the resulting phenotype was Nif$^+$Nod$^+$Fix$^+$, suggesting that the second copy of nifH was functional. Using probes from R. japonicum (Fuhrmann et al. 1985), sequences homologous to fixABC genes were detected. The fixBC region was localized in a 3.3 kb XhoI fragment adjacent to nifH copy 2. Homology to fixA was localized on the same PstI fragment as nifH copy 1.

Polypeptides synthesized by the wild type and insertion or deletion mutants were analyzed, by one- and two-dimension gel electrophoresis, after pulse-labeling of cells. Eighteen polypeptides which were not synthesized under conditions of ammonia assimilation and which might be nif specific were detected. Products corresponding to nifH, D and K were identified. Insertion in nifH copy 1 was polar on nifDK transcription; thus the three genes are likely part of the same transcription unit. Nitrogenase activity, in crude extracts of the nifE mutant and of the 4 kb BamHI deletion mutant, was restored by addition of ORS571 nitrogenase pure component 1, but not by component 2. This favors the hypothesis of mutations in nifEN-like products.

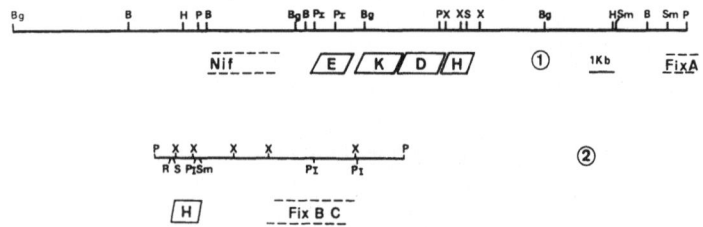

FIGURE 1. Physical map of the cloned nif regions. 1: nifHDKE cluster; 2: nifH copy 2; restriction sites: B: BamHI; Bg: BglII; H: HindIII; P: PstI; P$_I$: PvuI; S: SalI; Sm: SmaI; X: XhoI.
Norel F et al (1985) Mol. Gen. Genet. 199, 352-356.
Fuhrmann M et al (1985) Mol. Gen. Genet. 199, 315-322.

MOLECULAR GENETICS OF COMPETITION BETWEEN STRAINS OF RHIZOBIUM FOR NODULATION OF AFGHANISTAN PEAS.

D.N. DOWLING, U. SAMREY and W.J. BROUGHTON
MAX-PLANCK-INSTITUT FÜR ZÜCHTUNGSFORSCHUNG, D-5000 COLOGNE 30, FEDERAL
REPUBLIC OF GERMANY

The primitive pea cultivar Afghanistan can form nitrogen fixing nodules with R. leguminosarum strain TOM: this nodulation can be blocked in mixed inoculations by a European R. leguminosarum strain (PF$_2$), which is unable to nodulate this pea cultivar (Lie, 1978; Winarno, Lie, 1979). This nodulation blocking system provides a useful model in which to study the genetics of the bacterial component of competition (Broughton et al, 1980; 1982).

The nodulation blocking strain PF$_2$ has three large plasmids, the smallest of which (250 kb) hybridised to nif and nod genes from R. meliloti and was termed pSymPF$_2$. pSymPF$_2$ was labelled with the mobilizing transposon Tn5-Mob (Simon, 1984) and transferred into Rhizobium strains which did not block nodulation of cv. Afghanistan in competition tests (Dowling et al, 1985). Transconjugants with different combinations of PF$_2$ plasmids were screened for the ability to block the nodulation of TOM. These data suggest that nodulation blocking ability is encoded on the symbiotic plasmid pSymPF$_2$.

In order to clone the gene(s) involved in this competition process, a genomic clone bank of total PF$_2$ DNA was constructed in the wide host range cosmid pMMB33 (Frey et al, 1983). Cosmid clones which hybridised to a probe made from the plasmid fraction of PF$_2$ were mobilized by introduction of a Km^S R68.45 derivative into different Rhizobium strains and screened for nodulation blocking ability in competition tests with TOM. A number of cosmid clones were found which blocked nodulation of TOM on cv. Afghanistan; these clones will be characterised further.

REFERENCES

Broughton WJ, van Egeraat AWSM, and Lie TA (1980) Can. J. Microbiol. 26, 562-565.
Broughton WJ, Samrey U, and Bohlool BB (1982) Can. J. Microbiol. 28, 162-168
Dowling DN, Samrey U, and Broughton WJ (1985) In Magnien E and de Nettancourt D, eds, Genetic Engineering of Plants and Microorganisms Important for Agriculture, pp 56-57, Nijhoff/Junk, Dordrecht/Boston/Lancaster.
Frey J, Bagdasarian M, Feiss D, Franklin CH, and Deshusses J (1983) Gene 24, 299-308.
Lie AT (1978) Ann. Appl. Biol. 88, 462-465.
Simon R (1984) Mol. Gen. Genet. 196, 413-420.
Winarno R, and Lie TA (1979) Plant and Soil 51, 131-142.

COINOCULATION WITH SYMBIOTICALLY DEFECTIVE MUTANTS OF RHIZOBIUM MELILOTI

Ann M. Hirsch[1], Gretchen A. Kuldau[1], Shoshana Klein[2] and Ethan R. Signer[2]. Wellesley College, Wellesley, MA 02181, and Massachusetts Institute of Technology, Cambridge, MA 02139 U.S.A.

Fix[-] mutants are defined as making nodules that do not fix nitrogen. The purpose of this work is to describe the ultrastructural phenotypes specifically of Fix[-] nodules made after inoculation with certain symbiotically defective mutants of the alfalfa symbiont Rhizobium meliloti SU47, either singly or in pairwise combinations.

Mutants used were Nod[-] (Rm1126, nodC:IsRm1)(3); Nif[-] (Rm1354 nifA::Tn5) (7);(Rm1491 NifH::Tn5)(6); and Exo[-] (Rm5020 exoB355) (1).

Materials and Methods

Plant material, growing conditions, inoculations and acetylene reduction have been described previously (6). Nodules were surface sterilized (5) and squashed in LB agar containing 0.02% Calcofluor white (1,2). One-2 days after colony appearance, plates were illuminated with a hand-held ultraviolet lamp; colonies were counted and scored as "brights" or "darks."

For light and electron microscopy, nodules were prepared in Spurr's resin and examined as described previously (4).

Results and Discussion

In all single inoculations, nodules were Fix[-] after 6 weeks except for Rm 1126 which was Nod[-] (3). ExoB[-] nodules were small, white and bead-like and devoid of infection threads and intracellular bacteria. The few bacteria present were restricted to epidermal and subepidermal intercellular spaces (1).

In NodC[-] + NifA[-] or NodC[-] + NifH[-] coinoculations, there was some cooperation to produce Fix[+] nodules (ca. 20%). The majority were Fix[-] reflecting occupancy by the Nif[-] partner only. Cooperation of NifA[-]+ NifH[-] to produce a Fix[+] nodule was not observed. ExoB[-] + NifA[-] and ExoB[-] + NifH[-] coinoculations result in a majority of Fix[-] nodules. Both "brights" and "darks" were recovered from some nodules. In these nodules, two morphologically distinct bacteroid forms were observed within the same host cell. This suggests considerable cooperation for entry of the ExoB[-] mutant via root hairs and infection threads. However, by itself this would imply that the resulting nodules should be Fix[+], whereas in fact those nodules are Fix[-]. One explanation is that the ExoB defect is manifest not only at the early stage of entry via root hairs (1), but in addition at the later stage of differentiation into bacteroids that fix nitrogen.

References

1. Finan et al. 1985. Cell 40:869-877.
2. Leigh et al. 1985. PNAS. In press.
3. Long et al. 1981. Nature 298:485-488.
4. Hirsch et al. 1983. J. Bacteriol. 155:367-380.
5. Hirsch et al. 1984. J. Bacteriol. 158:1133-1143.
6. Meade et al. 1982. J. Bacteriol. 149:114-122.
7. Szeto et al. 1984. Cell 36 :1035-1043.

ON THE INHIBITION BY PJB5 OF NODULE INDUCTION BY
FAST-GROWING SOYBEAN RHIZOBIUM PRC205

KENNETH D. NADLER,
MICHIGAN STATE UNIVERSITY, EAST LANSING, MI 48824 USA

Genes in the symbiotic region of the Tn5-marked conjugative pea symbiosis plasmid pJB5 appear to inhibit nodulation of soybeans by the fast-growing soybean Rhizobium PRC205 (Nadler, 1984). PRC205pJB5 exconjugants poorly induce nodules on roots of 'Peking' soybeans whereas exconjugants containing a plasmid derived from pJB5 by deletion (of ca. 20 Mdal containing symbiosis genes) induce nodules as well as PRC205. We now report the isolation and symbiotic properties of derivatives of PRC205pJB5 either with deletions in pJB5 or cured of the plasmid. Such derivatives were isolated either as Kanamycin-sensitive strains (S-strains) following growth at elevated temperatures or as occupants (XN strains) of rare nodules induced by PRC205pJB5. Nine nod[+] derivatives induce nodules as well as PRC205; seven are sensitive to kanamycin and either have been cured of pJB5 or have suffered large deletions in this plasmid. Eleven nod[-] derivatives were isolated, of which nine retain pJB5 as indicated by their resistance to kanamycin and ability to act as high frequency donors of km[r]. Two exceptional nod[-] strains are sensitive to kanamycin but retain a pJB5-sized plasmid. It is likely that these two strains have merely deleted Tn5 but retained the pea symbiosis plasmid. Comparison of the properties of PRC205pJB5 and various cured derivatives confirms that genes encoded by the pea symbiosis plasmid inhibit nodule induction by the fast-growing Rhizobium PRC205.

*Supported by The MSU All University Research Initiation Grant and Michigan Soybean Committee.

GENETIC STUDY OF A LARGE PLASMID OF RHIZOBIUM MELILOTI

NICOLAS TORO/JOSE OLIVARES
DEPARTAMENTO DE MICROBIOLOGIA, ESTACION EXPERIMENTAL DEL ZAIDIN, CSIC,
18008-GRANADA, SPAIN.

INTRODUCTION

Genes involved in nodulation (nod) of legume roots and nitrogen fixation
(fix) are closely located in a of the large plasmids in the fast growing
Rhizobium strains. In R. meliloti the symbiotic genes are carried on a
very large plasmid (megaplasmid) of about 1000 MD (Banfalvi et al. 1981,
Rosenberg et al. 1981). Although other additional large plasmids in
the range of 90 to 250 MD may be also present, up to now no genes
directly related with the Nod and Fix phenotypes have been detected in
any of these plasmids. A genetic study on one of these plasmids has been
carried out as the aim of this work.

MATERIAL AND METHODS

The wild type strain R. meliloti GR4 and its cured derivatives GRO13 and
GRP4 were used. The plasmid pRmeGR4b (140 MD) was isolated and a specific
gene library was constructed using the broad host range vector pRK290.
This gene bank has been studied by colony hybridization, using the PstI
fragment of pID1 (nifH and D genes of R. meliloti Rm41) as probe, and
genetic complementation. The overlapping fragments have been used to
construct a first tentative EcoRI restriction map of pRmeGR4b. Plants
for symbiotic assays were grown as previously described (Olivares et al.
1980).

RESULTS AND DISCUSSION

The wild type strain R. meliloti GR4 harbours two megaplasmids of about
1000 MD and two large ones, pRmeGR4a and pRmeGR4b of 114 and 140 MD,
respectively. The study of cured derivatives of GR4 showed that none of
these two plasmids carry essential genes for nodulation of Medicago
sativa roots or nitrogen fixation. However, by genetic complementation
of a cured derivative of GR4 with the gene bank of pRmeGR4b, two regions
of this plasmid have been identified as involved in the infectivity of
the wild type strain. In addition, a 4.3 kb EcoRI fragment showing
homology to PstI fragment of pID1 is located adjacent to one of these
regions. Since no hybridization with pSA30 (nitrogenase genes of
Klebsiella pneumoniae) has been detected, the homology found might be
at level of the nifH gene promoter of R. meliloti. We can speculate
that this region might play a role in the control of the expression of
the adjacent region involved in the infectivity.

REFERENCES

Banfalvi Z et al (1981) Mol. Gen. Genet. 184, 318-325.
Olivares J et al (1980) Appl. Environ. Microbiol. 39, 967-970
Rosenberg C et al (1981) Mol. Gen. Genet. 184, 326-333.

AKNOWLEDGEMENT

This work has been supported by Comisión Asesora de Investigación Científica y Técnica, Grant Nº 1764/82

ANALYSIS OF A LEAF-CURL PHENONMENON IN PIGEONPEA (<u>CAJANUS</u>
<u>CAJAN</u> (L.) MILLSP.) INDUCED BY <u>RHIZOBIUM</u> NODULATION

[*]N.M. UPADHYAYA, [*]WILLIAM T. TUCKER, [**]J.V.D.K. KUMAR RAO AND [*]PETER J.
DART.
[*]Research School of Biological Sciences, Australian National University,
A.C.T. 2601, Australia. [**]International Crops Research Institute for
the Semi-Arid Tropics, Patancheru, Andhra Pradesh, India.

Nodulation of pigeonpea by 2 strains of fast-growing <u>Rhizobium</u> causes
pronounced curling and deformation of leaves (Kumar Rao et al.1984).
The symptoms appear first on the third developing trifoliate leaf bud
which bends outwards, followed by hypernasty, reduced leaf expansion,
suppression of apical dominance, sprouting of axial buds, and generally
stunted growth. These strains also effectively nodulate siratro and
<u>Desmodium</u> causing leaf-deformation. Following approach-grafting of a
normal plant and a plant showing leaf-curl, symptoms appeared on
developing shoots of the normal plant within 8 days of graft union,
above and below the graft. Shoots developing on the normal stock after
graft separation did not show symptoms, indicating that a continuous
supply of the inducing principle from the roots was necessary for the
production of symptoms. A spontaneous mutant of leaf-curl inducing
strain IC3342 resistant to streptomycin (Sm) and rifampicin (Rif) has
been used to study its competitiveness with other normal effective
strains. When mixed with the slow growing <u>Rhizobium</u> strain 32H1, the
leaf-curl strain formed only 9% of the nodules and with the fast growing
strain MNU1, only 11%. However, these plants developed leaf-curl
symptoms 23 days after sowing, the same time as plants inoculated with
the leaf-curl strain alone.
Strain IC3342 contains 2 mega-plasmids, one of which is a <u>Sym</u> plasmid,
established by hybridising with a <u>nif</u>HDK gene probe from the fast-
growing cowpea strain ANU240 and a <u>nod</u> gene probe from <u>R.trifolii</u> strain
ANU843. We are trying to cure the strain of plasmids and then to
reintroduce a mobilisable <u>Sym</u> plasmid from a non-leaf-curl strain to
establish the relationship between nodulation and leaf-curling. With
EcoRl-cleaved total genomic DNA, the <u>nif</u> gene probe hybridised to 2
fragments while the <u>nod</u> gene probe hybridised to 7 fragments.
Tn5 mutagenesis of the Sm[r]Rif[r] derivative, performed by mating <u>E.coli</u>
strain SM10 containing suicide vector pSUP1011, gave Km[r] colonies at a
frequency of 2.5×10^{-5} per recipient when the donor population was 7×10^6 cells/ml and that of recipient 4×10^5 cells/ml. Testing of these
presumptive Tn5 mutants for loss of nodulating and/or leaf-curling
ability is in progress. A test-tube screening system has been developed
using siratro as the test-plant.

Reference
Kumar Rao JVDK et al (1984) Soil Biol. Biochem. 16, 89-91.

RESTRICTION SITES IN THE NIFH,D,K GENE REGION OF ANABAENA
AZOLLAE ENDOSYMBIONTS FROM NINE DIFFERENT AZOLLA SPECIES

COHEN-BAZIRE, G.[1] and FRANCHE, C.[2]
1. Institut Pasteur, 28, rue du Dr. Roux,
   75724 Paris Cédex 15, France
2. ORSTOM, BP 1386, Dakar, Sénégal

The water fern Azolla contains a symbiotic heterocystous
cyanobacterium, Anabaena azollae, within specialized leaf
cavities. The six extant species of Azolla are divided into
two sections, Euazolla and Rhizosperma, according to the
reproductive structures. As the cyanophyte can supply the
association with its total nitrogen requirement by nitrogen
fixation, Azolla constitutes a potential nitrogen source in
agriculture.

Using probes carrying the nifH, nifD and nifK genes from
the free-living Anabaena PCC7120, we compared the restriction
sites within and around the nifH,D,K genes of DNA from vegetative cells of nine Anabaena azollae. The endosymbionts
were extracted from four New World species of Euazolla including A. caroliniana, A. filiculoides, A. mexicana and A.
microphylla, and five Old World species of Rhizosperma represented by two A. pinnata var. pinnata and three A. pinnata
var. imbricata.
The restriction sites of the nifK region differed among
members of Euazolla and Rhizosperma, but were identical for
all symbionts within each group. The HindIII and ECoRI sites
within nifD were conserved in all nine endosymbiotic strains.
On the contrary, the HindIII and ECoRI DNA fragments which
hybridized to the nifH probe were identical in the five
endosymbionts of A. pinnata, but slight differences were
observed among those of the four Euazolla species. Some
restriction sites in nifH were common to all members of both
groups.

Due to the difficulty of culturing A. azollae in vitro,
very little is known concerning the taxonomy of the symbionts
extracted from the different species of Azolla. Our data
suggest that there is a common ancestor A. azollae strain.
The divergence of the restriction sites in the nifH,D,K gene
region is in close relation with the geographical origin of
the ferns.

SYMBIOTIC GENES IN FRANKIA

DANIELA DRAKE[1]/JACK T. LEONARD[2]/ANN M. HIRSCH[1].
[1]WELLESLEY COLLEGE, WELLESLEY, MA 02181 USA/[2]HAMPSHIRE COLLEGE, AMHERST, MA 01002, USA.

The symbiosis that occurs between the actinorhizal endophyte Frankia and woody dicotyledons strongly resembles that between legumes and Rhizobium. Previous results indicate that symbiotic genes are conserved in Frankia. Homology to nifDK has been reported (1,5). We report that there is homology between cloned Rhizobium nodulation genes and Frankia DNA.

Materials and Methods

Frankia strain HFPCcI3 was obtained from Harvard Forest, Petersham, MA and grown in BAP (4). DNA was isolated after 7-10 days of growth. 1-5 ml of Frankia cells were suspended in TES and homogenized through a 23 guage needle. Lysozyme was added at 0.05 mg/ml and the cells were incubated overnight at 37 C. After centrifugation, SDS and RNase in TE were added to the spheroplasts. The lysate was incubated at 37 C for 30 min. and predigested pronase was added to the lysate. DNA was precipated with potassium acetate and isopropanol.

Bacterial strains were grown as described (3). Restriction digests of bacterial and Frankia DNA followed the manufacturer's directions. After electrophoresis, the DNA was transferred to filters (2). Probes of either nodAB or nodC were nick-translated and added to the hybridization mixtures. Washes were performed at either 65 or 37 C.

Results and Discussion

Like previous researchers (1,5), we have found homology between the nif probe and Frankia genomic DNA.

Using nodC as a probe gave relatively strong hybridization to EcoRl-restricted Frankia DNA but weaker hybridization to Sall-digested DNA. The hybridizations and washes occurred under conditions of high stringency. Probing with a nodAB probe gave less specific results at 65 C. Greater homology was observed with lower temperature washes.

We are in the process of constructing a genomic library of Frankia DNA to screen for sequences homologous to Rhizobium nodulation genes.

References

Lignon, JM and JP Nakas. 1984. 14th Steenbock Symposium of Nitrogen Fixation and CO_2 Metabolism. University of Wisconsin, Madison.
Maniatis, T, et al. 1982. Cold Spring Harbor Laboratory.
Meade et al, 1982. J. Bacteriol. 149:114-122.
Murry, MA, et al. 1984. Plant and Soil. 78:61-78.
Ruvkun, GB and FM Ausubel. 1980. Proc. Natl. Acad. Sci. USA, 77:191-195.

A STUDY OF RHIZOBIUM MELILOTI JJI MUTANTS IN INDOLE ACETIC ACID PRODUCTION.

IVAN HOOPER[*]/B. WHATELY/V.N. IYER
CARLETON UNIVERSITY, OTTAWA, ONTARIO, CANADA/[*]BOTANY DEPARTMENT,
UNIVERSITY OF WESTERN AUSTRALIA, NEDLANDS, WA 6009

Rhizobium meliloti Strain JJI has the genetic potential to produce a variety of phytohormones. The addition of tryptophan to JJI cultures results in the induction of the synthesis of indole acetic acid. In an attempt to elucidate the significance of rhizobial hormone production we have established a programme of obtaining Rhizobia mutants. A derivative (KN1001) of JJI without the cryptic 150 Md plasmid was manufactured for these studies. Five mutants of KN1001 resistant to 5-Methyl-Tryptrophan were selected (Whately et al., in preparation). Alternatively, using the suicidal vector pGS9, we constructed a bank of Tn5 mutants of KN1001. A colorimetric procedure using Salkowski's reagent enabled the detection of mutants deficient in indole acetic acid synthesis. Five MethylTryptophan resistant mutants did not give a positive reaction, and of 3000 Tn5 mutants screened four were indole acetic acid negative. Analysis of culture supernatants by GC-mass spectroscopy and HPLC techniques (Whately et al., in preparation) confirmed that, whilst KN1001 has the potential to produce indole acetic acid, the mutants produced only small amounts. The presence of a minor pathway for indole acetic acid synthesis is commonly found in those bacterial species which interact with plant systems.

The mutants were tested for symbiotic efficiency on Medicago sativa cv. Saranac. The results from these tests are complex. The 5-Methyl-Tryptophan resistant mutants were as effective as the parent. One of the Tn5 mutants indicated a normal nodulation pattern. However, the other three Tn5 mutants had difficulty producing nodules. When these three mutants did produce nodules, those nodulated plants were possibly not as effective as with the parent (KN1001). This was evident by a larger proportion of smaller plants, and an increase in the number of small nodules. An attempt was made to clarify the effects of culture growth phase and plant growth conditions on nodulation. It became apparent that changes in these factors were reflected in a complex manner by the nodulation results. Since these bacteria produce a complex mixture of auxins, then a more detailed study of these is required. It may become apparent that another complicating factor is the presence of an alternative nodulation mechanism.

INTRA- AND INTERGENERIC SIMILARITIES BETWEEN 23S RIBOSOMAL RNA CISTRONS FROM RHIZOBIUM AND RELATED BACTERIA

Brion D. W. Jarvis,[1] Moniek Gillis[2] and Josef DeLey[2]

Dept. of Microbiology & Genetics, Massey Univ., Palmerston North,
New Zealand[1] and Laboratory for Microbiology & Microbial Genetics,
Faculty of Science, State University, Ledeganckstraat 35,
B-9000 Gent, Belgium[2]

For many years, legume root nodule bacteria have been classified, identified and named according to the plant or group of plants on which they form nodules. This classification is based on one character determined by a small fraction of the genome. It is likely to mislead those who use it for comparative purposes because of its tendency to group together basically dissimilar strains and designate them as species. More reliable estimates of relationship can be obtained by comparing the whole genome or conserved sequences within it such as the ribosomal RNA cistrons.

We prepared [^{14}C]-labelled rRNA from the type strain Bradyrhizobium japonicum ATCC10324 and [^{3}H] rRNA from Rhizobium meliloti SU47 and R. loti NZP2037 labelled rRNA's from Agrobacterium tumefaciens. ICPB strain TT111 and Rhodopseudomonas palustris DSM130 were also used. These RNA's were hybridized under stringent conditions with filter-fixed DNA's from 24 strains of rhizobia representing nine previously identified DNA:DNA homology groups and with various strains whose relationship with Agrobacterium was already known. Two parameters were measured: $T_{m(e)}$, which is the temperature at which 50% of the hybrid is denatured, and the "percentage rRNA binding" which is the micrograms of labelled rRNA duplexed with 100 μg of filter-fixed DNA after ribonuclease treatment These parameters were used to draw rRNA similarity maps and the results were summarized in a $\Delta T_{m(e)}$ dendogram. The methods employed are described in detail by DeSmedt, LeLey (1975) and Gillis, DeLey (1980).

It was concluded that: (1) rRNA cistrons from the Rhizobium and Bradyrhizobium species studied resemble one another much less than do those of Rhizobium and other genera presently included in the family Rhizobiaceae (Agrobacterium and Phyllobacterium). (2) The genus Rhizobium contains at least three genetically distinct groups. Group I includes R. meliloti, R. fredii and R. leguminosarum. Group II is represented by R. loti and Group III by rhizobia from Galega sp. (3) The rRNA cistrons from species in Group I resemble those of Agrobacterium cluster 2 more closely than do those from species in Groups II or III. (4) Bradyrhizobium japonicum contains rRNA cistrons which have diverged from those of Rhodopseudomonas palustris to about the same extent as those of Rhizobium Group I and Agrobacterium cluster 2.

REFERENCES

DeSmedt J and DeLey J (1977) Int. J. Syst. Bact. 27, 222-240.
Gillis M and DeLey J (1980) Int. J. Syst. Bact. 30, 7-27.

R-PLASMID TRANSFER AND CHROMOSOMAL MOBILIZATION IN COWPEA RHIZOBIA

W. McLaughlin and M.H. Ahmad, Department of Biochemistry, University of the West Indies, Mona, Kingston 7, Jamaica

Genetic analysis of cowpea rhizobia is significant because of their ability to: (a) nodulate a broad host range (b) fix nitrogen ex. planta, and due to their wide distribution in tropical and subtropical soils.[1] The objective of this study was to determine the ability of two divergent cowpea rhizobia strains IRC256 and JRC23 to receive plasmids RP4 and R68.45 from E. coli. The ability of these plasmids to mobilize the chromosome of cowpea rhizobia was also examined. Chromosomal mobilization has never previously been reported in cowpea rhizobia.

MATERIALS AND METHODS: Donors used: E. coli J53 (RP4) (obtained from R. Dixon), E. coli 1230 (R68.45) (obtained from J.E. Beringer), cowpea rhizobia strains JRC23-SM20 (RP4), JRC23-SM20 (R68.45), IRC256-HA409 (RP4) and IRC256-HA409 (R68.45) (this laboratory). Recipients used: cowpea rhizobia strains JRC23-SM20 sm^r, IRC256-HA409 sm^r, JRC23-3185 rif^r ery^r met^- and E. coli JM6 rif^r (this laboratory).Tryptone yeast extract sodium chloride (TYS) medium[3] was used for E.coli, yeast extract mannitol (YEM) medium[4] or Mannitol Minimal (WMM) medium (yeast extract was replaced by $(NH4)_2SO4$ 1.0 g/1) was used for rhizobia. Selective medium for E. coli was TYS supplemented with kanamycin (200 ug/ml) and rifamicin (50ug/ml). Selective medium for rhizobia was YEM or WMM supplemented with kanamycin (200ug/ml), streptomycin (50ug/ml),erythromycin (200ug/ml), rifampicin (150ug/ml) and methionine (100ug/ml) as required. Filter matings on millipore membrane were done according to Beringer[2].

RESULTS AND CONCLUSIONS: R-plasmids RP4 and R68.45 were transferred from E. coli to two strains of cowpea rhizobia. The frequency of RP4 transfer in strains JRC23-SM20 and IRC256-HA409($1.0x10^{-2}$ and $4.7x10^{-2}$ per donor respectively) was higher than R68.45($6.38x10^{-6}$ and $1.66x10^{-5}$ per donor respectively).The plasmids were stably maintained in cowpea rhizobia as free living and after plant passage. The transconjugants expressed increased resistance to kanamycin and tetracyclin but not for ampicillin. Ampicillin resistance was re-expressed when plasmids were transferred from rhizobia to E.coli. Transfer of R-plasmids frequencies of RP4 and R68.45 in isogenic rhizobia strains was higher ($4.75x10^{-2}$ and $5.00x10^{-2}$ per donor respectively) than in non-isogenic strains ($8.7x10^{-5}$ and $5x10^{-6}$ per donor respectively).The ability of R-plasmids to mobilize chromosomal transfer in cowpea rhizobia was also examined. Mobilization of chromosomal markers Sm^r and met^+ by RP4 was more efficient ($1.34x10^{-6}$ and $1.58x10^{-6}$ per donor)than R68.45($1.74x10^{-7}$ and $1.06x10^{-7}$ per donor) for both the markers respectively. Plasmid RP4 could be used in mobilizing chromosomal genes transfer to develop a linkage map in cowpea rhizobia.

REFERENCES: 1. Ahmad, M.H. and McLaughlin, W. (1985) Adv. Biotech (in press) 2. Beringer, J.E. et al (1978) J. Gen. Microbiol. 104, 201-207. 3. Selvaraj, et al (1980) Mol. Gen. Genet. 178, 561-566. 4.Vincent, J.M. (1970) A manual for the practical study of root nodule Bact., Blackwell, Oxford.

CONSERVATION OF THE INSERTION SEQUENCE (IS66) HOMOLOGUE OF T-DNA AND VIR-DNA OF THE OCTOPINE TI-PLASMID IN RHIZOBIUM FREDII PLASMID DNA

NEELA RAMAKRISHNAN., R. K. PRAKASH., and ALAN G. ATHERLY. Dept. of Genetics, Iowa State University, Ames, IOWA-50011. U.S.A.

Hybridisation experiments were performed to determine if DNA sequences homologous to the right border region of TL-DNA of the octopine Ti-plasmid, pTiAch5, are conserved in different R. fredii strains, USDA 193,191,194, 201 and 206. A plasmid clone of EcoRI fragment 19a, a small fragment adjacent to the 'core' T-DNA, was used as the probe. It hybridised only to the plasmids carrying the symbiotic genes. Strain USDA 194 is an exception where the probe hybridised to the total DNA but not to any of the indigenous plasmids. Hybridisation to HindIII digested plasmid DNA of R. fredii strains suggested that the homology is confined to a 1.7kb fragment with a faint homology to an 8.8kb fragment. The 1.7kb HindIII fragment also hybridised to two regions of the vir-DNA of the octopine Ti-plasmid, pTiAch5. Recently, it has been reported that sequences homologous to an insertion element IS66 are present in the T-DNA region as well as in the virulence region of the Ti-plasmid, pTiB$_6$806. One copy of this element is present in the T-DNA region and two copies are located in the virulence region of the wild-type Ti-plasmid. Hybridisation studies proved that the 1.7kb HindIII fragment of R. fredii strain USDA 193 plasmid, homologous to T-DNA and the virulence region of the wild-type Ti-plasmid, is an IS66 homologue. Strong homology to the insertion sequence was also observed in regions adjacent to the nif genes of the strain USDA 193.

Total DNA digests of the different R. fredii strains hybridised to the 1.7kb HindIII fragment indicating the presence of at least one copy of IS66 homologue on the chromosome. Different strains differed in the number of copies of the element present on the chromosome. In R. fredii strain USDA 194, the 1.7kb HindIII fragment is present only on the chromosome and not on any of the plasmids present. The nod and nif genes of this strain homologous to the R. meliloti nod and nif probes are also located on the chromosome and not on the plasmids. There seems to be a close association of the insertion sequence homologue with the symbiotic genes in R. fredii strains.

TWO HIGHLY DIVERGENT SYMBIOTIC TYPES (SPECIES) COMPRISE BRADYRHIZOBIUM JAPONICUM

John Stanley, Gregory G. Brown and Desh Pal S. Verma, Biology Department, McGill University, 1205 Docteur Penfield Avenue, Montreal, CANADA H3A 1B1

There exist considerable physiological and symbiotic differences among Bradyrhizobium japonicum strains effectively nodulating soybean. Using specific gene probes, we have analyzed, for ten representative B. japonicum strains, the degree of DNA base substitution in and around symbiotically and constitutively expressed genes. Two genotypes were found. For the nifDH and nod genes, there is almost no sequence divergence within a group (less than 0.4%) but very high intergroup sequence divergence (greater than 16.5%). We termed these groups symbiotic types; sTI (whose nif gene organization is that of strain USDA 110) and sTII (previously undescribed). sTII strains are capable of high levels of in vitro nitrogen fixation in gluconate-glutamate media [1]. The fast-growing strains of R. japonicum are not related to either the sTI or sTII groups. Sequence divergence in and around the recA, glnA genes and two other cloned sequences further distinguished two sublines of sTI, between which symbiotic genes may have been transferred. Phenograms constructed by the unweighed group pair method depict the strain interrelationships. Here, Phenogram A is derived from nifDH and nod gene probes, while B is derived from recA, glnA and two other cloned DNAs. Breaks indicate no conservation of restriction fragments; and thus, a sequence divergence which is too great to be estimated. A range of symbiotic Tn5 mutants [2] of both genotypes (strains USDA 122 and 61A76) has been obtained in our laboratory. The highly divergent sTI and sTII groups, which are unlikely to have acquired nif or nod genes from a common origin or to give rise to a common descendant, can accurately be described as separate species, probably sharing host-specificity determinants. Thus, the definition of this species based on host-specificity is inadequate for molecular genetics [3].

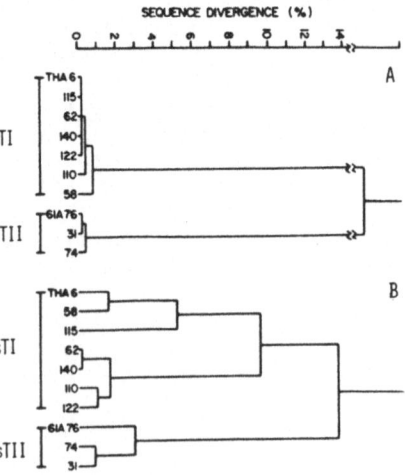

(1) Huber, T et al (1984) J Bacteriol 158, 1168-1171
(2) Rostas, K et al (1984) Mol Gen Genet 197, 230-235
(3) Stanley, J et al (1985) J Bact 163, 148-154

TRANSPOSON MUTAGENESIS OF *RHIZOBIUM* USING A DERIVATIVE OF Tn5 CONFERRING GENTAMICIN AND SPECTINOMYCIN RESISTANCE

C.L. WANG, J.R. SPOKES, M.J. WOODWARD* and P.R. HIRSCH
ROTHAMSTED EXPERIMENTAL STATION, HARPENDEN, HERTS., U.K. and
*OXFORD POLYTECHNIC, OXFORD, U.K.

Transposon mutagenesis using Tn5 has been an essential part in the development of genetic systems in many Gram-negative bacteria including members of the Rhizobiaceae (reviewed by Berg, Berg, 1983). Advantages of Tn5 compared with other transposons that can be used in *Rhizobium* include the ease with which the Tn5-determined kanamycin (Km) resistance can be selected, and the apparent randomness with which it inserts into chromosomal or plasmid DNA. Introduction of Tn5 into *Rhizobium* from *E.coli* has been mediated by the use of unstable suicide vectors: plasmids carrying Tn5 that can replicate in the *E.coli* donor but not in *Rhizobium* (Beringer *et al* 1978, Simon *et al* 1983). Km-r *Rhizobium* transconjugants from such crosses are found to contain Tn5 integrated into their DNA, and to have lost the vector plasmid.

A derivative of Tn5 in which the central region encoding kanamycin resistance has been replaced by a fragment from the IncP plasmid R1033 encoding resistance to gentamicin (Gm), spectinomycin (Sp) and a low level of streptomycin (Sm) has been constructed (Hirsch, Woodward, manuscript in preparation). This recombinant transposon, Tn5-GmSpSm, has been incorporated into an RP4::Mu cointegrate that is stably maintained in *E.coli* but not in *Rhizobium*, forming the suicide vector pXS102 (details to be published elsewhere).

Experiments using pXS102 have demonstrated that it can act as a suicide vector for the introduction of Tn5-GmSpSm from *E.coli* into *Rhizobium*, producing Gm-r Sp-r transconjugants that have lost the vector antibiotic resistance markers. Integration of Tn5-GmSpSm into both plasmid and chromosomal DNA of *Rhizobium* transconjugants has been detected.

The Gm resistance conferred by Tn5-GmSpSm inserted into *Rhizobium* plasmids has proved to be a useful marker in experiments to select for plasmid transfer between strains and also to monitor plasmid loss in curing experiments. In addition it has enabled investigation of the interaction between two copies of Tn5 in *Rhizobium*: pXS102 was transferred into a strain already carrying Tn5, and transposition of Tn5-GmSpSm was shown to be inhibited.

The suicide vector pXS102 carrying the recombinant transposon Tn5-GmSpSm appears to be a potentially important new tool for the genetic manipulation of *Rhizobium*.

REFERENCES

Berg DE, Berg CM (1983) Biotechnology 1, 417-435.
Beringer *et al* (1978) Nature 276, 633-634.
Simon R *et al* (1983) Biotechnology 1, 784-791.

THE GENETIC RELATIONSHIP BETWEEN RHIZOBIUM FREDII, GALEGA RHIZOBIA AND OTHER RHIZOBIUM AND BRADYRHIZOBIUM SPECIES

D. Neil Wedlock and Brion D. W. Jarvis

Dept. of Microbiology & Genetics, Massey Univ., Palmerston North, New Zealand

Groups of genetically related bacteria can be identified by quantitative DNA:DNA hybridization. These may be designated as species when their phenotypic characteristics are known. Six DNA:DNA homology groups have been identified among the fast-growing, acid-producing rhizobia (Crow et al. 1981; Jarvis 1983; Jarvis et al. 1980, 1982; Jordan 1984). The relationship of these DNA:DNA homology groups to R. fredii (fast-growing rhizobia which nodulate soybeans) and to fast-growing rhizobia which nodulate Galega officinalis or G. orientalis (goatsrue) is described below.

We used the following bacteria as sources of [^{32}P] labelled reference DNA: R. fredii strains USDA191 and USDA208; Galega rhizobia strains gal 1 and NW3; R. leguminosarum var. trifolii ATCC10004 and R. meliloti SU47. These DNAs were hybridized with DNA from 11 strains of R. fredii, 14 strains of Galega rhizobia and 18 strains representing known DNA:DNA homology groups from Rhizobium and Bradyrhizobium. The methods used were described by Jarvis et al. (1980).

DNA from the R. fredii reference strains USDA191 and USDA208 had mean relative percentage homologies at 65°C of 80 and 86%, respectively, with DNA from other R. fredii strains and 25 and 11%, respectively, with DNA from strains representing other Rhizobium and Bradyrhizobium DNA homology groups. Similarly, DNA from the Galega rhizobia reference strains gal 1 and NW3 had mean relative percentage homologies at 65°C of 79 and 85%, respectively, with DNA from other Galega rhizobia and 13 and 25%, respectively, with DNA from strains representing other Rhizobium and Bradyrhizobium DNA homology groups. These data were confirmed by hybridizations at 80°C and by melting point determinations on the hybrid DNAs formed. It was concluded that R. fredii and Galega rhizobia constituted two further DNA:DNA homology groups in the genus Rhizobium. A rapid identification method for these bacteria was devised which employed colony hybridization and in vitro labelled [^{32}P]-DNA from reference strains.

REFERENCES

Crow V L et al. (1981) Int. J. Syst. Bacteriol. 31, 152-172.
Jarvis B D W (1983) Current Microbiol. 8, 153-158.
Jarvis B D W et al. (1980) Int. J. Syst. Bacteriol. 30, 42-52.
Jarvis B D W et al. (1982) Int. J. Syst. Bacteriol. 32, 378-380.
Jordan D C (1984) In Kreig N R and Holt J C, eds, Bergey's Manual of Determinative Bacteriology, 9th Ed, pp 234-244, Williams & Wilkins, Baltimore.

Microbial Genetics Relevant to the Development
of Symbiotic Nitrogen Fixation

GARY S. DITTA, Univ. of California, San Diego, La Jolla, CA   92093

This session of the meeting was devoted to the vast array of
Rhizobium genes acting at early and intermediate stages of nodule
morphogenesis. There was much information that was new, much that was
exciting. For the past several years, the primary focus of Rhizobium
genetics has been to identify symbiotically essential genes and to sketch
the broad outline of their genetic organization. While this process is
still continuing, the emphasis is now shifting toward more detailed
analyses of the biochemical roles of the various gene products and toward
understanding how these gene products interact during symbiotic
differentiation. Such efforts are at an early stage, but given the
commitment by laboratories around the world to investigating the
Rhizobium-legume symbiosis, it is clear that the future holds great
promise indeed. Below I've attemped a very brief overview of the more
than 50 presentations offered in session 9.

Rhizobium genes necessary for nodule induction, nod genes, are
currently receiving the greatest attention by researchers. Relatively
few in number, nod genes have a plethora of effects on the Rhizobium-
legume interaction. They are involved in root hair curling, infection
thread growth, host range, and direct or indirect regulation of gene
expression. Some of the most exciting data presented at the meeting were
the findings by several laboratories that nod genes of fast-growing
Rhizobium species are induced in response to one or more soluble plant
factors. Preliminary evidence suggests the likely identification of nod-
specific transcriptional regulatory regions (i.e. promoters) involved in
this "signaling"of the bacteria by the plant. Thus far, the story seems
amazingly similar to the plant-mediated induction of A. tumefaciens Ti
plasmid vir genes first reported a year ago at the 2nd International
Symposium on the Molecular Genetics of the Bacteria-Plant Interaction.
One major difference, however, may be some specificity by Rhizobium for
plant substance(s) derived from legumes as opposed to non-legumes.

There were several reports in this session on the identification and
isolation of nod genes from slow-growing Rhizobium species. In B.
japonicum, certain nod genes (nodABC) have been identified near genes
involved in nitrogen fixation (nifA, fixA). Interestingly, it appears
that the nodD gene is not immediately adjacent to the nodABC genes,
suggesting that the arrangement of these so-called "common" nod genes may
differ somewhat from that found in fast-growing rhizobia. In Parasponia
Rhizobium ANU 289, a cowpea-like species, the nucleotide sequence for the
common nod region has revealed these genes to have substantial homology
to those from R. trifolii, and to have an organization identical to that
of fast-growing rhizobia. This is interesting in light of the ability of
Parasponia Rhizobium to infect the non-legume Parasponia (Ulmaceae), and
to do so by "crack entry" rather than infection thread invasion through
root hairs. Other slow-growing species for which nod studies were
reported include B. sp. (Lotus), and a cowpea group Rhizobium strain
capable of nodulating peanut and pigeonpea.

It has generally been anticipated that Rhizobium genes involved in the intermediate steps of nodule development, between nodule formation and the onset of nitrogen fixation, would be the most difficult to analyze in detail. It was therefore gratifying to see progress being made in this area. For example, a number of genetic loci have been identified that are involved in exopolysaccharide (eps) production, and mutants at these loci induce the formation of nodule like structures lacking bacteroids. The mutants are blocked at a very early stage of nodule invasion, prior to infection thread formation. Co-inoculation experiments reported at this meeting suggest the possibility that such mutants may also have blocks at later stages of bacteroid development. Results from two laboratories indicate a complex mode of regulation for eps genes, and it has been hypothesized that controls may be exerted during symbiosis to repress aspects of vegetative eps production. Another interesting set of symbiotically essential develomental genes was identified in R. meliloti that seem to be structural and functional homologues of chromosomal virulence (chv) genes in Agrobacterium tumefaciens. During the discussion session it was revealed by Dr. Gary Stacey that his laboratory has data in press showing that one of the chv loci in A. tumefaciens is involved in the production of cyclic β-1,2 glucans, a subclass of exopolysaccharides unique to Rhizobiaceae. An interesting angle on this is that chv mutants of A. tumefaciens are defective in their ability to attach to plant cells. In R. phaseoli, transposon mutants have been isolated that appear to lack the major lipopolysaccharide. Such mutants penetrate the root via infection thread formation, but such infection threads develop abnormally and soon abort.

In addition to those cases highlighted above, reports of ultrastructural analysis and/or DNA localization for other genes involved in the nodule differentiation pathway were also presented in this session, as were several instances of gene-specific competition phenomena.

ORGANIZATION AND REGULATION OF SYMBIOTIC NITROGEN FIXATION GENES FROM BRADYRHIZOBIUM JAPONICUM

H. HENNECKE, A. ALVAREZ-MORALES, M. BETANCOURT-ALVAREZ, S. EBELING,
M. FILSER, H.-M. FISCHER, M. GUBLER, M. HAHN, K. KALUZA, J.W. LAMB,
L. MEYER, B. REGENSBURGER, D. STUDER, AND J. WEBER

MIKROBIOLOGISCHES INSTITUT, EIDGENÖSSISCHE TECHNISCHE HOCHSCHULE,
ETH-ZENTRUM, UNIVERSITÄTSTRASSE 2, CH-8092 ZÜRICH, SWITZERLAND

1. INTRODUCTION

This article summarizes our recent data on the analysis of bacterial genes required for the formation of a nitrogen fixing soybean root nodule, with particular emphasis on nif and fix genes. The bacterium under study is the slow-growing Rhizobium japonicum (strain USDA 3I1b110). This species has been given a new generic name, Bradyrhizobium, to account for the many differences by which it is distinguished from the fast-growing rhizobia (Jordan 1984). The validity of this concept is supported by a phylogenetic analysis of a few Rhizobium species based on 16S rRNA oligonucleotide cataloguing (Hennecke et al. 1985). Figure 1 shows that the slow-growing Bradyrhizobium japonicum (strain 110) and R.lupini are well separated from the fast-growing R.meliloti and R.leguminosarum (by a S_{AB} value as low as 0.53); by this analysis all rhizobia fall into the α-subdivision of photosynthetic purple bacteria (Woese et al. 1984). Consequently, we will adopt the "Bradyrhizobium" terminology throughout this article to avoid confusion with the fast-growing R.japonicum strains (Keyser et al. 1982).

A few strains of the slow-growing rhizobia have been shown to be capable of inducing nitrogenase activity under free-living, microaerobic culture conditions (Hennecke 1981; and further references therein). It is thus possible to extract nif mRNA from such cultures, or to separate the total protein contents on two-dimensional gels, which has led to the identification of nitrogenase components (Scott et al. 1979; Hahn et al. 1984). The availability of such a system facilitates the analysis of mutant phenotypes, when it is of interest to know, for example, whether or not a presumptive regulatory mutant synthesizes nitrogenase proteins.

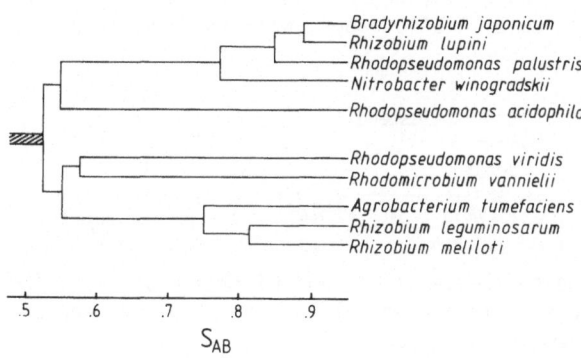

Fig. 1: Phylogenetic positions of Rhizobium species by 16S rRNA oligonucleotide cataloguing.

2. GENOME STRUCTURE OF B.JAPONICUM: PRESENCE OF NUMEROUS, DIFFERENT REPEATED SEQUENCES

In contrast to the fast-growing rhizobia, the slow-growing rhizobia have not been shown to contain symbiotic (Sym) plasmids. Consequently, it has been concluded that fix, nif, and nod genes are located on the chromosome. As will be seen in the following chapters there is now evidence that these genes may all be clustered rather than being scattered over the whole genome. This arrangement is further supported by an analysis of the genomic organization of newly detected repeated sequences (RSs), and by the mapping of large deletions which have been generated by homologous recombination between these repeats (Hahn et al. 1984; Kaluza et al. 1985a).

A 1126 bp sequence (RSα) has been found to be repeated 12 times in the B.japonicum genome. DNA sequencing of three RSα copies has revealed that they are almost identical (with only four nucleotide exchanges and one triplet deletion), and that they exhibit typical characteristics of prokaryotic insertion elements such as inverted repeats at its ends, potential target site duplications, and large open reading frames. Another repeated sequence (RSβ) is reiterated at least 6 times having an approximate length of 950 bp. One copy each of RSα and RSβ is located very close to the 5' side of the nifDK operon, but is not involved in its expression (Figs. 2 and 3). Deletion mutants have been obtained which have lost nifDK plus large stretches of flanking DNA on both sides. The largest deletion has lost 6 copies of RSα and 3 copies of RSβ. The endpoints of many deletions are within the RSs. Based on this analysis it has been possible to draw a map of the relative positions of these RSs (Fig. 2). Thus, it has become evident that many repeated sequences are clustered around some nif genes. As the deletion mutants grow well on minimal medium it is clear that the deleted genomic region does not carry genes that are absolutely essential for growth. Phenotypic analysis of the deletion mutants has shown that the region between RSβ2 and RSβ3 (Fig. 2) does not carry genes strictly essential for symbiosis, whereas the region between RSα12 and RSα3 (Fig. 2) is likely to contain a region involved in competitiveness and nodulation, but is not nodABC, nodF, or nodD (see also chapter 4).

In the course of hybridizing cloned regions back to total genomic DNA we have found further repeated sequences. The nature of these repeats is not yet known. These findings indicate that the analysis of the B.japonicum genome, particularly around the symbiotic genes, will turn out to be unpleasantly complicated.

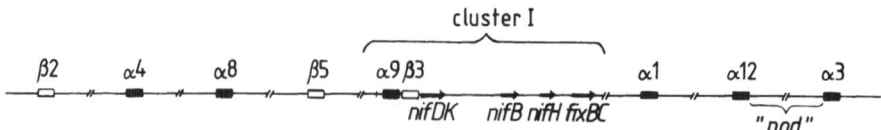

Fig. 2: Organization of repeated sequences around nif/fix cluster I. The relative order of the RSs was derived from deletion mapping, but the absolute distances between them are not known.

3. IDENTIFICATION AND ORGANIZATION OF NIF AND FIX GENES

To date we have identified nine DNA segments carrying genes (or potential genes) involved in symbiotic nitrogen fixation. These loci are organized in two clusters (Fig. 3). Genetic linkage between the two clusters has not yet been established. All the "genes" have been detected by inter-species hybridization, and for some of them additional information is available (mutations, sequence data, transcript and protein mapping) justifying their designation as "gene". It is possible that the two clusters harbor further nif or fix genes, but it is unlikely that they contain all essential nitrogen fixation genes because, after random Tn5 mutagenesis, we found several Fix⁻ mutants in which the mutations are located outside of these two clusters (see chapter 7). The following gives a more detailed account of what is known about the nif and fix genes shown in Fig. 3.

3.1. NifD, nifK, nifH.

The three nitrogenase genes, which are separated by 17 kb into two transcriptional units, have been most extensively characterized by the techniques of S1 mapping, protein expression, total sequencing, and mutational analysis. Both nifDK and nifH are transcribed from nif consensus promoters. NifK and nifH are immediately followed by potential transcription termination sites. The three genes are preceded by E.coli-like ribosome binding sites. An interesting observation was that the amino acid sequences of the MoFe protein's α and β subunits (the nifD and nifK gene products) share significant homology in the N-terminal third. This points to a possible evolutionary origin of nifD and nifK from a common ancestral gene. All these data have been published (Fuhrmann, Hennecke 1982; Kaluza et al. 1983; Hahn, Hennecke 1984; Fischer, Hennecke 1984; Kaluza, Hennecke 1984; Fuhrmann, Hennecke 1984; Thöny et al. 1985).

3.2. NifE.

A nifE-like region has been found downstream of nifK probably between the sites of two Fix⁺ Tn5 insertions (Fig. 3). By interspecies hybridization this region was found to be homologous to the Klebsiella pneumoniae nifE gene (Dixon 1984), the Sesbania Rhizobium nifE-like gene (Norel et al. 1985) and the R.meliloti fixE gene (Weber et al. 1985). Hence, the R.meliloti fixE gene may turn out to be a nifE-like gene.

3.3. NifB.

A nifB-like gene has been identified 11 kb downstream from the 3' end of nifK (Fuhrmann et al. 1985). It hybridizes specifically with nifB from K.pneumoniae (Dixon 1984) and fixZ (= nifB) of R.leguminosarum (Rossen et al. 1984). Partial DNA sequencing has confirmed the homology, and has revealed the direction of transcription as shown in Fig. 3. A strain was constructed with a 5 kb deletion (Fig. 3) affecting the 3' end of nifB. The deletion strain has a Fix⁻ phenotype, but this phenotype could arise from further nif/fix genes missing in the deleted region. Adams et al. (1984) reported on the location of a nifA-like gene at exactly the same

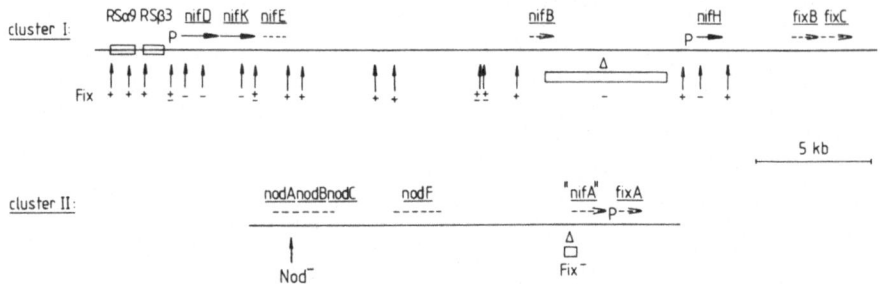

Fig. 3: Organization of nif, fix, and nod genes in B.japonicum. Tn5 mutations (arrows) and deletions (Δ) are indicated. p = promoter.

site where we found nifB. Even up- and downstream of nifB we did not obtain evidence for a possible nifA gene. (The Fix⁻ deletion mutant mentioned before (Fig. 3, cluster I) was able to synthesise the nifH gene product.)

3.4. The "nifA"-like gene.

In cluster II we have detected a gene region with homology to both nifA and ntrC of K.pneumoniae (Dixon 1984), and to a nifA-like regulatory gene of R.meliloti (Buikema et al. 1985; Weber et al. 1985). Sequencing of DNA from the central part of the gene showed 60 % homology to nifA, and 40 % homology to ntrC of K.pneumoniae. A deletion mutation was constructed affecting the 5' end of the gene (Fig. 3). The mutant strain was Fix⁻. Analysis of proteins synthesized by the mutant under microaerobic culture conditions showed that the nitrogenase Fe protein (the nifH gene product) and at least six further unidentified proteins were missing. Thus, it is likely that the product of this B.japonicum gene (tentatively called "nifA" gene) is functionally homologous to the K.pneumoniae nifA- or ntrC-specific transcriptional activators. (For further details see the poster summary of H.-M. Fischer et al. in this volume.)

3.5. FixA, fixB, fixC.

These genes have been detected via interspecies hybridization using the R.meliloti fixABC operon as probe (Fuhrmann et al. 1985). In contrast to R.meliloti (Pühler et al. 1984; Earl, Ausubel 1984) these genes are organized in two separate transcriptional units. FixB and fixC are located within gene cluster I (with the start of fixB being 3.2 kb downstream of the 3' end of nifH) whereas fixA is in cluster II immediately downstream of the nifA-like gene (Fig. 3). Partial DNA sequences from all three genes confirmed the homology and revealed the direction of transcription as shown in Fig. 3. FixA mRNA was extracted from soybean root nodules but could not be detected in aerobic B.japonicum culture. The fixA promoter was identified by S1 mapping and shown to fit with the nif consensus sequence.

4. IDENTIFICATION OF NOD GENES, AND LINKAGE TO NIF AND FIX GENES

R.phaseoli nod probes (Lamb et al. 1985) have been used initially to clone homologous regions from B.japonicum. Then the cloned regions were further identified using specific nodA, nodB, nodC, nodF, and nodD probes of R. leguminosarum (Downie et al. 1985; A. Downie, personal communication). NodABC- and nodF-homologous regions were thus detected in cluster II upstream of nifA and fixA (Fig. 3). The nodABC genes are located next to each other in the same way as in fast-growing rhizobia (Fisher et al. 1985). An insertion mutation has been constructed in the nodA-homologous region (Fig. 3). The mutant was unable to elicit nodules on soybean (Nod⁻). A nodD-specific hybridization was found with genomic DNA unlinked to clusters I and II.

5. STUDIES ON THE REGULATION OF NIF AND FIX PROMOTERS BY THE USE OF NIFH-, NIFD-, AND FIXA-LACZ FUSIONS

Translational fusions to lacZ were constructed with those genes for which the respective promoters had been mapped. Expression from these promoters was achieved in E.coli ET8000 (ntrA$^+$ ntrC$^+$) in the presence of plasmid pMC71A which determines constitutive synthesis of the K.pneumoniae nifA product (Table 1). Thus, both fix and nif genes may be controlled in a similar way. Activation of the fixA promoter is less pronounced in comparison to activation of nifH and nifD. This may reflect differences in the consensus promoter sequences, or may be due to the fact that expression was achieved by heterologous transcriptional activation factors. Experiments with nifD- and nifH-lacZ fusions have shown that the nifA-dependent activation was also dependent on the ntrA product (Alvarez-Morales, Hennecke 1985). However, replacement of the nifA protein by the ntrC product did not lead to activation (Table 1). This result is in contrast to the R.meliloti nifH promoter which is activated by either the ntrC or the nifA gene products (Sundaresan et al. 1983).

Fusion	β-Galactosidase activity		
	(1)	(2)	(3)
		nifA^C	ntrC^C
Bj nifH'-'lacZ	56	8516	71
Bj nifD'-'lacZ	89	14165	164
Bj fixA'-'lacZ	327	793	433

TABLE 1: Expression of nif- and fix-lacZ fusions in E. coli in the absence (1) or presence of constitutively synthesized nifA (2) or ntrC (3) proteins.

6. SITE-DIRECTED POINT AND DELETION MUTAGENESIS OF NIF PROMOTERS: FUNCTIONAL EVIDENCE FOR THE INVOLVEMENT OF THE NIF CONSENSUS SEQUENCE AND AN IMPORTANT UPSTREAM REGION FOR NIFA- PLUS NTRA-MEDIATED GENE ACTIVATION

In the B.japonicum nifH promoter a G to T transversion was introduced, by oligonucleotide mutagenesis, at position -25 of the nif consensus promoter sequence. It resulted in a 92 % decrease of promoter activity. Two other mutations (an A to G transition at -30, and a 63 bp deletion affecting the 5' untranslated nifH mRNA region) had little effect on expression (Kaluza et al. 1985b). Numerous deletions were constructed upstream of the nif consensus sequence in both the nifH and nifDK 5'-flanking regions. There are critical sequences up to position -200 which are required for maximal promoter activity. Lack of these sequences results in more than 94 % loss of promoter activity. (For further details see the poster summary of A. Alvarez-Morales et al. in this volume.)

7. THE ADVANTAGE OF B.JAPONICUM FOR THE ISOLATION OF MUTATIONS WHICH SPECIFICALLY AFFECT BACTEROID DEVELOPMENT AND REGULATION OF SYMBIOTIC NITROGEN FIXATION

After random Tn5 mutagenesis of the B.japonicum genome, kanamycin/strepto-mycin resistant strains were screened for Nod$^-$ and Nod$^+$Fix$^-$ phenotypes on soybean. Such mutants were then also tested for the ability to reduce ace-tylene under free-living, microaerobic culture conditions. In addition, these cultures were ^{35}S-labelled to examine the protein pattern by two-dimensional gel electrophoresis. Out of 17 potentially interesting mutants we here give phenotypic descriptions of two strains.

7.1. Mutant 3541.

This mutant is of interest because it has wild-type (Nif$^+$) activity in free-living culture whereas symbiotic nitrogen fixation is drastically reduced (by 95 %; Fix$^-$). The protein pattern is indistinguishable from the wild-type; nodulation is normal, and the nodules contain leghemoglobin. Electron microscopy has revealed that infected nodule cells contained underdeveloped bacteroids. In conclusion, the mutation (which is in a 12 kb EcoRI fragment somewhere outside of clusters I and II) does not direct-ly affect nitrogenase activity, but the mutant appears to be deficient in the development and/or maintenance of bacteroids.

7.2. Mutant 3160.

This mutant is Nif$^-$ in culture and Fix$^-$ in plants. Several proteins are missing including the nifH gene product; some of these proteins are also missing in the nifA deletion mutant (chapter 3.4.). Interestingly, a 15K protein is synthesized that is normally made only under aerobic conditions. Hence, the mutation may have affected a regulatory gene that is perhaps involved in respiratory control. Furthermore, the mutant elicits the pro-duction of poorly developed nodules ("bumps") and the plant cells in these bumps are completely uninfected. In conclusion, the mutation (which is in a 1.7 kb EcoRI fragment outside clusters I and II) indicates that there may be a common genetic regulatory element for the complex steps of nodu-le and bacteroid development, and nitrogen fixation.

The two mutants described before show that the free-living culture for assaying Nif activity has helped to differentiate particular phenotypes which, without this assay, might have been suspected to be concerned simply with nitrogen fixation (mutant 3541) or nodulation (mutant 3160). For both mutants, experiments are underway to prove that the observed single DNA insertions are the sole cause for the reported phenotypes.

8. REFERENCES

Adams TH et al. (1984) J.Bacteriol. 159, 857-862.
Alvarez-Morales A, Hennecke H (1985) Mol.Gen.Genet. 199, 306-314.
Buikema WJ et al. (1985) Nucleic Acids Res. 13, 4539-4555.
Dixon R (1984) J.Gen.Microbiol. 130, 2745-2755.
Downie JA et al. (1985) Mol.Gen.Genet. 198, 255-262.
Earl CD, Ausubel FM (1984) Proceedings of the 2nd International Symposium on The Molecular Genetics of the Bacteria-Plant Interaction, June 4-8, 1984, Cornell University, Ithaca, NY, abstr. no.6.
Fischer H-M, Hennecke H (1984) Mol.Gen.Genet. 196, 537-540.
Fisher RF et al. (1985) Appl.Environ.Microbiol. 49, 1432-1435.
Fuhrmann M, Hennecke H (1982) Mol.Gen.Genet. 187, 419-425.
Fuhrmann M, Hennecke H (1984) J.Bacteriol. 158, 1005-1011.
Fuhrmann M et al. (1985) Mol.Gen.Genet. 199, 315-322.
Hahn M, Hennecke H (1984) Mol.Gen.Genet. 193, 46-52.
Hahn M et al. (1984) Plant Mol.Biol. 3, 159-168.
Hennecke H (1981) In Bothe H and Trebst A, eds, Biology of Inorganic Nitrogen and Sulfur, pp. 309-316, Springer Verlag, Heidelberg.
Hennecke H et al. (1985) Arch.Microbiol., in press
Jordan DC (1984) In Krieg NR and Holt JG, eds, Bergey's Manual of Systematic Bacteriology, Vol.1, pp.234-256, Williams & Wilkins, Baltimore.
Kaluza K et al. (1983) J.Bacteriol. 155, 915-918.
Kaluza K, Hennecke H (1984) Mol.Gen.Genet. 196, 35-42.
Kaluza K et al. (1985a) J.Bacteriol. 162, 535-542.
Kaluza K et al. (1985b) FEBS Letters, in press.
Keyser HH et al. (1982) Science 215, 1631-1632.
Lamb JW et al. (1985) Gene 34, 235-241.
Norel F et al. (1985) Mol.Gen.Genet. 199, 352-356.
Pühler A et al. (1984) In Veeger C and Newton WE, eds, Advances in Nitrogen Fixation Research, pp.609-619, Nijhoff/Junk Publishers, The Hague.
Rossen L et al. (1984) Nucleic Acids Res. 12, 7123-7134.
Scott DB et al. (1979) Biochim.Biophys.Acta 565, 365-378.
Sundaresan V et al. (1983) Nature 301, 728-732.
Thöny B et al. (1985) Mol.Gen.Genet. 198, 441-448.
Weber G et al. (1985) In Szalay AA and Legocki RP, eds, Advances in Molecular Genetics of the Bacteria-Plant Interaction, pp. 13-15, Cornell University Publishers, Ithaca, NY.
Woese CR et al. (1984) Syst.Appl.Microbiol. 5, 315-326.

9. ACKNOWLEDGEMENTS

We thank S. Hitz, D. Reber, and C. Zugliani for expert technical assistance and H. Paul for typing the manuscript. We are very grateful to the following researchers for supplying us with strains, plasmids, or DNA sequence information: C. Earl, C. Elmerich, A. Downie, M. Drummond, G. Weber. Grants for this work were provided by the Swiss Federal Institute of Technology (ETH) and by the Agrigenetics Research Corporation. J.W.L. was supported by an EMBO long-term fellowship.

ORGANIZATION AND REGULATION OF Rhizobium meliloti AND
Parasponia Bradyrhizobium NITROGEN FIXATION GENES

Frederick M. Ausubel, William J. Buikema, Christopher D. Earl, John A.
Klingensmith, B. Tracy Nixon and Wynne W. Szeto

Department of Genetics, Harvard Medical School, and Department of Molecular
Biology, Massachusetts General Hospital, Boston, Massachusetts 02114 USA

A. INTRODUCTION

Most of what is known about the control of nif gene expression in
Rhizobium and Bradyrhizobium species has been obtained by analogy with the
more detailed knowledge of the regulation of nif genes in Klebsiella
pneumoniae.

K. pneumoniae nif genes are subject to two levels of positive
regulation in response to ammonia and oxygen. The first level of
regulation is nif-specific and is mediated by the products of the nifLA
operon. The nifA product is a transcriptional activator which is required
for the expression of all nif operons except its own (Buchanan-Wollaston et
al., 1981). The nifL product mediates O_2 repression of nifA activated
transcription (Hill et al., 1981; Merrick et al., 1982; Cannon, et al.,
1985). The second level of nif regulation is mediated by a centralized
system (the ntr system) which controls the expression of a variety of
nitrogen assimilatory genes in enteric bacteria. Under conditions of NH_4^+
starvation, the ntrC + ntrA products activate the nifLA operon (Ow,
Ausubel, 1983; Merrick, 1983) and also activate other genes involved in
nitrogen assimilation [e.g. hut (histidine utilization) and put (proline
utilization) (Magasanik, 1982)].

Recent data indicates that the nifA and ntrC genes are evolutionarily
related. First, as is the case for ntrC mediated activation, the ntrA
product is required in addition to the nifA product for activation of the
nifH promoter (Ow, Ausubel, 1983; Drummond et al., 1983; Sundaresan et al.,
1983a; Merrick, 1983). Second, in an ntrC deletion strain, the nifA
product substitutes for the ntrC product in activating several ntr
regulated promoters, including the nifA and glnA promoters (Ow, Ausubel,
1983; Merrick, 1983). Third, as illustrated in Fig. 1, nifA and ntrC
activated promoters share several common features characterized by the
consensus sequence 5'-TTGCA-3' at -15 to -10 (Ow et al., 1983; Sundaresan
et al., 1983a) and the sequence 5'-GGPyrPurPyrPur-3' at -25 to -20 (Beynon
et al., 1983). Finally, the K. pneumoniae nifA and ntrC gene products
share about 55% homology in a central conserved region of 200 amino acids
(Buikema et al., 1985).

Two key observations led to the conclusion that certain aspects of nif
control in Rhizobium species is very similar to nif control in K.
pneumoniae. First, nif promoter structure is highly conserved between
various Rhizobium species and K. pneumoniae suggesting that Rhizobium nif

A "nitrogen" regulated consensus promoters:

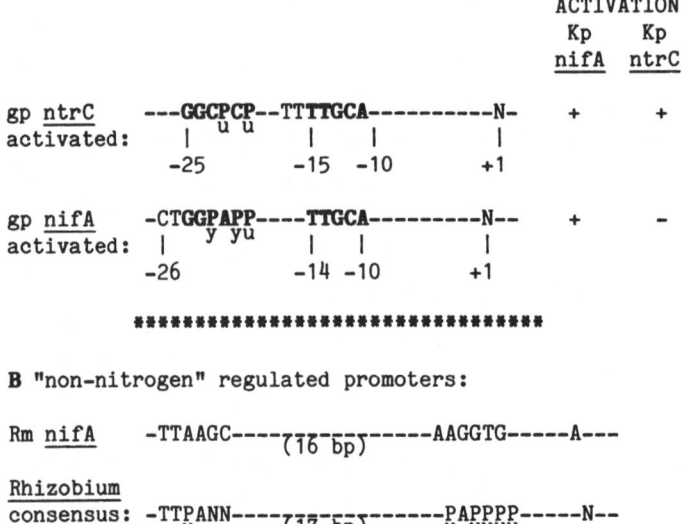

ACTIVATION
Kp nifA   Kp ntrC

gp ntrC activated:
```
   ---GGCPCP--TTTTGCA-----------N-
        u u       |    |        |
       -25      -15  -10       +1
```
\+   \+

gp nifA activated:
```
  -CTGGPAPP----TTGCA----------N--
      y yu      |    |        |
     -26      -14  -10       +1
```
\+   \-

****

B "non-nitrogen" regulated promoters:

Rm nifA   -TTAAGC----(16 bp)-----AAGGTG-----A---

Rhizobium consensus: -TTPANN----(17 bp)-----PAPPPP-----N--
                            u   u uuuu

E. coli consensus: -TTGACA----(17 bp)-----TATAAT-------N
```
    |                             |           |
   -35                          -10          +1
```

FIGURE 1. Consensus sequences for promoters activated by the K. pneumoniae nifA and ntrC gene products (A) and sequences for "non-nitrogen" promoters from Rhizobium species and E. coli (B). **A:** The bolded letters represent overlaps in the two consensus sequences. The ntrC consensus sequence was derived from the K. pneumoniae nifL (Ow et al., 1983; Drummond et al., 1983), the E. coli glnA (Reitzer, Magasanik, 1985), and the R. meliloti nifH (Sundaresan et al., 1983b) promoters and from several mutant derivatives of the K. pneumoniae nifH promoter which attained the ability to be activated by ntrC (Ow et al., 1985). The nifA consensus sequence was described earlier (Ausubel, 1984). **B.** The Rhizobium "non-nitrogen" consensus promoter was derived by Ronson and Astwood (this volume). The R. meliloti nifA promoter region is published (Buikema et al., 1985). The E. coli consensus promoter is from Hawley and McClure, 1983).

genes are regulated by homologues of the K. pneumoniae nifA and/or ntrC genes (for example, see Sundaresan et al., 1983b; Better et al., 1983; Adams, Chelm, 1984; Scott et al., 1983). Second, a R. meliloti nif regulatory gene which maps approximately 5.0 kb upstream of the nifHDK operon, (Szeto et al., 1984a) is approximately 50% homologous to both K. pneumoniae nifA and ntrC in the central 200 amino acid region that is also conserved between the two K. pneumoniae proteins (Buikema et al., 1985). Because the R. meliloti regulatory gene is located on the R. meliloti symbiotic megaplasmid and because the growth of strains carrying a mutation in the regulatory gene is not impaired on media containing proline, histidine, arginine, or glutamine as the sole nitrogen source, it is likely that the R. meliloti regulatory gene is a nif-specific regulatory gene like K. pneumoniae nifA rather than a general nitrogen regulatory gene such as ntrC (Szeto et al., 1984).

In K. pneumoniae, activation of the nif operons depends first upon activation of the nifLA promoter by the ntrC + ntrA products. There is no evidence that the same control circuitry operates in Rhizobium. Moreover, the mechanisms by which Rhizobium nifA-like regulatory genes are activated are unknown. Because nitrogen fixation in Rhizobium species occurs primarily in the symbiotic state, it is possible that activation of Rhizobium nifA-like genes is developmentally regulated and independent of a centralized nitrogen assmilation control system. One possibility is that Rhizobium nifA-like genes are activated by a signal from the host plant which acts through a Rhizobium symbiotic regulatory system which is responsible for activating many symbiotic genes -- for example, genes whose products are involved in the early steps in nodulation or in nodule construction and maintenance.

In this paper, we report several new findings concerning the organization and regulation of nif genes in R. meliloti. First, R. meliloti has an analogue of the K. pneumoniae nifB gene; as in K. pneumoniae, the R. meliloti nifB gene is located directly downstream of the nifA gene. Second, R. meliloti has an ntrC analogue as well as a nifA analogue; as expected, the ntrC analogue is involved in the derepression of nitrogen assimilation genes. Third, the R. meliloti ntrC gene is not required for the establishment of nitrogen fixing nodules. Fourth, although R. meliloti will not fix nitrogen ex planta, derepression of the nifH and nifA promoters occurs ex planta under nitrogen stress conditions; ntrC is required for this ex planta derepression. Finally, we also report some preliminary data concerning the organization and regulation of nif genes in the slow-growing Bradyrhizobium species which nodulates the non-legume Parasponia rigida. As in other Rhizobium and Bradyrhizobium species, nif genes in the Parasponia Bradyrhizobium species are clustered. However, our initial experiments suggest that the Parasponia Bradyrhizobium nif promoters may be less tightly regulated than the corresponding nif promoters in R. meliloti and K. pneumoniae.

B. RESULTS AND DISCUSSION

1. Organization of R. meliloti nif and fix Genes

In R. meliloti, a cluster of nif and fix genes has been mapped adjacent to the nifHDK operon (Ruvkun et al., 1982; Corbin et al., 1982, 1983; Forrai, et al., 1983; Puhler et al., 1984) (see Fig. 2).

A new addition to this map is a gene which is highly homologous to K. pneumoniae nifB, a gene required for the construction of the iron-molybdenum cofactor (FeMo-co) of nitrogenase. Transposon insertions immediately downstream of the nifA gene result in a Fix⁻ phenotype and DNA sequence analysis of this region initially showed homology to the amino-terminal sequence of the K. pneumoniae nifB gene (J. Beynon, F. Cannon, personal communication) and to the sequence of the R. leguminosarum "fixZ" gene (Rossen et al., 1984). Since only the amino-terminal sequence of the K. pneumoniae nifB gene was known, we sequenced the entire K. pneumoniae nifB gene as well as the nifQ gene downstream of it in order to compare these sequences to the putative R. meliloti nifB gene region. Based on

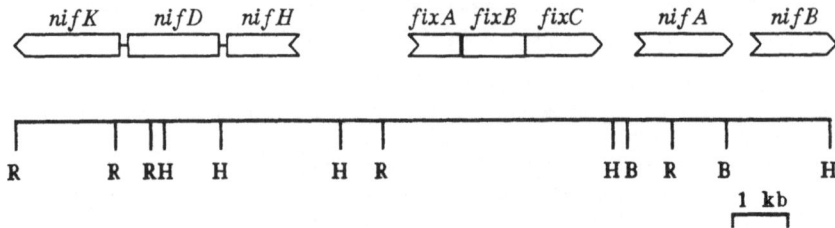

FIGURE 2. Genetic and physical map of the R. meliloti nif gene region. R = EcoRI, H = HindIII, B = BamHI.

---

these comparisons, ORFs sufficient to code for 54 kD, 51 kD, and 17 kD polypeptides were found within the R. meliloti nifB, the K. pneumoniae nifB and the K. pneumoniae fixQ gene regions respectively. The reported size for the K. pneumoniae nifB prouduct is 48 kD (Puhler et al., 1984). A transcriptional fusion of the E. coli lacZ promoter to the R. meliloti nifB gene was constructed. This construction was expressed in maxicells and synthesized a protein of approximately 50 kD in size. This corresponds well to 54 kD, the size predicted by the DNA sequence. The amino acid sequence of the R. meliloti nifB protein is 37% and 62% homologous to the nifB proteins of K. pneumoniae and R. leguminosarum respectively.

A third cluster of R. meliloti genes required for symbiotic nitrogen fixation is located between the nifHDK and nifAB regions on a single 4.5 kilobase HindIII fragment. There are three genes in this cluster which are generally referred to as the fixABC genes (Puhler et al., 1984). Their role in nitrogen fixation has not yet been determined. Genetic analysis showed that the fixABC genes comprise an operon (Ruvkun et al., 1982) and that the fixA promoter is activated by the R. meliloti nifA gene (Puhler et al., 1984; Szeto et al., 1984a) and contains the characteristic nif promoter consensus sequences described in Fig. 1 (Better et al., 1983).

Although the functions of the fixABC gene products are currently unknown, it is likely that the R. meliloti fixABC genes are analogous to three of the 17 K. pneumoniae nif genes. As an initial step in determining the functions and mode of regulation of these R. meliloti fix genes, the entire 4500 base pair HindIII fragment was sequenced using the shotgun dideoxynucleotide seqencing method. We found a putative promoter region with close homology to the R. meliloti nifH promoter and noted three open reading frames that code for polypeptides of molecular weights 31,146, 37,786, and 47,288 (see Fig. 2).

We are continuing our study of the fixABC operon by searching for physical homology between fixABC DNA and K. pneumoniae nif DNA but so far have not detected any homology at the DNA level by the Southern transfer and hybridization method. We have, however discovered homology to fixABC in other Rhizobium species. The strongest homology was detected to genomic DNA from the fast-growing species R. leguminosarum and R. trifolii; lesser homology was found with B. japonicum and the Patrasponia Bradyrhizobium.

2. The Roles of nifA and ntrC in the Regulation of R. meliloti nif Genes

Previous physical and genetic studies suggested the existence of an R. meliloti ntrC-like gene (Szeto et al., 1984b). To determine the respective roles of the R. meliloti nifA and ntrC genes in nif regulation, the R. meliloti ntrC-like gene was cloned and characterized. Using labelled E. coli ntrC DNA as a hybridization probe, Rhizobium ntrC-like DNA sequences were isolated from a genomic pHC79 cosmid library constructed previously by Buikema et al. (1983). Partial DNA sequence analysis demonstrated that the cloned R. meliloti DNA was indeed an analogue of E. coli ntrC. The R. meliloti ntrC gene was mutagenised with Tn5 and the mutated DNA recombined into the R. meliloti genome, resulting in the construction of three independent R. meliloti ntrC::Tn5 mutants.

Like ntrC mutants of enteric bacteria, R. meliloti ntrC::Tn5 strains grew poorly on minimal media containing arginine, proline or aspartate as the sole nitrogen source. However, unlike ntrC mutants of enteric bacteria, the R. meliloti ntrC mutant strains grew well on minimal medium containing low levels of glutamine as well as low levels of ammonia as the sole nitrogen source. Growth on low levels of ammonia requires the expression of glutamine synthetase (glnA), the synthesis of which in enteric bacteria is controlled by ntrC. This result suggests that the regulation of glutamine synthetase in R. meliloti differs from that in enteric bacteria.

One interesting feature of R. meliloti is that, despite the fact that nitrogenase activity cannot be detected ex planta, a nifH-lacZ fusion can be activated ex planta when R. meliloti is starved for NH_4^+ (Sundaresan et al., 1983c). In contrast to wild type, R. meliloti ntrC mutants failed to activate nifH-lacZ and fixA-lacZ fusions under conditions of nitrogen starvation ex planta. This indicates that the R. meliloti ntrC gene, like its counterpart in enteric bacteria, is activated when the cells are starved of fixed nitrogen and that the depression of the nifHDK and fixABC operons ex planta is mediated by the ntrC product.

In contrast to the complete failure to derepress the nifH and fixA promoters in ntrC mutants ex planta, nodules elicited by ntrC mutants only displayed a delayed fixation phenotype, i.e., the nodules were Fix^+ but did not develop the Fix^+ phenotype until 6 weeks after inoculation compared to 4 weeks for wild type. Therefore, the ntrC product does not appear to be absolutely required for nifA expression during symbiosis, although it does seem to play a role in the normal establishment of nitrogen fixation by R. meliloti bacteroids.

The results reported here indicate that unlike K. pneumoniae, the regulation of nif gene expression in R. meliloti bacteroids is not mediated by the central ntr system. This conclusion is supported by the observation that the R. meliloti nifA promoter shows no homology to the nif/ntr consensus promoter sequence (Buikema et al., 1985; Fig. 1). In fact, the nifA promoter is fairly homologous to a Rhizobium consensus promoter sequence for promoters which are not nitrogen regulated (Fig. 1; Ronson and Astwood, this volume). Currently, we are actively studying the regulation of the nifA promoter. One possibility is that the nifA promoter is activated by a symbiotic signal from the host plant.

3. Organization and Regulation of Parasponia Bradyrhizobium nif Genes

In contrast to K. pneumoniae and the fast-growing Rhizobium species, nifH is separated from nifDK in the slow-growing Rhizobium or "Bradyrhizobium" species examined to date (B. japonicum and the so-called Rhizobium "cowpea" miscellany). To determine whether this is also the case for Parasponia Bradyrhizobium strain Rp501, recombinant plasmids which contained nif genes were isolated from pLAFR1 (Friedman et al., 1982), pHC79 (Hohn, Collins, 1980) and pOCA7.9 (N. Olszewski, F. Ausubel, unpublished) cosmid clone banks by interspecies hybridization with Rhizobium cowpea nifH and nifD probes (Jagadish et al., 1985). Analysis of hybridizing DNA fragments from these libraries indicated that nifH is 25-30 kb from nifDK. A DNA fragment which hybridizes to the 5' end of R. meliloti nifB is located between these nitrogenase genes. DNA sequences homologous to the R. meliloti fixABC genes are also located near nifH in Rp501.

To study nifH regulation in the Parasponia Bradyrhizobium strain Rp501, the cloned nifH gene was sequenced using the Sanger di-deoxy method and by comparison with other Rhizobium nifH genes, the promoter region was identified and a translational fusion with E. coli lacZ was constructed. Preliminary results suggest that nifH derepression in free-living Parasponia Bradyrhizobium cells occurs by a different regulatory circuit than derepression in nodules. When placed in E. coli along with plasmids which cause constitutive production of E. coli ntrC or K. pneumoniae nifA products, the Parasponia Bradyrhizobium nifH promoter was expressed at a high constitutive level; this constitutive level was augmented about two fold by ntrC and nifA. A similar high level of constitutive expression was observed in Rp501 grown in rich medium. When placed under derepression conditions, this background level of expression increased three to five fold at the same time that acetylene reduction could be detected.

These results raise two questions. First, is there a constiutive level of nifH expression in Rp501; and second, does an ntrC and/or a nifA-like gene product augment such expression? To answer the first question, we are screening free-living Rp501 RNA for the presence of nifH. One explanation for the apparent constitutive transcription of nifH is the presence of a Rhizobium "non-nitrogen" promoter consensus sequence in the promoter region of the Rp501 nifH gene.

To determine whether an Rp501 ntrC-like gene is responsible for regulating the ex planta expression of nifH, we have used low stringency conditions of DNA hybridization to clone Rp501 DNA which has homology to the central conserved portion of ntrC from E. coli and K. pneumoniae ntrC genes. Sequence analysis indicates that the sequence cloned on the basis of homology does indeed code for an ntrC-like gene. We are now constructing ntrC mutants with Tn5 using a broad host range plasmid to introduce ntrC::Tn5 mutations into the Rp501 genome by marker exchange.

We have been unable to find Rp501 sequences homologous to the conserved, central portion of the nifA genes from K. pneumoniae and R. meliloti, even under low stringency hybridization conditions. In K. pneumoniae, R. meliloti and R. leguminosarum, a nifA gene is situated directly upstream of nifB. Mutational analysis of the DNA surrounding the Parasponia Rhizobium nifB gene may reveal a nifA-like gene.

C. REFERENCES

Adams TH and Chelm BK (1984) J. Mol. Appl. Genet. 2(4), 392-405.

Ausubel FM (1984) Regulation of nitrogen fixation genes. Cell 37, 5-6.

Better M, Lewis B, Corbin D, Ditta G and Helinski D (1983) Cell 35, 479-485.

Beynon JL, Cannon MC, Buchanan-Wollaston V and Cannon FC (1983) Cell 34, 665-671.

Buchanan-Wollaston V, Cannon MC, Beynon JL and Cannon FC (1981) Nature 294, 776-778.

Buikema WJ, Long SR, Brown SE, van den Bos RC, Earl CD and Ausubel FM (1983) J. Mol. Appl. Genet. 2, 249-260.

Buikema, WJ, Szeto WW, Lemley PV, Orme-Johnson WH and Ausubel FM (1985) Nucleic Acids Res. 13, 4539-4555.

Cannon, M, Hill, S, Kavanaugh, E and Cannon F (1985) Mol. Gen. Genet. 198, 198-206.

Corbin D, Barran L and Ditta G (1983) Proc. Natl. Acad. Sci. USA 80, 3005-3009.

Corbin D, Ditta G and Helinski DR (1982) J. Bacteriol. 149, 221-228.

Drummond M, Clements KJ, Merrick M and Dixon R (1983) Nature 301, 302-307.

Forrai T, Vincze E, Banfalvi Z, Kiss GB, Randhawa GS and Kondorosi A (1983) J. Bacteriol. 153, 635-543.

Friedman AM, Long SR, Brown SE, Buikema WJ and Ausubel FM (1982) Gene 18, 289-296.

Hawley DK and McClure W (1983) Nucl. Acids Res. 11, 2237-2255.

Hill, S, Kennedy, C, Kavanaugh, E, Goldberg RB, Hanau, R (1981) Nature 290, 424-426.

Hohn B and Collins J (1980) Gene 11, 291-298.

Jagadish MN, Yun AC, Noti J, Folkerts O and Szalay AA (1985) In Szalay AA and Legocki RP, eds, Advances in the Molecular Genetics of the Bacteria-Plant Interaction, pp 27-31, Cornell Univ. Publishers, Ithaca.

Magasanik B (1982) Ann. Rev. Genet. 16, 135-168.

Merrick M (1983) EMBO Journal 2, 39-44.

Merrick M, Hill, S, Hennecke, H, Hahn, M, Dixon, R and Kennedy C (1982) Mol. Gen. Genet. 185, 75-81.

Ow DW and Ausubel FM (1983) Nature 301, 307-313.

Ow DW, Sundaresan V, Rothstein D, Brown SE and Ausubel FM (1983) Proc. Natl. Acad. Sci. USA 80, 2524-2528.

Ow DW, Xiong Y, Gu Q and Shen SC (1985) J. Bacteriol. 161, 868-874.

Puhler A, Auilar MO, Hynes M, Muller R, Klipp W, Priefer U, Simon R and Weber G (1984) In Veeger, C and Newton WE, eds, Advances in Nitrogen Fixation Research, pp 609-619, Martinus Nijhoff/Junk, The Hague.

Reitzer LJ and Magasanik B (1985) Proc. Natl. Acad. Sci. USA 82, 1979-1983.

Rossen L, Ma Q-S, Mudd EA, Johnston AWB and Downie JA (1984) Nucleic Acids Res. 12, 7123-7134.

Ruvkun GB, Sundaresan V and Ausubel FM (1982) Cell 29, 551-559.

Scott KF, Rolfe BG and Shine J (1983) DNA 2, 149-155.

Sundaresan V, Ow DW, and Ausubel FM (1983a) Proc. Natl. Acad. Sci. USA 80, 4030-4034.

Sundaresan V, Jones JDG, Ow DW and Ausubel FM (1983b) Nature 301, 728-732.

Sundaresan V, Jones JDG, Ow DW and Ausubel FM (1983c) In Hamer D and Rosenberg M, eds, Gene Expression, UCLA Symposium on Molecular and Cellular Biology New Series, Vol. 8, pp 175-185, Alan R. Liss, New York.

Szeto WW, Zimmerman JL, Sundaresan V and Ausubel FM (1984a) Cell 36, 535-543.

Szeto WW, Zimmerman JL, Sundaresan V and Ausubel FM (1984b) In Davidson EH and Firtel RA, eds, Molecular Biology of Development, pp 611-617, Alan R. Liss, New York.

D. ACKNOWLEDGEMENTS

Unpublished experiments were funded by a grant from Hoechst A.G.

Rhizobium phaseoli: NITROGEN FIXATION GENES AND DNA REITERATION.

RAFAEL PALACIOS, MARGARITA FLORES, ESPERANZA MARTINEZ AND CARMEN QUINTO.
CENTRO DE INVESTIGACION SOBRE FIJACION DE NITROGENO, UNIVERSIDAD NACIONAL
AUTONOMA DE MEXICO. APARTADO POSTAL 565-A, CUERNAVACA, MORELOS. MEXICO.

INTRODUCTION

The bacteria belonging to the genus Rhizobium establish symbiosis with the
roots of legumes. The symbiotic process involves the differentiation of
both bacteria and plant cells. A specialized organ, the nodule, is formed in which nitrogen is fixed by the bacterial nitrogenase. The symbio
tic interaction is specific and Rhizobium "species" have been defined
according to the host they infect. This classification gathers into a
group different bacteria that may share only a certain nodulation ability.

R. phaseoli, the symbiont of Phaseolus vulgaris, common bean, has been
catalogued as a very heterogeneous group (Beynon, Josey, 1980; Roberts et
al, 1980; Catteau et al, 1982). We have found that R. phaseoli presents
a peculiar organization of nitrogen fixation gene sequences characterized
by the presence of DNA reiterations (Quinto et al, 1982). In the symbio
tic plasmid of R. phaseoli strain CFN-42 there are three non-contiguous
regions that contain nitrogen fixation structural genes. The complete
coding sequence of the nifH gene is found in the three regions. Interestingly, the three nifH genes are identical in both the coding sequence and
putative promoter regions (Quinto et al, 1985). This situation suggests
a high selective pressure to mantain these genes and their identity.
However, site directed insertion mutagenesis indicated that none of the
three nifH gene copies is indispensable for nitrogen fixation in symbiosis with Phaseolus vulgaris (Quinto et al, 1985). Moreover, some native
Rhizobium isolates do not present reiteration of nif genes and are able
to effectively nodulate P. vulgaris (Martínez et al, 1985). We have proposed that the biological significance of nif genes reiteration is related
to the general organization and dynamics of the Rhizobium genome (Palacios
et al, 1985).

ON THE NATURE OF Rhizobium phaseoli

Rhizobium phaseoli comprises a very heterogeneous group on the basis of
protein patterns (Roberts et al, 1980), plasmid profiles (Martínez, Palacios, 1982), antibiotic resistance (Beynon, Josey, 1980) and numeric
taxonomy (Catteau et al, 1984). To find out how general is the presence
of nif genes reiteration, a screening of 40 strains isolated from P. vulgaris nodules was performed. Such screening included strains from different geographical origins. Total DNA from each strain was digested
with a restriction endonuclease, subjected to agarose gel electrophoresis,
blotted and hybrized with a nifH gene specific probe from R. phaseoli
strain CFN-42. Only one of the 40 strains did not show reiteration of
nifH gene (Martínez et al, 1985). This non-reiterated strain, CIAT 899,

from Colombia, is alumminum and acid tolerant. In a screening of strains
from acid soils,we have isolated another non-reiterated strain from bean
nodules, strain UMR 1026 from Brazil.

An important difference between the "nif reiterated" and "non reiterated"
strains isolated from bean nodules is the host range of infection, strains
CIAT 899 and UMR 1026 were able to effectively nodulate Leucaena esculenta
while "nif reiterated" strains were specific for beans (Martínez et al,
1985).

The nodulation of Phaseolus vulgaris by a wide range of strains from tro
pical legumes has been reported (Lange, 1961), however there were no data
on nitrogen fixation. We isolated strains from different legumes and stu
died their capacity to nodulate and fix nitrogen in P. vulgaris. Some
strains isolated from legumes philogenetically distant from P. vulgaris
such as Leucaena leucocephala, Leucaena esculenta, Crotalaria pumila and
Dalea leporina were able to effectively nodulate P. vulgaris, (Martínez
et al, 1985). These broad host range strains did not present reiteration
of nif genes. In fact, a screening of Rhizobium strains from different
legumes for the presence of nif genes reiteration revealed that this
characteristic is present in strains from the genus Phaseolus and some
close relative genera of the Phaseolinae tribe (Martínez et al, 1985).
Data from other groups support this proposal. Strain ANU 240 has two nifH
copies (Morrison et al, 1983), and was originally isolated from Lab-lab
purpureus which belongs to the Phaseolinae tribe. Fast growing R.japonicum
have nif genes reiteration (Prakash and Atherly, 1984) and Glycine max,
its host, belongs also to the Phaseolinae tribe.

We suggest that there have been different evolutionary trends to become
R. phaseoli. The best fitted strains to effectively nodulate P. vulgaris
in nature are very specific for their host and have reiterated nitrogen
fixation genes as a common characteristic.

ON THE ORGANIZATION OF REITERATED NITROGEN FIXATION GENES IN Rhizobium
phaseoli.

We have studied the organization of nitrogen fixation gene sequences in
a "nif reiterated" R. phaseoli strain, CFN-42. This strain has six large
plasmids. One of these plasmids, p42-d, behaves as a "symbiotic plasmid"
since nodulation ability of the strain is completely lost when this plasmid is cured. Moreover, transconjugants of Agrobacterium tumefaciens con
taining p42-d are able to nodulate Phaseolus vulgaris. We found that
three different regions of this plasmid contain nitrogen fixation gene se
quences. Two of the regions (a and b) are 5 kb homologous and the third
one (c) is 1.3 kb homologous with the other two. The three regions contain the complete coding sequence of nifH gene, which is identical for
the three genes (Quinto et al, 1985). Hybridization experiments (Quinto
et al, unpublished data) indicate that regions a and b also contain genes
nifD and K downstream of nifH gene, while region c contains only nifH
gene (Figure 1).

In order to gain insight into the function of reiterated sequences, we
have constructed mutations by inserting antibiotic resistance sequences
that interrupt the different regions. We reported that mutant strains
carrying a single Km^r insertion in the coding sequence of nifH gene in

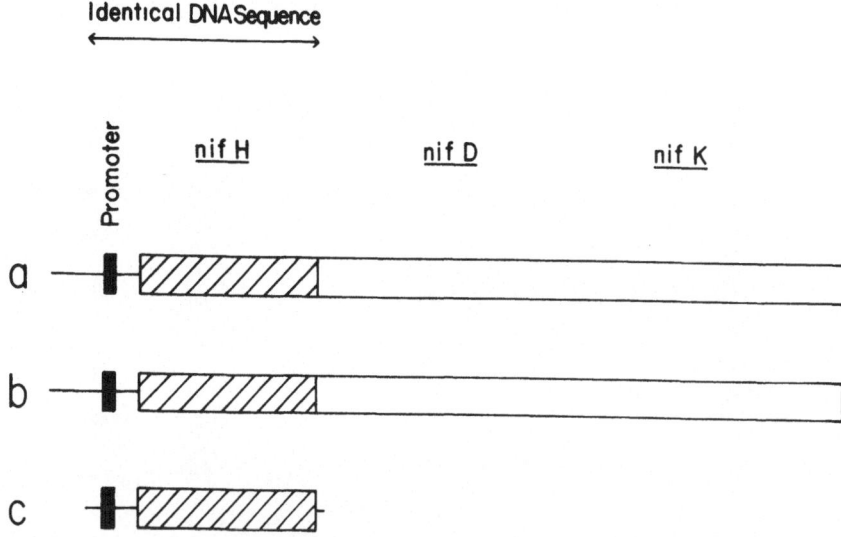

FIGURE 1. Schematic representation of the homology between nitrogenase structural genes in Rhizobium phaseoli strain CFN-42. These regions are non-tandemly distributed in the symbiotic plasmid (p42-d) and show perfect heteroduplex homology and conservation of the internal restriction sites.

any of the three regions are still able to effectively nodulate P. vulgaris (Quinto et al, 1985). We have also constructed strains carrying double mutations with different antibiotic resistance markers. Strains that have mutations in regions a and c or b and c are still able to fix nitrogen in symbiosis with P. vulgaris, while strains carrying mutations in both regions a and b are unable to fix nitrogen although they are still able to nodulate (Segovia et al, unpublished data). The position of the insertions and the symbiotic behaviour of the strains suggest that in regions a and b genes nifH, D and K are transcribed as a single unit starting in the promoter at the 5'end of gene nifH, and that either operon is sufficient to mantain nitrogen fixation during symbiosis with P. vulgaris. Moreover, strains carrying nifH-lacZ fusions suggest that both regions a and b are simultaneously expressed (Morett et al, unpublished data). These data support our conclusions that the reiteration of nif genes is not essential for their function.

The reiteration of nif genes has been reported in other organisms including Rhodopseudomonas capsulata (Scolnick,Haselkorn, 1984). Anabaena (Rice et al, 1982), Calotrix (Kallas et al, 1983) and other Rhizobia (see above). In R. meliloti (Better et al, 1983), R. trifolii (Scott et al, 1983; Watson et al, 1985) and R. leguminosarum (Watson et al, 1985) the DNA region that includes the promoter of the nifHDK operon is reiterated.

We have analyzed the 5'end of nif region a of R. phaseoli CFN-42 and found that it is reiterated five times in the genome. Three copies correspond to nif regions a, b and c in the symbiotic plasmid (p42-d), one copy (d) is also located in p42-d and the last one (e) is located in another plasmid (p42-a). The experiment presented in Figure 2 supports these conclusions.

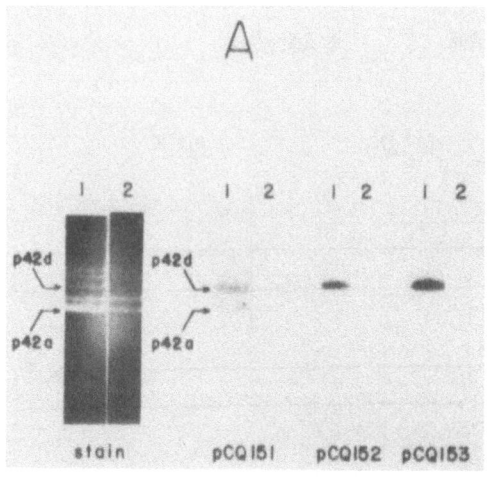
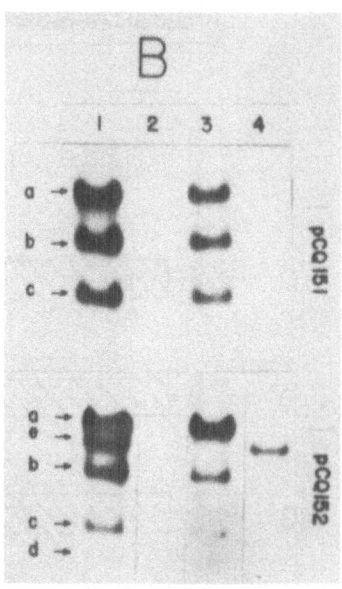

FIGURE 2. Identification of nif genes related regions in R. phaseoli strain CFN-42. Hybridization of blotted plasmid profiles (A) and BamHl total DNA digests (B) to P^{32} labeled inserts from different recombinant plasmids (see text) as indicated. Lanes 1, CFN-42. Lanes 2, CFN-2001, a derivative of CFN-42 cured of plasmids a and d. Lanes 3, GV3105 (p42-d::Tn5 mob), an A. tumefaciens transconjugant containing plasmid p42-d. Lanes 4, GV3105 (p42-a::Tn5), an A. tumefaciens transconjugant containing plasmid p42-a.

Nif region a was initially isolated as a 4.7 kb EcoRl insert from a pBR328 recombinant plasmid, pCQ15 (Quinto et al, 1982). Subclones of pCQ15 containing fragments of the nifH gene and or adjacent regions were constructed. pCQ151 contains a 1.1 kb EcoRl-Sall insert that extends from 1 kb upstream of the nifH gene initiation codon down to codon 29; pCQ152 contains a 273 bp Sall insert that extends from codon 29 down to codon 119 and pCQ153 has a 545 bp Sall insert that spans the rest of the coding region of nifH gene down to 9 base pairs after the termination codon. Inserts from these recombinant plasmids were used as hybridization probes to detect nifH-related sequences in plasmid profiles of R. phaseoli CFN-42 (Fig. 1A). All three probes hybridized with the 250 kb plasmid (p42-d); in addition pCQ151 also hybridized with p42-a, a 130 kb plasmid. There was no hybridization with CFN-2001, a strain cured of both plasmids. NifH-related DNA regions were identified by genome blot hybridization (Fig. 1B). pCQ151 hybridized with 5 fragments (a-e) of BamHl-digested DNA from strain CFN-42. The two inserts containing most of the coding sequence of the nifH gene (pCQ152 and pCQ153) revealed the same three fragments (regions a, b and c). Genome blots of A. tumefaciens transcon_ jugants containing either p42-d or p42-a revealed that regions a, b, c and d are located in p42-d while region e is located in p42-a (Fig. 1B).

Thus, in R. phaseoli strain CFN-42 there are five reiterated sequences related to nif structural genes or adjacent structures. Two regions (a and b) contain complete nifHDK operons, one (c) contains a complete nifH gene, and the other two (d and e) include regions upstream of the nifH gene a-d or a small portion of the 5'end of the coding sequence of the nifH gene.

It has been proposed that the repeats at the 5'end of nif structural genes could participate in gene expression during symbiosis (Better et al, 1983; Watson et al, 1985). On the other hand, it is interesting that some of the repeats in both R. meliloti (Better et al, 1983) and in R. trifolii contain part of the coding sequence of the nifH gene. It might be that reiteration of nif-related sequences is a common feature in Rhizobium. In certain groups such as R. phaseoli some reiterations are extended further downstream and cover complete genes. If such reiterations appeared only once during Rhizobium evolution, in some groups, part of them were lost while in others were conserved. Alternatively, reiterations could appear at different points in evolution or are part of a complex dynamic state of the Rhizobium genome.

ON THE GENERAL ORGANIZATION OF THE Rhizobium phaseoli GENOME.

DNA reiteration is a common feature of the eukaryotic genome. In contrast, few examples of DNA duplications have been found in prokaryotic organisms. An exception are the genomes of some archaebacteria such as Halobacterium halobium and Halobacterium volcanii which contain a very large number of families of repeated sequences whose members are dispersed on both chromosome and plasmid (Sapienza, Doolittle, 1982). Such repeated sequences participate in genome rearrangements that occur at high frequency (Sapienza et al, 1982).

We analized the degree of DNA reiteration in R. phaseoli. The total genome of strain CFN-42 was cloned in the EcoRl site of pBR329, seventy clones were selected at random and their recombinant plasmids were purified. These were used as hybridization probes with total genome Southern blots digested with EcoRl from two R. phaseoli "nif reiterated" strains: CFN-42 and CFN-285. It was surprising that in about half of the clones we detected evidence of DNA reiteration in one or both strains (Flores et al, unpublished data). The reiterated families contained different numbers of elements, from 2 to 15. The intensity of the hybridization bands suggested that some elements are short stretches of DNA while others could be several kilobases long. Some examples of reiterated families are presented in Figure 3.

These data show that the R. phaseoli genome has a high degree of DNA reiteration, nif-related sequences being one of the repeated families. If repeated families participate in genome rearrangements, and if such rearrangements are frequent, then our strains must be heterogeneous populations for such families. Preliminary experiments indicate that this is the case for at least one family examined in three different strains (Flores et al, unpublished observation). Moreover, Soberón et al (this volume) have detected genome rearrangements that involve nif genes.

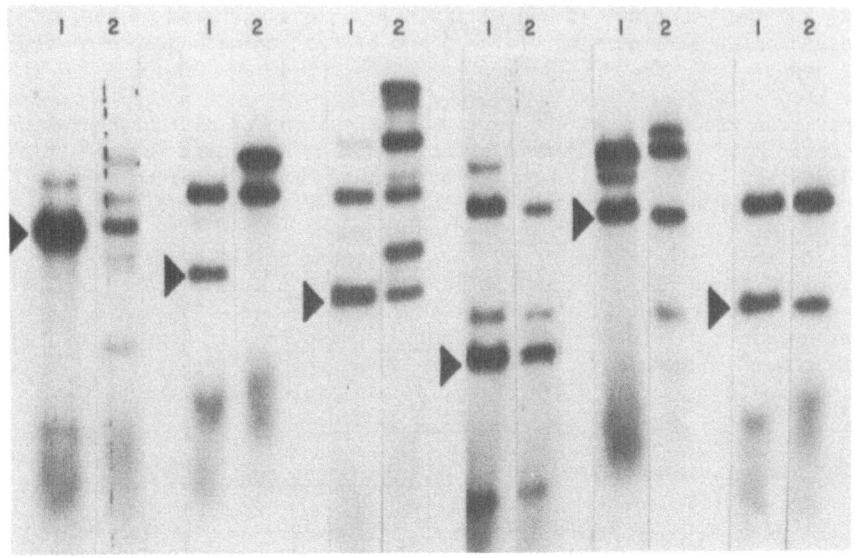

FIGURE 3. Genome Southern blots of strains CFN-42 (1) and CFN-285 (2) vs. different recombinant plasmids carrying single DNA fragments from strain CFN-42. Arrows indicate bands corresponding to the probe.

CONCLUSIONS AND PERSPECTIVES

An important goal in the field of nitrogen fixation is to understand and manipulate the genetic information that participates in symbiotic functions in Rhizobium. Our knowledge of the different genes that participate in nodulation and nitrogen fixation in different cross inoculation groups of Rhizobium is rapidly increasing. We can presume that in the near future a clear scheme of the genetic organization of symbiotic functions will emerge.

We think that a comprehensive knowledge of symbiotic genetic information must take into account that such information is immersed in the Rhizobium genome. In this context, the understanding of the general organization and dynamics of the Rhizobium genome is important.

Data presented here indicate that the R. phaseoli genome contains a large number of repeated families, nitrogen fixation gene sequences being one of such families. We have preliminary evidence that repeated families are not a peculiarity of R. phaseoli but a more general phenomenon in Rhizobium and also in Agrobacterium. As opposed to the eukaryotic genome, in particular the plant genome, a high degree of DNA reiteration has been found only a few cases in bacteria (Sapienza, Doolittle, 1982). As expected, this genomic structure has been related with a high frequency of rearrangements (Sapienza et al, 1982). There is evidence that genome rearrangements are indeed frequent in Rhizobium (Soberón et al, this volume; Flores et al, unpublished observations). The presence of a highly dynamic genome in Rhizobium might be related with the possibility of adaptation to a changing environment in both the free-living and the symbiotic state. It could also be related to a better fitness with the particular host the bacteria comes in contact with. These propositions must be taken as questions that can be experimentaly adressed.

ACKNOWLEDGEMENTS.

This work was performed with the technical assistance of Rosa María Ocampo and Virginia Quinto. The authors are grateful to Lorenzo Segovia for critically reading the manuscript. Partial financial support for this research was provided by the U.S. National Academy of Science/National Research Council by means of a grant from the U.S. Agency for International Development, and by Grants from the Consejo Nacional de Ciencia y Tecnología.

REFERENCES

Beynon JL, Josey DP (1980) J. Gen. Microbiol. 118, 437-442.
Catteau M, Khanaka H, Segrand MD and Guillaume J (1984) In Veeger C and Newton WE, eds, Advances in Nitrogen Fixation Research, p 330, Martinus Nijhoff/Junk, The Hague.
Kallas T, Rebiere MC, Rippka R and Tandeau de Marsac N (1983) J. Bacteriol. 155, 427-431.
Lange RT (1961) J. Gen. Microbiol. 26, 351-359.
Martínez E. Pardo MA, Palacios R and Cevallos MA (1985) J. Gen. Microbiol., in press.
Martínez E, Palacios R (1984) In Veeger C and Newton WE eds, Advances in Nitrogen Fixation Research, p 60, Martinus Nijhoff/Junk, The Hague.
Morrison NA, Nah CY, Trinick MJ, Shine J and Rolfe B (1983) J. Bacteriol. 153, 527-531.
Palacios R, Quinto C, Leemans J, Flores M, De la Vega H and Martínez E. (1985). In van Vloten-Doting L, Groots G and Hall t, eds. Molecular Form and Function of the Plant Genome, pp 455-465, Plenum Publishing Corporation.
Prakash RK, Atherly AG (1984) J. Bacteriol. 160, 785-787.
Quinto C, De la Vega H, Flores M, Fernández L, Ballado T, Soberón G, and Palacios R (1982). Nature (London) 299, 724-726.
Quinto C, De la Vega H, Flores M, Leemans J, Cevallos MA, Pardo MA, Azpiroz R, Girard ML, Calva E, and Palacios R (1985). Proc. Natl. Acad. Sci. USA 82, 1170-1174.
Rice D, Mazur BJ and Haselkorn R (1982). J. Biol. Chem. 257, 13157-13163.
Roberts G, Leps WT, Silver LE and Brill WJ (1980) Appl. Envir. Microbiol. 39, 414-422.
Sapienza C, Doolittle WF (1982). Nature (London) 295, 384-389.
Sapienza C, Rose MR and Doolittle FW (1982). Nature (London) 299, 182-185.
Scolnick PA, Haselkorn R (1984) Nature (London) 307, 389-292.

FUNCTIONAL ANALYSIS OF THE <u>BRADYRHIZOBIUM JAPONICUM NIFH</u>, <u>NIFD</u> AND <u>FIXA</u> PROMOTER REGION

A. ALVAREZ-MORALES/M. BETANCOURT-ALVAREZ/M. GUBLER/K. KALUZA/H. HENNECKE.
MIKROBIOLOGISCHES INSTITUT, EIDGENÖSSISCHE TECHNISCHE HOCHSCHULE,
ETH-ZENTRUM, UNIVERSITÄTSTRASSE 2, CH-8092, ZÜRICH, SWITZERLAND

B. japonicum nifH'-, nifD'- and fixA'-'lacZ translational fusions were constructed and used to analyse the effect of the K. pneumoniae nifA or ntrC regulatory products on transcription from the B. japonicum promoters. The 3 promoters responded positively to the nifA product, however, no effect was observed with the ntrC protein. Activation of fixA was only 2-fold whereas that of the nif promoters was greater than 100-fold. From the nifH- and nifD-lacZ fusion plasmids a series of 5' deletions, random (with Bal31) or defined (using restriction sites), were constructed. A region of at least 70bp upstream the -100 position was identified as essential to obtain wild-type levels of nifA-mediated activity (fig.1). Furthermore, a 67bp internal deletion (from -110 to -43) in the nifH promoter suggested that the 70bp region contains one or more specific sequences which need to be in a fixed position to obtain wild-type activation (Table 1). A G to T transversion affecting the nifH promoter (TGGC .. TTGCA to TTGC .. TTGCA) did not completely abolished nifA-mediated activation but resulted in 92% decrease in promoter activity.

Table 1.

nifD				nifH		
Plasmid	Activity %	End point		Plasmid	Activity %	End point
pRJ1008	100.0	-428		pRJ1009	100.0	-700
pRJ1021	104.4	-165		prJ1070	1.2	(-110 to -43)
pRJ1012	37.9	-132		pRJ1071	1.0	(-144 to -43)
pRJ1014	6.6	-101		pRJ1072	3.1	- 43
pRJ1017	5.8	- 65		pRJ1073	4.2	-110

Fig. 1 Effect of 5' deletions in nifA-mediated activation of the B. japonicum nifH promoter.

P=promoter, β-Galactosidase activity in Miller units.

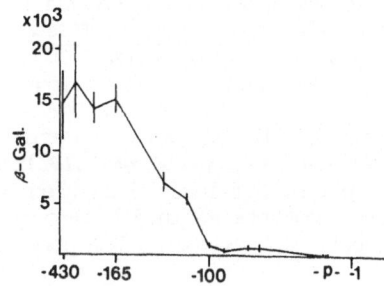

VECTOR-INSERTION ANALYSIS OF RHIZOBIUM sp. ORS571 N$_2$ FIXATION GENE LOCI.

ROBERT G. K. DONALD, DAVID W. NEES, CHRISTOPHER K. RAYMOND, ALBERT I. LOROCH, and ROBERT A. LUDWIG.

Department of Biology, Thimann Laboratories, University of California, Santa Cruz, CA. 95064, USA.

INTRODUCTION

The Vector-insertion (Vi) mutagenesis/cloning technique allows the direct recombinant DNA cloning of any genes for which mutants can be selected or screened [DONALD et al., J.Bacteriol. 162,317-323(1985)]. As employed here, Vi mutants carry plasmid/genome cointegrates between the Tn5-containing plasmid pVP2021 and the N$_2$ fixation genes of the Rhizobium sp. ORS571 genome. Vi mutants form via IS50R-mediated, plasmid/genome cointegration events and not direct Tn5 transpositions. Because ORS571 Vi mutants are stable, genomic VP2021-derived DNA sequences constitute a targeted cloning vector. For recombinant DNA cloning, total genomic DNA is first isolated, is digested with an appropriate restriction endonuclease, is ligated under dilute conditions, and is used to directly transform E.coli. Recombinant DNA plasmids are obtained; no exogenous vector is employed. Recombinant plasmids may be used to physically map the site of the original Vi mutation and to obtain by DNA hybridization the corresponding ORS571 wild-type genomic regions under study.

RESULTS

65 independent, Rhizobium sp. ORS571 N$_2$ fixation-defective (Nif$^-$) Vi mutants were selected, cloned as recombinant DNA plasmids, and mapped to the ORS571 genome. Recombinant Nif::Vi plasmids so obtained were used as DNA hybridization probes to isolate homologous recombinant phages from a genomic library of ORS571 constructed in λEMBL3. Genomic maps were drawn for three ORS571 Nif gene-loci. DNA from λNif phages comprising all three genomic Nif-loci was subcloned in broad host-range plasmids that stably replicate in ORS571. Plasmid Nif subclones were introduced into ORS571 strains whose Nif:Vi mutants had been physically mapped. Nif$^+$ genetic complementation tests were conducted. 45 Nif::Vi mutants in genomic Nif-locus 1 defined two gene clusters separated by 8 kb of DNA. Of these, 36 Nif::Vi mutants mapped to a 7 kb DNA segment that showed limited DNA homology with Klebsiella pneumoniae nifHDKNE. This ORS571 DNA segment encodes at least two Nif operons. 9 Nif::Vi mutants mapped to a 1.5 kb DNA segment and showed homology with both K. pneumoniae and R. meliloti nifA. This 1.5 kb DNA segment encodes a separate Nif operon. 15 Nif::Vi mutants defined Nif-locus 2, mapped to a 3.5 kb DNA segment, and showed DNA homology with the R. meliloti P2 fix-operon. Nif-locus 2 carries a second nifH (nifH2) gene. 4 Nif::Vi mutants defined Nif-locus 3, mapped to a 2 kb DNA segment, and showed no DNA homology with Nif-loci 1 and 2.

LOCALIZATION AND PRELIMINARY CHARACTERIZATION OF A NIFA-LIKE GENE IN BRADYRHIZOBIUM JAPONICUM

HM. Fischer, D. Reber and H. Hennecke

Mikrobiologisches Institut, Eidgenössische Technische Hochschule, ETH-Zentrum, Universitätsstrasse 2, CH-8092 Zürich, Switzerland

By interspecies hybridization using Klebsiella pneumoniae nifA and ntrC and a Rhizobium meliloti nifA-like gene (1) as probes a nifA-like gene was localized in the soybean symbiont B. japonicum. A 0.8 kb XhoI fragment ending about 1 kb upstream of the previously localized fixA-like gene (2) showed maximal hybridization to all three probes used (fig.1). By partial sequencing this XhoI fragment and comparison of the deduced amino acid sequence to the ones of the genes used as probes it was found that the nifA-like gene has the same orientation as the fixA-like gene and assuming a similar length as the K. pneumoniae nifA gene its approximate location could be determined (fig.1). Of the 56 determined amino acids 25 (47%) were identical in all sequences compared.

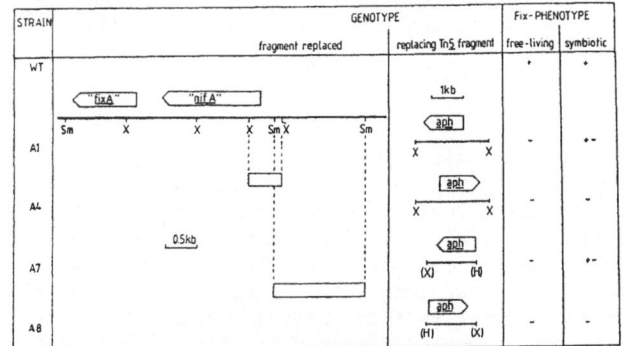

Fig. 1:
Physical map of the B. japonicum fixA-nifA region and its functional analysis by site directed replacement mutagenesis.
H: HindIII, Sm: SmaI, X: XhoI, aph: aminoglycosidephosphotransferase.

The nifA-like gene was further analyzed functionally by site directed replacement mutagenesis as shown in fig. 1. Under free-living conditions the Fix-activity of all mutants was below the level of detection whereas in symbiosis only the mutants A1 and A7 which carry the aph gene of the inserted Tn5 fragment in the same orientation as the nifA-like gene showed Fix-activity. This might be due to a read out from the aph promotor. Nodules from A7 infected plants were indistinguishable from wildtype nodules. In nodules originating from A1 the visible leghemoglobin was drastically reduced and nodules from plants infected with A4 or A8 were brown to black and showed necrotic deformations. Analysis of total protein extracts by two dimensional gel electrophoresis showed that beside expression of nifH the expression of at least six other proteins is affected by the mutations.

References

1: Buikema WJ et al (1985) Nucl. Acid Res 13, 4539-4555.
2: Fuhrmann M et al (1985) Mol Gen Genet 199, 315-322.

REITERATION OF DNA SEQUENCES IN RHIZOBIUM PHASEOLI

MARGARITA FLORES, VICTOR GONZALEZ AND RAFAEL PALACIOS.
CENTRO DE INVESTIGACION SOBRE FIJACION DE NITROGENO, UNIVERSIDAD NACIONAL
AUTONOMA DE MEXICO. APARTADO POSTAL 565-A, CUERNAVACA, MORELOS. MEXICO.

INTRODUCTION

We have reported (Quinto et al, 1982) that nitrogen fixation gene sequences are reiterated in Rhizobium phaseoli. However, such reiteration does not seem to be indispensable to establish an effective symbiosis with P. vulgaris (Quinto et al, 1985; Palacios et al, this volume). To better understand the significance of nif genes reiteration, we decided to analize the overall frequency of DNA reiteration in the R. phaseoli genome.

PROCEDURE

Total DNA of strain CFN-42 was cloned in the EcoRl site of pBR329. Clones were selected at random and recombinant plasmids were purified and used as hybridization probes against total genome Southern blots of two R. phaseoli strains: CFN-42 and CFN-285. The following criteria were used as evidence for reiteration: in the homologous strain, CFN-42, the presence of more hybridization bands than those (usually one) present in the probe; in strain CFN-285, the presence of more bands than those that could be explained by the presence of the probe as a contiguous DNA stretch.

RESULTS

We have analyzed 63 recombinant plasmids that contain inserts from strain CFN-42, as hybridization probes against total genome Southerns from two strains. 33 recombinant plasmids revealed repeated families in either one or both strains. The intensity of the hybridization bands suggested that in some cases repeated DNA sequences are short DNA stretches while in other cases could be several kilobases long. Reiterations were revealed by probes from the chromosome as well as from different plasmids of strain CFN-42. Some of the probes that revealed repeated families were hybridized against other Rhizobium strains and against Agrobacterium tumefaciens. Some of them revealed DNA reiteration.

DISCUSSION

Our data indicate that the R. phaseoli genome has a large degree of DNA reiteration. This complex organization is probably a general feature in Rhizobium and Agrobacterium, as some preliminar data suggests. As opposed to the eukaryotic genome, few cases of a high frequency of DNA reiteration have been reported. In such cases, reiteration has been related to the presence of genome rearrangements (Sapienza et al, 1982). There is evidence that genome rearrangements do occur in R. phaseoli (Soberón et al, this volume; Flores et al, unpublished data).

REFERENCES

Quinto C et al (1982) Nature (London) 299, 724-726.
Quinto C et al (1985) Proc. Natl. Acad. Sci. USA 82, 1170-1174.
Sapienza C et al (1982). Nature (London) 299, 182-185.

HOST-SPECIFIC EXPRESSION OF SYMBIOTIC NITROGEN-FIXATION GENES IN FAST-GROWING RHIZOBIUM SPECIES

SIIRI E. IISMAA[1,2], JOHN M. WATSON[2] and PETER R. SCHOFIELD[2]
(1) CENTRE FOR RECOMBINANT DNA RESEARCH, R.S.B.S., AUSTRALIAN NATIONAL UNIVERSITY and (2) CSIRO, DIVISION OF PLANT INDUSTRY, CANBERRA, A.C.T. 2601, AUSTRALIA.

Reiterated symbiotic gene promoter sequences have been described in the fast-growing Rhizobium species R. meliloti (Better et al. 1983), R. trifolii (Watson, Schofield 1985), R. leguminosarum (Watson et al. 1985) and R. phaseoli (Quinto et al. 1985). These sequences constitute reiterations of the nifHDK promoter regions (from about -160 to + 20 bp) and, in some copies the conservation extends through the leader sequence and into the nifH coding region.

Under stringent hybridization conditions, an R. trifolii repeated sequence-specific probe does not hybridize to DNA from other fast-growing species of Rhizobium, indicating that such sequences are species specific. Ruvkun and Ausubel (1980) showed that the nitrogenase structural genes (nifHDK) of diverse nitrogen-fixing microorganisms are highly conserved. The divergence of the nifHDK promoter regions among the fast-growing Rhizobium species, coupled with the narrow host range of effectiveness of these species, suggest that the repeated promoter sequences may be the sites at which host-specific expression of the nifHDK (and other symbiotic genes) is mediated.

Computer alignment of four copies each of the repeated promoter sequences from R. trifolii and R. meliloti reveals two domains which are conserved between the species. These domains are separated by a third domain which is conserved within each species (Schofield, Watson 1985). The highly conserved nature of these promoter regions (extending to ca -160 bp), within a given species, is consistent with coordinate expression of the symbiotic genes (nif and fix). The existence of these domains suggests that the mechanism of activation of the symbiotic genes may be complex.

One of the conserved domains, revealed by the computer alignment, contains the nif consensus -10 and -20 sequences which have been found to precede the transcription initiation points of all characterized nif genes (Beynon et al. 1983; Ow et al. 1983). These sequences are possible sites at which RNA polymerase binds. We propose a model in which the postulated binding of RNA polymerase to the nif consensus promoter sequences requires the prior interaction of both general and species-specific regulatory gene products at appropriate sites within the promoter region. According to this model, expression of the nifHDK (and other symbiotic genes), in fast-growing Rhizobium species, requires the presence of the appropriate species-specific domain.

Better M et al. (1983) Cell 35, 479-485.
Beynon J et al. (1983) Cell 34, 665-671.
Ow DW et al. (1983) Proc Natl Acad Sci 80, 2524-2528.
Quinto C et al. (1985) Proc Natl Acad Sci 82, 1170-1174
Ruvkun GB, Ausubel FM (1980) Proc Natl Acad Sci 77, 191-195.
Schofield PR, Watson JM (1985) Nucleic Acids Res 13, 3407-3418.
Watson JM, Schofield PR (1985) Mol Gen Genet 199, 279-289.
Watson JM et al. (1985) In Szalay AA and Ausubel FM, eds. Molecular Genetics of the Bacteria-Plant Interaction, pp 22-26, Cornell University Publishers, New York.

TRANSCRIPTIONAL ORGANIZATION OF A RHIZOBIUM MELILOTI NIF/FIX GENE CLUSTER DOWNSTREAM FROM P2 - OVERLAPPING TRANSCRIPTION

CHOONG-HYUN KIM, GARY DITTA and DONALD R. HELINSKI, Univ. of California, San Diego, La Jolla, CA 92093

A 14 kb nif/fix gene cluster of Rhizobium meliloti contains a number of genes necessary for symbiotic nitrogen fixation with alfalfa, including the nitrogenase structural genes (nifHDK), a regulatory gene (nifA), and genes of unspecified function (fixABC). Studies have shown that there are at least three symbiotically essential transcription units in this cluster (1,2). Promoter regions controlling nifHDK (P1) and fixABC (P2), have been well characterized by DNA sequencing and S1 nuclease mapping (3). A third transcription unit containing nifA is downstream from the fixABC genes and is transcribed in the same direction as fixABC (4). Details of transcriptional organization downstream from nifA are lacking. Experiments reported here were undertaken to study gene expression in the P2/PnifA region of R. meliloti.

Two approaches were used to examine promoter activities in R. meliloti. In one case, promoter-containing DNA fragments were fused to the E. coli lacZ gene in pGD499, a broad host range promoter probe vector, and β-galactosidase activities were monitored in bacteroids and broth-grown cells. These data were used to ascertain relative individual promoter strengths. It was found that the three promoters, P1, P2 and PnifA, are all highly active in bacteroids. P2 has activity comparable to that of P1, while PnifA activity is approximately half that of P1 or P2. Little or no activity was observed for these promoters in vegetatively grown cells. Possible weak promoter activity was observed in a 0.7 kb BglII-HindIII fragment downstream of nifA.

A second strategy involved introducing CATn5, a promoter probe fragment carrying the chloramphenicol acetyltransferase (CAT) gene cartridge and the neomycin resistance gene of Tn5, downstream of P2 at various identical locations in plasmid constructs containing or lacking the P2 promoter. CAT activity was then monitored from these constructs in bacteroids from 5 week old alfalfa nodules. These data allowed us to determine the contribution of P2 to nifA gene expression. It was found that 93% of fixABC and 57% of nifA transcription originates from P2 in bacteroids. Since PnifA is highly active in bacteroids (53% of P2), and since over half of nifA transcription (57%) is in fact derived from P2, it must further be concluded that there is only modest termination (30%) of P2 transcription between fixC and nifA. Thus, nifA expression in alfalfa nodules depends both on transcription originating from its own promoter, PnifA, and on significant levels of read-through transcription originating from P2, located 4.0 kb upstream.

(1) Corbin et al (1983) Proc. Natl. Acad. Sci., USA 80, 3005-3009.
(2) Ruvkun et al (1982) Cell 29, 551-559.
(3) Better et al (1983) Cell 35, 479-485.
(4) Szeto et al (1984) Cell 36, 1035-1043.

ORGANIZATION AND FUNCTIONAL ANALYSIS OF <u>RHIZOBIUM</u> <u>LEGUMINOSARUM</u> PRE SYM-GENES.

R. Klein Lankhorst, J. Hontelez, W. van de Greef, A. van Kammen and R. van den Bos.
Department of Molecular Biology, Agricultural University, Wageningen, The Netherlands.

In <u>R.leguminosarum</u> PRE the genes involved in symbiotic nitrogen fixation are located on 400 kb sym-plasmid. To elucidate the organization of sym-genes on this plasmid we have constructed libraries of sym-plasmid DNA in phage λ-EMBL3 and cosmid vector pJB8. The libraries were screened by hybridization with probes specific for the nitrogenase genes (<u>nif</u> HDK), <u>nif</u> A and nodulation (<u>nod</u>) genes. By chromosome walking a 100 kb region containing the <u>nif</u> K, D and H sequences and a 50 kb region containing the <u>nif</u> A and <u>nod</u> genes have been mapped. The <u>nod</u>-genes are situated about 10 kb from <u>nif</u> A. The distance between the <u>nif</u> HDK region and <u>nif</u> A has not been determined yet, but appears to be greater than 75 kb.

To investigate the possible involvement of a membrane bound dehydroge-nase in nitrogen fixation we compared membranes isolated from free-living <u>R.leguminosarum</u> PRE and bacteroids. In an <u>in</u> <u>vitro</u> dehydrogenase assay a bacteroid specific membrane-bound dehydrogenase was found which appeared to be NADH-dependent. The expression of this dehydrogenase gene is not regulated by <u>nif</u> A since activity can still be detected in <u>nif</u> A$^-$ mutant bacteroids. Analysis of the dehydrogenase complex by SDS-PAGE revealed two proteins with apparent molecular weights of 28 and 29 kd which were present in much larger amounts in bacteroid than in bacterial membranes.

In order to detect genes involved in synthesis of the FeMo-cofactor of nitrogenase EPR spectroscopy might be used to detect the presence or absence of this cofactor in mutant bacteroids. In preliminary experiments we studied the feasibility of obtaining EPR spectra with intact wild type and mutant bacteroids, which were compared with the spectrum obtained with purified <u>Azotobacter</u> <u>vinelandii</u> nitrogenase. These spectra showed that the FeMo-cofactor can be determined in intact bacteroids by whole cell EPR-spectroscopy and that this technique thus can be used to assign <u>R.leguminosarum</u> gene functions involved in FeMo-cofactor synthesis.

ISOLATION AND CHARACTERIZATION OF THE GENES FROM
FRANKIA THAT CODE FOR THE COMPONENT I PROTEINS OF NITROGENASE

JAMES M. LIGON AND JAMES P. NAKAS
COLLEGE OF ENVIRONMENTAL SCIENCE AND FORESTRY, SYRACUSE, NY 13210, USA

Genomic DNA was isolated from a Frankia alder root nodule isolate and was used to construct a gene library in the cosmid vector pHC79. Clones from this gene library were screened by in situ colony hybridization with a ^{32}P-labeled nitrogenase gene (nifHDK) fragment from Klebsiella pneumoniae to identify clones containing the Frankia genes that code for the component I and II proteins of nitrogenase. A clone with a plasmid containing 49 kilobases (kb) of Frankia DNA, including the nifD and nifK genes, was identified. Attempts to demonstrate hybridization of nifH genes derived from K. pneumoniae and Azotobacter vinelandii to this plasmid were negative, whereas these fragments hybridized to a 7kb Pst I fragment in the Frankia genomic DNA. Restriction mapping of the nifDK region has shown that the nifD and nifK genes are contiguous. These results demonstrate that the nifDK genes and the nifH gene are distantly separated in Frankia. Nucleic acid sequence analysis of the nifD and nifK genes has been initiated and the DNA sequence data, although incomplete at this time, compares favorably to the DNA sequences of the same genes from other diazotrophic species.

TRANSCRIPTION OF <u>RHIZOBIUM</u> <u>MELILOTI</u> <u>NIF</u> PROMOTERS BY A <u>RPOB</u> MUTANT RNA POLYMERASE.

BRENT L. NIELSEN, PHIL KNIGHT, AND LYLE R. BROWN. OREGON STATE UNIVERSITY, CORVALLIS, OR 97331 U.S.A.

INTRODUCTION

Control of <u>nif</u> gene expression in <u>R. meliloti</u> is similar to <u>nif</u> gene control in <u>Klebsiella pneumoniae</u>. In <u>K. penumoniae</u>, expression of <u>nif</u> genes is integrated with the control of the general nitrogen regulatory system (<u>ntr</u>). Primary control involves activation of the <u>nifLA</u> operon by <u>ntrA</u> (<u>glnF</u>) proteins. The <u>nifA</u> protein then functions with <u>ntrA</u> protein as co-activators of <u>nifHDK</u> gene expression (Magasanik, 1982). We established an <u>in vitro</u> transcription system to directly test the specific requirements for <u>nif</u> expression in <u>R. meliloti</u>.

METHODS

RNA polymerase was extracted and purified from <u>R. meliloti</u> HYP-9 (Nielsen and Brown, 1985), a rifampicin-resistant <u>rpoB</u> RNA polymerase mutant, and RM41 (wild-type). RNA polymerase was used in conjunction with reconstituted <u>K. pneumoniae</u> <u>nifA</u> protein or 55,000 to 60,000 m.w. bacteroid protein extract to construct an <u>in vitro</u> system for analysis of transcription from <u>R. meliloti</u> <u>nifHDK</u> and <u>nif</u>-like promoters. <u>In vitro</u> transcripts from purified <u>nif</u> gene templates were analyzed on formaldehyde-denaturing gels.

RESULTS

We show that a 55,000 to 60,000 m.w. bacteroid protein extract has the functional characteristics of a "<u>nifA</u>" protein in our <u>in vitro</u> system. The bacteroid extract stimulates specific transcription by <u>R. meliloti</u> wild-type polymerase from <u>nif</u> and <u>nif</u>-like promoters, and is dependent upon the presence of the upstream -180 to -80 conserved promoter region for this stimulation. It is unlikely that our <u>R. meliloti</u> bacteriod fraction contains a <u>ntrA</u> protein unless the protein is much smaller than analogous proteins (75,000 to 85,000 m.w.) from other species. The <u>in vitro</u> system does not require <u>ntrA</u> as a co-activator for <u>nif</u> gene expression.

Both run-off transcription assays and Southern hybridization experiments showed that the unique mutant RNA polymerase purified from HYP-9 specifically transcribed linear and circular <u>nif</u> promoter containing templates <u>in vitro</u>. Neither the <u>ntrA</u> protein nor the <u>nifA</u> protein were required for this transcription.

Magasanik B (1982) Ann. Rev. Genet. 16, 135-168.
Nielsen BL and Brown LR (1985) J. Bacteriol. 162, 645-650.

ORGANIZATION AND CHARACTERIZATION OF NITROGEN FIXATION GENES FROM BRADYRHIZOBIUM JAPONICUM STRAIN I110

JOHN D. NOTI/ARTHUR N. TURKEN/ALADAR A. SZALAY
BOYCE THOMPSON INSTITUTE FOR PLANT RESEARCH, CORNELL UNIVERSITY, TOWER ROAD, ITHACA, NY 14853, USA

Physical mapping of a 39 kilobase pair (kb) segment of DNA from Bradyrhizobium japonicum strain I110 which contains all three structural genes (nif) for nitrogenase showed that nifH is located 17.2 kb downstream of the nifDK operon. With northern blot analysis of bacteroid RNA we found that the nifDK and nifH transcripts were approximately 3000 and 900 nucleotides, respectively, in length. Since the nifD,K and H genes are 1545, 1554 and 882 nucleotides in length, respectively, this shows that only these three genes comprise the two operons.

A total of 95 independent Tn5 insertions within this 39 kb region were obtained in Escherichia coli and transferred to the wild type strain I110 by marker exchange. Individual transconjugants containing a Tn5 insertion were inoculated onto Glycine max cv. Wilkin (soybeans) and analyzed for their effect on symbiotic nitrogen fixation. The 95 Tn5 insertions distributed throughout this 39 kb region defined the boundaries of these two operons and revealed three additional regions essential for nitrogen fixation. Fix genes were located 9 kb upstream and 1.5 kb downstream of the nifDK operon. The DNA downstream of the nifDK operon from strain I110 showed very strong hybridization to the regions downstream of nifK in Bradyrhizobium sp. (Vigna) strain IRc 78 and in B. japonicum strains USDA 123 and RCR 3407. An essential fix gene which is part of the nifDK operon is located in this region in strain IRc 78 (unpublished data).

A cluster of fix genes spanning 4 kb of DNA was located 4.5 kb upstream of nifH. Insertions of Tn5 in this latter region resulted in transconjugants that formed small nodules which expressed less than 9% of the wild type nitrogenase activity. DNA from this latter region hybridized with the nif regulatory gene from R. meliloti. Transconjugants with Tn5 insertions in this homologous region formed nodules which had reduced levels (less than 7% of wild type) of the nifDK and nifH transcripts. This finding implicates this region in the regulation of the nif genes. Recent results (Morales and Hennecke, 1985) that demonstrate K. pneumoniae nifA-mediated activation in E. coli of the nifD and nifH promoters of B. japonicum further supports the conclusion that a nifA-like regulatory gene is present in the slow-growers. Although both the nifH and nifDK operons were coordinately activated by the K. pneumoniae nifA gene product (Morales and Hennecke, 1985), our results are congruent with other reports and show that the levels of the nifH transcripts were higher than the nifDK transcripts in nodules formed by the wild-type strain.

Reference

Morales, A.A. and Hennecke, H. (1985) Mol. Gen. Genet. 199, 306-314.

GENETIC INSTABILITY OF RHIZOBIUM PHASEOLI

G. SOBERON-CHAVEZ/R. NAJERA/L. CASTREJON
CENTRO DE INVESTIGACION SOBRE FIJACION DE NITROGENO, UNAM, CUERNAVACA, MOR. MEXICO.

Reiterated sequences are not common among bacteria, but their presence have been described in *Streptomyces sp.* (2) *Halobacterium halobium* (5) and *Pseudomonas psyrangae pv phaseolicola* (6). The presence of reiterated sequences may be related to the genetic variability of some of these bacteria (1, 3).

The structural genes of the nitrogenase are reiterated in *Rhizobium phaseoli* (4). We report here that it is common to isolate *Rhizobium phaseoli* strains that loose frequently their symbiotic phenotype when culture at 37°C for 5 days. Fifty four out of a hundred strains studied loose their nodule forming ability at a frequency of 25% or higher.

Using hybridization procedures, we showed that in one unstable strain different genetic rearrangements occur that modify the possition of *nif* genes and affect the nodulation ability of the bacteria.

These rearrangements are may be related to the presence of reiterated sequences in *Rhizobium phaseoli*, and we think that this plasticity makes this bacteria a highly adaptable symbiont.

It is dificult to explain the high frequency in wich unstable *R. phaseoli* strains are isolated, may be the instability represents some selective advantage to the bacteria in the soil. We will determine the competitivness of different unstable and stable strains.

REFERENCES:

1.- Altenbuchner J. and J. Cullum (1984) Mol. Gen. Genet. 195:134-138.
2.- Ono H. *et al* (1982) Mol. Gen. Genet. 186: 106-110.
3.- Pfeifer F. *et al* (1981) J. Bacteriol. 145: 375-381.
4.- Quinto C *et al* (1982) Nature (London) 299: 724-726.
5.- Sapienza C. and F. Doolitle (1982) Nature (London) 245: 384-389.
6.- Szabo L. J. and D. Mills (1984) J. Bacteriol 157: 821-827.

A MOLECULAR ANALYSIS OF THE nif CLUSTER IN THE PARASPONIA RHIZOBIUM
STRAIN ANU289

JEREMY J. WEINMAN[1,2], PETER M. GRESSHOFF[2] and KIERAN F. SCOTT[1], Centre
for Recombinant DNA Research, Research School of Biological Sciences[1]
and Department of Botany, Faculty of Science[2], Australian National
University, Canberra, ACT 2601, Australia

Parasponia Rhizobium strain ANU289 effectively nodulates a wide range
of tropical legumes and the non-legume tree Parasponia (Trinick, 1973).
To allow the genetic requirements for the expression of nitrogen fixation
in this strain to be determined we have isolated and characterized a
number of genes involved in this process.

The nitrogenase structural proteins are encoded on two separate operons, nifDK and nifH, located 21kb apart. DNA sequencing has predicted
the primary amino acid structure of all three structural genes (Scott et
al., 1983; Weinman et al., 1984). We have identified restriction fragments in addition to those containing the structural genes which are
transcribed in the nodule; in particular, a 10kb region 3' to the nifH
gene. By using cloned R. meliloti fix genes (kindly supplied by H.
Reilander) as hybridization probes we have identified 3.5kb 3' to nifH
a 2.5kb PstI fragment and (further 3') a 1.95kb PstI fragment containing
sequences homologous to fixB and fixC respectively. We have also located
sequences on an unlinked lambda clone homologous to a region 3' to the
R. meliloti fixD gene required for a fix[+] response (Szeto et al., 1984).

The promoter regions for the ANU289 nifH and nifDK transcripts have
been determined after S1 nuclease mapping experiments (Scott et al.,
1983; Weinman et al., 1984). Comparison of these regions reveals a
consensus promoter sequence (see below) which can also be observed
prior to other nif genes in other Rhizobium strains (Better et al.,
1984; Sundaresan et al., 1983; Scott et al., 1983; Adams, Chelm, 1984;
Kaluza, Hennecke, 1984; Fuhrmann, Hennecke, 1984) and is similar to
that preceeding the K. pneumoniae nif operons (Beynon et al., 1983).
Homology of this consensus to sequences prior to genes not regulated
by nifA or the related ntrC; viz xylABC in Pseudomonas putida (Inouye
et al., 1984) and the heat shock induced transcript of rpoD in E. coli
(Taylor et al., 1984).

Rhizobium consensus	YTGGCAYC....TTGCA/T
rpoD (HS)	CTGcCACc .. TTGaA
xylABC	aTGGCATC....TTGCT

suggests that Rhizobium nif activation is not unique but rather a
specialized adaption of a broader mechanism. Further, the homology
to the rpoD (HS) promoter region may imply that nif promoter activation
also makes use of a variant sigma factor.

REFERENCES
Trinick MJ (1973) Nature 244, 459-460.
Scott KF et al (1983) DNA 2, 141-148.
Weinman JJ (1984) Nucl. Acids Res. 12, 8329-8344.
Szeto WW et al (1984) Cell 36, 1035-1043.
Better M et al (1983) Cell 35, 479-485; Sundaresan V et al (1983) Nature
301, 728-732.
Adams TH and Chelm BK (1984) J. Mol. App. Genet. 2, 392-405.
Kaluza K and Hennecke H (1984) Mol. Gen. Genet. 196, 35-42.
Fuhrmann M and Hennecke H (1984) J. Bacteriol. 158, 1005-1011.
Beynon J et al, (1983) Cell 34, 665-671.
Inouye S et al (1984) PNAS (USA 81, 1688-1691.
Taylor WE et al (1984) Cell 38, 371-381.

A MODEL OF NUTRIENT EXCHANGE IN THE RHIZOBIUM-LEGUME SYMBIOSIS

MICHAEL L. KAHN[#*]/JENNIFER KRAUS[*]/JOHN E. SOMERVILLE[#]
DEPARTMENT OF MICROBIOLOGY[#] AND PROGRAM IN GENETICS AND CELL BIOLOGY[*]
WASHINGTON STATE UNIVERSITY, PULLMAN, WA 99164-4340, USA

Rhizobium bacteria and leguminous plants can establish a symbiotic relationship in which photosynthetic products made by the plant supply energy used by Rhizobium to reduce atmospheric dinitrogen (Bergerson, 1982; Verma and Long, 1983). Reduced nitrogen is returned to the plant to complete the exchange. Details of the interaction are not completely understood but it is clear that the relationship has required considerable adaptation by both symbiotic partners. For instance, the low oxygen tension needed to stabilize nitrogenase is preserved by leghemoglobin, a plant-derived protein produced only in the nodule. Unusual forms of glutamine synthetase are found in both the plants and the bacteria (Cullimore et al. 1983; Darrow and Knotts, 1977).

This paper analyzes the exchange of nutrients between Rhizobium and its host plant and asks how particular types of exchange might help maintain the symbiotic interaction. Reviews of nitrogen and carbon metabolism in the nodule (e.g., Rawsthorne et al, 1980; Emerich et al, 1983) have concluded that assimilation of fixed nitrogen occurs in the plant cytosol and that at least some reactions in the bacterial citric acid cycle are essential for fixation. However, the precise flow of energy from carbohydrate to nitrogenase is not clear, nor is the path taken by nitrogen after its reduction by nitrogenase. Many enzymes capable of metabolizing reduced nitrogen have been detected in nodules (Robertson and Farnden, 1979), but it is difficult to determine the importance of each enzyme because many alternate metabolic pathways exist. It has generally been assumed that enzymes that catalyze reversible reactions act to assimilate ammonia. We present here a model that questions this assumption. We argue that the model can account for some unusual properties of the symbiotic forms of Rhizobium.

Free-living, nitrogen-fixing bacteria reduce nitrogen when the availability of fixed nitrogen limits their growth and other conditions such as oxygen tension and mineral availability are satisfied (Roberts and Brill, 1981; Magasanik, 1982). They respond to nitrogen stress by inducing nitrogen assimilatory enzymes like glutamine synthetase (GS) and glutamate synthase (GOGAT), catabolic enzymes that can release ammonia from poor nitrogen sources, and high-affinity transport systems for nitrogen-containing compounds in addition to enzymes related to nitrogen fixation. Transcriptional and post-translational controls inhibit nitrogen fixation when a good source of fixed nitrogen becomes available. This is "reasonable" behavior given the high cost of reducing N_2.

Rhizobium bacteroids that are fixing nitrogen do not have these characteristics. GS and GOGAT activities are low in effective bacteroids (Werner et al, 1980; Brown and Dilworth, 1975). Bacteroids do not express the high-affinity ammonia uptake system that is induced by nitrogen limitation in free-living R. leguminosarum (O'Hara et al, 1985). Most striking is that bacteroids export fixed nitrogen instead of importing it (Bergerson and Turner, 1967).

Why does Rhizobium continue to reduce nitrogen even though it appears to have a sufficient supply? One might argue that Rhizobium exports fixed nitrogen to allow the plant to grow better so that it can supply more carbohydrate to the bacteria. This is unlikely. Bacteria that do not fix nitrogen will benefit from the improved health of the plant but will be exempt from the energy costs of contributing to it. They should then be able to replicate faster and eventually dominate the interaction. The plant must protect itself from this possibility in order to keep mutualism from becoming parasitism. During infection the plant can reject bacteria that do not have the proper surface chemistry or produce the right hormones but some type of control may also exist in the nodule. Bacteroids are enclosed within the plant-derived peribacteroid membrane. This membrane could regulate the nutrients available to the bacteroid, the waste products that the plant will accept in return and the bacteroid's oxygen and mineral supply in a way completely determined by the plant.

How can the plant manipulate the interaction to maximize the fixed nitrogen it receives? We propose that, by feeding the bacteroid a nitrogen (N)-containing compound, the plant can obtain fixed nitrogen from the interaction with less risk of being exploited. We also propose that, if an N-containing compound is used to feed the bacteroid, the unusual pattern of fixing nitrogen and exporting ammonia could be part of a bacterial strategy to increase the carbon and energy supplied to it by the plant. It has been suggested previously that amino acids may be important in bacteroid metabolism (O'Gara and Shanmugam, 1976; Tubb, 1976). We develop this idea below and explore its consequences.

Our speculations have focused primarily on glutamate catabolism but can be generalized to include other N-containing compounds. In the model diagrammed in Figure 1, glutamate is fed to the bacteroid where it is catabolized to yield ammonia or an amino acid, energy and carbon. The waste products are returned to the plant with no loss of the fixed nitrogen and the plant uses the nitrogen to regenerate glutamate and complete the upper cycle. If a certain quantity of energy must be transferred into the bacteroid in order to reduce a unit of nitrogen, glutamate can be as efficient an energy source as a carbohydrate or organic acid since the NADH used to make glutamate can be recovered during its catabolism. Because N-containing compounds are available to the bacteroid, we expect its nitrogen assimilatory enzymes to be repressed. The waste products will include nitrogen and we therefore expect the bacteroid to export nitrogen. These properties of the model are in agreement with the observed properties of bacteroids.

Built into the model is the idea that fixed nitrogen is used to carry carbon and energy. As a result of this, the bacteroid has a direct interest in the nitrogen status of the plant since the more fixed nitrogen the plant contains, the more N-containing compound can be delivered to the bacteroid. The upper cycle in Figure 1 benefits only the bacteroid since all of the nitrogen is being recycled. However, if the plant removes some of the ammonia in order to support its own growth, an "appropriate" bacterial response would be to produce more ammonia in order to replenish the fixed nitrogen that serves as a carrier of organic acids. Nitrogen fixation can therefore be thought of as the bacteroid's solution to a problem of carbon and energy availability rather than one of nitrogen availability. In a recent review, Magasanik (1982) divides nitrogen-regulated proteins into two classes, a class that functions only in nitrogen assimilation and a class that can also be used to supply the cell with

FIGURE 1. Proposed nutrient flow between plant and bacteroid. Vertical lines indicate the peribacteroid membrane (PBM) and bacteroid membrane (BM).

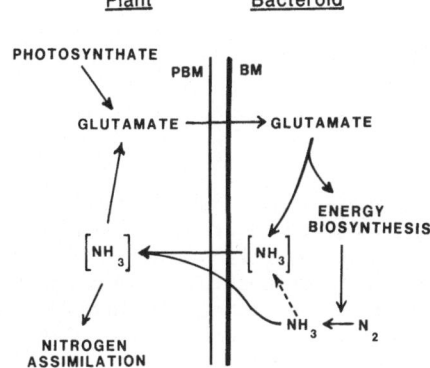

carbon and energy. He considers <u>Klebsiella</u> nitrogenase to be in the first class and enzymes like histidase or arginase to be in the second. We suggest that the <u>Rhizobium</u> <u>nif</u> proteins are regulated like enzymes of the second class. We do not exclude a role for nitrogen regulatory proteins such as <u>nifA</u> or <u>ntrC</u> but suggest that the <u>Rhizobium</u> <u>nif</u> genes may also respond to a set of effector molecules that are different from those used to activate <u>Klebsiella</u> nitrogenase. Consistent with this idea, the conserved <u>nif</u> promoter sequences are longer in <u>R. meliloti</u> than in <u>Klebsiella</u> and are comparable to the length of the control regions of several class 2 genes from enteric bacteria (Ow et al, 1983). Moreover, the region of the <u>R. meliloti</u> <u>nifH</u> promoter that is stimulated by the <u>Klebsiella</u> <u>nifA</u> protein is not the same region as that needed for nitrogenase expression in the nodule (Better et al, 1985).

In one important respect, rhizobia behave as the hypothesis predicts they should. A source of fixed nitrogen is generally needed to induce <u>Rhizobium</u> to fix nitrogen ex planta and maximum activity is found when this nitrogen source is an amino acid such as glutamate or aspartate (Rao et al, 1978). Under these growth conditions the nitrogen fixed by the bacteria is not assimilated, ammonia is exported into the medium and the bacteria do not grow (O'Gara and Shanmugam, 1976). Only simple molecules are required to induce this syndrome. This implies that the plant does not need to interfere with the bacteroid's metabolism using specific inhibitors or complicated inducers in order to establish nitrogen fixation.

We have focused our attention on glutamate because glutamate is a crucial compound in the synthesis of all amino acids and is also important in ammonia assimilation. Glutamate is an unusual effector for <u>Rhizobium</u>-adding 0.1% glutamate to cultures of <u>R. japonicum</u> growing in gluconate and ammonia significantly inhibits their growth (Upchurch and Elkan, 1978) but <u>Rhizobium</u> 32H1 grows much faster with glutamate present (Ludwig, 1978). Glutamate can also lead to the development of pleimorphic cells (Kaneshiro and Kurtzman, 1982). Glutamate is often needed to induce nitrogen fixation by ex planta cultures of <u>Rhizobium</u>. Glutamate catabolism is accomplished by ammonia excretion (O'Gara and Shanmugam, 1976; Tubb, 1976; this lab).

Biochemical and genetic arguments suggest that glutamate is available to the bacteroids. Glutamate in bacteroids is rapidly labeled by pulses of $^{14}CO_2$ and $^{13}N_2$ (Schubert and Coker, 1981) which indicates both that glutamate is present and that it has a rapid turnover. If glutamate is present under conditions where ammonia is also available we expect a decreased need for GS. As described above, bacteroid GS levels are low. Since GOGAT mutants of R. meliloti are Fix[+] (Kondorosi et al, 1977), glutamate synthesis by the GS/GOGAT pathway is not needed for an effective symbiosis. It has been argued that glutamate synthesis by GDH is unlikely (Kondorosi et al, 1977). We suggest that glutamate is imported by the bacteroid to satisfy its needs.

Glutamate can be catabolized in a large number of different ways (Figure 2) and there are many ways of including glutamate in the metabolic flow needed to support symbiotic nitrogen fixation. One difficulty in sorting out these possibilities is that many of the enzymes potentially involved can have either a catabolic or anabolic role and are subject to allosteric control. Some of the diversity in the literature may reflect the use of different pathways by different plant-bacteria combinations. We consider some of these alternatives below.

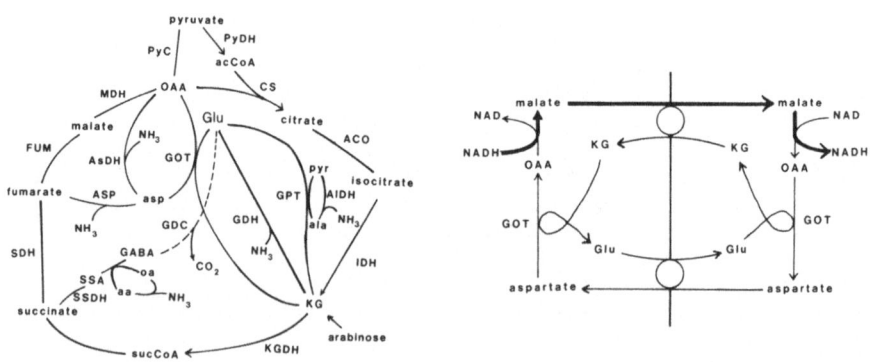

FIGURE 2. (left) Pathways of glutamate degradation. Abbreviations are: oxaloacetic acid (OAA), α-ketoglutarate (KG), glutamate (Glu), aspartate (asp), pyruvate (pyr), alanine (ala),γ-aminobutyric acid (GABA), succinate semialdehyde (SSA), amino acid (aa), organic acid (oa), pyruvate carboxylase (PyC), pyruvate dehydrogenase (PyDH), citrate synthase (CS), aconitase (ACO), isocitrate dehydrogenase (IDH), α-ketoglutarate dehydrogenase (KGDH), glutamate dehydrogenase (GDH), glutamate-pyruvate aminotransferase (GPT), glutamate-oxaloacetate aminotransferase (GOT), glutamate decarboxylase (GDC), alanine dehydrogenase (AlDH), aspartate dehydrogenase (AsDH), succinate dehydrogenase (SDH), malate dehydrogenase (MDH), aspartase (ASP), succinate semialdehyde dehydrogenase (SSDH) and fumarase (FUM).

FIGURE 3. (right) The malate-aspartate shuttle.

One method of catabolizing glutamate is to remove the ammonia to give α-ketoglutarate. This can be done directly using glutamate dehydrogenase (GDH), or by transamination of oxaloacetate to give aspartate, or by trans-amination of pyruvate to give alanine. Other α-keto acids could also be transaminated. Aspartate can be deaminated by aspartase or by aspartate dehydrogenase to release ammonia; alanine can be deaminated by alanine dehydrogenase. These deaminations will regenerate the organic acid acceptor for subsequent use in a cycle. All of the reactions that release ammonia are reversible. Efficient operation of these pathways thus requires that the end products be removed. This can most easily be done by excreting ammonia from the cell and by catabolizing the organic acids further. The α-ketoglutarate can be catabolized through the citric acid cycle to yield GTP, reducing equivalents and organic acids to be used in transamination reactions or biosynthesis. The organic acids used in transamination could also be acquired from the plant under certain circumstances.

These possibilities are consistent with many observations in the literature. The low level of NADP-dependent isocitrate dehydrogenase (IDH) in R. meliloti bacteroids (Henson et al, 1982) and the decline in IDH level during maturation of R. japonicum bacteroids (Karr et al, 1984) indicate that the plant may be supplying α-ketoglutarate to the nodule. We find low levels of IDH in cells grown on glutamate. An α-ketoglutarate dehydrogenase mutant of R. meliloti is ineffective (Duncan and Fraenkel, 1979), which indicates that metabolism of α-ketoglutarate is important to nodule function. Mutants that lack malate dehydrogeanse (J.K. Waters and D.W. Emerich, personal communication) and succinate dehydrogenase (Gardiol et al, 1982) are also ineffective. Significant activities of aspartase, asparaginase, glutaminase, aspartate dehydrogenase, alanine dehydrogenase, glutamate-oxaloacetate amino transferase (GOT) and NAD-GDH have all been detected in bacteroids. A mutant of R. japonicum that is GOT forms ineffective nodules that are capable of reducing acetylene (Zlotnikov et al, 1984). It was speculated that this mutant has a defect in transferring fixed nitrogen to the plant. Alanine, aspartate and glutamate are rapidly labeled in pulse labeling experiments.

Glutamate can also be degraded by decarboxylation to yield γ-amino butyric acid (GABA). The amino group is then removed by transamination of an α-keto organic acid to give succinate semialdehyde (SSA). SSA can be oxidized to give succinate or reduced to give 4-hydroxybutyrate. The latter could be metabolized to give poly-3-hydroxybutyrate. Large amounts of GABA can be found in some nodules (Freney and Gibson, 1975) and this amount depends on the bacterial strain forming the nodules. Because some of these reactions are irreversible, detection of the appropriate enzymes would be evidence for glutamate catabolism.

A third possibility involves the malate-aspartate shuttle (Figure 3). In this pathway malate and glutamate are imported into the bacteroid in exchange for α-ketoglutarate and aspartate, respectively. Malate is oxidized within the bacteroid to give OAA which is then transaminated by glutamate to give aspartate. α-ketoglutarate is transaminated in the plant by aspartate to yield glutamate. The net result is the transfer of NADH into the bacteroid without the transfer of carbon. This pathway would not support bacteroid growth but would still support nitrogen fixation. It represents the model of Figure 1 in its purest form since the amino group here is used only as a carrier of organic acids. If the limiting reaction was the export of aspartate, the bacteroid would have a large pool of N-containing compounds but would have an energy supply dependent on generating still more fixed nitrogen.

The malate-aspartate shuttle is used to transfer reducing equivalents into mitochondria (Meijer and Van Dam, 1974) and should not require novel behavior from the plant. The needed transport proteins would be made by the plant but would be located in the peribacteroid membrane instead of the inner membrane of the mitochondrion. It has been suggested that an ancestor of Rhizobium might have been the precursor to mitochondria (Baltscheffsky and Baltscheffsky, 1981). If so, this pathway might be a remnant of the interaction between a protoeukaryote and a nitrogen-fixing endosymbiont that has persisted after the endosymbiont lost its nitrogen-fixing ability.

More than one pathway could operate during the maturation of the nodule. Initially the exchange appears to be most to the bacteria's advantage since the bacteroids increase in number and the nodule enlarges. Later, as the plant withdraws nitrogen from the pool, the bacteroid would produce more ammonia to replenish it. If the plant continues to remove nitrogen, this situation could progress until most of the bacteroid's energy was directed into nitrogen fixation.

One difference between plants that make ureides and those that make amides could result from differences in bacterial catabolism. In the pathways above, glutamate catabolism returned dicarboxylic acids, pyruvate or the cognate amino acids to the plant. If catabolism proceeds further to generate glyoxalate or its transamination product, glycine, the plant would have a supply of purine precursors. Glycine is incorporated directly into purines and it can also be metabolized to give methylene tetrahydro-folate (MTF), another precursor of purine biosynthesis. Purines are precursors of ureides. The relatively low activity of plant serine hydroxy-methyltransferase, an enzyme that makes glycine and MTF from serine, is consistent with this possibility (Reynolds et al, 1982).

In some Rhizobium symbioses, the ^{15}N concentration is significantly higher in the nodules than in the plant (Shearer et al, 1982). Since nitrogen is fixed to give ammonia and free ammonia is not involved in cycles that use only transamination reactions, there is a possibility that two pools of fixed nitrogen can exist. Nitrogen used for bacteroid bio-synthesis will predominantly come from the amino acid pool.

We have isolated a number of the Tn5 induced mutants that affect glutamate synthesis or degradation. Five mutants defective in glutamate catabolism (Gca-) are Fix$^-$. These also have defects in proline catabolism but grow normally on mannitol-ammonia or arabinose-ammonia medium and give an alkaline reaction on mannitol-ammonia-glutamate medium. Other Gca-Fix$^-$ mutants have been isolated that have defects in arabinose or dicar-boxylic acid utilization. We have not yet been able to determine the en-zymatic defect in these mutants.

We propose that an amino acid such as glutamate carries carbon and energy into the bacteroid. This arrangement allows the plant to obtain nitrogen from the interaction in any way that assures that it will not be exploited. It may seem counterintuitive for the plant to release fixed nitrogen to the bacteria but the fixed nitrogen is not lost and can be recycled. The proposal suggests how fixing nitrogen will be advantageous to the bacteroid even though it is supplied with amino acids. Our model reduces the symbiotic exchange of carbon for nitrogen to the molecular level by suggesting that the medium of exchange is a compound that con-tains both. We believe that similar strategies may operate in other nitro-gen-fixing symbioses.

REFERENCES

Baltscheffsky, H and M Baltscheffsky (1981) in "Mitochondria and Micro-somes", CP Lee et al, eds., Addison-Wesley, pp 519-540.

Bergerson, FJ (1982) "Root Nodules of Legumes: Structure and Functions" Research Studies Press.

Bergerson, FJ and GL Turner (1967) Biochem. Biophys. Acta 141:507-515.

Better, M et al (1985) in "Adv. in Molecular Genetics of Plant-Bacteria Interactions, AA Szalay and R Legocki, eds. Cornell Press, pp 7-9.

Brown, CM and MJ Dilworth (1975) J. Gen. Microbiol. 86:39-48.

Cullimore, JV et al (1983) Planta 157:245-253.

Darrow, RA and RR Knotts (1977) Biochem. Biophys. Res. Comm. 78:554.

Duncan, MJ and DG Fraenkel (1979) J. Bacteriol. 137:415-419.

Emerich, DW et al (1983) in "Nitrogen Fixation" WJ Broughton, ed., Clarendon Press pp. 213-244.

Freney, JR and AH Gibson (1975) Aust. J. Plant Physiol. 2:663-668.

Gardiol, A et al (1982) J. Bateriol. 151:1621-1623.

Henson, CA et al (1982) Plant and Cell Physiol. 23:227-235.

Kaneshiro, T and MA Kurzman (1982) J. Appl. Bacteriol. 52:210-207

Karr, DB et al (1984) Plant Physiol. 75:1158-1162.

Kondorosi, A et al (1977) Mol. Gen. Genet. 151:221-226.

Ludwig, RA (1978) J. Bacteriol. 135:114-123.

Magasanik, B (1982) Ann. Rev. Genet. 16:135-168.

Meijer, AJ and K Van Dam (1974) Biochem. Biophys. Acta 346:213-244.

O'Gara, F and K. Shanmugam (1976) Biochem. Biophys. Acta 437:313-321.

O'Hara, GW et al (1985) J. Gen. Microbiol. 131:757-764.

Ow, D et al (1983) Proc. Nat. Acad. Sci. (USA) 80:2524-2528.

Rao, VR et al (1978) Biochem. Biophys. Res. Comm. 81:224-231.

Rawsthorne, S et al (1980) Phytochemistry 19:341-355.

Reynolds, PHS et al (1982) Plant Physiol. 69:1334-1338.

Roberts, GP and WJ Brill (1981) Ann. Rev. Microbiol. 35:207-235.

Robertson, JG and KJF Farnden (1979) in "Biochemistry of Plants", PK Stumpf and EE Conn, eds, 5:65-113.

Schubert, KR and GT Coker (1981) Adv. Chem. 197:317-339.

Shearer, G et al (1982) Plant Physiol. 70:465-468.

Tubb, RS (1976) Appl. Environ. Microbiol. 32:483-488.

Upchurch, RG and GH Elkan (1978) Biochem. Biophys. Acta 538:244-248.

Verman, DPS and S Long (1983) Int. Rev. Cytology Suppl. 14:211-245.

Werner, D et al (1980) Planta 147:320-329.

Zlotnikov, KM et al (1984) Dokl. Akad. Nauk. SSSR Micro. 275:189-192.

This work was supported by grants from the Competitive Research Grants Office of the U.S.Department of Agriculture and by a National Science Foundation predoctoral fellowship to J. Kraus. We thank L. Moore for helpful comments.

GENES INVOLVED IN THE CARBON METABOLISM OF BACTEROIDS

Clive W. Ronson and Patricia M. Astwood
Grasslands Division, DSIR, Palmerston North, New Zealand.

INTRODUCTION

Nitrogen fixation by the Rhizobium-legume symbiosis may be limited by nodule carbon or oxygen supply (reviewed by Stacey, Upchurch, 1984). In either case, rhizobial energy production pathways must be efficiently coupled to nitrogenase to optimise symbiotic nitrogen fixation. Bacteroids within plant nodule cells are enclosed within membranes of plant origin, and are dependent on the plant for provision of carbon/energy substrates. The host plant also requires carbon substrates to assimilate the ammonia fixed by bacteroids. Hence knowledge of what carbon sources are supplied to bacteroids is essential in order to understand the regulatory mechanisms by which plant and bacteroid carbon metabolism is co-ordinated. In addition, it is of interest to compare the regulation of rhizobial carbon catabolism genes in the free-living and symbiotic states, and to determine whether such genes respond to the symbiotic regulatory system proposed to regulate symbiotic gene expression (Ausubel, 1984).

ORGANIC ACIDS ARE SUPPLIED TO BACTEROIDS

To assess the physiological significance of rhizobial pathways of carbohydrate catabolism for symbiotic nitrogen fixation, mutants defective in several steps of sugar catabolism have been isolated from different species (eg. Arias et al., 1979; Ronson, Primrose, 1979; Duncan, 1981; Glenn et al., 1984; Cervenansky, Arias, 1984). Bacteroids formed by R.leguminosarum mutants defective in various enzymes also lacked the enzyme, suggesting that the same enzymes are formed in the free-living and symbiotic states (Glenn et al., 1984). In general, such mutants form effective nodules, indicating that sugars are not essential substrates for bacteroid nitrogen fixation. Mutants with defects in the tricarboxylic acid cycle form ineffective nodules (Duncan, Fraenkel, 1979; Gardiol et al., 1982; Finan et al., 1983). However this finding does not prove that the TCA cycle is required for symbiotic nitrogen fixation as the mutants were severely impaired for growth on several carbon sources. A mutant of R.leguminosarum characterised as defective in pyruvate dehydrogenase formed effective nodules, implying that a complete TCA cycle may not be required. However the mutant grew normally on succinate, indicating that it must either form acetyl-CoA by another pathway or utilise an incomplete TCA cycle (Glenn et al., 1984).

Biochemical observations that succinate, fumarate, and malate were transported by isolated bacteroids and stimulated nitrogen fixation (reviewed by Ronson et al., 1981) suggested that the C4-dicarboxylates may play an important role in the symbiosis. This was confirmed by the finding that mutants of R. trifolii and R.leguminosarum defective in C4-dicarboxylate

transport (<u>dct</u> mutants) but otherwise unimpaired in growth
formed ineffective nodules (Ronson et al., 1981; Finan et
al.,1981, 1983; Arwas et al.,1985). The requirement for
C4-dicarboxylates seems to relate specifically to nitrogen
fixation since nodules formed by <u>dct</u> mutants contain plant
cells filled with bacteroids of morphology similar to
wild-type. Hence rhizobia within plant cells probably utilise
carbon sources other than C4-dicarboxylates to support growth
and division (Ronson et al., 1981; Finan et al., 1983). This
suggestion is supported by observations that a mutant of
<u>R.trifolii</u> in which the C4-dicarboxylate transport system was
hypersensitive to catabolite repression formed ineffective
nodules (Ronson et al., 1981), and that an <u>R.leguminosarum</u>
mutant defective in gluconeogenesis formed effective nodules
(McKay et al., 1985). The gluconeogenic enzymes were
repressed by 0.4mM sugar in wild-type free-living bacteria
but were formed in bacteroids, suggesting that bacteroids may
have access to only low concentrations of sugars (McKay et
al., 1985). While the symbiotic role of C4-dicarboxylates
seems likely to be to provide energy substrates for nitrogen
fixation, other roles such as precursors for heme
biosynthesis cannot be excluded. The pathways by which
symbiotic rhizobia metabolise C4-dicarboxylates have not been
defined but may differ from those used by free-living
rhizobia for growth, as mutants unable to use succinate as
sole carbon source while free-living yet able to fix nitrogen
symbiotically have been isolated (Ronson, unpub. data).

GENETIC ORGANISATION AND REGULATION OF <u>DCT</u> GENES FROM <u>R.LEGUMINOSARUM</u>

Because of the importance of the C4-dicarboxylate
transport system for symbiotic nitrogen fixation, we have
begun a study of the <u>dct</u> genes from <u>R.leguminosarum</u>. Initial
studies, using a combination of genetic complementation,
sub-cloning, and transposon mutagenesis, identified three
loci <u>dct</u>A, <u>dct</u>B, and <u>dct</u>C within a 5.0 kb region of DNA (Fig.
1). The A and B loci were defined by their ability to
complement structural and regulatory mutants respectively,
while the C locus was identified by its ability to complement
B mutants when expressed from a vector promoter (Ronson et
al., 1984). To establish the phenotype of mutants in the C
locus, and to determine whether the DNA fragment contained
further <u>dct</u> loci, we have recombined several transposon Tn<u>5</u>
insertions from the cloned fragment into the genome, using
the plasmid incompatibility method (Ruvkun, Ausubel, 1981).
These experiments (Fig. 1) revealed another gene <u>dct</u>D between
<u>dct</u>B and <u>dct</u>C, mutants in which displayed a clean succinate-
negative phenotype similar to <u>dct</u>A mutants, not the leaky
phenotype characteristic of <u>dct</u>B mutants. Complementation
experiments showed that D mutants, unlike B mutants, were not
complemented by plasmids expressing the A or C loci
constitutively, confirming that D was a separate locus. The
site-directed mutagenesis also revealed that <u>dct</u>C mutants
grew on succinate plates at wild-type rates, while mutants in
the A and B loci showed clean and leaky phenotypes respectively, as expected. However when inoculated onto pea plants,

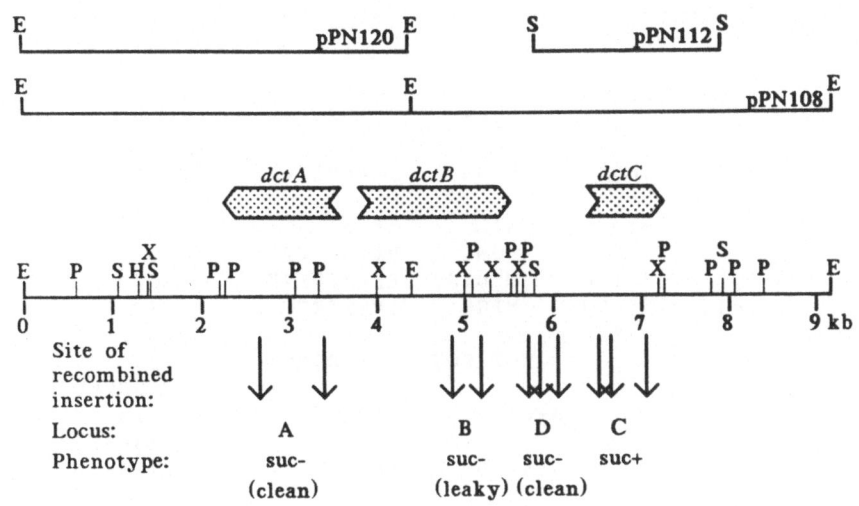

Figure 1. Map of dct loci (Ronson et al., 1984), sites of transposon insertions recombined into genome, and phenotypes of resultant mutants. Plasmids pPN108, pPN112, and pPN120 are also shown.

only the A mutants formed ineffective nodules (no acetylene reduction); mutants in the B and D loci formed partially-effective nodules (acetylene reduction 8% of wild-type rates), while C mutants formed effective nodules (acetylene reduction >50% of wild-type).

To study the regulation of the dct genes, we have isolated transcriptional fusions of the genes to the E.coli lacZ gene. The fusions were created by transposon mutagenesis of cloned DNA fragments with Tn3-HoHol, a modified version of Tn3 that contains lacZ coding sequences inserted next to the left inverted repeat of Tn3 (Stachel et al., 1985). Results obtained with fusions into pPN108 (Ronson et al., 1984) which contains the entire dct regulon (Fig. 1) showed that the dctA gene was induced at least 10-fold by succinate, with maximal activity reached within 3 h. The dctB, C and D genes were expressed at a low (15-20 units) constitutive level (not shown). To further analyse the regulation of the dctA gene, fusions were isolated in pPN120, a pLAFR1-based plasmid with the 4.4kb EcoR1 fragment (Fig. 1) containing the dctA gene and the 5' portion of the dctB gene only. Assay of the fusion in the wild-type and various dct mutant backgrounds (Table 1) showed that the fusion was not inducible in dctB or dctD backgrounds, and was inducible to about 60% of wild-type level in a dctC background. The fusion was expressed constitutively in a dctA background, and in a wild-type strain containing pPN112 (Ronson et al., 1984) , a plasmid that expresses dctC constitutively from a vector promoter. Similar results were obtained when the strains were assayed in nodules (Table 1). Overall, the results suggest that

Table 1. Activity of dctA::lacZ fusion in various backgrounds.

Back-ground	B-Gal activity[a]		Nodule activity	Symbiosis
	Glucose[b]	Succinate[b]		
WILD-TYPE	4-9	200-318	HIGH	EFFECTIVE
dctB	0	7-17	NIL	8% of WT
dctD	0	7	LOW	8% of WT
dctC	3	88-120	MEDIUM	EFFECTIVE
dctA	150-180	450-600	HIGH	INEFFECT.
pPN112 in WT	118	217	NOT DONE	NOT DONE

a)units calculated according to Miller (1972).
b)sole carbon source used for growth.

```
U  C
A   C
C   A
UA
GC
CG
CG
CG
GC
GC
AGCGUUUCUCUUUUC
```
Figure 2. DctA transcriptional terminator.

regulation of the dctA locus is complex: the B and D gene products may be positive regulatory elements required for A gene expression (or D may encode a product required for inducer to reach its site of action), while the dctC product may modulate A gene expression to a lesser extent. However, constitutive expression of the C gene product can activate the A gene, and alleviate the requirement for the B (but not D) product for growth. Finally, the A gene seems to be under negative autogenous control. The gene fusion experiments gave no evidence of any symbiosis-specific regulation of dct gene expression, but did confirm that the dctA gene is strongly expressed under symbiotic and free-living conditions.

NUCLEOTIDE SEQUENCE OF THE DCTA GENE.
    To gain insight into the promoter structure of a gene that is expressed in both free-living and symbiotic rhizobia, and to aid studies of the expression of the dctA gene, we have sequenced both strands of the 3kb HindIII-EcoRI fragment containing the dctA gene and 5' terminus of the dctB gene (Fig. 1). Two open reading frames (ORFs) were identified, one of 1332 nucleotides which encoded the A gene product and one of 597 nucleotides which encoded the N terminal portion of the B gene product. The ORFs correlated well with earlier genetic data (Ronson et al., 1984), and there were no other ORFs in either direction in the A gene region, or in the 800 bp downstream. However about 40bp downstream of the A termination codon there was an inverted repeat which when transcribed would form a very stable stem-loop structure (Fig. 2) characteristic of a rho-independent transcriptional terminator. Hence the dctA locus probably consists of one gene encoding a protein of 444 amino acids or 49kdal. The dctA product is highly hydrophobic with 68% apolar residues, indicating it is embedded in the membrane. Predictions of secondary structure (Parry, Ronson, unpub.) suggest that the protein contains 12 α-helical transmembrane shafts arranged perpendicular to the membrane surface, a structure similar to that proposed for the E.coli lactose permease (Foster et al.,

1983).

The region between the <u>dct</u>A and B coding regions (Fig. 3) is 223bp and 49%GC, compared to 61-63%GC for the coding

```
     1                   _____>  <__  __  ____  _____
        CATTTTTCT GCACGAAACG CAAATGGATT TGTGCGGATT TCCGCATTGC
        GTAAAAAGA CGTGCTTTGC GTTTACCTAA ACACGCCTAA AGGCGTAACG
DctB<Met       RBS?
     50
        TTAGTTAGTT GTTAGCAGTC TCGTAAATTT TTCATTAATA AATTCAATCG
        AATCAATCAA CAATCGACAG AGCATTTAAA AAGTAATTAT TTAAGTTAGC

    100                                  C TGGCACG      TTGCG
        GTTGGTTGGC GACTTAAAA       GCG A      AAGG AGGTGGCAAC
        CAACCAACCG CTGAATTTTG ACCGTGCCGC TAACGCTTCC TCCACCGTTG

    150      ____  __  ____ >  <___  ____  _____  _____>
        AACGGCTGAG CTGTTGGACT TGAAGCGAAC GGCTCGGGAG GCCGGAGTTC
        TTGCCGACTC GACAACCTGA ACTTCGCTTG CCGAGCCCTC CGGCCTCAAG

    200  <_____               RBS        Met>DctA
        GTTCCGGACG AGCCACTAGG AGGACATCAT G
        CAAGGCCTGC TCGGTGATCC TCCTGTAGTA C
```

Figure 3. Nucleotide sequence of region between <u>dct</u>B and <u>dct</u>A coding regions. Dyad symmetries are marked by arrows and the AT-rich region by a dashed line. Potential ribosome binding sites and promoters discussed in the text are high-lighted. The "<u>nif</u>A-regulated" promoter is also super-scripted.

regions. The A gene contains a good ribosome-binding site, the B gene a less-so one. Other features of the sequence outlined in Fig. 3 include a number of regions of imperfect dyad symmetry that regulatory molecules might recognize, and a particularly AT-rich (25/27bp) region. The transcription start sites of the <u>dct</u> genes are currently being determined; however comparison of the sequence 5' to <u>dct</u>A with published promoter sequences suggests several possibilities (Fig. 4). The sequence contains a good match to the consensus sequence for constitutively expressed <u>E.coli</u> promoters (Fig. 4a); however the <u>dct</u>A gene is not expressed in <u>E.coli</u> and positively-regulated <u>E.coli</u> genes show a poor match to the consensus. Hence the significance of the sequence remains unknown. The <u>dct</u>A sequence also contains a good match to the "<u>nif</u>A-regulated" consensus sequence characteristic of <u>nif</u> gene promoters (Ausubel, 1984)(Fig. 4b). Finally the <u>dct</u>A sequence has a match to a consensus derived from promoters of the <u>R.meliloti</u> ALA synthetase (Leong et al., 1985) and <u>nif</u>A Buikema et al., 1985) genes (Fig. 4c). Matches to the consensus can also observed 5' to the <u>dct</u>B gene, and <u>rec</u>A and <u>nod</u>A genes from <u>R.meliloti</u> (Fig. 4d; Ronson, Buikema, unpub. observations), suggesting that the consensus may represent a class of <u>Rhizobium</u> "non-nitrogen" promoters. The consensus has an <u>E.coli</u>-like -35 sequence and a purine-rich -10 sequence.

Of particular interest is the positioning of the <u>dct</u>A

A) E.coli consensus promoter (constitutive):
TTGACa   17b   TATAAT 5-8b cat

  dctA sequence (starting at bp61):

```
**  **         *            ****
TTAGCa        17b          ATTAAT 5b  CAA
```
B) "nifA-regulated" consensus promoter:
CTGGCACP 4b TTGCA   10-12b N
       u

  dctA sequence (starting at bp119):

```
******** *  ****
CTGGCACG 4b TTGCG      10b  CAA
```

C) R.meliloti "non-nitrogen" promoters:
i) ALA synthetase P1 (Leong et al., 1985)
TTGACC   17b   AAGAAA 5b TGCCA
ii) ALA synthetase P2 (Leong et al., 1985)
TTGACT   17b   GAATGA 4b ACA
iii) nifA (Buikema et al., 1985)
TTAAGC   16b   AAGGTG 3bCCA
Consensus:
TTPANN   17b.   PAPPPP4-5b CA
   u   u uuuu

dctA sequence (starting at bp113):

```
****          *            ******
TTAAAA        17b          AAGGAG 4b CA
```

D) Other matches to "non-nitrogen" consensus:
a)dctB (this paper)
TTGAAT   17b   ACGAGA 7b CA
b)R.meliloti nodA (Torok et al., 1984)
TTAATG   17b   GAGAAA 5b CA
c)R.meliloti recA (Buikema, unpub.)
TTGACG   17b   AACAAA 7b CA

Figure 4. Comparison of dctA sequence with published
promoter sequences, and derivation of a consensus Rhizobium
"non-nitrogen" promoter sequence. Transcription start-points
that have been determined are high-lighted.

"nifA-regulated" promoter with respect to the "non-nitrogen"
promoter (Fig. 3). Their juxtaposition suggests that
transcription might start at the same nucleotide irrespective
of which promoter is used. The presence of the "nifA-
regulated" promoter suggests that dctA gene expression may be
selectively enhanced co-ordinately with nif gene expression
during symbiosis, through the action of nifA (Szeto et al.,
1984). To test this hypothesis, we are currently determining
the transcription start points of the dctA gene, and
investigating whether nifA influences dctA expression.

FURTHER DISCUSSION
    The presence of the "nifA-regulated" promoter sequence 5'
to the dct structural gene suggests a mechanism by which dct

and <u>nif</u> genes may be co-ordinately activated in symbiosis,
although <u>nif</u> genes map on symbiotic plasmids and <u>dct</u> genes on
the chromosome (near <u>his</u> in <u>R.meliloti</u>; Finan, pers. comm.).
Other genes involved in bacteroid carbon metabolism (eg those
encoding cytochromes) have yet to be identified, and it will
be of interest to determine whether they are regulated in a
similar manner. Another neglected area is the possible role
of catabolites and oxygen in the regulation of rhizobial
energy metabolism genes. Of particular interest are the
observations that mutants of <u>B.japonicum</u> in which hydrogenase
synthesis was insensitve to oxygen and carbon substrate
repression (Hup^c mutants), expressed both hydrogenase and
ribulose bisphosphate carboxylase constitutively while free-
living, but only hydrogenase when grown symbiotically. In
addition, the Hup^c mutants formed 5-fold more hydrogenase
than wild-type in nodules, suggesting that hydrogenase may be
subject to catabolite repression in nodules (Merberg, Maier,
1983, 1984; Merberg et al., 1983).
    Reduced levels of C4-dicarboxylate transport in
bacteroids formed by mutants of <u>R.leguminosarum</u> correlated
with reduced acetylene reduction levels (Finan et al., 1983).
It is therefore of interest to determine if enhanced
expression of <u>dct</u> genes can cause enhanced nitrogen fixation
rates. With the identification of <u>dct</u>A promoter sequences, it
may be possible to isolate such mutants by oligonucleotide-
directed site-specific mutagenesis.

References
Arias A et al (1979) J. Bacteriol. 137, 409-414.
Arwas R et al (1985) J. Gen. Microbiol. 131, in press.
Ausubel FM (1984) In Shapiro L and Losick R, eds, Microbial
Development, pp275-298, CSHL, New York.
Buikema WJ et al (1985) Nuc. Acids Res. 13, 4539-4555.
Cervenansky C and Arias A (1984) J. Bacteriol.160, 1027-1030.
Duncan M (1981) J. Gen. Microbiol. 122, 61-67.
Duncan M and Fraenkel DG (1979) J. Bacteriol. 137, 415-419.
Finan TM et al (1981) J. Bacteriol. 148, 193-202.
Finan TM et al (1983) J. Bacteriol. 154, 1403-1413.
Foster DL et al (1983) J. Biol. Chem. 258, 31-34.
Gardiol A et al (1982) J. Bacteriol. 151, 1621-1623.
Glenn AR et al (1984) J. Gen. Microbiol. 130, 239-245.
Leong SA et al (1985) Nuc. Acids Res., in press.
McKay IA et al (1985) J. Gen. Microbiol. 131, in press.
Merberg D and Maier RJ (1983) Science 220, 1064-1065.
Merberg D and Maier RJ (1984) J. Bacteriol. 160,448-450.
Merberg D et al (1983) J. Bacteriol. 156, 1236-1242.
Miller JM (1972) Experiments in Microbial Genetics, Cold
Spring Harbor Laboratory, New York.
Ronson CW and Primrose SB (1979) J. Gen. Microbiol.112,77-88.
Ronson CW et al (1981) Proc.Natl.Acad.Sci.USA 78, 4284-4288.
Ronson CW et al (1984) J. Bacteriol. 160, 903-909.
Ruvkun GB and Ausubel FM (1981) Nature 289, 85-89.
Stacey G and Upchurch RG (1984) Trends in Biotech. 2, 65-70.
Stachel SE et al (1985) EMBO J. 4, 891-898.
Szeto WW et al (1984) Cell 36, 1035-1043.
Torok I et al (1984) Nuc. Acids Res. 12, 9509-9524.

CHARACTERIZATION, SIGNIFICANCE AND TRANSFER OF HYDROGEN UPTAKE GENES
FROM RHIZOBIUM JAPONICUM

G.R. Lambert, A.R. Harker, M. Zuber, D.A. Dalton, F.J. Hanus, S.A.
Russell, and H.J. Evans, Laboratory for Nitrogen Fixation Research,
Oregon State University, Corvallis, OR   97331, U.S.A.

Introduction

Nitrogenases from different sources exhibit ATP-dependent formation
of both NH_4^+ and H_2. Many N_2-fixing microorganisms are able to recycle
the nitrogenase-mediated H_2 using a membrane-bound hydrogenase, thereby
recovering some of the energy utilized in the formation of H_2. The
physiology, biochemistry and energetics of H_2 recycling have been
extensively reviewed (Dixon, 1978; Robson, Postgate, 1980; Adams et al.,
1981; Eisbrenner, Evans, 1983; Evans et al., 1985).

Improvement of legume productivity by increasing the nitrogen-fixing
capability of Rhizobium species is a major research objective. Since
direct transfer of nif genes has not realized this goal, an alternative
strategy is to isolate and transfer the genetic information for other
traits which improve nitrogen fixing ability. In terms of H_2 evolution,
attempts to mutate nitrogenase so that it will produce NH_4^+ but not H_2
may be futile, since H_2 evolution appears to be inherent in the nitrogenase mechanism (Chatt, 1980; Simpson, Burris, 1984). Assuming that H_2
recycling capability can improve the efficiency of nitrogen fixation, it
is desirable to transfer hup genes to commercially important Rhizobium
strains which lack this capability. Recent advances in Rhizobium
genetics in terms of the range of techniques available make such a
genetic study of H_2 oxidation timely. In addition to the potential
agricultural application of this work, an understanding of the
organization and regulation of expression of hup genes in both the free-
living bacteria and nodules is of considerable interest as an example of
a symbiotically expressed enzyme activity in Rhizobium.

Recent developments in understanding of the biochemistry of nitrogen
fixation and advantages conferred by hydrogen oxidation in R. japonicum
in plants and for the free-living bacteria in soil will be discussed in
this paper. In addition, progress in isolation, transfer and expression
of hup genes both in free-living and symbiotic conditions in diverse
species of Rhizobium will be summarized.

The Biochemistry of H_2 Oxidation

The process of hydrogen oxidation involves the activation of H_2 by
hydrogenase and electron transport to O_2. The biochemical characterization of this complex series of reactions is incomplete. Further
efforts to define the biochemical processes involved in H_2 oxidation are
essential for the characterization of the genetic determinants of the H_2
oxidation system.

The membrane-bound hydrogenase from Rhizobium japonicum has been
purified from autotrophically cultured cells (Harker et al., 1984) and
bacteroids (Arp, 1985). Hydrogenase from both sources is composed of
two subunits with molecular weights near 60,000 and 30,000. Purified
hydrogenase contains Ni as demonstrated by direct analysis (Arp, 1985)
and by [63]Ni labelling (Harker et al., 1984). Stults et al. (1984)
detected [63]Ni labelled hydrogenase from a hydrogenase uptake constitutive mutant of R. japonicum by two-dimensional electrophoresis.

Subunit analysis of the labelled polypeptide yielded only one polypeptide (M_R 67,000 (Stults et al., 1984). Purified preparations of hydrogenase contained no detectable cytochrome b_{559}, although the presence of this cytochrome in R. japonicum cells has been correlated with hydrogenase activity (Eisbrenner et al., 1983).

The process of electron transport from H_2 to O_2 is complex and involves carriers in normal respiratory pathways. A flavoprotein (O'Brian, Maier, 1983) and b- and c-type cytochromes (Eisbrenner, Evans, 1982a; O'Brian, Maier, 1983) have been implicated in H_2 oxidation. The involvement of ubiquinone has been indicated by the inhibition of H_2 uptake by inhibitors of ubiquinone oxidation-reduction (Eisbrenner, Evans, 1982a) and by direct reduction of ubiquinone by H_2 during electron transport (O'Brian, Maier, 1985b).

The proximal acceptor of electrons from reduced hydrogenase has not been identified. What is believed to be a unique cytochome b_{559} has been detected in Hup bacteroids (Eisbrenner, Evans, 1982b) and chemolithotrophically cultured cells (Eisbrenner et al., 1982) of R. japonicum. These authors demonstrated a positive correlation (r=0.98) between hydrogenase activity and the concentration of this cytochrome. Cytochrome b_{559} was not detected in Hup⁻ point mutants or Hup⁻ wildtype strains. Recent unpublished work in our laboratory has demonstrated the presence of this H_2 reducible cytochrome in membranes isolated from chemolithotrophically grown cells. H_2-dependent reduction of cytochrome b_{559} occurs in the presence of cyanide. The physiological significance of this component has not been established.

O'Brian and Maier (1985a) demonstrated that constitutive mutants of R. japonicum contain elevated levels of cytochrome-o. They have further shown that under heterotrophic growth conditions the hydrogenase expressed in these mutants is unable to reduce any cytochrome in the presence of HQNO (O'Brian, Maier, 1985b), implying direct reduction of ubiquinone by hydrogenase. The failure to detect an H_2 reducible cytochrome b_{559} in a heterotrophically cultured mutant which expresses hydrogenase activity constitutively or in a Hup⁺ strain derepressed for hydrogenase expression, does not exclude the possible role of such a cytochrome in chemolithotrophically cultured cells or bacteroid forms of Hup⁺ wild-type strains of R. japonicum.

Beneficial Effects of Hydrogen Oxidation

The potential benefits of an efficient H_2 oxidation system to N_2-fixing organisms were outlined by Dixon (1972) and have been discussed in a review by Eisbrenner and Evans (1983) and by others. In addition to recovery of some of the energy expended during nitrogenase-dependent H_2 evolution and protection of nitrogenase from O_2 damage, Dixon (1972) considered the possibility that removal of H_2 by oxidation might prevent H_2 inhibition of nitrogenase. More recently, Dixon and Blunden (1983) have observed that the apparent fraction of nitrogenase electrons allocated to N_2 reduction in pea nodules was inversely related to nitrogenase activity under a variety of experimental conditions. They concluded that their observations could be explained by an inhibition by H_2 of N_2 fixation.

Minamisawa et al. (1983) have reported that the rate of transport of fixed N from roots of soybean plants nodulated by Hup⁺ strains was greater than that of plants nodulated by Hup⁻ strains and proposed that an operative H_2 recycling system affected the balance of carbon utilization and the nitrogen assimilation process in nodules. More recently Minamisawa et al. (1984) have made the interesting observation that nodules formed by several Hup^+ R. japonicum strains accumulated more than ten-fold higher concentrations of a compound X (possibly a polyamine) than nodules formed on soybean plants by Hup^- R. japonicum strains. This information indicates that the functioning of a H_2 oxidation system within nodules may affect a much broader spectrum of nodule metabolic processes than was initially imagined.

In experiments with Rhizobium ORS 571 which nodulates Sesbania, Stam et al. (1984) have found that the addition of H_2 to succinate limited N_2-fixing cultures of this bacterium resulted in an increase in the molar growth yield (Y succinate) from 27 to 35 and a slight reduction in the molar growth yield based on O_2 consumption (YO_2). Therefore, less energy per mole of substrate utilized was derived from H_2 oxidation than from substrate oxidation. Furthermore, DeVries et al. (1984) have concluded that only one of two sites of oxidative phosphorylation is utilized when H_2 is oxidized, an observation that is consistent with the decreased energy yields when H_2 instead of succinate is utilized.

It seems obvious that H_2 oxidation has the potential to benefit N_2 fixing organisms in several different ways. The system is so complex that conclusions about the advantages of H_2 recycling in vivo must be based on carefully planned growth experiments. Evans et al. (1984) have discussed the necessity to conduct legume growth trials with Hup^+ and Hup^- strains that are isogenic with the exception of the H_2 oxidation characteristic. Also, they have argued that plants in evaluation trials should be grown to maturity to provide the possibility for compounding of any benefits from H_2 recycling during the logarithmic phase of the accumulation of fixed N. In an experiment with soybeans, where plants were grown to maturity in one-meter tiles with 10 replicates of each treatment, the mean increases in the N contents of plants inoculated with a Hup^+ R. japonicum strain compared with data from plants inoculated with an otherwise isogenic Hup^- strain were: 8.6% for seed N and 11% for total N. These differences were significant at the 0.02 and 0.03 levels, respectively and are considered realistic measurements of the advantages of use of an R. japonicum strain capable of H_2 recycling.

Another experiment was conducted with the objective of evaluating any beneficial effect of the capability for chemoautotrophic growth in R. japonicum on the survival of R. japonicum in soil. As shown in Figure 1, the addition of 1% H_2 over soil inoculated with a Hup^+ strain of R. japonicum had a dramatic effect, increasing the numbers of R. japonicum from 1×10^7 to about 2×10^8. The difference persisted during most of the 160 day incubation period. The addition of H_2 to Hup^- R. japonicum in soil had no significant effect on numbers of cells in the soil during the incubation period. These results support the argument that possession of a capacity for autotrophic growth with H_2 as the energy source could benefit Rhizobium japonicum survival in soils.

Isolation and Characterization of hup Genes

Since hydrogen recycling capability is a desirable trait in a species of Rhizobium, as far as legume productivity is concerned, it is desirable to transfer hydrogen uptake (hup) genes to species of Rhizobium which lack this ability. Such gene transfer presupposes the isolation of all hup genes from a strain with an efficient H_2 recycling system. Considerable progress has been made in the fast-growing species where hydrogen recycling ability is plasmid encoded (Brewin et al, 1980; De Jong et al., 1982; Kagan, Brewin, 1985; Behki et al, 1985). In

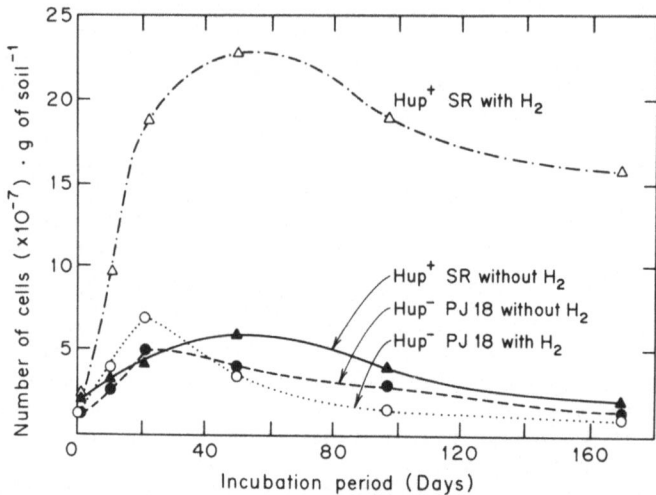

Fig. 1. The effect of H$_2$ on the survival of Hup$^+$ R. japonicum SR and Hup$^-$ mutant PJ18 in sterile soil. Three g samples of gamma irradiated soil adjusted to field capacity were placed in a series of test tubes. Each tube in one series was inoculated with 10^7 R. japonicum SR and a second series inoculated with 10^7 R. japonicum PJ18. One-half of each series containing either Hup$^+$ R. japonicum or Hup$^-$ PJ18 in sealed jars was treated with 5% CO$_2$, 1% O$_2$ and the remainder N$_2$ while the other half of each of the series was treated with 5% CO$_2$, 1% O$_2$, 1% H$_2$ and the remainder N$_2$. The jars with inoculated tubes of soil were incubated at 21° C at night and 27° C during the day. Samples were removed and numbers from R. japonicum counted at times indicated. Results plotted are means of three replicate samples of soil.

contrast, in the slow-growing species several R. japonicum Hup$^+$ strains showed no detectable plasmid DNA (Cantrell et al., 1983). Consequently, a gene bank of R. japonicum USDA122DES was constructed in a cosmid pLAFR1 (Cantrell et al., 1982). The gene bank was transferred into a Hup$^-$ R. japonicum point mutant (PJ17nal) by conjugation and cosmids which contained hup genes were identified by a rapid colony screening method. Hup$^+$ R. japonicum colonies can be distinguished visually by their localized decolorization of filter paper impregnated with methylene blue in the presence of respiratory inhibitors (Haugland et al., 1983). Eleven cosmids which complemented the Hup$^-$ mutant PJ17 showed 3 DNA fragments in common after digestion with EcoR1 (Cantrell et al., 1983). One of these cosmids, pHU1, was subjected to transposon Tn5 mutagenesis in order to determine the extent of hup-specific DNA. The Hup phenotype conferred by each Tn5 insertion was determined after integration of these mutated genes into the R. japonicum USDA122DES chromosome. The results indicate that hup-specific sequences spanned about 16 kb of DNA within pHU1 (Haugland et al., 1984). Complementation studies indicated the existence of at least two transcriptional units. Despite the extent of hup-specific DNA within pHU1, several lines of evidence suggested that additional hup specific DNA remained to be isolated. pHU1 complements only 2 out of 6 Hup$^-$ mutants, although it returns hup activity to 3 of these mutants at low frequency (Haugland et al.,1984). Acquisition of pHU1 by Hup$^-$ wild-type R. japonicum and R.

meliloti strains did not confer Hup activity in the free-living state. Cantrell et al. (1985) isolated an additional hup cosmid from the gene bank by complementation of one of the Tn5 generated Hup⁻ mutants. One of the new cosmids isolated, pHU52, is similar to pHU1 except that it possesses a 5.5 kb EcoR1 fragment at the righthand end. By contrast with pHU1, pHU52 complemented all but one of the Hup⁻ mutants and is therefore a promising cosmid for intraspecies transfer of hydrogenase activity (Cantrell et al, 1985). Recently, cosmids which contain both nif and hup genes have been isolated from an R. japonicum USDA110 gene bank (Hom et al., 1985). Whether these cosmids are homologous to pHU1 remains to be determined.

Transfer and Expression of Cosmid-Borne R. japonicum hup Genes in the Free-Living State

Acquisition of pHU52 by 4 Hup⁻ wild-type R. japonicum strains, USDA 16, USDA 117, USDA 138, and USDA 144, conferred upon these strains the ability to perform the oxyhydrogen reaction after derepression in the free-living state (Lambert et al., 1985). Likewise, three Hup⁻ R. meliloti strains and R. leguminosarum 128C53 were converted to a Hup⁺ phenotype after acquiring pHU52 by conjugation. Hydrogen oxidation in these transconjugants was coupled to ATP formation as demonstrated by the ability of these strains to grow autotrophically using hydrogen gas as the sole energy source (Lambert et al., 1985). In these experiments, the cosmid pHU52 is maintained in the cells by antibiotic selection pressure in the case of heterotrophic cultures, and by its essentiality for growth in the case of autotrophic cultures.

Symbiotic Expression of Cosmid-Borne hup Genes

Although studies with free-living bacteria have indicated that Hup activity can be transferred across species boundaries, the important question from the viewpoint of biological nitrogen fixation is whether these cosmid borne hup genes are expressed in nodules. Several reports indicate that the expression of Hup activity in nodules can depend upon the particular plant host (Keyser et al., 1982; Bedmar et al., 1983) and it was therefore of interest to determine whether Hup activity encoded by pHU52 would be expressed not only in soybean, but also in alfalfa and clover nodules. Using dense bacteroid suspensions and methylene blue as the electron acceptor we have detected low levels of Hup activity in nodules formed by R. japonicum USDA 138 (pHU52), R. meliloti 102F28 (pHU52) and R. trifolii SU794 (pHU52) (Table 1). The low levels of Hup activity are mostly due to the instability of the cosmid as reflected by the low levels of tetracycline resistance in bacteria isolated from the nodules (Table 1). Nevertheless, the data does indicate that conditions in soybean, alfalfa and clover nodules are conducive to the expression of the R. japonicum cosmid-borne hup genes. The instability of pHU52 varies greatly between different strains. In R. japonicum USDA 123 Spc, a Hup⁻ strain, 20-25% of bacteria isolated from nodules formed by USDA 123 Spc(pHU52) retain the tetracycline resistance marker, and in these nodules 70% of the nitrogenase-mediated hydrogen formation is recycled.

Further Genetic Research

In order to produce genetically engineered Hup⁺ strains capable of recycling all nitrogenase mediated H_2, the cosmid-borne hup genes must

Table 1. Expression of cosmid-borne hup genes in legume nodules

Inoculum	Host	H_2 uptake rate[1] of bacteroids (pmol.min^{-1}.mg protein^{-1})	Fraction of cells showing tetracycline resistance[2]
R. japonicum USDA138 (Hup$^-$)	Glycine max (cv. williams)	<20[3]	0
R. japonicum USDA138 (pHU52)	Glycine max cv. williams	168	3.9×10^{-4}
R. meliloti 102F28 (Hup$^-$)	Medicago sativa (cv. Vernal)	<20	0
R. meliloti 102F28 (pHU52)	Medicago sativa (cv. Vernal)	241	2.2×10^{-3}
R. trifolii SU794 (Hup$^-$)	Trifolium incarnatum	<20	0
R. trifolii SU794 (pHU52)	Trifolium incarnatum	925	6.8×10^{-3}

1. Determined amperometrically using methylene blue as the electron acceptor.
2. All tetracycline resistant cells were Hup$^+$ as determined by the methylene blue colony assay (Haugland et al., 1983).
3. Data are means from 3 replicates from soybean and clover and from 1 replicate for alfalfa. Soybean plants were harvested after 25 days growth, alfalfa after 38 days and clover plants after 45 days.

be stabilized in their new environment. This could be accomplished using new generations of more stable plasmid vectors (Hom et al., 1985), or by using an integrating plasmid to deliver these hup genes into the chromosome. An understanding of how the hup genes are turned on and off at the molecular level is essential in order to successfully transfer and stabilize these genes in Rhizobium spp. lacking the hydrogen uptake genes. To this end, experiments have been designed to identify the structural genes for the 60 kd and 30 kd polypeptide components of the hydrogen uptake enzyme (see Zuber et al., this conference).

Summary

The membrane-bound hydrogenase from Rhizobium japonicum is composed of two subunits with molecular weights near 60,000 and 30,000 and contains nickel. The concentration of a cytochrome b_{559} correlates with hydrogenase activity under different conditions. Hydrogenase activity is a desirable trait in R. japonicum in terms of legume productivity and ability to grow in soil under certain conditions.

Of a number of hup cosmids isolated, only pHU52 is able to confer Hup activity and autotrophic growth capability to diverse Hup$^-$ slow- and fast-growing Rhizobium species in the free-living state. The transferred <u>hup</u> cosmid is also expressed in alfalfa, clover and soybean nodules, the level of expression depending on the stability of the cosmid in the strain of <u>Rhizobium</u> used.

Acknowledgements

We thank Sheri Woods-Haas for typing the manuscript. This work was supported by the U.S. Department of Agriculture Competitive Research Grant 82 CRCR-1-1073, the National Science Foundation Research Grant PCM81-18148, the Agrigenetics Corporation, the Murdock Trust and the Oregon Agricultural Experiment Station from which this is Technical Paper no. 7623.

References

Adams MWW et al (1981) Biochim. Biophys. Acta 594, 105-176.
Arp DJ (1985) Arch. Biochem. Biophys. 237, 504-512.
Bedmar EJ et al (1983) Plant Physiol. 72, 1011-1015.
Behki et al (1985) Arch. Microbiol. 140, 352-357.
Brewin NJ et al (1980) Nature 288, 77-79.
Cantrell MA et al (1982) Arch. Microbiol. 131, 102-106.
Cantrell MA et al (1983) Proc. Natl. Acad. Sci. USA 80, 181-185.
Cantrell MA et al (1985) In Szalay AA and Legocki RP, eds, Advances in the Molecular Genetics of the Bacteria-Plant Interaction, pp. 64-66, Cornell University Press, New York.
Chatt J (1980) In Stewart WDP and Gallon JR, eds, Nitrogen Fixation, pp. 1-17, Academic Press, London.
DeJong TM et al (1982). J. Gen. Microbiol. 128, 1829-1838.
DeVries W et al (1984) Antonie van Leewenhoek, 50, 505-524.
Dixon ROD (1972) Arch. Microbiol. 85, 193-201.
Dixon ROD (1978) Biochimie, 60, 233-236.
Dixon ROD (1983) Plant and Soil 75, 131-138.
Dixon ROD and Blunden EAG (1983) Plant and Soil 75, 131-138.
Eisbrenner G and Evans HJ (1982a) J. Bacteriol. 149, 1005-1012.
Eisbrenner G and Evans HJ (1982b) Plant Physiol. 70, 1667-1672.
Eisbrenner G and Evans HJ (1983) Ann. Rev. Plant Physiol. 34, 105-136.
Eisbrenner G et al (1982) Arch. Microbiol. 132, 230-235.
Evans HJ et al (1984) World Soybean Research Conference III, Westview Press, Boulder CO (in press).
Evans HJ et al (1985) In Ludden PW and Burns JE, eds, Proceedings of the Fourteenth Steenbock Symposium, pp 1-11, Elsevier, New York.
Harker AR et al (1984) J. Bacteriol. 159, 850-856.
Haugland RA et al (1983) Appl. Environ. Microbiol. 45, 892-897.
Haugland RA et al (1984) J. Bacteriol. 159,1006-1012.
Hom SSM et al (1985) J. Bacteriol. 161, 882-887.
Kagan SA and Brewin NJ (1985) J. Gen. Microbiol. 131, 1141-1148.
Keyser HH et al (1982) Plant Physiol. 70, 1626-1630.
Lambert GR et al (1985) Proc. Natl. Acad. Sci. USA 82, 3232-3236.
Minamisawa K et al (1983) Soil Sci. Plant Nutr. 29, 85-92.
Minamisawa K et al (1984) Soil Sci. Plant Nutr. 30, 435-444.
O'Brian MR and Maier RJ (1982) J. Bacteriol. 152,422-430.
O'Brian MR and Maier RJ (1983) J. Bacteriol. 155, 481-487.
O'Brian MR and Maier RJ (1985a) J. Bacteriol. 161, 507-514.
O'Brian MR and Maier RJ (1985b) J. Bacteriol. 161, 775-777.
Robson RL and Postgate JR (1980) Ann. Rev. Microbiol. 3, 83-207.
Simpson FB and Burris RH (1984) Science 224, 1095-1097.
Stam H et al (1984) Arch. Microbiol. 139, 53-60.
Stults LW et al (1984) J. Bacteriol. 159,153-158.

ISOLATION, CHARACTERIZATION AND EXPRESSION OF
BRADYRHIZOBIUM JAPONICUM nif AND GLUTAMINE SYNTHETASE GENES

B.K. Chelm, T.A. Carlson and T.H. Adams
DOE-Plant Research Laboratory, Michigan State University, East Lansing,
MI 48824 USA

Bradyrhizobium japonicum has two species of glutamine synthetase (GS) designated GSI and GSII. The gene encoding GSI, *gln*A, has been isolated and characterized. GSI is both homologous and analogous to the glutamine synthetase found in other bacteria and is not significantly regulated by transcriptional control; its regulation being mostly posttranslational. The enzyme GSII is unique to bacteria of the Rhizobiaceae family and is highly regulated in response to changes in nitrogen source, carbon source, oxygen, and symbiotic development. We have set out to clone the gene encoding GSII. The enzyme GSII was purified 27 fold from crude extracts of *B. japonicum* USDA 110 using AffiGel Blue affinity chromotography and DEAE ion exchange HPLC as the major fractionation steps. The amino acid sequence of the amino terminus was determined by automated gas phase Edman degradation. The amino terminal sequence is Met-Thr-Lsy-Tyr-Lys-Leu-Glu-Tyr-Ile-Trp-Leu-Asp-Gly-Tyr-Thr-Pro-Thr-Pro-Asn-Leu-. Using the amino acid sequence and the genetic code as a guide, we synthesized a mixture of oligo-nucleotides that would hybridize to a 17 base region of the gene encoding GSII. Screens of genomic Southerns and a genomic cosmid library have identified one region of the *B. japonicum* chromosome. This isolated region is now being further analyzed for the presence of a gene encoding GSII.

We have been studying *nif* genes from *B. japonicum*. Three transcription units have been studied in greatest detail; *nif*DK, *nif*B and *nif*H, organiz-ed in that order within a region of about 22 kbp. Promoters for these three transcription units have been mapped by S1 protection analyses and show a very high degree of homology in the region of -34 to +1 nucleotides relative to the start of transcription. The consensus sequence for this region is 5' t-agaCC-tGGCAtGcc-GTTGCtaAt-ctcgg-a 3', where the capitol letters represent bases which are conserved in all three promoters and lower case letters represent bases found in only two of the three. Expression studies have been carried out on these three transcription units. All three are of course expressed in bacteria isolated from nodules. They are also induced in free-living cultures grown under oxygen-limiting conditions, however, the inducibility differs somewhat between the differ-ent promoters and is currently being further defined.

B. japonicum contains 5 or 6 gene regions with significant homology to *K. pneumoniae nif*A which encodes a positive regulator of *nif* gene express-ion. One of these, Hnal, has the least homology and is located upstream of *nif*B. A *B. japonicum* strain has been constructed in which this gene contains a deletion mutation. This mutant strain incites nodules of wild-type appearance which can reduce acetylene, however the plants appear no healthier than uninoculated plants. The molecular detail of this defect-ive symbiosis is being further characterized. A second *nif*A-homologous gene, Hna2, which has greater than 60% homology to the *K. pneumoniae nif*A over about 600 bp, has also been mutated in *B. japonicum* yielding a strain with no detectable phenotype. *B. japonicum* strains mutated in the other *nif*A homologies are currently being constructed in the search for the true *nif*A gene.

CLONING AND CHARACTERIZATION OF GLUTAMINE SYNTHETASE GENES FROM RHIZOBIUM MELILOTI 1021, R. SESBANIA ORS571 AND AGROBACTERIUM TUMEFACIENS C58

Frans J. de Bruijn[*], Silvia Rossbach and Jeff Schell

Max Planck Institut, 5000 Köln 30, West Germany.

Introduction and Results.

Rhizobia and agrobacteria are well known for their symbiotic/parasitic interactions with plants. Both types of interactions, wether rhizobial nitrogen fixation or agrobacterial opine catabolism in crown galls, involve synthesis, catabolism and assimilation of nitrogenous compounds. Glutamine synthetase (GS) plays an essential role in nitrogen assimilation in both rhizobia and agrobacteria, which have been shown to have two distinct forms of the enzyme (GSI,II). We would like to examine how the glnA (GSI,II) genes are regulated in relation to rhizobial nitrogen fixation (nif) and agrobacterial nopaline (arginine) catabolism (noc, arc, aut) genes in the free living vs. symbiotic (parasitic) state, comparing this to Klebsiella pneumoniae, where nif, glnA and aut genes have been shown to be coordinately regulated by a central nitrogen regulation (ntr) system (de Bruijn, Ausubel 1981;1983; Merrick 1983). Moreover we would like to determine if indeed GSI or II plays a role in nif regulation in rhizobia as suggested (Ludwig, Signer 1977; Kondorosi et al. 1977). R. sesbania ORS571 was included in the compaitive study because of its unique capacity to be a diazotroph (Dreyfus, Elmerich 1983), making a proper comparison of nif and glnA regulation in free living vs. symbiotic state possible.

Clone banks of Rm1021, RsORS571 and AtC58 were constructed and used to isolate the GSI structural gene by complementation of a Gln⁻ E. coli. These glnAI cosmids encode an essentially heat stable GS enzyme activity and hybridize strongly to oneanother and to the glnA(GSI) gene from Rm104A14 (Sommerville, Kahn 1983). The AtC58 glnAI region encodes a 57kd protein in maxicells (= expected M. wt. of GSI).

From the Rm1021 cosmid bank a second region was identified, also able to complement Gln⁻ E. coli strains, which carries no homology to the glnAI regions, to Rm104A14 glnA(GSI) or the K. pneumoniae glnA gene (de Bruijn, Ausubel 1981). It encodes an essentially heat labile GS (biosynthetic) enzyme activity and carries a 4.5kb long, complex operon which synthesizes among others a 35kd protein (= expected M.wt. of GSII). Tn5 insertions in the glnAII region of Rm1021 give rise to a Gln⁺ Nod⁺ Fix⁺ phenotype, suggesting that the presumptive GSII region plays no role in nif regulation. The Rm1021 glnAII region was found to be highly conserved and used to isolate the analogous regions from RsORS571 and AtC58. The AtC58 glnAII region was found to be able to complement Gln⁻ E. coli strains, like its Rm1021 counterpart, but only after strong selection during which minor rearrangements were induced in the cloned (promoter?) region.

References.

de Bruijn FJ and Ausubel FM (1981) Mol. Gen. Genet. 183, 289-297.

de Bruijn FJ and Ausubel FM (1983) Mol. Gen. Genet. 192, 342-353.

Dreyfus BL, Elmerich C and Dommergues YR (1983) Appl. Envir. Microb.45,711

Kondorosi A, Svab Z, Kiss GB and Dixon RA (1977) Mol. Gen. Genet.151, 221

Ludwig RA and Signer ER (1977) Nature 267, 245-248

Merrick M (1983) EMBO Journal 2, 39-44

Somerville JE and Kahn ML (1983) J. Bacteriol. 156, 168-178

CLONING OF GENES INVOLVED IN GLUTAMINE SYNTHESIS IN RHIZOBIUM PHASEOLI

GUADALUPE ESPIN, SOLEDAD MORENO AND MARINA WILD.
CENTRO DE INVESTIGACION SOBRE FIJACION DE NITROGENO. UNIVERSIDAD NACIONAL
AUTONOMA DE MEXICO.

Bacteria of the genus Rhizobium posses two forms of glutamine synthetase
GS: GSI similar to that of enteric bacteria, and the heat labile GSII
specific to the Rhizobeacea. In a previous study, we described the isolation of two R.phaseoli mutants CFN2012 and CFN2017 that lost GSII transferase activity (1), further studies of these mutants have shown that they
retained GSII biosynthetic activity(J.Mora personal communication). To
further study genes involved in glutamine synthesis, we have cloned the
wild type sequences mutated in CFN2012 and CFN2017 as well as three other
sequences that complemented the glutamine auxotrophy of UNF1811, a glnA
mutant of Klebsiella pneumoniae.
Plasmid p2012 carries the wild type sequences mutated in strains CFN2012
and CFN2017, It was isolated from a cosmid bank of the wild type strain
CFN42 constructed in vector pSUP205. p2012 restored GSII transferase activity of CFN2012 and CFN 2017 but did not confered UNF1811 the ability to
grow in the absence of glutamine.
Two cosmids p240 and p320 able to complement the glutamine auxotrophy of
UNF1811 were also isolated. We also cloned a 12 kb BamHI fragment of
strain CFN42 that showed homology to pFB6164 a plasmid carrying a glnA
gene from R.meliloti (2), pMW2 containing this fragment confered to UNF
1811 the ability to grow in the absence of glutamine.
GS biosynthetic and transferase activities of the UNF1811 derivatives
harbouring pMW2, p240 or p320 were determined: UNF1811 pMW2 showed high
levels of GS biosynthetic activity, but extremely low levels of transferase, in contrast UNF1811 p320 showed high levels of the transferase activity and low levels of the biosynthetic. UNF1811 p240 had low levels of
both activities. None of the activities detected in the three strains
were heat labile.
Strains CFN2012 and CFN2017 carry mutations leading to the lost of GSII
transferase activity, the fact that plasmid p2012 carrying the wild type
sequences mutated in these strains did not complement strain UNF1811, together with the finding that these mutants still have GS biosynthetic
activity suggest that mutations in CFN2012 and CFN2017 are not in the
structural gene for GSII.
The ability of UNF1811 to grow in the absence of glutamine when carries
plasmids pMW2, p240 or p320 indicates that we have cloned a gene o genes
coding for GS. The heat stability of the GS activities in these strains
may indicate that the three plasmids carry a gene for GSI, if this is the
case the differences in the levels of transferase and biosynthetic activities remain to be explained. We may also have cloned the genes for both
GSI and GSII, the phenotype of UNF1811 pMW2 resembles that of mutants
CFN2012 and CFN2017 in their lack of transferase activity, suggesting
that pMW2 may carry a gene for GSII, the heat stability of the GS activity in pMW2 remains also to be explained.

REFERENCES

1. Morett et al (1985) Mol. Gen. Genet. in press
2. de Bruijn et al (1984) In Veeger C and Newton WE, eds, Advances in
Nitrogen Fixation Research, Nijhoff/Junk Pudoc, Wageningen p

BACTERIAL CYCLIC AMP AND HEME IN THE
BRADYRHIZOBIUM JAPONICUM/SOYBEAN SYMBIOSIS

Mary Lou Guerinot and Barry K. Chelm. DOE Plant Research Laboratory and
the Department of Microbiology and Public Health, Michigan State
University, East Lansing, MI 48824

Regulation of gene expression in the Bradyrhizobium japonicum/soybean
symbiosis is likely to involve the exchange of specific chemical signals
between the symbiotic partners. We are currently investigating the possibilities that cyclic AMP and/or heme may be effectors for integrating
plant and bacterial metabolism. To facilitate this analysis, we have
constructed mutants of Bradyrhizobium japonicum strain USDA 110 in which
the DNA sequences encoding enzymes necessary for either cyclic AMP synthesis or heme synthesis have been deleted by a gene-directed
mutagenesis technique. The mutant in cyclic AMP synthesis, MLG 10, was
constructed using a previously isolated fragment of bradyrhizobial DNA
which had been shown to encode an adenylate cyclase activity.
Surprisingly, cyclic AMP could still be detected in cultured cells of
this mutant strain. The possibility that this bacterium has isozymes of
adenylate cyclase is being investigated. No phenotype has yet been
detected for this mutant, either in free-living culture or on plants.
To construct the mutant in heme biosynthesis, we first isolated the gene
encoding 5-aminolevulinic acid synthase (hemA) from a cosmid library of
B. japonicum DNA using a fragment of the R. meliloti hemA gene as a
hybridization probe. The B. japonicum mutant strain, MLG 1, can no
longer carry out the first step in heme biosynthesis. It has no
5-aminolevulinic acid synthase activity and is unable to grow in minimal
medium unless 5-aminolevulinic acid is added. Despite this auxotrophy,
the nodules incited by the mutant strain appear normal, contain heme and
are capable of high levels of acetylene reduction. Bacterial extracts
prepared from these nodules have no detectable ALAS activity, in accord
with the free-living phenotype of this strain. As previous studies have
suggested that the formation of the heme moeity of leghemoglobin is
exclusively a function of the bacterial symbiont, the apparent rescue of
the B. japonicum hemA mutant by the soybean plant suggests that the
regulation of the synthesis of the heme moeity itself as well as the
role heme plays as an effector may be more complex than previously
envisioned.

ALTERED NITROGENASE REGULATION IN NITRATE REDUCTASE OVERPRODUCING
MUTANTS OF PARASPONIA RHIZOBIUM STRAIN ANU289

SUSAN M. HOWITT/KIERAN F. SCOTT*/PETER M. GRESSHOFF.
BOTANY DEPARTMENT,*CENTRE FOR RECOMBINANT DNA RESEARCH, AUSTRALIAN
NATIONAL UNIVERSITY, PO BOX 4, CANBERRA, ACT, 2601 AUSTRALIA

We have attempted to study nitrogen control in Parasponia
Rhizobium strain ANU289 by using the ammonia analogue, methylamine.
Methylamine represses several enzymes involved in nitrogen assimilation
in a manner similar to ammonia (Howitt, Gresshoff, 1985). Nitrate
reductase was one such enzyme. Transposon mutants in which nitrate
reductase was insensitive to repression by methylamine were isolated.
These showed higher levels of nitrate reductase than the wild type in
both the presence and absence of ammonia. The mutants did not appear
to be affected in general nitrogen control - other enzymes examined
(glutamine synthetases I and II, histidase and an ammonia carrier)
were present at the same levels as the wild type. However, in agar
cultures, in the absence of ammonia, the mutants showed higher
than normal nitrogenase activity. Ammonia completely repressed
nitrogenase in both the mutants and the wild type. We are continuing
to investigate nitrogenase regulation in these mutants.

REFERENCES
Howitt SM and Gresshoff PM (1985) J. Gen. Microbiol. 131, 1433-1440.

COSMID CLONING OF DNA FROM RHIZOBIUM SP WR1001 AND IDENTIFICATION OF THE GENE CODING FOR GLUTAMATE SYNTHASE

SUI-SHENG T. HUA AND DAVID A. LAWRENCE
WESTERN REGIONAL RESEARCH CENTER, ARS-USDA, ALBANY, CA 94710, USA

1. INTRODUCTION

The glutamate synthase (GOGAT) of salt-tolerant Rhizobium sp WR1001 has apparently evolved to possess some unique properties on adaptation to saline environments. Of particular interest are the facts that the NADPH-dependent GOGAT has an alkaline pH optimum of 8.2-8.3 and its activity is no longer inhibited by high concentrations of glutamate up to 300mM (Hua et al, 1984). Since glutamate dehydrogenase (GDH) is negligible in R. sp WR1001, GOGAT may be particularly important for ammonia assimilation and glutamate production under salt stress (Hua et al, 1982). To understand further the nature of the glutamate synthase gene in R. sp WR1001 and to establish its relationship to salt tolerance, nitrogen metabolism and nitrogen fixation, we have cloned the gene for molecular genetic analysis.

2. PROCEDURE

R. sp WR1001 DNA was extracted by phenol and purified by cesium chloride-ethidium bromide density gradient. Purified DNA was partially digested by the restriction endonuclease EcoRl and size fractionated on a sucrose gradient (Ditta et al, 1980). DNA fragments in the size range of 10-20 Kb were cloned onto the cosmid vector pLAFR1 and transfected into E. Coli DH1 (Friedman et al, 1982).

3. RESULTS

A total of 1,500 colonies were pooled and stored as the genomic library of R. sp WR1001. Selected colonies were analyzed and 90% of them contained fragments of WR1001 DNA. The cloned gene library was mass-conjugated into Salmonella tyhpimurium JB2112 (Δ glt) in a triparental mating using DH1 (pRK2013) as a source of mobilizing plasmid. Transconjugants of JB2112 which appeared to be Glt+, were purified. The GOGAT (glt) gene of WR1001 is within an 8.3 Kb DNA insert in the hybrid plasmid pHL1. A physical map of pHL1 was constructed using restriction endonucleases EcoR1, Pstl, HindIII, BglII and BamH1. Expression of The cloned glt gene of WR1001 in JB2112 results in an enzyme with similar properties and Km values to that from WR1001. The glt gene hs been subcloned onto a multicopy plasmid, pBR325, to facilitate enzyme purification and other molecular genetic studies.

4. REFERENCES

Ditta G et al (1980) Proc Natl Acad Sci USA 77, 7347-7351.
Friedman AM et al (1982) Gene 18, 289-296.
Hua SST et al (1982) Applied and Environmental Microbiology 44, 135-140.
Hua SST et al (1984) In Veeger C and Newton WE, eds, Advances in Nitrogen Fixation Research, pp 263 Matinus Nijhoff/Junk, The Hague.

CHARACTERIZATION OF AUXOTROPHS OF RHIZOBIUM MELILOTI 104A14 AND CLONING OF THE GENES FOR CARBAMYLPHOSPHATE SYNTHETASE

TOM K. KERPPOLA/MICHAEL L. KAHN*
DEPARTMENTS OF BIOCHEMISTRY AND MICROBIOLOGY*, WASHINGTON STATE UNIVERSITY, PULLMAN, WA 99164-4340, USA

R. meliloti 104A14 was mutagenized with nitrous acid and treated with penicillin to enrich for auxotrophs. Auxotrophic mutants were tested for growth on different amino acids, nucleosides, nucleotide bases and vitamins, and classified by their growth factor requirements. Mutations in ornithine transcarbamylase, argininosuccinate synthetase, and serine or glycine biosynthesis do not affect the symbiosis with alfalfa. Mutations in carbamylphosphate synthetase (CS), ornithine, pyrimidine, purine, asparagine, leucine, tyrosine, and methionine biosynthesis lead to ineffective symbiosis as do mutations in one-carbon metabolism. Complementation of the auxotrophic mutants by a P2 cosmid gene bank of R. meliloti 104A14 was used to clone the genes affected by many of these mutations. Effectiveness was restored in prototrophic revertants and in mutant strains that carried complementing plasmids.

Some arginine auxotrophs require a pyrimidine and either citrulline or arginine for growth and cannot make carbamylphosphate, a key nitrogen metabolite. These presumably have a defect in CS. Based on the ability of plasmids that complement the CS mutants to hybridize to E. coli carA and carB DNA, we have named the R. meliloti genes carA and carB. We believe that they correspond to the light and heavy subunits of the enzyme. Unlike the E. coli genes, the Rhizobium genes were not linked to each other and therefore are not organized in an operon. Three complementation groups were determined by the cross-complementation behavior of the CS mutants and plasmids. The carA group was complemented by the pTK11 series of plasmids. The two complementation groups in carB overlap since the pTK12 series of plasmids complemented all of the carB mutants but the pTK13 plasmids complemented only a subset of these. Some of the pTK12 plasmids do not have any Rhizobium DNA in common with the pTK13 plasmids.

We consider intragenic complementation to be the best explanation for this complementation pattern. In E. coli, the carB region codes for only one protein but this protein has a very large internal duplication (Nyunoya and Lusty, 1983). Although E. coli CS can form oligomeric associations, the monomer is active and is sensitive to allosteric effectors. These observations are consistent with a model in which the enzyme has two similar domains that act cooperatively. The two domains on separate polypeptides might interact to restore activity. Intragenic complementation could also occur if intermediates, like carbamate or carbonyl phosphate, were released from one mutant protein and metabolized by a protein with a complementary defect.

REFERENCES

Nyunoya, H. and C.J. Lusty (1983) Proc. Nat.Acad.Sci. (USA) 80, 4629.

MOLECULAR GENETICS OF c-AMP SYNTHESIS IN RHIZOBIUM MELILOTI

F. O'GARA, R. LATHIGRA, M. O'REGAN, B. KIELY and B. BOESTEN,
MICROBIOLOGY DEPARTMENT, UNIVERSITY COLLEGE, CORK, IRELAND.

INTRODUCTION

The formation of nitrogen fixing root nodules between Rhizobium spp.
and its respective host legume depends on the differential expression of
genes in both symbiotic partners. The signals controlling gene expression in Rhizobium during this process are not well understood. In other
Gram negative bacteria cAMP is an important effector molecule
controlling gene expression (1, 2). cAMP is detectable in R. meliloti
and the recent isolation of a DNA fragment encoding adenyl cyclase
sequences (3) provides the framework to elucidate the role of this
nucleotide in Rhizobium.

RESULTS AND DISCUSSION

The R. meliloti adenyl cyclase gene was encoded on a 5.27 Kb BglII
fragment cloned in pRK290. Two large subfragments (A and B) are
produced in double digests of this 5.27 Kb fragment with BglII and
BamHI restriction endonucleases. Subcloning experiments in pBR322
indicated that the cya gene is located on the 2.2 Kb BamHI fragment
(i.e. B-fragment) and depends on the tetracycline promoter of pBR322 for
expression. These results suggest that the cya promoter is located on
the 2.6 Kb BglII - BamHI A fragment. This was confirmed by a number of
Tn5 insertions in the 5.27 Kb BglII fragment that resulted in a Cya
negative phenotype and some of these mapped to the BamHI end of
fragment A. Cya related promoter activity on fragment A was
demonstrated by the construction of gene fusions with MudI (Ap^r lac)
phage and the promoter probe vector pMC1403 (4). The results indicate
that the cya related promoter and translation initiation sites are
recognised in an E. coli background and that transcription proceeds from
the BglII to the BamHI site on fragment A. 'In vivo' transcription-
translation experiments in E. coli maxicells demonstrated the presence
of a 28 K polypeptide when the B fragment was expressed from the
chloramphenicol acetyltransferase promoter in plasmid pACYC184 and
complemented Δcya mutations. This product was not observed in the
non-complementing reverse orientation. Results from DNA hybridization
experiments indicated that although both A and B fragments are highly
conserved in R. meliloti strains no evidence of significant sequence
homology was detected in other rhizobial species or E. coli under the
conditions employed. This may indicate that there are differences in
the adenyl cyclase protein among Rhizobium species.

REFERENCES

1. Botsford JL (1981) Microbiol. Rev. 45: 620-642
2. Ullman A and Danchin A (1983) In : Greingard P and Robison GA, eds
    Adv. Cyclic Nucleotide Res. 15, Raven Press, New York pp. 1-53
3. Kiely B and O'Gara F (1983) Mol. Gen. Genet. 192: 230-234
4. Casadaban et al (1980) J. Bacteriol. 143: 971-980

SYMBIOTICALLY DEFECTIVE HISTIDINE AUXOTROPHS OF BRADYRHIZOBIUM JAPONICUM

Prakash Sista, Katalin Rostas, Michael Sadowsky and Desh Pal S. Verma, Centre for Plant Molecular Biology, Biology Department, McGill University, Montreal, CANADA H3A 1B1.

Symbiotically defective auxotrophic mutants have been isolated from several fast-growing rhizobia (1); however, little information is available about the effects of general metabolic mutations on symbiotic functions of the slow-growing rhizobia. Mutagenesis procedures have been successful in inducing defined mutations in metabolic and symbiotic functions in Bradyrhizobium japonicum (2,3,4). This report describes the symbiotic defect of histidine auxotrophs of B. japonicum USDA 122.

Four histidine requiring auxotrophs of B. japonicum USDA 122 were isolated by random Tn5 mutagenesis. The mutants grew on minimal medium supplemented with L-histidine or L-histidinol but failed to grow with L-histidinol phosphate. Reversion to prototrophy occured at a frequency of 10^{-7} on minimal medium without antibiotics but prototrophs could not be isolated from minimal medium containing 100 ug/ml kanamycin and streptomycin. Two of the four mutants were symbiotically competent, while the other two were defective and failed to form nodules on Glycine max cvs. Lee and Peking and on G. soja. The prototrophic revertants were capable of forming effective symbioses on all soybean cultivars examined. When histidine was supplied to the plant growth medium, both nodulation defective mutants formed effective symbioses. Occasionally, the nodulation defective mutants could form effective nodules on histidine unamended plants. Three classes of bacteria were isolated from such nodules: class 1 was kanamycin-resistant auxotrophs; class 2 was kanamycin-sensitive prototrophs and class 3 was kanamycin-sensitive auxotrophs. Our results suggest that two Tn5 insertion mutations in B. japonicum USDA 122 leading to histidine auxotrophy affect nodulation in some way. These mutations are in regions that show no homology to any of the Rhizobium meliloti "common" nodulation genes.

(1) Schwinghamer EA (1974) In Hardy RWF, ed, Dinitrogen Fixation, vol. 3, pp 577-622, John Wiley and Sons, New York.
(2) Wells SE and Kuykendall DL (1983) J. Bact. 156, 1356-1358.
(3) Hom SSM et al (1984) J. Bact. 159, 335-340.
(4) Rostas K et al (1984) Mol. Gen. Genet. 197, 230-235.

MALATE DEHYDROGENASE-DEFICIENT MUTANTS OF RHIZOBIUM JAPONICUM

JAMES K. WATERS/GLENN G. PRESTON/RONG-TI LIANG/DAVID W. EMERICH.
DEPARTMENT OF BIOCHEMISTRY, UNIVERSITY OF MISSOURI, COLUMBIA, MO 65211, USA

The supply of photosynthate to the bacteroid endophyte is thought to be a major limitation to nitrogen fixation (Havelka, 1975). The translocated photosynthate is mainly composed of sugars but the bacteroids have been found to have little glucose-, fructose-, sucrose-, lactose-, or ribose- dependent oxygen consumption which indicates bacteroids are unable to metabolize these sugars efficiently (Glenn et al 1981). The pentose phosphate pathway and phosphofructokinase are not present in Rhizobium japonicum bacteroids (Reibach et al, 1983). Rhizobium japonicum bacteroids do possess a functional Enter-Doudoroff pathway and a functional tricarboxylic acid cycle (Kurz et al, 1977).

The concentration of organic acids (succinate, fumarate, malate) in legume nodules has been measured in the mM range (Stumpf et al, 1981; Reibach et al, 1983). When these organic acids are added to suspensions of bacteroids, oxidative respiration and nitrogen fixation (acetylene reduction) are stimulated significantly (Glenn et al, 1981; Mulongoy et al, 1977; Keele et al, 1969; Argueso et al, 1979). Malate is likely to be a main source of the C-4 dicarboxylic supply since it is found in high concentration in many symbiotic legume nodule relationships (Devries, 1980; Antonie et al, 1978; Stumpf et al, 1979).

Methods: Malate dehydrogenase-deficient mutants of Rhizobium japonicum 3I1b-143 were isolated after Tn5 mutagensis. Generation of mutants was accomplished by conjugation of the R. japonicum rif^R nal^r recipient with the E. coli SM10 donor containing the plasmid SUP-2021. Two methods were used for selecting for presumptive mutants (1) absence of growth on malate media and (2) the inability of colonies to reduce the dye nitrotetrazolium blue.

Approximately 10% of the 2,000 Tn5-generated mutants screened resulted in a presumptive malate dehydrogenase-negative phenotype. Of these presumptives, 150 were grown to late log phase in a liquid mannitol/tryptone media and assayed for enzyme activity. Mutants were grouped into three classes: (i) normal malate dehydrogenase (presumptive uptake mutants); (ii) deficient malate dehydrogenase; and (iii) elevated malate dehydrogenase activity. Representatives from each group were inoculated onto soybeans and assayed for in planta acetylene reduction and bacteroid malate dehydrogenase activity.

Results: Mutants of Rhizobium japonicum were isolated which had reduced levels of malate dehydrogenase activity. In addition, a few isolates possessed elevated levels of this enzyme activity. These isolates also had elevated levels of acetylene reduction activity. When the malate dehydrogenase-deficient mutants were used as inoculum onto soybeans, the isolated bacteroids were found to have levels of malate dehydrogenase similar to those from the wild type strain.

Argueso T. et al (1979) Arch. of Microbiology 121, 199-206.
Devries E. et al (1980) Plant Science 20, 119-123.
Glenn A. et al (1981) J. of Gen. Microbiology 126, 243-247.
Havelka, U.D. (1975) Symbiotic Nitrogen Fixation Volume 1, 421-439.
Kurz W. et al (1977) (Can. J. of Microbiology 23, 1197-1200.
Reibach P. et al (1983) Plant Physiology 72, 634.
Stumpf D. et al (1979) Anal Biochem 95, 311-315.
Stumpf D. et al (1981) Plant Physiology 68, 989-991.

HUP ACTIVITY DETERMINED BY PRL6JI DOES NOT INCREASE N2 FIXATION

S.D. Cunningham*, Y. Kapulnik*, S.A. Kagan#, N.J. Brewin#, D.A. Phillips*

* Department of Agronomy, Univ. California, Davis, CA 95616, U.S.A.
# John Innes Institute, Norwich NR4 7UH, U.K.

Comparisons of random collections of Hup^+ and Hup^- Rhizobium isolates have shown that in the case of R. japonicum (1) and mungbean rhizobia (2) Hup^+ strains averaged more N_2 fixation than Hup^- bacteria. Analagous tests with R. leguminosarum showed little symbiotic advantage for Hup^+ phenotypes relative to Hup^- strains (3,4). However, in previous collaborative studies our research groups used plasmid pIJ1008 (= pVW5JI/pRL6JI) to convert Hup^- R. leguminosarum strains into Hup^+ phenotypes with greater N_2 fixation capability (5). In those experiments pRL6JI, but not pVW5JI, carried Nod^+, Fix^+ and Hup^+ markers which produced 31 to 128% increases in N_2 fixation compared to $Nod^+Fix^+Hup^-$ recipient strains. The present study reports the results of a direct test for benefits associated with Hup activity conferred by pRL6JI.

Tn5-mob insertions into pRL6JI were used to produce Hup^- isolines from the Hup^+ R. leguminosarum 3855 (6). Six different Hup^- mutants that resulted from single insertions of Tn5-mob, a Hup^+ control strain 518 (3855 pRL6JI::Tn5-mob), and 3855 were tested for symbiotic performance in three cultivars of pea (Pisum sativum L. cvs. Feltham First, Alaska, and JI1205) and one line of vetch (Vicia benghalensis L.). Plants were grown with N-free nutrient solution under environmentally and microbiologically controlled conditions. Total dry weight, Kjeldahl N content, C_2H_2 reduction, H_2 evolution, and $^3H_2$ incorporation were determined after the flowering stage. Bacterial isolates from nodules were identical to inoculant strains in all cases tested.

The Kjeldahl N data showed clearly that in every case at least one Hup^- mutant fixed as much N_2 as the control strain 518. Data from other parameters also were consistent with the interpretation that the Hup activity conferred by pRL6JI did not increase N_2 fixation. Because the test conditions were identical to those under which pRL6JI produced large increases in N_2 fixation by Nod^+Fix^+ rhizobia on Alaska peas (5), these results indicate that pRL6JI carries some trait other than the Hup^+ phenotype which can markedly improve symbiotic performance.

REFERENCES

1. Albrecht SL et al (1979) Science 203, 1255-1257.
2. Pahwa K and Dogra RC (1981) Arch. Microbiol. 129, 380-383.
3. Nelson LM (1983) Appl. Environ. Microbiol. 45, 856-861.
4. Truelsen TA and Wyndaele R (1984) Physiol. Plant. 62, 45-50.
5. DeJong TM et al (1982) J. Gen. Microbiol. 128, 1829-1838.
6. Kagan SA and Brewin NJ (1985) J. Gen. Microbiol. 131, 1141-1147.

FREE-LIVING AND SYMBIOTIC EXPRESSION OF RHIZOBIUM JAPONICUM HYDROGENASE GENES IN SLOW- AND FAST-GROWING RHIZOBIUM SPECIES

GRANT R. LAMBERT, F. JOE HANUS, ALAN R. HARKER, MICHAEL A. CANTRELL, STERLING A. RUSSELL, and HAROLD J. EVANS, LABORATORY FOR NITROGEN FIXATION RESEARCH, OREGON STATE UNIVERSITY, CORVALLIS, OR   97331, U.S.A.

A cosmid pHU52 isolated from a Rhizobium japonicum gene bank by complementation of a Tn5-induced Hup$^-$ mutant, confers hydrogen uptake (Hup) activity and autotrophic growth capability upon Hup$^-$ slow- and fast-growing Rhizobium (Lambert et al., 1985). These experiments used bacteria derepressed for hydrogenase activity in the free-living state, where antibiotic selection ensures the retention of the cosmid. However, pHU52, a pLAFR1 derivative is unstable in nodules, such that bacteria isolated from nodules show tetracycline resistance at a frequency typically in the range 10^{-4} to 10^{-2}. We have detected Hup activity (ranging from 40 to 2,000 pmol.min^{-1}.mg protein^{-1}) in bacteroid suspensions from nodules in the case of R. japonicum USDA138(pHU52), R. meliloti 102F28(pHU52) and R. trifolii SU794(pHU52). These results indicate that the conditions in alfalfa and clover nodules are appropriate for the expression of R. japonicum hydrogenase genes. However, further genetic manipulation is necessary to increase the stability of these genes in their new hosts.

The regulation of Hup activity appears to be different in free-living and symbiotic R. japonicum, suggesting that the genetic determinants required for symbiotic and free-living Hup activity may also differ. pHU1 confers Hup activity in soybean and alfalfa nodules but not in free-living bacteria. Likewise, both subunits of the uptake hydrogenase have been detected immunologically in bacteroids carrying pHU1 or pHU52 and in free-living cells carrying pHu52. Free-living transconjugants carrying pHU1 contain no detectable hydrogenase polypeptides. Based on the comparative maps of pHU1 and pHU52 and the different Hup phenotype of several Tn5 insertions in the R. japonicum chromosome when determined in the free-living state and in nodules, it is suggested that the 5.5 kb EcoR1 fragment unique to pHU52 contains part of a gene or genes necessary only for free-living expression of Hup activity.

The stability of pHU1 and pHU52 in plant tests is dependent on the Rhizobium strain used. For R. japonicum USDA123Spc (Hup$^-$), the USDA123Spc (pHU1) or USDA123Spc(pHU52) transconjugants can recycle 70% of nitrogenase-mediated H_2 formation in nodules, resulting in an increase in relative efficiency of transfer of electrons to nitrogen via nitrogenase from 0.64 to 0.90.

References

Lambert, G.R., Cantrell, M.A., Hanus, F.J., Russell, S.A., Haddad, K.R. and Evans, H.J. (1985) Proc. Natl. Acad. Sci. 82: 3232-3236.

GENETIC ANALYSIS OF PLASMID pIJ1008 FROM RHIZOBIUM LEGUMINOSARUM

S.S.MANIAN, P.GRÖNGER, M.O'CONNELL and A.PÜHLER
LEHRSTUHL FÜR GENETIK, UNIVERSITÄT BIELEFELD, 4800 BIELEFELD, F.R.G.

INTRODUCTION

The recombinant plasmid pIJ1008 (pRL4JI::Tn5, pRL6JI) is self-transmissible and carries the genetic determinants for Hup activity and symbiosis. Detailed information regarding the genetic organisation of the plasmid is lacking. A genetic analysis of the plasmid has been initiated in our laboratory. Preliminary results on the organisation of several fix genes encoded by the plasmid are presented.

METHODS

A gene bank of the plasmid was constructed in the vector pSUP205 as follows; a) construction of a cosmid gene bank of R.meliloti 2011 carrying pIJ1008, b) transfer of the clones individually into R.leguminosarum strains carrying pIJ1008, pRL6JI or pVW5JI and c) screening of exconjugants for clones in which the cosmids have been stably maintained as a result of cointegrate formation.

RESULTS AND DISCUSSION

One cosmid clone (Cos 4) contained a 5.45 kb EcoR1 fragment which hybridised to the R.meliloti fixABC region and a 1.8 kb EcoR1 fragment which hybridised to the R.meliloti fixD gene. Heteroduplex studies between the 5.45 kb EcoR1 fragment of R.leguminosarum and the R.meliloti fixABC region revealed two stretches of homology. A long stretch (3.7 kb) of homology was observed over the entire fixABC region and a very short stretch (0.12 kb) of homology was observed in the 5'-terminal portion of the fixD gene. As the R.meliloti DNA fragment used in these studies did not contain the entire fixD gene, the extent of homology in the 3'-terminal portion of the gene could not be ascertained. Preliminary results from restriction analysis of Cos 4 place the 5.45 kb and 1.8 kb EcoR1 fragments adjacent to one another. Taken together the data suggest that the organisation of the fixABC region in both R.meliloti and R.leguminosarum may be similar.

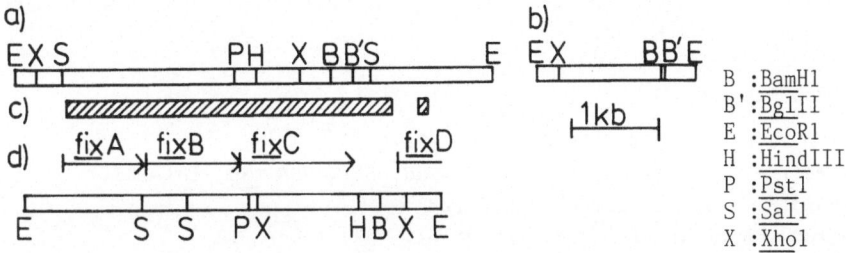

Fig. 1. The restriction maps of the 5.45 kb (a) and the 1.8 kb (b) EcoR1 fragments. The organisation of the R. meliloti DNA fragment (d) used for the hybridisation studies and the extent of homology (c) between the two species are also indicated.

PLASMID LOCATION OF HUP GENES IN STRAINS OF RHIZOBIUM LEGUMINOSARUM

TOMAS RUIZ-ARGUESO,ANTONIO LEYVA,JOSE PALACIOS,TERESA MOZO, BELEN CAMPOS
DEPARTMENT OF MICROBIOLOGY, ETS INGENIEROS AGRONOMOS,28040-MADRID,SPAIN

Legume root nodules produced by most strains of Rhizobium evolve hydrogen gas as an obligate by-product of the nitrogenase reaction. Some strains of R.leguminosarum possess an active H_2-uptake (Hup) system permitting H_2 to be recycled in nodules, and are believed to be more energy-efficient symbionts than Hup$^-$ strains.There is evidence indicating that Hup determinants are plasmid born in a Hup$^+$ strain of R. leguminosarum (Brewin et al 1980) and in newly isolated strains from pea nodules (Seifert et al 1983). To investigate if this characteristic is a general trait in R.leguminosarum and, particularly in strains that,possessing a highly effective H_2-uptake system,are potential donors for transfering the H_2-recycling capability to Hup$^-$ Rhizobium species, we have examined 5 Hup$^+$ and 6 Hup$^-$ strains of R. leguminosarum for plasmid content and location of specific nif and hup genes.

Plasmids were resolved on agarose gels by the procedure of Eckhardt 1978. The gels were dried and hybridized to R.meliloti nifHD DNA and to DNA from pCM7 that contains hup-specific DNA sequences from R.japonicum strain 122DES (Haugland et al 1984). To prepare pCM7, the 5 kb Eco R1 fragment of pHU1 (Cantrell et al 1983) was subcloned into vector pUC13. A common feature of both Hup$^+$ and Hup$^-$ strains was the presence of 3 to 6 large plasmids of MW ranging from 60 to higher than 500 Mdal.Only one plasmid in each strain showed hybridization to the R.meliloti nif probe (Sym plasmid). The MW of the Sym plasmids varied between 140 and 200 Mdal in 10 out of the 11 strains examined. In one strain (UPM-Ve6) a megaplasmid of MW higher than 480 Mdal was identified as the nif-containing plasmid.The Sym plasmids of all the Hup$^+$ strains but not the Sym plasmids of any of the Hup$^-$ strains were found to hybridize as well to the hup-specific R.japonicum DNA probe.

Total DNA from the studied strains of R.leguminosarum was Eco R1 restricted, separated on agarose gels, blotted and hybridized to the 5 kb Eco R1 fragment from pCM7. All the Hup$^+$ strains contained 2 common Eco R1 fragments of 2.2 and 1.6 kb that strongly hybridized to the R.japonicum hup-specific DNA. These two fragments belong to the Sym-hup plasmid DNA since they were not present in total DNA digests from Hup$^-$ or pSym-cured derivatives of Hup$^+$ strains.

REFERENCES

Brewin NJ et al (1980) Nature 288, 77-79
Cantrell et al (1983) Proc. Natl. Acad. Sci. USA 80, 181-185
Eckhardt T (1978) Plasmid 1, 584-588
Haugland RA et al (1984) J. Bacteriol. 159, 1006-1012
Seifert BL et al (1983) In Veeger C and Newton WE, eds, Advances in Nitrogen Fixation Research, p 721, Nijhoff/Junk, The Hague

MOBILIZATION OF SYM-HUP PLASMIDS FROM STRAINS OF RHIZOBIUM LEGUMINOSARUM

TOMAS RUIZ-ARGUESO, JOSE M. PALACIOS, ANTONIO LEYVA AND ANA VILLA
DEPARTMENT OF MICROBIOLOGY, ETS INGENIEROS AGRONOMOS, 28040-MADRID,SPAIN

Some strains of Rhizobium leguminosarum and R. japonicum (Hup[+] strains) specify an oxygen-dependent hydrogen oxidation system that recycles H_2 produced by nitrogenase in legume nodules. Since this H_2-recycling capability seems to be beneficial for the Rhizobium-legume symbiosis (1),the transfer of genes responsible for the H_2-uptake system (hup genes) to several species of Rhizobium that lack this phenotype may improve their symbiotic performance.

There is genetic and physical evidence that at least some of the hup genes are located on the symbiotic plasmid in Hup[+] strains of R. leguminosarum (2,3,Ruiz-Argüeso,these proceedings).Three of such Hup[+] strains,128C53, 128C56 and 128C23 possess Sym-hup plasmids of 175,165 and 184 Mdal,respectively.To transfer the H_2-recycling capability from these strains to other Rhizobium,we inserted into their Sym-hup plasmids a selectable marker and a mobilization element by mating each of the strains with E.coli S17-1 containing a Tn5-mob system as part of the suicide plasmid pSUP5011 constructed by Simon et al (4).Kan[r] colonies from each cross were pooled and the resulting cell suspension was conjugated "en masse" with a non-infective Sym-cured strain of R. leguminosarum (128C53 str[r]) using the helper plasmid RK2013.Str[r]kan[r] colonies appeared at a frequency of 10^{-6}-10^{-7} per recipient. The plasmids present in 25 str[r] kan[r] clones per cross,were separated on Eckhardt gels and hybridized to R. meliloti nif HD and pSUP5011 DNA probes.Two, 4 and 5 transconjugants from the crosses with strains 128C53,128C56 and 128 C23 respectively,acquired a new plasmid of a size similar to the Sym-hup plasmid of the corresponding donor strain.In all these transconjugants the acquired plasmid hybridized to both nif and pSUP5011 DNAs.

Transconjugans were used as inocula for pea seeds,and in all cases the transferred pSym-hup::Tn5-mob conferred to the recipient strain 128C53.4str[r], the capacity to effectively nodulate peas.Bacteroids from pea nodules produced by transconjugants containing the Tn5-mob tagged pSym-hup plasmid of strains 128C53 and 128C56 showed a capacity to oxidize H_2.No hydrogenase was observed however,in bacteroids from transconjugants containing the Sym-hup plasmid of strain 128C23 although the plasmid was mantained in the transconjugants after passing through the plant.

REFERENCES

1 Evans HJ et al (1985) In Ludden PW and Burris JE, eds,Nitrogen Fixation and CO_2 Metabolism, pp 3-11,Elsvier, New York.
2 Brewin NJ et al (1980) Nature 288, 77-79.
3 Seifert BL et al (1983) In Veeger C and Newton WE,eds, Advances in Nitrogen Fixation Research p 721,Nijhoff/Junk, The Hague.
4 Simon R et al (1983) In Pühler A,ed, Molecular Genetics of the Bacteria-Plant Interaction pp 98-106, Springer-Verlag, Berlin.

MOLECULAR CLONING OF UPTAKE HYDROGENASE GENES OF RHIZOBIUM LEGUMINOSARUM B10

HANS-VOLKER TICHY, PETRA RICHTER AND WOLFGANG LOTZ
INSTITUT FÜR MIKROBIOLOGIE UND BIOCHEMIE, UNIVERSITÄT
ERLANGEN-NÜRNBERG, EGERLANDSTR. 7, D-8520 ERLANGEN, FRG.

The Rhizobium leguminosarum strain B10 has been isolated from pea root nodules. It fixes nitrogen in symbiosis with pea (Pisum sativum) and vetch (Vicia angustifolia). In addition, this strain expresses an uptake hydrogenase in pea root nodules (Tichy, Lotz 1985). A cosmid gene bank of R. leguminosarum B10 total DNA in pMMB34 (Frey et al. 1983) has been constructed using the cloning method of Ish-Horowicz and Burke (1981). From this gene bank we have isolated the plasmid pRlB505 which carries sequences homologous to R. japonicum hup genes. The hybridization probe used to screen the gene bank was a subclone of pHU1 (Cantrell et al. 1983) containing the 6Kb HindIII-fragment. In addition to this fragment of pHU1, all EcoRI-fragments of the pHU1 insert, except the smallest one, hybridized to pRlB505, the two large EcoRI fragments strongly, the three other EcoRI fragments weakly. Restriction enzyme cleavage sites have been mapped on pRlB505 for BamHI, HindIII, SstI, EcoRI, SalI and BglII by analyzing subcloned fragments from a partial HindIII digest, single and double digestions (only for mapping the two BglII sites) and the hybridization of pHU1 subclones with digested pRlB505 DNA. Two HindIII-subclones of pRlB505 have been used to characterize R. leguminosarum strains of different origin (Nelson et al. 1985). Presently, the cloned R. leguminosarum hup-DNA is analyzed by site-directed-mutagenesis. Tn5-insertions have been isolated on three pRlB505-subfragments cloned in pACYC184. Since these subclones could not be mobilized into strain B10, cointegrates of the Tn5-tagged plasmids with pRK290 have been constructed. These cointegrates could be mobilized by pRK2013 into R. leguminosarum B10. Transconjugants were grown over about fifty generations and single colonies isolated which had lost the plasmid but retained Tn5. Two Tn5 insertion mutants have been isolated at present, one of these, B10-1201, carries Tn5 inserted 2.5 Kb from the right end of the DNA cloned in pRlB505. Bacteroids of this strain were tested for methylene blue reduction (Tichy, Lotz 1985) and found to be Hup$^+$. Therefore, this insertion is probably located outside the R. leguminosarum B10 hup-region, provided that the homology between R. japonicum and R. leguminosarum hup-DNA extends to the R. japonicum DNA fragment cloned in pHU52 but not in pHU1. In this fragment one border of the R. japonicum hup-region is located (Lambert et al. 1985). The other mutant, B10-1151, carries a Tn5 insertion in the region of pRlB505 showing low homology with R. japonicum hup-DNA. Preliminary results indicate, that this mutation causes a Hup$^-$ phenotype.

REFERENCES

Cantrell MA et al (1983) Proc. Natl. Acad. Sci. U.S.A. 80, 181.
Frey J et al (1983) Gene 24, 299.
Ish-Horowicz D and Burke JF (1981) Nucleic Acid Res. 9, 2989.
Lambert GR et al (1985) Proc. Natl. Acad. Sci. U.S.A. 82, 3232.
Nelson LM et al (1985) FEMS Microbiol. Lett. (in press).
Tichy HV and Lotz W (1985) FEMS Microbiol. Lett. 27, 107.

EXPRESSION OF RHIZOBIUM JAPONICUM GENES FROM A HUP-COMPLEMENTING COSMID IN ESCHERICHIA COLI

MOHAMMED ZUBER/ALAN R. HARKER/HAROLD J. EVANS
LABORATORY FOR NITROGEN FIXATION RESEARCH, OREGON STATE UNIVERSITY,
CORVALLIS, OR 97331, USA

A cosmid, pHU52, isolated from a Rhizobium japonicum gene bank confers chemoautotrophy and also complements hydrogen uptake activity in Hup strains of Rhizobium (Lambert et al., 1985). Recently, Harker et al (1984) have purified uptake hydrogenase polypeptides of 60,000 and 30,000 mol. weight from a Hup R. japonicum strain and also raised antibodies against them. As an effort toward the identification of the structural genes of these two polypeptides, we have attempted to express these genes in E. coli following the "maxicell" technique of Sancar et al (1979). We have constructed a new expression vector pMZ545 wherein appropriate DNA fragments can be cloned into one or more of the four unique cloning sites placed downstream to the E. coli promoter for the lac operon, Plac. An added advantage of this vector is that cloning of DNA fragments results only in an operon fusion but not a gene fusion. Subclones of various overlapping fragments from the Rhizobium insert DNA of pHU52 constructed in pMZ545, when used for "maxicell" experiments, clearly showed the expression of R. japonicum genes in E. coli under the conditions used to turn on the Plac promoter (glucose was replaced by fructose as the carbon source). The 6.1 kb BglII fragment (Fig. 1) expressed three proteins of about 40,000, 30,000 and 18,000 mol. weights, in either orientation relative to the transcription initiation from the Plac promoter. Similarly, the 5.5 kb EcoRI fragment (Fig. 1) showed the expression of three proteins of mol. weights of about 45,000, 35,000 and 22,000, irrespective of its orientation. The possibility still remains that one or two of these proteins could be truncated polypeptides because of the presence of translational stops in all three reading frames downstream to the cloning sites in pMZ545. None of these proteins cross react with antibodies raised against 60 kd and 30 kd polypeptide components of uptake hydrogenase. Whether a more sensitive ELISA assay would help detect the expression of 60kd and 30kd polypeptides in E. coli remains to be investigated. Expression of R. japonicum genes from 6.1 kb BglII and 5.5 kb EcoRI fragments of pHU52 in E. coli in either orientation relative to transcription initiation at the Plac promoter on the vector pMZ545 clearly suggests the presence of Rhizobium promoters on these fragments and their active functioning in E. coli.

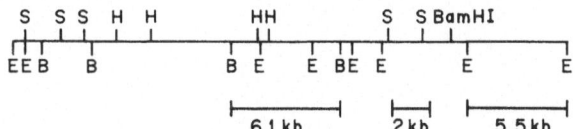

Fig. 1. Restriction map of pHU52. Abbreviations used: kb, kilobase pairs; H, S, B, and E represent the respective restriction sites: HindIII, SmaI, BglII, EcoRl

References
Harker, A. et al (1984) J. Bacteriol. 159, 850-856.
Lambert, G.R. et al (1985) PNAS 82, 3232-3236.
Sancar, A. et al (1979) J. Bacteriol. 137, 692-693.

POSTER DISCUSSION SUMMARY

MICROBIAL GENETICS RELEVANT TO NODULE FUNCTION

Convenor: N.J. BREWIN,
JOHN INNES INSTITUTE, NORWICH NR47UH, U.K.

The discussion theme was microbial genes that affect nodule function. The posters were grouped into 3 major areas and the findings summarized below.

1. MOLECULAR GENETICS OF FIX GENES

A wide range of studies was reported on bacterial genes affecting N_2 fixation within the nodule (fix genes). These genes include nifH (nitrogenase component II), nifDK (nitrogenase components Ia and b), nifA and ntrC (positive regulators for nif promotors), nifB, and various ill-defined fix genes. Map positions have been assigned by cosmid cloning. Gene functions have been identified by: hybridization to corresponding cloned genes from Klebsiella pneumoniae; DNA sequence comparisons; complementation analysis; and S1 nuclease mapping. The control of gene expression has been extensively examined using gene fusion systems to β-galactosidase or the chloramphenicol. (CAT) gene cartridge. Gene promoters have been analyzed by DNA sequencing, deletion analysis, and localized point mutagenesis. In fast-growing Rhizobium species, the promoter regions of nifHDK are reiterated 5-6 times on the symbiotic plasmid at positions which control the expression of other genes expressed in the symbiotic state. Within these reiterated promotor sequences are sequence domains common to all Rhizobium species and other domains that are species-specific. These observations have led to a model for coordinated host-specific expression of symbiotic nitrogen-fixation genes.

2. AMMONIA ASSIMILATION AND CARBON METABOLISM

Glutamine synthetase of Rhizobium (and Agrobacterium) exists in two forms, and several posters reported the isolation of the corresponding genes. Several different techniques were used for these and other genes of intermediary metabolism, but a convenient method of isolation from a cosmid gene bank was to select for suppression of well-characterized mutations in E. coli. Using this strategy, a gene encoding δ-amino levulinic acid synthetase was isolated from B. japonicum. Surprisingly, when the corresponding chromosomal gene was mutagenized in B. japonicum, the phenotype was Fix$^+$, implying either that the bacterium was synthesizing heme by an alternative route or that the plant was compensating for deficiencies in bacterial heme biosynthesis. New mutants were also reported for adenyl cyclase, arginine biosynthesis, histidine biosynthesis, dicarboxylic acid transport and malate dehydrogenase, illustrating the range of interests in this sphere of intermediary metabolism.

3. HYDROGENASE

A cosmid clone, pHU52, isolated from a R. japonicum gene bank, confers chemoautotrophy and hydrogen-uptake activity on Hup$^-$ strains of R. japonicum and on fast-growing strains of Rhizobium. This cosmid clone encodes two polypeptides of 60K and 30K molecular weight. Hydrogenase genes showing DNA homology to those of R. japonicum have been shown to be present on the symbiotic plasmids of some strains of R. leguminosarum. Mutants of R. leguminosarum that lacked hydrogenase activity have been isolated and used in plant growth tests. The results indicated that hydrogen recycling within pea nodules did not confer any enhancement of symbiotic N_2 fixation.

RECOGNITION AND INFECTION
IN SYMBIOSIS

INVITED PAPERS

POSTER SUMMARIES:

(i)   Competition and attachment.
(ii)   Host-induced phenomena.
(iii)   Compositional variations.
(iv)   Actinorhizal systems.

DISCUSSION GROUP SUMMARY

RECOGNITION AND INFECTION OF CLOVER ROOT HAIRS BY RHIZOBIUM TRIFOLII

FRANK B. DAZZO, RAWLE I. HOLLINGSWORTH, JOHN E. SHERWOOD, MIKIKO ABE, ESTELLE M. HRABAK, ALICIA E. GARDIOL, H. STUART PANKRATZ, KATHRYN B. SMITH, AND HONG YANG
DEPARTMENT OF MICROBIOLOGY, MICHIGAN STATE UNIVERSITY, EAST LANSING, MI 48824, U.S.A.

1. INTRODUCTION

The infection of clover roots by the nitrogen-fixing bacterium, Rhizobium trifolii, involves interaction of both symbionts and is host specific. The host specificity expressed during root hair infection occurs before formation of the infection thread (Li, Hubbell 1969) and constitutes a model for cellular recognition between procaryotes and eucaryotes. A white clover glycoprotein lectin called trifoliin A (ca. 53 kD) accumulates on the surface of root hairs at the growing tip and appears to function as a cell recognition molecule (Dazzo et al. 1978; Dazzo, Truchet 1983). Therefore, our studies of recognition in this symbiosis have focused on trifoliin A on the root and complementary saccharide receptors on the bacterial symbiont. In this paper, we review the current understanding of interactions between these cell surface components as affected by symbiotic genes on the nodulation plasmid of R. trifolii.

2. BIOSYNTHESIS OF TRIFOLIIN A AS AFFECTED BY NITRATE SUPPLY

In vivo biosynthesis of trifoliin A in white clover roots has been demonstrated by incorporation of labelled amino acids into protein immunoprecipitable with homologous anti-trifoliin A during the heterotrophic stage of seedling development (Sherwood et al. 1984a). Labelled trifoliin A was detected in the roots (but not the hypocotyl or cotyledons) after 2 hr, and was elutable from the root surface by the hapten 2-deoxy-D-glucose after 6-8 hr. Much of the labelled lectin was in root exudate. Excess nitrate supply did not repress trifoliin A sythesis. However, 10-30 fold less labelled lectin accumulated on the root surface and the R. trifolii-binding ability of trifoliin A in the root exudate was reduced 97%. These results (Sherwood et al. 1984a) explain why R. trifolii 0403 and anti-trifoliin A bind less to root hairs of seedlings grown with nitrate levels which prevent infection (Dazzo, Brill, 1978), and why lower levels of "active" trifoliin A (capable of binding to R. trifolii) are detected in the root exudate of clover grown with 15 mM nitrate solutions (Dazzo, Hrabak 1981).

3. POLYSACCHARIDE RECEPTORS ON R. TRIFOLII WHICH BIND TRIFOLIIN A

R. trifolii grown in defined B3 medium produces polysaccharide receptors for trifoliin A which change with culture age. They are the CPS on agar surfaces (Dazzo et al. 1978; Sherwood et al. 1984b; Abe et al. 1984) and LPS in broth culture (Hrabak et al. 1981). Trifoliin A binds CPS from 5-day old plate cultures better than CPS at other

culture ages. Extracellular acidic polysaccharide (EPS) of the same
5-day old cultures (which remains in the supernatant after
centrifugation of cells) does not bind trifoliin A (Abe et al. 1984).
We treated the CPS and EPS with bacteriophage-borne β-lyase and
analyzed the glycosyl components of the isolated oligosaccharides by
GC/MS and the non-carbohydrate substituents by ^{1}H-NMR (Hollingsworth
et al. 1984a). The results indicated that these polysaccharides
consisted of glucose, galactose, and glucuronic acid, and were
substituted with ester-linked acetate, ketal-linked pyruvate, and
ether-linked 3-hydroxybutyrate (Hollingsworth et al. 1984a;
Hollingsworth et al. 1984b; Abe et al. 1984). The neutral glycosyl
composition of either polysaccharide did not vary with culture age
(Sherwood et al. 1984b; Hollingsworth et al. 1984a; Abe et al. 1984)
but the degree of acetate, pyruvate, and 3-hydroxybutyrate
substitutions (moles of substituent per mole of oligosaccharide) in
the CPS did change with culture age and was higher than in the EPS of
the same 5 day-old culture. Hence, the acidic CPS and EPS of
plate-grown cultures are not identical molecules. The binding of
trifoliin A to the CPS oligosaccharide fragments reflects the
age-dependent level of binding of the native CPS of R. trifolii 0403
in plate culture (Abe et al. 1984). Partial removal of acetate or
pyruvate substitutions of the oligosaccharide fragments results in a
reduction of trifoliin A binding. Thus, the binding of trifoliin A
appears to be sensitive to small changes in the degree of
noncarbohydrate substitutions in the CPS.

Binding of trifoliin A to R. trifolii 0403 LPS is minimal during
mid-exponential phase and optimal in early stationary phase (Hrabak et
al. 1981). Combined GC/MS and ^{1}H-NMR reveal increases in several
glycosyl components of R. trifolii 0403 LPS isolated at early
stationary phase, including quinovosamine, 2-O-methylfucose,
2-O-methylrhamnose, and N-methyl-3-amino-3,6-dideoxygalactose (Hrabak
et al. 1981; and this publication). β-linked quinovosamine is an
effective hapten of trifoliin A-R. trifolii interactions (Hrabak et
al. 1981). However, since small amounts of quinovosamine are present
in LPS of mid-exponential phase cultures, the mere presence of this
haptenic sugar in the LPS does not automatically mean it will bind
trifoliin A (Hrabak et al. 1981).

The effect of culture age on polysaccharide chemistry accounts for
many of the inconsistencies among laboratories testing the
lectin-recognition hypothesis. These changes provide clues to the
preferred minimal binding site of the lectin and may reflect bacterial
regulation of recognition in the Rhizobium-legume symbiosis (Hrabak et
al. 1981). The efficiency of the R. trifolii inoculum in infection
(clover root hair infections per constant inoculum size) varies with
culture age and parallels the transient appearance of trifoliin A
receptors on the bacteria in broth culture (Hrabak et al. 1981).

4. PHASES OF R. TRIFOLII ATTACHMENT TO CLOVER ROOT HAIRS

We have examined the time-course and the orientation of attachment of
R. trifolii to root hairs on white clover seedlings inoculated with
the subpopulation of encapsulated bacteria which bind trifoliin A
uniformly (Dazzo et al. 1984a). Attachment is a dynamic sequential
process, beginning with specific reversible interactions involving

trifoliin A (Phase 1), followed by irreversible interactions involving extracellular microfibrils (Phase 2). A similar sequence of reversible and irreversible phases of attachment of <u>Agrobacterium</u> <u>tumefaciens</u> to plant cells has been proposed (Matthysse et al. 1981).

Phase 1 attachment can be subdivided into 3 steps when 10^7 to 4×10^8 cells are inoculated per seedling under defined slide-culture conditions (Dazzo et al. 1984a). Phase 1A occurs within minutes and is recognized as clumping of cells in random orientation at root hair tips. Phase 1B occurs within the next 4 hrs and involves erosion of the capsule of unattached cells by enzymes in root exudate. During this time, trifoliin A in root exudate binds to the bacteria. In contrast to the 1A clumping, Phase 1C involves the attachment of single cells to the root hair. 1C attachment is randomly oriented for the first few hr after inoculation and predominantly polar after 4 hr. Granular, electron-dense aggregates (called "Truchet particles") and trifoliin A can be detected at the interface between the bacteria and the root hair surface in 1A and 1C attachments after several hours.

The enzyme(s) in clover root exudate which alter the capsular polysaccharide of <u>R. trifolii</u> during Phase 1B are antigenically unrelated to trifoliin A (Dazzo et al. 1982). Immunoelectrophoresis (Dazzo et al. 1982) and dialysis studies (Solheim, Fjelheeim 1983) suggest that the capsular polysaccharides of <u>R. trifolii</u> are cleaved into smaller, dialyzable fragments. Clover root exudate and extract are more active in degrading the capsular polysaccharide of <u>R. trifolii</u> than of <u>R. meliloti</u> or <u>R. leguminosarum</u> (Dazzo et al. 1982; Solheim, Fjelheim 1984). Bhuvaneswari and Solheim (1985) have proposed that these enzyme-mediated cleavages result in oligosaccharide products which induce clover root hair branching.

Within 4 hr, <u>R. trifolii</u> assumes on most of the clover root hairs a pattern of attachment combining 1A clumping at the tip and 1C polar attachments along the side of the same root hair. This attachment pattern is symbiont-specific and 2-deoxy-D-glucose inhibitable, and is present on approximately 93% of the infected root hairs examined 4 d after inoculation with <u>R. trifolii</u> 0403 (Dazzo et al. 1984a). Perhaps this orientation provides the optimal distribution of bacteria for marked curling and successful infection of the root hair (van Batenburg et al. 1984; Ervin, Hubbell 1985).

The hapten 2-deoxy-D-glucose can inhibit the attachment of trifoliin A-binding <u>R. trifolii</u> cells to clover root hairs (Dazzo et al. 1976; Zurkowski 1980; Dazzo et al. 1984a). However, bacterial attachment occurring in the absence of this hapten becomes progressively less dissociable, suggesting a Phase 2 mechanism of adhesion following Phase 1 (Dazzo et al. 1984a). Under defined slide culture conditions (Dazzo et al. 1984a), cells are in Phase 2 adhesion after 12 hr incubation with seedlings. These cells remain attached to clover root hairs when exposed to the shear forces of high-speed vortexing. At Phase 2, extracellular microfibrils associated with the bacteria attached to the root hair surface can be seen by SEM. Although the nature of these microfibrils is unknown, possibilities include bundles of cellulose and/or fimbriae.

5. INVOLVEMENT OF pSym NODULATION GENES on TRIFOLIIN A RECEPTORS, SURFACE POLYSACCHARIDE CHEMISTRY, AND ATTACHMENT TO CLOVER ROOT HAIRS

R. trifolii genes important for expression of trifoliin A receptors on the cell surface have been transferred to and expressed in Azotobacter vinelandii (Bishop et al. 1977) and pTi-cured Agrobacterium tumefaciens (Dazzo et al. 1982). Genes for incorporation of quinovosamine into the R. trifolii LPS and for the ability of R. trifolii to attach to clover root hairs in a hapten-inhibitable manner are located on the symbiosis plasmid (Russa et al. 1982; Zurkowski 1980). Following is a summary of new genetic evidence for the importance of certain nodulation genes on the Sym plasmid of R. trifolii in the binding of cells to trifoliin A, the chemistry of the trifoliin A-binding surface polysaccharides, and the specific attachment of R. trifolii to clover root hairs.

Analysis of Tn5-induced mutations in the Sym plasmid of R. trifolii ANU 843 (from B. Rolfe, Australian National University) has identified at least three adjacent regions affecting nodulation of white clover roots (B. Rolfe and M. Dordjevic, personal communication). These regions are designated as the hair-curling region I (Hac), the "superHac" region II, and the host-specificity region III (Hsp). Hac- mutants in region I are Inf- Nod-. Region II and III mutants induce a greater frequency of marked root hair curling but are significantly impaired in root hair infection and nodulation. In addition, region III mutants have gained the ability to infect pea roots. A pSym-cured mutant strain 845 (hac- inf- nod-) was also available for analysis.

The binding of trifoliin A to these mutant strains has been examined by fluorescence microscopy after "ex planta" growth on B3 agar plates or "in situ" in the clover root environment using previously described methods (Dazzo et al. 1982; Sherwood et al. 1984b). In the "ex planta" assay, wild type R. trifolii 843 bound trifoliin A transiently, and was optimal in 5 day-old cultures. The binding of trifoliin A was ca. 95% less for Hac- region I mutant 851 and essentially negative for pSym- mutant 845.

The "in situ" binding of trifoliin A was examined for the wild type and several mutant strains after 16 hr incubation with white clover seedlings in slide cultures. Significantly fewer cells of mutants in each of the three nodulation regions (except strain 246 discussed below) bound trifoliin A than did wild type 843 in the root environment. Consistent with R. trifolii 0403 (Dazzo et al. 1982), binding of trifoliin A to strain 843 in the root environment was uniform on some cells and polar on other cells. In contrast, none of the mutants except strain 246 bound trifoliin A at one cell pole.

The pattern of in situ binding of trifoliin A to strain 246 (Tn5 in Hac region I) is similar to wild type and is consistent with its inf+ nod+ phenotype on white clover. Genetic analysis (M. Dorjdevic, personal communication) suggests that strain 246 has a Tn5 inserted between genes in Hac region I. However, we have found that root hair infection is somewhat delayed and less frequent with this strain on white clover, and therefore, it is not equivalent symbiotically to wild type strain 843.

^{1}H-NMR analysis of oligosaccharide fragments of the CPS of plate-grown cultures indicates that certain Hac- region I mutants are altered in acetate and pyruvate substitutions. In addition, the CPS of pSym-strain 845 is depolymerized by β-lyase PD-I enzyme with a Vmax higher than that of wild type CPS. Thus, certain essential nodulation genes on the Sym plasmid affect CPS chemistry when grown on B3 medium. ^{1}H-NMR analysis of LPS from some of Barry Rolfe's strains provided by R. Carlson (Eastern Illinois University) and purified further by us indicates that the LPS from a Hac- region I mutant (strain 851) differs from the LPS of wild type strain 843 in the N-acetyl region. The LPS of a region II mutant differs in the O-acetyl region and the ratios of fatty acids. These results indicate that certain nodulation genes on the Sym plasmid affect LPS chemistry.

We utilized light and scanning electron microscopy (Sherwood et al. 1984b; Dazzo et al. 1984a) to compare the attachment of wild type and mutant strains to white clover root hairs. These methods are not complicated by the false overestimates of bacterial attachment to root hairs based on plate counts or scintillation counting of radiolabelled bacteria. Our findings, relative to wild type 843, are summarized as follows: (i) in a 2 hr beaker assay, certain region I and III mutants attach in fewer numbers to root hairs 200 μm in length; (ii) in a 4 hr slide culture assay, fewer root hairs at all stages of development have the combination of 1A plus 1C attachment pattern for certain region I, II, and III mutants; (iii) in a vortexing experiment after 12 hr incubation in slide cultures, fewer root hairs at all stages of development have Phase 2 firm adhesions with certain regions I, II, and III mutants. SEM examination of seedlings after 4 d incubation in slide cultures indicate that, in contrast with wild type, extracellular microfibrils are not found associated with attached bacteria representing certain region I Hac- mutants, and are less abundant for certain region III Hsp- mutants. Consistent with results using R. trifolii 0403 (Dazzo et al. 1984a), 2-deoxy-D-glucose specifically inhibited the pattern of attachment of the wild type strain 843 to white clover root hairs during the first 4 hr. Pattern 1A plus 1C was most inhibited, pattern 1A only was intermediate in inhibition, and pattern 1C was least inhibited. This indicates that trifoliin A-polysaccharide interactions contribute greatest to the 1A plus 1C pattern, intermediate to the 1A only pattern, and least to the 1C pattern during the first 4 hr. Consistent with this interpretation, the attachment of pSym- strain 845 (which does not bind trifoliin A) was predominantly 1C only and was not inhibited by 2-deoxyglucose. Therefore attachment by strain 845 was accomplished by nonspecific mechanisms.

Results of these studies provide strong evidence that trifoliin A-surface polysaccharide interactions are involved in recognition and infection of clover root hairs by R. trifolii. The Sym plasmid carries genes essential for expression of trifoliin A receptors on the surface of R. trifolii. Tn5 insertions in certain essential nodulation genes on the Sym plasmid of R. trifolii affect binding of trifoliin A, CPS and LPS chemistry, and root hair attachment. Certain essential nod genes on pSym are required for trifoliin A-mediated attachment to clover root hairs during Phase 1 and for production of extracellular microfibrils associated with Phase 2 adhesion. Host specificity genes affect the accumulation of these microfibrils. The Hac nod D gene affects the chemistry of both CPS and LPS (as exemplified by strain 851) and is expressed ex planta while growing on B3 medium.

6. ALTERNATIVE LECTIN-RECOGNITION "SIGNAL" HYPOTHESIS

One of the mechanisms of attachment of R. trifolii to clover root hairs during the first 4 hr after inoculation is specific and involves an interaction between trifoliin A and polysaccharide receptors. This interaction is viewed as an early recognition event of this symbiosis. Specific attachment is hapten-reversible, symbiont-specific, characteristic of the combined 1A plus 1C pattern on the same root hair under controlled conditions, and requires nodulation genes on the Sym plasmid of R. trifolii. However, cells which do not bind trifoliin A, including wild type R. trifolii at certain culture ages, certain non-infective mutants of R. trifolii (e.g., pSym- cured mutants), and heterologous rhizobia have an alternate mechanism of attachment to clover root hairs (Dazzo et al. 1976; Badenoch-Jones et al. 1985) which is unaffected by 2-deoxy-D-glucose and therefore non-specific. These results indicate that attachment to root hairs PER SE cannot be the key determinant of host specificity in this symbiosis (Dazzo & Truchet 1983). The results with the mutants do, however, elevate the importance of lectin-mediated attachment to successful root hair infection. Because of these alternate, nonspecific mechanisms, we have focused on an alternative "signal" hypothesis involving root lectin and its saccharide receptors in symbiont recognition. This hypothesis, first proposed by W. Kamberger (1979), states that attachment is followed by symbiont-specific lectin-polysaccharide interactions which produce a signal (gene expression?) that triggers successful infection of the receptive root hair.

7. INFECTION-RELATED BIOLOGICAL ACTIVITIES OF TRIFOLIIN A-BINDING POLYSACCHARIDES OF RHIZOBIUM TRIFOLII

We are testing Kamberger's "signal" hypothesis by studying the effect of trifoliin A-binding polysaccharides from R. trifolii on clover root hair infection. Trifoliin A-binding CPS and LPS from R. trifolii 0403 can bind to clover root hairs (Dazzo, Brill 1977; Dazzo et al. 1984b). Minute quantities (0.1-5 µg) of these polysaccharides enhance root hair infection of white clover by R. trifolii 0403 and inhibit infection at higher quantities (Dazzo et al. 1984b; Abe et al. 1984). The trifoliin A-binding oligosaccharide fragments of the R. trifolii 0403 EPS and CPS, and LPS from serologically unrelated strains of R. trifolii also display this biological activity. In contrast, root hair infection is not enhanced with non-lectin binding native EPS of 5 day-old cultures or the LPS of mid-exponentially growing cells of R. trifolii 0403, or with LPS from R. meliloti 102F28, Escherichia coli 0127 B:8, or Salmonella typhimurium W. Thus, the trifoliin A-binding ability and infection-related biological activities of these R. trifolii polysaccharides in this low concentration range are correlated. Stimulation of root hair infection by R. trifolii LPS is restricted to the region of the seedling root present at the time of exposure to the polysaccharide, occurs by pretreatment of clover seedlings for as little as 1 min (optimal after 1-2 hr), and is inhibited by combined nitrogen. There is a distinct increase in accumulation of electron-dense granules at tips of clover root hairs treated with the trifoliin A-binding LPS of R. trifolii 0403.

These studies support the hypothesis that specific
lectin-polysaccharide interactions trigger an early signal which
modulates infection of clover root hairs by R. trifolii. Further
description of surface polysaccharide changes of R. trifolii mutants
are presented in poster abstracts from this laboratory.

8. REFERENCES

Abe M et al. (1984) J. Bacteriol. 160, 517-520.
Badenoch-Jones J et al. (1985) Appl. Environ. Microbiol. 49,
1511-1520.
Bhuvaneswari TV and Solheim B (1985) Physiol. Plant. 63, 25-34.
Bishop PE et al (1977) Science 198, 938-940.
Dazzo FB and Brill WJ (1977) Appl. Environ. Microbiol. 33, 132-136.
Dazzo FB and Brill WJ (1978) Plant Physiol. 62, 18-21.
Dazzo FB and Hrabak EM (1981) J. Supramol. Struct. Cell. Biochem. 16,
133-138.
Dazzo FB and Truchet GL (1983) J. Membrane Biol. 73, 1-16.
Dazzo FB et al (1976) Appl. Environ. Microbiol. 32, 166-177.
Dazzo FB et al (1983) Proc. Amer. Soc. Microbiol. (Abstr. K9, p 178).
Dazzo FB et al (1978) Biochim. Biophys. Acta 536, 276-286.
Dazzo FB et al (1979) Curr. Microbiol. 2, 15-20.
Dazzo FB et al (1982) Appl. Environ. Microbiol. 44, 478-490.
Dazzo FB et al (1984a) Appl. Environ. Microbiol. 48, 1140-1150.
Dazzo FB et al (1984b) In Veeger C and Newton WE, eds, Advances in
Nitrogen Fixation Research, p 413, Martinus Nijhoff/Junk, The Hague.
Erwin SE and Hubbell DH (1985) 49, 61-68.
Hollingsworth RI et al (1984a) J. Bacteriol. 160, 510-516.
Hollingsworth RI et al (1984b) Carbohydr. Res. 144, C7-C11.
Hrabak EM et al (1981) J. Bacteriol. 148, 697-711.
Kamberger W (1979) FEMS Microbiol. Lett. 6, 361-365.
Li D and Hubbell DH (1969) Can. J. Microbiol. 15, 1133-1142.
Matthysse AG et al (1981) J. Bacteriol. 145, 583-595.
Russa R et al (1982) FEMS Microbiol. Lett. 13, 161-165.
Sherwood JE et al (1984a) Planta 162, 540-547.
Sherwood JE et al (1984b). J. Bacteriol. 159, 145-152.
Solheim BJ and Fjelheim KE (1984) Physiol. Plant. 62, 11-17.
van Batenburg FD et al (1984) Physiol. Plant. 59, 363-372.
Zurkowski W (1980) Microbios 7, 27-32.

9. ACKNOWLEDGEMENTS

Portions of this work were supported by NIH Grant 1 RO1 GM34331-01,
USDA Competitive Grant 85-CRCR-1-1627, and the Michigan Agricultural
Experimental Station. This work is a collaboration with B. Rolfe. The
authors thank H. Sadoff, K. Nadler, D. Hubbell, and W. Brill for
helpful suggestions.

RECOGNITION AND INFECTION BY SLOW-GROWING RHIZOBIA

WOLFGANG D. BAUER, T.V. BHUVANESWARI, HARRY E. CALVERT, IAN J. LAW,
NASIR S.A. MALIK and STEPHEN J. VESPER.
Battelle-Kettering Research Laboratory, Yellow Springs, Ohio   45387.

ATTACHMENT OF RHIZOBIA TO ROOTS

A field isolate of R. japonicum has recently been isolated that may
help to answer the question of whether or not rhizobia need to attach to
the host root surface in some particular way in order to infect and
nodulate. This field isolate, designated 1007, was obtained from a
large nodule in the crown region of a soybean plant from a local field
(S.J. Vesper, T.V. Bhuvaneswari, W.D. Bauer, in preparation).

Initial observations have indicated that 1007 is unusual in its
attachment capabilities. In some experiments, attachment of this strain
was determined by incubating a diluted culture for 15 min with soybean
root segments excised from the infectible zone of 3-day-old seedlings,
rinsing to remove unattached or loosely attached bacteria, sonicating to
release firmly attached bacteria and plate counting of the released
bacteria as described (Vesper, Bauer 1985). In other experiments,
strain 1007 was incubated with intact roots for 6 h, gently rinsed and
examined under the microscope for attachment to root hairs.

As shown in Table 1, strain 1007 appears to be a very poor attacher
relative to strain 110. If it attaches at all, 1007 does not appear to
attach firmly enough to withstand gentle rinsing, whereas cells of 110
attach firmly enough to withstand a jet of water from a squirt bottle or
vigorous shaking in water suspension.

COMPARISON OF ATTACHMENT CAPABILITIES		
	Strain 110	Strain 1007
Firm attachment to root segments in 15 min	2700 bacteria/ segment	3 bacteria/ segment
Firm attachment to root hairs in 6 hours	133 bacteria/ root	0 bacteria/ root

Table 1

Figure 1

Given these differences in attachment capability, we wanted to know
whether there were corresponding differences between 1007 and 110 in
their ability to infect and nodulate the host. To determine the
relative nodulating ability of these two strains, roots of seedlings in
growth pouches were inoculated with a series of dilutions of the
bacterial cultures. The bacteria were squirted onto the root surface,
the infectible zones of the roots were then marked, and the nodules
which formed within this zone counted a week later (Bhuvaneswari et al.,
1980). The results of these experiments are shown in Fig. 1.

The results in Figure 1 indicate that 1007 nodulates just as efficiently, cell for cell, as 110, despite its apparent inability to attach firmly to the root surface. To us, this was a rather unexpected result and raises a real question in our minds concerning the need for any kind of firm attachment in the infection process.

These results have led us to wonder whether or not there are looser kinds of associations between rhizobia and roots that might be important, weak kinds of attachments that could not withstand rinsing. One can imagine three different classes of bacteria in the film of water on the root surface: firmly attached cells which remain associated with the root despite vigorous rinsing; loosely attached cells which are tied to the root surface in some way, but not strongly enough to withstand vigorous rinsing; and free cells, cells with no significant ties to the root surface.

We have devised a relatively simple way to distinguish the bacteria in each of these three classes (Vesper, Bauer 1985). A relatively small number of rhizobia are incubated with a large number of roots or root segments. The suspension is gently agitated so that every bacterial cell has a good chance to come in quick contact with a root surface and then to attach if it can. The number of _free_ bacteria at any given time is measured by withdrawing and plating replicate aliquots of the bacterial suspension. At the end of the experiment, the roots are removed, rinsed vigorously in a large volume of water to remove unattached and loosely attached bacteria, then subjected to homogenization or sonic vibration to release the firmly attached bacteria. The number of _firmly_ attached cells is measured by plating the bacteria recovered from the roots after rinsing. Since any bacteria associated with the root surfaces will disappear from the suspension of free bacteria, the total number of _loosely plus firmly_ attached bacteria can be estimated by subtracting the number of bacteria that remain free at the end of the experiment from the original number of free bacteria that one started with. The number of _loosely_ adhering bacteria is simply the difference between the total number of attached bacteria minus the number of firmly attached bacteria.

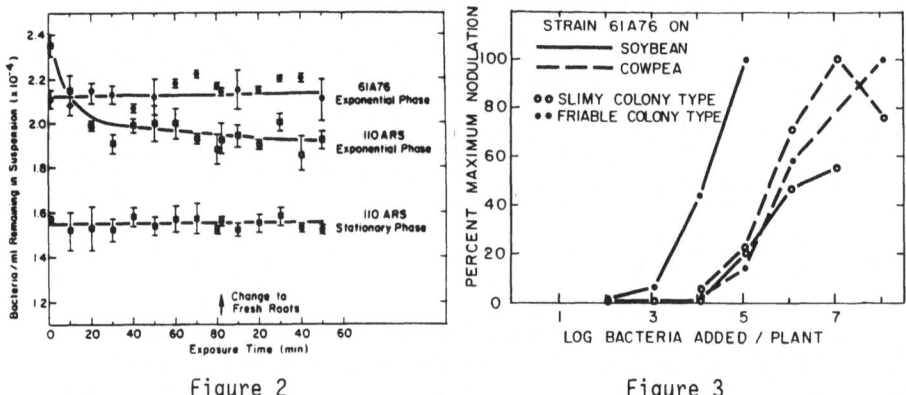

Figure 2   Figure 3

We have found the results of experiments like this really very surprising. With exponential phase cultures of strain 110, there was a rapid decline in the number of free bacteria (Fig. 2). This corresponds, of course, to a rapid increase in the number of firmly plus loosely attached bacteria. There was little further change in the number of adhering bacteria after approximately 20-30 minutes. Only about 15-20% of the bacteria in these suspensions disappeared from suspension and became detectably associated with the roots.

There are several alternatives to explain why the majority of the bacteria failed to associate with the roots. Perhaps only a few special sites for attachment were present on the roots, and these were quickly saturated. Perhaps a rapid equilibrium was established between bacteria associating and disassociating with the roots. Or perhaps the unattached bacteria lacked necessary attachment structures. To test these possibilities, we removed the original set of root segments from the bacterial suspension after 80 min and replaced them with a fresh set of root segments. There was no further reduction in the number of free bacteria (Fig. 2), indicating that few bacteria attached to the fresh set of roots.

This result and others (Vesper, Bauer 1985) appear to rule out the possibility of attachment site saturation and the possibility that an equilibrium had been established. We are left with the conclusion the bacteria remaining free in suspension did not attach because they didn't have the necessary equipment to do so.

In the experiments with exponential phase cultures of strain 110 (Fig. 2), it appears that there are three distinct subpopulations; i) the firmly attaching subpopulation comprising about 4% of the cells; ii) the loosely attaching subpopulation comprising about 12% of the cells; and iii) the non-attaching subpopulation comprising about 86% of the cells. Growth phase and strain can markedly affect the relative sizes of these subpopulations (Fig. 2). The genetic mechanisms that determine the subpopulation sizes remain to be identified.

As yet we know nothing about the substances or structures that are involved in loose adhesion of rhizobia to the root surface. It is possible that binding of host root lectin to receptor polysaccharides of the R. japonicum capsules could contribute to this loose kind of attachment. However, we should note that 50-80% of the cells in these exponential phase cultures have lectin-binding capsules. This does not correlate well with the 12% subpopulation capable of loose attachment. Firm attachment of R. japonicum appears to involve pili present on the surfaces of a small subpopulation of cells (S.J. Vesper, W.D. Bauer, manuscript submitted).

HOST RANGE AND SPECIFICITY

Our studies in regard to host range and specificity are in a rather early stage. What we offer here are just two observations in search of molecular and genetic explanations.

Typically, host range or specificity will be evidenced by having some Rhizobium isolates that nodulate only one of two potential hosts, other isolates that nodulate only the other of the two potential hosts, and perhaps some isolates that nodulate both of the potential hosts. Surprisingly, however, every strain of rhizobia which we know to nodulate soybean also nodulates cowpea. We don't know yet why this special relationship exists, but we are trying to find out. One possibility is that the soybean-cowpea isolates are just the same as the cowpea-specific isolates except that they have an extra tool or two that allows them to get into soybean as well as cowpea. Alternatively, it is also possible that the "cowpea specific" strains of rhizobia have all the tools needed to nodulate both soybean and cowpea, but that soybean plants reject these isolates as potential pathogens because they are wearing the wrong clothes. Mutant screening experiments are in progress to decide between these alternatives.

Since many Rhizobium isolates can nodulate both cowpea and soybean plants, does this mean that they do so by the same mechanisms? To all outward appearances, this appears quite possible. However, preliminary studies have revealed important quantitative differences in response to common isolates.

Different dosages of the same Rhizobium strain were used to inoculate both soybean and cowpea plants. This strain, 61A76, normally has two colony types, a slimy type and a friable type. All of the progeny of the friable colony type are also the friable type, but about 10% of the progeny of the slimy type are the friable type. So we can't inoculate with a pure culture of the slimy colony type, but actually add a 90:10 mixture of slimy and friable.

Results of experiments with this strain are shown in Figure 3. There are three kinds of host response differences that deserve mention.

1. The first is that it takes approximately 100 times as many cells of the slimy colony type to generate a half-maximal response in cowpea as it does in soybean. It is as though cowpea plants have built into them a 100-fold higher threshold for the rhizobia to climb over in activating crucial host responses. We don't understand the basis for this threshold.

2. The second point, not obvious from Fig. 3, is that even though it may take 100 times as many Rhizobium cells to cross the response threshold, the nodulation response in cowpea plants is much greater once the threshold is crossed. Maximum nodulation density is roughly 10 times higher in cowpea than in soybean. We don't understand the basis for this difference in nodule density.

3. The third point is that the two colony types of strain 61A76 have very different nodulation efficiencies on soybean, but essentially the same efficiencies on cowpea. It seems reasonable to suppose that the different efficiencies of the two colony types on soybean may be related to the amount of a limiting signal substance, with the slimy colony type producing more of this substance and therefore having a higher cell-for-cell level of efficiency. If this is true, then the equal response to the two colony type by cowpea implies that the amount of substance limiting to soybean is irrelevant to cowpea, perhaps because cowpea responds to a different signal substance.

As we shall see in the final section of this presentation, there is more direct evidence to indicate that rhizobia may have to supply different signals to nodulate different hosts.

We hope that the studies briefly indicated here will stimulate others to clarify the molecular and genetic bases of overlapping host ranges and host response thresholds. It is also hoped that these demonstrations of 10 and 100-fold response differences will help others to recognize that there are important quantitative differences that could easily be confused with qualitative host specificity differences.

INDUCTION OF CORTICAL CELL DIVISIONS

We have recently learned that the first one or two rounds of Rhizobium-induced cell divisions in soybean roots are complete within the first 24 h after inoculation, before penetration and infection thread formation (Calvert et al., 1984). This discovery has initiated attempts in our laboratory to track down the substances from rhizobia that induce these cortical cell divisions. There are some good hints to start with. It is known that cytokinins are effective inducers of plant cell division. Exogenous kinetin has been reported to induce cortical divisions in pea (Torrey, 1961). We know from our own observations (N.S.A. Malik, H.E. Calvert, and W.D. Bauer, unpublished) that exogenous benzyladenine induces cortical divisions in soybean, cowpea and alfalfa. It has also been reported that rhizobia synthesize cytokinins (Giannattasio, Coppola 1969; Phillips, Torrey 1970, 1972; Wang et al., 1982). So there are rough indications to suggest that rhizobia probably secrete cytokinins and that these probably induce host cortical cell divisions. To our knowledge, however, no one has isolated the inducer of cortical cell divisions from rhizobia, determined whether the substance is a typical cytokinin, or investigated the host specificity of its activity. This is what we are attempting to do.

All of our initial attempts to isolate an inducer activity from Rhizobium culture filtrates have failed (N.S.A. Malik, W.D. Bauer, unpublished). Studies by Bhagwat and Thomas (1982) and later studies by Halverson and Stacey (1984, 1985) indicated that substances in host root exudates of cowpea and soybean can stimulate the homologous rhizobia to nodulate more rapidly. It thus seemed possible that growth on host root exudate might be required in order to stimulate Rhizobium synthesis and secretion of the cortical cell division inducing factor. Bacteria were therefore grown in the presence of the host root and then culture filtrates from these bacteria were assayed by serial sectioning methods for inducer activity. Again the results were negative.

There are many possible reasons for these negative results. However, we began to wonder if perhaps the cortical division inducer was not secreted by the bacteria until the bacteria had begun to penetrate into the root. In order to test this possibility, we interposed a Milipore or Nucleopore membrane in between the bacteria and the host root, preventing penetration of the root by the bacteria, but allowing substances to diffuse freely between the two. The ability of bacteria in membrane patches to induce cortical cell divisions was assayed by complete serial sectioning of the root segments lying under the patches. This is a laborious technique, but one which reveals even a single induced cell division. A number of contamination tests were included to ensure that the observed cortical cell divisions were not caused by bacteria that leaked onto root. Lack of contamination was confirmed by the lack of infection threads and nodules in these experiments.

The patch experiments clearly showed that the homologous bacteria can induce cell divisions in the host root cortex even though the bacteria remain separated from the root surface. We believe these experiments provide reliable evidence that rhizobia secrete some diffusible substance that induces cell divisions in the host root cortex.

Serial sectioning methods have also been used to examine soybean roots inoculated with several different heterologous rhizobia. There was no evidence that the heterologous species induced any cortical cell divisions. I think you will agree that this is a rather surprising result. To the best of our knowledge, no one has suggested that the induction of cortical cell divisions would be a likely determinant of host specificity.

While these serial sectioning studies have provided clear evidence of a host specific response, at least in soybean, they do not tell us where the specificity resides. Perhaps the specificity resides in the inducing substances secreted by the rhizobia. Alternatively, as suggested by the earlier studies by Bhagwat and Thomas (1982) and by Halverson and Stacey (1984, 1985), the specificity may reside in the ability of substances in host root exudates to stimulate only homologous rhizobia to initiate infections.

Experiments were designed to decide between these two alternatives. We reasoned that if a soybean root and an alfalfa root were grown side-by-side and exposed to a Rhizobium japonicum strain that nodulated only soybean, then these bacteria would induce cell divisions only in the soybean root if the Rhizobium inducer was specific. However, if the soybean root stimulated the R. japonicum strain to secrete a non-specific inducer, then both the soybean root and the adjacent alfalfa would respond. To ensure that we were dealing only with diffusible inducing factors in our experiments, the rhizobia were kept in Nucleopore patches like those mentioned previously.

The results of these experiments (Table 2) are proving quite interesting. With only one exception, our present results indicate that rhizobia secrete inducers of cortical cell divisions that affect responses in their host, but not in non-host plants. From these results, it appears that the specificity of this response may reside in the inducer substances secreted by rhizobia. This makes us all the more anxious to purify and characterize these substances.

RHIZOBIUM INDUCTION OF CORTICAL CELL DIVISIONS

Strain of Rhizobium	Plants paired under Nucleopore patches		
	soybean & alfalfa	soybean & cowpea	cowpea & alfalfa
R. jap. 100	soybean + alfalfa -	soybean + cowpea -	cowpea + alfalfa -
R. spp. 32H1	soybean - alfalfa -	soybean - cowpea +	cowpea + alfalfa -
R. mel. 1021	soybean - alfalfa +	soybean - cowpea -	cowpea - alfalfa +

Table 2

We would like to end this discussion with a note of caution. Our most recent analyses have revealed that a few of the alfalfa roots in these experiments did respond slightly to the cowpea-specific strain 32H1. So, before jumping to any strong conclusions about the host specificity of these inducers, it will be important to isolate the substances and determine whether different plants respond specifically to different substances, or alternatively whether different plants respond to the same substances but with substantially different activation thresholds.

ACKNOWLEDGMENTS

The studies reported here have been supported in part by grants
83-CRCR-1-1283 and 84-CRCR-1-1508 from the USDA Competitive Grants
Office, DE-FG02-84ER13211 from the DOE Office of Basic Energy Science,
and PCM-8309137 from NSF. We gratefully acknowledge the find technical
assistance of J. Lakomski, M. Pence and S. Parker and the help of D.
Patten and S. Dunbar in preparation of the manuscript and figures.

REFERENCES

Bhagwat AA and Thomas J (1982) Appl. Environ. Microbiol. 43, 800-805.
Bhuvaneswari TV et al (1980) Plant Physiol. 66, 1027-1031.
Calvert HE et al (1984) Can. J. Bot. 62, 2375-2384.
Giannattasio M and Coppola S (1969) Frank. Giorn. Bot. Ital. 103, 11-17.
Halverson LJ and Stacey G (1984) Plant Physiol. 74, 84-89.
Halverson LJ and Stacey G (1985) Plant Physiol. 77, 621-625.
Torrey JG (1961) Exp. Cell Res. 23, 281-299.
Vesper SJ and Bauer WD (1985) Symbiosis, in press.

INVOLVEMENT OF ANTIBIOTIC PRODUCTION IN COMPETITIVENESS OF RHIZOBIUM LEGUMINOSARUM BV. TRIFOLII STRAIN T24.

Terese M. Barta and Eric W. Triplett, University of California, Riverside, CA 92521 U.S.A.

We have shown that antibiotic production is involved in the competitiveness of Rhizobium leguminosarum bv. trifolii strain T24. This ineffective strain was discovered by Schwinghamer, who also showed that strain T24 produces an antibiotic and is very competitive against other R. leguminosarum bv. trifolii strains for nodulation of clover roots. The strain T24 antibiotic is bacteriostatic toward R. leguminosarum bvs. trifolii and phaseoli, and R. fredii. Inhibition of R. leguminosarum bv. viceae strains is variable. The R. meliloti, Agrobacterium, and Bradyrhizobium strains tested were not affected. The antibiotic is produced constitutively on both rich medium and minimal medium. Clover plants were inoculated with varying concentrations of ineffective strain T24, in combination with effective USDA strain 2046. Shoot fresh weight and observations of plant health were used to assess competitive ability of nodule bacteria, similar to methods used by Berestetskii et al (1983). Stunted yellow plants were observed within four weeks of inoculation in those treatments in which there was a predominance of ineffective symbiosis with strain T24. Fresh weights of the shoots were also significantly lower than those of plants inoculated with only the effective strain. The effective stain was mixed in ratios of 1:10 and 1:1 (effective:ineffective) with a transposon Tn5 mutant of strain T24 lacking antibiotic production. Average plant fresh weight was significantly higher than in the 1:10 and 1:1 combinations with wild-type T24. A spontaneous mutant of strain 2046 resistant to the antibiotic, when co-inoculated in a 1:1 ratio with strain T24, produced healthy green plants with significantly higher fresh weights than plants inoculated with the sensitive parent strain in the same ratio with strain T24. These data demonstrate that antibiotic production is important in the competitiveness expressed by strain T24.

REFERENCES

Berestetskii, OA, Novikova, AT, and Knyazeva, VL. 1983. Simple method for assessing the competitive ability of nodule bacteria. Mikrobiologiya 52:651-657.

Schwinghamer, EA, and Belkengren, RP. 1968. Inhibition of rhizobia by a strain of Rhizobium trifolii: some properties of the antibiotic and of the strain. Arch. Mikrobiol. 64:130-145.

WHAT ATTRACTS RHIZOBIUM MELILOTI TO LOCALIZED SITES ON ALFALFA ROOTS?

Kostia Bergman, Rose C. Larosilière, and Melissa Maina.
Northeastern University, Boston, MA 02115

Previous work from this laboratory has shown that rhizobia are attracted to localized sites on the surface of legume roots. We have used *Rhizobium meliloti* strain SU47 and mutants derived from it for physiological analysis of this behavior. Non-motile mutants have completely lost their ability to swim because of defective synthesis (Fla⁻) or rotation (Mot⁻) of flagella. As expected these mutants are not attracted to root sites. Generally non-chemotactic mutants (Che⁻) are motile but have lost responses to a wide variety of attractants because of defective communication between specific receptors and the flagella motor. NR4300(*che-6*), a Che⁻ strain that swims smoothly but never tumbles is not attracted to root sites. Our expectation was that all Che⁻ mutants would be similarly defective however, NR3000(*che-1*) and NR4000(*che-3*) which do not respond to any of the known attractants tested are still attracted to root sites. The strong implication from the existence of these different types of Che⁻ mutants is that the wild type has two chemotaxis systems - one for locating potential food molecules and one for locating an appropriate root site. The amount of convergence between the two pathways and indeed whether they both exist in all of the individuals of a culture or only in specialized cells is presently unclear.

Before attempting to purify the unknown attractants from alfalfa roots we developed a reliable but quick assay of specific chemotaxis. Single colonies of rhizobia are stabbed into soft agar swarm plates containing minimal media and allowed to grow for 24h. Filter paper squares inoculated with droplets (8µl) of known or unknown attractants are placed 1cm from the colonies. A positive response, which is indicated by a flare of bacteria directed toward the filter paper, can be read in 24h. Using a single plate, interesting (infection site?) attractants that attract the mutants as well as the wild type (SU47) can be differentiated from common attractants (potential foods including amino acids and sugars) which attract only SU47.

Extracts from alfalfa root tips contain interesting attractant activity. It is stable to autoclaving but destoyed by hydrolysis with 6N HCl and by modification with dansyl chloride. In HPLC analysis using a reverse phase column in the presence of dodecyl-sulfate as a strong ion pair we have separated the activity into two fractions. Fraction A2 attracts NR3000, NR4000 and SU47 while Fraction A1 attracts only NR3000 and SU47. Both fractions contain more than one peptide as well as other substances. Further physiological and chemical analysis will require scale-up of the isolation procedure.

ATTACHMENT OF NONMOTILE AND NONCHEMOTACTIC MUTANTS OF RHIZOBIUM MELILOTI TO ALFALFA ROOTS

G. CAETANO ANOLLES, L.G. WALL, A.T. DE MICHELI, E.M. MACCHI* and
G. FAVELUKES
QUIMICA BIOLOGICA I, FACULTAD DE CIENCIAS EXACTAS, UNIVERSIDAD NACIONAL
DE LA PLATA AND *INIFTA, 1900-LA PLATA, ARGENTINA

Motility and chemotaxis by Rhizobium meliloti - though not absolute requirements for nodule formation in alfalfa - contribute to the competitive ability of strains to nodulate the host (Ames, Bergman, 1982; Gulash et al. 1984). We have studied the influence of motile behavior of derivatives of R. meliloti L5-30 on their early attachment to alfalfa roots during preinfection, and their ability to compete with wild-type for attachment.

METHODS

Behavioral mutants were isolated by enriching for bacteria which did not migrate in semisoft agar (Napoli, Albersheim, 1980). Motility and chemotaxis were characterized by direct L.M. and by Adler's capillary tube method (Adler, 1973); flagella were examined by T.E.M. and L.M. (Gray's staining). Bacterial attachment (4 hour) to 5-day old seedlings was quantitatively assayed as described elsewhere (G.C.A.,G.F., submitted). The time-course of nodule formation was studied on 5-day old seedlings inoculated with 10^5 rhizobia/ml and grown in Fåhraeus mineral solution.

RESULTS AND DISCUSSION

Spontaneous mutants: nonmotile without (Fla⁻), or with flagella (Mot⁻), and motile but generally nonchemotactic (Che⁻),were able to nodulate alfalfa as did parent L5-30. Mutants showed attachment to specific as well as nonspecific sites on alfalfa roots (G.C.A, G.F., submitted); their overall attachment was considerably smaller (up to 50 times). Host-specific attachment of L5-30 increased with time to reach a maximum after 6 hours of incubation. Compared with the parent strain, the time-rate of host-specific attachment of strain LP101 (Fla⁻), LP206 (Mot⁻) and LP302 (Che⁻) was decreased 50 or more times. Their ability to compete with wild-type R. meliloti for 4 hour attachment (G.C.A., G.F., submitted) was markedly diminished at intermediate competitor concentrations (10^4-10^6/ml); higher than 10^6/ml gave full competition, as L5-30. Nodulation by these mutants was delayed compared with L5-30.

These results indicate that: a. the absence of flagella, motility and chemotaxis does not prevent the expression of symbiont recognition during attachment, and b. the competitive advantage for nodulation conferred by motility and chemotaxis is already expressed during the early process of attachment to roots.

Adler, P. (1973) J. Gen. Microbiol. 74, 77-91.
Ames, P., Bergman, K. (1982) J. Bacteriol. 148, 728-729.
Gulash, M. et al. (1984) Appl. Environ. Microbiol. 48, 149-152.
Napoli, C.A., and Albersheim, P. (1980) J. Bacteriol. 141, 979-980.

Supported by CICBA, CONICET, and SECYT (Argentina).

USE OF A FLUORESCENT STAIN TO CHARACTERIZE ALFALFA NODULE INITIATION

MARK DUDLEY AND SHARON LONG
DEPARTMENT OF BIOLOGICAL SCIENCES, STANFORD UNIVERSITY, STANFORD CA 94305,
U.S.A

The inoculation of alfalfa (<u>Medicago</u> <u>sativa</u>) with <u>Rhizobium</u> <u>meliloti</u> induces <u>de</u> <u>novo</u> formation of the symbiotic nitrogen fixing organ, the root nodule. The events in meristematic nodule morphogenesis include the dedifferentiation and proliferation of cortical cells, infection thread formation and penetration into the root cortex, bacterial release into host cells, and cellular differentiation. The earliest events have been characterized in pea (Newcomb et al, 1979), while later morphogenesis has been studied in alfalfa (Hirsch et al, 1982; Vance et al, 1980). We have developed a fluorescent staining technique and used it to study early events in alfalfa nodule development. Nodules or root segments were fixed embedded in JB-4 resin, and sectioned into 1-6 micron sections. These sections were stained with a combination of DAPI (4'-6-damidino-2-phenylindole) at 6 ug/ml and AO (acridine orange) at 25 ug/ml in PBS, destained 3 times in PBS, then viewed with UV epifluorescent illumination using a UG-1 exciter filter, a "U" dichroic mirror, and a 420nm barrier filter. To facilitate analysis of early plant responses, a modification of the Turgeon and Bauer spot inoculation technique was used.

The first cells to undergo mitosis were observed at 21 hours after inoculation dividing anticlinally in the root inner cortex. By 24 hours a distinct region of the cortex displayed active cell division, enlarged interphase nuclei with prominent nucleoli, brightly staining cytoplasm, and cytoplasmic strands across the cell vacuole. At 48 hours infection threads were often visible in epidermal and hypodermal cells outward from activated regions, which displayed anticlinal and periclinal divisions. Incipient nodules 3 days after inoculation showed branched infection threads throughout the activated region. Cells in the center of these pre-emergent nodules were isodiametric and filled with cytoplasm and numerous small vacuoles, and showed no bacteria were not observed in the cytoplasm. Differentiation of vascular bundles and xylem vessels, and release of bacteria into host cells was evident in samples harvested 4.5 days after inoculation. Mitotic activity was confined to cells in the apical dome of the nodule, and the characteristic pattern of a nodule meristem was established.

The staining technique discussed here is complementary to ultrastructural analysis of morphology, and facilitates the visualization of infection threads, bacteroids within host cells, and overall structural organization in semi-thin sections. We plan to use this technique further to characterize the phenotypes of mutants involved in nodule differentiation or morphogenesis.

REFERENCES

Hirsch AM et al (1982) J. of Bacteriol. 151, 411-419.
Newcomb W et al (1979) Can. J. Bot. 57, 2603-2616.
Turgeon BG, Bauer WD (1983) Protoplasma 115, 122-128.
Vance CP et al (1980) Physiological Plant Pathology 17, 167-173.

ACKNOWLEDGEMENTS
We are grateful for support of M.D. by the McKnight Foundation and for a Shell Research Foundation award to S.L.

RHIZOBIUM MELILOTI SPECIFICALLY ATTACHED TO ALFALFA ROOTS, OBLIGATORY PRECURSOR TOWARDS NODULATION

G. FAVELUKES and G. CAETANO ANOLLES

QUIMICA BIOLOGICA I, FACULTAD DE CIENCIAS EXACTAS, UNIVERSIDAD NACIONAL DE LA PLATA, 1900-LA PLATA, ARGENTINA

Initial attachment of rhizobia to legume roots, a very early event in the process of infection and nodule formation, has been found to show already the host-symbiont specificity of nodulation (Hsn) in clover (Dazzo et al. 1976), in peas (Kato et al. 1980), and in soybeans (Stacey et al. 1980), and is generally regarded as an early step for symbiotic recognition. In the alfalfa system, early specific binding (Esb) of Rhizobium meliloti to the root surface has been recently demonstrated (G.C.A., G.F., submitted), in a study of the effects of competitor bacteria on the attachment of an infective R. meliloti "indicator" strain. While saturating levels of heterologous bacteria decreased it to a limited extent, homologous rhizobia abolished it completely. The difference was taken to represent the occupation and blockage of symbiont-specific binding sites on the roots, only by the homologous rhizobia. Here we address the question of the significance of early specific binding to roots, for infection and nodulation in alfalfa. We have done so by studying the root-binding properties of three non-nodulating, Tn-5 induced mutants of R. meliloti 1021 (Meade et al. 1982; Hirsch et al. 1982) and their effects upon the nodulation by the parent strain.

The degree of attachment of R. meliloti L5-30-1 indicator strain (in low concentration) to alfalfa roots at 4 hours in Fåhraeus solution was determined in the presence of different competitor bacteria. The extent of nodulation of alfalfa seedlings by an R. meliloti infective strain in the presence of noninfective competitor strains was axenically assayed in vermiculite supported nitrogen-free mineral solution in closed jars.

All three nod⁻ mutants 1027, 1028 and 1126 were able to attach to alfalfa roots. Competitors 1126 and - to a lesser extent - 1027 inhibited completely attachment of an R. meliloti indicator strain. Competitor 1028 in turn caused a limited inhibition. Thus, while 1126 and 1027 conserve the parental ability to bind to specific root sites (Esb$^+$), 1028 phenotype is Esb$^-$. The extent of nodulation by parental strain 1021 was markedly inhibited by Esb$^+$ competitors 1126 and 1027. This effect appears to require specific binding of the competitor, since Esb$^-$ mutant 1028 was not inhibitory. We conclude that: a. early binding of rhizobia to symbiont-specific root sites is an obligatory step during preinfection, and b. only specifically bound rhizobia progress towards infection and nodulation.

Dazzo, F.B. et al. (1976) Appl. Environ. Microbiol. 32, 168-171.
Hirsch, A.M. et al. (1982) J. Bacteriol. 151, 411-419.
Kato, G. et al. (1980) Agric. Biol. Chem. 44, 2843-2855.
Meade, H.M. et al. (1982) J. Bacteriol. 149, 114-122.
Stacey, G. et al. (1980) Plant Physiol. 66, 609-614.

Acknowledgments: We are indebted to Drs. W.J. Buikema and F.M. Ausubel for nod⁻ mutants. Supported by CICBA, CONICET, and SECYT (Argentina).

ATTACHMENT OF RHIZOBIUM LEGUMINOSARUM 248 TO PEA ROOT HAIRS.

JAN W.KIJNE, GERRIT SMIT, CLARA L.DIAZ and BEN J.J.LUGTENBERG,
DEPT. OF PLANT MOLECULAR BIOLOGY,BOTANICAL LABORATORY,
UNIVERSITY OF LEIDEN, NONNENSTEEG 3, 2311VJ LEIDEN, THE NETHERLANDS.

1. INTRODUCTION

Rhizobium-root hair attachment-inhibition experiments with lectin haptenic monosaccharides have led to seemingly conflicting results and to
opposite opinions about the lectin-recognition hypothesis. This situation
possibly is caused by the use of different rhizobial culture media and
by quantification of attachment to non-essential parts of the roots.

2. MATERIAL AND METHODS

R. leguminosarum 248 was batch-cultured in TY-medium (tryptone-yeast
extract, Beringer 1974) or YM-medium (yeast extract-mannitol, Zurkowski,Lorkiewicz 1978, with 150 instead of 40 mg/l $CaCl_2$). Lateral roots
of 6-8 days old pea seedlings (Pisum sativum cv. Rondo, coarse gravel,
N-free Raggio medium) were incubated in a rhizobial suspension (1-2 x
10^8 cells/ml in 25 mM phosphate buffer). After various incubation periods
at room T, 100-150 root hairs in the developing root hair zone were
classified into four categories (LM phase contrast, 400x), ranging from
- (no attachment) to ++ (abundant tip attachment).

3. RESULTS

The optimal pH for attachment was 7.5 for rhizobia cultured in either
medium. Optimal attachment of TY-rhizobia was found at late log/station-
ary phase, which in this medium is specified by carbon-limitation. YM-
rhizobia showed two attachment optima, at OD_{620} 0.5 and, especially,
0.9; the culture conditions specifying these growth phases are not yet
known. In both media optimal attachment is accompanied by autagglutin-
ation of the rhizobia. YM-rhizobia adhered quicker to pea root hairs
(25% ++ root hairs in 30 min) than TY-rhizobia (5% ++ root hairs in 30
min). Attachment of TY-rhizobia was not inhibited by monosaccharides
(50 mM, including pea lectin haptens), but was almost completely inhibited by 0.1 M NaCl. On the contrary, attachment of YM-rhizobia was
specifically reduced in the presence of pea lectin haptens; inhibition-
activity of haptens was linearly correlated with lectin affinity. NaCl-
addition up to 0.2 M resulted only in a 50% reduction of attachment.

4. CONCLUSION

The results point at two different root hair attachment mechanisms for
R. leguminosarum 248, determined by the culture conditions. As for the
lectin-recognition hypothesis: are the pea rhizosphere conditions TY-
like or YM-like?

5. REFERENCES

Beringer, JE (1974) J. gen. Microbiol. 84, 188-198.
Zurkowski, W and Lorkiewicz, Z (1978) Gen. Res. 32, 311-314.

COMPETITION BETWEEN STRAINS OF RHIZOBIUM TRIFOLII
FOR NODULATION OF WHITE CLOVER

P.M. STEPHENS, J.E. COOPER and A.J. HOLDING.
The Queen's University of Belfast, Newforge Lane, Belfast,
BT9 5PX,Northern Ireland.

Nodulation in white clover occurs in two distinct phases:
an initial transient infectibility that is restricted to
the developing root hair zone (DRH) and the no root hair
zone (NRH), and a delayed, induced susceptibility of mature
clover root hair cells (Bhuvaneswari et al., 1981).

A method was developed for determining the location of
infectible host cells on the roots of white clover in a
free-flowing rooting solution system. Nodulation profiles
over a period of 10 days were obtained for 3 strains of
Rhizobium trifolii (P3 (rif), IDL (ery) and 1186 (spec))
at pH 6.7 and pH 5.0 in N-free rooting solution at $18^{o}C$.

For all 3 strains the majority of nodules first appeared
in the NRH zone at both pH values. Delayed inoculation
indicated that this region became progressively less
susceptible to infection over a period of 12 hours. In
single culture, strains differed significantly with regard
to the percentage of plants nodulated in the NRH zone
after 10 days at both pH values. There was a positive
correlation between the ability of strains to form nodules
in the NRH zone in single culture and their competitive
ability in mixed culture. This relationship was maintained
or amplified when solution pH was lowered from 6.7 to 5.0.

REFERENCE

T.V. Bhuvaneswari, A.V. Bhagwat and W.D. Bauer (1984)
Plant Physiology 68, 1144-1149

MEMBRANE POTENTIAL CHANGES IN SOYBEAN ROOTS
INOCULATED WITH INFECTIVE AND NONINFECTIVE RHIZOBIA

T. ÉRSEK, A. NOVACKY, S.G. PUEPPKE
UNIVERSITY OF MISSOURI, DEPT. OF PLANT PATHOLOGY, COLUMBIA, MO 65211

1. INTRODUCTION

Transmembrane potential difference (E_m) is a sensitive indicator of the status of plant cell membranes (Novacky 1983). We conducted a comparative study of the influence of Nod⁻ R. meliloti 102F51 and Nod⁺ R. japonicum 74 on E_m of root cells of soybean cv. McCall. Our objectives were to determine whether (i) membranes of soybean roots respond to Nod⁻ strains and (ii) electrophysiology has potential usefulness for analysis of mutants.

2. MATERIALS AND METHODS

Seedlings were dip-inoculated with 5×10^8 rhizobia ml^{-1} and grown in plastic pouches. Segments extending an equal distance on each side of the initial root tip were excised, mounted on holders, aged in an aerated perfusion solution and analyzed (Novacky, Ullrich-Eberius 1982).

3. RESULTS

Nod⁻ cells, though inducing no visible signs, alter the E_m of cortical cells to a greater extent than do Nod⁺ cells (Table 1). The passive (diffusion, E_D) rather than the functional (pump, E_P) properties of plasmalemma are affected at 1 day. The reduction of E_m is transient. The lack of increase in E_P with age of R. meliloti-inoculated tissue, unlike in the the the compatible system and control, indicates that the link between cellular metabolism and membrane transport may also be affected.

TABLE 1. $E_m - E_D = E_P$ values (mV) in cortical cells of soybean roots.

Days	Control	R. japonicum	R. meliloti
1	149 − 116 = 33	138 − 112 = 26	131 − 98 = 33
4	145 − 108 = 37	150 − 112 = 38	143 − 107 = 36
7	149 − 107 = 42	150 − 106 = 44	137 − 102 = 35

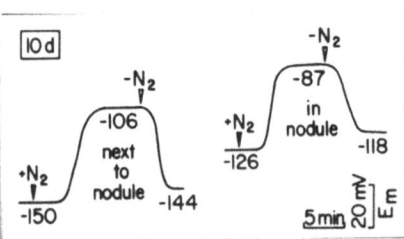

FIGURE 1. Comparison of E_m in nodule cells and cortical cells next to a nodule at 10 days.

The E_m of nodule cells is considerably lower than that of adjacent root cells (Fig. 1). The decreased E_m is apparently due to changes in the lipid matrix and/or ion channels of the membrane. In summary, changes in E_m seem to be not an immediate result of previously reported bacterial adsorption (Pueppke 1984) and they do not require the formation of infection threads. According to our results electrophysiology appears to be a useful technique for analysis of symbiotically defective mutants, too.

4. REFERENCES

Novacky A (1983) In Callow JA, ed. Biochemical Plant Pathology, pp 347-366, John Wiley & Son Ltd.
Novacky A and Ullrich-Eberius CI (1982) Physiol. Plant Path. 21, 237-249.
Pueppke SG (1984) Plant Physiol. 75, 924-928.

LECTIN INDUCTION OF NODULATION COMPETENCY IN BRADYRHIZOBIUM JAPONICUM

L. J. HALVERSON, D. L. GERHOLD, S. AUGER AND G. STACEY. DEPARTMENT OF MICROBIOLOGY, UNIVERSITY OF TENNESSEE, KNOXVILLE, TN USA.

To investigate the recognition process involved in the Bradyrhizobium japonicum soybean association, a mutant of B. japonicum (strain HS111) defective in nodule initiation was utilized. Nodulation in soybeans is developmentally restricted to the area between the root tip (RT) and smallest emerging root hair (SERH). The position of the uppermost nodule from the RT mark, made at the time of inoculation, enables one to infer the relative rate at which nodulation is initiated. Previously, we reported that pretreatment of mutant strain HS111 in soybean root exudate (RE), soybean lectin (SBL) or a galactose-specific root protein reversed its delayed nodulation mutant phenotype. The following data will demonstrate that lectin also induces nodulation competency in the wild type B. japonicum strain USDA110.

Strain USDA110 when inoculated on soybean formed nodules with an average distance of the uppermost nodule being approximately 2.3 ± 0.3 mm above the RT mark and 18 mm below the RT mark, respectively. Inoculation of strain USDA110 at a cell concentration of 10^4 cells/seedling exhibits a delay in nodule initiation similar to mutant strain HS111. The average distance of the uppermost nodule from the RT mark for the wild type at a cell concentration of 10^4 cells/seedling is approximately 12 mm below the RT mark. Strain USDA110 pretreated in soybean RE or SBL prior to inoculation at a concentration of 10^4 cells/seedling does not exhibit the pattern of delayed nodule initiation. The average distance of the uppermost nodule from the RT mark following SBL or RE pretreatment is approximatley 1 mm above the RT mark.

To determine if lectin induces RNA or protein synthesis, the effect of lectin on the nodulation characteristics of strain USDA110 was tested in the presence of antibiotics. USDA110 cells were pretreated in minimal media containing 10 ug/ml SBL and 50 ug/ml chloramphenicol or 25 ug/ml rifampicin. The data in Table I show that both these antibiotics inhibit the nodulation enhancing characteristics of SBL.

Table 1. Effect of antibiotics on the nodulation enhancing property of SBL.

Treatment	Ave. distance of Uppermost nod. from RT	% Nod. only below RT
USDA110 MM	-15.1 ± 2.9	73
USDA110 SBL	$- 0.3 \pm 1.0$	33
USDA110 MM-chl	-10.5 ± 3.3	64
USDA110 SBL-chl	-10.4 ± 2.8	69
USDA110 MM-rif	-11.6 ± 2.1	75
USDA110 SBL-rif	-14.8 ± 3.1	75

MM = Minimal Media; SBL = soybean lectin; chl = chloramphenicol; rif = rifampicin.

NODULE SUPPRESSION IN SOYBEANS PRODUCED BY SUPEROPTIMAL INOCULA

Stephen T. Takats, Temple University, Philadelphia, PA 19122, USA

Double serial inocula of <u>Rhizobium japonicum</u> 110 were applied to primary roots of Pride 216 soybean seedlings in plastic growth pouches, following the protocol of Pierce & Bauer (1983), Plant Physiol. 73, 286-290. Primary inocula were of optimal size for nodulation, 10^5 ml^{-1}. Secondary inocula varied in size from 0 to 10^{10} ml^{-1}, and were applied 10 hours after the primary. Superoptimal secondary inocula resulted in the suppression of nodules in the basal region of the root, whereas optimal or suboptimal secondary inocula had little effect (see figure below; RT_1 and RT_2 indicate the positions of the root tips at the times of the two inoculations.) Suppression was dependent on the size of the superoptimal secondary inoculum and did not occur with UV-killed bacteria. The loss of nodules occurred in the region of the root which would normally have responded to the primary inoculum; the nodules which appeared apically could have originated from either or both inocula. These results are inconsistent with those obtained by Pierce & Bauer; from their results, no effect from a secondary should have been observed following an optimal primary inoculum. Apical suppression of nodulation by the establishment of prior nodules on the root is well known for legumes. It is likely that the basal suppression reported here depends not only on the size of the secondary inoculum but on the short time between the applications of the two inocula. The basis for the suppression is not yet understood. It may be due to an interaction on the root surface between the invasive bacteria in the primary inoculum and superficial bacteria in the secondary inoculum.

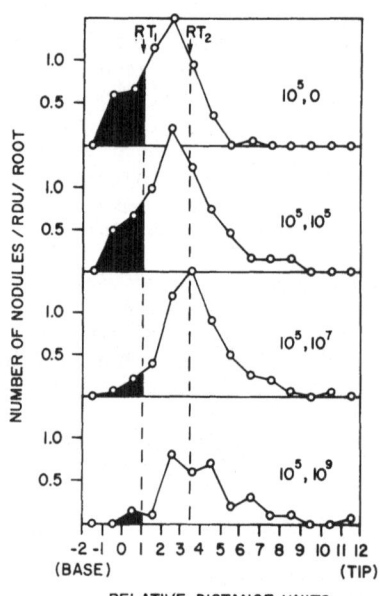

INCORPORATION OF LIPOPOLYSACCHARIDE ANTIGENS FROM RHIZOBIUM BACTEROIDS INTO THE PERIBACTEROID MEMBRANE OF PEA ROOT NODULES

NJ BREWIN[1], DJ BRADLEY[2], EA WOOD[1], B WELLS[1], AP LARKINS[2], G GALFRE[2], GW BUTCHER[2]

1. JOHN INNES INSTITUTE, COLNEY LANE, NORWICH NR4 7UH, UK
2. MONOCLONAL ANTIBODY CENTRE, AFRC INSTITUTE OF ANIMAL PHYSIOLOGY, BABRAHAM, CAMBRIDGE CB2 4AT

Rat monoclonal antibodies that react with Rhizobium lipopoly-saccharide (LPS) from bacteroids were used to examine the specificity, size heterogeneity and cell distribution of LPS antigens. Two lines of evidence indicated that LPS from bacteroids was physically associated with the plant-derived peribacteroid membrane. Peribacteroid material, isolated from pea root nodule homogenates by differential centrifugation, contained about 1% of the bacterial LPS antigen. In a sandwich immunoassay an immobilised monoclonal antibody, AFRC MAC 57, was used to bind LPS antigen from this material. Subsequently, binding of peribacteroid membrane was detected using biotinylated AFRC MAC 64 as an immunochemical marker for this membrane (Brewin et al., 1985). The physical association of MAC 57 antigen (LPS) and MAC 64 antigen (peribacteroid membrane glycoprotein) was disrupted by treatment with detergent (0.5% NP40) which did not inhibit the binding of individual monoclonal antibodies to their respective antigens. The binding of peribacteroid membrane to immobilised MAC 57 could also be inhibited competitively by the addition of purified LPS derived by phenol extraction from free-living cultures of R. leguminosarum.

In another series of experiments pea nodules were fixed in glutar-aldehyde, embedded for electron microscopy and thin sections were treated with MAC 57 and stained with immunogold. In addition to LPS antigen being detected in the cell wall of bacteroids, it was also present in small amounts in the peribacteroid membrane.

Although LPS derived from free-living cultures of Rhizobium was very heterogeneous, resulting in a ladder of immuno-staining bands after SDS gel electrophoresis, the LPS from bacteroids was predominantly of the fast-migrating form. However, growth of Rhizobium in culture medium containing low concentrations of $CaCl_2$ resulted in a size distribution of LPS similar to that observed for bacteroids. In addition, there was enhanced release of LPS antigen into the extracellular medium. These changes may be associated with the inhibition of capsular and extracellular polysaccharide synthesis under conditions of low Ca^{++}, resulting in the exposure of naked bacterial outer membrane.

On the basis of these observations, it is proposed that the bacteroid outer membrane may 'bleb off', traverse the peribacteroid space and fuse with the peribacteriod membrane. In this way, material from the bacterial periplasmic space may be transferred to the plant cytoplasm.

REFERENCES

Brewin NJ et al., (1985) EMBO Journal 4, 605-611.

ANALYSIS OF THE EXTRACELLULAR POLYSACCHARIDES (EPSs) FROM RHIZOBIUM
MUTANTS ALTERED IN THEIR SYMBIOTIC PLASMIDS (pSYM).

R.W. CARLSON/L. CUMMINS/F. HUSTMEYER
CHEMISTRY DEPARTMENT, EASTERN ILLINOIS UNIVERSITY, CHARLESTON, ILL.
61920

1. INTRODUCTION

Considerable effort has been directed towards determining the role of
Rhizobium polysaccharides in symbiosis, one of these polysaccharides
being the acidic EPSs. This report presents results from the analysis
of mutants of R. trifolii and R. leguminosarum which have been altered
in their pSyms. Mutants derived from R. trifolii ANU843:
ANU851Nod-Hac- has a Tn5 insert in nod region I of pSym, ANU2252Nod-Hac-
also has a Tn5 insert in nod region I of pSym, ANU871Nod-Hac- has a
40-50kb deletion of pSym which includes regions I and II of pSym,
ANU845Nod-Hac- is missing the pSym, ANU845pRT150Nod-Hac++ is derived
from ANU845 and contains region I of pSym, ANU845pBR1ANNod+Fix+ is
derived from ANU 845 and contains another pSym. Mutants derived from R.
trifolii ANU 794: ANU437Nod+Fix-Muc- contains a Tn5 insert which is not
in pSym, ANU437RP4+Nod+Fix+Muc+ is derived from ANU437 and contains a
4kb piece of DNA from the wild-type strain. Mutants derived from R.
leguminosarum 128C53sm(r)rif(r): ANU54Nod-Muc- (also known as
128C53sm(r)rif(r)exo(-1)) is missing pSym, ANU54pBR1AN Nod+(on
clover)Fix- contains an R. trifolii pSym, ANU54pJB5J1 Nod+(on pea)Fix-
contains an R. leguminosarum pSym. All mutants from Dr. B. Rolfe.

2. PROCEDURE

EPSs were precipitated from culture supernatants by EtOH. EPSs from R.
trifolii strains were purified by precipitation with CTAB. EPSs from R.
leguminosarum strains and R. trifolii ANU437 were purified by Sepharose
4B chromatography (in EDTA/TEA=10mM/30mM). Neutral sugars were
quantitated by GC analysis of their alditol acetates. Glycosyl linkages
were determined by methylation and GC/MS analysis (NIH Regional Center,
Wash. Univ., St. Louis, Mo.). Acyl groups, pyruvate, uronic acids, and
KDO were determined by colorimetric assays. Antisera was prepared from
white New Zealand rabbits. LPSs and EPSs were analyzed by SDS-PAGE.
Immunoblots were made by transfer to nitrocellulose, incubation with
antisera and visualization with goat anti-rabbit peroxidase conjugate.

3. RESULTS

EPSs from ANU843 and its mutants all have the same glycosyl and pyruvyl
linkages. Acidic EPS could not be detected in any R. leguminosarum
mutants derived from strain 128C53sm(r)rif(r), including the mutants
with pSyms. Immunochemical tests suggest that all three R.
leguminosarum mutants excrete LPS, and that the mutants containing pSyms
excrete additional antigens. These antigens require further
investigation. Mutant ANU437 produces 0.9% of the parental acidic EPS.
It also excretes LPS. Strain ANU437RP4+, which contains a 4kb piece of
DNA from the wild-type strain, is corrected in its ability to fix
nitrogen and produce acidic EPS.

4. CONCLUSIONS

Results from ANU843 mutants suggest that pSym does not dictate EPS
glycosyl or pyruvyl arrangements. R. leguminosarum results support this
conclusion. In addition the results suggest that EPS is not required
for early symbiotic events. Results for the ANU794 mutants support the
idea that EPS is not required for early events of symbiosis, but suggest
requirement for events, subsequent to initial infection, which lead to
proper nitrogen fixation.

ALTERATIONS IN CAPSULAR POLYSACCHARIDE OF TN5-INDUCED MUTANTS IN THE
RHIZOBIUM TRIFOLII NODULATION REGION.

A.E. GARDIOL, R.I. HOLLINGSWORTH, F.B. DAZZO, M.A. DJORDJEVIC, AND B.G.
ROLFE, MICHIGAN STATE UNIVERSITY, EAST LANSING, MI 48824 U.S.A., AND
AUSTRALIAN NATIONAL UNIVERSITY, CANBERRA CITY, AUSTRALIA.

Tn5 mutants of R. trifolii 843 in three nodulation regions (I,II,III)
(Djordjevic et al. 1985, submitted) of the sym plasmid were examined.
The objective of this work was to define if these strains had alterations
in their capsular polysaccharide (CPS) and lectin-binding ability. We
also studied the enzymatic incorporation of pyruvate into CPS of R. trifolii 843, 0403 rif and mutant strains.
CPS was isolated and a glucuronic acid-specific lyase (PD-I) was used to
hydrolyze the CPS into its oligosaccharide repeating units (OS) (Higashi
et al. 1978) (Hollingsworth et al. 1984). PD-I was used for kinetic
study of CPS depolymerization rates and non-carbohydrate substitutions of
OS were measured by ^{1}H-NMR (Hollingsworth et al. 1984). In vitro incorporation of pyruvate into CPS was measured using EDTA-treated cells(Tolmasky et al. 1980). Lectin-binding ability was examined by indirect immunofluorescence (Dazzo et al. 1982).

RESULTS AND DISCUSSION CPS from mutants in region I(Hac) showed different
rates of depolymerization than wild type CPS. OS from mutants in this region also had different levels of acetic and pyruvic acid substitutions.
In contrast, CPS depolymerization rates as well as the level of non-
carbohydrate substitutions were unchanged by mutations in regions II and
III. Mutant strains had a significantly lower ability to bind the clover
lectin. An in vitro assay to measure the enzymatic incorporation of pyruvate into R. trifolii CPS was developed using EDTA-treated cells, UDP-
Glucose, UDP-Galactose, UDP-Glucuronic acid, and $[1-^{14}C]$PEP as pyruvate
donor. Pyruvylation occurred at the lipid-bound intermediate stage from
$[1-^{14}C]$PEP. Incorporation of radioactivity into glycoconjugates soluble
in organic solvent (CPS pyruvyl transferase activity) was measured for
two wild type R. trifolii 843, 0403 rif and mutants having more or less
pyruvate in the CPS. The levels of this activity in two mutants of R.
trifolii 843 within region I were significantly different. R. trifolii
251 with a single Tn5 in the sym plasmid and more pyruvate in the CPS
than wild type R. trifolii 0403 rif, had higher CPS pyruvyl transferase
activity. We conclude that Tn5 mutations in certain nod genes on the
sym plasmid affect the levels of non-carbohydrate substitutions in the
CPS of R. trifolii and the lectin binding ability of mutant strains.
Functions related to polysaccharide synthesis may be encoded in the
sym plasmid.

REFERENCES Dazzo FB et al (1982) Appl. Environ. Microbiol. 44:478-490.
Higashi S et al (1978) J. Gen. Appl. Microbiol. 24:143-163.
Hollingsworth RI et al (1984) J. Bacteriol. 160:510-516.
Tolmasky M et al (1980) Arch. Biochem. Biophys. 203:358-364.

STRUCTURAL STUDIES ON THE LIPOPOLYSACCHARIDE OF RHIZOBIUM TRIFOLII 0403.

RAWLE I. HOLLINGSWORTH, ESTELLE M. HRABAK AND FRANK B. DAZZO
DEPARTMENT OF MICROBIOLOGY, MICHIGAN STATE UNIVERSITY, EAST LANSING,
MI 48824-1101 USA

INTRODUCTION

In order to better understand i) the phenomenon of variation of biological activity of lipopolysaccharide (LPS) with growth phase and ii) the effect of mutations in the nod(ulation) genes of rhizobia on the structures of their surface polysaccharides, it is necessary to have a good understanding of the structure and composition of these polysaccharides. This information is essential in determining the role of surface polysaccharides in the Rhizobium-legume symbiosis.

METHODS

Cells were grown in BIII defined medium at 30 C to either mid-exponential or early stationary phases and the LPS isolated as described (Hrabak et al., 1981). Lipid A was cleaved from the LPS by hydrolysis with 1% acetic acid at 100 C for 1 hr or by 0.1 M HCl at 100 C for 15 min, followed by chloroform extraction. Carbohydrates were analyzed as alditol acetates by GC/MS and by ^{1}H-NMR spectroscopy of the aqueous phase from the lipid A extraction. Fatty acids were determined by GC/MS after conversion to methyl esters with methanolic HCl. Standards of 2-0-methylrhamnose, 2-0-methylfucose, 2-0-methylquinovose, N-methyl-3-amino-3,6-dideoxygalactose and N-methyl-3-amino-3,6-dideoxyglucose were prepared by direct synthesis.

RESULTS

2-0-Methylrhamnose, 2-0-methylfucose, N-methyl-3-amino-3,6-dideoxy-galactose and glucuronic acid (confirmed by deuterium labelling) were confirmed as components of LPS from both mid-exponential and early stationary phases. With the exception of glucuronic acid, all the sugars mentioned were present in larger proportions in LPS from the later growth phase. Other sugars known to be present in LPS of R. trifolii 0403 are rhamnose, fucose galactose, glucose, quinovosamine, heptose and glucosamine. The lipids present were mostly 3-hydroxy fatty acids, which were linked (apparently through 3-hydroxy hexa- and/or octa-decanoic acids) to a novel monosaccharide component. The lipid A was shown to be radically different to the glucosamine disaccharide model accepted for the Enterobacteriaceae. No phosphorus was detected by either colorimetric or ^{31}P-NMR studies. Infrared, ^{1}H-NMR and methylation studies with diazomethane all confirmed that a large proportion of the hydroxy fatty acids were esterified at the 3-hydroxy groups with other fatty acids. The picture of the lipid A which emerges is one of a complex of cross-linked fatty acids attached to a monosaccharide unit through 3-hydroxy hexadecanoic acid and/or 3-hydroxyoctadecanoic acid through N-acyl and O-acyl linkages.

REFERENCES

Hrabak, EM et al (1981) J. Bacteriol. 148, 697-711.

COMPOSITIONAL VARIATION IN LIPIDS FROM LIPOPOLYSACCHARIDE OF RHIZOBIUM TRIFOLII 843 AND NOD⁻ MUTANTS.

ESTELLE M. HRABAK, RAWLE I. HOLLINGSWORTH, FRANK B. DAZZO, MICHAEL A. DJORDJEVIC, AND BARRY G. ROLFE
DEPARTMENT OF MICROBIOLOGY, MICHIGAN STATE UNIVERSITY, EAST LANSING, MICHIGAN 48824 USA AND GENETICS DEPARTMENT, AUSTRALIAN NATIONAL UNIVERSITY, CANBERRA, AUSTRALIA

Lipopolysaccharide (LPS) was isolated from wild type Rhizobium trifolii 843 (at mid-exponential or early stationary growth phase) and from Tn5-induced mutants in nodulation regions I, II and III of the Sym(biotic) plasmid (at early stationary phase). Identification of the fatty acids was accomplished by conversion to methyl esters by methanolysis, followed by GC/MS. The typical fatty acids were hexa-, and octa-decanoate, 3-hydroxy tetra-, hexa-, and octa-decanoate, 2-methyl hexa- and octa-decanoate, 3-hydroxy hexadecanoate(ante-iso form), and octadecan-9-enoate. The major fatty acids were the 3-hydroxy fatty acids, with 3-hydroxy tetradecanoate being the predominant typical fatty acid in all LPSs examined. LPS from R. trifolii 262 (region II mutant) had significantly more 3-hydroxy hexadecanoate, hexa- and octa-decanoate than wild type. Mid-exponential phase wild type 843 LPS had more hexa- and octa-decanoate than LPS from early stationary phase. Two fatty acids (2-methyl hexa- and octa-decanoate) were detected in small amounts and have not been previously reported in Rhizobium. A structure, derived from 4-keto-3-methyl hexadecanoate, was proposed for two isomeric non-typical fatty acids which are formed during methanolysis of the LPSs.

Lipid A, containing all the fatty acids from the intact LPS, was separated from the rest of the molecule by 1% acetic acid hydrolysis, followed by chloroform extraction. Methylation of this lipid A for 2 hr released mostly hexa- and octa-decanoate, but after 12 hr hydrolysis, the major fatty acids were the 3-hydroxy acids. This indicates that the 3-hydroxy fatty acids are either N-acyl(amide)-linked or buried in the internal structure of the lipid A, and therefore not as susceptible to methanolysis.

Mild saponification of LPS released primarily 3-hydroxy tetradecanoate, hexa- and octa-decanoate, indicating that these fatty acids are ester-linked. The fatty acids which remain attached to the lipid A, 3-hydroxy hexa- and octa-decanoate, are presumably amide-linked to other components in the LPS. ^{1}H-NMR spectroscopy of the lipid A indicated the presence of a carbohydrate moiety which is very unstable and decomposes rapidly during attempts to separate it from the lipids by hydrolysis.

The aqueous phase from lipid A isolation was examined by ^{1}H-NMR and signals corresponding to O-methyl, N-methyl, O-acetyl, N-acetyl and 6-deoxy functions were identified in all samples. R. trifolii 851 (region I mutant) had significantly more N-acetyl groups than wild type 843. R. trifolii 262 (region II mutant) had a lower ratio of O-methyl to N-methyl groups than wild type. Water-soluble material from mid-exponential phase 843 appeared to contain considerably less signals in the acetyl region than the early stationary phase 843, although this may not be significant because of the extreme lability of these groups.

R. trifolii 843 and 851 LPS were assayed colorimetrically for phosphorus. No P was detected in wild type LPS, but 851 contained appreciable amounts of P, which was determined by ^{31}P-NMR to be bound as a monoester.

ISOLATION AND CHARACTERIZATION OF THE OUTER MEMBRANE OF RHIZOBIUM SP. 127E15, ITS MUTANTS, AND REVERTANTS

THOMAS B. MAY, PETER P. WONG, AND JAMES A. GUIKEMA
Kansas State University, Manhattan, KS 66506 U.S.A.

The goal of this study was to characterize the outer membrane (OM) of Rhizobium sp. 127E15, its mutants, and revertants in order to investigate the role of OM in salt tolerance and its effect on nodulation. 127E15 induces effective nodules on lima beans. Ineffective mutants and their revertants have been obtained (Wong, Guikema, 1984). Properties of the mutants suggest that a surface alteration prevents nitrogen fixation but allows nodule formation. Salt tolerance was also lost with the mutation.

The OM was isolated by (1) French Pressure Cell disruption, (2) differential centrifugation to remove unbroken cells and (3) sucrose density gradient centrifugation. Mg^{+2} reduced yield 20-fold by causing OM to aggregate with unbroken cells. This was unexpected since many isolation protocols use Mg^{+2} to activate DNase and for selective solubilization of cytoplasmic membrane by Triton X-100 (Schnaitman, 1971). Sodium lauroyl sarcosine was found to be a useful replacement for Triton X-100 (this work, Filip et al., 1973).

Mutant OM differed from wildtype in two significant ways. First, mutants show an additional 32 KDa peptide. 127E15 has 12-15 proteins, of which 5 are major OM peptides. Topology probed with proteases shows that these proteins are protected from digestion, perhaps by lipopolysaccharide (LPS) (Rosenbusch, 1974). Only the 32 KDa mutant band and a 22 KDa peptide were cleaved. Second, wildtype and mutants differed in the appearance of a shaded region when gels were silver stained. This region from wildtype also stained with Periodic Acid Schiff Base (PAS), which stains sugar residues. A low molecular weight mutant band (at dye front) also stained with PAS. The revertants regained the 127E15 profile except the PAS-stained region migrated faster. This suggests that (1) PAS-stained material is similar for wildtype, mutants, and revertants and (2) that it is LPS or glycoprotein. In addition, the shaded area appeared to slow the migration of 127E15 major proteins. Migration of ovalbumin was slowed when mixed with 127E15 OM, and these proteins migrated at a similar position to mutants and revertants if bands were excised and run on a second gel. We propose that the shaded region is LPS, since it is a major OM constituent, and that loss of LPS is related to loss of salt tolerance and effective nitrogen fixation. LPS also appears to alter protein migration such that differences in protein profiles may be artifacts.

REFERENCES

Filip CP et al (1973) J. Bacteriol. 115, 717-722.
Rosenbusch JP (1974) J. Biol. Chem. 249, 8019-8029.
Schnaitman CA (1971) J. Bacteriol. 108, 545-552.
Wong PP and Guikema JA (1984) In Veeger C and Newton WE, eds., Advances in Nitrogen Fixation Research, p. 274, Martinus Nijhoff/Junk, The Hague.

FINE STRUCTURE OF EARLY STAGES IN NODULATION OF <u>ALNUS RUBRA</u> BY <u>FRANKIA</u>

A.M. BERRY[1] AND M.E. McCULLY[2]. [1]DEPT. OF ENVIRONMENTAL HORTICULTURE, UNIV. OF CALIFORNIA, DAVIS, CA 95616, USA; [2]BIOLOGY DEPARTMENT, CARLETON UNIVERSITY, OTTAWA, ONT. CANADA K1S5B6

The initial invasion of <u>Alnus</u> roots by <u>Frankia</u> occurs typically through a folded region in a deformed root hair. Electron micrographs of roots at the prenodule stage demonstrate the subcellular changes associated with the infection process, at the root hair infection site and within the prenodule, prior to nodule lobe expansion.

The <u>Frankia</u> hypha is the infective agent throughout the infection process, passing first through the root hair wall at a deeply folded region of the deformed hair. The hypha may branch at the infection site, but no specialized adhesive structures such as pili or capsular material were noted. A distinctive host-derived mucilage layer specific to the infected hair surrounds the <u>Frankia</u> hyphae in the rhizoplane.

At the site of <u>Frankia</u> penetration, the primary cell wall of the host shows evidence of fibrillar disorganization, without obvious abrupt discontinuity. Secondary wall lamellae are also deposited in the infected hair, an anomalous condition for <u>Alnus</u> root hairs. Wall ingrowths characteristic of transfer-cell wall are localized at the zone of initial infection. Subsequently, the <u>Frankia</u> hyphae follow a pathway of symplastic infection through the root hair base and into the subjacent cortical cells. Within the hair and in the young prenodule, <u>Frankia</u> hyphae are surrounded by an apparently host-derived wall encapsulation, deposited outside the host plasmalemma. In the infected hair, this encapsulation appears to be continuous with the transfer-cell wall proliferation. Microtubular arrays in the host cytoplasm parallel the longitudinal axis of the hair, which is also the axis of <u>Frankia</u> growth toward the root interior. Thus wall proliferation and increased plasmalemmal surface area are two major structural features which characterize the host root hair response to infection.

<u>Frankia</u> passes through the root hair base into a succession of recently-divided cortical cells. In the infection pathway the cortical cells expand greatly, and cell-to-cell passage of <u>Frankia</u> occurs through the newly-expanded walls. Golgi bodies are frequent in the early stages of cortical infection, probably an indication that new wall or encapsulating material is synthesized. The encapsulation layer is continuous with the invaded cell wall, suggesting that the new wall is deposited or redeposited at the time of <u>Frankia</u> passage.

INTERCELLULAR PENETRATION - A NOVEL MECHANISM OF INFECTION IN
ACTINORHIZAL PLANTS

IAIN M. MILLER and DWIGHT D. BAKER,
BATTELLE-KETTERING LABORATORY, 150 E. SOUTH COLLEGE STREET,
YELLOW SPRINGS, OHIO 45387-0268 USA

The initiation and development of root nodules in the actinorhizal
family Elaeagnaceae departs radically from the well-documented root hair
infection (RHI) mechanism which occurs in other actinorhizal families.
The mode of entry in Elaeagnaceous genera is by direct intercellular
penetration (IP) of the epidermis and apoplastic colonization of the root
cortex. This is in sharp contrast to entry via root hair infection in
which the actinomycete enters deformed and branched root hairs by
digesting through the cell wall. In IP, colonization of the cortex by
the endophyte induces host cortical cells to secrete a darkly staining
material into the intercellular spaces which serves as a substrate
through which the filamentous bacterium grows. The presence of this
darkly staining material makes it possible to see the infection sites at
the light microscope level although electron microscopy is necessary to
visualize the Frankia filaments. Several infection sites may be seen
surrounding a single nodule and we suggest that a single nodule might
contain multiple strains due to this phenomenon.

Formation of a prenodule in the root cortex below the site of entry,
such as is found during nodule development in Alnus and Myrica, does not
occur in Elaeagnus or Hippophae. In fact, there is a complete lack of
cell division in the root cortex. In both RHI and IP, a primary nodule
primordium is initiated from the root pericycle at a very early stage in
the infection process, well in advance of the colonizing Frankia endophyte. The developing nodule primordium in Elaeagnus and Hippophae is
surrounded by a heavily tanninized protoperiderm which the endophyte
easily traverses in its intercellular growth toward the nodule. Once
inside the cortex of the primary nodule primordium, the endophyte
ramifies throughout the cortex in the small intercellular spaces at
multiple cell junctions and in the middle lamellae. Actual invasion of
cortical cells occurs when a side branch of the endophyte is initiated
and grows through the host cell wall. During this cell invasion the host
cell membrane remains intact and the advancing Frankia is encapsulated
with a fibrous cell wall like material deposited by host cell secretory
vesicles. Once inside the host cell, the endophyte may differentiate
the specialized vesicles and begin nitrogen fixation. However, incom-
patible symbioses often occur in the Elaeagnaceae and vesicles or
nitrogen fixation may not be observed. In summary, it is important to
note that in IP the growth of Frankia from the rhizoplane to the cortical
cells of the nodule is entirely intercellular, whereas in RHI the
pathway from rhizoplane to nodule cortex is always intracellular.

DISCUSSION SESSION ON RECOGNITION AND INFECTION

RUSSELL W. CARLSON, EASTERN ILLINOIS UNIVERSITY, CHEMISTRY DEPARTMENT, CHARLESTON, ILL. 61920

The following is an attempt to summarize the discussion on "Recognition and Infection" (session 6) held on Thursday, August 8. The readers are referred to each participant's summary for details of their work. There are also many other interesting reports in session 6 which were not covered in the discussion session.

Thomas May (working with Peter Wong), Biology Department, Kansas State University. Mr. May summarized his work on the analysis of the outer membranes (OM) of Rhizobium sp. 127E17 (symbiont of lima bean) mutants. These mutants are Nod+Fix-. His results indicate the presence of an additional OM protein (34 kd) in the mutants which is absent in both the parent and revertant strains. In addition his results suggest alterations in the LPSs; the mutants seem to be lacking a higher molecular weight form of the LPS. Since the presence of the higher molecular weight LPS seems to alter the OM protein banding pattern it is not certain whether the change in the protein banding pattern of the mutant is a real change in OM protein or an apparent change caused by the LPS alteration. Questions were raised concerning treatments by which the LPS and OM proteins could be separated prior to PAGE analysis, e.g. EDTA treatment. Some of these methods have been attempted without success. The LPSs are being purified and will be further analyzed. In addition parental LPS can be added to the OM proteins from the mutants in order to see if the LPS will alter the protein banding pattern of the mutants.

Dr. Tibor Ersek (working with Dr. Steve Pueppke), University of Missouri, Department of Plant Pathology. Dr. Ersek summarized his work in which the effect on the membrane potential of soybean root cells by infective (R. japonicum) and non-infective (R. meliloti) rhizobia is measured. It is found that the non-infective rhizobia alter the membrane potential to a greater extent than infective bacteria. This effect seems to be long lasting (up to 7 days). Several questions were raised: 1. What is the location of the electrode? My impression from Dr. Ersek's answer, as well as from the abstract, is that the electrode was placed in the infectible zone of the soybean root. 2. Which ion transport was affected? The answer to this question is not known. The abstract states that 100 fold increases in Na^+ magnify the effect. 3. Why was R. meliloti chosen as the non-infective strain? This experiment was done to establish a useful technique for the analysis of symbiotically defective mutants. Thus an absolutely negative (non-infective) strain was used. 4. Has a supernodulating strain been tested? A supernodulating strain was not examined.

Dr. N.J. Brewin, John Innes Institute. Using monoclonal antibodies to the bacteroid LPS and to a peribacteroid membrane glycoprotein, work was done which shows that LPS is transferred from the bacteroid to the peribacteroid membrane. Furthermore the work suggests the bacteroid LPS is lacking the higher molecular weight form of the LPS, i.e. possibly the O-antigen. A model was given in which LPS in the form of a vesicle is released from the bacteroid and fused into the peribacteroid membrane. It was suggested that the results would be more convincing if verified by using additional monoclonal antibodies to other immunogenic sites on the LPS as well as to other peribacteroid proteins. The function of this transfer process is not known. Dr. Brewin stated that the results suggest a way in which bacteroid material could be transferred to the plant cell.

Dr. Rawle Hollingsworth (working with Dr. Frank Dazzo), Michigan State University, Department of Microbiology and Public Health. Dr. Hollingsworth has been comparing the CPS and LPS structures from Tn5 generated mutants of R. trifolii. The CPSs were hydrolyzed to single repeating units by a phage enzyme. Both the CPS repeating units and LPSs were compared by nmr analysis. For the CPSs from several mutants, quantitative changes in the substituent groups were noticed; in pyruvyl, acetyl and β–hydroxybutyric acid groups; e.g. decreases in strain ANU851 (a nod D mutant), and increases in strain ANU252 (a nod A mutant). The major change in the LPSs seems to be the presence of phosphate in the LPS from strain ANU851. Phosphate is not detected in the LPS from the parent strain. Further work is in progress to determine if these changes are functionally related to the phenotypic changes of these mutants.

Dr. Jan Kijne, University of Leiden. Dr. Kijne presented data which suggest that lectin–mediated attachment of R. leguminosarum to pea root hairs is dependent on the media in which the bacteria are grown. Attachment by bacteria grown in yeast extract/mannitol media was inhibited by lectin haptens while attachment by bacteria grown in tryptone/yeast extract was not inhibited by haptens but was correlated with the amount of piliated cells. Dr. Kijne pointed out the need to evaluate the growth conditions present in the rhizosphere in order to properly study possible recognition phenomena, such as lectin-binding.

Dr. Gary Stacey, University of Tennessee, Department of Microbiology. Dr. Stacey summarized his work on the effect of root exudate and soybean lectins on nodulation by wild-type R. japonicum 3Ilb110 and delayed nodulator, HS111. Pretreatment of HS111 with root exudate or lectin resulted in normal nodulation. Pretreatment of 3Ilb110 resulted in enhanced nodulation. Experiments with chloramphenicol and rifampicin suggest that the effect of lectin and root exudate requires protein synthesis. This work also points out the importance of rhizosphere factors, in this case possibly lectin, in nodulation. The work also suggests that these factors induce the expression of genes in the bacteria which are required for nodulation. Other work (see session 9) at this conference has shown that certain of the "common nod" gene products are induced by a small molecular weight molecule released in root exudate.

PHYSIOLOGY OF
NITROGEN-FIXING SYMBIOSES

INVITED PAPERS

POSTER SUMMARIES:

(i)   Structural studies.
(ii)   Factors affecting nodulation.
(iii)   Nitrogen metabolism.
(iv)   Carbon metabolism.
(v)   Oxygen transport and effects.
(vi)   Hydrogen uptake and evolution.
(vii)   Actinorhizal systems.
(viii) Associative relationships.

DISCUSSION GROUP SUMMARY

CARBON METABOLISM IN LEGUME NODULES

JOHN G. STREETER and SEPPO O. SALMINEN
DEPARTMENT OF AGRONOMY
OHIO STATE UNIVERSITY (OARDC), WOOSTER, OH 44691, USA.

1. INTRODUCTION

This paper reviews recent efforts to understand carbon metabolism in legume nodules. Because of space limitations the review is limited, with a few exceptions, to papers published within the past 2 years (1983-85). Also, because of the emphasis on nodules, papers dealing with carbon metabolism in cultured Rhizobium have, for the most part, been omitted. Within the past few years, papers on the identification and metabolism of carbon compounds in actinorhizal nodules have begun to appear, but this work is also not reviewed here. Several topics relevant to this subject have recently been reviewed in another symposium volume (Ludden, Burris, 1985).

Many estimates of the carbon requirement for N_2 fixation by legume nodules have been published. Many workers have considered only the cost for the conversion of N_2 to NH_4^+. A recent report by Rainbird et al (1984) includes all probable costs for the operation of legume nodules and suggests an overall cost of about 12 g carbohydrate per g N fixed (Table 1). This cost is substantial and one driving force for work on carbon metabolism is the hope that this requirement might be reduced.

TABLE 1. Estimate of the carbohydrate requirement for N_2 fixation by Glycine max nodules. (Adapted from Rainbird et al, 1984).

Component	Basis for estimate	g carbohydrate/ g N fixed
N_2 fixation + H_2 evolution	respiration	7.29 [a]
nodule "maintenance"	respiration	2.68 [a]
assimilation, transport of NH_4^+	calculation	1.86
growth	calculation	0.26
Total		12.1 [b]

[a] A very similar estimate of these two component costs has recently been reported by Heytler, Hardy (1984) using a novel calorimetry procedure.
[b] A very similar estimate for total carbon cost (12.4) has recently been calculated by Gordon et al (1985) from entirely different data sets.

In all legumes examined thus far, sucrose is the major carbohydrate translocated from shoots to nodules. Sucrose is also a major carbon compound in all nodules examined. However, legume nodules contain high concentrations of some "unusual" carbon compounds and these will be discussed briefly before considering carbon metabolism.

2. CYCLITOLS, α,α-TREHALOSE, and MALONATE.

Cyclitols have 6-carbon ring structures (hexa-hydroxy-cyclohexanes) typified by myo-inositol, a common constituent of all plant cells. Various cyclitols have long been known to be major carbohydrates in legumes, but only within the past 10 years have they been shown to be major carbohydrates in nodules. In some nodules (e.g. soybean), cyclitols may comprise the bulk of the mono-and disaccharide pool. In recent studies of Lupinus angustifolius nodules, pinitol concentration was greater than sucrose in all samples (J.G. Streeter, unpublished).

Cyclitol composition of nodules of a few legumes is given in Table 2. Note that cyclitol accumulation is not confined to nodules formed by fast- or slow-growing rhizobia or to ureide- or amide-exporting nodules. It is likely that some cyclitols are synthesized in nodules and not just accumulated from other plant organs. In soybeans for example, chiro-inositol is essentially absent in other plant parts. However, work on the metabolism of these compounds in nodules has not been reported.

TABLE 2.   Cyclitols in legume nodules[a].

Legume	myo-inositol	D-chiro inositol	D-pinitol[b]	Ononitol[c]	O-methyl-scyllo inositol
Glycine max	++++	++++	++	+	–
Trifolium repens	+	+++	+++	–	–
Pisum sativum	tr	–	–	++++	++
Arachis hypogea	tr	–	++	++	–
Pueraria thunbergiana	tr	tr	+	–	–
Albizia julibrissin	tr	tr	+	–	–
Lupinus angustifolius	tr	tr	++++	–	–
Medicago sativa	tr	–	+++	+	–
Phaseolus vulgaris	tr	–	–	–	–

[a] Data are from Streeter (1985a, earlier publications, and unpublished data), Davis, Nordin (1983), Phillips et al (1984), and Skot, Egsgaard (1984). tr = trace.
[b] 3-0-methyl-D-chiro-inositol.
[c] 4-0-methyl-myo-inositol.

The role of cyclitols in nodules is unknown. In studies where labeled CO_2 is supplied to plants, the cyclitols are not readily labeled in nodules (section 4), suggesting that they are metabolically inert relative to other carbon compounds. Perhaps they play an osmotic role as has been suggested for other cyclitol-containing tissues (Ford, 1984). Suggestions regarding the function of cyclitols in nodules must take into account the apparent absence of cyclitols in Phaseolus vulgaris nodules (Table 2). While their function remains a mystery, the cyclitols are major carbon compounds in many legume nodules and their presence should not be ignored by workers in this field. Binder and Haddon (1984) have recently published a major contribution to the analytical chemistry of these compounds.

Although there are no conclusive reports of the synthesis of α,α-trehalose in higher plants, trehalose may accumulate to high concentrations in nodules. Trehalose concentration varies widely in nodules depending on Rhizobium strain and there is now good evidence that this glucose-glucose disaccharide is synthesized in bacteroids (Reibach, Streeter, 1983; Streeter, 1985b). Trehalose is accumulated by all Rhizobium species in culture with concentrations reaching as high as 3 or 4% of dry weight (Streeter, 1985b). Although a role for trehalose has not been established, the sum of evidence presently available suggests that the compound may be important in the free-living portion of the life cycle of Rhizobium (Streeter, 1985b).

Malonate has been known for several years as a major dicarboxylic acid in many legume nodules. Recent studies have shown that, like the cyclitols, this compound is not appreciably labeled following CO_2-labeling of shoots (section 4). Thus, malonate is yet another carbon compound for which no role in nodules is known. Uptake and oxidation of malonate by R. japonicum bacteroids is slow relative to cultured bacteria and malonate does not significantly reduce uptake and metabolism of succinate, pyruvate, or α-ketoglutarate by bacteroids (Werner et al, 1982).

3. ENZYMES OF CARBON METABOLISM

Recent reports which document the presence of carbon-metabolizing enzymes in legume nodules are summarized in Table 3. The list should not be regarded as complete because it does not reflect the earlier literature. Nevertheless, it indicates improved recent progress in elucidating what reactions are possible and will serve to introduce the reader to this literature.

Most of the recent reports have dealt with Glycine max and these reports permit some tentative conclusions for soybean nodules: (a) the presence of the pentose phosphate pathway in R. japonicum bacteroids is questionable; (b) utilization of glucose via glycolysis in bacteroids is probably limited because of the low activity of P-fructokinase; (c) no mechanism has yet been reported for the hydrolysis of sucrose in bacteroids. Combined, these observations suggest limited capability for direct utilization of carbohydrates in bacteroids and, coupled with the presence of relatively high activity of TCA cycle enzymes in bacteroids, are consistent with the view of dicarboxylic acids as the principal source or reducing equivalents for bacteroids (section 5).

4. STUDIES WITH INTACT PLANTS AND NODULES

4.1. Labeling patterns in nodules after supplying shoots with labeled CO_2

Several major studies in this subject area have been reported recently, all of them with soybean (Gordon et al, 1985; Kouchi et al, 1985; Kouchi, Nakaji, 1985; Kouchi, Yoneyama, 1984; Reibach, Streeter, 1983). The following general conclusions have emerged from these studies: (a) Carbon enters the nodules almost exclusively as sucrose; (b) Most (80%) of the carbon entering the nodule is lost through respiration or is exported to other plant parts; (c) Some major carbon pools (cyclitols,

TABLE 3.   Recent reports on enzymes of carbon metabolism in legume
            nodules; B = bacteroid, C = cytosol, DH = dehydrogenase.

Reference & [legume]	Enzyme	Typical Specific activity
		μmole/mg prot x min
Bassarab et al (1984)	α-mannosidase (B&C)	0.90 (B), 8.1 (C)
[Glycine max]	α-galactosidase (C)	3.6
	α-glucosidase (B&C)	1.1 (B), 1.5 (C)
Henson et al (1982)	NAD-isocitrate DH (B&C)	0.50 (B), 0.54 (C)
[Medicago sativa]	NADP-isocitrate DH (B&C)	0.33 (B), 0.95 (C)
Henson, Collins (1984)	invertase (C)	0.10
[Medicago sativa]	hexokinase (C)	0.18
	glucose-6P DH (C)	0.03
	NADP-isocitrate DH (C)	1.8
	malate DH (C)	2.1
	amylase (C)	0.18
Karr et al (1984)	β-ketothiolase (B)	2.4
[Glycine max]	pyruvate DH complex (B)	0.04
	fumarase (B)	0.22
	malate DH (B)	4.4
	hydroxybutyrate DH (B)	0.25
	NADP-isocitrate DH (B)	0.36
	acetoacetate-succinyl-CoA transferase (B)	0.08
Copeland, Morell (1985)	hexokinase (C)	a
[Glycine max]	fructokinase (C)	a
Morell, Copeland (1984)	alkaline invertase (C)	10.4[a]
[Glycine max]		
Morell, Copeland (1985)	sucrose synthase (C)	15.1[a]
[Glycine max]		
Reibach, Streeter (1983)	P-glucomutase (B&C)	0.05 (B), 0.06 (C)
[Glycine max]	hexokinase (glucose)(B&C)	0.04 (B), 0.02 (C)
	hexokinase (fructose) (B&C)	0.007(B), 0.04 (C)
	P-glucose isomerase (B&C)	0.27(B), 0.91 (C)
	P-fructokinase (B&C)	b , 0.25 (C)
	fructose-1,6-bisP-aldolase (B&C)	0.006(B), 0.006(C)
	glucose-6P-DH (B&C)	0.008(B), 0.07 (C)
	NADP-6P-gluconate DH (B&C)	0 (B), 0.11 (C)
	NAD-6P-gluconate DH (B&C)	0.04 (B), 0 (C)
Streeter (1982)	invertase, pH 5.4 (B&C)	0 (B), 0.007(C)
[Glycine max]	invertase, pH 7.8 (B&C)	0 (B), 0.11 (C)
	maltase (B&C)	0.03 (B), 0.008(C)
	trehalase, pH 3.8 (B&C)[c]	0.04 (B), 0.02 (C)
	trehalase, pH 6.6 (B&C)[c]	0.06 (B), 0.02 (C)
Tajima, Yamamoto (1984)	glucose-6P-DH (C)	0.06
[Glycine max]	NAD-6P-gluconate DH (C)	0
	NADP-6P-gluconate DH (C)	0.06
Vance, Stade (1984)	PEP carboxylase (C)	42.3[a]
[Medicago sativa]		

TABLE 3.   (Continued)

Reference & [legume]	Enzyme	Typical Specific activity
		µmole/mg prot x min
Vance et al (1983)		
[Glycine max]	PEP carboxylase (C)	0.49
[Vigna angularis]	PEP carboxylase (C)	0.41
[Phaseolus vulgaris]	PEP carboxylase (C)	0.49
[Medicago sativa]	PEP carboxylase (C)	0.30
[Lotus corniculatus]	PEP carboxylase (C)	0.64
Deroche et al (1983)		
[Phaseolus vulgaris]	PEP carboxylase (C)	0.19[d]
[Medicago sativa]	PEP carboxylase (C)	0.15[d]
[Pisum sativum]	PEP carboxylase (C)	0.03[d]
[Glycine max]	PEP carboxylase (C)	0.22[d]
Waters et al (1985)		
[Glycine max]	malate DH (B)	1270[a]

[a] Report on purification and properties; activity, where given, is for purified enzyme.

[b] Enzyme activity originally reported as zero but recent work indicates a trace (0.01) of activity in bacteroids.

[c] Enzymes of trehalose synthesis and breakdown in soybean nodules are described in a separate report (Salminen, Streeter) in this volume.

[d] Data reported as activity/g fresh wt; protein concentration of 15 mg/g fresh wt was assumed for these calculations.

malonate) in nodules are only weakly labeled; (d) Labeled C rapidly appears in organic acids (mostly malate) in the cytosol and subsequently in amino acids; (e) Most of the radioactivity recovered in bacteroids was in the neutral sugars. If one assumes that point (e) indicates rapid turnover of organic acids relative to sugars, the overall results of these studies are consistent with the view that 4-carbon dicarboxylic acids are the principal source of reducing equivalents for bacteroids.

The recent studies of Kouchi are noteworthy because of the $^{13}CO_2$ steady-state labeling technique used and the thoroughness of the analytical chemistry. The most recent results of this group suggest that a large proportion of organic acids in soybean nodules are physically separated from the major site of respiration (Kouchi, Nakaji, 1985; Kouchi et al, 1985). Another interesting use of isotopes is the use of ^{13}N and ^{11}C to provide estimates of actual flow rates of C and N in phloem and xylem of nodulated alfalfa (Caldwell et al, 1984).

4.2.   Recycling of CO_2 via PEP carboxylase

Respiration and N_2-fixing activities of nodules are very closely coupled (Kanamori et al, 1984), and it has been estimated that 32% of the re-spired CO_2 is re-assimilated via PEP carboxylase (Rainbird et al, 1984). While this recycled carbon makes no contribution to the supply of reduc-ing equivalents in nodules, it may provide substrate for bacteroids, acids required for NH_4^+ assimilation, and carboxylate required to balance cation transport (Gadal, 1983).

There is recent evidence to indicate that the latter function of CO_2 assimilation is the major function in ureide-exporting nodules (Vance et al, 1985). With alfalfa and birdsfoot trefoil nodules, which export amino acids, about 80% of the ^{14}C in xylem exudate following labeling of nodules with $^{14}CO_2$ was in amino acids (Maxwell et al, 1984). However, radioactivity in the xylem exudate from soybean and adzuki bean (ureide exporters) was found mainly 70 to 87%) in the organic acid fraction (Vance et al, 1985).

5. CARBON SOURCES FOR BACTEROIDS

Several mutants lacking the ability to absorb malate and/or succinate have been isolated from several Rhizobium species. All of these mutants, to our knowledge, form ineffective symbioses, and this work will be reviewed elsewhere in this volume by C. Ronson et al. In addition, mutants lacking glucokinase, fructokinase, pyruvate dehydrogenase and glucose-6P-dehydrogenase have recently been described, and all form effective nodules (Cervenansky, Arias, 1984; Glenn et al, 1984). A recent report by McKay et al (1985) indicates that gluconeogenic capability is also not required for the formation of an effective symbiosis.

R. japonicum bacteroids absorb dicarboxylic acids by an active (energy) requiring) mechanism but they absorbed carbohydrates (including pinitol) only slowly and passively (Reibach, Streeter, 1984). Although the fast-growing species of Rhizobium generally absorb and oxidize sugars readily in culture, bacteroids exhibit little or no sugar uptake while actively absorbing dicarboxylic acids (de Vries et al, 1982; Saroso et al, 1984). Although typical respiration rates have been reported for R. legumino-sarum bacteroids supplied with very high fructose concentrations (560 mM) (Hooymans, Logman, 1984), the physiological relevance of this result is not clear.

The working hypothesis adopted by most workers in this area for the past several years has been that dicarboxylic acids are the principal source of reducing equivalents for bacteroids. At this point in time, there seems to be little evidence to refute the hypothesis while evidence supporting the hypothesis continues to grow. However, unless there are limited transport activities in the peri-bacteroid membrane, it seems likely that bacteroids are exposed to a wide range of carbon compounds and it may be unwise to expect bacteroids to be dependent on a single source of reducing equivalents.

6. NEEDS FOR FURTHER WORK

a) Given the probability that organic acids are very important in nodule function, more effort on the quantitative analysis and metabolism of these compounds under various conditions is needed.
b) The localization of carbon pools and enzymes of carbon metabolism in nodules is an almost untouched area. Efforts to localize metabolites and enzymes need to go beyond bacteroid/cytosol separations and to consider tissues, types of host cells, organelles, and the peri-bacteroid space.
c) Rapid progress is being made in the isolation and purification of intact plant membranes and the characterization of specific transport activities localized therein. However, only rarely has this technology been applied to nodules (e.g. Mellor et al, 1984). Much more effort on the composition and function of nodule membranes is needed.

d) The abundance of cyclitols, α,α-trehalose, and malonate in many nodules would seem to warrant additional work on the possible role of these compounds in nodule function.

7. ACKNOWLEDGEMENT

We thank A. Glenn, L. Copeland, H. Kouchi, D. Emerich and C. Vance for sharing in press manuscripts prior to publication.

8. REFERENCES

Bassarab S et al (1984) Physiol. Plant Path. 24, 9-16.
Binder RG and Haddon WF (1984) Carbohyd. Res. 129, 21-32.
Cervenansky C and Arias A (1984) J. Bact. 160, 1027-1030.
Caldwell CD et al (1984) J. Exptl. Bot. 35, 431-443.
Copeland L and Morell MK (1985) Plant Physiol. (in press).
Davis LC and Nordin P (1983) Plant Physiol. 72, 1051-1055.
Deroche M-E et al (1983) Physiol. Veg. 21, 1075-1081.
Ford CW (1984) Phytochem. 23, 1007-1015.
Gadal P (1983) Physiol. Veg. 21, 1069-1074.
Glenn AR et al (1984) J. Gen. Microbiol. 130, 239-245.
Gordon AJ et al (1985) J. Exptl. Bot. 36, 756-759.
Henson CA et al (1982) Plant Cell Physiol. 23, 227-235.
Henson CA and Collins M (1984) Crop Sci. 24, 727-732.
Heytler PG and Hardy RWF (1984) Plant Physiol. 75, 304-310.
Hooymans JJM and Logman GJJ (1984) J. Plant Physiol. 116, 273-277.
Kanamori T et al (1984) Soil Sci. Plant Nutr. 30, 231-237.
Karr DB et al (1984) Plant Physiol. 75, 1158-1162.
Kouchi H et al (1985) Ann. Bot. (in press).
Kouchi H and Nakaji K (1985)Soil Sci. Plant Nutr. (in press).
Kouchi H and Yoneyama T (1984) Ann. Bot. 53, 883-896.
Ludden PW and Burris JE (1985) Nitrogen Fixation and CO_2 Metabolism, Elsevier, New York.
Maxwell CA et al (1984) Crop Sci. 24, 257-264.
McKay IA et al (1985) J. Gen. Microbiol. (in press).
Mellor RB et al (1985) Z. Naturforsch.40c, 73-79.
Morell MK and Copeland L (1984) Plant Physiol. 74, 1030-1034.
Morell MK and Copeland L (1985) Plant Physiol. 78, 149-154.
Phillips DV et al (1984) J. Agric. Food Chem. 32, 1289-1291.
Rainbird RM et al (1984) Plant Physiol. 75, 49-53.
Reibach PH and Streeter JG (1983) Plant Physiol. 72, 634-640.
Reibach PH and Streeter JG (1984) J. Bact. 159, 47-52.
Saroso S et al (1984) J. Gen. Microbiol. 130, 1809-1814.
Skot L and Egsgaard H (1984) Planta 161, 32-36.
Streeter JG (1982) Planta 155, 112-115.
Streeter JG (1985a) Phytochem. 24, 174-176.
Streeter JG (1985b) J. Bact. (in press).
Tajima S and Yamamoto Y (1984) Soil Sci. Plant Nutr. 30, 85-94.
Vance CP et al (1983) Plant Physiol. 72, 469-473.
Vance CP and Stade S (1984) Plant Physiol. 75, 261-264.
Vance CP et al (1985) Plant Physiol. (in press).
de Vries GE et al (1982) J. Bact. 149,872-879.
Waters JK et al (1985) Biochem. (in press).
Werner D et al (1982) Z. Naturforsch.37c, 921-926.

FACTORS LIMITING N$_2$ FIXATION BY THE LEGUME-RHIZOBIUM SYMBIOSIS

FRANK R. MINCHIN, JOHN E. SHEEHY and JOHN F. WITTY
The Animal and Grassland Research Institute, Hurley, Maidenhead, Berks., SL6 5LR, U.K. and Rothamsted Experimental Station, Harpenden, Herts., AL5 2JQ, U.K.

1. INTRODUCTION

It is not our intention to produce yet another review of the numerous publications concerning factors limiting N$_2$ fixation but rather to recommend novel directions for the future. The need for a new impetus was perceived in a review of environmental factors by Sprent et al (1983) where it was noted that many alternative hypotheses were available which postulated either a direct effect of these factors on the nodule system or an indirect effect acting via the host plant. Such conflicts must be resolved if research into limiting factors is to contribute to the improvement of nitrogen fixation by legume crops. However, before progress can be made it is necessary to improve the accuracy of techniques which measure nitrogen fixation and related processes, and to focus greater attention on the role of oxygen supply.

2. MEASUREMENT TECHNIQUES

2.1. The Acetylene Reduction Assay

During the past 18 years the acetylene reduction assay has been the mainstay of research into factors which limit N$_2$ fixation. The use of (open) flow-through gas systems, which enable simultaneous measurements of acetylene reduction and respiration, has revealed a major error in this assay as it is normally used with nodule systems. This error is associated with a rapid decline in nitrogenase activity caused by the addition of acetylene (Minchin et al, 1983). Furthermore, we have shown with soyabean that the use of disturbed nodule systems produce misleading results and the 'classic' acetylene reduction assay cannot even be used for comparative purposes (Minchin et al, poster summary, these proceedings). Therefore, the validity of results obtained and conclusions drawn from experiments in which the conventional acetylene reduction assay was used must be questioned.

Other users of flow-through systems have not reported an acetylene-induced decline with soyabean (Mederski, Streeter, 1977; Patterson et al, 1983; Layzell et al, 1984). The reasons for this discrepancy are not apparent, but it may be that within any species not all symbiotic combinations are sensitive to acetylene. With sainfoin we have found that the occurrence of an acetylene-induced decline in nitrogenase activity depends on the strain of Rhizobium (unpublished data). Therefore, sensitivity to acetylene must be checked using a flow-through system. The use of such a system also provides an accurate determination of nitrogenase activity by undisturbed, intact plants (as the maximum rate of acetylene reduction (Minchin et al, 1983) and permits other investigations into the basic mechanisms of nitrogen fixation.

2.2 Relationships Between Respiration and Nitrogen Fixation

The understanding of factors which limit N_2 fixation can be increased by an analysis of respiration associated with this process, in terms of both the allocation of carbohydrate to nitrogenase activity and the efficiency of respiratory carbohydrate consumption in providing reductants and ATP. In the past, techniques to determine carbohydrate allocation have been destructive and incorporated errors associated with the use of disturbed systems (e.g. detached nodules) and techniques designed to determine the efficiency of nitrogenase-linked respiration have often been problematical (Minchin et al, 1981). These objections can be overcome by a technique employing flow-through systems and the fact that acetylene reduction and respiration are linearly related when both processes are being reduced by exposure to acetylene (or reductions in external oxygen concentrations, see below). The intercept of this linear relationship on the respiration axis, at zero nitrogenase activity, allows for the separation of total root respiration into nitrogenase-linked and growth and maintenance components. The slope of the relationship measures the efficiency of carbohydrate use with either intact roots or detached nodules (Witty et al, 1983). The value of this approach has been demonstrated in studies on the response of white clover to nitrate (Fig 1a) and variation in the efficiency of pea nodules formed by genetically engineered rhizobia (Fig 1b).

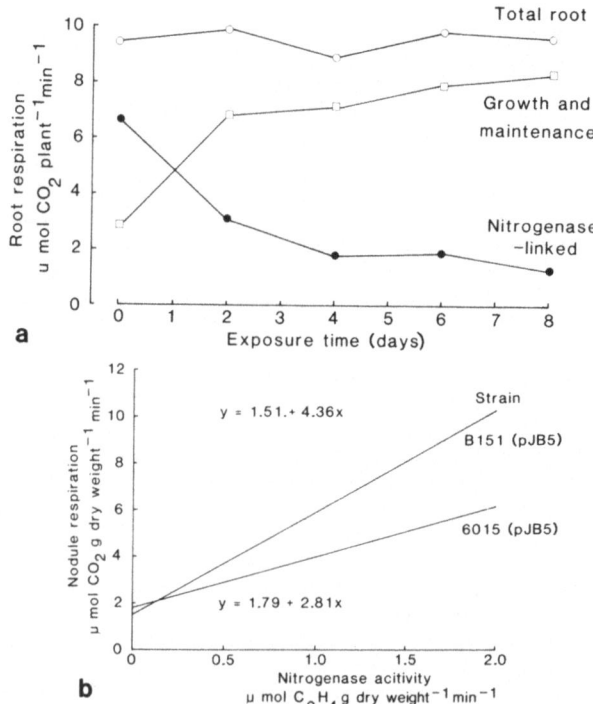

FIGURE 1. Examples of the use of flow-through systems for measurements of (a) the separation of root respiration for white clover exposed to 330 ppm NO_3-N, and (b) the efficiency of pea nodules formed by genetically engineered rhizobia. Data for (a) Minchin et al, unpublished, data for (b) Skot et al, unpublished.

3.   THE INVOLVEMENT OF OXYGEN

3.1. The Role of a Variable Diffusion Barrier in Oxygen Protection by
      Legume Nodules

The use of flow-through systems has also led to the discovery that
nodules can protect their nitrogenase by a combination of respiratory
oxygen consumption and controlled variation of oxygen diffusion (Sheehy
et al, 1983). Reduced oxygen diffusion due to gross morphological
disturbances in the cortex of lenticular nodules has previously been
suggested as the cause of reduced nitrogenase activity in water stressed
or detached nodules (Parkhurst, Sprent, 1975; Ralston, Imsande, 1982).
However, we have now established that reversible changes in diffusion
resistance can occur with undisturbed nodule systems of both lenticular
and non-lenticular types. Protection of nitrogenase against an external
concentration of 80%, in the absence of any marked increase in nodule
respiration, is explained by this controlled resistance (Sheehy et al,
1983).

A gradual increase in the resistance also explains the decline in nodule
activity in the presence of acetylene or when nitrogen is replaced by
argon; both treatments which prevent ammonium production by nitrogenase.
Similar concomitant declines in nitrogenase activity and respiration can
be induced by reductions in external oxygen concentration (Witty et al,
1983), whilst the magnitude of the acetylene-induced decline is less in
nodules where the diffusion resistance has already been increased by
prior exposure of root systems to 80% external oxygen (Witty et al,
1984). Errors in the acetylene reduction assay related to the use of
disturbed systems (Minchin et al, poster summary, these proceedings) can
also be explained in terms of this disturbance increasing the diffusion
resistance of the nodules.

At present the nature of the variable resistance is unknown, but it
probably involves variations in the length of a water-filled section of
the diffusion pathway. Such a zone could be cell layers devoid of
intercellular air spaces (Tjepkema, 1984) or water within narrow
intercellular pores (Sheehy et al, 1985). Regardless of its nature the
presence of a variable diffusion resistance gives nodules the potential
to overcome complications in oxygen supply arising from variations in
individual nodule size; changes in surface area to volume ratio can be
matched by changes in the diffusion resistance. However, oxygen
diffusion in the network of intercellular spaces within the bacteroid
zone may become limiting if the nodule radius becomes too large.
Nevertheless, for an individual plant the morphological feature of the
nodule system which ultimately limits the rate of N_2 fixation will be the
weight of operational bacteroids.

3.2. Oxygen as a Limiting Factor

The normal function of the diffusion resistance can be envisaged as the
regulation of oxygen supply, so that the respiratory activity of the
nodule is matched to its carbohydrate supply. However, an imbalance in
oxygen and carbohydrate supplies could result in bacteroid function
becoming oxygen limited. Knowledge of the interactions between
acetylene, nodule disturbance and the regulation of oxygen supply now
make it possible to assess the apparent contradictions in the literature

regarding oxygen as a limiting factor. All experiments dealing with this topic are inconclusive. Marked increases in activity of detached nodules exposed to high external oxygen concentrations (e.g. Bergersen, 1970; Mague, Burris, 1972) can be re-interpreted as a removal of the oxygen stress caused by increases in the diffusion resistance following nodule detachment. Similar increases for intact root systems exposed to high oxygen concentrations (Minchin et al, unpublished) can be interpreted as removal of the oxygen stress caused by acetylene-induced increases in the diffusion resistance. Conversely, the absence of any significant increase in nitrogenase activity of undisturbed nodule systems for oxygen concentrations between 20 and 40% (Criswell et al, 1977; Ralston, Imsande, 1982) does not mean that oxygen supply is not limiting. The lack of increase could be due to nitrogenase damage resulting from inoperation of the oxygen protection mechanism, because of either an insufficient carbohydrate supply or a poor response by the diffusion resistance. It cannot be ascertained whether oxygen limiting conditions would prevail if the oxygen protection system was functional.

With these difficulties in using nodule systems whose diffusion resistance has been increased by disturbance and/or exposure to acetylene it is difficult to envisage an experiment which will provide direct evidence for or against oxygen limitation. However, an indirect approach can be made through the use of mathematical models employing the known parameters of soyabean nodule morphology and biochemistry and the physics of gaseous diffusion (Sheehy et al, 1985; Sheehy, Bergersen, in preparation). One conclusion to be drawn from these models is that because of the need for nitrogenase protection the soyabean nodule functions under a degree of oxygen limitation. Despite the presence of leghaemoglobin a small oxygen concentration gradient will exist across an infected cell. Thus, if oxygen supply was regulated to allow the bacteroids at the edge of infected cells to fix at optimum rates those towards the centre of the cells would operate at below optimum rates. An increase in oxygen supply to the central bacteroids would result in a supra-optimal oxygen concentration at the outer bacteroids with consequent damage to their nitrogenase (Sheehy et al, in preparation).

Given these constraints on nodule functioning the question still remains as to whether nodule activity is ultimately limited by carbohydrate supply or oxygen supply. With carbohydrate limitation, N_2 fixation would respond positively to increases in carbohydrate supply with an actual rate of activity lower than the potential rate if all bacteroids were operating with an optimal oxygen supply. With oxygen limitation, N_2 fixation would only respond to increases in carbohydrate supply up to a certain level, after which activity would be limited by oxygen supply and independent of carbohydrate supply. At present, data required to directly answer this question are absent. However, experiments involving long-term exposure of pea and soyabean plants to high, atmospheric or low oxygen concentrations have produced no significant differences in rates of plant growth or nitrogen fixation (Table 1). Such results show that the oxygen diffusion resistance can be varied and suggest that the N_2 fixation rate may be controlled by carbohydrate supply. Essentially similar conclusions were reached by Criswell et al (1976) from short-term experiments involving variations in oxygen. However, from their data it appears that soyabean nodules may take up to 24 hours to recover from exposure to 6% O_2, a recovery which can now be interpreted as a reduction of the oxygen diffusion resistance. Changes in diffusion resistance on this time scale could not cope with short-term increases in carbohydrate supply (e.g. during a period of high irradiance). Such ephemeral oxygen-limiting situations could account for the high seasonal mean concentrations of apparently surplus carbohydrates, especially polyhydroxybutryrate, found in soyabean nodules (Wong, Evans, 1971; Streeter, 1981).

TABLE 1. Effect of a 28d exposure to sub or supra-atmospheric O_2 concentrations on vegetative growth and N_2 fixation of soyabean and pea

	Soyabean		Pea	
% O_2	Total plant D.Wt. (g)	Total fixed N (mg)	Total plant D.Wt. (g)	Total fixed N (mg)
10	6.10	155	4.90	163
21	6.37	153	4.81	157
30	6.83	148	4.99	158
s.e.	0.473	11.1	0.388	12.5

Data of Witty, Giller (unpublished)

3.3 Oxygen and Limiting Environmental Stress Factors

The protection of nitrogenase from oxygen damage through the operation of a variable diffusion resistance should also be taken into account when considering environmental stress factors which limit nitrogen fixation. Variations in oxygen supply have previously been considered in relation to water stress and waterlogging (Sprent, 1969; Schwinghamer et al, 1970; Pankhurst, Sprent, 1975; Denison et al, 1983). The data of Pankhurst and Sprent is entirely consistent with an increase in diffusion resistance in response to increased water stress of nodules. However, possible interactions with oxygen protection mechanisms have not been considered in relation to such factors as temperature, defoliation and nitrate.

Our experiments with flow-through systems have demonstrated the involvement of nodule resistance changes in controlling nitrogenase-linked respiration during a diurnal $25/15^{\circ}C$ temperature change in controlled environment cabinets. Indeed, such changes must occur to prevent oxygen damage resulting from reduced carbohydrate use at the lower temperature. Of greater significance is the finding that different symbioses react to the temperature change at different rates and to different extents; the reaction of pea being greater than white clover (Fig 2a). Conversely, the speed of change in root respiration due to the response of the diffusion resistance to defoliation (total shoot removal) is greatest in white clover, least in pea and intermediate in soyabean (Fig 2b). It is tempting to speculate that these reaction speeds may be related to the adaptation of these species to temperature and defoliation.

Experiments involving exposure of white clover to 330 ppm NO_3-N have shown that both nitrogenase activity and associated respiration decline markedly during an 8 day exposure period. If the nodules are to maintain oxygen protection during this period it is necessary for the diffusion resistance to increase, causing a reduction in the response to acetylene

(Witty et al, 1984). However, throughout the 8 day exposure period plants displayed a greater than 60% reduction in nitrogenase activity in response to acetylene, suggesting that the diffusion resistance barrier was not increased. A cessation of ammonium production from nitrogenase is known to cause a change in resistance (Minchin et al, 1983) and it is possible that ammonium produced by nitrate reduction in the root and/or nodule cytosol could interfer with control of the resistance, resulting in oxygen damage of the nitrogenase. Although this hypothesis is highly speculative it is included as an example of the type of novel approach that can be made when the oxygen protection mechanism is considered.

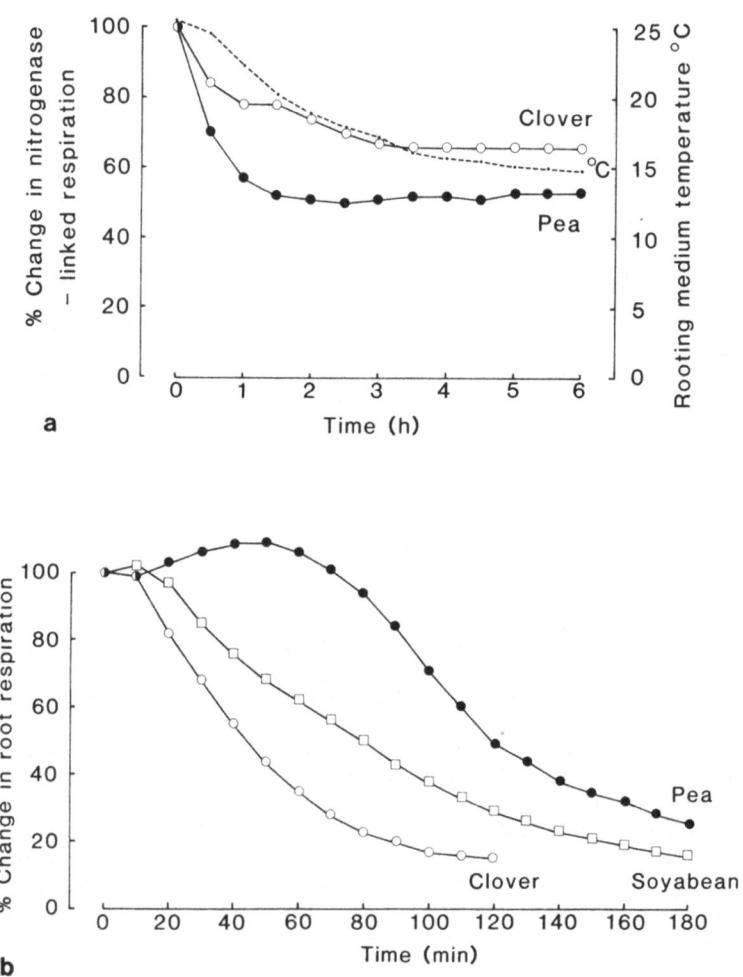

FIGURE 2. Reactions of different symbioses to (a) a diurnal temperature change from 25 to 15°C, and (b) total shoot removal. Data from Minchin et al, unpublished.

4. CONCLUSIONS

The rate at which a legume plant can fix nitrogen must be considered as a function of oxygen supply to the nodules as well as carbohydrate supply, efficiency of carbohydrate use and weight of operational bacteroids within the nodule system. At present the interrelationships between these 4 components and the relative effects of environmental stress on them are not well understood. Further research is required in these areas if the study of limiting factors is to have an input into legume improvement. The use of flow-through systems in this research is essential; the use of the 'classic' acetylene reduction assay should be limited to detecting the existence of nitrogenase activity.

5. REFERENCES

Bergersen FJ (1970) Aust. J. biol. Sci. 23, 1015-1025.
Criswell JG et al (1976) Plant Physiol. 58, 622-625.
Criswell JG et al (1977) Crop Sci. 17, 39-44.
Denison RF et al (1983) Plant Physiol. 73, 648-651.
Layzell DB et al (1984) Can. J. Bot. 62, 965-971.
Mague TH and Burris RH (1972) New Phytol. 71, 275-286.
Mederski HJ and Streeter JG (1977) Plant Physiol. 59, 1076-1081.
Minchin FR et al (1981) Plant, Cell and Environ. 4, 5-26.
Minchin FR et al (1983) J. exp. Bot. 34, 641-649.
Pankhurst CE and Sprent JI (1975) J. exp. Bot. 26, 287-304.
Patterson et al (1983) Plant Physiol. 70, 695-700.
Ralston EJ and Imsande J (1982) J. exp. Bot. 33, 208-214.
Schwinghamer EH et al (1970) Plant Soil 33, 192-212.
Sheehy JE et al (1983) Ann. Bot. 52, 565-571.
Sheehy JE et al (1985) Ann. Bot. 55, 549-562.
Sprent JI (1969) Planta 88, 372-375.
Sprent JI et al (1983) In Jones DG and Davies DR, eds, Temperate Legumes: Physiology, Genetics and Nodulation, pp 269-317, Pitman, London.
Streeter JG (1981) Ann. Bot. 48, 441-450.
Tjepkema JD (1984) In Veeger C and Newton WE, eds, Advances in Nitrogen Fixation Research, pp 467-473, Nijhoff/Junk-Pudoc, The Hague.
Witty JF et al (1983) J. exp. Bot. 34, 951-963.
Witty JF et al (1984) Ann. Bot. 53, 13-20.
Wong PP and Evans HJ (1971) Plant Physiol. 47, 750-755.

6. ACKNOWLEDGEMENTS

The Animal and Grassland Research Institute and Rothamsted Experimental Station are financed through the Agricultural and Food Research Council.

THE SITE OF NITROGENASE IN FRANKIA IN FREE-LIVING CULTURE AND IN SYMBIOSIS

JOHN G. TORREY, HARVARD FOREST, HARVARD UNIVERSITY, PETERSHAM, MA 01366

Actinorhizal plants comprise a diverse array of woody dicotyledonous species whose roots are susceptible to infection by the filamentous soil bacterium Frankia of the Actinomycetales. The present tally of host plants includes in excess of 200 species distributed among twenty-three genera in eight families (Moiroud and Gianinazzi-Pearson 1984). The filamentous bacterium infects susceptible plants either via root hair deformation and invasion (Callaham et al. 1979, Berry et al. 1985) or through root epidermal penetration via intercellular spaces and cortical cell penetration (Miller, Baker 1985). Within the host cortical cells produced from proliferating modified lateral roots which form the multilobed nodule (Bowes et al. 1977), the invading bacterium differentiates branched filaments which typically end in terminal swellings called vesicles. Within the vesicles the prokaryotic enzyme nitrogenase is formed, catalyzing the reduction of dinitrogen to ammonia which is assimilated into the metabolism of both the endophyte and its symbiotic host.

Although earlier anatomical descriptions of actinorhizal root nodules existed (cf. Bond 1967), the definitive demonstration of the actinomycetous nature of Frankia came from ultrastructural studies by Becking et al. (1964) who described the filamentous character of the endophyte and the complexities of vesicular structure in Alnus glutinosa. The definitive description of sporulation by Frankia in nodules came from the later study by van Dijk and Merkus (1976). With the successful isolation and culture of Frankia from root nodules of Comptonia peregrina by Callaham et al. (1978) it was possible to study these structural features of Frankia outside of the host (Newcomb et al. 1979) and to begin to analyze the comparative physiology and biochemistry of the bacterium in the plant symbiosis and in the free-living state.

When grown in a complex liquid medium containing yeast extract (Callaham et al. 1978, Baker, Torrey 1979, Lalonde, Calvert 1979), Frankia proliferates as a filamentous mat consisting of branched septate hyphae 0.5-1.5 μm in diameter. Usually in response to culture in defined nutrient medium lacking fixed nitrogen compounds, short branched filaments of Frankia develop into swollen terminal vesicles averaging 3.0-5.0 μm in diameter. Concomitant with vesicle formation one can demonstrate the onset of acetylene reduction activity (Tjepkema et al. 1980, 1981) or $^{15}N_2$ incorporation (Torrey et al. 1981). The ontogeny of vesicle formation, the structural nature of the vesicle both in the host plant and in the cultured organism and the physiology and biochemistry of vesicles have been the subject for intense study over the past decade.

In host plants, Frankia vesicles assume various shapes (Fig. 1). In nodules of Alnus and Elaeagnus the spherical vesicles are arrayed around the periphery of infected cortical cells of the nodule lobes (Lalonde, Knowles 1975a, Baker et al. 1980). Vesicles in Ceanothus and Chamaebatia are similarly peripheral but somewhat pear shaped (Strand, Laetsch 1977, Newcomb, Heisey 1984). In Myrica and Comptonia the vesicles are club-shaped or clavate with the swollen ends oriented toward the cell periphery (Newcomb et al. 1978, Benson, Eveleigh 1979, Schwintzer et al 1982). Datisca and Coriaria root nodules show Frankia vesicles that are elongate and club-shaped but oriented toward a central vacuole occupying each infected cell (Newcomb, Pankhurst 1982, Calvert et al. 1979). In the

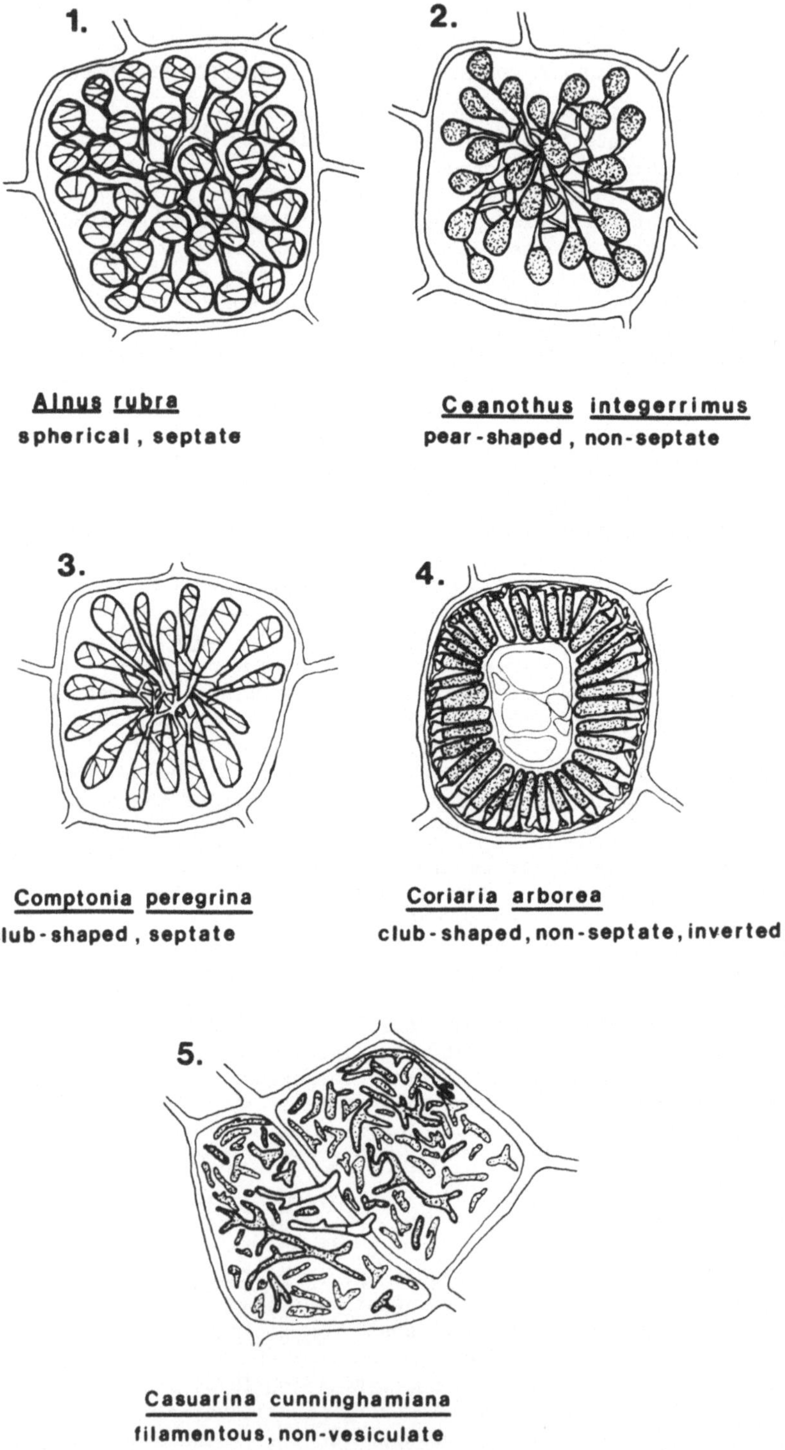

294

1.

2.

Alnus rubra
spherical, septate

Ceanothus integerrimus
pear-shaped, non-septate

3.

4.

Comptonia peregrina
club-shaped, septate

Coriaria arborea
club-shaped, non-septate, inverted

5.

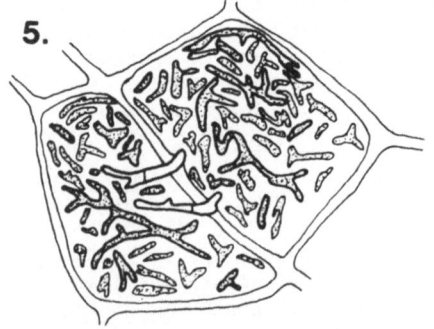

Casuarina cunninghamiana
filamentous, non-vesiculate

FIG.1. Diagrams of Frankia-infected root nodule cells of 5 hosts (see text).

genera of the Casuarinaceae including Casuarina and Allocasuarina, no vesicles are seen in actively fixing nodules (Torrey 1976, Zhang, Torrey
1985). The filamentous Frankia is differentiated into two types, the
thickened invasive filamentous strands and short non-septate terminal
endings showing little or no terminal swelling (Berg 1983).

Within the host plant, beginning at the invasion site and extending
throughout all the infected cells of the host nodules, Frankia is surrounded by a polysaccharide, probably pectic, layer (Lalonde, Knowles 1975b),
produced by the host plant cytoplasm in response to the presence of the
invading microorganism. This encapsulation produced by host dictyosomes
and ER serves to separate the Frankia from the host cell and is bordered
throughout by the host plasmalemma. Both in the nodule and in the cultured state, Frankia vesicles can be shown to possess another specialized
layer called the vesicle envelope which consists of a multilaminate structure, evident in freeze-fracture preparations, as 12-15 or more closely
packed and continuous thin layers surrounding the vesicle itself and extending along the stem to the basal septum where the vesicle is attached
to the bacterial filament (Torrey, Callaham 1982, Lalonde, Devoe 1976,
Newcomb et al. 1986). The vesicle envelope has special properties that
facilitate vesicle function.

As in all prokaryotes, the enzyme nitrogenase in Frankia is oxygen-
labile (Benson et al. 1979). Special protection of the enzyme is provided
in a variety of ways. In culture Frankia can be shown to possess acetylene reduction activity over a range of oxygen concentrations (Tjepkema
et al. 1981, Murry et al. 1984a). In vitro the enzyme is inactive in the
absence of oxygen, dependent as it is for the aerobic generation within
the bacterium of ATP and reductant. Frankia grown in the absence of fixed
nitrogen forms vesicles which show increasing acetylene reduction activity
at pO_2 values up to normal atmosphere (O_2 = 20%), showing reduced activity
when 30% O_2 or greater is reached. At 40% O_2 acetylene reduction activity
is markedly reduced. One hypothesis is that the multilaminate vesicle envelope provides a physical barrier to inward diffusion of oxygen, thereby
protecting the nitrogenase within the vesicle from denaturation by O_2.
Kinetic studies of respiration in cultured Frankia support the idea of a
physical barrier to O_2 diffusion into vesicles (Murry et al. 1984a).

In pure culture under aerobic conditions, all Frankia strains studied
to date show the presence of spherical-shaped vesicles, regardless of the
appearance of the filaments within the nodule. Even Frankia isolates from
Casuarina such as HFPCcI3 form typical vesicles in culture. The host
cells within the nodule, however, may modify both the occurrence and the
form of the vesicles as well as their function.

From our studies of Frankia in Alnus rubra, a temperate timber tree
from Oregon, one can characterize the most typical (most studied) relationship between host and endophyte. Infection involves root hair deformation and root hair cell wall penetration (Berry, Torrey 1983, Berry et
al. 1985). The nodule with multiple lobes is coralloid. Vesicles formed
in cortical cells of nodule lobes are spherical and peripherally arrayed
(Fig. 1-1). The organism isolated and grown in culture (Berry, Torrey
1979) grows readily in yeast extract medium and in defined media (Murry
et al. 1984b). With mechanical stirring and air-sparging, HFPArI3 grows
well on defined medium with propionate as carbon and energy source and
NH_4Cl as nitrogen source. Doubling times of the filamentous growth are
24-48 hr at 28°C. Such cultures can be homogenized, transferred to

medium lacking combined nitrogen and induced to form vesicles in abundance (Tjepkema et al. 1980, 1981, Murry et al. 1984b). A study of the
ontogeny of vesicles of ArI3 shows first a lag in vesicle induction followed by the development of provesicles which lack acetylene-reducing activity, and then the differentiation of phase-bright, well developed a-
cetylene-reducing vesicles (Fontaine et al. 1984). Free-living dinitrogen
fixation occurs at rates sufficient to sustain growth of the organism when
provided adequate carbon substrates under fully aerobic conditions (Murry
et al. 1984b).

A contrast can be made with the behavior of Frankia in Casuarina
cunninghamiana, a tropical fuel-wood tree native to Australia. In culture
Frankia HFPCcI3, isolated from root nodules of C. cunninghamiana (Zhang
et al. 1984) looks and behaves much like ArI3 (for a detailed comparison
see Zhang et al. 1985). The organism grows as a filamentous mat in defined medium, can be induced to form numerous large spherical vesicles by
withholding combined nitrogen and shows high acetylene-reducing activity
as long as carbon substrates are adequate. Freeze-substituted material
of cultured CcI3 prepared for electron microscopy (Lancelle et al. 1985)
shows complex multiseptate terminal vesicles and a series of interesting
novel ultrastructural features whose functions have yet to be elucidated.
The ultrastructural details of the laminated vesicle envelope remain to
be determined in transmission electron microscopy as the envelope itself
is difficult to fix for sectioning, possibly because of its lipid character (Torrey, Callaham 1982).

The striking difference between Alnus rubra and Casuarina cunninghamiana is seen, not in the two cultured microorganisms grown in isolation but in the remarkably different expression of the two Frankias within their respective hosts. Frankia HFPCcI3 used for inoculation of seedlings of C. cunninghamiana grown at 33°C produces nodules within 3 weeks.
The nodules are formed following root hair infection by rapid proliferation of multiple lateral roots induced at the infection site (Torrey
1976). The apical end of each nodule lobe is terminated by an elongate
determinate nodule root which tends to grow vertically upward. The
Frankia proliferates throughout cortical cells in the swollen basal lobes
of the nodule and, without development of terminal vesicles (Fig. 1-5),
begins to show acetylene-reducing activity with the endophyte remaining
in the filamentous state (Tyson, Silver 1979, Berg 1983).

It was apparent that in Casuarina root nodules the oxygen-protection
for nitrogenase provided in Alnus by the vesicle envelope was not functioning in the Casuarina symbiosis and that some other mechanism for oxygen-protection had evolved. This problem has been explored from different
points of view. Berg (1983) produced structural evidence that infected
host cells in Casuarina root nodules had specially modified cell walls,
either lignified or suberized or both, the latter potentially providing
a barrier to free oxygen diffusion into the infected cells. Tjepkema
(1979, 1983a) had shown in Alnus and in Myrica root nodules that intercellular air spaces allowed aeration of nodule tissue up to the infected
cells. The presumption would be that suberized cell walls would lead to
reduced O_2 access and therefore lowered O_2 in the infected cells. Berg
pointed out the occurrence of modified mitochondria within infected cells
reminescent of oxygen-deprived mitochondria in other systems. Tjepkema
(1983b) has followed the lead of Davenport (1960) in the study of the occurrence of hemoglobin-like compounds in root nodules of actinorhizal
plants. Casuarina root nodules can be shown to contain the highest levels

of these compounds found among the species studied. The suggestion is that these pigments might serve in a way analagous to leghemoglobins in root nodules of legumes to modulate the low O_2 levels necessary to allow nitrogenase to function.

With these ideas in mind Murry et al. (1985) grew the Frankia isolate CcI3 from Casuarina in defined medium, produced nitrogenase-induction by fixed nitrogen deprivation under controlled atmospheres in which the oxygen levels were varied from zero up to atmospheric concentrations. Under anaerobic conditions no growth occurred and no acetylene reduction by the culture could be demonstrated. At 20% O_2 in the gas phase above the liquid medium, normal spherical vesicles differentiated and good acetylene-reduction activity was observed. At $\sim$ 0.3% O_2 the cultured CcI3 cells grew slowly, remained filamentous but could be shown to have low acetylene-reducing activity. Presumably this low oxygen environment in vitro simulates the situation in the Casuarina root nodule in which, by virtue of specialized wall modification and perhaps other physiological factors, a reduced pO_2 is achieved in the infected cell. Low O_2 levels suppress Frankia vesicle formation and allow the enzyme nitrogenase to be synthesized within what appear under phase optics to be unmodified terminal filaments of the endophyte.

The symbiosis in Myrica gale and Comptonia may represent an intermediate situation between Alnus and Casuarina. The terminal vesicles within the root nodule vary in shape considerably from overtly enlarged club-shaped terminal endings to only slightly enlarged elongate terminal filaments (Fig. 1-3). The state of the vesicle envelope within these nodules has not been determined. As in Casuarina, the cell walls of infected Myrica root nodules show specialization (VandenBosch, Torrey 1985) and an intermediate level of hemoglobin-like compounds has been reported (Tjepkema 1983b). Atmospheric levels of O_2 are available within the nodule to the infected cell (Tjepkema 1983a) and even in waterlogged sites, aeration of the nodule through air passages within the nodule roots allows the nodule to undergo aerobic respiration (Tjepkema 1978) and would necessitate some special modification of infected cells to achieve O_2 protection of the nitrogenase within the Frankia filaments and vesicles. The interesting and peculiar orientation of the elongate vesicles in Datisca and Coriaria (Fig. 1-4) suggests yet another modification which may have evolved to achieve the end of O_2 protection of the endophyte nitrogenase. This association which has been well described (Calvert et al. 1979, Newcomb, Parkhurst, 1982) remains to be explored in functional terms.

In the cases described above, it is clear that the host cells and tissues exert a remarkable degree of control over the expression of the invading microorganism. This host control is part of the symbiotic relationship which keeps the invader under control and allows the symbiosis to work to the benefit of the host while allowing the endophytic symbiont a site and substrates to function. The evident control of host over endophyte expression is especially seen in cross inoculations, that is, the circumstances created when one inoculates a host plant of one species, genus or family with an infective microsymbiont isolated from a different host species, genus or family. We still know relatively little about the interactions involved which allow cross-inoculations to occur although, as isolations and culture of Frankia increase, we learn more about these possibilities (see, for example, cross-inoculations among members of the families Betulaceae and Myricaceae summarized by VandenBosch, Torrey 1983). In one such example, first described by Lalonde (1979), it was

shown that inoculation of <u>Alnus glutinosa</u> (of the family Betulaceae) with
the <u>Frankia</u> strain HFPCpI1 isolated from <u>Comptonia peregrina</u> (of the family Myricaceae) led to the formation of spherical vesicles within the effective root nodules which developed rather than the club-shaped vesicles
which CpI1 forms in its own host <u>Comptonia</u>. A careful study has yet to
be made as to whether this cross-inoculation results in altered nitrogenase activity. Examples of cross-inoculations which result in modified,
usually reduced, acetylene reduction activity have been reported recently
by several research groups (Dillon, Baker 1982, VandenBosch, Torrey 1984,
1985).

Vesicles that develop in culture are typically spherical, separated
from the vegetative filament by a basal septum, and surrounded by a
thickened envelope which is visible in phase or Nomarski optics (Fontaine
et al. 1984). Such vesicles possess internal septations that develop
during ontogeny, reaching most elaborate development before senescence.

In root nodules of different types (cf. Fig. 1) internal vesicle
septations are readily observed in ultrastructural studies in some nodules
and are lacking in others. Thus, for example, internal vesicular septation is lacking in nodules such as <u>Ceanothus</u> (Fig. 1-2) and similar
nodules, e.g., <u>Chamaebatia</u>, <u>Dryas</u> and <u>Purshia</u>. Septations are also lacking in vesicles of <u>Coriaria</u> and <u>Datisca</u> and nodules show an inverted arrangement of these vesicles (Fig. 1-4).

The role of septations within vesicles is not clear. In cultured
<u>Frankia</u> prepared by freeze-substitution for transmission electron micros-
copy (Lancelle et al. 1985), internal vesicular membranes are multilaminar
and frequently show ribosomes associated on both sides of the outermost
layers, suggesting high metabolic activity. Other ultrastructural features associated with septa membranes include mesosomes, tubules and mi-
crofilaments. Further study will be needed to understand structure-func-
tion relationships.

Unfortunately, high quality preservation for ultrastructural analysis
of vesicles in nodules is much more difficult than with free-living cultured <u>Frankia</u>. In so far as the available evidence goes, vesicles in
some circumstances can function in dinitrogen-fixation in the absence of
vesicle septation. What physiological circumstances lead to non-septate
vesicles remain to be determined.

REFERENCES

Baker D et al (1980) Can J Microbiol 26, 1072-1089.
Baker D and Torrey JG (1979) In Gordon JC et al. eds. Symbiotic Nitro-
gen Fixation in the Management of Temperate Forests, pp. 38-56. Oregon
State Univ. Corvallis.
Becking et al (1964) Ant v Leeuwenhock J Microbiol and Serol 30, 343-376.
Benson et al (1979) Science 205, 688-689.
Benson DR and Eveleigh DE (1979) Bot. Gaz (Suppl.) 140, S15-S21.
Berg RH (1983) Can J Bot 61, 2910-2918.
Berry et al (1985) Can J Bot (in press)
Berry AM and Torrey JG (1979) In Gordon JC et al, eds. Nitrogen Fixation
in the Management of Temperate Forests, pp 69-83. Oregon State Univ.
Corvallis.
Berry AM and Torrey JG (1983) Can J Bot 61, 2863-2876.
Bond G (1967) Ann Rev Plant Physiol 18, 107-126.
Bowes B et al (1977) Amer J Bot 64, 516-525.

Callaham D et al (1978)   Science   199, 899-902.
Callaham D et al (1979)   Bot Gaz   140 (Suppl.) S51-S59.
Calvert HE et al (1979)   In Gordon JC et al. eds. Nitrogen Fixation in the
Management of Temperate Forests, pp 474-475. Oregon State Univ. Corvallis.
Davenport HE (1960)   Nature   186, 653-654.
Dijk C van and Merkus E   (1976)   New Phytol   77, 73-91.
Dillon JT and Baker D   (1982)   New Phytol   92, 205-219.
Fontaine MS et al (1984)   J Bact   160, 921-927.
Lalonde M (1979)   Bot Gaz   140 (Suppl.)   S35-S43.
Lalonde M and Calvert HE (1979)   In Gordon JC et al. eds. Nitrogen Fixa-
tion in the Management of Temperate Forests, pp 95-110, Oregon State
Univ. Corvallis.
Lalonde M and Devoe IW (1976)   Physiol Plant Path   8, 123-129.
Lalonde M and Knowles R (1975a)   Can J Microbiol   21, 1058-1080.
Lalonde M and Knowles R (1975b)   Can J Bot   53, 1951-1971.
Lancelle, SA et al   (1985)   Protoplasma (in press).
Miller IM and Baker D   (1985)   Protoplasma (in press).
Moiroud A and Gianinazzi-Pearson V (1984)   In Verma DPS and Holm T eds.
Genes Involved in Microbe-Plant Interactions, pp 205-223, Springer Verlag,
New York.
Murry M et al (1984a)   Arch Microbiol   139, 162-166.
Murry M et al (1984b)   Plant Soil   78, 61-78.
Murry M et al (1985)   Can J Microbiol   (in press).
Newcomb W et al   (1986)   Can J Bot   (in press).
Newcomb W et al (1979)   Bot Gaz   140 (Suppl.) S22-S34.
Newcomb W and Heisey RM   (1984)   Can J Bot   62, 1697-1707.
Newcomb W and Pankhurst CE   (1982)   N Zealand J Bot 20, 93-103.
Newcomb W et al   (1978)   Can J Bot   56, 502-531.
Schwintzer CR et al   (1982)   Can J Bot   60, 746-767.
Strand R and Laetsch WM   (1977)   Protoplasma   93, 165-178.
Tjepkema JD   (1978)   Can J Bot   56, 1365-1371.
Tjepkema JD   (1979)   In Gordon JC et al, eds. Nitrogen Fixation in the
Management of Temperate Forests, pp 175-186, Oregon State Univ.
Corvallis.
Tjepkema JD   (1983a)   Amer J Bot   70, 59-63.
Tjepkema JD   (1983b)   Can J Bot   61, 2924-2929.
Tjepkema JD et al   (1980)   Nature   287, 633-635.
Tjepkema JD et al   (1981)   Can J Microbiol   27, 815-823.
Torrey JG   (1976)   Amer J Bot   63, 335-344.
Torrey JG and Callaham D   (1982)   Can J Microbiol   28, 749-757.
Torrey JG et al   (1981)   Plant Physiol   68, 983-984.
Tyson JG and Silver WS   (1979)   Bot Gaz   140 (Suppl.) S44-S48.
VandenBosch KA and Torrey JG   (1983)   Can J Bot   61, 2898-2909.
VandenBosch KA and Torrey JG   (1984)   Plant Physiol   76, 556-560.
VandenBosch KA and Torrey JG   (1985)   Amer J Bot   72, 99-108.
Zhang Z et al   (1984)   Plant and Soil   78, 79-90.
Zhang Z et al   (1985)   Plant and Soil   (in press).
Zhang Z and Torrey JG (1985)   Ann Bot   (in press).

ACKNOWLEDGEMENTS

The author expresses his indebtedness to his research colleagues and
support staff. The figure was drawn by Mary Lopez. Research summarized
here was supported in part by Department of Energy Research Grant DE-FG
02-84ER-13198 and by the Maria Moors Cabot Foundation for Botanical Research, Harvard University.

USE OF ^{13}N TO STUDY N$_2$ FIXATION AND ASSIMILATION BY CYANOBACTERIAL-LOWER PLANT ASSOCIATIONS

J.C. MEEKS, C.S. ENDERLIN, C.M. JOSEPH, N. STEINBERG, AND Y.M. WEEDEN.
DEPARTMENT OF BACTERIOLOGY, UNIVERSITY OF CALIFORNIA, DAVIS, CA 95616
USA

1. INTRODUCTION

Symbiotic associations involving nitrogen-fixing cyanobacteria and eucaryotes are of general interest because, in a manner analogous to interactions between rhizobia and legumes, the cyanobacteria provide fixed nitrogen for growth of the associations (Stewart et al. 1980). We are interested in identifying the chemical interactions between partners of these associations that influence cyanobacterial growth, heterocyst differentiation, and N$_2$ fixation as well as other metabolic processes. We reasoned that such a study must utilize relatively rapidly growing material in pure culture that is also amenable to experimental manipulation. Accordingly, we established pure cultures of gametophyte tissue of the bryophyte Anthoceros punctatus L. for growth submerged in liquid medium and defined conditions for reconstitution of the association with various wild-type and mutant strains or species of Nostoc (Enderlin, Meeks, 1983). These experimental manipulations are not currently achievable with the Azolla-Anabaena association (Peters, Calvert, 1983).

In this report, we will summarize the results of our initial physiological studies directed toward identification of the primary route(s) for assimilation of ammonium by Anthoceros and Nostoc and determination of the fraction of dinitrogen-derived ammonium assimilated by associated Nostoc. We have used the radioactive isotope of nitrogen as [^{13}N]N$_2$ and ^{13}NH$_4^+$ as a tracer in these in vivo and in situ experiments. A number of parallel experiments have also been done with Azolla-Anabaena and isolated Anabaena in collaboration with Dr. G.A. Peters, Battelle-Kettering Res. Lab., Yellow Springs, OH USA.

2. LABELING ^{13}N

A flow scheme summarizing the sequential processes in generation of ^{13}N and synthesis of different molecular species is given in Fig. 1. The ^{13}N is generated at the University of California, Davis, Crocker Nuclear Laboratory cyclotron by the ^{16}O(p,α)^{13}N nuclear reaction using glass distilled water as substrate. The 60 ml recirculating target system was described by Parks and Krohn (1978). Typically about 0.5 Ci (ca. 1.9 X 10^{10} Bq) of ^{13}N, primarily as ^{13}NO$_3^-$ with trace amounts of ^{13}NO$_2^-$ and ^{13}NH$_4^+$, is generated in a 20 min irradiation. The radioactive anions are concentrated by HPLC onto an anion exchange resin (SAX), eluted with phosphate buffer and collected in the desired volume as described (Chasko, Thayer, 1981). The concentrated ^{13}NO$_3^-$ solution can be used directly or transferred to a reaction flask containing saturated NaOH and Devarda's alloy for reduction to ammonia. The ^{13}NH$_3$

1. Generation: H_2O $\xrightarrow[\text{20MeV, }20\mu A]{^{16}O(p,a)\,^{13}N}$ $^{13}NO_3^- \ggg {}^{13}NO_2^- > {}^{13}NH_4^+$

2. Concentration: 60 ml $^{13}NO_3^-$ $\xrightarrow[\text{SAX Column}]{\text{HPLC on}}$ ≤1ml up to 3ml $^{13}NO_3^-$

3. Reduction : $^{13}NO_3^-$ $\xrightarrow[\text{65°C, Saturated NaOH}]{\text{Devarda's Alloy(Cu/Al/Zn)}}$ $^{13}NH_3$

4. Oxidation : $^{13}NH_4^+$ $\xrightarrow[\text{1.5}\mu\text{mol }^{14}NH_4^+]{\text{Na/KOBr}}$ $[^{13}N]\,N_2$

Fig. 1. Flow scheme of processing of ^{13}N. Details in text.

is vacuum distilled and trapped in 5 mM buffer (Mops or Mes, initial pH about 3.0) if it is to be used directly in an assimilation experiment (Meeks et al., 1983). The $^{13}NH_3$ is trapped in 0.35 M formic acid if it is to be converted to $[^{13}N]N_2$. Routinely about 150 mCi of $^{13}NH_4^+$ is recovered in a 3 ml volume. The oxidation of $^{13}NH_4^+$ plus carrier $^{14}NH_4^+$ to $[^{13}N]N_2$ is catalyzed by hypobromite (Burris, 1972). The reaction is done under vacuum, with the $[^{13}N]N_2$ swept from the liquid by gentle sparging with ultra-pure CO_2, water vapor is removed by passage over anhydrous $CaSO_4$ and carrier CO_2 is trapped in liquid N_2 (Meeks et al. 1985). The $[^{13}N]N_2$ is collected, compressed and transferred using a Toeppler pump in a manner similar to that described by Austin et al. (1975). An average of 0.93 μmol of dinitrogen and 50 mCi of ^{13}N are recovered in the 3-ml sample vial at the start of incubation.

The conditions for culture of Anthoceros-Nostoc, isolation of symbiotic Nostoc and incubation of associated tissues or Nostoc are detailed in Enderlin and Meeks (1983) and Meeks et al. (1983; 1985). The assimilatory reactions are initiated by addition of the radionuclide and terminated by mixing or homogenization of the samples in approx. 80% methanol. The methanolic extracts are processed either for distillation to recover NH_4^+ or amide-nitrogen, or for thin-layer electrophoresis on 0.1 mm thick layers of cellulose to separate amino acids, essentially as described by Wolk et al. (1976). The ^{13}N-labeled amino acids are detected by radioelectrophoretogram scanning and the radioactivity quantitated by time-corrected integration of peaks in the radioscans (Meeks, 1981). Unincorporated $[^{13}N]N_2$-derived $^{13}NH_4^+$ is also determined by the difference in time-corrected total radioactivity between pre- and post-electrophoresis of the thin-layer plate (Meeks et al., 1985).

3. PATHWAYS OF ASSIMILATION OF EXOGENOUS AND $[^{13}N]N_2$-DERIVED $^{13}NH_4^+$ BY ANTHOCEROS-NOSTOC.

When Anthoceros-Nostoc was incubated for up to 5 min with $^{13}NH_4^+$ or 10 min with $[^{13}N]N_2$, 3 radioactive amino acids were detected; glutamine and glutamate accumulated the highest radioactivity, with alanine being a minor product. Glutamine and glutamate could have been formed in parallel from ammonium by glutamine synthetase (GS) and glutamate dehydrogenase (GDH) or in a sequential manner through the activities of

GS and glutamate synthase (GOGAT) (the glutamate synthase cycle; Miflin, Lea, 1980), or by both routes.

Four experimental approaches are commonly used with radiotracers to identify the operation of a metabolic sequence: (1) the rate or kinetics of incorporation of radiolabel into specific products; (2) the transfer of radiolabel between products formed during short time periods (pulse) when followed by continued incubation with excess nonradioactive substrate (chase); (3) the effect of specific enzyme inhibitors on the formation of radiolabeled products; and (4) dilution of radiolabeled end product by excess unlabeled substrate or presumed metabolic inter-mediates. We have applied these four experimental approaches to the assimilation of exogenous (Meeks et al., 1983) and [^{13}N]N$_2$-derived ^{13}NH$_4^+$ (Meeks et al., 1985) by Anthoceros-Nostoc. Based on the results summarized below, we conclude that ammonium assimilation occurs primarily by the glutamate synthase cycle with little contribution by GDH.

When ^{13}NH$_4^+$ was the nitrogen source, more than 90% of the radio-activity extracted with methanol was associated with glutamine and less than 10% with glutamate after 5 s of incubation (Fig. 2). Total radio-activity in both compounds increased in a linear manner for up to 5 min, but the fractions of radioactivity recovered in glutamine and glutamate did not markedly change with increasing incubation time. More than 90% of the ^{13}N in glutamine was initially in the amide position, decreasing only to 86% after 5 min of incubation. The kinetics of labeling of glutamine and glutamate by Anthoceros-Nostoc supplied with [^{13}N]N$_2$, however, showed a typical precursor (glutamine)-product (glutamate) relationship: 95% of the ^{13}N recovered in the organic fraction after 30 s of incubation was associated with glutamine, declining to 70% by 120 s, while the fraction of radioactivity in glutamate correspondingly increased (Fig. 2).

The transfer of ^{13}N between glutamine and glutamate was not readily observed when 10, 30, 60, or 120 s pulses of ^{13}NH$_4^+$ and [^{13}N]N$_2$ were followed by chase periods of up to 10 min with 2.5 mM NH$_4^+$ or air. While kinetic and pulse-chase experiments with ^{13}NH$_4^+$ and [^{13}N]N$_2$ were consistent with the formation of glutamine by GS, they yielded equivocal results with respect to the route of glutamate formation.

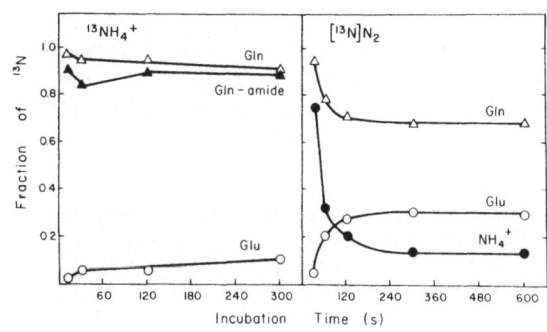

Fig. 2. Time course of the fractions of ^{13}N recovered in glutamine, glutamate and NH$_4^+$ when Anthoceros-Nostoc was incubated with ^{13}NH$_4^+$ and [^{13}N]N$_2$. The fraction of [^{13}N]N$_2$-derived ^{13}NH$_4^+$ was based on total ^{13}N recovered, all others were on total ^{13}N in the organic fraction. Values are means of 3-22 experiments.

The analogues methionine sulfoximine (MSX) and diazo-oxonorleucine (DON) were used in attempts to discriminate between GOGAT and GDH as the major enzyme catalyzing the formation of glutamate from exogenous or $[^{13}N]N_2$-derived $^{13}NH_4^+$. MSX is an analogue of glutamate that irreversibly inhibits glutamine synthetase (Ronzio et al., 1969). If glutamate is synthesized primarily from ammonium by GDH, MSX should have little effect on its radiolabeling. Conversely, if formed from glutamine by GOGAT, inhibition of the synthesis of glutamate by MSX should be similar to that of glutamine. In the presence of 0.25 to 1.0 mM MSX, the formations of ^{13}N-labeled glutamine and glutamate were inhibited by more than 90% when Anthoceros-Nostoc was supplied with $^{13}NH_4^+$ or $[^{13}N]N_2$ (Table).

Table. Effect of analogues and carrier ammonium on the formation of glutamine and glutamate from $^{13}NH_4^+$ and $[^{13}N]N_2$ by Anthoceros-Nostoc

| Addition | cpm in compound/g FW | | | |
| | per nCi $^{13}NH_4^+$ added | | per μCi $[^{13}N]N_2$ added | |
	Glu	Gln	Glu	Gln
None	473 ± 93	4044 ± 1162	15 ± 2	37 ± 5
MSX	16 ± 4 (3)	67 ± 40 (2)	1 ± .5 (7)	3 ± .5 (8)
DON	78 ± 9 (16)	2696 ± 531 (66)	2 ± .1 (13)	24 ± 2 (65)
NH_4^+	13 ± 4 (3)	164 ± 22 (4)	-	-

Additions were DON and NH_4^+ at 1 mM, MSX at 1 mM for $^{13}NH_4^+$ and 0.25 mM for $[^{13}N]N_2$. Incubation times prior to exposure to $[^{13}N]N_2$ were 120 min and for $^{13}NH_4^+$ were; DON = 60 min, MSX = 120 min and NH_4^+ = 0.5 min. Incorporation times were 2 min for $^{13}NH_4^+$ and 10 min for $[^{13}N]N_2$. ( ) = % of no additions.

DON, an analogue of glutamine, irreversibly inhibits glutamine-amide transfer reactions (Miflin, Lea, 1980). The presence of DON should inhibit the formation of $[^{13}N]$glutamate by GOGAT, but should have no effect on that catalyzed by GDH. The ^{13}N recovered in glutamine should be greater than, or equal to, that in the absence of DON unless the exposure time to DON results in depletion of the pool of glutamate prior to introduction of radiolabeled substrate. When Anthoceros-Nostoc was incubated with 1 mM DON for 60 min or longer prior to the introduction of $^{13}NH_4^+$ or $[^{13}N]N_2$, the subsequent radiolabeling of glutamate was inhibited 84 to 87% and that of glutamine by less than 35% (Table). The amounts of $[^{13}N]$glutamine formed were distinctly dependent upon the prior incubation time with DON (Fig. 3); ^{13}N accumulated in glutamine after short periods (< 30 min) before declining after progressively longer exposures.

Assuming that in both plants and cyanobacteria the affinity of GS and GDH for ammonium differs (Miflin, Lea, 1980), then lowering the specific activity of $^{13}NH_4^+$ by addition of carrier 1 mM NH_4^+ should result in: (a) a differential decrease in the radiolabeling of glutamate and glutamine if they are formed in parallel by GS and GDH; or (b) a proportional decline if they are formed sequentially in the glutamate

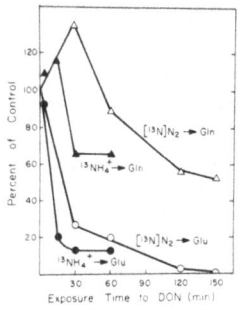

Fig. 3. Effect of various incubation times in 1 mM DON on subsequent radiolabeling of glutamine and glutamate by Anthoceros-Nostoc supplied with $^{13}NH_4^+$ and $[^{13}N]N_2$. Values are means of 3 to 5 experiments.

synthase cycle. Addition of carrier NH_4^+ resulted in a proportional decline in the ^{13}N-labeling of glutamine and glutamate (Table).

Thus, the results of inhibitor and radiolabel dilution experiments are consistent with assimilation of exogenous and N_2-derived NH_4^+ primarily by the glutamate synthase cycle in Anthoceros-Nostoc. In similar experiments with Azolla-Anabaena we found no indications that GDH has a biosynthetic role in the assimilation of ammonium.

4.   ASSIMILATION OF AMMONIUM BY ASSOCIATED NOSTOC.

Digitonin was used to distinguish assimilation of exogenous $^{13}NH_4^+$ by associated Nostoc from that by Anthoceros (Meeks et al., 1983); digitonin increases the membrane permeability of eucaryotes more extensively than that of procaryotes. In the presence of 0.05% (w/v) digitonin, assimilation of $^{13}NH_4^+$ by Nostoc alone was unaffected, but that by Anthoceros-Nostoc was inhibited more than 99%. These data indicate that symbiotic Nostoc incorporates an insignificant amount (< 1%) of the exogenous ammonium assimilated by Anthoceros-Nostoc and, by default, that Anthoceros, grown under symbiotically-associated conditions, must assimilate ammonium primarily by the glutamate synthase cycle (see section 3).

When immediately isolated from Anthoceros, Nostoc incorporated exogenous $^{13}NH_4^+$ into glutamine and glutamate and the incorporations were inhibited more than 93% in the presence of 100 µM MSX (Meeks et al., 1985). The amount of ^{13}N assimilated per unit protein by symbiotically-grown Nostoc was about 5-fold lower than that by Nostoc grown in the absence of Anthoceros; control experiments showed that the lower rate of assimilation was not an artifact of the procedure for rapid isolation of Nostoc.

The lower assimilation of exogenous $^{13}NH_4^+$ and the lower rate of GS activity in vitro (See Joseph, Meeks, these proceedings) suggest that symbiotic Nostoc may not assimilate all of its N_2-derived NH_4^+. Experiments to directly quantitate excretion of $[^{13}N]N_2$-derived $^{13}NH_4^+$ by isolated symbiotic Nostoc were hindered because the relative rates of acetylene reduction and $[^{13}N]N_2$ fixation decline to less than 30% of the control rates of intact tissue when Anthoceros-Nostoc was treated for up to 60 min with digitonin (Fig. 4). Therefore, the amount of N_2-derived NH_4^+ excreted by Nostoc under associated conditions is substantially underestimated when assayed apart from Anthoceros even though the

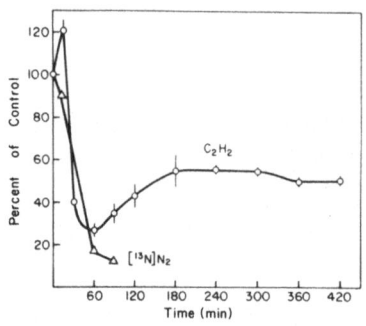

Fig. 4. Effect of various incubation times in 0.05% (w/v) digitonin on acetylene reduction and $[^{13}N]N_2$ fixation by Anthoceros-Nostoc. Values are means of 3 to 4 experiments with $[^{13}N]N_2$ reflect total radioactivity recovered in methanolic extracts.

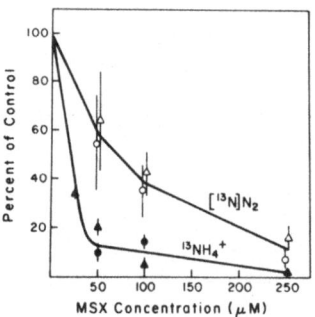

Fig. 5. Effect of various MSX concentrations on the incorporation of $[^{13}N]N_2$ and $^{13}NH_4^+$ by Anthoceros associations reconstituted with wild-type (circles) or MSX-resistant (triangles) strains of Nostoc. MSX was added 120 min prior to addition of ^{13}N. Values are means of 3 experiments.

reduction of acetylene recovers to about 50% of the control values after about 3 h of continual exposure to digitonin. Consequently, the rate of nitrogenase-dependent ammonium production by isolated Nostoc is artifactually lowered to more closely parallel the potential rate of its assimilation.

To circumvent the problems with isolated Nostoc, Anthoceros associations reconstituted with either the wild-type or a strain of Nostoc selected as resistant to 50 µM MSX were used to estimate the fraction of N_2-derived NH_4^+ assimilated in situ by symbiotic Nostoc (Fig. 5; Meeks et al., 1985). The formation of ^{13}N-labeled glutamine and glutamate in the absence of MSX was the primary contol in these experiments; the secondary control was the wild-type association in the presence of MSX. If symbiotic Nostoc assimilates all of its N_2-derived NH_4^+, the radiolabeling of glutamine and glutamate by the association with the MSX-resistant strain should be the same in the presence or absence of MSX. Conversely, if symbiotic Nostoc assimilates none of its N_2-derived NH_4^+, the presence of MSX should inhibit assimilation equally in associations with wild-type or MSX-resistant Nostoc; in this case only MSX sensitive Anthoceros would be active in ammonium assimilation. The formations of ^{13}N-labeled glutamine and glutamate from $^{13}NH_4^+$ and $[^{13}N]N_2$ in the presence of various MSX concentrations were inhibited by nearly the same extent in Anthoceros associations with either MSX-resistant or wild-type strains of Nostoc (Fig. 5). These results indicate that symbiotic Nostoc assimilates only a minor fraction of its N_2-derived NH_4^+ and that ammonium, not organic nitrogen, must be the compound made available to <u>Anthoceros</u> tissue. The amount of N_2-derived NH_4^+ assimilated by symbiotic <u>Nostoc</u> was calculated to be on the order of 10% of the total fixed (Meeks et al., 1985).

In contrast to symbiotic <u>Nostoc</u>, rapidly isolated <u>Anabaena</u> from <u>Azolla</u> retains its relatively high rate of acetylene reduction (Peters, Calvert, 1983). We examined the cellular and extracellular distribution of radiolabel after incubating isolated <u>Anabaena</u> with $[^{13}N]N_2$ for 10 min (Meeks et al., 1985b); 38% of the fixed nitrogen was in an extra-cellular fraction and 99% of that fraction was $^{13}NH_4^+$. The intra-cellular fraction contained ^{13}N-labeled glutamine, glutamate and alanine. The formations of radiolabeled glutamine and glutamate were inhibited by MSX and DON.

Symbiotically-associated cyanobacteria, therefore, assimilate N_2-derived NH_4^+ by the glutamate synthase cycle, although N_2 fixation and NH_4^+ assimilation are not tightly coupled. The extent of uncoupling by repression or inhibition of the assimilatory cycle and the mechanism(s) of regulation may not be the same in different associations (Stewart et al., 1980; Orr, Haselkorn, 1982; Joseph, Meeks, these proceedings).

5. REFERENCES

Austin SM et al (1975) Nucl. Inst. Methods 126, 373-379.
Burris RH (1972) Methods Enzymol. 126, 415-431.
Chasko JH and Thayer JR (1981) Int. J. Appl. Radiat. Isotopes 32, 645-649.
Enderlin CS and Meeks JC (1983) Planta 158, 157-165.
Meeks JC (1981) In Root JW and Krohn KA, eds, Rec. Adv. Chem. Ser. No. 197: Short-lived Radionuclides in Chemistry and Biology, pp 269-294, Am. Chem. Soc. Wash. DC.
Meeks JC et al (1983) Planta 158, 384-391.
Meeks JC et al (1985) Planta 164, 406-414.
Meeks JC et al (1985b) Arch. Microbiol. 142, in press.
Miflin BJ and Lea PJ (1980) In Miflin BJ, ed, The Biochemistry of Plants, vol. 5: Amino Acids and Derivatives, pp 169-202, Academic Press, NY, London.
Orr J and Haselkorn R (1982) J. Bacteriol. 152, 626-635.
Parks NJ and Krohn KA (1978) Int. J. Appl. Radiat. Isotopes 29, 754-756.
Peters GA and Calvert HE (1983) In Goff LJ, ed, Algal Symbiosis, pp 109-145, Cambridge Univ. Press, Cambridge.
Ronzio R et al (1969) Biochemistry 8, 1066-1075.
Stewart WDP et al (1980) In Stewart WDP and Gallon JR, eds, Nitrogen Fixation, pp 239-277, Academic Press, NY, London.
Wolk CP et al (1976) J. Biol. Chem. 251, 5027-5034.

6. ACKNOWLEDGEMENTS

The work reported herein was supported in part by USDA-CRGO Grants No. 70-59-2063-1-1-276-1 and 83-CRCR-1-1295 and NSF Grant PCM 79-04136. We thank Lewanna Archer for formatting the manuscript.

ULTRASTRUCTURE OF SOYBEAN NODULES INOCULATED WITH
ELECTRON TRANSPORT MUTANTS OF RHIZOBIUM JAPONICUM 61A76

MEHRESHAN T. EL MOKADEM
BOTANY DEPARTMENT, WOMEN'S COLLEGE, AIN SHAMS UNIVERSITY, HELIOPOLIS,
CAIRO, EGYPT

The differentiation and developmental changes of cells and organelles
of soybean (Glycine max cv. Williams) nodules, infected with Rhizobium
japonicum 61A76 wild type or electron transport mutant strains, were
studied by transmission electron microscopy. The three 61A76 mutant
strains, namely Na_1, Na_3, and Na_0, although totally deficient in C- and
aa_3-type cytochromes, were symbiotically infective but formed ineffective
nodules. Seedlings were maintained in Leonard jars in sterilized sand on
a N-free nutrient solution. 11 days after inoculation, cells in both effective and ineffective nodules have the features of typical meristimatic
cells. The three mutant strains provoked different and unusual responses
in the cytoplasm of plant cells. A feature induced by strain Na_1 was
electron-dense cytoplasm; cell vacuoles were not common. The most remarkable structural changes occurred in host mitochondria, which appeared
elongated, disrupted or dumbbell-shaped; mitochondrion-amyloplast associations were recorded. The cytoplasm in cells of nodules induced by strain
Na_3 was heterogeneous; variously sized vesicles and vacuoles were scattered throughout the cytoplasm. Fewer abnormalities were observed in
nodules induced by strain Na_0.

35-45 days after inoculation, major differences were observed in
bacteroid development between wild type and mutant strains. In ineffective nodules, bacterial cells were either still enclosed within intercellular spaces, enveloped in masses of polysaccharide-like material or
trapped in large infection thread vesicles surrounded by massive cell
walls. Holl (1975) and Newcomb et al (1977) showed that ineffectiveness
is associated with incomplete bacteroid development. The failure of rhizobia to be released from different envelopes (Pankhurst, 1977) could
make these mutants more sensitive to host cells (Pankhurst, 1974) and
lead to unsuccessful nodule development.

Bacteroids from mutant strains, senesced and deteriorated much sooner
than wild-type. The host cells survive this degeneration and most are
still alive 35-45 days after inoculation. This indicates that the auto-
lysis process is "selective" in that it is more damaging to the rhizobia
than to the host (Basset et al, 1977a, b).

REFERENCES

Bassett B et al (1977a) Can.J.Microbiol. 23; 573-582
Bassett B et al (1977b) Can.J.Microbiol. 23; 873-883.
Holl FB (1975) Euphytica 24; 767-770.
Mackenzie CR and Jordan DC (1974) Con.J.Bat., 20; 755-759.
Newcomb W et al (1977) Can.J.Bot. 55; 1891-1907.
Pankhurst CE (1974) J.Gen.Microbiol. 82; 405-413.
Pankhurst CE (1977) Can.J.Microbiol. 23; 1026-1033.

INVOLVEMENT OF MICROBODIES IN PEANUT-RHIZOBIUM SYMBIOSIS.

S. Hameed and A. K. Bal
Department of Biology,
Memorial University of Newfoundland,
St. John's, Newfoundland, Canada A1B 3X9.

Microbodies, present in uninfected cells interspersed among the infected ones in soybean root nodules, have been implicated in ureide metabolism and export of recently fixed nitrogen (Newcomb et al. 1985). In peanut, infected cells that form the core of the nodules, are not interspersed by uninfected cells (Sen et al. in preparation). Unlike soybean, microbodies are found in the infected cells at all stages of development-early stages of infection to maturation. Cytochemical and biochemical studies of microbodies in different stages of the peanut nodule are presented.

Nodules of different developmental stages were assayed for nitrogenase (acetylene reduction) activity following the method of Hardy et al. (1968). Catalase (marker enzyme) activity (Beer, Sizer 1952) was measured in both the bacteroid and host cytosol fraction. Samples of nodules fixed in aldehyde (Karnovsky 1965) were processed for ultrastructural studies. Sections of nodules were subjected to 3,3'diaminobenzedine (DAB) reaction in the presence of hydrogen peroxide (H_2O_2) at pH 9 in propanediol buffer, prior to osmium tetroxide treatment (Hayat 1974). A set of controls were run by adding 3 amino,1,2,4-triazole (AT), potassium cyanide (KCN) in the incubation mixture; control reactions were also carried out in the absence of DAB and H_2O_2.

A good correlation between catalase activity and nitrogenase activity was observed, both reaching a peak at 35 day old nodules. Osmiophilic electron dense reaction product was detected in typical microbodies bound by a membrane, showing a core of denser nucleoid-like structure. In some cases the membrane showed discontinuity and in other forms the central core of dense structure was lined by an electron transluscent layer within the membrane. Microbodies were found in close contact with the peribacteriod membrane and instances of membrane fusion could be seen. Connections of microbodies with endoplasmic reticulum resulted in pear shaped structures. A DAB positive structure has been recognized in contact with the outer membrane of the bacteroid; this structure often caused bulging of the peribacteroid membrane. Lipid bodies (oleosomes) were also found in contact with the peribacteroid membrane.

The above studies strongly suggest intimate involvement of the enzymes of microbodies in peanut nodule interaction between host and the bacteroid.

Beers RF, Sizer, LW (1952) J. Biol. Chem. 195, 133-140.
Hardy RWF et al. (1968) Plant Physiol. 43, 1185-1207
Karnovsky MJ (1965) J. Cell Biol. 27, 137a-138a.
Newcomb EH et al. (1985) Protoplasma 125, 1-12.
Hayat MA (1974) Electron Microscopy of Enzymes 2, pp 12-16 Van Nostrand Reinhold, New York.

STRUCTURAL FEATURES OF NORTH AMERICAN SOYBEAN ROOT NODULES FORMED BY TWO DIFFERENT STRAINS OF RHIZOBIUM FREDII

S. SHANTHARAM, N. DUTEAU, R. K. PRAKASH and A. G. ATHERLY, Department of Genetics, Iowa State University, Ames, Iowa, U.S.A. 50011.

The nodulation of five commercial cultivars of Glycine max L. by Rhizobium fredii strains, USDA 191 and 193 were studied using transmission electron microscopy. R. fredii USDA 193 produced effective symbiosis with cv. Peking and Virginia but not with cv. Evans, Hill, Rampage and Williams. The ultrastructure of root nodules of cv. Peking and Virginia were normal with several bacteroids enclosed in each peribacteroid membrane envelope and they were rich in poly-B-hydroxybutyrate granules and showed the presence of leghemoglobin. In the rest of the commercial cultivars, the root nodules were whitish (lack of leghemoglobin) and showed poor degree of infection both in terms of the number of host cells infected and in the number of bacteroids enclosed per membrane envelope; there were enlarging vacuoles in the infected cells and contained intact starch grains; the peribacteroid membrane envelopes were disintegrated and depending on the cultivar, the bacteroids showed varying degrees of plasmolysis and degeneration.

Fortuitously, due to repeated sub–culturing of R. fredii 193 in the laboratory medium without the plant selection pressure, a spontaneous Nif⁻ derivative was encountered. This mutant produced ineffective root nodules on cv. Peking where the bacteroids showed plasmolysis and degeneration. The leghemoglobin was not visually detectable.

In contrast, the root nodules formed by the infection of R. fredii USDA 191 on north American cultivars of soybean were effective and showed normal cytology with the exception of cv. Rampage where the peribacteroid membrane envelopes were broken down to form small membranous vesicles.

It is believed that the above information will augment the efforts to understand the interactions of the fast–growing microsymbionts with soybean.

We would like to acknowledge the technical help of Mrs. Anita Cody during electron microscopy. This investigation was supported by a research grant 59-219-0-1-494-0 from the U.S. Department of Agriculture and funds from Land O'Lakes Corporation, Minneapolis, Minnesota.

ULTRASTRUCTURAL LOCALIZATION OF NODULE-SPECIFIC
URICASE IN SOYBEAN NODULES

KATHRYN A. VANDENBOSCH and ELDON H. NEWCOMB. Department of Botany,
University of Wisconsin, Madison, WI 53706.

During legume root nodule development, the plant encodes a group
of nodule-specific proteins, the nodulins, that are not found in other
plant organs. One of these proteins, nodulin-35 from soybean, has
been identified as a subunit of uricase (Bergmann, et al., 1983),
which is involved in the conversion of purines to ureides in the
metabolism of recently fixed nitrogen. In soybeans and closely
related legumes, ureides are the principle forms of nitrogen that are
translocated from the nodule to other parts of the plant. In this
study, we have employed immunogold labeling to localize
nodule-specific uricase in soybean nodules.

Sections of resin-embedded nodules were first incubated in a
polyclonal antibody preparation made against nodulin-35. They were
then transferred to a suspension of protein-A coated gold particles,
which thus provide a marker for uricase. Glutaraldehyde-fixed, osmium
post-fixed specimens, when treated with periodate prior to labeling,
showed excellent ultrastructural preservation, good preservation of
antigenicity and very low background labeling.

Mature nodules showed pronounced labeling for uricase in enlarged
peroxisomes in uninfected cells of the central infected zone of the
nodule, and in no other subcellular compartment. Small, presumptive
peroxisomes found in bacteroid-containing cells did not exhibit
specific labeling. These results confirm that the conversion of urate
to allantoin occurs in the peroxisomes of uninfected cells.

Developing nodules were examined for the first appearance of
uricase. Uricase was first detectable in uninfected cells coincident
with the release of Rhizobium bacteroids from infection threads in
adjacent infected cells. Gold particles, indicating the presence of
uricase, occurred over young, developing peroxisomes in these cells.
The plant cytoplasm and endoplasmic reticulum remained unlabeled. In
older nodules, peroxisomes were more heavily labeled, indicating an
increase in uricase concentration. In summary, uricase appears before
the nodule has fully differentiated and increases in concentration as
the nodule matures.

Reference:
Bergmann, H. et al., 1983. EMBO J. 2: 2333-2339.

Acknowledgement:
We thank D.P.S. Verma for antibody preparations used in this study.

SOYBEAN ROOT RESPONSE TO SYMBIOTIC INFECTION. GLYCEOLLIN I ACCUMULATION
IN AN INEFFECTIVE TYPE OF SOYBEAN NODULES WITH AN EARLY LOSS OF THE
PERIBACTEROID MEMBRANE

D. WERNER*, R.B. MELLOR*, M.G. HAHN** and H. GRISEBACH**
* Fachbereich Biologie, Botanisches Institut der Philipps-Universität,
  Lahnberge, D - 3550 Marburg/L, FRG
** Institut für Biologie II/Biochemie der Pflanzen, Universität Frei-
  burg, Schänzlestr. 1, D - 7800 Freiburg, FRG

As a very general host response in Glycine max as modified by Rhizobium mutants compartmentation by the peribacteroid membrane is analysed in: Type A as an effective wild type, Type B with an early loss of
the peribacteroid membrane (PBM) with stable bacteroids, Type C with a
degeneration of bacteroids and peribacteroid membranes at the same time
and Type D with a lysis of bacteroids and fusion of lytic compartments,
however only in the vicinity of the host cell nucleus. Type B nodules
infected with Rhizobium japonicum 61-A-24 are biochemically characterized by a reduction of glutamine synthetase activity to 10 to 30 %
compared to the wild-type nodules (type A), a reduction of the total
pool of the soluble amino acids to less than 1/3 and by a continuous
increase of nodule number (more than 150 per plant) compared to a constant number type A nodules (about 20 to 30 nodules per plant) (1, 2).
A glyceollin I accumulation of about 6000 pmol x mg dry weight^{-1}, a
tenfold increase above control root tissue was found in type B nodules
from Glycine max which had been infected with a fix$^-$ strain (61-A-24)
of Rhizobium japonicum. In nodules infected with one other ineffective
(fix$^-$) strain of Rhizobium japonicum (RH 31-Marburg) or with two
fix$^+$ strains of Rhizobium japonicum (61-A-101 and USDA 110), no increase in glyceollin I concentrations above control values was found
at either 20 d or 34 d after infection. Nodules infected with Rhizobium japonicum 61-A-24 are distinguished by an early loss of the peribacteroid membrane in the infected host cell, whilst the bacteroids
themselves remain stable (3).

REFERENCES

(1) Werner D et al (1980) Planta 147, 320-329.
(2) Werner D et al (1984) Planta 162, 8-16.
(3) Werner D et al (1985) Z. Naturforsch 40c, 179-181

BIOGENESIS AND FUNCTION OF THE PERIBACTEROID MEMBRANE IN NODULES

D. WERNER, R.B. MELLOR, E. MÖRSCHEL, S. BASSARAB, T. CHRISTENSEN and M. RÖHM, Fachbereich Biologie, Botanisches Institut der Philipps-Universität, D - 3550 Marburg-L, FRG

A. Using mutants of Rhizobium japonicum, we have demonstrated that the biogenesis and stability of the peribacteroid membrane formed by the membrane-building organelles of the host cell (ER, Golgy Body) is affected by the symbiont. Comparing nodules with a stable peribacteroid membrane until senescence, three other PPM developments could be characterized:
a) Early loss of the peribacteroid membranes with stable bacteroids in the host cell cytoplasm (1);
b) Lysis of bacteroids in fusionating peribacteroid membranes (2): This development is observed only in the vicinity of the host cell nucleus, indicating a cellular gradient from the nucleus to the periphery of the host cell in this respect;
c) Degenerating peribacteroid membranes as well as bacteroids in early stages of nodule development (2).

B. In wild type nodules of Glycine max with stable peribacteroid membranes besides the constitutive choline kinase activity (CK I) with a K_M of about 150 µM, a second choline kinase (CK II) was separated with a K_M of 81 µM. This choline kinase has also a slightly different molecular weight of 60 Kd compared to 58-59 Kd for CK I, by using non-denaturing and SDS-gel electrophoresis no subunit structure for both enzyme forms was found. The isoelectric point of CK I is 8.1 and for CK II 8.5. Two dimensional mapping of whole cytoplasm from nodule and root tissue showed that CK II is absent in root tissue and also in nodules with an early loss of the peribacteroid membrane.

C. The particle density of the peribacteroid membrane in Glycine max cv. Mapple Arrow was determined to be between 2700 to 2800 $P/\mu m^2$ (Fig. 1). This is significangly more than in the outer membrane of chloroplasts (660 P) but somewhat less than in the inner membrane of chloroplasts (4200 P).

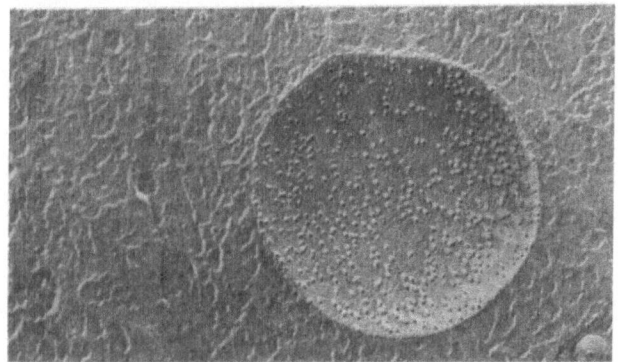

Fig. 1: Particle density of the peribacteroid membrane of
Glycine max cv Mapple Arrow ⊢————⊣   0.2 µm

ASPECTS OF THE PHYSIOLOGY OF COMBINED-N INHIBITION OF SOYBEAN NODULATION

ALLAN R. J. EAGLESHAM and ADRIENNE K. LA FAVRE
BOYCE THOMPSON INSTITUTE, ITHACA, NY 14853, USA.

1. INTRODUCTION

Although mineral N inhibition of symbiotic N_2-fixation by legumes has been a subject of study over many years, the mechanisms involved are poorly understood, even at the level of the whole-plant physiology. The interaction of the environment on the mineral N:symbiotic N balance has received relatively little attention despite possibly important implications for yield potential of the agriculturally important legumes. Here we present a brief report of efforts to elucidate how mineral-N nutrition of root and shoot separately influence nodulation by soybean.

2. MATERIALS AND METHODS

2.1. Expt. 1. The Effects of Elevated Temperatures on Early Nodulation by Soybean. In order to examine patterns of combined-N (KNO_3) assimilation and their relationship to nodulation without complication from symbiotic-N assimilation, seedlings (cv. Wilkin, inoculated at planting) were harvested at 7,10,13 or 16 days after planting (DAP). Seedlings were grown in sand culture, 4 pot^{-1}, 45 or 90mg N pot^{-1} added at planting and with a temperature regime of 30/25°C or 41/25°C.

2.2. Expt. 2. The Effects of High Levels of Applied NO_3 on Nodulation of a Second Root on the Same Plant. The cv. Harosoy forms adventitious root initials on the lower stem, making it possible to produce doubly rooted plants within 4 weeks of planting. Plants were grown in sand culture with KNO_3 applied to the original root at levels sufficient to delay initiation of, or completely inhibit, nodulation of that root, viz. 500mg N at planting or 250-500 ppm N continuously supplied. The second root was fed with N-free nutrient solution. Both roots were inoculated at 28 DAP and plants harvested 10-26 days later.

3. RESULTS

3.1. Expt. 1. Reduction in nodulation by the warm-treated 45mg N plants was mainly a consequence of reduced growth potential. In contrast, the 90mg N plants showed a nodulation reduction relatively more severe than the loss of growth potential. This could not be explained in terms of temperature effects on N-assimilation or of residual unabsorbed N in the pot. High levels of unassimalated NO_3 were found in the shoots, but not in the roots of the 90mg N, warm-treated plants.

3.2. Expt. 2. N applied to the original root did not inhibit nodulation on the second root. Profuse nodulation only on the second root was explainable in terms of high levels of NO_3 and Kjeldahl-N in the original root, and low levels in the shoot.

4. CONCLUSION

The root and shoot independently influence soybean nodulation. When one component is relatively low in N status, nodulation is precluded or reduced if the other component is relatively high in N status.

Acknowledgements   We are grateful to Allied Corporation for funding and to John Austin and Barbara Goldman for technical assistance.

NODULATION OF BEAN (PHASEOLUS VULGARIS L. SAVI.) BY RHIZOBIA FROM
ALFALFA NODULES

B. D. Eardly[1,2], D. B. Hannaway[1], and P. J. Bottomley[1]
Oregon State University[1] and Boyce Thompson Institute[2]

Alfalfa (Medicago sativa L.) is rarely nodulated by Rhizobium species from outside the Medicago cross-inoculation group (Vincent, 1970). Ineffective nodulation of this host has typically been attributed to R. meliloti strains which lack the intrinsic capability to develop an efficient symbiosis (Burton, 1972).

An investigation of a population of indigenous alfalfa nodule isolates from a moderately acid silt loam soil revealed that all isolates were symbiotically ineffective, and that many of the isolates contained diverse, and often atypical characteristics for R. meliloti. For example, most of the isolates studied had a relatively slow mean generation time in broth culture (4-10 h) independent of carbon source and vitamin supplementation. Isolates also lacked 39°C temperature tolerance, and many failed to markedly acidify yeast extract mannitol culture media. Since these characteristics are atypical for Rhizobium meliloti, infectiveness testing of selected isolates was conducted, using representative hosts of the various legume cross-inoculation groups. Results revealed that all isolates tested were able to nodulate both common bean and alfalfa, irrespective of host cultivar.

To assess the possibility of contamination of the inoculant cultures, an alfalfa nodule reisolate (191R) was selected for further scrutiny. Twenty single colonies from this reisolate were obtained by serial dilution. The promiscuous infectivity of each of the 20 subcultures was confirmed using both alfalfa and common bean as plant hosts. Additionally, further reisolations of strain 191R were made from several nodules from both host species, and these were used to confirm the reciprocal infectivity suggested by earlier results.

Preliminary evidence indicates that in comparison to effective commercial strains and uninoculated controls, both acetylene reduction activity and shoot dry matter production were very low in alfalfa inoculated with strain 191R, and intermediate in common bean inoculated with the same isolate. The partial effectiveness exhibited by strain 191R with common bean, together with the diversity of cultural characteristics of the group of native isolates, suggests that they are more closely related to R. leguminosarum biovar phaseoli than R. meliloti. The results of these experiments provide an alternate explanation for the question of ineffective nodulation in alfalfa, i.e. nodulation by promiscuous native Rhizobium species.

Burton, J.C. (1972) Agronomy 15:229-246.
Vincent, J.M. (1970) IBP Handb. no. 15. Blackwell Sci., Oxford.

EFFECT OF PHOSPHORUS NUTRITION ON SYMBIOTIC DINITROGEN FIXATION

EFFECT OF PHOSPHORUS NUTRITION ON SYMBIOTIC
DINITROGEN FIXATION IN SOYBEAN PLANTS

DANIEL W. ISRAEL, USDA AND SOIL SCIENCE DEPARTMENT, NORTH CAROLINA STATE
UNIVERSITY, RALEIGH, NC 27695-7619 USA

Soybean (Glycine max L. Merr.) plants inoculated with
Bradyrhizobium japonicum strain USDA 110 were grown in Perlite and
supplied nutrient solution containing 0.5 mM KH_2PO_4 and 2.0 mM KNO_3
until 24 days after transplanting. The 2.0 mM KNO_3 did not inhibit
nodule initiation and growth during this period. After day 24, plants
were supplied minus N nutrient solutions containing 0.01, 0.1, 0.5,
1.0, and 2.0 mM H_2PO_4. On day 53 photosynthetic measurements were
made and on day 54 plants were harvested for measurement of nodulation, nitrogenase activity, dry weight, and tissue content of N and P.
Phosphorus treatments were replicated four times.

This range of nutrient solution P concentrations resulted in
plants with leaflet total P concentrations ranging from 0.08 to 0.43%
of dry weight. Improving plant P status from limiting to sufficient
levels resulted in concomitant increases in 1) N concentration in all
plant tissues, 2) whole plant dry weight and N accumulation, 3) nodule
weight and number per plant and weight per nodule, 4) whole plant N to
whole plant dry weight ratios, 5) nodule dry weight to whole plant dry
weight ratios, 6) specific-acetylene-reduction activity of nodules,
and 7) whole plant leaf area and photosynthetic rates per unit leaf
area (CER's).

The increasing whole plant N to whole plant dry weight ratios with
increased P supply indicated greater response of the symbiotic-
dinitrogen-fixation process than of overall plant growth to improved P
nutrition. This enhancement of symbiotic dinitrogen fixation resulted
from increases in nodule initiation, growth, and function (specific-
acetylene-reduction activity). It was concluded that P has a specific
role or roles in the symbiotic-dinitrogen-fixation process of soybean
plants apart from its influence on overall growth of the host plant.

EFFECTS OF SOIL ACIDITY AND SOIL BUFFER CAPACITY ON THE GROWTH AND NITROGEN FIXATION OF BROAD BEANS (<u>VICIA</u> <u>FABA</u> L.)

ERIKA JOST/KONRAD MENGEL
INSTITUTE OF PLANT NUTRITION, JUSTUS LIEBIG UNIVERSITY, GIESSEN,
SUDANLAGE 6, D-6300 GIESSEN

The effect of soil acidity and soil buffer capacity on growth and nitrogen fixation of broad beans was examined in two pot experiments.

Plant seedlings inoculated with a suspension of Rhizobium leguminosarum were grown in pots on a loamy soil. The pH of the soil was adjusted to pH 4.5, 5.2, 6.0 and 7.2 with $Ca(OH)_2$. The rates of nitrogen fixed during the vegetation period were tested by the acetylene reduction technique using intact plants in intervals of 7 to 10 days. Plants grown under extreme acid conditions fixed only 30 % of the total nitrogen fixed by plants growing in a neutral medium (pH 7.2). Nodulation at pH 4.5 was not totally repressed but strongly delayed.

The influence of proton secretion by plant roots on the nodulation process was tested in a second pot experiment. Broad bean plants were grown in eight soils adjusted to the same pH (7.0) but differing in buffer capacity (clay, organic matter, carbonate content). Every 4 week from sowing to harvest nitrogen fixation was measured using the same technique. Until the stage of early pod filling the development of nitrogen fixation was quite the same for all soils. At a later stage, however, plants grown in soils with high clay content showed a distinct higher nitrogen fixation than plants grown in weakly buffered soils. No correlation was found between the decrease in pH and the decrease in nitrogenase activity. Soils which release relatively high amounts on exchangeable NH_4^+ during the early pod filling stage were characterized by high nitrogenase activity.

A STUDY ON CORRELATION BETWEEN LEGHEMOGLOBIN
AND NITROGEN CONTENT IN CHICKPEA (CICER ARIETINUM L.)

PANDURANG L. PATIL/PRABHAKAR H. RASAL.
MAHATMA PHULE AGRICULTURAL UNIVERSITY, COLLEGE OF AGRICULTURE, PUNE-411
005 INDIA.

The investigations were undertaken to study the nodulation pattern,
leghemoglobin and nitrogen content and its correlation with the age of
some promising Cv. of chickpea. At the initial stage of the crop growth,
effective nodules were more than ineffective ones. Results reveal that
more number of nodules were observed in a Cv. PG-5 at the 60th day followed by N-31, PG-2, N-59 and Chaffa. The size of nodules of a Cv. PG-5
was bigger than other cultivars at the flowering stage of the crop.
Similarly, dry weight of nodules/plant was also more in a Cv. PG-5 as
compared to the rest of the cultivars. The content of leghemoglobin and
nitrogen at the initial stage of the crop was less. However, they were
maximum at the flowering stage of the crop. At the maturity of the crop,
the leghemoglobin content was reduced. Similarly, the nitrogen fixation
was also reduced to a very low level. In general, it was found that
there was a good correlation between leghemoglobin content of nodules
and nitrogen fixation by a bacterium at all stages of the crop in most
of the cultivars.

SOWING DEPTH CAN DETERMINE NODULATION PATTERN IN CHICKPEA.

O.P. RUPELA, M.R. SUDARSHANA AND J.A. THOMPSON. ICRISAT, PATANCHERU, P. O. 502 324, A.P., INDIA.

Two cultivars of chickpea (Cicer arietinum L.) namely Annigeri, an average nodulator, and K 850, a superior nodulator, were sown under receding soil moisture conditions at 5 and 10 cm depth in a heavy black soil (Vertisol) having about 10^4 chickpea rhizobia per g dry soil. Deeper sown chickpeas generally had lower total nodule numbers, nodule mass and acetylene reduction (AR) activity, although differences were not always significant at each sampling time of 37, 47 and 59 days. These parameters were usually greater for K 850 than for Annigeri. Plants sown at 10 cm formed epicotyl roots and nodules but those shown at 5 cm did not. Epicotyl nodules formed up to 40% of total nodule number and mass and accounted for up to 50% of AR activity. Formation of epicotyl nodules seems to depend largely on moisture availability in this zone during the early stages of plant growth. This pattern of nodulation as affected by sowing depth may be typical of Vertisols as about 90% of nodules form in the top 15 cm. Nodules are rarely seen in 30-45 cm zone while chickpea Rhizobium and roots can invariably be seen in 90-120 cm zone.

EFFECTS OF NITRATE LEVEL AND SOIL TEXTURE ON SOYBEAN ROOT GROWTH AND NODULATION

J. A. STONE and B. R. BUTTERY
AGRICULTURE CANADA RESEARCH STATION, HARROW, ONTARIO, CANADA

Observing root growth at a glass or acrylic plastic-soil interface is a widely used research technique because it allows direct observation of roots without destruction, reducing replication and labor (Bohm, 1979). It has been shown that the growth medium used will have a substantial impact on how a root system develops (MacKey, 1973). The objective of the present study was to determine the effect of nirate level and soil texture on soybean (Glycine max (L.) Merr.) root growth and nodulation. 'Amcor' and 'Kentland' soybean cultivars were grown in 0.0, 10.0, and 40.0% mixtures of perlite and Brookston clay loam (clayey, mixed, mesic, Typic Haplaqualls) and watered with nutrient solutions containing 0.0 or 6.0 mM nitrate. All soil mixtures were supplied with Rhizobium japonicum US 110. Corresponding experiments were carried out in the field and growth room. In the growth room experiment, plants were grown until 21 d after emergence in 0.076 m I.D. by 0.92 m long acrylic tubes immersed in 23°C constant temperature water baths located in a controlled environment room having a 16 h photoperiod providing 160 uE $m^{-2}s^{-1}$ at plant level. In the field experiment, plants were grown until 53 d after emergence in 0.076 m I.D. by 1.83 m acrylic tubes installed in the ground at 15° from vertical. Response variables measured were the rate of taproot extension, root counts at the acrylic-soil interface, and top, root, and nodule dry weight. Although it was hoped that we would be able to directly observe nodule development on the roots at the soil tube interface, this proved to be virtually impossible, because few nodules formed on the surface roots, and these were difficult to see.

Increasing the proportion of perlite increased rates of root extension, cumulative root counts, and top dry weights in the field and growth room experiments. However, the soil mixture had no effect on nodule dry weight at either location, or on root dry weight in the growth room. Nitrate suppressed nodule development and increased top dry weight but had no effect on the rate of taproot extension. This suggests that nodulation itself had no effect on the rate of extension. Nitrate increased cumulative root counts and root dry weights in the field test, but decreased the same characters in the growth room test. This amomalous effect of nitrate may be related to the different light levels experienced by plants in the different environments.

REFERENCES

Bohm W (1979) Methods of Studying Root Systems, Springer-Verlag, New York
MacKey J (1973) Proc. 4th Int. Wheat Genet. Symp., Missouri Agric. Exp. Stn. 1973:827-842.

IMMUNO-GOLD LOCALIZATION OF GLUTAMINE SYNTHETASE
IN FREE-LIVING CYANOBACTERIA

B. BERGMAN, P. LINDBLAD AND A. PETTERSSON.   INST. OF PHYSIOL. BOTANY,
UNIVERSITY OF UPPSALA, BOX 540, S-751 21 UPPSALA, SWEDEN

In nitrogen-fixing cyanobacteria, glutamine synthetase (GS; EC 6.3.1.
2) is the primary NH_4^+-incorporating enzyme. By using an immuno-gold
technique, coupled to transmission electron microscopy, a detailed cellular and subcellular localization of GS in some free-living cyanobacteria
can be presented. The rabbit-antiserum used was that raised towards GS
purified from Anabaena 7120 (Orr and Haselkorn, 1982). The following results were obtained: In Anabaena cylindrica (1403/2a) GS was present in
all three cell types: heterocysts (see Figure), vegetative cells and
akinetes. The specific gold label was always more pronounced in heterocysts compared to vegetative cells and showed a uniform distribution in
the cytoplasm of all three cell types. No specific label was associated
with subcellular inclusions such as membranes, carboxysomes, polyphos-
phate granules (Bergman et al, 1985). Similar results were obtained
using Anabaena azollae (Newton 2B) and Nostoc 73102. When omitting the
rabbit-anti-GS-antiserum, virtually no label was observed.

REFERENCES

Bergman, B., Lindblad, P., Pettersson, A., Renstrom, E.R. Tiberg, E.
(1985) Planta (in press).
Orr, J. and Haselkorn, R. (1982) J. Bact. 152:626.

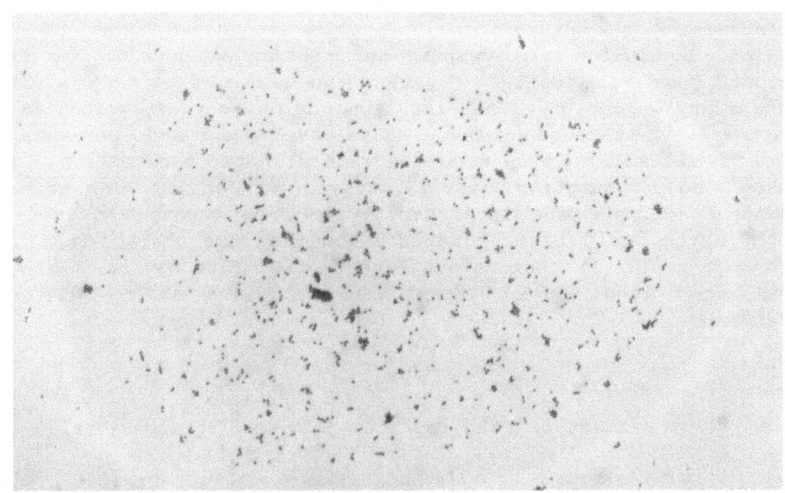

GS/gold labelled heterocyst of Anabaena cylindrica (not post-stained).

PURIFICATION AND DEVELOPMENTAL EXPRESSION OF THE NODULE SPECIFIC URICASE FROM PHASEOLUS VULGARIA L.

JEAN MARC BONNEVILLE, FRANCISCO CAMPOS, JAIME PADILLA AND FEDERICO SANCHEZ.
CENTRO DE INVESTIGACION SOBRE FIJACION DE NITROGENO, UNIVERSIDAD NACIO-NAL AUTONOMA DE MEXICO. APARTADO POSTAL 565-A, CUERNAVACA, MORELOS. MEXICO.

Tropical legumes such as soybean and bean transport most of the symbio-tically fixed nitrogen as ureides. Uricase (EC. 1.7.3.3) from soybean nodules (Bergmann et al, 1983) is a tissue specific protein and a good developmental marker for the differentiation of uninfected cells. Phaseolus vulgaris (mexican bean) nodule uricase was purified to homo-geneity from 4 week old nodules. The purified enzyme presents only one band on SDS-PAGE of about 33 K daltons of MW. The enzyme purification factor was 50X, which indicates that uricase is aprox. 2% of the total soluble protein from the nodule. Specific antisera was used to follow the appearance and antigen accumulation during the nodule development. Uricase was first antigenically detected in 8 days old nodules, 2 days before the onset of nitrogen fixation. Polysomal mRNA from nodules was translated in a reticulocyte lysate and the in vitro translation products were immunoprecipitated with the uricase specific antibodies. This enzyme's in vitro translation product has the same MW as the in vivo product. Our data indicates that uricase mRNA appears early in the nodule development and that the onset of nitrogen fixation is not re-quired for its induction in Phaseolus vulgaris.

REFERENCES

Bergmann H et al (1983) EMBO J. 2, 2333-2339

FIXATION OF NITROGEN AND CARBON BY LEGUME STEM NODULES

BERTRAND D. EARDLY and ALLAN R. J. EAGLESHAM
BOYCE THOMPSON INSTITUTE FOR PLANT RESEARCH AT CORNELL UNIVERSITY,
ITHACA, NY 14853, U.S.A.

The stem nodule system is a novel alternative for examining physiological aspects of the Rhizobium-legume relationship.

After inundation, nodules form on the submerged stems of some species of Aeschynomene and Sesbania. Inoculation of dry stems of these species also results in the development of nodules, visible within several days (Eaglesham and Szalay, 1983; Eaglesham et al., 1984). When fully developed, stem nodules are dark green in external appearance. The red-pigmented bacteroid zone is centrally located within the nodule and is surrounded by a cortex of chloroplast-containing cells. Energy sink and energy source are therefore in close juxtaposition; chloroplasts are occasionally seen in recently infected Aeschynomene cells, but not within mature bacteriod-zone cells.

Because the stem nodule is borne aerially and the bacteroid zone is surrounded by photosynthetic tissue, it may be expected that N_2-fixing activity would be inhibited by excess oxygen, at least in the daytime. However, high specific acetylene reducing activities (ARAs) were observed in the light (up to 448 umoles/g nod/h [Eaglesham and Szalay, 1983]) with A. scabra stem nodules, and when placed in the dark ARA immediately decreased. Increasing oxygen supply to A. indica and S. rostrata stem nodules in the dark resulted in increased ARA. Defoliation of plants bearing root and stem nodules led to large reductions in root ARA, less so in S. rostrata stem ARA, and had no effect on A. indica ARA.

The stem nodule is a combination of photosynthetic and N_2-fixing tissues which are functionally well integrated; the indications are that photosynthetically-derived-O_2 diffusion into the bacteroid zone maximizes N_2-fixing activity, and that in A. indica the nodule is an energy sufficient unit.

Eaglesham, A.R.J., A.A. Szalay (1983) Plant Science Letters, 29, 265-272.
Eaglesham, A.R.J. et al. (1984) In: Advances in Nitrogen Fixation Research, p. 501 (Veeger, C. and W.E. Newton, eds.), The Hague: Nijhoff/Junk.

BIOLOGICAL N$_2$ FIXATION AND ^{15}N MINERAL ABSORPTION BY BEAN (PHASEOLUS VULGARIS L.)^2 CULTIVARS.

SONIA M. FONSECA; SIU MUI T. SAITO; PAULO C.O. TRIVELIN
CENTRO DE ENERGIA NUCLEAR NA AGRICULTURA-CENA/USP.
AV. CENTENÁRIO, S/N?, CAIXA POSTAL 96. PIRACICABA, S.P. BRASIL.

The ability of fixing N$_2$ by Phaseolus vulgaris bean cultivars was evaluated in the greenhouse using ^{15}N isotopic dilution methodology and having wheat as non-fixing control plant. Four cultivars were selected under previous field and greenhouse tests. Cultivars were Carioca, Negro Argel, Rio Tibagi and 2234; the latter is being utilized by CNPAF-EMBRAPA in associated cultivation of maize-bean experiments. Pots with 5 kg of TRE soil having 4 plants each, were used to evaluate the fixed N translocated to the seed during the growth cycle of the cultivars. Experimental design was N rates x cultivars factorial and 4 replications totalling 80 pots. Labelled fertilizer - (^{15}NH$_4$)$_2$SO$_4$ was applied at the following rates: 0, 12, 24 and 48 mg N/kg soil. In general, the increased responses to applied N fertilizer showed bean dependence on exogenous N sources. Low rates of N (12 ppm) were sufficient to increase total N level by 40,45,67 and 140% on Negro Argel, Rio Tibagi, 2234 and Carioca cultivars, respectively. A highly N mineral responsive cultivar was 2234, that showed increased amounts of total N due to N addition, in detriment to biological nitrogen fixation. On the opposite, Carioca showed high N$_2$ fixation efficiency when it was grown under the presence of 24 ppm of N (24 mg N/kg soil). Amounts of N accumulated in the seed demonstrate that rates of (NH$_4$)$_2$SO$_4$ higher than 24 mg N/kg soil should not be recommended for these cultivars.

EFFECT OF SALINITY ON GROWTH OF AZOLLA MEXICANA ON N_2 AND FIXED N

GORDON J. JOHNSON

DEPARTMENT OF BIOLOGY, UNIVERSITY OF NEW MEXICO, ALBUQUERQUE, NM 87131 USA

The Azolla-Anabaena symbiosis is a highly productive N_2-fixing system with significant agricultural applications (Peters, 1978). Here, the salt tolerance of Azolla, under either obligatory N_2-fixing or nitrate-fed conditions is evaluated. Azolla mexicana was collected near Alburquerque, New Mexico, USA, and grown on N-free medium (Johnson et al, 1966). When nitrate-fed, 2 mM $Ca(NO_3)_2$ was added and $CaSO_4$ omitted. Environmental chamber conditions were 16 hr photoperiod (195 $\mu Em^{-2}sec^{-1}$, 25C) and 8 hr dark period (22C).

10 mM NaCl resulted in stimulation of Azolla growth under N_2-fixing conditions compared to the control (Table). At 50 mM NaCl, growth decreased slightly. At 100 mM, growth decreased to 37% of control. At 150 mM, growth was severely depressed. N_2 fixation (acetylene reduction) decreased only slightly up to 100 mM NaCl, but was undetectable at 150 mM NaCl. Azolla yields were reduced when nitrate grown even though harvested after 28 days rather than the 16 days used for plants without fixed nitrogen. Dry weight of nitrate-supplied Azolla was not affected by NaCl up to 100 mM, but at 150 mM, it decreased to 56% of control. The number of Anabaena cells in fronds of nitrate-supplied Azolla decreased similarly (Table). Thus, symbiotic Anabaena and N_2-fixation are more sensitive to NaCl inhibition than growth of the nitrate-fed host fern.

TABLE. Effect of NaCl on Azolla/Anabaena on N-Free and NO_3^- Media

NaCl (mM)	0	10	50	100	150	200	250
N_2							
mg[1]	54 + 5	74 + 18	46 + 10	20 + 7	9 + 1	5 + 1	-
ARA[2]	4312 ± 416	3303 ± 397	2978 ± 659	3016 ± 297	0	0	-
NO_3^-							
mg[1]	19 + 5	22 + 2	19 + 4	18 + 3	11 + 1	4 + 1	3 + 1
Cells[2]	100	99	85	80	15	19	21

[1]Dry weight. [2]Acetylene reduction activity in nmoles $(g\ fr\ wt)^{-1}hr^{-1}$.
[3]Anabaena cell count as % control $(g^{-1}\ fr\ wt)$.

Johnson, CV et al (1966) Plant Physiol. 41, 852-855.
Peters, GA et al (1978) BioScience 28, 852-855.

Support by the New Mexico Water Resources Research Institute and the technical assistance of Clinton Ready and Miles Prices is gratefully acknowledged.

GLUTAMINE SYNTHETASE SPECIFIC ACTIVITY AND PROTEIN OF NOSTOC SP. IN SYMBIOTIC ASSOCIATION WITH ANTHOCEROS

CECILLIA M. JOSEPH AND JOHN C. MEEKS
DEPARTMENT OF BACTERIOLOGY, UNIVERSITY OF CALIFORNIA, DAVIS, CA 95616
USA

The heterocyst frequency of Nostoc in symbiotic association with the bryophyte Anthoceros approaches 45% of the total cyanobacterial cells compared to 3-8% heterocysts in free-living, N_2-grown Nostoc. Addition of exogenous NH_4^+ to N_2-grown Anthoceros-Nostoc results in a gradual decrease in the frequency of heterocysts in associated Nostoc (to less than 3% of the total cells) and in the rate of N_2 fixation (acetylene reduction) (Enderlin, Meeks, 1983). We have examined the assimilation of nitrogen through in-situ, in-vivo, and in-vitro measurements of activities by symbiotically-associated Nostoc in comparison to Nostoc grown by itself under N_2 fixing conditions. Herein we present data on the in-vitro activity of glutamine synthetase (GS), the primary ammonium assimilating enzyme, and the amount of GS antigen present in Nostoc from associations grown with and without combined nitrogen.

Colonies of symbiotic Nostoc were isolated by enzyme digestion of Anthoceros tissue followed by gradient centrifugation (Meeks et al, 1985). The in-vitro activity of GS was monitored by the non-physiological transferase assay (Stadtman et al., 1970). The specific activities of GS from Nostoc in N_2- and NH_4^+-grown Anthoceros associations were 34% and 26%, respectively, that of free-living, N_2-grown Nostoc (0.85 units/mg protein). These values correlate well with the relative rates of in-vivo $^{13}NH_4^+$ assimilation (Meeks et al., 1985). Based on an enzyme linked immunosorbant assay, the amounts of GS antigen in symbiotic N_2- and NH_4^+-grown Nostoc were 74% and 65%, respectively, the amount in free-living, N_2-grown Nostoc (68 μg GS/mg protein). Similar values of GS antigen were obtained by immunoelectrophoresis.

The above data indicate that the level of GS activity and the amount of GS protein in symbiotic Nostoc do not correlate with heterocyst frequency or the rate of nitrogen fixation. These data also imply that the regulation of GS is by post-translational mechanisms in symbiotic Nostoc. This is in contrast to the Anabaena symbiont of Azolla caroliniana where immunological data imply regulation of the synthesis of GS (Orr, Haselkorn, 1982; Stewart et al., 1980).

REFERENCES

Enderlin CS and Meeks JC (1983) Planta 158, 157-165.
Meeks JC et al (1985) Planta 164, 406-414.
Orr J and Haselkorn R (1982) J. Bacteriol. 152, 626-635.
Stadtman et al (1970) Adv. Enzyme Regul. 8, 99-118.
Stewart WDP et al (1980) In Stewart WDP and Gallon JR, eds, Nitrogen Fixation, pp 239-277, Academic Press, NY, London.

IMMUNO-GOLD LOCALIZATION OF GLUTAMINE SYNTHETASE
IN SYMBIOTIC CYANOBACTERIA

P. LINDBLAD AND B. BERGMAN, INSTITUTE OF PHYSIOLOGICAL BOTANY, UNIVERSITY
OF UPPSALA, BOX 540, S-751 21 UPPSALA, SWEDEN

A reduced glutamine synthetase (GS, EC 6.3.1.2) activity due to a
low GS antigen level has been observed in the Azolla-Anabaena symbiosis
(Orr and Haselkorn, 1982; Stewart et al, 1983) and in the Peltigera-
Nostoc symbiosis (Stewart et al, 1983). By using an immuno-gold techni-
que, coupled to transmission electron microscopy, a detailed cellular
and subcellular localization of GS in the Azolla-Anabaena symbiosis and
in a cycad Zamia-Nostoc symbiosis can be presented. The rabbit-anti-
GS-antiserum used was raised towards GS purified from filaments of
Anabaena 7120 (Orr and Haselkorn, 1982). When omitting the rabbit-anti-
GS-antiserum, virtually no label was observed.

In Anabaena in Azolla caroliniana, the pattern of distribution of
GS in heterocysts was different as compared to that found in free-living
cyanobacteria (Bergman et al, 1985). GS was mainly observed in the cell
wall and in the polar nodules. In the vegetative cells, the number of
GS antigenic sites were reduced when compared to that found in vegetative
cells of free-living cyanobacteria.

In Nostoc of actively growing coralloid roots of Zamia skinneri,
GS was present in heterocysts and vegetative cells. The intensity of
the gold label and the pattern of distribution were similar to that found
in free-living, nitrogen-fixing cyanobacteria (Bergman et al, 1985).

REFERENCES

Bergman, B., Lindblad, P., Pettersson, A., Renstrom, E. & Tiberg, E.
(1985) Planta (in press).
Orr, J. & Haselkorn, R. (1982) J. Bact. 152:626.
Stewart, W.D.P., Rowell, P. & Rai, A.N. (1983) Ann. Microbiol. 134B:205.

TRANSLOCATION AND METABOLISM OF GLYCINE BETAINE BY YOUNG SALT-STRESSED NODULATED ALFALFA PLANTS

DANIEL LE RUDULIER, THEOPHILE BERNARD, and JEAN-ALAIN POCARD, LAB. DE MICROBIOLOGIE ET PHYSIOLOGIE DES SYMBIOSES, UNIVERSITE DE RENNES I, 35042 RENNES CEDEX, FRANCE

We have previously demonstrated that the addition of glycine betaine (10mM) to the growing medium partially restores the nitrogen fixation activity of young NaCl-stressed nodulated *Medicago sativa* L. plants (cv Europe). To investigate the role of this betaine, [methyl-^{14}C] glycine betaine and [1,2-^{14}C] glycine betaine were fed to the roots of stressed (0.2 M NaCl) or non-stressed young plants (31 days) inoculated with *Rhizobium meliloti*, strain 102F34. A 40 % reduction of the uptake of ^{14}C glycine betaine was observed in stressed plants. After an incubation time of 6 h, 96 and 4 % of the labelling were found, respectively, in the ethanol soluble products and the insoluble fraction, in both stressed and unstressed plants. During a metabolization phase of 48 or 96 h, no loss of ^{14}C was observed in the gas phase but a small increase of the insoluble fraction was seen (9 %). Catabolism of glycine betaine occured very slowly in any organ except in nodules of unstressed plants. The distribution pattern of ^{14}C among the various plant organs showed a ^{14}C glycine betaine mobilization with a transport of betaine in the nodules of stressed plants. Accumulation of free glycine betaine in these nodules might contribue to maintain a higher water status and to protect nitrogenase activity.

This work was supported by a grant from the Ministère de l'Education Nationale, "Mission des Biotechnologies".

AMMONIUM NUTRITION OF CLOVER IN A SPLIT-ROOT SYSTEM.

PAUL H S REYNOLDS AND MICHAEL J BOLAND
APPLIED BIOCHEMISTRY DIVISION   DSIR   PALMERSTON NORTH   NEW ZEALAND

In the legume nodule NH_4^+ serves as N source. It is incorporated into asparagine or allantoin and made available for the plant's N nutrition. In contrast to the symbiotic supply of NH_4^+, direct supply to roots can cause toxicity problems and interfere with K^+ uptake (Raven, Smith 1976). These problems can be overcome by exposing only a part of the root system to a steady flow of buffered NH_4^+. Further, this steady supply of NH_4^+ to one portion of the roots mimics the constant production of NH_4^+ in the N-fixing nodule. This work asks if roots thus separated and exposed to NH_4^+ act as 'nodules' and assimilate N in the same way as nodules.

Methods. Ramets from a single plant of Trifolium repens cv Grasslands Huia, were rooted for 14 d in pumice and set out for hydroponic growth in 1 ℓ foil-covered beakers. Four treatment groups were set up at 18 d as follows: A. N-free nutrient, 4 µg/ml tetracycline (to suppress nodulation); B. N-free nutrient, inoculated with R. trifolii NZP560; C. 'split', (N-free nutrient); D. +N nutrient. Nutrient was modified NCSU (Pritchard et al. 1984), changed every 48 h. In the N-free solution NH_4NO_3 was omitted and KNO_3 replaced by K_2SO_4. At day 22 about 1/3 of the roots of plants in group C was directed into a container through which was circulated 4 mM NH_4^+ MES pH 6.8, 1 mM $CaCl_2$ at 15 ml/h (complete turnover in 12 h). Plants were grown for another 30 d in the growth chamber with 12 h daylength at 25/20 C.

Results and Discussion. The four treatment groups and the results from the various enzyme assays and analyses are shown below:

Treatment group	dry wt g	total N gN/g	protein mg/g	asparagine µmol/g	asparagine /plant	GS nmol/min/gfwt	GOGAT nmol/min/gfwt	K+ mg/g
A. tops	0.16	0.07						
roots	0.10	0.32						
B. tops	2.31	0.31						
roots	0.57	0.42		56	6.4			
C. tops	3.32	0.28						34
-N roots	0.35	0.34	1.6	25	1.8	29	18	49
+N roots	0.57	0.76	3.5	28	3.2	124	36	17
D. tops	4.88	0.30						38
roots	1.57	0.59	2.6	5	1.5	121	47	21

There were no nodules in A or D. Excellent nodulation was seen in B. The few nodules in the -N roots of C developed late and were small and pale. These data show the production of healthy plants by growth in a split-root system with NH_4^+ as sole N source. NH_4^+ toxicity problems were avoided by circulating buffered NH_4^+ to the split-root, and competition with K^+ uptake was avoided by exposure of only part of the root to NH_4^+. GS and GOGAT levels were 4-fold higher on a g fresh wt basis in the part of the split root supplied with NH_4^+. Asparagine levels in the roots of the split-root plants (5 µmol) were comparable to the levels in nodulated roots (6.4 µmol) and considerably higher than in the roots of +N plants (1.5 µmol). The +N split root appeared to act as a 'simulated nodule'. This model may have implications in the ammonia nutrition of non-leguminous plants.

Pritchard, M.W. et al. (1984) Plant Soil 81, 389-402.
Raven J.A., Smith F.A. (1976) New Phytol. 76, 415-431.

ENZYMES OF UREIDE BIOSYNTHESIS IN SOYBEAN NODULES:
EFFECT OF NITRATE TREATMENT

Kathryn A. Schuller, David A. Day and Peter M. Gresshoff
Botany Department, The Australian National University,
Canberra. ACT., 2601

The effect of nitrate treatment on nitrogen fixation and the
activities of the nodule cytoplasmic ureide synthesizing enzymes was
investigated in soybeans. In order to eliminate the inhibitory effect
of nitrate treatment on nodule development, nitrate (10mM) was
supplied daily to well nodulated plants which had been grown in pots
watered with a nitrogen-free nutrient solution for 6-8 weeks prior to
the commencement of the treatment. Nitrogen fixation was assayed
in situ (nodulated roots) and in anaerobically isolated bacteroids.
Three different techniques, acetylene reduction, $^{15}N_2$ incorporation and
xylem ureide content gave similar results with respect to the effect
of nitrate treatment on in situ nitrogen fixation. Nodule cytoplasmic
enzymes were assayed in vitro and activities expressed on a g nodule
fresh weight basis. Despite a 50-70% inhibition of in situ acetylene
reduction there was no effect of a 2-day nitrate treatment on either
acetylene reduction by isolated bacteroids or in vitro activity of
glutamine synthetase (GS), glutamine oxoglutarate aminotransferase
(GOGAT), xanthine dehydrogenase, uricase or allantoinase. After 7
days of nitrate treatment acetylene reduction by isolated bacteroids
was inhibited 100% (totally dependent on endogenous bacteroid
substrates) and 79% (5mM malate added to supplement endogenous
substrates). GOGAT activity was inhibited 43%. However, there was
still no effect on any of the other nodule cytoplasmic enzymes. This
corresponded with a 94% inhibition of in situ acetylene reduction.
During the second week of nitrate treatment a decline in xanthine
dehydrogenase activity and GS activity was accompanied by a decline in
nodule leghaemoglobin content and fresh weight. Neither of these
latter two parameters was affected during the first week of nitrate
treatment. Nodule soluble protein content was not affected by nitrate
throughout the treatment period. In conclusion then, nitrate
treatment fails to induce a decline in the net synthesis of
nitrogenase and the ureide synthesizing enzymes, at least in the short
term. Consequently, this cannot explain the decline in in situ
nitrogen fixation upon nitrate treatment of soybean plants.

EFFECT OF NITRATE ON THE ENZYME LEVELS OF <u>RHIZOBIUM JAPONICUM</u> BACTEROIDS

Mark T. Smith, Rexford R. Hays, Janet L. Clawitter and David W. Emerich
Department of Biochemistry, University of Missouri, Columbia, MO 65211

The effects of nitrate on the physiology and morphological development of the symbiosis between <u>Rhizobium</u> and leguminous plants has been reported by many investigators. However, there are few reports, with the obvious exception of nitrogenase, in which the change in a particular enzyme has been monitored following the addition of nitrate. We have followed the change in activity of the following enzymes as a function of plant age after nitrate additions: nitrogenase, malate dehydrogenase, isocitrate dehydrogenase, pyruvate dehydrogenase, fumarase, hydroxybutyrate dehydrogenase, β-keto-thiolase, acylketothiolase, acetoacetate-succinyl-CoA thiolase, glutamine synthetase, aspartate-oxaloacetate aminotransferase, α-keto-glutarate dehydrogenase, glucose-6-phosphate dehydrogenase, and glyceraldehyde-3-phosphate dehydrogenase. Nodule weights and leghemoglobin contents also were measured.

As observed previously, by many other investigators, addition of nitrate decreases nitrogenase activity after an initial, transient increase. Nitrate dramatically decreased nodule weights but elevated the leghemoglobin content.

The specific activities of organic acid metabolic enzymes decrease in the presence of nitrate. The specific activity of the glycolytic enzymes and ammonia assimilation/shuttling enzymes increase when nitrate is added. No consistent pattern was observed for the three poly-3-hydroxybutyrate metabolic enzymes included in the study. There was no change observed in the specific activity of acylketothiolase.

Enzyme	Per Cent Change in Specific Activity of the following Enzymes in the Presence of Nitrate*	
	Day 15	Day 20
Nitrogenase	-30	-75
Malate Dehydrogenase	-30	-75
Fumarase	-60	-75
Glucose-6-phosphate dehydrogenase	+75	+55
Glyceraldehyde-3-phosphate dehydrogenase	+70	+200
Hydroxybutyrate dehydrogenase	-55	-80

*10mM nitrate added to plant growth nutrient solution on day 12

ASPARTATE AND ASPARAGINE BIOSYNTHESIS IN ALFALFA NODULES.

SIEGLINDE S. SNAPP AND CARROLL P. VANCE, DEPARTMENT OF AGRONOMY AND PLANT
GENETICS, UNIVERSITY OF MINNESOTA, AND USDA-ARS, SAINT PAUL, MINNESOTA
55108, USA

Asparagine biosynthesis and transport have a pivotal role in assimilation and utilization of symbiotically fixed nitrogen. Asparagine is the
predominant amino acid in nodules and xylem sap of alfalfa, birdsfoot
trefoil, and lupine. Although ureides are the major nitrogen transport
product in soybeans, substantial quantities of asparagine have been
detected in soybean nodules and xylem sap. Indirect evidence suggests
the pathway for biosynthesis of asparagine involves aspartate as a substrate, however little direct conversion of aspartate to asparagine has
been demonstrated. The objectives of this research were to: a) demonstrate rapid direct conversion of ^{14}C-aspartate to ^{14}C-asparagine in alfalfa nodules; b) assess if aminooxyacetate (AOA) blocks aspartate carbon
flow into TCA cycle acids and the effect on asparagine biosynthesis; and
c) assess the effects of methionine sulfoximine (MSO) and azaserine (AZA)
on asparagine biosynthesis.
Within 20 minutes ^{14}C-asparagine was readily synthesized from ^{14}C-
aspartate in both control and inhibitor treated alfalfa nodules. AOA, an
inhibitor of aminotransferase activity, significantly reduced the flow of
carbon from aspartate into TCA cycle acids and decreased ^{14}CO$_2$ evolution
by 99%. Concomitantly AOA treated nodules showed a 15-fold increase in
conversion of ^{14}C-aspartate to ^{14}C-asparagine. MSO and AZA, ammonia
assimilation inhibitors, stimulated short term (20 minutes) conversion of
aspartate to asparagine. However, after 60 minutes MSO and AZA appear
to block asparagine biosynthesis. Malate, succinate, and fumarate were
the predominant organic acids labeled from both ^{14}C-aspartate and ^{14}CO$_2$.
Short term exposure of excised alfalfa nodules to AOA, MSO, and AZA had
little effect on nodule CO$_2$ fixation. By contrast, over a 2 hour period
radioactivity in nodules and xylem sap of nodulated roots exposed to
^{14}CO$_2$ was reduced 35 to 80% by exposure to inhibitors. Inhibitor treatment also reduced the total amino acid content of xylem sap by 50 to 80%.
Radioactivity in xylem sap aspartate and asparagine was substantially
reduced by inhibitor treatment.
The data demonstrate glutamine dependent asparagine biosynthesis
from aspartate in alfalfa nodules. Aspartate aminotransferase activity
readily converted aspartate carbon into TCA cycle acids. The conversion
of aspartate to organic acids may have contributed to difficulties previously reported in using ^{14}C-aspartate as a precursor for ^{14}C-aspara-
gine. Nodule CO$_2$ fixation in alfalfa appeared to be dependent upon a
continued assimilation of ammonia. Transport of asparagine and aspartate
in the xylem sap was found to be dependent upon uninterrupted aminotransferase activity and uninhibited ammonia assimilation.

PRODUCTION, CHARACTERIZATION, AND APPLICATION OF MONOCLONAL ANTIBODIES SPECIFIC FOR SOYBEAN NODULE XANTHINE DEHYDROGENASE

ERIC W. TRIPLETT/CRAIG R. LENDING/DAVID J. GUMPF/CARE F. WARE
DEPARTMENT OF PLANT PATHOLOGY AND DIVISION OF BIOMEDICAL SCIENCES,
UNIVERSITY OF CALIFORNIA, RIVERSIDE, CA 92521 USA

To ensure antibody specificity in subsequent studies, monoclonal antibodies were produced against soybean nodule xanthine dehydrogenase (XDH), an enzyme involved in ureide synthesis.

Production and screening of the monoclonal antibodies. Female BALB/c mice were immunized subcutaneously twice with purified XDH, and were boosted intravenously three days prior to fusion. Hybridomas were initially screened using solid phase sandwich ELISA. Fourteen positive cell lines, stable for antibody production, were subcloned by limiting dilution. Seven subclones were selected for production of acites fluid and tested for their ability to immunoprecipitate XDH. An indirect immunoprecipitation assay using rat anti-mouse kappa chain monoclonal antibody covalently linked to Sepharose-CL4B was used. After incubation in cell culture supernatant, the beads were incubated with XDH and, following centrifugation, the XDH activity remaining in the supernatant was measured. The immunoprecipitated protein was released from the beads and analyzed using polyacrylamide gel electrophoresis.

Characterization of the monoclonal antibodies. Specificity of each hybridoma clone for XDH was demonstrated by purifying XDH to homogeneity from a nodule crude extract by immunoaffinity chromatography using monoclonal antiboby covalently bound to Sepharose-CL4B beads. Each clone was found to be in the IgG_1 subclass. Competitive ELISAs demonstrated that two of the seven clones shared the same epitope while the remaining five clones bound to unique sites on the protein.

Applications of the monoclonal antibodies. The monoclonal antibodies have been used for the large scale purification of XDH and the determination of the organ and species specificity of soybean nodule XDH. Using immunoaffinity chromatography, 2 mg of XDH was purified to homogeneity from 200 g of soybean nodules in seven hours. Organ specificity of xanthine dehydrogenase in soybean was measured by ELISA. Crude extracts of nodules, roots, stems, and leaves cross-reacted with all seven monoclonal antibodies. These data demonstrate that XDH is not nodule specific. Species specificity of each monoclonal antibody was also examined. A double sandwich ELISA was used to determine the variability among the seven clones for their binding to nodule crude extracts of six legume species. Among the seven clones, six different binding specificities were observed. These results correspond well with the epitope determination data which showed that the seven clones bind to six different binding sites on the enzyme. The two clones which were found to bind to the same epitope also showed the same binding pattern to the nodule extracts from the six legume species. Another clone was found to bind only to soybean nodule extracts.

ENZYMATIC DEGRADATION OF ALLANTOATE IN SOYBEANS

RODNEY G. WINKLER/JOSEPH P. POLACCO, DALE G. BLEVINS, KRYSTYNA LUKASZEWSKA/
DOUGLAS D. RANDALL
PLANT PHYSIOLOGY AND BIOCHEMISTRY GROUP AND UNIVERSITY OF MISSOURI-
COLUMBIA, MO 65211, USA.

The ureides, allantoin and allantoate are important nitrogen transport compounds in soybeans as well as several other tropical legumes. Although soybeans represent the majority of nitrogen utilized by soybeans under N_2-fixing conditions, virtually nothing is known about their catabolism. The objective of this research was to elucidate the first step of the allantoate catabolic pathway.

Crude extracts of developing seed coats incubated with allantoate (1mm Mn^{2+}. pH 8.75) yielded labile glyoxylate derivatives (glyoxylate + ureidoglycolate + ureidoglycine), NH_3, CO_2 and low levels of urea (< 10% of allantoate degradation) after 1h at $30°C$. Low levels of urea production suggest allantoate is not degraded by an amidinohydrolase, which would catalyze the production of equimolar urea and ureidoglycolate. The activity of partially purified extracts was similar to that of the crude extract. NH_3, $^{14}CO_2$ and glyoxylate + its labile derivative were released in a 2:1:1 ratio. The enzymatic release of products was inhibited by 10 mM EDTA, 10 mM acetohydroxamate (a urease inhibitor and metal chelator), 10 mM borate but not by the potent urease inhibitor phenyl phosphordiamidate.

HPLC analysis of deproteinized extracts revealed the enzyme dependent production of a novel peak that will be referred to as a peak D. Peak D was the first product detected by HPLC, glyoxylate was detected at later time points. Peak D did not comigrate with allantoin, allantoate, ureidoglycolate, glyoxylate, glycolate, oxamate, oxalurate, glycine, or urea. When BME was varied in the reaction mixture, glyoxylate production was unaffected, but the level of peak D was positively correlated with the level of BME in the reaction mixture, suggesting that peak D may be an adduct of BME and a reactive intermediate. We hypothesized that peak D was 2-ethanolthio 2'-ureido acetic acid ($NH_2CONHCHCO_2HSCH_2CH_2OH$) and synthesized, and identified this compound as outlined in Winkler et al. (1). Purified Peak D and the synthesized compound were identical by UV and chromatographic comparison.

Similar results were obtained with partially purified leaf extracts. NH_3 release was independent of urease and the level of urea production was much less than stiochiometric (< 5%). Acetohydroxamate was an inhibitor of allantoate breakdown in leaves exhibiting a K_i of 5 mM for NH_3 and glyoxylate production indicating that it is not a specific urease inhibitor as has been assumed by some workers. Our results are not consistent with allantoate amidinohydrolase, although it is generally assumed to be the allantoate degrading enzyme of higher plants. Our results, however, indicate that allantoate amidohydrolase is the allantoate degrading activity of soybeans.

(1) Winkler, R.G., et al (1985). Plant Physiol (In Press).

SUCROSE BREAKDOWN IN SOYBEAN NODULES

LES COPELAND AND MATTHEW MORELL
DEPT AGRICULTURAL CHEM, UNIVERSITY OF SYDNEY, N.S.W., AUSTRALIA 2006

INTRODUCTION

Symbiotic nitrogen fixation in root nodules of legumes is dependent on the supply of carbohydrates from the host plant. Investigations in this laboratory are aimed at describing how sucrose, the main carbohydrate translocated into the nodules, is utilized to provide nutrients for the bacteroids, energy for nitrogenase, and carbon skeletons for the assimilation of fixed NH_4^+. The initial step in the breakdown of sucrose in soybean nodules is cleavage in the plant cytoplasm by either alkaline invertase (EC 3.2.1.26) or sucrose synthase (EC 2.4.1.13). The kinetic properties of these enzymes have recently been described (Morell, Copeland 1984, 1985). This report describes the enzymes in the plant cytosolic fraction of soybean nodules which are capable of phosphorylating the glucose and fructose produced from sucrose. Soybean nodules infected with Rhizobium japonicum strain CB 1809 were used in this study.

RESULTS

DEAE-cellulose chromatography has shown that the plant cytosolic fraction of soybean nodules contained two enzymes which phosphorylate hexoses - a hexokinase (EC 2.7.1.1) and a fructokinase (EC 2.7.1.4). There was only a small amount of hexokinase in comparison to fructokinase. Hexokinase displayed typical Michaelis-Menten kinetics. The enzyme had a high affinity for glucose (Km 0.075 mM) and a relatively low affinity for fructose (Km 2.5 mM). The kinetic parameters indicate that this enzyme would phosphorylate mainly glucose in vivo. Fructokinase was specific for fructose (Km apparent 0.077 mM). Substrate inhibition was observed at fructose concentrations greater than 0.4 mM at pH 8.2 but not at pH 6.6. Fructokinase required K^+ ions for maximum activity.

CONCLUSIONS

When sucrose is cleaved in the plant cytoplasm of soybean nodules infected with R. japonicum CB 1809 the fructose produced is likely to be metabolized further in the same subcellular compartment, following phosphorylation by a specific fructokinase, rather than be taken by the bacteroids. The low concentration of fructose in soybean nodules (Streeter 1980) suggests that the activity of fructokinase is adequate to meet the requirement for fructose phosphorylation. The bacteroids isolated from these nodules have very little fructose phosphorylating activity (Morell, Copeland 1984). However, glucose may be available for uptake by the bacteroids of these nodules, due to the limited potential for the phosphorylation of glucose in the plant cytoplasm. Glucose phosphorylating activity is present in the bacteroids of these nodules (Morell, Copeland 1984) and also soybean nodules infected with R. japonicum USDA 110 or 138 (Reibach, Streeter 1983).

REFERENCES

Morell MK, L Copeland 1984 Plant Physiol 74,1030-1034
Morell MK, L Copeland 1985 Plant Physiol 78,149-154
Reibach PH, JG Streeter 1983 Plant Physiol 72,634-640
Streeter JG 1980 Plant Physiol 66,471-476

ENZYMES OF CARBON METABOLISM AS INDICATORS OF THE CARBON NUTRITION OF NODULE BACTEROIDS

A.R. GLENN, I.A. McKAY, S. SAROSO, R. ARWAS & M.J. DILWORTH
NITROGEN FIXATION RESEARCH GROUP, SCHOOL OF ENVIRONMENTAL & LIFE SCIENCES
MURDOCH UNIVERSITY, MURDOCH, WESTERN AUSTRALIA 6150

We have previously examined the physiology of rhizobial bacteroids by investigating their transport properties (Dilworth & Glenn 1985) and the symbiotic properties of sugar (Glenn *et al*. 1984) or organic acid (Arwas *et al*. 1985) mutants. We have now investigated the gluconeogenic enzymes in *Rhizobium leguminosarum* MNF 3841 and the sugar catabolic enzymes in cowpea *Rhizobium* NGR 234 which are regulated by the available carbon sources.

In *R. leguminosarum* MNF 3841 gluconeogenesis takes place via phosphoenol pyruvate carboxykinase (PEPCK) and fructose 1,6-bisphosphate aldolase. In chemostat culture PEPCK is completely repressed when cells are grown on sucrose and derepressed when grown on fumarate. Addition of sucrose (0.4 mM) to a fumarate chemostat completely inhibits synthesis of PEPCK. A Tn5 induced PEPCK$^-$ mutant (MNF 3085) fails to grow on a range of simple non-sugar compounds but is Nod$^+$Fix$^+$ on peas. Pea bacteroids of MNF 3841 isolated on a Percoll gradient contain detectable PEPCK which can be shown by nucleotide specificity to be bacterial in origin. It is not found in bacteroids of the PEPCK$^-$ mutant. These data suggest that pea bacteroids receive sufficient sugar to compensate for the gluconeogenic defect in MNF 3085, but insufficient to completely repress synthesis of PEPCK in bacteroids of the wild-type.

In cowpea *Rhizobium* NGR 234, invertase, fructokinase, glucose 6-phosphate dehydrogenase and the Entner-Doudoroff enzymes are present in sugar-grown cells, but are barely detectable in succinate-grown cells. Isolated snake-bean bacteroids of NGR 234 contain very low activities of these four enzymes. In the free-living form C_4 dicarboxylates exert some repressive effect (about 50%) on the induction of these sugar catabolic enzymes. These data suggest that the sugar catabolic enzymes, unlike the dicarboxylate transport system (Saroso *et al*. 1984) are not induced in snakebean bacteroids of NGR 234.

Data from both species of *Rhizobium* indicate that relatively small quantities of sugars are received by bacteroids. We suggest that the peribacteroid membrane is relatively impermeable to sugars and so dictates the carbon source(s) available to the bacteroid.

Dilworth M.J. & Glenn, A.R. (1985). In: Ludden, P.W. & Burris, J.E. eds. "Nitrogen Fixation and CO$_2$ Metabolism" pp 53-61. Elsevier, New York.

Glenn, A.R. *et al*. (1984). J. gen. Microbiol. 130, 239-245.

Arwas, R. *et al*. (1985). J. gen. Microbiol. in press.

Saroso, S. *et al*. (1984). J. gen. Microbiol. 130, 1809-1814.

DARK CARBON DIOXIDE FIXATION AND RE-EVOLUTION IN NODULATED SOYBEAN

B.J. King, D.B. Layzell and D.T. Canvin, Dept. of Biology, Queen's University, Kingston, Ont., Canada K7L 3N6

INTRODUCTION

Several roles have been proposed for dark CO_2 fixation in soybean root nodules, including: (1) organic and amino acid production for xylem transport, (2) production of intermediates in the ureide pathway, and (3) organic acid production for bacteroid C metabolism. The purpose of this study was to examine the pathway and role of dark CO_2 fixation in soybean nodules.

METHODS

The plants used were vegetative stage Harosoy 63 X USDA 16, a Hup^- _Rhizobium_ strain. Dark CO_2 fixation was measured in intact plants at saturating $[CO_2]$ by feeding $^{14}CO_2$ for 2 to 60 min to nodulated roots. The label from a 2 min pulse was followed through the organic and amino acid fractions. Other aspects of C and N metabolism were considered, including respiration, C and N accumulation in nodule dry matter, and the C and N requirement for transport of organic acids, ureides and amino acids.

RESULTS AND DISCUSSION

The rate of CO_2 fixation measured following a 2 min $^{14}CO_2$ feed was 102 μmol $gDW^{-1}h^{-1}$, and was equivalent to 14% of net respiration. The total C requirement for organic acids and amino acids accounted for only 19% of the observed dark CO_2 fixation rate. Therefore, the role of the excess dark fixation was investigated.

Pulse-chase experiments and varying the $^{14}CO_2$ feed length showed that 75 to 92% of CO_2 fixed during a 2 min pulse was re-released within 1 h. Fractionation demonstrated that 67% of the original label was in organic acids, primarily malate. Treatment of the malate fraction with malic enzyme demonstrated the presence of label in C-4, consistent with PEP carboxylase fixation.

Loss of label in malate, citrate and fumarate over the chase period accounted for most of the observed decline. Amino acids, including aspartate, glutamine and glycine, showed a less pronounced decline. No evidence was seen for the presence of label in ureides or intermediates, possibly due to very rapid or very slow turnover of the ureide pathway.

The role of dark CO_2 fixation in regard to N_2 fixation was then examined. Replacing N_2 with Ar had no effect on dark CO_2 fixation. Treatment with 100% O_2 resulted in a 54% inhibition of dark CO_2 fixation. It was concluded that PEP carboxylase fixation may be associated with nitrogenase activity.

Based on these results, dark CO_2 fixation by PEP carboxylase may play a major role in nodules by providing organic acids for respiration in support of nitrogenase activity.

ACKNOWLEDGEMENTS

Supported by NSERC (Can.) grants (DBL, DTC) and scholarship (BJK) and a DSS (Agric. Can.) contract. Travel supported by the School of Graduate Studies and Research and the Faculty of Arts and Science, Queen's Univ.

DYNAMICS OF CARBON USE IN SOYBEAN ROOT NODULES: A C-13 KINETIC STUDY

HIROSHI KOUCHI, TADAKATSU YONEYAMA AND JUNJI ISHIZUKA
DEPARTMENT OF APPLIED PHYSIOLOGY
NATIONAL INSTITUTE OF AGROBIOLOGICAL RESOURCES
YATABE-MACHI, TSUKUBA, IBARAKI, 305 JAPAN

Metabolism of translocated photoassimilates in legume root nodules is not yet fully understood. Isolated bacteroids from nodules are capable of utilizing wide varieties of carbon substrates particularly the TCA cycle intermediates to support their respiration and/or dinitrogen fixing activity (Peterson & LaRue, 1981), but little is known about the differentiation and in vivo interrelationship of carbon metabolism between plant cell cytosol and bacteroids. We have presented some quatitative data on the metabolism and utilization of photoassimilated carbon in soybean nodules by using a steady-state $^{13}CO_2$ labelling technique.

Well-nodulated, water-cultured soybean plants (40-55 days old) were fed $^{13}CO_2$ with a constant specific activity (Kouchi & Yoneyama, 1984). Time-course of respiratory $^{13}CO_2$ evolution from the nodules of nodulated roots was determined (Kouchi et al., 1985), and compared with the kinetics of ^{13}C-labelling of some metabolites in plant cell cytosol and bacteroids isolated from the nodules.

It was demonstrated that nodule respiration was strongly dependent on "recently assimilated carbon", since the respired CO_2 from the nodules was rapidly labelled with ^{13}C to reach an isotopic equilibrium at the level of 80-90% of labelled carbon within 4hr $^{13}CO_2$ feeding. The levels and time-course pattern of labelling of sucrose, which was the most highly labelled compound in the nodules, were in good agreement with those of respired CO_2. A very close agreement of labelling patterns of sucrose between plant cell cytosol and bacteroids was observed. This suggests that sucrose is freely diffusible across the bacteroid membrane and not metabolically compartmented between these two fractions.

The levels of labelling of organic acids in the nodules were considerably lower than that of respired CO_2. Organic acids of TCA cycle intermediates, especially succinate exhibited significantly lower levels of labelling in the bacteroids than in the cytosol, but the time-course patterns of the labelling were in parallel between the two fractions. A large amount of 3-hydroxybutyrate was found in the bacteroids but it exhibited consistently low labelling rate. These findings suggest that the bacteroids contain relatively large inactive pools of organic acids and the active pools of TCA cycle intermediates in the bacteroids are very small.

REFERENCES
Kouchi H and Yoneyama T (1984) Ann. Bot. 53, 875-882, 883-896
Kouchi H et al. (1985) Ann. Bot. 56, in press.
Peterson JB and LaRue TA (1981) Plant Physiol. 68, 489-493

MATHEMATICAL ASSESSMENT OF SYMBIOTIC N_2 FIXATION EFFICIENCY IN DETERMINATE NODULES.

EDWARD L. McCOY and LARRY BOERSMA
OREGON STATE UNIVERSITY, CORVALLIS, OR   97331 U.S.A.

1. INTRODUCTION

The carbon use efficiency of symbiotic N_2 fixation is not completely understood on the nodule level due to the complexity and coordination of the reaction processes. A mathematical model is formulated to investigate the fluxes of carbon, nitrogen, hydrogen and energy in a soybean nodule. The model examines nitrogen metabolism in relation to 1) carbon supply, 2) energy yield from respiration, 3) relative efficiency of N_2 fixation, and 4) hydrogenase activity.

2. PROCEDURE

The model structure describes the major metabolic pathways for carbon, nitrogen and hydrogen flux in the determinate nodule. Hexose entering the host cell passes into the TCA cycle for host respiration, NH_4^+ assimilation, and bacteroid uptake of organic acids. The energy ($2 e^- = 3$ ATP, maximum 6.3 ATP/mole C) from host respiration participates in NH_4^+ assimilation. The CO_2 respired by the host is assimilated by dark CO_2 fixation or exported to the atmosphere. The energy from bacteroid respiration (maximum 4.8 ATP/mole C) participates in reduction of N_2 or H^+. The CO_2 respired by the bacteroid enters the host CO_2 pool. N_2 is reduced by nitrogenase (requiring 21 ATP/mole N_2), and passed to the host cytoplasm as NH_4^+ for assimilation in conjunction with the host TCA pool. The assimilated N is exported as amides and ureides. Bacteroid H^+ is reduced by nitrogenase (requiring 3.5 ATP/mole H^+) and is lost to the atmosphere or oxidized by hydrogenase (yielding 3 ATP/mole H_2). The model is formulated as 19 linear, algebraic equations for the steady state flux of carbon, nitrogen and hydrogen in the nodule. The model equations result from a combination of mass conservation, energy transduction, and empirical relationships based on experimental observations. Data of the surface areas and volumes of the nodule, vascular tissue, host, and bacteroid cells are from Bergersen (1982).

3. RESULTS AND DISCUSSION

Theoretical analysis on the efficiency of N_2 fixation indicate: 1) N_2 uptake decreases with carbon supply, and vanishes at a carbon supply corresponding to basal respiration of 18.9 μmole CO_2 g^{-1} fresh wt. h^{-1}. 2) Increased N_2/H_2 ratio of nitrogenase increases N_2 uptake at all carbon supply rates. 3) Decreased energy yield from respiration decreases N_2 uptake and H_2 evolution but slightly increases CO_2 evolution. 4) 65.7-78.4% of the carbon supply is evolved as CO_2 depending on the N_2/H_2 ratio and the % maximum energy yield from respiration. 5) Hydrogenase increases N_2 uptake 5.08% at a N_2/H_2 ratio = 2.0, 9.70% at a N_2/H_2 ratio = 1.0, and 17.15% at a N_2/H_2 ratio = 0.5.

4. REFERENCES

Bergersen FJ (1982) Root Nodules of Legumes: Structure and Function, Research Studies Press/Wiley, Chichester.

FRUCTOSE-DEPENDENT DIFFERENTIATION, N_2 FIXATION, AND GROWTH IN THE CYANOBIONT ANABAENA AZOLLAE

Anat Rosen, Hanna Arad, and Elisha Tel-Or
Department of Agricultural Botany, The Hebrew University of Jerusalem,
Rehovot 76100, Israel

Axenic cultures of Anabaena azollae were found earlier to preferentially take up and utilize low concentrations of fructose for enhanced nitrogenase activity (Tel-Or et al, 1984). The low concentration of 1 mM fructose was insufficient to enhance growth in the light or in the dark, and the introduction of 10-20 mM fructose to the growth medium enhanced myxotrophic growth in the light, and in the light in the presence of DCMU (10^{-5}M) as well as heterotrophic growth in the dark. Sucrose also enhanced growth at the same concentrations, while glucose was not effective. The fructose-enhanced cell growth was characterized by a sequence of changes: glycogen accumulation, heterocysts differentation, glycogen breakdown, induced dehydrogenase and nitrogenase activity, and faster cell growth. A fructose-grown batch culture showed that the fructose taken up by the cells was converted to glycogen which consisted of 20-40 percent of the cell's dry weight. The glycogen was broken down only when the frequency of heterocysts increased to a maximum of 18 percent after four days of growth. The drastic breakdown of glycogen was accompanied by an 8 fold increase in glucose-6-P dehydrogenase activity, approaching 150 nmole.NADPH.mg prot^{-1}.min^{-1}. Nitrogenase activity was enhanced 7-10 fold to a rate of 150 umoles aceylene reduced. gr dry weight^{-1} hr^{-1}, which enhanced cell growth. The biomass of the fructose grown cells was three to four times higher than of the control cells.

It is suggested that the pattern of fructose supported nitrogenase activity in the isolated cyanbiont Anabaena azollae is similar to its development in situ in the leaf cavity of the host fern Azolla, where the host provides the phycobiont with fructose and sucrose.

References

Tel-Or E et al (1984) In Veeger C and Newton WE, eds, Advances in Nitrogen Fixation Research, pp 461-465, Martinus Nijhoff/Junk, The Hague.

ENZYMES OF TREHALOSE METABOLISM IN SOYBEAN NODULES

SEPPO O. SALMINEN and JOHN G. STREETER
DEPARTMENT OF AGRONOMY
OHIO STATE UNIVERSITY (OARDC), WOOSTER, OH 44691 USA.

Trehalose is one of the major carbohydrates found in soybean nodules. It accumulates in all species of Rhizobium tested. The pathway for its synthesis has been established in yeast: UTP + glucose-1-P $\overset{\rightarrow}{1}$ UDPG + glucose-6-P $\overset{\rightarrow}{2}$ trehalose-6-P $\overset{\rightarrow}{3}$ trehalose. Enzymes metabolizing trehalose have been demonstrated in various organisms.

Nodules from soybean (Glycine max (L.) Merr, cv. Beeson 80) inoculated with R. japonicum strain USDA 110 (low trehalose accumulator) or 61A76 (high trehalose) were macerated and fractionated into I. cell debris, II. bacteroid soluble protein, III. bacteroid fragments, and IV. cytosol. The non-particulate fractions, II and IV, were gel-filtered. The fractions were then assayed for enzymes of trehalose metabolism. All activities are reported on a per minute per gram nodule fresh wt. basis.

UDPG pyrophosphorylase (1) activity was high: .94 µmoles in fr. II and 14 µmoles in fr. IV in extracts of nodules formed by USDA 110. Similar high activities were seen in 61A76. Trehalose-6-P synthetase (2), the key biosynthetic enzyme, was confined to bacteroids, mostly to fr. II, where 61A76 (high trehalose) had activity of 7.2 nmoles relative to 3.9 nmoles in USDA 110. 61A76 had also a higher trehalose phosphatase (3) activity (103 nmoles) than USDA 110 (32 nmoles) in fr. II, whereas similar rates were seen in fr. IV, 86 and 95 nmoles, respectively).

Trehalase, the major catabolic enzyme, converting trehalose to two glucoses, had two pH optima, 3.8 and 6.6. A significant amount of activity was found in fr. I, USDA 110 had a rate of 284 nmoles, whereas the rates in frs. II and IV were 631 and 1960 nmoles, respectively at pH 6.6. 61A76 showed a similar rate in fr. I, but fr. II had only 63% and fr. IV 83% of the corresponding rates seen with USDA 110. Trehalose phosphorylase, converting trehalose to glucose + β-glucose-1-P was present in frs. II (63 and 68 nmoles) and IV (35 and 151 nmoles) in USDA 110 and 61A76. Phosphotrehalase, converting trehalose-6-P to glucose + glucose-6-P had activities of 4.1 nmoles in fr. II and 7.8 nmoles in fr. IV with similar rates in 61A76.

The data indicate that trehalose is synthesized in the bacteroids, whereas it is utilized in both the bacteroid and host tissues. The high trehalose strain 61A76 had higher trehalose-6-P synthetase activity than the low trehalose strain USDA 110.

The high activities of the catabolic enzymes relative to trehalose-6-P synthetase would seem to require separation of trehalose and the catabolic enzymes for trehalose accumulation to occur.

EFFECTS OF BLOCKING PHOTOSYNTHATE TO SOYBEAN ROOT NODULES

CHARLES SLOGER
USDA, ARS, NITROGEN FIXATION AND SOYBEAN GENETICS LABORATORY
BELTSVILLE, MD. 20705 USA

Soybean nodules are dependent upon a supply of photosynthate for growth and N_2 fixation. The objective was to begin to elucidate the responses of carbohydrates, H_2 uptake activity, respiration and nitrogenase activity of nodules to a sudden cutoff in photosynthate supply.

MATERIALS AND METHODS.

Soybean, Glycine max (L.) Merr. cv. Bonus was grown with N-free nutrient in sterile vermiculite inoculated with Bradyrhizobium japonicum USDA 110 (Hup$^+$ phenotype). Plants were grown in a greenhouse at least 45 days. The transport of photosynthate to nodules was blocked by steam girdling the the stem for 1 minute. Nitrogenase activity was measured with 10 minute acetylene reduction assays of detached roots. Nodules were homogenized in 0.05 M phosphate buffer pH 7.6 at 40°C, filtered through cheesecloth and centrifuged 10,000xg for 10 minutes. The cytosolic and bacteroidal fractions were extracted with 80% (v/v) ethanol (Streeter, Bosler, 1976). Trimethylsilyl derivatives of sugars and organic acids were formed using N,0-bis(trimethylsilyl) trifluoroacetamide. Gas chromatography was according to Streeter, Bosler (1976). The rate of endogenous O_2 and H_2 uptake by bacteriods was determined ampero-metrically (Hanus et al., 1980). Ureides in cytosol were determined according to van Berkum, Sloger (1983).

RESULTS AND DISCUSSION.

The sucrose concentration in the cytosolic fraction of the nodules decreased 34% in the first hour and 73% in 2 hours after blocking trans-port of photosynthate. The rate of nitrogenase activity decreased 56% in the first hour and 79% in 2 hours. The concentration of sucrose was 40x higher in the cytosol than in the bacteroids. The sucrose content decreased rapidly in the bacteroids. The concentration of myo-inositol, glucose and malic acid in the cytosol and bacteroids did not change much in 2 hours. Endogenous H_2 uptake activity, and O_2 uptake by isolated bacteroids and ureide content in the cytosol were relatively uneffected by the cutoff in photosynthate supply.

The results indicate that the N_2 fixing process is more responsive than H_2 uptake or bacteroid respiration to a cutoff in photosynthate to nodules. Nitrogenase activity in soybean nodules is closely related to the supply of sucrose in nodules.

REFERENCES.

van Berkum P, Sloger C (1983) Plant Physiol. 73:511-513.
Hanus FJ, Carter KR, Evans HJ (1980) Methods Enzymol. 69:731-739.
Streeter JG, Bosler ME (1976) Plant Sci. Letters. 7:321-329.

ORGANIC ACID METABOLISM OF SOYBEAN BACTEROIDS UNDER LOW
OXYGEN CONCENTRATION

SHIGEYUKI TAJIMA,
LAB. BIOCHEM., FACULTY OF AGRIC., KAGAWA UNIV., MIKI-CHO,
KITA-GUN, KAGAWA 761-07, JAPAN

1. INTRODUCTION

Functioning TCA cycle are reported to present in bacteroids.
However it might not be operative under low oxygen concentration. Inside
of the nodules is believed to be anaerobic, and leghemoglobin is necessary
for bacteroid respiration. The presence·of anaerobic carbon metabolism
was already reported. The purpose of our research is to elucidate how
organic acids are metabolized in the bacteroids and generate ATP and
reducing power for the nitrogenase system under low oxygen concentration.

2. MATERIAL & METHODS

Soybean plants were grown by water culture as already reported
(Tajima and Yamamoto 1977).
The method for bacteroid isolation was basically of Emerich et al
(1979). The O_2 concentration around the mortar and pestle was determined
by an O_2 analyzer. The value was always below 0.05%.
The purification of the leghemoglobin was performed as Bergersen
and Turner reported (1967).
Organic acid determination was performed by a HPLC system (Carboxylic
acid analyser, Seishin Pharmaceutical Co., LTD.).

3. RESULTS

In 2% oxygen concentration, an optimum condition for bacteroid
acetylene reduction activity, the organic acid degradation by bacteroids
was very slow, and both lactate (or β-hydroxybutyrate) and acetate were
accumulated in the incubation system.
In the presence of leghemoglobin in the incubation system the
organic acid degradation by bacteroids was accelarated extensively even
in 2% oxygen concentration, and the formation of lactate (or β-hydroxy-
butyrate) and acetate were negligible. Addition of acetylene to the
incubation system slightly inhibited the organic acid degradation.
2,3-^{14}C-succinate was incubated with nodule bacteroids under 1%
oxygen concentration. Time course of $^{14}CO_2$ formation showed that the
succinate degradation was very slow at high succinate concentration.
For rapid degradation of such high (ca. 4mM) concentration of succinate
the presence of leghemoglobin in the incubation system was necessary.
The radioactivity in the bacteroid fraction was mainly detected in
various organic acids of TCA cycle.

4. REFERENCES

Bergersen, FJ and GL Turner (1967) Biochim. Biophys. Acta 141, 507-515
Emerich, DW et al (1979) J. Bacteriol. 137, 153-160
Tajima, S and Y Yamamoto (1977) Plant Cell Physiol. 18, 247-253

LOW TEMPERATURE INHIBITION OF N_2 FIXATION AND C AND N PARTITIONING IN SOYBEAN

Walsh K.B. and D.B. Layzell, Dept. of Biology, Queen's
University, Kingston, Ont. K7L 3N6, Canada

1. INTRODUCTION

Low soil temperatures during vegetative growth of soybeans are known to limit plant growth and yield under Canadian agricultural conditions. However, little is known of the physiological basis for the observed inhibition. In this study, detailed analyses of C and N metabolism were carried out using N_2 fixing soybeans which were exposed to root temperatures of either 15 or $25^{\circ}C$ for a period in late vegetative growth. From these results, a mechanism was proposed to account for the inhibitory effects of low root temperature on plant growth in N_2 fixing soybeans.

2. MATERIALS AND METHODS

Five week old soybean (cv. Harosoy 63) inoculated with Rhizobium japonicum USDA 16 (a Hup⁻ strain) were subject to a root chilling treatment of $15^{\circ}C$ for an 11 day period. Measurements were made of gas exchange (CO_2, H_2, H_2O, C_2H_2 reduction), plant growth (dry weight increment, leaf area) and %C, %N and starch content of various plant organs. These data were used to reconstruct and compare the pattern of C and N partitioning under the two temperature treatments.

3. RESULTS and DISCUSSION

Despite an increase in the RE of nitrogenase in the $15^{\circ}C$ treatment, N_2 fixation was immediately inhibited to 80% of that in the $25^{\circ}C$ plants. The subsequent changes in the pattern of C and N partitioning resulted in N limitation of shoot apical growth. Fig. 1 outlines possible cause and effect relationships which may account for the inhibition of plant growth at $15^{\circ}C$.

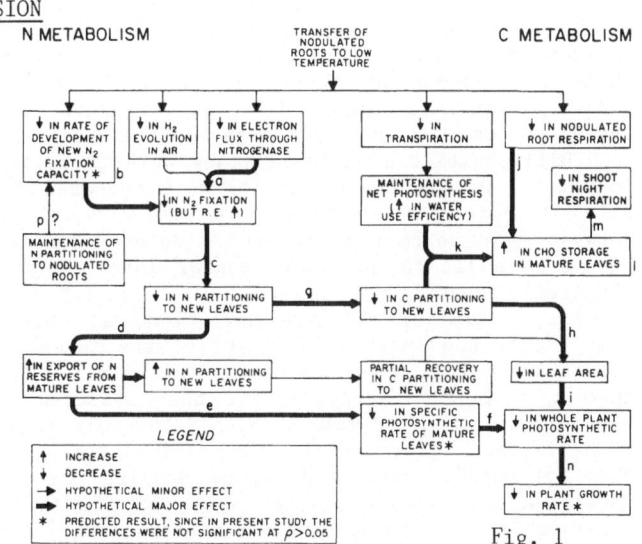

Fig. 1

It was concluded that tolerance to low root temperature in N_2 fixing soybeans will involve the development of a greater capacity for primary assimilation of nitrogen at low root temperatures.

4. ACKNOWLEDGEMENTS

Supported by a NSERC (Canada) grant (DBL) and a DSS (Agric. Canada) contract and travel supported by the School of Graduate Studies and Research and the Faculty of Arts and Science, Queen's University.

DETERMINING GAS DIFFUSIVITIES IN NODULES FROM
LAG PHASES IN ACETYLENE REDUCTION

LAWRENCE C. DAVIS, LARRY E. ERICKSON, AND G. TRAVIS JONES
Kansas State University, Manhattan, Kansas 66506, U.S.A.

Diffusion of gases through legume nodules is of obvious importance for nitrogen fixation. We have applied engineering techniques to develop a model for diffusion with reaction in the nodule. It is treated as a sphere with an inert outer layer, a reactive layer and an inert core. The dimensions of each may be altered to match real cases. When both inward and outward diffusion are involved, as with measuring ethylene production from acetylene, no analytical solution to the mass flow equations is possible. Instead, a numerical simulation is used, with dimensionless parameters. For a sphere, there is a characteristic lag time dependent on the square of the radius of the sphere and the diffusivity of the gas through the sphere. Product accumulation is expressed as the fraction of external material converted.

The kinetic properties of the enzyme are treated by two additional dimensionless parameters β and α. The first (β) reflects the extent of enzyme saturation and the second (α) reflects the specific activity of the active tissue. Lag times depend on both of these with low substrate giving a longer lag and high specific activity a shorter lag. The shortest possible lag is about 0.1 of the characteristic time (Dt/r^2) while the longest is about 0.3. Thickness of the inert outer layer is very important and outward diffusion of product may make a major contribution to observed lags with high substrate levels.

Diffusivities used for fitting were those of acetylene and ethylene in water. Dimensions of nodules were measured by microscopy and reasonable estimates of enzyme parameters were obtained from steady-state assays. For vetch nodules, the simulated dimensions of the nodule that give a best fit to the experimental data are somewhat smaller than the physically measured dimensions (0.7 vs 1.0 mm radius). With larger nodules, a similar relative deviation was observed. This would imply that either the inert outer layer or the inner active zone has a greater permittivity than water or that irregularities in nodule dimensions give them a smaller effective radius than measured. The difference between observed and expected values is only two-fold so we can probably not make a distinction between these two possibilities. We are forced to conclude that for vetch, sesban and lupine nodules there is no evidence for air passageways through the outer zone, nor is the need to invoke a highly impermeable layer to account for restricted gas diffusion.

Trials with a non-legume (Eleagnus) indicated a significantly greater porosity than expected on the basis of physical dimensions. The estimated effective radius was less than 0.5 mm for a nodule 1.6 mm thick and 1 cm in diameter. The absence of a distinctive zone containing leghemoglobin in these nodules made it difficult to apply a model assuming discrete inert and active layers.

ACKNOWLEDGEMENT. Supported by a USDA grant 83-CRCR-1-1301 to (L.C.D.) and the Kansas State Agricultural and Engineering Experiment Stations.

CONSERVATION OF SOYBEAN LEGHEMOGLOBIN STRUCTURES

WILLIAM H. FUCHSMAN*/REID G. PALMER
*CHEMISTRY DEPARTMENT, OBERLIN COLLEGE, OBERLIN, OH 44074, USA/IOWA STATE
UNIVERSITY, AMES, IA 50011, USA.

Soybean leghemoglobin (Lb) is heterogeneous, consisting of four major Lb's (a, c_1, c_2, c_3) so well as minor Lb's. The four major Lb's are separate gene products. This report addresses three related questions: Are Lb's from different soybean cultivars and plant introductions the same? Is soybean Lb heterogeneity functional? And are discrepancies in published amino acid sequences for soybean Lb's due to cultivar differences?

Major soybean Lb's were compared by isoelectric focusing under conditions capable of discriminating proteins with pI differences of 0.01 pH unit (Fuchsman, Appleby, 1979; Fuchsman, 1983). Lb's from 69 cultivated soybean (Glycine max) and 18 wild soybean (Glycine soja) cultivars and plant introductions were compared. The cultivars and plant introductions were diverse in maturity group, in geographical source, and in seedling root fluorescence (Delannay, Palmer, 1982) phenotype and genotype (a property that has not been selected in breeding programs). All cultivars and plant introductions produced Lb's a, c_1, c_2 and c_3, and each of the four major Lb's exhibited a pI that was independent of the source cultivar or plant introduction.

Therefore, soybean Lb heterogeneity is conserved in a genetically diverse selection of G. max and G. soja cultivars and plant introductions. By the criterion of less than 0.01 pH unit variation in pI, the structures of soybean Lb's a, c_1, c_2 and c_3 also are conserved in G. max and G. soja. The results provide circumstantial evidence that soybean Lb heterogeneity is functional, since conservation of a trait in an otherwise diverse population suggests that the trait is functional.

Published amino sequences of soybean Lb's (Ellfolk, Sievers, 1974; Hurrell, Leach, 1977; Sievers et al., 1978; Hyldig-Nielson et al., 1982; Wiborg et al., 1982) contain discrepancies, some of which involve charged amino acids. The comparative results reported here, which include results for many of the soybean cultivars used in sequence studies, show that at least some of the reported discrepancies cannot be due to differences in Lb structures.

REFERENCES
Delanney X and Palmer RG (1982) Crop Sci. 22, 278-281.
Ellfolk N and Sievers G (1974) Acta Chem. Scand. B28, 1245-1246.
Fuchsman WH (1983) Crop Sci. 23, 165-166.
Fuchsman WH and Appleby CA (1979) Biochim. Biophys. Acta 579, 314-324.
Hurrell JGR and Leach SJ (1977) FEBS Lett. 80, 23-26.
Hydig-Nielsen JJ et al (1982) Nucl. Acids Res. 10, 689-701.
Sievers G et al (1978) Acta Chem. Scand. B32, 380-386.
Wiborg O et al (1982) Nucl. Acids Res. 10, 3487-3494.

CYTOCHROME COMPOSITION OF <u>RHIZOBIUM JAPONICUM</u> BACTEROIDS

DONALD L. KEISTER AND SARAH S. MARSH, Nitrogen Fixation & Soybean
Genetics Lab., ARS, Beltsville, MD 20705 and Battelle-Kettering Research
Lab., Yellow Springs, OH 45387.

Recently it has become clear that there is a large degree of diversity
between <u>R. japonicum</u> strains based on DNA homology (Hollis et al.,1981)
and previously we have identified some phenotypic characteristics of the
different DNA homology groups (Keister et al., 1984). Characteristics of
DNA Homology group II (e.g. 61A76, USDA117, USDA83, USDA86) included high
<u>ex planta</u> nitrogenase activity, high extracellular polysaccharide (EPS)
production, the major acidic EPS is composed of rhamnose and 4-0-methyl-
glucuronic acid (Huber et al., 1984), small colony morphology on agar,
high intrinsic antibiotic resistance (Kuykendall, personal commun.) and
production of rhizobitoxin (Devine and Weber, 1977). In contrast,
characteristics of DNA homology group I and Ia are low or negative <u>ex
planta</u> nitrogenase activity, low EPS production, acidic EPS composed of
mannose, glucose, galactose and 4-0-methylgalacturonic acid, and low
intrinsic antibiotic resistance. Preliminary studies revealed another
possible phenotype, retention of cytochrome aa_3 in bacteroids by strains
of DNA homology Group II (Keister, et al, 1984).

Appleby (1969) in his classical studies on the electron transport system
of <u>R. japonicum</u> 505 bacteroids found no cytochrome aa_3 or o although
these are the major aerobic oxidases of cultured cells (Appleby, 1969b).
In contrast we found that bacteroids of strain 61A76 retained both of
these cytochromes (Keister et al., 1983). In order to determine whether
our results with 61A76 or Appleby's results with strain 505 were representative of <u>R. japonicum</u>, we analyzed bacteroids from 44-50 day old crown
nodules from field grown plants (Keister et al., 1983) inoculated with 17
different strains. Strains of DNA homology group I (10324, USDA38,USDA58,
D193, 5633) and group Ia (8°, 61A50, USDA62, 110, 140) contained very low
or no cytochrome aa_3 while strains of DNA homology group II (61A76, USDA
strains 29,83,86,117 and 130) all retained significant amounts of this
oxidase. Only three strains were devoid by cytochrome o (505, D193,
USDA58). Thus retention of cytochrome aa_3 by bacteroids appears to be
phenotypic of group II. Studies are in progress to determine whether aa_3
and o are functional in bacteroids which retain them.

REFERENCES

Appleby CA (1969a) Biochim. Biophys. Acta 172, 71-87.
Appleby CA (1969b) Biochim. Biophys. Acta 172, 88-105.
Devine TE and Weber DF (1977) Euphytica 26, 527-535.
Hollis AB, Kloos WE and Elkan GE (1981) J. Gen. Microbiol. 123, 215-222.
Keister DL, Marsh SS and El Mokadem MT (1983) Plant Physiol. 71, 194-196.
Keister DL, et al. (1984) In Veeger C and Newton WE, eds, Advances in
Nitrogen Fixation Research, p. 264.

These studies were supported by grants 81-CRCR-1-0653 and 81-CRCR-1-0772
from the USDA.

FORMS OF LEGHEMOGLOBIN IN INTACT LEGUME NODULES ATTACHED TO ROOTS

ROBERT V. KLUCAS, KEUK-KI LEE, AND LINDA SHEARMAN
DEPARTMENT OF AGRICULTURAL BIOCHEMISTRY, UNIVERSITY OF NEBRASKA,
LINCOLN, NEBRASKA, 68583-0718

Leghemoglobin (LB) was assessed in intact, whole yellow blossom sweet clover and soybean nodules using a spectrophotometric technique. The technique was valuable not only for evaluating the role of LB in legume nodules, but also as a method for probing the effects of various treatments on nodules using LB as an internal marker. Excellent spectra of the smaller type of indeterminant nodules on sweet clover as well as alfalfa, birdsfoot trefoil, and red clover were obtained. Less definitive spectra but nevertheless interpretable spectra were obtained with larger soybean nodules. Rates of acetylene reduction (AR) of the spectrally-monitored nodule were measured during some of the treatments. Comparisons of spectra of whole nodules that were attached to intact plants and exposed sequentially to atmospheres of air, 100% N_2, 100% O_2, and then air revealed that ferrous LB was the dominate form of LB in young, active nodules. In situ, ferrous LB was reversibly oxygenated. In young sweet clover nodules, less than 5% of the ferrous LB was oxygenated in air. However, older nodules and older (proximal versus distal) sections of sweet clover nodules possessed more oxygenated ferrous LB than younger nodules. Thus far, ferric LB has not been detected in any intact nodule under normal or chemically-treated conditions. Nodulated sweet clover roots that were immersed in nicotinate-containing nutrient solution below the zone of nodulation transported nicotinate to nodules within 20 min as shown by the formation of the distinctive nicotinate complex with ferrous LB. In air, this complex was the dominate form of LB for at least one day after treatment with nicotinate, but under 100% O_2, the nicotinate was reversibly displaced by O_2. Formation of the ferrous LB nicotinate complex was demonstrable in soybean nodules after exposing nodulated roots to nicotinate but spectra were less definitive. The effect of nicotinate on AR by sweet clover or soybean nodules was ambiguous. In most experiments, treatment of nodulated roots with nicotinate resulted in a decline in AR, but the decline in AR may be due to secondary effects other than the specific binding of nicotinate to LB. With either sweet clover or soybean in air, removal of the upper portion of the plant resulted in a large decrease in AR within 4 hours. Concomitant with the decline in AR was an increase in ferrous LB and a decrease in oxygenated ferrous LB.

FURTHER ERRORS IN THE USE OF THE ACETYLENE REDUCTION ASSAY WITH LEGUMES

FRANK R. MINCHIN, JOHN, E. SHEEHY AND JOHN F. WITTY
The Animal and Grassland Research Institute, Hurley, Maidenhead, Berks.
SL6 5LR, U.K. and Rothamsted Experimental Station, Harpenden, Herts. AL5
2JQ, U.K.

Previous work with open, flow-through gas systems using intact whole
plants (attached root systems) has demonstrated an acetylene-induced
decline in the nitrogenase activity of legumes (Minchin et al, 1983).
This decline was shown to reflect a reduction in O_2 supply to bacteroids
resulting from an increase in the O_2 diffusion resistance of the nodules
(Witty et al, 1984). This poster illustrates further errors associated
with the conventional use of detopped/disturbed root systems in
comparative acetylene reduction assays.

With 'attached' root systems of soyabean the time course of ethylene
production shows an increase during a 4-5 minute mixing period, a maximum
activity period of 1-2 minutes and then a marked acetylene-induced
decline over the next 50-60 minutes. Relative to this ethylene
production the time course of a detopped/disturbed system (shoot removed,
rooting medium shaken off) shows a reduced maximum rate of activity and
little response to acetylene. This is evidence that the nodular O_2
diffusion resistance increases in response to detopping and disturbance
prior to exposure to acetylene.

When 'attached' root systems are used to compare nitrogenase activity at
both 25 and 15°C the two time courses of ethylene production are very
similar to those obtained in the above comparison of 'attached' versus
detopped/disturbed roots. Therefore, the nodular O_2 diffusion resistance
of 'attached' roots is increased in response to a 10°C temperature
decrease. However, when detopped/disturbed roots are used to make the
same temperature comparison they show reduced activity relative to the
'attached' root and the acetylene-induced decline in nitrogenase activity
is greater at 15°C. Thus, detopping and disturbance increases the O_2
diffusion resistance of the nodules at both temperatures, but the effect
is greater at 25°C.

When the acetylene reduction technique is used as a comparative assay
with 'attached' roots in flow-through systems the maximum rate of
activity declines by approximately 40% in response to a decrease in
temperature from 25 to 15°C. But, when the same comparative assay is
performed with detopped/disturbed root systems there is an apparent
increase, of approximately 20%, in response to the temperature decrease.
This unusual result can be explained by disturbance induced changes in
the nodules' O_2 diffusion resistance.

CONCLUSION
The conventional acetylene reduction assay for nitrogenase activity, as
performed with detopped/disturbed root systems, should not be used even
as a comparative assay. The only accurate method is to use 'attached'
roots in a flow-through gas system.

REFERENCES
Minchin FR et al (1983) J. exp. Bot. 34, 641-649.
Witty JF et al (1984) Ann. Bot. 53, 13-20.

RESPIRATION AND OXIDATIVE PHOSPHORYLATION OF MITOCHONDRIA FROM NODULES OF COWPEA (VIGNA UNGUICULATA L.)

STEPHEN RAWSTHORNE and THOMAS A. LARUE
BOYCE THOMPSON INSTITUTE, ITHACA, NY 14853, U.S.A.

In the nodule the free O_2 concentration is about 10nM and an O_2 limitation to mitochondrial function is likely.

Mitochondria were extracted and separated from bacteroids under microaerobic conditions. When incubated aerobically (20-220 μM O_2) with malate and ADP they showed immediate O_2 uptake (100-140 nmol min^{-1} mg $protein^{-1}$). Respiratory control ratios (RCR) were between 5 and 8, and ADP/O and ATP/O ratios were between 2.3 and 2.6. Succinate produced greater O_2 uptake (150-250 nmol min^{-1} mg $protein^{-1}$) although RCR's and ADP/O or ATP/O ratios were expectedly smaller (2.7-3.0 and 1.4-1.6, respectively). These activities are typical of mitochondria from aerobic plant tissues. We did not see the low activities reported for mitochondria isolated from coleoptiles of anaerobically-germinated monocot seeds (Vartapetian BB et al., 1978; RA Kennedy, pers. comm.).

Microaerobic respiration was assayed by monitoring the deoxygenation spectra of leghemoglobin in sealed cuvettes with no gas phase. With either malate or succinate, saturating or 'primer' quantities of ADP produced continuous state III, or state IV activities which declined with decreasing O_2 concentration (150-5nM). At 10nM O_2, state III respiration rates were, respectively, only 7-10% and 5-6% of aerobic rates for malate and succinate. Aerobic and microaerobic oxygen uptake was inhibited 98% by KCN. This rules out the alternative pathway of respiration.

ATP/O ratios were calculated during declines in O_2 concentration from 150-5nM. In this range, ATP/O ratios were smaller for malate (1.22) and succinate (0.47) suggesting that oxidative phosphorylation becomes inefficient with decreased O_2 concentration. In addition, state III respiration rates intersect with state IV rates between 9-15nM O_2, suggesting that below this range there is no longer an increase in electron flux in the presence of ADP; i.e. oxidative phosphorylation is not occurring.

Mitochondria at 10nM O_2 may therefore produce only small amounts of ATP. Nevertheless, at O_2 concentrations only 2-3 fold larger, oxidative phosphorylation is more effective, respiration rates are larger, and so ATP synthesis is likely to be greater. In a recent model of oxygen flux in nodule cells (Sheehy JE et al., 1985) the predicted free O_2 concentration at the periphery of the infected cell was 20-26nM. This may explain why mitochondria are observed around the periphery of mature infected cells (e.g., Newcomb EH et al., 1985).

References
Newcomb EH et al. (1985) Protoplasma 125, 1-12.
Sheehy JE et al. (1985) Ann. Bot. 55, 549-562.
Vartapetian BB et al. (1978) In Hook D and Crawford RMM, eds, Plant Life in Anaerobic Environments.

Acknowledgements: Supported by US Dept. of Energy, #82ER12066.

ROLE OF NODULE SIZE ON THE NITROGENASE ACTIVITY IN COWPEA

DIPANKAR SEN and R.W. WEAVER
Department of Soil and Crop Sciences
Texas A & M University, College Station, TX 77843, U.S.A.

The size of nodules on the root system of cowpea (Vigna unguiculata (L.)Walp.) vary considerably. We examined the biological importance of nodule size in determining the specific activity of nodules in fixing N_2. Possibly nodules with thinner cortices and higher surface/volume ratios would have higher specific activities due to better O_2 availability to their bacteroids. The relationship of nodule size with other relevant nodule parameters, such as quantity of bacteroids, size of the bacteroidal zone, and thickness of nodule cortex was also examined.

Nodules were collected from cowpea plants (var. California Blackeye) one month after inoculation with an effective strain (32H1) of Rhizobium and grown under laboratory conditions (16 hr. daylength at room temperature with PAR of 325 μ einstein . m^{-2}. sec^{-1}). Nodules were graded by sizes of 2 mm, 3 mm, and 5 mm ($\pm$ 0.1 mm) diameters and their acetylene reduction (AR) activities were measured under 20, 30, and 40% O_2 after 12 min. exposure to C_2H_2.

AR activities of nodules of different sizes were not significantly different at atmospheric O_2 concentration (Table 1). Increasing O_2 to 30% greatly enhanced activities of nodules in the smallest and the largest classes. Increasing O_2 to 40% resulted in further enhancement of activity only for nodules in the middle sized class.

Table 1. AR by cowpea nodules of different sizes at different O_2 tensions.

Average nodule diameter mm	AR activities at O_2 conc. of		
	20%	30%	40%
2.0	86.0 a*	268.8 c	157.2 ab
3.0	84.7 a	132.3 ab	218.2 bc
5.0	70.5 a	262.6 c	91.6 a

*Values denoted with the same letter are not significantly different at the 5% level by Waller-Duncan test.

The interaction of the bacteroids and their nodular environment is complex. With an increase in the nodule diameter from 2 to 5 mm, in spite of an increase in the thickness of nodule cortex from an average of 0.32 to 0.54 mm, the proportion of the bacteroidal zone increased from 32 to 48% of the nodule volume. The number of bacteroids apparently decreased slightly (15.5 x 10^{10} to 11.1 x 10^{10}. g nodule^{-1}) and resulted in a decrease in the number of bacteroids per unit volume of the bacteroidal zone. This may reduce competition among bacteroids for available substrates and O_2 leading to increased nitrogenase activity. However, as the nodule diameter increased from 2 to 5 mm , the surface/volume ratio decreased from 3.0 to 1.2 mm^{-1} . This, accompanied by the concurrent increase in the thickness of nodule cortex may reduce diffusion of gases resulting in a decrease in nitrgenase activity.

The results show that the smaller nodules were not more active than larger nodules in nitrogen fixation at the atmospheric O_2 concentration.

EFFECT OF OXYGEN PRESSURE ON NITRITE
ACCUMULATION INSIDE SOYBEAN AND ALFALFA NODULES

JEAN-FRANCOIS SOUSSANA, MARIE-ODILE HECKMANN, RACHIDA WAKAIM, JEAN-JACQUES DREVON, LOUIS SALSAC, Laboratoire de Recherche sur les Symbiotes des Racines, INRA, 9 Place Viala, 34060 Montpellier Cedex, France

A supply of nitrate higher than 3 mM represses soybean nodule nitrogenase activity (Jatimiliansky et al., 1982). The observation that nitrite, a product of nitrate reduction, affects leghaemoglobin (Rigaud, Puppo, 1977) and the component II of nitrogenase (Trinchant, Rigaud, 1982) prompted us to study the nitrate and nitrite metabolism in soybean and alfalfa nodules.

The nitrate reductase (NR) capacities (in vitro activities) of bacteroids and root nodule cytosol were comparable and varied little during life cycle, except during and after maturation of pods when the bacteroidal NR capacity increased. Both activities were always significantly higher than that of the host root.

The nitrite reductase (NiR) activity was detected inside nodules. The cytosol NiR capacity was higher than that of roots which was higher than that of bacteroids. The NiR capacity inside nodules was in the same range of magnitude as the NR capacity.

The nodule concentration of NO_2^- was estimated to vary from 0.6 mM until pod filling, to 1.2 mM after this stage, which was much higher than the nitrite concentration in roots. Increasing the external O_2 pressure on excised soybean or alfalfa nodules resulted in a significant decrease of the nodule NO_2^- (and also ethanol) content (Fig. 1).

Such concentration of NO_2^- inside nodules could inhibit nitrogenase (see Rigaud et al., 1973). It appears to be due to the microaerobiosis inside nodules; in plant roots indeed, the NiR activity is decreased under low O_2 tension because of a reduction of the entry of glucose 6P in the plastids (Dry et al., 1981). The increase in nodule NO_2^- pool may also be due, at least partially, to dissimilatory NR activity; this activity is accompanied in free-living Rhizobium under anaerobiosis by an increase in NO_2^- pool in bacterial cytoplasm (O'Hara, Daniel, 1985).

References

Dry I et al (1981) Planta 152, 234-238
Jatimiliansky JR et al (1982) Physiol. Veg. 20, 407-422.
O'Hara GW and Daniel RH (1985) Soil Biol. Biochem. 17, 1-9.
Rigaud J et al (1973) J. Gen. Microbiol. 77, 137-144.
Rigaud J and Puppo A (1977) Biochim. Biophys. Acta 497, 702-706.
Trinchant JC and Rigaud J (1982) Appl. Environ. Microbiol. 44, 1385-1388.

HEMOGLOBINS WHICH SUPPLY OXYGEN TO INTRACELLULAR PROKARYOTIC SYMBIONTS

JONATHAN B. WITTENBERG/BEATRICE A. WITTENBERG/MICHAEL TRINICK/QUENTIN H.
GIBSON/ANTHONY I. FLEMING/DIDIER BOGUSZ/CYRIL A. APPLEBY
ALBERT EINSTEIN COLLEGE OF MEDICINE, NEW YORK, NEW YORK 10461, USA

Hemoglobin is found within the symbiont-harboring cells of the nitrogen-
fixing associations: legume/Rhizobium, Parasponia/Rhizobium, and woody
dicot/Frankia. Hemoglobin is also found, ocassionally at very great
concentration, within the cytoplasm of symbiont-harboring cells of the
carbon-fixing associations between bacteria and certain bivalve molluscs.
We ask: Are the ligand-binding properties of each hemoglobin adapted to
the particular (and different) intracellular environment of its symbiosis? Or, can we discern a set of properties common to all? On the basis
of limited data so far accumulated, we discern a common pattern among the
plant hemoglobins. Very great oxygen affinity is achieved by very rapid
combination with oxygen together with a modest rate of dissociation.
This was not anticipated since the spectrally apparent terminal oxidase
systems of soybean, Sesbania, and Parasponia bacteroids differ (Appleby,
1984) and free-living Rhizobia from Sesbania fix nitrogen and grow at
very much higher oxygen pressure than other cultured Rhizobia (Gebhardt
et al. 1984).

In contrast, the one clam hemoglobin (Lucina pectinata) so far examined
achieves high oxygen affinity by very rapid combination with oxygen in
the face of rapid dissociation.

Oxygen Reactions of Some Hemoglobins at 20°

	Combination $k'(M^{-1} s^{-1} \times 10^{-6})$	Dissociation $k(s^{-1})$	Equilibrium K_{diss} (nM)	$p1/2$ (torr)
SYMBIONT-HARBORING PLANTS				
Soybean	116	5.55	47	0.026
Sesbania	250[a]	7.35	30	0.02
Parasponia	165	14.8	89	0.05
Casuarina	45	5.5	120	0.07
SYMBIONT-HARBORING CLAM		.		
Lucina	155[a]	55	360[a]	0.2[b]
NON-SYMBIOTIC TISSUE				
Horse Myoglobin	14	11	770	0.7
Busycon (mollusc)[c]	48	71	1500	1.2

a. estimated; b. Read (1962); c. Schreiber and Parkhurst (1984)

REFERENCES

Appleby CA (1984) Ann. Rev. Plant. Physiol. 35, 433-478.

Gebhardt C, Turner GL, Gibson AH, Dreyfus BL, Bergersen FJ (1984) J.
Gen Microbiol. 130, 843-848.

Read KRH (1962) Biol. Bull. (Woods Hole) 123, 605-617.

Schreiber JK, Parkhurst LJ (1984) Comp. Biochem. Physiol. 78A, 129-135.

Supported by Research Grants PCM 83-03698, PCM 84-16016, DMB 08454 from
the National Sciences Foundation and HL 19299 and GM 14276 from the
USPHS. JBW is a Career Awardee, HL 1-K6-733, of the USPHS, NHLBI.

DIRECT EVIDENCE FOR A VARIABLE BARRIER TO DIFFUSION INTO LEGUME NODULES

JOHN F. WITTY, LEIF SKØT and NIELS P. REVSBECH
ROTHAMSTED EXPERIMENTAL STATION and AARHUS UNIVERSITY, DENMARK.

Previous studies show that nitrogenase activity is limited by the rate of O_2 diffusion into the nodule and indirect evidence shows that in many symbioses nodules can rapidly alter their resistance to gaseous diffusion (Minchin et al 1985, Witty et al 1984). Using O_2 specific micro-electrodes (Revsbech, Ward 1983) we have demonstrated directly the operation of this barrier. The electrode consists of an inner gold cathode surrounded by an aqueous electrolyte and shielded from the environment by an oxygen selective membrane within the 4 μm tip. Unlike the naked platinum electrode used in nodule studies by Tjepkema and Yocum (1974) this electrode measures quantitatively free molecular O_2 and not redox potential in solution.

Within functional nodules of pea (*Pisum sativum*) and french bean (*Phaseolus vulgaris*) the oxygen concentration sensed by the electrode in the bacteroid containing region was less than 1 μM. In a number of experiments the electrode tip was advanced through the cortex to a position just inside the infected tissue and then ambient O_2 concentration was increased to 40%. In 9 out of 21 experiments the oxygen remained low and unchanged and in 4 out of 21 cases it increased irreversibly to near ambient value. In the remaining experiments the O_2 concentration at the electrode tip increased (in 1 to 2.5 min) and then decreased to its former concentration. We postulate that when ambient O_2 concentration was raised flux into the nodule exceeded respiratory consumption and O_2 concentration increased. We ascribe the subsequent decrease in concentration to the operation of the barrier which increased in resistance so that respiration could once more accommodate flux. It is likely that apparent differences in behaviour are due to the exact positioning of the electrode tip and the respiratory capacity of the infected tissue.

Within nodules of french bean formed by an ineffective strain of *Rhizobium*, O_2 concentrations ranged from 60-100 μM and these concentrations followed those of the gas phase. Work is continuing on the method of operation and implications of the barrier.

REFERENCES

Minchin FR et al (1985) Ann. Bot. 55, 53-60.
Tjepkema JD and Yocum CS (1974) Planta 119, 351-360.
Witty JF et al (1984) Ann. Bot. 53, 13-20.
Revsbech NP and Ward DM (1983) Appl. Environ. Microbiol. 45, 755-759.

IMMUNOLOGICAL COMPARISON OF H$_2$-OXIDIZING HYDROGENASES

Daniel Arp, Lillian McCollum, Charles Doyle and Lance Seefeldt, Dept. of Biochemistry, University of California, Riverside, CA   92521, U.S.A.

Hydrogenases are involved in improving the efficiency of N$_2$ fixation in both free-living and symbiotic aerobic microorganisms. Recently, the hydrogenase of Rhizobium japonicum has been purified to homogeneity from both bacteroids (Arp, 1985) and free-living bacteria (Harker et al., 1984). This enzyme is similar to the membrane-bound hydrogenases from the hydrogen oxidizing bacteria Alcaligenes eutrophus and Alcaligenes latus. We have examined the extent of this homology (Arp et al., 1985). Given the diversity of the catalytic and molecular properties of hydrogenases in general, it is useful to define groups of hydrogenases with similar properties. We have purified the hydrogenases from the three microorganisms above as well as that of Azotobacter vinelandii (Seefeldt and Arp, this volume). The migrations of these proteins on native and SDS-PAGE have confirmed the similarities in subunit composition and molecular weight (Arp et al., 1985). The amino acid compositions of the proteins also indicated substantial similarity. The immunological cross reactivity was compared with antibody raised in a rabbit against the purified hydrogenases of R. japonicum and A. eutrophus. All four hydrogenases cross-reacted with each antibody when tested with the ELISA technique. The antibody raised against R. japonicum hydrogenase inhibited the activity of three of the four hydrogenases. These experiments have now been extended to an analysis of each subunit of the hydrogenase. The subunits of R. japonicum hydrogenase were separated via SDS-PAGE, excised from the gel, and used to raise antibody in a rabbit. The specificity of each antibody towards only its respective subunit was confirmed with a Western blot (data not shown). The results of an ELISA with each of these antibodies and the four hydrogenases are presented in Figure 1. As with the antibody directed against the holoenzyme, cross reaction was evident with all four hydrogenases with either of the subunit antibodies. This indicates that the antigenic cross reactivity is not confined to one or the other subunit. We were unable to demonstrate any inhibition of H$_2$ oxidation coupled to methylene blue reduction with the subunit antibodies.

Arp DJ (1985) Arch. Biochem. Biophys. 237, 504-512.
Harker AR et al (1985) J. Bacteriol. 159, 840-856.
Arp DJ et al (1985) J. Bacteriol. 163, 15-20.

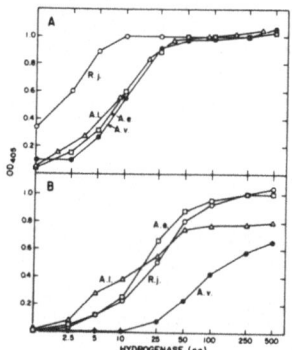

Figure 1. Comparison of the cross-reactivities of hydrogenases with an ELISA. Shown is the reaction of antisera raised against the large subunit (A) or the small subunit (B) of R. j. hydrogenase with purified hydrogenases from R. japonicum (O), A. latus (Δ), A. vinelandii (●), and A. eutrophus (□). For details of the procedure see Arp et al., 1985.

RELATIONS AMONG NITROGENASE, HYDROGENASE AND NITRATE-REDUCTASE ACTIVI-
TIES IN RHIZOBIUM LUPINI

M.A. CHAMBER; S. BAUTISTA

S.I.A., Apartado 13, San José de La Rinconada, Sevilla, SPAIN

1.- INTRODUCTION
Lupinus sp. are considered as a promising grain legumes in temperate cli-
mas, because of their high contents of protein, being adapted to grow in
acid and sandy soils.

2.- MATERIALS AND METHODS
Two species of Lupinus (Albus and Luteus) planted in Leonard jars and ino
culated by eight different strains of R. lupini were grown along 16 weeks,
measuring nodulation per plant, reduction of C_2H_2, Hydrogen evolution (R.
Efficiencies), H_2 uptake, dry weight and N contents of plants, and NO_3-
reductase in whole nodules and their bacteroids.

3.- RESULTS AND DISCUSSION

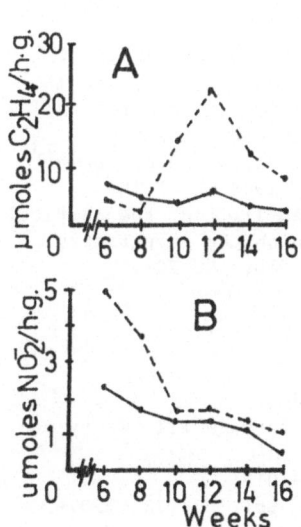

The Acetylene Total and Specific Activities
(Fig. A) reached their maxima at the beginning
of pod formation (plants 12 weeks old); five
out of the eight strains could be considered
as efficient ones (over 4 umoles C_2H_4/hxg.no-
dule). The ratios of H_2 evolved by nitrogenase were positively and significantly correlated to the A.T.A. and A.S.A. values, however,
the relative efficiencies (R.E.) hardly changed along the period under study. The Luteus
species (dotted lines in the Figures) got values higher than Albus for all of the parameters after blooming, except for nodules number per plant and R.E. One strain (IM-438) ob
tained an average R.E. over 0.99 and meaning
ful H_2 uptakes (2.9 umoles/hxmg.protein). The
data from NO_3-reductase in whole nodules (Fig.
B) showed a rapid decline at first, being held
for most of the growing period, and almost
disappearing at the end; there were significant correlations to NO_3-reduc
tase in their bacteroids suspensions.

4.- REFERENCES
Eisbrenner G., Evans H.J. (1983) Ann.Rev.Plant Physiol, 34, 105-136
Stephens B.B., Neyra C.A. (1983) Plant Physiol., 71, 731-735

INFLUENCE OF HYDROGENASE ON THE GROWTH OF GLYCINE AND VIGNA SPP.

JEAN-JACQUES DREVON, VIPIN CHANDRA KALIA, MARIE-ODILE HECKMANN, LOUIS SALSAC, Laboratoire de Recherche sur les Symbiotes des Racines, INRA, 9 Place Viala, 34060 Montpellier Cedex, France

Legume-root nodules evolve H_2 except those harbouring Hup[+] Rhizobium strains (Schubert, Evans, 1976). Hydrogen recycling capability has been postulated to be beneficial for symbiotic nitrogen fixation (Dixon, 1976).

The comparison of Hup[+] (PJ17-1) and Hup[-] (PJ17) isogenic mutants of R. japonicum in symbiosis with soybean shows a lower consumption of carbon substrates for the Hup+ strain (Drevon et al., 1982).

A further evaluation of the effect of this enzyme on the symbiosis has been conducted by comparing the growth of legumes nodulated by these two mutants.

In greenhouse tests after 35 days, there was no significant difference in growth of V. radiata, V. unguiculata and G. max. c.v. Hodgson inoculated with PJ17 and PJ17-1. But with G. usuriensis, the growth was significantly higher with PJ17 than with PJ17-1. In longer term experiments, the dry weight of soybean c.v. Hodgson, grown in liquid aerated medium, was significantly higher with the Hup[-] mutant.

In field test although the yield was slightly higher with PJ17, the difference from the Hup[+] symbiosis was not significant.

By contrast, the growth of soybean is higher with PJ18-HR (a complementation of the Hup[-] mutant PJ18 with the hup DNA isolated from the parental strain) than with PJ18 (Evans et al., 1984). However, the PJ18 mutation may affect other functions in addition to the hydrogenase (see Drevon et al., 1982; Haugland et al., 1984).

The lower growth of soybean inoculated with the Hup[+] strain in our experiments may be due to the preferential use of H_2 (shown by the lower CO_2 evolution in the Hup[-] mutant). This could be detrimental to nitrogenase activity as (1) the P/O ratio of H_2 as a respiratory substrate may be less than that of carbon substrates, which has been shown recently in Rhizobium ORS 571 (Stam et al., 1984); (2) The bacteroid respiration inside nodules may be oxygen-limited, the critical oxygen pressure (external pressure of O_2 under which the respiratory rate decreases) of excised soybean nodules being close to 70%.

References

Dixon ROD (1976) Nature 263, 173.
Drevon JJ et al (1982) Plant Physiol. 70, 1341-1346.
Evans HJ et al (1984) World Soybean Research Conference III, Westview Press, Boulder (in press).
Haugland RA et al (1984) J. Bacteriol. 159, 1006-1012.
Schubert KR and Evans HJ (1976) Proc. Natl. Acad. Sci. USA 73, 1207-1271.
Stam H et al (1984) Arch. Microbiol. 139, 53-60.

IMMUNOLOGICAL DETECTION OF TWO HYDROGENASE SUBUNITS IN TRANSCONJUGANTS OF HUP⁻ STRAINS OF RHIZOBIUM CARRYING PHU52 AND IN HUP⁺ STRAINS OF E. COLI

A.R. HARKER/G.R. LAMBERT/F.J. HANUS/M. ZUBER/H.J. EVANS
LABORATORY FOR NITROGEN FIXATION RESEARCH, OREGON STATE UNIVERSITY,
CORVALLIS, OR 97331 USA

The uptake hydrogenase of Rhizobium japonicum as purified in our laboratory is a nickel containing enzyme composed of two subunit polypeptides of M_r 60,000 and 30,000 (Harker et al. 1984). Although similar results have been recently reported for the hydrogenase isolated from bacteroids of soybean nodules (Arp 1985), the biochemical significance of the previously unreported 30kD subunit is unclear. The Hup-complementing cosmid pHU52 confers hydrogenase activity on Hup⁻ wildtype strains of Rhizobium when derepressed in the free-living state (Lambert et al. 1985). The purpose of this present work was to determine whether the acquisition of Hup specific cosmids by Hup⁻ recipients of Rhizobium could be correlated with the expression of Hup activity and the synthesis of the hydrogenase subunits.

The antisera produced against the hydrogenase polypeptides exhibited strong cross-reaction with a common contaminating polypeptide on nitrocellulose blots of SDS gels of crude extracts. This non-specific cross-reaction precluded detection of the hydrogenase subunits in crude extracts. Hydrogenase-specific antibodies were purified by separate passage of each serum through a column containing purified hydrogenase covalently bound to sepharose. This procedure eliminated cross-reaction with all proteins in crude extracts except the appropriate hydrogenase subunits.

Eight Hup⁻ recipient strains and 17 transconjugant strains carrying pHU1, pHU52, pHU53, or the pLAFR1 vector, were screened for Hup activity and the presence of immunologically detectable hydrogenase subunits. Only derepressed transconjugant strains carrying the cosmid pHU52 expressed Hup activity. The level of activity in the transconjugants ranged from 3% to 38% of that in derepressed Hup⁺ R. japonicum USDA 122DES. Both subunits of the enzyme were detected in crude extracts of all strains of Rhizobium carrying pHU52, in approximate proportion to the measured activity. No detectable cross-reaction was observed in Hup⁻ recipient strains, transconjugants carrying pHU1, pHU53, or pLAFR1, nor in any strain that was not derepressed or derepressed in the absence of H_2.

Concurrent screening of Escherichia coli HB101 (pHU52) revealed immunological homology betweeen the uptake hydrogenase of several strains of E. coli and R. japonicum hydrogenase. E. coli MBM7061 lacked Hup activity when grown anaerobically or aerobically and failed to cross-react with antibodies against R. japonicum hydrogenase. Anaerobically cultured E. coli MBM7061 expresses formate hydrogen-lyase activity indicating that the hydrogenases involved in hydrogen uptake and formate dependent hydrogen evolution are biochemically distinct.

Arp DJ (1985) Arch. Biochem. Biophys. 237, 504-512.
Harker AR et al (1984) J. Bacteriol. 159, 850-856.
Lambert GR et al (1985) Proc. Natl. Acad. Sci. U.S.A. 82, 3232-3236.

USING H_2 EVOLUTION IN $AR:O_2$ AS A MEASURE OF ELECTRON FLOW THROUGH NITROGENASE

Layzell D.B., G.E. Weagle and D.T. Canvin, Dept.of Biology, Queen's University, Kingston, Ont., K7L 3N6, Canada

INTRODUCTION
In whole plant studies, the use of H_2 evolution in $Ar:O_2$ as a measure of electron flow through nitrogenase has not been fully char-acterized. In this study, changes in H_2 evolution were examined in five legume symbioses following transfers between $N_2:O_2$ and $Ar:O_2$ atmospheres.

METHODS
Five legume species were inoculated with effective Hup$^-$ Rhizobium strains and grown N-free in silica sand. At mid vegetative growth, the pots were connected to an open gas exchange system capable of continuously monitoring H_2 (Layzell et al 1984) and CO_2 evolution under gas phases of N_2 or Ar in 21% O_2. Alternatively, 10% C_2H_2 was used to obtain a measure of the C_2H_4 production rate.

RESULTS and DISCUSSION
Following transfer to an $Ar:O_2$ atmosphere, the rate of H_2 evolution from intact legume nodules did not immediately increase to a new steady-state rate. Rather, highly reproducible fluctuations were observed, and these were characteristic of each symbiosis examined. Since the acetylene reduction rate of soybean or lupin did stabilize following exposure to 10% C_2H_2, the instability of H_2 evolution in $Ar:O_2$ seem to be a different phenomenon from the C_2H_2-induced decline reported previously (Minchin et al,1983).
In soybean, lupin, pea and alfalfa, exposure to an $Ar:O_2$ atmosphere inhibited the subsequent rate of H_2 evolution in $N_2:O_2$ to 58 to 88 % of the inital rate in $N_2:O_2$. This suppressed rate recovered within 30 to 60 min following reexposure to $N_2:O_2$.
CO_2 evolution of nodulated roots also declined following transfer to an $Ar:O_2$ atmosphere. The magnitude of the decline was similar to that expected from the theoretical cost of NH_3 assimilation and N transport (1.0 CO_2 evolved / NH_3). In soybean, the fluctuations in H_2 evolution under $Ar:O_2$ were reflected as smaller variations in CO_2 evolution, suggesting some relationship to energy metabolism.
These fluctuations are not easily explained with our present understanding of nodular N_2 fixation. However, they may reflect aspects of in vivo nitrogenase regulation which are not yet fully appreciated. In addition, the instability of H_2 evolution in $Ar:O_2$ may result in inaccuracies in estimates of electron flow through nitrogenase.

REFERENCES
Layzell DB et al (1984) Plant Physiol. 75, 585-585.
Minchin FR et al (1983) J. Exp. Bot. 34, 156-162.

ACKNOWLEDGEMENTS
Supported by a NSERC (Canada) grant (DBL) and a DSS (Agric. Canada) contract and travel supported by the School of Graduated Studies and Research and the Faculty of Arts and Science, Queen's University.

EXAMINATION OF HUP EXPRESSION BY SOYBEAN BRADYRHIZOBIA BELONGING TO DIFFERENT SEROGROUP PHENOTYPES

Peter van Berkum, Harold H. Keyser, Deane F. Weber
USDA-ARS, BARC-W, Beltsville, MD 20705, U.S.A.

1. INTRODUCTION

Eisbrenner,Evans (1983) reported that 25% or less soybean bradyrhizobia oxidize H_2. Hup appears to be predominantly associated with strains of the 110 and 122 serogroups. Keyser et al. (1984) reported a low frequency of Hup[+] phenotypes among other serogroups. However, the Hup characterization used free-living cultures under conditions for expression of ex planta nitrogenase activity and tritium exchange. Emerich et al. (1980) reported no measurable isotope exchange with bradyrhizobial Hup, and Robson, Postgate (1980) suggested the presence of nitrogenase activity to cause an apparent exchange reaction. Therefore, we have used amperometry with bacteria and bacteroids to evaluate selected isolates for Hup.

2. PROCEDURE

The origin maintainance and purity checks of cultures, preparation of inoculum, plant culture, and C_2H_2 reduction were as described by van Berkum, Keyser (1985). H_2 was determined by gas chromatography and amperometry (Hanus et al. 1980). Chemoautotrophic growth was determined in liquid medium (Hanus et al. 1979). Cell suspensions in buffer of heterotrophically grown bacteria were subjected to Hup measurement according to Maier (1981).

3. RESULTS

Fifty two isolates of eight serogroups were selected according to their reported Hup phenotype. All were effective, evolving H2 except for the Hup+ 110 and 122 serogroups. Eleven strains (21%) were Hup[+] as bacteroids and bacteria, although 33 (63%) should have been Hup[+]. Our analysis identified wild-type Hup[+] 6,31 and 123 serogroup phenotypes, and wild-type Hup[-] 110 and 122 serogroup phenotypes. Chemoautotrophic growth was observed by the Hup[+] 6 and 31 isolates, but not by the Hup[+] 123 strains.

4. CONCLUSIONS

Our results show that most of the Hup[+] survey isolates did not oxidize H_2 as bacteroids or as cell suspensions in buffer. The disparity in the results between the two studies may indicate a lower incidence of strains in soils across the USA capable of forming Hup[+] symbioses.

5. REFERENCES

Eisbrenner G and Evans HJ (1983) Ann Rev Plant Physiol 34,105-136
Emerich DW et al. (1980) Plant Physiol 66,1061-1066
Hanus FJ et al. (1979) Proc Natl Acad Sci USDA 76,1788-1792
Hanus FJ et al. (1980. Methods Enzymol 69,731-739
Keyser HH et al. (1984) Appl Environ Microbiol 47,613-615
Maier RJ (1981) J Bacteriol 145,533-540
Robson RL and Postgate JR (1980) Ann Rev Microbiol 34,183-207
van Berkum P and Keyser HH (1985) Appl Environ Microbiol 49,772-777

6. ACKNOWLEDGEMENTS

We wish to express our gratitude to Dr R Hauck for the gift of an amperometric chamber, and to Dr R Maier and Mr F Moshiri for useful discussions concerning Hup measurement.

NITROGENASE ACTIVITY AND CARBON SUPPLY IN *FRANKIA* AFTER ADDITION
OF AMMONIUM TO THE HOST PLANT *ALNUS INCANA*

KERSTIN HUSS-DANELL
DEPT OF PLANT PHYSIOLOGY, UNIVERSITY OF UMEÅ, S-901 87 UMEÅ, SWEDEN

One clone of *Alnus incana* (L.) Moench was grown in symbiosis with a local source of *Frankia* or with *Frankia* ArI4. Seven to nine weeks old plants were given 20 mM NH_4Cl (20 mM KCl=control) for three days. On the first day nitrogenase activity of intact plants decreased to about 80 % of initial rate and then further to about 25 and 10 % on the second and third days, respectively. Control plants were not affected. The hypothesis that decreased nitrogenase activity was due to shortage of energy (assimilates) in *Frankia* was tested in two ways. (i) Measurements of nitrogenase activity in root nodule homogenates (in vitro measurements) indicated loss of active nitrogenase rather than shortage of energy in *Frankia*. (ii) Shoots were exposed to $^{14}CO_2$ and translocation of ^{14}C to *Frankia* vesicles in root nodules was studied. *Frankia* vesicles from ammonium treated plants contained about half as much ^{14}C as those of control plants during all three days studied. It is suggested that reduced supply of carbon to (part of) *Frankia* vesicles in the root nodules may have caused reduced protein synthesis, including synthesis of nitrogenase. However, any changes in nitrogen metabolism of the nodules should not be excluded as a possible explanation for the observed effects.

TISSUE CULTURE OF TWO ACTINORHIZAL SPECIES:
COWANIA STANSBURIANA AND COWANIA SUBINTEGRA

JUDY L. JAKOBEK/RALPH A. BACKHAUS/JEAN C. STUTZ
DIVISION OF AGRICULTURE, ARIZONA STATE UNIVERSITY, TEMPE, AZ 85287 USA

Cowania stansburiana Torr. and Cowania subintegra Kearney are
actinorhizal members of the family Rosaceae (Righetti, Munns 1980 and
Perry et al. 1985). Recently there has been interest concerning the
possibility of expanding nitrogen fixation capability to economically
important non-nitrogen fixing plants. One of the most promising
techniques through which this goal may be realized is through somatic
hybridization by protoplast fusion of cells from nitrogen fixing and
non-nitrogen fixing plants. Cowania stansburiana and C. subintegra may
be good candidates for somatic hybridization because they are nitrogen
fixing plants which are related to many agriculturally important
members of the Rosaceae. Essential to the success of this procedure is
the development of a cell and callus culture system and demonstration
of the totipotentiality of the cultured cells. Presented here are the
first applications of cell and callus culture techniques to
actinorhizal members of the Rosaceae.

SUMMARY OF RESULTS

Callus was initiated when tissue explants from various sources,
including shoot proliferation cultures (Jakobek et al. 1985) or newly
germinated seedlings, were transferred to medium containing NAA.
Shoots regenerated from callus transferred to media with 0, 0.1, or 0.3
NAA. Habituated cell lines were also obtained which grew indefinitely
on hormone free medium, and several lines regenerated shoots and
somatic embryoids. Cell suspension cultures were initiated from
callus, habituated and NAA maintained, as well as from 2 mm pieces of
seedling explants. Calli were regenerated from suspension cultures
plated onto solidified medium, but growth was limited. Suspensions of
habituated cells produced somatic embryoids in vitro. The system
developed makes possible culture of cells in vitro and demonstrates
their totipotency.

REFERENCES
Jakobek JL et al (1985) Micropropagation of Rosaceous Actinorhizal
Species: Cowania stansburiana and Cowania subintegra. In prep.
Perry CB et al (1985) Proc. 6th Inter. Symposium on Nitrogen Fixation.
In press.
Righetti TL and Munns DN (1980) Plant Physiol. 65, 411-412.

ACKNOWLEDGEMENTS: This research has been supported by NSF grant PCM-
840078.

IN VITRO MICROGRAFTING OF ACTINORHIZAL DESERT SHRUBS

NANCY E. KYLE/DR. TIMOTHY L. RIGHETTI
DEPARTMENT OF HORTICULTURE, OREGON STATE UNIVERSITY, CORVALLIS, OR 97331

Several genera of the Rosaceae are capable of fixing nitrogen in symbiosis with Frankia. Many commercial crops are in the Rosaceae, but do not fix nitrogen. All of these tree crops are grafted in efforts to combine desirable characteristics of rootstock and scion. The possibility exists of introducing nitrogen-fixing capability into these crops by grafting them onto nitrogen-fixing rootstocks. Although efficiency differs, the rosaceous shrubs Cowania mexicana, Purshia glandulosa and P. tridentata nodulate and fix nitrogen. The closely related species, Fallugia paradoxa, does not. An in vitro micrografting procedure has been developed to obtain all possible scion-rootstock combinations of these four species. Although considerable barriers to future commercial applications exist, techniques for elucidating the physiology of scion-rootstock interactions with regard to nitrogen fixing capability and efficiency in rosaceous genera have been developed (Kyle 1984). Grafted plants can be transferred to soil in the greenhouse with better than a 40% success rate.

Nodulation in the Rosaceae is slow. Seedlings generally take 3 months to nodulate in the greenhouse (Dalton, Zobel 1977). Nodulation trials of micrografted plants are still underway. To date, we have produced nodules on 3 plants with P. tridentata scions, 1 on a C. mexicana and 2 on P. glandulosa rootstocks. These plants are green and vigorous compared to non-nodulated plants growing in the same containers. It is not surprising that these plants should nodulate, since both graft partners nodulate on their own. However, the nodules on these plants indicate that micrografting per se does not inhibit nodulation. Further nodulation studies are in progress.

Dalton DA and Zobel DB (1977) Plant and Soil 48, 57-80.
Kyle NE (1984) Micrografting between N-Fixing and Non-N-Fixing Genera
    of the Rosaceae. M.S. Thesis, Arizona State University

ASPARAGINE AND GLUTAMINE AS NITROGEN SOURCES CONTROLLING DEVELOPMENT AND
NITROGENASE ACTIVITY OF <u>FRANKIA</u> SP. STRAIN HFP CpI1

HAYES LAMONT, JOHN G. TORREY, and PAT YOUNG

CABOT FOUNDATION, HARVARD UNIVERSITY FOREST,
PETERSHAM, MA 01366, USA

<u>Frankia</u> sp. strain HFP CpI1, a N_2-fixing actinomycete from the root
nodules of <u>Comptonia</u> <u>peregrina</u> (sweet fern), was cultured in shaken
liquid media containing mineral nutrients, an organic buffer system,
biotin, 5 mM propionate, and 5 mM L-amino acid from among arginine,
aspartate, asparagine, citrulline, glutamate, and glutamine. Harvested
17 days after inoculation, cultures were assayed for nitrogenase activity (reduction of C_2H_2), soluble cell protein, concentration of vesicles,
and concentration of sporangiospores releasable by sonication.

Both asparagine and glutamine, and none of the other four amino
acids, supported protein yields of about 17 µg ml^{-1} (6.1 doublings),
equal to the yield with 5 mM NH_4Cl. Sporangia and releasable sporangiospores were much less abundant in the glutamine culture than in any
other culture with protein yield exceeding 3 µg ml^{-1}. The only cultures
having nitrogenase activity were a control without combined N, the arginine culture, and the citrulline culture; these were rich in vesicles
and yielded 6 to 9 µg protein ml^{-1}. The glutamine culture had few vesicles, but the asparagine culture, despite its lack of nitrogenase activity, had them in abundance.

We retested asparagine and glutamine at 2.5 mM in a time-course experiment, harvesting after 4, 7, 11, and 17 days of growth. The glutamine
cultures had zero nitrogenase activity at all harvest times; so also did
the control NH_4^+ cultures. The asparagine cultures, however, reduced
acetylene as early as day 4, after which activity rose to a peak of 1.53
nmol mg^{-1} min^{-1} at about day 7 and then fell, being close to zero by day
17. The asparagine growth curve was exponential (k = 0.24 day^{-1})
through all harvest times despite the extreme variation of nitrogenase
activity. As in the first experiment, control cultures lacking combined
N displayed higher and more persistent nitrogenase activity, and slower
growth, than did the asparagine cultures.

Our results show that CpI1 can use L-glutamine as a N source and that
glutamine, like NH_4^+, suppresses vesicle formation and biosynthesis of
active nitrogenase in CpI1. The suppressive effect is consistent with
the view of some researchers that glutamine is a more direct regulator
of nitrogenase biosynthesis than NH_4^+. Our results also show that
glutamine favors growth of vegetative hyphae with few sporangiospores.

From our evidence, it appears likely that CpI1 utilizes L-asparagine.
Derepression of one or more enzymes of asparagine metabolism could have
led to a gradual increase of intracellular glutamine concentration such
that nitrogenase was eventually inhibited and/or repressed. Furthermore, given that the C:N ratio (molar) of the asparagine medium was only
3.5 in the first experiment and 5.0 in the second, net production of
NH_4^+ may have gradually raised the intracellular concentrations of NH_4^+
and of glutamine to levels sufficient to inhibit and/or repress nitrogenase.

PURIFICATION AND PROPERTIES OF TREHALASE IN FRANKIA ARI3

MARY F. LOPEZ[1]/JOHN G. TORREY[2]
[1]DEPARTMENT OF BOTANY, UNIVERSITY OF MASSACHUSETTS, AMHERST, MA 01003, USA
[2]CABOT FOUNDATION, HARVARD UNIVERSITY, PETERSHAM, MA 01366, USA

Trehalase was purified from cultures of Frankia strain ArI3 grown
on media with or without NH_4Cl. The purified enzyme was specific for
trehalose, exhibited a broad pH optimum of pH 4.5 to 5.3 and had a Km
for trehalose of 4.2 mM. The trehalase was inhibited in vitro completely
by sucrose, glucose and mannose and partially by mannitol and sorbitol.
In addition to the specific trehalase, a mixture of non-specific α and β-
glucosidases which exhibited some activity with α,α-trehalose as a substrate were also partially purified in Frankia extracts made form nitrogen-fixing cells. These enzymes were not detected in the purifications
of crude extracts made from non-nitrogen-fixing cells (grown on media
supplemented with NH_4Cl). Trehalase activity in crude extracts increased
over time when cells were induced to fix nitrogen, and the maximum specific activity of trehalase from nitrogen-fixing cultures was 4 times the
maximum activity from non-fixing cultures. Trehalase activity was also
examined in crude extracts made from Frankia vesicle clusters isolated
from Alnus rubra nitrogen-fixing nodules infected with ArI3. The maximum activity of trehalase in these clusters was 6-7 times greater than
in the nitrogen-fixing pure cultures of ArI3 and 26-33 times greater
than the non-fixing pure cultures.

VARIATION IN NITROGEN FIXATION AMONG POPULATIONS OF FRANKIA AND CEANOTHUS IN ACTINORHIZAL ASSOCIATION.

DAVID L. NELSON AND CARLOS LOPEZ, USDA FOREST SERVICE, INTERMOUNTAIN RESEARCH STATION, SHRUB SCIENCES LABORATORY, PROVO, UT 84601 U.S.A.

Improvement of western United States wildland shrubs is required by increased demand for disturbed land reclamation and modification and improvement of livestock range and wildlife habitat. Selection of superior nitrogen fixation characteristics for actinorhizal shrubs could improve plant productivity in wildland ecosystems. Experimental evidence supports existence of variable nitrogen-fixing activity of pure-culture isolates of Frankia. The objective of this study was to demonstrate possible variation in nitrogen fixation rate of actinorhizal associations among natural populations of Frankia and Ceanothus.

Ceanothus velutinus seed and soil collected from five sites in Idaho and Washington were used in reciprocal combination in greenhouse culture to form actinorhizal associations. Nitrate was initially added to the test soil to encourage uniform seedling establishment. Thereafter, nitrate was withheld and all other essential nutrients added periodically. A completely randomized split-plot experimental design was used with up to 33 plants per accession per soil source. At one year plant age, soil was washed from roots, and plants with nodules intact were tested. Nitrogen fixation rate was measured by acetylene reduction and gas chromatography.

Original soils ranged from pH 5.8 to 6.8 and amended test mixtures varied about 0.8 pH units. Sulfur, nitrate and phosphate levels were variable and typical for mountainous soils. Vesicular-arbuscular mycorrhizae occurred on 63.0 to 98.2% of the plants. More than 90% of the plants became nodulated in all soils except one. There was a significant difference in the nitrogen fixation rate of nodulated plants among different soil sources but no significant difference between plant accessions. There appeared to be significant interaction between Ceanothus accessions and soil source for acetylene reduction rate. Number of nodules per plant, nodule weight, plant weight, and root-shoot weight ratio had no clear relationship to acetylene reduction rate.

In this study variability in nitrogen fixation rate of actinorhizal associations appeared to be more a function of Frankia, the endophyte, than Ceanothus, the host. An approximate four-fold difference in acetylene reduction rate among actinorhizal populations derived from the various soil sources suggests a potential for improvement. For use in wildland settings, it should be more biologically sound to select for superior nitrogen fixing Frankia populations than for superior pure culture strains.

RESPIRATION BY SYMBIOTIC *FRANKIA* IN ROOT NODULE HOMOGENATES PREPARED FROM *ALNUS INCANA*

PER-ÅKE SELANDER AND KERSTIN HUSS-DANELL
DEPT. OF PLANT PHYSIOLOGY, UNIVERSITY OF UMEÅ, S-901 87 UMEÅ, SWEDEN

Respiration by the actinomycete *Frankia*, from root nodules of *Alnus incana* (L.) Moench, was measured with an oxygen electrode. Endophyte vesicles were prepared anaerobically from root nodule homogenates by a filtration procedure. O_2-uptake measurements were made in a hypotonic medium to decrease the possibility of respiration by contaminating host cell mitochondria. The medium also contained KCl and EDTA to decrease the possibility for contaminating enzymes on the vesicle surface. Respiration was defined as oxygen consumption sensitive to inhibitors of electron transport, cyanide or antimycin a. A number of different sugars, amino acids, carboxylic acids and lipids were tested as carbon sources. Respiration was supported by three different substrates; NADH, glucose-6--phosphate together with NAD and finally malate together with NAD and glutamate. As the vesicle membranes might be "leaky", further investigations are needed before the physiological significance of these effects can be evaluated.

BIOMASS PRODUCTION AND NITROGEN UTILIZATION BY *ALNUS INCANA* WHEN GROWN ON N_2 OR NH_4^+.

ANITA SELLSTEDT AND KERSTIN HUSS-DANELL
PLANT PHYSIOLOGY, UNIVERSITY OF UMEÅ, S-901 87   UMEÅ, SWEDEN

Cuttings of one clone of alder were rooted and divided into two groups. One group was inoculated with *Frankia* inoculum and fixed atmospheric nitrogen. The other group was not inoculated but received ammonium at the same rate as the first group fixed their nitrogen. Nitrogen fixation was calculated from frequent measurements of acetylene reduction and hydrogen evolution. This calculation is relevant as, this *Frankia* inoculum did not show any detectable uptake of hydrogen.

Alders fixing nitrogen developed longer shoots, larger leaf areas, more biomass, and contained more nitrogen than alders receiving ammonium.

Alders fed with ammonium excreted part of the nitrogen in an organic form while alders inoculated with *Frankia* did not excrete any detectable amount of nitrogen.

It seems that the energy demand of nitrogen fixation is not so high that biomass production in alders is retarded. It is concluded that alders fixing nitrogen produced more biomass partly due to a better utilization of their nitrogen.

HEME CONTENT AND DIFFUSION LIMITATION OF RESPIRATION AND NITROGENASE ACTIVITY IN NODULES OF CASUARINA CUNNINGHAMIANA

J. D. TJEPKEMA and M. A. MURRY[+]
DEPARTMENT OF BOTANY, UNIVERSITY OF MAINE, ORONO, ME 04469, U.S.A.
[+]CABOT FOUNDATION, HARVARD UNIVERSITY, PETERSHAM, MA 01366, U.S.A.

INTRODUCTION We compared the sensitivity to O_2 of Frankia isolate CcI3, a Casuarina isolate, in symbiosis within Casuarina nodules with previous results for its sensitivity to O_2 in culture (Murry et al., 1985). We also compared the heme content with that of other actinorhizal nodules.

In Casuarina nodules the walls of the infected cells contain a suberin-like material that may restrict O_2 diffusion into the cell (Berg, 1983). In contrast to other actinorhizal plants Frankia does not form vesicles in Casuarina nodules. Vesicles probably protect nitrogenase from O_2 inactivation. In culture, Frankia isolate CcI3 does not form vesicles under microaerobic conditions but does form them under aerobic conditions.

RESULTS AND DISCUSSION Atmospheric values of pO_2 were required to maximize C_2H_2 reduction rates in nodules induced by Frankia strain CcI3, whereas maximum activity in avesicular cultures occurred only at very low pO_2. This result is evidence for a diffusion barrier in the nodule, such as suberized cell walls, that greatly restricts O_2 diffusion to the endophyte.

Table 1. Heme contents of nodules and suberization of host cells.

Species	Total heme (nmol g^{-1} fresh wt)	CO-reactive heme (nmol g^{-1} fresh wt)	C_2H_4 (μmol $h^{-1}g^{-1}$ fresh wt)	Suber- ization
Casuarina cunninghamiana	97 + 3	44 + 5	24.3 + 2.0	+
Myrica gale	107 + 19	26 + 7	38.5 + 3.2	+
Alnus rubra	28 + 2	3.2 + 0.3	46.4 + 3.1	−
Datisca glomerata	15 + 1	0	19.4 + 2.2	−

Suberization was correlated with high concentrations of total and CO-reactive heme, which might facilitate O_2 transport within the diffusion barrier. In genera where there is no suberization and relatively little heme there may be little restriction of O_2 diffusion to the endophyte. In this case the vesicle wall of the endophyte may be the primary means of protecting nitrogenase from excessive O_2.

Berg RH (1983) Can. J. Bot. 61, 2910-2918.
Murry MA et al (1985) Can. J. Microbiol., in press.

INHIBITION OF UPTAKE HYDROGENASE BY ACETYLENE IN ACTINORHIZAL NODULES

LAWRENCE J. WINSHIP and KENDALL J. MARTIN
HAMPSHIRE COLLEGE, AMHERST, MA 01002   USA

Introduction

The reduction of dinitrogen to ammonia by nitrogenase is normally accompanied by the production of hydrogen gas. However, no hydrogen is evolved by healthy, nitrogen-fixing actinorhizal nodules, due to the activity of an uptake hydrogenase (Hup). (Eisbrenner, Evans 1983) In nodules of Alnus rubra infected by Frankia strain ArI3 the uptake hydrogenase was irreversably inactivated by acetylene (Winship, Tjepkema 1984). so that large amounts of hydrogen were evolved when the nodules were returned to an acetylene-free atmosphere. While this type of inhibition has been reported for other nitrogenase-containing systems, its occurance in symbiotic associations varies greatly (Arp 1985). The objective of this study was to screen actinorhizal symbioses for acetylene-sensitive uptake hydrogenase.

Materials and Methods

Nodules from plants grown in the greenhouse and in the field were excised and placed in plastic 20 ml syringes. The syringes were flushed with the appropriate assay gas mixture, then an initial sample was injected into a Carle Model III gas chromatograph equipped with dual columns (Porapak T and MS 5A) and dual thermal conductivity detectors, able to measure carbon dioxide, ethylene, acetylene, hydrogen, oxygen, and nitrogen concentrations. After incubation at 24C for 30 minutes, a final sample was taken and the syringe flushed with the next assay gas.

The sequence of gas mixtures used for each nodule sample was;
1. air, 2. argon/oxygen (79:21), 3. argon/oxygen/acetylene (69:21:10), 4. argon/oxygen (79:21), and 5. air. Rates were calculated from the differences in peak height, calibrated with known gas mixtures.

Results

Two patterns emerged. I. Nodules of Alnus rubra x ArI3, of Alnus rugosa x soil, and of Eleagnus angustifolia infected with both soil and Frankia strain WgCCi17 showed the expected increase in hydrogen evolution after exposure to acetylene (in gas mixtures 4 and 5). II. Nodules of Datisca glomerata and of Ceanothus americana, both inoculated with soil, showed no change in hydrogen evolution following incubation with 10% v/v acetylene. While specific activities of nodule samples varied, the patterns remained consistent throughout the survey.

References

Arp DJ (1985) in Ludden PW and Burris JE, eds, Nitrogen Fixation and CO$_2$ Metabolism, p 121.
Eisbrenner G and Evans HJ (1983) Ann. Rev. Pl. Physiol. 34, 105.
Winship LJ and Tjepkema JD (1984) Can. J. Bot. 62, 1602.

ASSOCIATIVE N₂-FIXATION BETWEEN <u>KLEBSIELLA</u> SP. AND RICE (<u>ORYZA</u> <u>SATIVA</u> <u>L</u>.)

JARIYA BOONJAWAT AND JIRAPORN LIMPANANONT, CHULALONGKORN UNIVERSITY, BANGKOK 10500, THAILAND

R15 and R17 are heterotrophic bacteria isolated from the rhizosphere of tropical rice grown in acid and arid soils of Thailand through the use of acetylene reduction(AR) assay(Boonjawat et al,1983). Both belong to the Genus <u>Klebsiella</u> (A. Choonhahirun,J. Boonjawat,unpublished). The question, whether these bacteria enter or merely attach to the root surface has led us to stain the living bacteria in acridine orange dye(1:10,000) and follow their adhesion on rice root. <u>Klebsiella oxytoca</u> 1301 was used as control for positive association, and <u>E. coli</u> K12 for negative association. Rice seeds used are Hua-cho-chi-momor(HCCMM),IR42,IR58 from IRRI, and RD5 from the Department of Agriculture, Thailand. Inoculation of 2.7×10^7 bacterial cells per 3 rice seedlings on day 7th after germination resulting in clumping of K. <u>oxytoca</u> 1301,R15 and R17 on the root surface at 2h, and formation of micronodules at 36h after inoculation. Observation under SEM revealed the presence of numerous spherical structures with uniformity in size(10-15 M^{-6} in diameter) as shown in <u>Fig.1</u>. The cross-section (<u>Fig.2</u>) showed aggregated bacteria packed in membrane bound micronodule, that strongly associated with the root cell. Concurrent AR assay showed that significant C_2H_2 formation in the head space of 100 1^{-3} tube at the concentration higher than 4 Mol^{-6} 1^{-3} could be detected at 3 days after inoculation only when high density of micronodules were observed. R17 is the best colonizer followed by R15 and K. <u>oxytoca</u> as inoculated into RD5,HCCMM, IR58 and IR42 seedlings with respect to acetylene reduction activity.

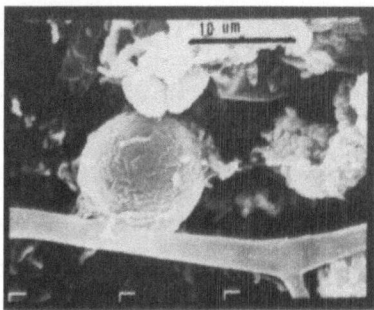 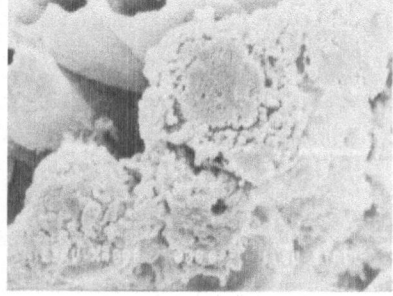

Figure 1. Micrnodule formation.   Figure 2. Cross-section of the
                                              micronodule.

REFERENCE
Boonjawat et al (1983) In Vacharotayan S and Takai Y, eds,
Comparative Ecological Study on Nitrogen Economy of Paddy Soil between
Tropical and Temperate Regions, pp 135-145,
NODAI Research Institute, Japan.

ACKNOWLEDGEMENTS
The authors wish to express their gratitude to all who assisted them in preparing this manuscript.

NITROGEN-FIXATION AND DENITROFICATION BY AZOSPIRILLUM GROWN IN FREE-LIVING CULTURES AND IN ASSOCIATION WITH WHEAT

H. Bothe, W. Zimmer, G. Danneberg, A. Kronenberg, G. Neuer
Botanical Institute, The University of Cologne, Gyrhofstr. 15, D-5000 Köln 41, F.R.G.

Introduction

Azospirillum can live in association with grasses and may enhance plant growth by providing products of N_2-fixation and/or phytohormones to the plants. Azospirillum has also been described to perform denitrification. We have recently performed model experiments with wheat-Azospirillum associations (Neuer et al. Arch. Microbiol. 141, 364, 1985). For this, Azospirillum and germinated wheat seeds were grown for a week in semi-solid agar medium containing solely inorganic salts. The association was then assayed for N_2-fixation (C_2H_2-reduction) and denitrification (N_2O-formation) activities. These results will be summarized here. In addition, some biochemical and physiological properties of denitrification by Azospirillum will be presented.

Results

A. Properties of the wheat-Azospirillum association

The data published in detail in Neuer et al. (Arch. Microbiol. 141, 364, 1985) can be summarized as follows:

a) C_2H_2-reduction and N_2O-formation activities are strictly dependent on the presence of both plants and Azospirillum.

b) Both activities can readily be demonstrated 3-5 h after the removal of air out of the assay flasks (actual concentration of $O_2 = 0.5 - 2\%$ in the gas phase of the vessels).

c) The Azospirillum strain used determines the rates of N_2O-formation and C_2H_2-reduction in the association.

d) The number of plants, the pH, the concentration of C_2H_2 and particularly the temperature are critical for optimal activities in these experiments.

B. Physiological and biochemical properties of denitrification by Azospirillum

Azospirillum grows anaerobically with nitrate as sole respiratory electron acceptor. Nitrate is converted to either nitrite, N_2O or N_2, depending on the strain used. Previous experiments (Zimmer et al., Arch. Microbiol. 138, 206, 1984 and Stephan et al. Arch. Microbiol. 138, 212, 1984) have shown that Azospirillum can also grow with nitrite or nitrous oxide as respiratory electron acceptor. Growth with nitrate or nitrite suppresses nitrogenase biosynthesis. In contrast, Azospirillum synthesizes nitrogenase when growing anaerobically with N_2O. The molar growth yields are comparable for the growth with nitrite and O_2, but are approx. 2/3 lower when nitrate is used as the respiratory electron acceptor (Zimmer and Bothe, in prep.). Dissimilatory nitrite reductase is a cytochrome c,d containing enzyme. The biosynthesis of nitrous oxide reductase requires Cu (G. Danneberg, W. Zimmer, H. Bothe, In: Azospirillum III, W. Klingmüller (ed), Springer, Berlin, in press).

Conclusion

Azospirillum is probably the most versatile organism with respect to the nitrogen metabolism. The model experiments show that Azospirillum performs either N_2-fixation or denitrification, depending on the amount of O_2 and nitrate available. Similar events may occur in nature particularly under tropical conditions.

THE CONTRIBUTION OF SYMBIOTIC NITROGEN FIXATION
TO THE NITROGEN BUDGET OF SHIPWORMS

BARBARA J. KAMICKER/SCOTT GALLAGER/JOHN B. WATERBURY
WOODS HOLE OCEANOGRAPHIC INSTITUTION, WOODS HOLD, MA 02543 USA

A newly discovered bacterium is the first microorganism known to degrade cellulose and to fix nitrogen. The bacterium lives symbiotically in the gland of Deshayes in the gills of teredinid molluscs (shipworms). It has been isolated in pure culture and will grow in a simple mineral medium containing cellulose and no source of combined nitrogen. The bacterium's unique combination of physiological properties almost certainly plays an important role in the metabolism of shipworms. We measured the acetylene reduction activity of the bacteria in shipworms to determine the contribution of symbiotic nitrogen fixation to the nitrogen budget of the host.

The acetylene reduction assay was done on intact animals in wood or on gills that were removed from the animals and homogenized in sterile seawater. We calculated the amount of nitrogen fixed from acetylene reduced using the ratio of 3 moles of N_2 fixed per mole of acetylene reduced. Because shipworms have no net excretion of nitrogenous compounds, we were able to calculate how much of the fixed nitrogen could be incorporated into the host.

Animals that did not receive phytoplankton supplements were smaller and fixed nitrogen at about 1/10 the rate of animals that received phytoplankton. As the shipworms grew, phytoplankton enhanced the rate of nitrogen fixation in 23 and 28 day old animals. However, at 39 days post-settlement, the rates of nitrogen fixation were the same regardless of the animals' diet. Shipworms grown in wood in flowing coastal seawater fix less nitrogen per mg dry weight (dw) as they grow. This affects how much fixed nitrogen is contributed to the nitrogen budget of the shipworm. The animals may derive 15-22% of their nitrogen requirement from symbiotic nitrogen fixation. The nitrogen in phytoplankton and dissolved organic matter probably supply the rest of the nitrogen requirement. Shipworms grown in wood in filtered, autoclaved seawater fix nitrogen at about the same rate for 6 months until they become nutrient limited and die. They are able to supply most of their nitrogen requirement from symbiotic nitrogen fixation.

Shipworms fix nitrogen at different rates during growth which appears to be related to the availability of phytoplankton and to their physiological condition. Phytoplankton are used by very young animals to help them grow quickly and to establish an effective symbiosis. Even though phytoplankton-fed older animals grow larger, they fix less nitrogen than shipworms that did not receive a phytoplankton supplement. Shipworms can reach sexual maturity when they use wood as their sole nutrient source for up to 9 months by relying heavily on the bacterial symbiont to degrade cellulose and to fix nitrogen.

We have shown that the bacterium may supply 15-22% of the animal's nitrogen requirement under natural growth conditions. Under stressed conditions, i.e., using wood as the sole food source, the symbiont can supply nearly all of the nitrogen needs of the shipworm. Thus, it appears that the shipworm bacterium is the first known nitrogen-fixing symbiont in an animal to contribute a significant amount of combined nitrogen to its host.

POSTER DISCUSSION 5 PHYSIOLOGY OF NITROGEN-FIXING SYMBIOSIS

Robert Klucas, Department of Agricultural Biochemistry, University of Nebraska; Lincoln, NE 68583-0718

The posters in this session were truly international in that they represented research from 17 counties and covered a wide range of symbiotic systems involving bacteria, actinorhiza and cyanobacteria. Because of the large number (78) of posters, it was not possible to discuss each poster separately so posters were grouped into topics such as (1) structure and nodulins (2) nitrogen metabolism (3) carbohydrate metabolism (4) factors affecting nodulation (5) hydrogen evolution and uptake (6) properties of the actinorhizal system (7) oxygen transport and effects and (8) associative symbiotic relationships. Six individuals were invited to present synopses of research in their laboratories.

Dr. Birgitta Bergman of the University of Uppsala described research on cycads which are gymnosperms harboring nitrogen-fixing cyanobacteria (Nostoc) in a special type of root. The cyanobacterial filaments are localized intercellularly in a distinct zone along the root. Cross sections of these roots revealed that heterocyst frequency increased gradually toward the older parts of the root and that nitrogenase and glutamine synthetase activities were higher at the growing tip and lower in older parts of the roots. Apparently, the cyanobacteria in the cycad-Nostoc symbiosis show characteristics similar to free-living cyanobacteria or to typical symbiotic cyanobacteria depending on the developmental stage of the symbiosis.

Dr. Paul Reynolds of DSIR, New Zealand discussed the enzymology of carbon supply for purine biosynthesis in the synthesis of ureides in soybean nodules. Phosphoserine aminotransferase (PSAT) was detected in soybean nodules and increased in specific activity during nodule development. This enzyme is proposed to be involved in a pathway that moves carbon via phosphoglycerate dehydrogenase and serine hydroxymethyltransferase for purine synthesis. PSAT was localized in the proplastid fraction which is the site of other enzymes involved in purine synthesis.

Dr. D. B. Layzell of Queens University presented evidence which indicates that a major role for dark CO_2 fixation in soybean nodules is to provide C skeletons for bacteroid metabolism in support of N_2 fixation. He also reported that unexpected fluctuation occurred in H_2 evolution rates when legume nodules were exposed to an Ar: O_2 atmosphere. These fluctuations and instabilities may cause inaccuracies in estimates of election flow through nitrogenase when H_2 evolution in Ar: O_2 is used.

Professor D. Werner of the University of Marburg discussed research in his laboratory on the biogenesis and function of the peribacteroid membrane in nodules. He presented evidence that the peribacteroid membrane is a physiological barrier and that a second form of choline kinase is induced only in peribacteroid membrane-building nodules. This choline kinase may be a nodulin. The particle density of peribacteroid membranes was investigated by Professor Werner using a freeze-factoring technique.

Dr. J. Tjepkema of the University of Maine summarized his work on hemoglobins in actinorhizal nodules. Hemoglobin appears to be totally absent from Datisca glomerate nodules and total heme is low which leads to the conclusion that hemoglobin is not an absolute requirement in actinorhizal nodules. Myrica gale and Casuarina cunninghamiana nodules do possess hemoglobin which may function as in legume nodules for the transport of O_2 within a diffusion barrier. The function of hemoglobin in Alnus nodules is not known because no diffusion barrier has been identified except for the cytoplasm of the infected cells.

Dr. J. Wittenberg discussed the kinetics of hemoglobins which supply oxygen to intracellular prokarotic symbionts. On the basis of data from hemoglobins isolated from legume, Parasponia and actinorhiza nodules as well as a hemoglobin from certain bivalve mollusc, it appears that the very high oxygen affinity of these hemoglobins is achieved by very rapid combination with oxygen coupled with a modest rate of dissociation.

Dr. John Witty of Rothamsted Experimental Station presented direct evidence for a variable barrier to gas diffusion into functional pea and french bean nodules. Oxygen concentrations were measured within a nodule with an O_2-specific microelectrode inserted into the nodule. In air, the O_2 concentration in the nodule was below 1 μM, but when the nodule was exposed to an elevated level of O_2, the O_2 concentration in the nodule transiently increased and then returned to the former concentration.

PHYSIOLOGY AND ECOLOGY OF
NITROGEN-FIXING MICROORGANISMS

INVITED PAPERS

POSTER SUMMARIES:

(i)	Rhizobium - competition.
(ii)	Rhizobium - survival.
(iii)	Rhizobium - general.
(iv)	Associative bacteria.
(v)	Cyanobacteria.
(vi)	Frankia.
(vii)	Miscellaneous.

DISCUSSION GROUP SUMMARY

RECENT ADVANCES IN THE ECOLOGY OF RHIZOBIUM

EDWIN L. SCHMIDT and FRANCOISE M. ROBERT
UNIVERSITY OF MINNESOTA, ST. PAUL, MN 55108 U.S.A.

1.   INTRODUCTION

Microbial ecology deals with the interactions between microorganisms and their environment. These interactions regulate the biochemical transformations carried out by microorganisms in nature and
hence are of great theoretical and practical importance. Information
on the ecology of those microorganisms with the ability to transform
atmospheric nitrogen into biologically useful nitrogen is particularly
critical to the goal of enhancing biological nitrogen fixation. This
review will concentrate on the ecology of the nitrogen-fixing soil
bacteria of the genus Rhizobium prior to its entry into the root of a
crop legume host plant.

2.   DIRECT APPROACHES TO THE ECOLOGY OF RHIZOBIA

The ecology of rhizobia in the soil, and especially in the rhizosphere, is clearly crucial to the establishment and performance of the
legume symbiosis, but is extremely difficult to study. Two major
approaches to the free-living ecology of rhizobia have emerged: the
use of immunofluorescence and the use of antibiotic resistance markers.

2.1 Immunofluorescence (IF). IF was first applied to Rhizobium ecol-
ogy in 1968. The technique is a general one, applicable to autecological study of virtually any bacterium that will elicit a distinctive
immune response. The autecological capability of IF is particularly
useful for study of rhizobia since most ecological questions deal with
some individual strain judged to be of special interest. The use of IF
for direct study of rhizobia in the soil and rhizosphere has expanded
in recent years with greater availability and improved performance of
fluorescence microscopes. Extraction procedures to separate bacteria
from soil particles prior to IF examination have been improved
(Kingsley, Bohlool 1981), but specific protocols are not applicable to
all soils. Use of monoclonal antibodies for IF may be expected to
enhance substantially the strain specificity attainable.

2.2 Antibiotic Resistance Markers. Strains with mutations which result
in antibiotic resistance provide the possibility of autecological study
of that strain in soil. Soil dilution plating on media containing
various antibacterial and antifungal agents has been used to enumerate
spontaneous antibiotic resistant mutants of Rhizobium (Pena-Cabriales,
Alexander 1983). To be useful in ecological studies, antibiotic resistance markers must be stable, unaltered in symbiotic and ecological
attributes, and not shared by indigenous soil bacteria. A recent aute-

cological application of antibiotic resistance/selective plating attained impressive detection sensitivity in the range of 30 rhizobia g^{-1} of soil (Bushby 1981). Robert and Schmidt (unpublished) combined a selective antibiotic medium with an immunoenzymatic assay to detect an antibiotic resistant mutant of R. japonicum serogroup 138 at 50 cells/ g^{-1} of soil. Immunoenzymatic confirmation of 138 colonies was necessary since colonies of other soil bacteria appeared in the plates. Moreover, attempts to apply the technique to a mutant of serogroup 123 with different antibiotic markers was unsuccessful because of high background counts of other bacteria in the same soil.

3.   INDIRECT APPROACHES TO THE ECOLOGY OF RHIZOBIA

In contrast to the few techniques that allow a direct approach to the ecology of rhizobia, there is a large and growing number of techniques for the indirect approach through nodule analysis. Ecological questions that can be addressed indirectly center on the identification of a Rhizobium that has attained its nodule niche. Identification may be achieved by analysis of the nodule directly or by isolation of the Rhizobium and subsequent analysis of the pure culture isolate.

3.1 Strain Identification by Direct Examination of Nodule Contents. Identification of rhizobia directly from nodule preparation without isolation or subculture was first accomplished by the application of serology in the form of agglutination. Increasingly, agglutination is being supplanted by other serological techniques that offer greater resolution and reliability. Immunodiffusion assays using bacteroids directly as antigens have been reported as feasible and sensitive, but have been little used. Most nodule typing for Rhizobium strain currently is by IF or by ELISA (Enzyme-linked immunosorbent assay). Strain identification by IF examination of nodules was used successfully in ecological studies of R. trifolii (Roughley et al. 1976), R. japonicum (Moawad et al. 1984), R. phaseoli (Robert, Schmidt 1983), R. leguminosarum (May, Bohlool 1983), and chick pea rhizobia (Kingsley, Bohlool 1983). Large scale nodule typing by all serological techniques has been simplified since the finding (Somasegaran, et al. 1983) that nodules may be oven-dried for long-term storage. IF techniques are particularly useful in the detection of mixed infections in nodules (May, Bohlool 1983; Moawad et al. 1984). ELISA is capable of serotyping large numbers of nodules, and providing highly specific and highly sensitive strain identification without the need for microscopy. The reliability of ELISA for nodule serotyping has been demonstrated with both fast- and slow-growing strains (Rice et al. 1984). Recent improvements in the technique include use of a fluorescent substrate (Morley, Jones 1980) and substitution of beta galactosidase for alkaline phosphatase (Martensson et al. 1984). Enhancement of sensitivity and a decrease in nonspecific adsorption was attained by replacing goat anti-rabbit antibody with Protein A (Kishinevsky, Maoz 1983). The precision of strain identification by both IF and ELISA will be greatly enhanced by the availability of monoclonal antibodies.

3.2 Strain Identification of Pure Cultures. Serological techniques are reliable for identification of strains in nodules only if there is no cross-reaction with other strains, or if a cross-reaction can be removed by adsorption. Numerous alternatives to serology are available to identify the strain of interest in ecological studies once it is isolated from the nodule. The more commonly used and most promising of these will be mentioned.

Sometimes rhizobia that develop stable resistance to fairly high levels of antibiotics can be used as already noted for direct ecological studies in soil. Antibiotic resistant mutants are more readily detected in the nodule than in soil by plating on antibiotic selective media. Advantage is taken of this in ecological studies to evaluate the success of an antibiotic-marked inoculant strain in competition studies (McLoughlin et al. 1984). In addition many rhizobia are naturally resistant to low concentrations of various antibiotics and since this resistance varies with the antibiotic and with the strain, a given strain may express a distinctive pattern of antibiotic resistance. Thus, intrinsic antibiotic resistance (IAR) has been proposed and used for strain identification in studies with ecological implications (Kingsley, Bohlool 1983). Fluorescent antibodies produced against strains with distinct IAR patterns proved to be highly specific (B.B. Bohlool, personal communication). IAR has the major advantage over the use of mutants in eliminating concerns of altered symbiotic or free-living biological properties of the mutant.

Strain typing by means of bacteriophage sensitivity has been used for many years for both slow- and fast-growing rhizobia, but patterns of sensitivity may become quite complex. Lesley (1982) for example, detected as many as 80 phage groups in a collection of R. meliloti on the basis of patterns of reaction obtained with 15 phages.

Gel electrophoresis for the differentiation of whole cell proteins of Rhizobium has been proposed for detailed characterization or R. japonicum strains of ecological interest (Noel, Brill 1980). The method has high discriminatory power but further work is necessary to evaluate effects of culture medium composition on protein patterns. Gel electrophoresis of DNA digested by restriction endonucleases has been shown to give patterns which allow genetic typing of Rhizobium strains (Mielenz et al. 1979). This method may be expected to come into increasing use in ecological studies of Rhizobium.

4.   ECOLOGY OF RHIZOBIA AS FREE-LIVING BACTERIA IN SOIL.

Rhizobia have a selective niche in nature, the legume root nodule, but apparently need not depend on that niche for survival. Instead, evidence is accumulating that rhizobia introduced into a soil join the soil community and maintain a population base often independent of any opportunity to attain their selective niche. While co-existing with other soil biota, they must compete for substrate and space, respond to the environmental features of their soil microsites, and profit if possible from the advent of an appropriate legume host root into that microsite. Why and how certain of these indigenous soil rhizobia accomplish the transition from soil to legume bacteria while others do not is not known, but their behavior as free-living bacteria must be a major determinant of success in nodulation.

4.1 Diversity of Indigenous Soil Rhizobia. Evidence increasingly points to the serological diversity of rhizobia indigenous to field soils. Direct enumeration in soil by IF indicated that several serogroups of R. japonicum coexisted in similar numbers at a particular field site despite the dominance of serogroup 123 in the nodules of soybean (Moawad et al. 1984). Analysis of nodule content, albeit biased in favor of the most competitive strains, has also provided insight into the distribution of specific strains of Rhizobium. Using complementary methods for nodule strain analysis Dughri and Bottomley (1984) noted serogroup diversity among indigenous R. trifolii and further documented diversity within serogroups of interest. Substantial diversity also occurs within indigenous R. japonicum serogroup 123 (E.L. Schmidt, unpublished). Strain diversity must be given greater consideration in studies of the indigenous rhizobia.

4.2 Effects of Soil Factors on Rhizobium Populations. Soil acidity has been identified as a factor influencing Rhizobium survival and nodulation depending on the species of Rhizobium. Cowpea rhizobia are tolerant (Bushby 1981), whereas R. meliloti are relatively sensitive to soil acidity (Lowendorf, Alexander 1983). Strains of R. meliloti selected for resistance to acidity in culture did not fare better than acid sensitive strains when inoculated into soil (Lowendorf, Alexander 1983). Extreme soil acidity is often linked to excess free aluminum, but Al sensitivity was found to have little effect on the survival of cowpea rhizobia in Al-rich acid soils (Hartel, Alexander 1983).

Rhizobia as free-living bacteria must exploit some substrates in soil in competition with other soil biota. The nature of these substrates remains unknown, but recent autecological studies have approached this question. Various organic amendments had no effect on the growth of several rhizobia in nonsterile soil, however 1% mannitol additions allowed proliferation of 2 antibiotic-marked strains of slow-growing rhizobia but not 2 strains of fast-growers. Antibiotics together with 0.1% mannitol presumably reduced competition from other bacteria and allowed growth of all 4 strains in the nonsterile soil (Pena-Cabriales, Alexander 1983).

Survival of rhizobia as a function of soil temperature and desiccation continues to be of ecological concern. Many studies still use sterilized soils, but as the data of Boonkerd and Weaver (1982) point out, survival in sterilized soil provides no basis for extrapolation to natural soil.

A recently developed split-root growth system was used to study salinity stress and rhizobial inoculation applied to one-half of soybean root systems (Singleton, Bohlool 1984). It was observed that processes involved in nodule initiation were extremely sensitive to low salt concentrations. This was probably due to sensitivity of root infection sites since rhizobial survival and root colonization were unaffected. The split-root technique promises to be a very valuable tool for ecological study of rhizosphere interactions in that the plant may be maintained under relatively unstressed growth conditions.

5. BEHAVIOR OF RHIZOBIA IN THE RHIZOSPHERE

In their transition from indigenous soil bacteria to legume nodule bacteria, rhizobia first become rhizosphere bacteria. The use of autecological techniques in the past decade has provided evidence that soil rhizobia are adept at becoming rhizosphere rhizobia. At least for some, and perhaps for all, the rhizosphere need not be that of a host legume , or even a non-host legume, but also can be a non-legume. And the host rhizosphere may be colonized as densely by strains which do not achieve nodule occupancy as those that do.

5.1 Rhizobia in Non-Legume Rhizospheres. Stimulation of strains of various species of Rhizobium occurred in the rhizospheres of oats, corn, and wheat. The intensity of the response depended on plant cultivar and Rhizobium strain (Reyes, Schmidt 1979; Pena Cabriales, Alexander 1983; Moawad et al. 1984). In some cases rhizosphere stimulation was lower in the non-host than in the host (Moawad et al. 1984), but in other instances as for R. japonicum in certain oat cultivars (Reyes, Schmidt 1979; Pena-Cabriales, Alexander 1983), and R. phaseoli in certain corn varieties (Pena-Cabriales, Alexander 1983) the rhizosphere effects were greater than with host plants.

5.2 Non-Host Legume Rhizospheres. Legume rhizospheres attract rhizobia other than strains which are potential nodulators. Both R. leguminosarum and R. japonicum were found at higher densities in the rhizospheres of peas than in those of soybeans at early stages (1-2d) whereas the reverse was true at later stages (3-35d) (Bohlool et al. 1984). A comparison of a strain of R. japonicum with one of R. phaseoli showed that the rhizosphere effect of the former was greater in both kidney bean and red clover rhizospheres (Pena-Cabriales, Alexander 1983). Further indication that rhizosphere stimulation depends on both plant and Rhizobium strain and is not related to host legume was reported by Robert and Schmidt (1983). Populations of R. phaseoli were higher in soybean rhizospheres than were R. japonicum populations in Phaseolus vulgaris. In addition an aggressive inoculant strain of R. phaseoli developed rhizosphere/soil ratios with soybean that were 10-20 times higher than an indigenous strain of R. phaseoli.

5.3 Rhizobia in Host Rhizospheres. Earlier concepts of the legume root eliciting a strong selective rhizosphere response on the part of its homologous rhizobia under the impetus of a "specific stimulation" have not been borne out. Studies with both soybeans and beans (P. vulgaris) showed that rhizosphere responses were slight for homologous rhizobia during early growth stages when nodulation was initiated, not selective for homologous rhizobia, and not significantly different from non-host rhizospheres (Reyes, Schmidt 1979). Marked rhizosphere effects with rhizobial densities of 10^7 or greater g^{-1} rhizosphere soil were encountered only in mature plants from flowering through nodule decay (Bushby 1981, 1984; Robert, Schmidt 1983; Moawad et al. (1984). Rhizobia which did not accomplish nodulation were as good rhizosphere bacteria as those that did nodulate (Bushby 1984; Moawad et al. 1984). Rhizobia are obviously good rhizosphere bacteria but still represent a minority of the total rhizosphere bacteria even in host rhizospheres (Moawad et al. 1984).

6. COMPETITION AS A PRE-INFECTION EVENT

The most pressing problems in the ecology of Rhizobium center on the matter of competition, precipitated by the fact that certain indigenous soil strains regularly outcompete both other indigenous strains and more desirable inoculant strains for nodule occupancy. The attributes that contribute to the success of an indigenous strain are unknown but must be elucidated if the potential of the most effective nitrogen fixing rhizobia is to be exploited.

6.1 The Successful Inoculant. In some instances inoculant strains have been competitive, and were able not only to replace the indigenous populations in the nodules but also become established in the soil. Two strains of R. leguminosarum introduced on pelleted seeds into a field in Hawaii containing indigenous strains of the same species completely supplanted the resident population in the nodules of lentil for two consecutive seasons. The resident population of lentil rhizobia however was low, -- only about 100 cells g^{-1} soil (May, Bohlool 1983). In Mississippi soils with low indigenous populations of R. trifolii (less than 1000 g^{-1}) all five inoculant strains tested proved to be highly successful in replacing the native strains in red clover nodules (Materon, Hagedorn 1982). R. phaseoli strain Viking 1 was highly competitive as an inoculant of P. vulgaris in a field where previously the indigenous population had virtually excluded inoculant strain QA1062 from nodulation. The complete dominance of the Viking 1 carried over the year following inoculation, when it was present at or below the level of the resident population (Robert, Schmidt 1983). Success of an inoculant strain was found to vary with soil type. Two strains which were very successful in an oxisol were less so in a mollisol, whereas a third strain, not competitive in the oxisol, occupied most nodules in the mollisol (Moawad, Bohlool 1984).

6.2 The Highly Competitive Indigenous Rhizobium. Inoculants are widely applied in soybean growing areas of midwestern U.S., but inoculant strains are rarely found in the nodules. The dominant strains in the nodules belong to R. japonicum serogroups 123 derived from the indigenous soil populations (Moawad et al. 1984). Considerable attention has been given to this phenomenon and has led to consideration of a number of possible bases for the competitiveness of the successful indigenous Rhizobium.

Rhizosphere response was studied in Minnesota soybean field soils to examine time and intensity of soybean rhizosphere colonization by three indigenous R. japonicum serogroups. All three serogroups (110, 123, 138) coexisted in the soil in similar densities at the time of planting, and all three showed similar weak rhizosphere responses in 1 to 3 week old plants (Moawad et al. 1984). In most instances rhizosphere effects for 123 were even lower than for competitors, yet virtually all the nodules that developed were of serogroup 123. In order to determine if serogroup 123 was the first to colonize the rhizosphere or to out-number competitors at an even earlier growth stage, colonization of seedling radicles was followed from germination to emergence (Robert, Schmidt 1985). Again serogroup 123 appeared to have no advantage in rhizosphere population dynamics.

First exposure to a strain nevertheless has been shown to influence nodulation dramatically under some conditions. Inoculation of pea cv Afghanistan with a mixture of nodulating (TOM) and non-nodulating (PF$_2$) strains of R. leguminosarum completely inhibited nodulation. However the nodulation ability of TOM was completely restored when this strain was inoculated 3 days prior to inoculation with PF$_2$ (Winarno, Lie 1979). The outcome of competition between certain strains of R. japonicum could be altered by pre-exposure of the roots of soybean to one strain before inoculation with a second (Kosslak et al. 1983; Kosslak, Bohlool 1984). When USDA strains 110 and 138 were inoculated together on soybean seedlings grown in vermiculite, 110 dominated in the nodules. However, prior exposure to poorly competitive strain 138 for as little as 24 h favored nodule occupancy by 138. Results of these studies (Kosslak et al. 1983) and of subsequent split-root experiments (Singleton, Bohlool 1984) indicate that interactions which occur in the rhizosphere during very early plant growth are critical in determining the outcome of competition among strains of R. japonicum.

Among other attributes of possible relevance to the success of an indigenous strain, only soil pH will be mentioned. Ability of R. japonicum serogroup 135 to displace serogroup 123 in high pH midwestern soils is relatively well known. Recent work by Dughri and Bottomley (1984) has shown that soil pH markedly changed the outcome of competition among indigenous R. trifolii for nodulation of subterraneum clover and that differences were expressed at the serotype or serotype variant level as well as serogroup level.

7. CONCLUSION

New and improved methodologies have accelerated progress and expanded interest in the ecology of Rhizobium. The autecological capability of some of these techniques has made it possible to address biological events that lead to the symbiotic relationship and determine its nature. The critical areas of Rhizobium behavior in the soil, and especially of Rhizobium-plant interactions in the rhizosphere, are now receiving attention but need additional emphasis.

8. REFERENCES

Bohlool BB et al. (1984) In Veeger C and Newton WE, eds, Advances in Nitrogen Fixation Research, pp 287-293, Martinus Nijhoff/Junk, The Hague.
Boonkerd N, Weaver RW (1982) Appl. Environ, Microbiol. 43, 585-589.
Bushby HVA (1981) Siol Biol. Biochem. 13, 241-245.
Bushby HVA (1984) Soil Biol. Biochem. 16, 635-641.
Dughri MH, Bottomley PJ (1984) Soil Biol. Biochem. 16, 405-411.
Hartel PG, Alexander M (1983) Soil Sci. Soc. Am. J. 47, 502-506.
Kingsley MT, Bohlool BB (1981) Appl. Environ, Microbiol. 42, 241-148.
Kingsley MT, Bohlool BB (1983) Can. J. Microbiol. 29, 518-526.
Kishinevsky B, Maoz A (1983) Current Microbiol. 9, 45-49.
Kosslak RM et al. (1983) Appl. Environ. Microbiol. 46, 870-873.
Kosslak RM, Bohlool BB (1984) Plant Physiol. 75, 125-130.
Lesley SM (1982) Can. J. Microbiol. 28, 180-189.
Lowendorf HS, Alexander M (1983) Soil Sci. Soc. Am. J. 47, 935-938.
Martensson AM et al. (1984) J. Gen. Microbiol. 130, 247-253.
Materon LA, Hagedorn C (1982) Appl. Environ. Microbiol. 44, 1096-1101.
May SN, Bohlool BB (1983) Appl. Environ. Microbiol. 45, 960-965.
McLoughlin TJ et al. (1984) J. Appl. Bacteriol. 56, 131-135.
Mielenz JR et al. (1979) Can. J. Microbiol. 25, 803-807.
Moawad HA, Bohlool BB (1984) Appl. Environ. Microbiol. 48, 5-9.
Moawad HA et al. (1984) Appl. Environ. Microbiol. 47, 607-612.
Morley SJ, Jones DG (1980) J. Appl. Bacteriol. 49, 103-109.
Noel KD, Brill WJ (1980) Appl. Environ. Microbiol. 40, 931-938.
Pena-Cabriales JJ, Alexander M (1983) Soil Sci. Soc. Am. J. 47, 241-245.
Reyes VG, Schmidt EL (1979) Appl. Environ. Microbiol. 37, 854-858.
Rice WA et al. (1984) Can. J. Microbiol. 30, 1187-1190.
Robert FM, Schmidt EL (1983) Appl. Environ. Microbiol. 45, 550-556.
Robert FM, Schmidt EL (1985) Soil Biol. Biochem. 17, in press.
Roughley RJ et al. (1976) Soil Biol. Biochem. 8, 403-407.
Singleton PW, Bohlool BB (1984) Plant Physiol. 74, 72-76.
Somasegaran P et al. (1983) J. Appl. Bacteriol. 55, 253-261.
Winarno R, Lie TA (1979) Plant Soil 51, 135-142.

PHYSIOLOGICAL AND BIOCHEMICAL ASPECTS OF N_2-FIXING CYANOBACTERIA

PETER ROWELL, NIGEL W. KERBY, ALLAN J. DARLING and WILLIAM D.P. STEWART
AFRC RESEARCH GROUP ON CYANOBACTERIA AND DEPARTMENT OF BIOLOGICAL
SCIENCES, UNIVERSITY OF DUNDEE, DUNDEE DD1 4HN, UK.

1. INTRODUCTION

Various aspects of the physiology and biochemistry of cyanobacteria have
recently been reviewed (see Stewart, 1980; Hawkesford et al., 1983;
Bothe et al., 1984; Houchins, 1985; Stewart et al., 1985). Here we discuss
several aspects of recent interest to our laboratory.

NH_4^+, the product of N_2-fixation is assimilated via the glutamine synthetase
(GS)/glutamate synthase pathway in cyanobacteria (see Stewart, 1980).
Products of NH_4^+ assimilation may be involved in the regulation of
nitrogenase (Stewart, Rowell, 1975) and recent work on cyanobacteria
(Machray, Stewart, 1985) suggests a nitrogen regulatory system in close
proximity to the *gln*A gene, as in *Klebsiella* (Espin et al., 1982).
However, the mechanisms by which exogenous NH_4^+ is taken up and the ways
in which products of its assimilation are involved in regulation of
nitrogenase have yet to be fully elucidated.

The isolation of mutant strains deficient in GS is important for several
reasons. Firstly, to study the role of GS in nitrogen metabolism;
secondly, to study the role of GS in the regulation of nitrogenase and
thirdly, to obtain mutant strains for the photoproduction of NH_4^+.

Several approaches have been used to obtain GS deficient mutants and
these have included the selection of glutamine auxotrophs, methionine
sulphoximine (MSX)-resistant strains, and strains resistant to the
ammonium analogues methylammonium ($CH_3NH_3^+$) and ethylenediamine (1,2-
diaminoethane, EDA). The use of EDA (Polukhina et al., 1982) appears to
be particularly successful and has been used in our laboratory. However,
the reasons for its success were unknown. We have recently characterized
the transport and metabolism of EDA and compared it with the transport and
metabolism of $CH_3NH_3^+$, which is widely used to study NH_4^+ transport systems
(see Kleiner, 1981; Rai et al., 1984). These studies have provided
information on the mechanism of NH_4^+ transport and its subsequent metabolism via GS, and we have obtained preliminary evidence using an immuno-
cytochemical technique for the intracellular localization of GS which
relates to its role in NH_4^+ uptake and assimilation. The localization of
GS is compared with that of certain other key enzymes of nitrogen and
photosynthetic carbon metabolism.

2. AMMONIUM TRANSPORT AND METABOLISM

2.1. The uptake and accumulation of $CH_3NH_3^+$ by N_2-fixing cyanobacteria

$CH_3NH_3^+$ and NH_4^+ appear to share a common transport system in various cyanobacteria and the kinetics of $^{14}CH_3NH_3^+$ uptake by cyanobacteria are usually
biphasic (Rai et al., 1984; Boussiba et al., 1984; Kerby et al., 1985a).
Fig. 1 shows the kinetics of uptake of $^{14}CH_3NH_3^+$ at pH 7.0 and pH 9.0 for
N_2-fixing *Anabaena variabilis* and *Anabaena cylindrica*. Data for *Nostoc*
CAN (not presented) are similar to those for *Anabaena variabilis* (Kerby
et al., 1985a). The initial rapid phase of uptake is attributed to
uptake via a $CH_3NH_3^+$ (NH_4^+) transport system at pH 7.0 and probably to
passive diffusion of uncharged CH_3NH_2 and trapping by protonation at

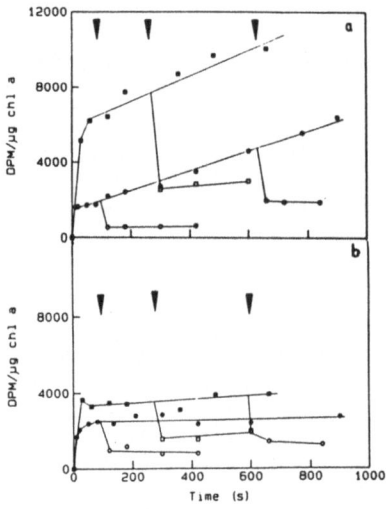
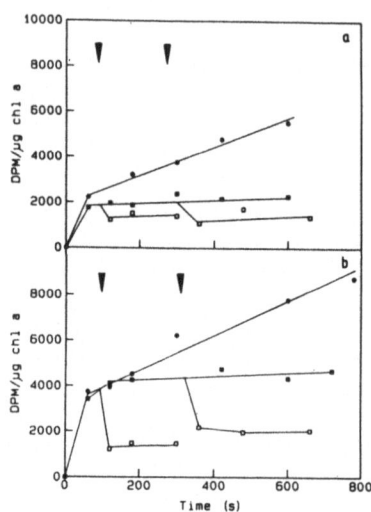

Fig. 1.   Fig. 2.

Fig. 1. CH$_3$NH$_3^+$ uptake by (a) *A. variabilis* Kutz (ATCC 29413), and (b)
A. cylindrica (CCAP 1403/2a) which were grown in continuous culture in
N-free medium. Cells were harvested, washed and resuspended in either
10 mol m^{-3} Hepes NaOH (pH 7.0) or 10 mol m^{-3} Tricine NaOH (pH 9.0).
30 mmol m^{-3} ^{14}CH$_3$NH$_3^+$ was added (specific activity 10 kBq cm^{-3}) at t = 0,
and the incorporation of ^{14}C was determined over the time period shown
(see Rai et al., 1984; Kerby et al., 1985a). 3 mol m^{-3} NH$_4^+$ was added at
the times indicated (▼). (●) pH 7.0, (■) pH 9.0, (O) pH 7.0 + NH$_4^+$
and (□) pH 9.0 + NH$_4^+$.

Fig. 2. Effect of MSX on CH$_3$NH$_3^+$ uptake by *A. variabilis* at (a) pH 7.0
and (b) pH 9.0. Experimental details are as given in Fig. 1.
50 mmol m^{-3} MSX was added and, after incubating for 2 h, the cells were
washed and resuspended in fresh medium in the absence of MSX. (●)
control, (■) MSX-treated and (□) MSX-treated with 3 mol m^{-3} NH$_4^+$ added
at the time indicated (▼).

pH 9.0. This initial phase of uptake was greater at pH 9.0 than at
pH 7.0 and was followed by a second slower phase of uptake, although the
second phase was barely detectable in *A. cylindrica*. Addition of
3 mol m^{-3} NH$_4^+$ specifically displaced a large proportion of the ^{14}CH$_3$NH$_3^+$
taken up by the cells at both pH values and prevented further uptake.
In contrast, Na$^+$ (up to 100 mol m^{-3}) displaced little or no ^{14}CH$_3$NH$_3^+$ and
did not prevent further uptake.

2.2. The metabolism of ^{14}CH$_3$NH$_3^+$ and accumulation of γ-methylglutamine

Following incubation of *A. variabilis* with ^{14}CH$_3$NH$_3^+$ for 6 h, negligible
radioactivity was incorporated into proteins but most was associated with
an amino acid peak which was not detected in untreated cells (Rai et al.,
1984). This component was tentatively identified as γ-methylglutamine,
a product of CH$_3$NH$_3^+$ metabolism as in other microorganisms (see Kleiner,
1981). Table 1 shows that accumulation of γ-methylglutamine occurred in
both *A. variabilis* and *Nostoc* CAN (data not shown) at pH 7.0 and pH 9.0
but that virtually no accumulation was detected in *A. cylindrica*.

These data are in agreement with the finding that in *A. cylindrica* the second phase of $^{14}CH_3NH_3^+$ uptake is virtually undetectable and suggest that sustained uptake of $^{14}CH_3NH_3^+$ is dependent on its metabolism.

When *A. variabilis* was treated with MSX to inactivate GS, then washed to remove any NH_4^+ produced (Stewart, Rowell, 1975), the second phase of $^{14}CH_3NH_3^+$ uptake was abolished but the initial rapid phase was unaffected (Fig. 2). Thus MSX inhibited metabolism of $^{14}CH_3NH_3^+$ by GS but did not directly affect the initial rapid uptake. This is in agreement with data for *Anacystis nidulans* (Boussiba et al., 1984; Boussiba, Gibson, 1985); it does not support the view (Turpin et al., 1984) that MSX inhibits the NH_4^+ transport system *per se*.

The data obtained indicate that cyanobacteria possess an active NH_4^+ ($CH_3NH_3^+$) transport system which may be important for scavenging low concentrations of external NH_4^+ at low pH values. Additionally, since biological membranes are permeable to NH_3, the transport system may be important for the retention of NH_4^+ synthesised internally by processes such as N_2 fixation (Raven, 1980; Castorph, Kleiner, 1984). The usefulness of $CH_3NH_3^+$ for study of NH_4^+ transport systems is limited by the fact that it is usually rapidly metabolised. The fact that metabolism of $CH_3NH_3^+$ was virtually undetectable in *Anabaena cylindrica* suggests that this cyanobacterium may be suitable for the further study of the transport system without requiring the use of mutant strains deficient in GS or the use of inhibitors of GS activity such as MSX.

2.3. Uptake and assimilation of EDA

Resistance to EDA has been used for the selection of mutant strains of cyanobacteria which are partially deficient in GS (Sakhurieva et al., 1982; Stewart et al., 1985) and are effective NH_4^+-liberating mutants with derepressed nitrogenase activity (Polukhina et al., 1982). Uptake of ^{14}C-EDA (Kerby et al., 1985b) occurred at pH 9.0 with an initial rapid uptake followed by a second slower phase of uptake. 200 mmol m^{-3} NH_4^+ and 10 mol m^{-3} Na$^+$ each displaced a small proportion of the radioactivity at pH 9.0 but did not inhibit further uptake indicating that NH_4^+ did not specifically inhibit EDA uptake or its subsequent metabolism. At pH 7.0, there was an initial high accumulation of label, with little subsequent uptake. The initial uptake, at pH 7.0, is attributable to non-specific binding of EDA since it can be almost totally prevented in the presence of 100 mol m^{-3} Na$^+$. Thus the uptake of EDA appears to be via diffusion in response to a pH gradient, and not via an active transport system such as that involved in the uptake of $CH_3NH_3^+$ (NH_4^+).

2.4. Metabolism of EDA

As with $CH_3NH_3^+$, almost all of the label from ^{14}C-EDA taken up at pH 9.0 was incorporated into the ethanol-soluble fraction and amino acid analysis revealed that virtually all of the radioactivity was associated with a single peak. This component, which was not formed at pH 7.0, or at pH 9.0 in cells pretreated with MSX to inactivate GS (Table 2), was tentatively identified as aminoethylglutamine. The initial rapid uptake of ^{14}C-EDA at pH 9.0 was not prevented by MSX pretreatment of the cells (Kerby et al., 1985b). Thus, as with $CH_3NH_3^+$, sustained uptake of EDA is dependent on metabolism via GS. Such data suggest that the successful isolation of mutants of *A. variabilis* deficient in GS (Polukhina et al., 1982) by selecting for resistance to the toxic effects of EDA may be due to the fact that unwanted transport mutants will not be obtained, and that EDA is metabolised via GS to a product which appears to be toxic.

TABLE 1. Free amino acid pools of *A. variabilis* (Av) and *A. cylindrica* (Ac) in the presence or absence of added CH_3NH_3Cl.

Organism	Medium pH	$\pm\ CH_3NH_3Cl$	nmol amino acid $\mu gchla^{-1}$		
			Glu	Gln	MeGln
Av	7	−	0.63	0.18	0
Av	9	−	0.72	0.12	0
Av	7	+	0.24	0.18	0.68
Av	9	+	0.13	0.11	1.15
Ac	7	−	0.59	0.92	0
Ac	9	−	0.69	1.71	0
Ac	7	+	0.61	0.69	0
Ac	9	+	0.70	1.54	0.12

Cells were suspended in either 10 mol m^{-3} Hepes (pH 7.0) or 10 mol m^{-3} Tricine (pH 9.0) buffers and equilibrated for 30 min prior to addition of 30 mmol m^{-3} CH_3NH_3Cl. After a further 30 min incubation at 25°C and at a photon flux density of 75 μmol m^{-2} s^{-1} cells were filtered, washed with 100 mol m^{-3} NaCl to remove non-specifically bound CH_3NH_3Cl and free pool amino acids extracted and analysed as previously described (see Kerby et al., 1985a). Only data for glutamate, glutamine and γ-methylglutamine (MeGln) are shown.

TABLE 2. Free amino acid pools of *A. variabilis* in the presence or absence of added EDA (30 mmol m^{-3}), with or without pretreatment with 50 mmol m^{-3} MSX for 2.5 h.

pH of medium	$\pm$ MSX	$\pm$ EDA	nmol amino acid $\mu gchla^{-1}$		
			glu	gln	X
9	−	−	0.72	0.12	0
7	−	+	0.52	0.24	0
8	−	+	0.41	0.13	0.12
9	−	+	0.31	0.11	0.62
9	+	−	0.90	0	0
9	+	+	0.82	0	0

For other experimental details see Table 1 and Kerby et al. (1985a,b). X represents the component tentatively identified as amino-ethylglutamine.

3. ENZYME LOCALIZATION

Studies on cyanobacterial metabolism have involved the separation of heterocyst and vegetative cell fractions in order to determine the distribution of key enzymes. Problems associated with such studies include assessing the structural integrity of isolated heterocysts and the contamination of such preparations by vegetative cells. In addition, little information on the intracellular location of components is obtained.

The use of specific antisera against key enzymes of cyanobacterial metabolism (see Murry et al. (1984) for localization of nitrogenase in heterocysts using ferritin labelled antibodies), which can be labelled with colloidal gold (Horisberger, 1983), offers a direct means by which these proteins can be localized in thin sections of whole filaments. We have localized several proteins (Fig. 3).

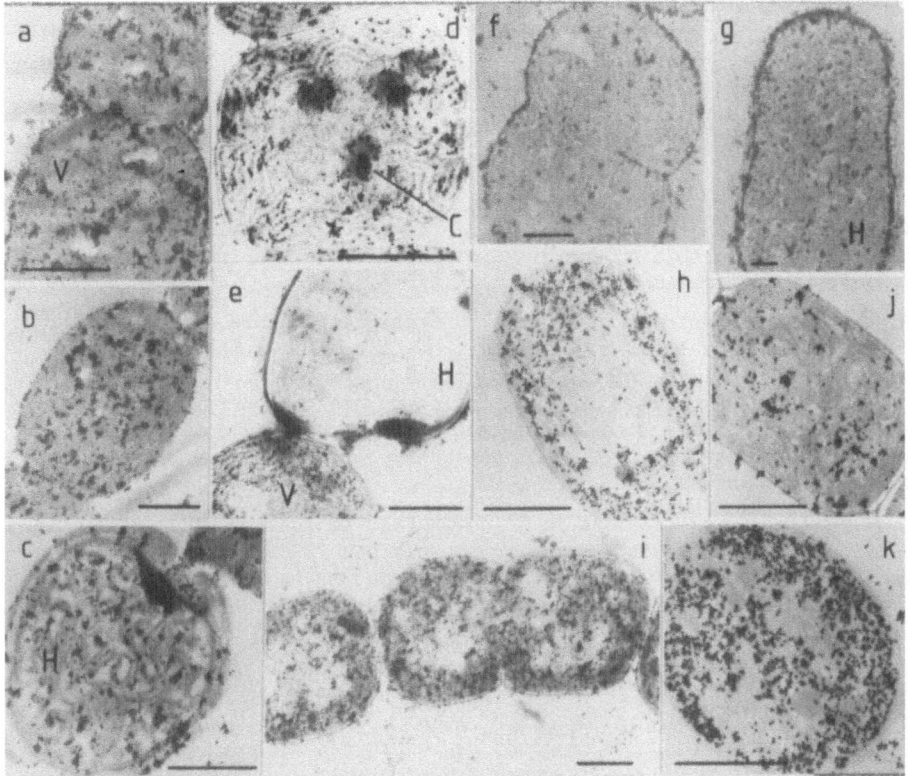

Fig. 3. Localization of GS (a-c) RuBisCO (d,e) FNR (f,g) and thioredoxin
(h-k) in *A. cylindrica* (a,d-j) and *A. variabilis* (b,c,k) using antisera
against purified GS and FNR from *A. cylindrica*, RuBisCO from *Microcystis*
PCC 7820 and thioredoxin from *A. cylindrica* (j) and *A. variabilis* (h,i,k)
in conjunction with colloidal gold conjugated goat anti-rabbit IgG
(Cossar et al., 1985a,b). Bar indicates 1 μm. H, heterocyst; V,
vegetative cell and C, carboxysome.

3.1. Glutamine synthetase, the enzyme responsible for primary NH_4^+
assimilation, was localized in *A. cylindrica* and *A. variabilis* using
antiserum against *A. cylindrica* GS (Fig. 3a-c). The data confirm that GS
is present in both heterocysts and vegetative cells but in *A. cylindrica*
labelling was greatest at the periphery of the vegetative cells,
suggesting that GS may be located in close proximity to the sites of NH_4^+
transport (see above) and this is consistent with the hypothesis that NH_4^+
transport and assimilation are tightly coupled (see section 2.2.).

3.2. Ferredoxin-$NADP^+$ oxidoreductase (FNR), which is generally considered
to be involved in electron transfer from NADPH to nitrogenase (see Apte et
al., 1978; Houchins, 1985; Schrautemeier et al., 1985), was localized in
A. cylindrica and *A. variabilis* using antiserum against enzyme purified
from *A. cylindrica*. Vegetative cells and heterocysts of *A. cylindrica*
and *A. variabilis* were extensively labelled, particularly at the
periphery of the cells (Fig. 3f,g). In neither case was there any clear
association with the chromatoplasm which is surprising in view of the
role of FNR in photosynthetic electron transfer. However, since

cyanobacteria have multiple molecular forms of FNR (Rowell et al., 1981) (which show NADPH-dependent diaphorase, ferredoxin-dependent cytochrome c reductase and transhydrogenase activity) it is possible that the antiserum does not react with all forms of the enzyme.

Clearly, such a location for an enzyme with FNR activity may be important in terms of the physiological function of this enzyme, as an apparent association with the plasmalemma is consistent with the possibility that this is a site of NADPH oxidation and respiratory electron transfer (Nitschmann, Peschek, 1985). This finding is also consistent with our previous finding that an energised plasmalemma is required for sustained N_2ase activity (see Hawkesford et al., 1983).

In the above experiments, using *A. cylindrica*, antisera against GS and FNR were found to label the periphery of cells in particular. It is unlikely that this represents a non-specific association of the antisera because there was little or no labelling at the periphery of the cells when pre-immune sera were substituted for the specific antisera or when antisera against ribulose 1,5-bisphosphate carboxylase/oxygenase (RuBisCO) and thioredoxin were used (see below).

3.3. RuBisCO was localized in *A. cylindrica* (Cossar et al., 1985a) using antiserum against RuBisCO from *Microcystis* PCC 7820 (Fig. 3d,e). The data confirm that RuBisCO, the key carboxylating enzyme of the Calvin cycle, is absent from mature heterocysts and that carboxysomes are sites of RuBisCO accumulation (see Codd, Marsden, 1984) and also suggest an association with the photosynthetic thylakoids. In stationary phase cultures labelling was particularly associated with carboxysomes, whereas in continuous cultures there was extensive labelling of the chromato-plasm which increased with increasing growth rate.

3.4. Thioredoxin was localized in *A. cylindrica* and *A. variabilis* using antisera raised against the purified protein from each cyanobacterium (Cossar et al., 1985b). The antiserum against thioredoxin from *A. cylindrica*, reacted only with *A. cylindrica* thioredoxin, which was located mainly in the nucleoplasm of vegetative cells and showed no association with the chromatoplasm (Fig. 3j). This is consistent with an involvement in processes such as ribonucleotide reduction (Laurent et al., 1964; Whittaker, Gleason, 1984). In contrast, the antiserum against *A. variabilis* thioredoxin labelled vegetative cells of both cyanobacteria (Fig. 3h,i,k) but, in the case of *A. cylindrica* (Fig. 3h,i) labelling was associated mainly with the chromatoplasm, suggesting the presence of a second thioredoxin or thioredoxin-like protein. Using either antiserum, the labelling of heterocysts was substantially lower than that of vegetative cells. Thioredoxin(s) may be involved in, for example, the light-dependent activation of certain Calvin cycle enzymes (see Buchanan, 1984; Ip et al., 1984; Whittaker, Gleason, 1984). Of particular interest is the potential role of thioredoxin in the light-dependent deactivation of glucose-6-phosphate dehydrogenase of vegetative cells (Cossar et al., 1984; Udvardy et al., 1984) which may be important for inhibition of the oxidative pentose phosphate pathway in the light. However, the decreased level of thioredoxin(s) in heterocysts may be related to the fact that glucose-6-phosphate dehydrogenase activity remains high in heterocysts in the light (Stewart et al., 1985) thus maintaining the generation of NADPH as a source of reductant for nitrogenase (Apte et al., 1978; Houchins, 1985).

4. REFERENCES

Apte SK, Rowell P and Stewart WDP (1978) Proc. Roy. Soc. Lond. B 200, 1-25.
Bothe H, Nelles H, Hager K-P, Papen H and Neuer G (1984) In Veeger C and
Newton WE, eds, Advances in Nitrogen Fixation Research, pp 199-210,
Martinus Nijhoff/Junk, The Hague.
Boussiba S, Dilling W and Gibson J (1984) J. Bacteriol. 160, 204-210.
Boussiba S and Gibson J (1985) FEBS Lett. 180, 13-16.
Buchanan BB (1984) Bioscience 34, 378-383.
Castorph H and Kleiner D (1984) Arch. Microbiol. 139, 245-247.
Codd GA and Marsden WJN (1984) Biol. Rev. 59, 389-422.
Cossar JD, Rowell P and Stewart WDP (1984) J. Gen. Microbiol. 130, 991-998.
Cossar JD, Rowell P, Darling AJ, Murray S, Codd, GA and Stewart WDP (1985a)
FEMS Microbiol. Lett. 25, 65-68.
Cossar JD, Darling AJ, Ip SM, Rowell P and Stewart WDP (1985b) J. Gen.
Microbiol. - in press.
Hawkesford MJ, Rowell P and Stewart WDP (1983) In Papageorgiou GC and
Packer L, eds, Photosynthetic Prokaryotes: Cell Differentiation and
Function, pp 199-218, Elsevier Biomedical, New York.
Horisberger MC (1983) Trends in Biochemical Sciences 8, 395-397.
Houchins JP (1985) In Ludden PW and Burris JE, eds, Nitrogen Fixation and
CO_2 Metabolism, pp 261-268, Elsevier, New York.
Ip SM, Rowell P, Aitken A and Stewart WDP (1984) Eur. J. Biochem. 141,
497-504.
Kerby NW, Rowell P and Stewart WDP (1985a) Arch. Microbiol. - in press.
Kerby NW, Rowell P and Stewart WDP (1985b) Arch. Microbiol. 141, 244-245.
Kleiner D (1981) Biochim. Biophys. Acta 639, 41-52.
Laurent TC, Moore EC and Reichard P (1964) J. Biol. Chem. 239, 3436-3444.
Machray GC and Stewart WDP (1985) Proc. Roy. Soc. Edin. 85B, 239-252.
Murry MA, Hallenbeck PC and Benemann JR (1984) Arch. Microbiol. 137, 194-
199.
Nitschmann WH and Peschek GA (1985) Arch. Microbiol. 141, 330-336.
Polukhina LE, Sakhurieva GN and Shestakov SV (1982) Microbiology 51, 90-95.
Rai AN, Rowell P and Stewart WDP (1984) Arch. Microbiol. 137, 241-246.
Raven JA (1980) Advances in Microbial Physiology 21, 47-226.
Rowell P, Diez J, Apte SK and Stewart WDP (1981) Biochim. Biophys. Acta
657, 507-516.
Sakhurieva GN, Polukhina LE and Shestakov SV (1982) Microbiology 51,
308-312.
Schrautemeier B, Bohme H and Boger P (1985) Biochim. Biophys. Acta 807,
147-154.
Stewart WDP (1980) Ann. Rev. Microbiol. 34, 497-536.
Stewart WDP and Rowell P (1975) Biochem. Biophys. Res. Commun. 65, 846-857.
Stewart WDP, Rowell P, Cossar JD and Kerby NW (1985) In Ludden PW and
Burris JE, eds, Nitrogen Fixation and CO_2 Metabolism, pp 269-279,
Elsevier, New York.
Turpin DH, Edie SA and Canvin DT (1984) Plant Physiol. 74, 701-704.
Udvardy J, Borbely G, Juhasz A and Farkas GL (1984) J. Bacteriol. 157,
681-683.
Whittaker MM and Gleason FK (1984) J. Biol. Chem. 259, 14088-14093.

ACKNOWLEDGEMENTS

This work was supported by the AFRC, SERC, NERC and the Royal Society.

EFFECTS OF PHOSPHORUS AND LIME UPON NODULATION BY INDIGENOUS SEROGROUPS OF RHIZOBIUM TRIFOLII

ANGELA S. ALMENDRAS[1] AND PETER J. BOTTOMLEY,[1,2]
DEPARTMENTS OF SOIL SCIENCE[1] AND MICROBIOLOGY[2], OREGON STATE UNIVERSITY, CORVALLIS, OR 97331 U.S.A.

Introduction. Serogroup 6 was the dominant and serogroup 36 was a minor nodule occupant of non-inoculated subclover (Trifolium subterraneum L.) cv. 'Mt. Barker' grown in unlimed acid soil (pH 4.8 ± 0.1). However, the situation was reversed when such soil was amended with $CaCO_3$ (Dughri and Bottomley, 1983). Experiments were conducted to determine: i) if other liming materials, with or without calcium, would have the same effect, ii) if the application of phosphate – considered to become more available as a result of liming – would change the composition of nodule occupants and, iii) if the rate of phosphorus (P) applied would influence the extent of nodule occupancy of serogroup 36.

Results and Discussion. Occupancy by serogroups 6 and 36 in the nodules of subclover and direct enumeration of these serogroups in soil prior to planting and after harvest were determined by immunofluorescence using serogroup specific FITC-IgG conjugants.

Amending acid soil (pH 5.1 ± 0.1, 5 mg extractable P Kg^{-1}), with either $CaCO_3$, $Ca(OH)_2$, MgO or K_2CO_3 increased significantly the % nodule occupancy by serogroup 36 (33-52%). In all cases serogroup 6, the dominant nodule occupant in unamended soil, became a minor nodule occupant as a result of liming irregardless of the lime source. A phosphorus application (25 mg P Kg^{-1}) to either unlimed soil or soil amended with $Ca(OH)_2$ resulted in an increase of the % nodules occupied by serogroup 36. Although the addition of P alone had no significant effect on the numbers of nodules occupied by serogroup 6, the presence of P with lime prevented the decline in nodule occupancy by serogroup 6 observed in treatments receiving lime alone. Application of P rates greater than 25 mg P Kg^{-1} had no further effect on increasing the number of nodules occupied by serogroup 36.

Although the population sizes of both 6 and 36 increased during the incubation period before planting in both unlimed and limed soil, no differential proliferation was observed. The numbers of serogroup 36 were consistently higher than serogroup 6 in both soil treatments. Numbers measured after nodule harvest were not differentially affected by lime or phosphorous application either. In summation, there was no correlation between the population sizes of both serogroups in the soil and the changes in nodule occupancy observed in response to soil treatments.

Reference

Dughri, M. H. and P. J. Bottomley. 1983. Appl. Environ. Microbiol. 46:1207-1213.

SEROLOGY AND PHYSIOLOGY OF NATIVE BRADYRHIZOBIUM
JAPONICUM POPULATIONS IN IOWA SOILS

R. K. BERG, JR. and T. E. LOYNACHAN. Iowa State University, Ames, IA
50011. USA

Physical, chemical, and biological factors that might influence native
rhizobial populations were evaluated in 12 soils from major soybean-
producing regions of Iowa. Soil traits examined included pH, organic
matter, available P and K, extractable Zn and S, and soil phosphatase
activity. Parent material, slope, aspect, landscape position, texture,
inoculation history, and other management practices were also considered.
Rhizobial populations at each site were estimated by most probable
number. Agglutination reactions were used to identify serogroup distribution on 100 nodules from each soil by using antisera prepared against
USDA strains 142, 138, 135, 123, 122, 110, 94, 76, 31, and 6.
Individual isolates from each soil were used to monitor tolerance to pH
(4.5 and 8.5), temperature (5 and 35°C), and salt (1% NaCl). Relative
isolate symbiotic efficiency was measured by biomass production, nodulation, and specific nitrogenase activity (C_2H_2 reduction) in a growth
chamber.

The soils examined were mostly Udolls and Aquolls having silty clay
loam or loam textures formed from loess and glacial till. They varied
from slightly acid (6.1) to moderately alkaline (7.8), with organic
matter contents ranging from 1.6 to 6.4%. Native populations measured
in August and September on soybeans in their reproductive growth stages
ranged from 5.34 to 7.55 log rhizobia g^{-1} soil.

The majority of the nodules typed (77 to 96%) reacted with at least one
of the antisera, and mixed reactions with more than one antiserum per
nodule were common. The dominate rhizobial component in each soil
belonged to serogroup 123. Some nodules gave differential agglutination
with antisera prepared from USDA strains 142 and 122 even though these
serotypes belong to the same serogroup. In some soils, nodules having
mixed reactions involving 142 were nearly as common as those having
mixed reactions with 123. Common mixed reactions observed were 123/122,
123/76, 142/123, and 142/123/122. Nodules with 123/76 and 123/122(142)
reactions appeared to be antigenically similar to USDA strains 127 and
129, respectively. Relatively low levels of 123/110, 123/6, and 142/122
were also found.

Individual isolates capable of growing on acid plates were more common
than those growing on alkaline plates. The 5°C temperature appeared
more favorable for isolate growth, but a few isolates grew at 35°C. A
wide range of salt tolerance was seen, with between 18 to 67% of soil
isolates being tolerant to 1% NaCl. Nodule number, mass, and specific
nitrogenase activity were nearly two times greater in some soils than
in others.

ACKNOWLEDGMENTS

The authors wish to acknowledge the financial assistance provided
by Allied Corporation and to thank Dr. Robert M. Zablotowicz for his
technical assistance and advice.

ECOLOGICAL STUDIES OF RHIZOBIUM JAPONICUM:
RELATIONSHIPS BETWEEN INOCULUM ESTABLISHMENT AND NODULE
OCCUPANCY IN COMPETITIVE AND NON-COMPETITIVE SOILS

J. BROCKWELL/R.J. ROUGHLEY/D.F. HERRIDGE
CSIRO, Canberra, A.C.T. 2601/ Dep. Agric., Gosford, N.S.W. 2250/
Dep. Agric., Tamworth, N.S.W. 2340, Australia.

Soil populations of rhizobia often interfere with attempts to introduce new strains by legume inoculation. A three-year study to define this problem as it applied to soybean cropping was conducted on a vertisol initially free of R. japonicum. Related strains of rhizobia (CB1809, CB1809str, CB1809strspc), distinguishable by differential resistance to streptomycin and spectinomycin, were used in three successive years to inoculate Bragg soybeans. The same seed beds were used for each crop so that 2nd- and 3rd-year inocula met with competition from background populations of rhizobia introduced previously. Uninoculated control treatments were included at each sowing.

In year 1, the CB1809 inoculant colonized rhizospheres, formed nodules and established in the soil. A 200-fold increase in CB1809 in the soil during the fallow period between crops, attributed to release of bacteria from disintegrating nodules, ensured a large background population for future experimentation. At the same time, a few CB1809 cells contaminated uninoculated soil thus providing a small background population for subsequent work. In years 2 and 3, rhizosphere populations of background CB1809 increased greatly between days 14 and 42 while populations of inoculant rhizobia remained static or declined. In year 3, CB1809 dominated other strains in colonizing the rhizosphere in all treatments even when the background population of CB1809 was small.

In any year, the strain first used as inoculant formed the majority of nodules. In year 2 in the presence of a large background population of CB1809, the CB1809str inoculant was overwhelmed and all nodules were formed by CB1809; in the presence of a small background population of CB1809, all nodules formed by day 42 were due to CB1809str inoculant but some CB1809 nodules appeared later. In the treatment inoculated in year 2 with CB1809str against a small background population of CB1809, CB1809 formed a higher proportion of nodules in year 3 than it had in year 2. The proportion of nodules formed by inoculant strains was consistently greater than their relative representation in the rhizosphere. This was ascribed to specific placement of inoculant in that zone of the soil where infection foci first form on roots.

It was concluded that a strain of rhizobia first introduced into a soil previously free of R. japonicum 'adapted' to the environment by its rapid development of a very large, well-dispersed population and this background dominated inocula introduced subsequently by numerical superiority. However, the results indicated that background rhizobia were most susceptible to competition from inoculation rhizobia shortly after germination of the host. This period should be the particular focus of inoculation strategy.

SOIL DISTRIBUTION, CULTIVAR EFFECTS, AND COMPETITIVENESS OF A
RHIZOBIUM MELILOTI SEROGROUP.

M. D. BUSSE,[1] M. B. JENKINS,[2] AND P. J. BOTTOMLEY,[1,2] DEPARTMENTS OF
SOIL SCIENCE[1] AND MICROBIOLOGY,[2] OREGON STATE UNIVERSITY, CORVALLIS,
OREGON 97331 U.S.A.

Recent evidence has been obtained which shows the heterogeneity
of an indigenous soil population of Rhizobium meliloti. One
serogroup within that diverse population was found to be the dominant
nodule occupant on non-inoculated alfalfa (Jenkins and Bottomley,
1985).

To enhance the understanding of the potential ecological and
agronomic significance of these findings we examined further both the
distribution of serogroup 31 in a larger field plot adjacent to the
original and the effects of host cultivar on nodule occupancy and
symbiotic effectiveness.

Nodules were sampled from replicate plots of 'Anchor', 'Vernal',
and 'Saranac' alfalfa planted to a large variety trial covering 0.2
ha. Purified isolates were compared by whole cell, somatic
agglutination tests and antibiotic-resistance patterns. Serological
data confirmed the extensive distribution of serogroup 31 throughout
the field site since isolates were identified from plants in 11 out
of 12 plot areas. Within the 11 plots, nodule occupancy ranged from
11 to 75%, with a mean of 45%. A significant cultivar effect was
identified: average nodule occupancy by 31 was greater on 'Vernal'
(60%) compared to 'Anchor' (28%) or 'Saranac' (35%). Although
variation in nodule occupancy within cultivars was found, it was not
consistent across all cultivars.

Antibiotic-resistance patterns were compared to determine if
there were additional isolates which occupied a high percentage of
nodules. Ten distinct patterns were found among the remaining
isolates which failed to agglutinate with antiserum 31. Although two
of the patterns comprised the largest percentage of isolates among
the cultivars, these values were substantially less than those found
for serogroup 31. Therefore, in addition to the observed cultivar
effect, members of serogroup 31 were the most dominant nodule
occupants from the indigenous population on all varieties.

Seedlings of the three cultivars were grown under greenhouse
conditions to compare the symbiotic effectiveness of an isolate of
31. From a comparison of the shoot dry weight of inoculated versus
nitrate supplemented plants, the effectiveness was found to be
significantly higher on 'Vernal' compared to 'Anchor' seedlings.

It can be inferred that members of serogroup 31 expressed a
combination of both persistent nodulating and superior effectiveness
traits on 'Vernal' plants far stronger than on the other two
cultivars. The expanded characterization of serogroup 31 presented
in this study further supports its value as a tool for nodulation and
competition studies.

Jenkins, M. B., and P. J. Bottomley. (1985). Soil Sci. Soc.
Am. J. 49, 326-328.

NODULATION OF DIFFERENT CULTIVARS OF <u>TRIFOLIUM</u> <u>SUBTERRANEUM</u> L.
BY INDIGENOUS <u>RHIZOBIUM</u> <u>TRIFOLII</u>

DAVID H. DEMEZAS[1] AND PETER J. BOTTOMLEY,[1,2] DEPARTMENTS OF
MICROBIOLOGY[1] AND SOIL SCIENCE[2], OREGON STATE UNIVERSITY, CORVALLIS
OR 97331 U.S.A.

I.   Introduction. Effects of subclover cultivars on the competition
for nodulation between indigenous R. trifolii have recently been
observed with plants grown in soil (Dughri and Bottomley, 1984). The
possibility that indigenous strains of R. trifolii might proliferate
differentially in rhizosphere soil and account for competitive
advantage in nodulation by these strains on different cultivars has
not been examined.
II.   Results and Discussion. Subclover cvs. 'Mt. Barker' and
'Woogenellup' were sown into Whobrey soil. Nodules collected after 8
weeks and the occupants' identity determined using a serogroup
specific fluorescent antibody conjugants (FAB). An apparent cultivar
effect on the composition of nodule occupants was observed.
Serogroup 1-01 occupied 25% of the nodules on 'Mt. Barker' but 5% of
the nodules on 'Woogenellup". Both serogroups 1-01 and 2-01 occupied
a similar percentage of nodules on 'Mt. Barker' and the latter
occupied a similar percentage on both cultivars.
     Using a modified procedure of Kingsley and Bohlool (1981) for
extracting rhizobia from soil, the rhizosphere populations of
serogroups 1-01 and 2-01 were followed from the time of sowing of
seeds to the appearance of nodules (14d) using FAB. There was no
apparent rhizosphere stimulation of the numbers of either serogroup
1-01 or 2-01 on either cultivar when these data were compared to the
unplanted soil. The population densities of serogroups 1-01 and
2-01 in the rhizosphere were similar and averaged 0.32×10^6 and 0.59
$\times 10^6$ per gram of soil respectively over the course of the
experiment. The average soil populations of the two individual
serogroups in the unplanted control did not differ from the
rhizosphere average population densities. Serogroup 2-01 population
was more stable and consistently higher than that of serogroup 1-01
over the time course of the experiment.
     Nodule occupancy was also determined in the latter experiment at
4 and 12 weeks after sowing. Serogroup 2-01 occupied approximately
50% of the 4 week nodules of both cultivars. In contrast, serogroup
1-01 occupied a lower percentage on both 'Mt. Barker' (7.1%) and on
'Woogenellup' (16.1%). On the former cultivar the percentage nodule
occupancy by serogroup 1-01 increased to 29.1% of the nodules at 12
weeks, but did not change on 'Woogenellup' (15.7%).

Dughri, M. H., and P. J. Bottomley (1984) Soil Biol. Biochem. 16,
405-411.
Kingsley, M. T., and B. B. Bohlool (1981) Appl. Environ. Microbiol.
42, 241-248.

DIVERSITY OF AN INDIGENOUS POPULATION OF RHIZOBIUM MELILOTI

A. HARTMANN and N. AMARGER
I.N.R.A., Microbiologie des sols
B.V. 1540, Dijon 21034, France

In order to determine the composition of the Rhizobium meliloti population in nodules formed on alfalfa, isolates from root nodules of uninoculated field grown plants of Medicago sativa, cv Europe, were characterized with respect to indigenous plasmids, phage sensitivity and symbiotic effectiveness on alfalfa.

Five sites were chosen in an alfalfa field, from each site five plants were selected, from each plant five root nodules were picked up and one Rhizobium strain was isolated from each nodule, which gave a total of 125 isolates.

The plasmid profile of each isolate was obtained using the "in gel" lysis technique of Eckhardt modified by Rosenberg et al (1982). The molecular mass was estimated by comparisons with standard plasmids of known molecular mass. The phage typing system described by Lesley (1982) was used to group the isolates according to their phage susceptibility. The symbiotic effectiveness was estimated by measuring the dry matter produced by alfalfa seedlings inoculated with R. meliloti isolates and grown on a nitrogen free medium for five weeks.

A megaplasmid (M.M. > 650 kb) was detected in all the isolates but one. Besides this megaplasmid, plasmids of molecular mass ranging approximately from 45 to 450 kb were also detected, their number varying from 0 to 4. On the basis of plasmid number and size 25 different groups could be differentiated. Almost two thirds of the isolates presented one plasmid besides the megaplasmid, the most important group represented 40 isolates which had one plasmid of molecular mass from 215 to 250 kb.

The differential susceptibility of the isolates to the 13 phages tested allowed to divide the isolates into 40 groups. There was no correspondence between the groups obtained by phage typing and those obtained by plasmid profile. The class grouping the 40 isolates with one plasmid of molecular mass 215-240 kb could be subdived into 17 distinct groups by phage typing. By combining the two types of criteria, the larger group observed was formed of 17 isolates.

The effectiveness of 80 % of the isolates was not significantly different from the effectiveness of the standard strain, one isolate was ineffective, one was poorly effective (20 % of the standard), the effectiveness of the other isolates was between 60 an 80 % of the effectiveness of the standard.

The results indicate that the field population studied was composed of different strains of R. meliloti with no dominant type. Plasmid profiles are reproducible and can be used to distinguish among different isolates of a species of Rhizobium.

References
Lesley SM (1982) Can. J. Microbiol. 28, 180-189.
Rosenberg C. et al. (1982) J. Bact. 150, 402-406.

COMPETITIVE GROWTH OF SLOW GROWING RHIZOBIUM JAPONICUM AGAINST FAST
GROWING ENTEROBACTER AND PSEUDOMONAS SPECIES AT LOW CONCENTRATIONS OF
SUCCINATE AND OTHER SUBSTRATES IN DIALYSIS CULTURE

C. HUMBECK[1], H. THIERFELDER[1], P.M. GRESSHOFF[2] and D. WERNER[1]
[1] Fachbereich Biologie, Botanisches Institut der Philipps-Universität,
Lahnberge, D - 3550 Marburg/L, FRG
[2] Department of Botany, Australian National University, Canberra

A cultivation system with simultaneous growth of six bacterial cultures
in separate bags in dialysis culture was developed. In a medium with
no added carbon source (one half concentrated Hoagland solution, water
deionized and distilled), cell number of Rhizobium japonicum increased
during a 7 day period by a factor of 35, whereas the number of Enterobacter aerogenes cells decreased to one half. With a concentration of
100 nM succinate as an additional carbon source in the inflow, Rhizobium japonicum 61-A-101 cell number increased by a factor of 50 during
a 8 day period, whereas cell number of Enterobacter cloacae NCTC 10005
only doubled and of Enterobacter aerogenes NCTC 10006 decreased. At
10 mM concentration of succinate in the inflow, doubling time of the
two Enterobacter strains was about 12 h, compared to about 24 h for
the Rhizobium japonicum strain. Varying the succinate concentration
from 10 mM to 100 nM in the inflow, Rhizobium japonicum 61-A-101 sur-
passed the Enterobacter aerogenes strains in the growth rate between
1 mM and 100 µm succinate in the inflowing medium. Growing in competition with two Pseudomonas strains, Rhizobium japonicum RH 31 Marburg
(fix⁻) did overgrow also Pseudomonas fluorescens, was however out-
grown by Pseudomonas putida. In utilizing low concentrations of a
[14]C labelled organic acid (malonate), three strains of Rhizobium japonicum left 2-4 times smaller amounts of ^{14}C in the medium than two
species of Pseudomonas and two species of Arthrobacter. Succinate uptake in Rhizobium japonicum 61-A-101 has a biphasic kinetic, indicating
the presence of a high and low affinity uptake system. The apparent K_M
data are 2.4 µM (high affinity system) and 172 µM (low affinity system).

REFERENCES

Humbeck C et al (1985) Arch. Microbiol. (in press)

COMPETITION FOR NODULATION BY TWO BRADYRHIZOBIUM JAPONICUM STRAINS ON SOYBEANS GROWN UNDER DIFFERENT ENVIRONMENTAL CONDITIONS

JEFFREY S. LA FAVRE, ADRIENNE K. LA FAVRE and ALLAN R. J. EAGLESHAM
BOYCE THOMPSON INSTITUTE, ITHACA, NY 14853, USA.

1. INTRODUCTION

Introduction of new Rhizobium strains into the legume crop ecosystem is often hampered by rhizobia indigenous to the soil. The indigenous rhizobia compete with the inoculum strain for nodulation of the legume host, often to near exclusion of the inoculum strain. The mechanisms underlying competition need to be elucidated so that introduction of superior rhizobia can be facilitated in the future. Here we report the results of preliminary experiments describing the competitive interaction between two strains of Bradyrhizobium japonicum.

2. MATERIALS AND METHODS

Bradyrhizobium japonicum strains USDA 76 and RCR 3410 were used in this study. Various mixtures of these two strains were used to inoculate Wilkin soybeans grown in sand culture in an environmentally controlled light room. The percentage of nodules occupied by each strain was determined by typing nodules on antibiotic media, the strains being identified by differences in intrinsic antibiotic resistance and colony morphology.

3. RESULTS

In the first experiment soybeans were harvested at five sequential time periods (19 - 46 days after planting). The percentage of nodules occupied by USDA 76 dropped significantly with increasing plant age. A subsequent experiment suggested that the above results were not due to differential proliferation and/or survival through time of the two strains. Examination of the sand solution revealed that it changed pH and ionic strength with time, however further experiments suggested that pH and ionic strength did not contribute directly to changes in nodule occupancy by the two strains. In the next experiment, plants inoculated at planting were compared with plants inoculated 21 days after planting. Soybeans inoculated at planting had 36% of their nodules occupied by USDA 76 while plants inoculated 21 days after planting had 10% of their nodules occupied by USDA 76 (a significant difference by ANOVA $P = 0.01$). In the final experiment all plants were inoculated 29 days after planting but grown at three different light intensities. Nodule occupancy by USDA 76 was low for all treatments but the higher light treatment, which resulted in greater root growth, had a higher percentage of nodule occupancy by USDA 76.

4. CONCLUSIONS

The competitive ability of USDA 76 appears to be greater on young and rapidly growing roots whereas that of RCR 3410 is greater on older root systems. We speculate that USDA 76 is more competitive at nodulating roots when the rate of root tip extension is greater and further experiments are being conducted to address this possibility

OSMOREGULATION IN <u>RHIZOBIUM</u> <u>MELILOTI</u>:   ROLE OF GLUTAMIC ACID

JAMES BOTSFORD, VIRGINIA SALAS, TRAVIS FISHER. New Mexico State University.

When <u>R.</u> <u>meliloti</u> grows in a completely defined medium with 250 mM NaCl, glutamic acid is made in excess (Arch. Microbiol. 137:124). A similar effect is seen when cells grow with inhibitory concentrations of KCl, polyethylene glycol, or sucrose indicating the effect is due to high osmolarity and not to sodium. The role that glutamic acid plays is uncertain. The addition of glutamate to the growth medium does not make the cells more resistant to inhibition of growth by NaCl. Cells will use glutamate as a sole source of nitrogen indicating that it is transported and metabolized.

It has been proposed that glutamate is accumulated by gram negative bacteria growing in high NaCl to act as a counter ion for the potassium that accumulates under these conditions. This hypothesis was tested. The intracellular concentration of potassium and glutamate was determined in cells grown in minimal defined medium with inhibitory concentrations of NaCl, KCl, polyethylene glycol (PEG) or sucrose. Potassium was determined with the atomic absorption spectrometer. Free glutamate was determined either with by amino acid analysis or enzymically using glutamic dehydrogenase. The concentration of potassium was found to vary 4 fold at most. The concentration of glutamate varied more than 20 fold. There was no correlation between the intracellular concentrations of potassium and glutamate. Ammonium, like glutamate, was found to increase in response to higher osmolarity. However, there was no stoichiometric relationship between the two parameters.

The mechanism to account for excess glutamate production is uncertain. However, time course studies indicate that when cells are shifted from regular medium to one of high osmolarity, glutamate is accumulates rapidly in the absence of de novo protein synthesis. A clone from a <u>R.</u> <u>meliloti</u> gene library on pRK290 that complements a <u>gdh</u> <u>glt</u> glutmate auxotroph of <u>Escherichia</u> <u>coli</u> has been found. In the heterologous system, glutamate is produced in excess when the host grows with inhibitory concentrations of NaCl.

At this point, the data suggest that conditions of high osmolarity cause changes in the cell such that glutamate synthase is activated. It is not certain if glutamate plays an active role in the cell's response to high osmolarity.

AMERICAN SESBANIA YIELD FAST-GROWING, FREE-LIVING, NITROGEN-FIXERS

SUSAN J. BROWN, JAMES E. URBAN, EVERETT ROSEY AND LAWRENCE C. DAVIS.
Div. of Biology and Dept. of Biochemistry, Kansas State University,
Manhattan, Kansas 66506.

Nodule and Rhizobium-like soil bacteria associated with Sesbania
vesicaria (from a dry, upland area of Fayette Co., Texas) and S.
drumondi (from a wet, lowland area) were isolated for comparison with
each other and with the S. rostrata stem nodule isolate, ORS571 (1).
Large (1) and small (s) colony variants were isolated from upland (U)
and lowland (L) soil samples. Us, Ul, Ls, and Ll were differentiated
by carbon and nitrogen source preferences. KSB001 (from S. vesicaria
nodules) appeared indentical to KSB003 (from S. vesicaria seeds) in
all traits examined. KSB002 (from S. drummondi nodules) grew more
slowly than the Upland strains, doubling in yeast extract medium in
2.5 hr. The Upland strains resembled ORS571, doubling within 1.3 hr.
KSB002, ORS571, and the Upland strains produced an alkaline pH change.

KSB001, Us, and ORS571 showed very similar cytochrome profiles on
SDS-PAGGE, although their protein profiles and EcoRl restriction fragment patterns suggest they are different strains. The Lowland strains
each showed differences in cytochrome and protein profiles, as well as
in EcoRl fragment patterns.

ORS571 grew on succinate-minimal medium supplimented with ammonium
chloride but not on mannitol-minimal medium containing glutamate. The
newly isolated strains preferred mannitol-minimal medium. All newly
isolated strains grew in nitrogen-free minimal medium solidified with
0.75% Bacto agar. Growth appeared too vigorous to be supported solely
by nitrogen scavenging, yet acetylene reduction activity was observed
only for ORS571 and KSB002 under the conditions examined.

ORS571 effectively nodulated S. vesicaria, forming root nodules but
not stem nodules. ORS571 did not nodulate S. drummondi. KSB003 and Us
also appeared to effectively nodulate S. vesicaria.

1. Dreyfus et. al., Appl. and Environ. Microbiol. 45:711-713, 1983.

COMPETITION AND SURVIVAL OF <u>RHIZOBIUM</u> <u>PHASEOLI</u> IN WATER STRESS BEAN PLANTS*.

D. ESPINOSA-VICTORIA, R. FERRERA-CERRATO, A. LARQUE-SAAVEDRA and R. LEPIZ-ILDEFONSO. Colegio de Postgraduados, Chapingo, Mex., Mexico, and Instituto Nacional de Investigaciones Agricolas.

There are numerous reports on the effect of water stress on plants and microorganisms. However, little is known about the effect of drought on the symbiosis <u>Rhizobium</u> <u>phaseoli</u> – <u>Phaseolus</u> <u>vulgaris</u> L. A study of three different stages of development (vegetative, flowering and pod filling) in <u>P</u>. <u>vulgaris</u> – <u>Rhizobium</u> symbiosis, under drought conditions, is reported here.

Two antibiotic resistant strains of <u>Rhizobium</u> <u>phaseoli</u> (CPMEX1$_{Str150}$ and CPMEX 22$_{Spe200}$) were studied separately on its survival and competitiveness when developed in the rhizosphere of bean (<u>Phaseolus</u> <u>vulgaris</u> L.) cv. Negro 150 G3. The experiment was carried out under greenhouse conditions and the plants developed in pots with non sterile soil. Irrigated and water stress treatments were compared at vegetative, flowering and pod filling stages of development. Water stress plants reached transpiration rates of 1 µg H_2O cm^{-2} sec^{-1} with a leaf resistance of 30 sec cm^{-1}. Soil water potential was below -15 bars. During the vegetative stage the number of viable rhizobia diminished in both water conditions. However, during flowering or pod filling stages in the irrigated pots there was a significant increment in viable rhizobia as compared with the water stress ones. Under well watered conditions for the three stages of development the highest percentages of nodulation were found with CPMEX22$_{Spe200}$ (39, 37 and 26%, respectively). Under water stress the same strain was the most successful one while CPMEX 1$_{Str150}$ gave lower nodule yield during the three stages of development equivalent to 13, 10 and 9%. Nitrogenase activity was at its highest during flowering (200 µmol of ethylene per plant h^{-1}) when inoculated with CPMEX22$_{Spe200}$ under well water regimens (Table 1).

Table 1. Nitrogenase activity* of nodules with or without water stress during vegetative or flowering stage of development in <u>Phaseolus</u> <u>vulgaris</u>.

		µmol C_2H_4 . h^{-1} per plant	
TREATMENT (Strain-Humidity level)		Vegetative $\overline{X}$	Flowering $\overline{X}$
CPMEX1 str150	irrigated	56.971	158.440
CPMEX1 str150	droughted	0.274	39.333
CPMEX1 spe200	irrigated	60.524	200.165
CPMEX22 spe200	droughted	0.289	12.804

*By acetylene reduction technique.

This work was supported in part by Grant The Bean/Cowpea CRSP, Title XII project.

CARBON SUBSTRATE REGULATION OF SUCCINATE METABOLISM
BY FREE-LIVING BRADYRHIZOBIUM JAPONICUM

Suzanne M. Ferrenbach and Joe Eugene Lepo
Microbial Physiology Laboratories
Department of Biology
The University of Mississippi
University, MS 38677 U.S.A.

The utilization of dicarboxylic acids by rhizobia may be crucial to the establishment of effective nitrogen-fixing Rhizobium-legume symbioses. However, the regulation of the mechanisms of dicarboxylate transport and oxidation in rhizobia are poorly characterized. Thus, we have monitored the expression of succinate transport, succinate dehydrogenase, and intracellular cyclic AMP in batch cultures of free-living Bradyrhizobium japonicum, strain USDA 136, grown in defined media containing various substrates as the sole source of carbon. This study differs in methodology from those done previously in that cells were "adapted" to each substrate examined by two consecutive passages to late log-phase growth on defined media containing that substrate; this procedure ensured stable physiological response to a given carbon substrate. Strain USDA 136 grew best on pyruvate, those carbon substrates entering the TCA cycle via conversion to pyruvate, and α-ketoglutarate. In general, growth was poor on the TCA cycle intermediates and those substrates entering the cycle via the Embden-Meyerhoff-Parnas, and Entner-Doudoroff pathways. Growth rate and maximum optical density reached were dissimilar for any two substrates tested. Expression of the succinate uptake (dicarboxylate transport, Dct) system was inducible: rates of succinate uptake were 7-fold higher in succinate-grown cells than in arabinose-grown cells. Cells grown on the other substrates tested exhibited rates of uptake intermediate between these two. Better growth on a particular substrate was not always associated with an increase in the rate of succinate uptake. Mid-log phase rates of succinate uptake on dual-carbon substrate media containing arabinose and succinate were similar to that seen in cells grown on succinate alone. Addition of arabinose or glucose to mid-log phase succinate-grown cells also had no significant effect on the rate of succinate uptake. Succinate dehydrogenase was expressed constitutively with the mid-log phase levels being essentially the same regardless of the substrate(s) employed for growth. Cyclic AMP did not appear to regulate either succinate transport or succinate dehydrogenase; there was no correlation between these activities and intracellular cyclic AMP during growth on any of the substrates tested. In addition, diauxic growth was not observed during experiments in which cells were grown on dual-carbon substrate media; and, the addition of exogenous cyclic AMP to succinate- or arabinose-grown cells had essentially no effect on the rates of succinate uptake. On the other hand, the addition of cyclic AMP to cultures growing in the presence of both succinate and arabinose stimulated growth rates by 50%, a result suggesting relief of catabolite repression of arabinose utilization by succinate.

ISOLATION AND CHARACTERIZATION OF A PHAGE AND
ITS MUTANT SPECIFIC FOR RHIZOBIUM JAPONICUM USDA 117

Fawzy M. Hashem and J. Scott Angle
University of Maryland, College Park, MD   20742, U.S.A.

1. INTRODUCTION
Rhizobiophage are often detected in soils cropped to legumes. The
presence of rhizobiophage in fields suggests that they could play an
important role in selection, propagation or elimination of Rhizobium
genotypes in nature (Vincent, 1977). Bacteriophage have been
isoloated for most Rhizobium species including R. lupini (Lotz and
Mayer 1972), R. trifolii (Atkin 1973), R. leguminosarum (Ley et al.
1972) and R. japonicum (Stacey et al 1984). The current study
examines the isolation and characterization of a phage and its
mutant, both of which are specific for R. japonicum USDA 117.

2. PROCEDURE
Twenty-three soybean Rhizobium strains, which represent a wide range
of serotypes, were examined as hosts for phage isolation. A phage
was isolated from the rhizosphere soil of soybean plants. A mutant
of this phage was obtained during long-term incubation in soil which
contained the original phage and R. japonicum USDA 117.   The
physiological characterisitics of the phage were examined in
adsorption and one-step growth experiments according to Adams (1959).

3. RESULTS
A phage and its mutant were isolated and found to specifically infect
R. japonicum USDA 117. The primary distinction between the two phage
was the plaque size produced on the host. The original phage and the
mutant produced plaques with maximum sizes of 2.2 and $11.4mm^2$,
respectively.   Both phage exhibited similar morphologies with a
hexagonal head 60 nm in diameter. The mutant phage was adsorbed
faster to the host than the original phage. Characterization of both
phage indicated that the burst size, rise period and generation time
were 100 PFU/ml, 12 min. and 80 min. for the original isolate and 210
PFU/ml, 6 min., and 70 min. for the mutant phage.

4. CONCLUSION
The isolation and characterization of two phage specific for R.
japonicum USDA 117 was reported. Both phage were adsorbed relatively
quickly to their host and exhibited rapid generation time. The two
phage, however, were quiet dissimiller when growth rates were
compared. The mutant phage infects its host much more quickly than
the original phage. This suggests that the pathogenicity of the
mutant phage may have been altered.

5. REFERENCES
Adams MH (1959) Bacteriophage. Interscience, NY
Atkin GH (1973) J. Virol. 12, 149-156
Ley AN et al. (1972) Can. J. Microbiol. 18, 375-384
Lotz W and Mayer F (1972) Can. J. Microbiol. 18, 1271-1274
Stacey G et al. (1984) Appl. Environ. Microbiol. 48, 68-72
Vincent JM (1977) In A Treatise on Dinitrogen Fixation 3, 277-300

RHIZOBIAL ECOLOGY OF THE WOODY LEGUME, PROSOPIS GLANDULOSA,
IN THE SONORAN DESERT.

M. B. JENKINS, R. A. VIRGINIA AND W. M. JARRELL
SOIL & ENVIRONMENTAL SCIENCES, AND ECOSYSTEMS RESEARCH GROUP
UNIVERSITY OF CALIFORNIA, RIVERSIDE, 92521, U.S.A.

Near Harper's Well in the California Sonoran Desert, mesquite (Prosopis glandulosa var. torreyana) grows as a phreatophyte, developing two lateral root systems, one at the surface (0–0.6m) and one in the sub-saturated (phreatic) soil above groundwater. These root systems are separated by a dry intermediate zone (0.6–3.9m). Although the natural ^{15}N abundance of mesquite tissue at Harper's Well indicates that the trees fix N_2 (Shearer et al., 1983), few field root nodules have been observed. The objectives of this study were to estimate rhizobial soil population densities of the two root zones the soil environment of which are contrasting, and collect and characterize isolates from the indigenous population of mesquite-nodulating rhizobia, in order to determine if surface and phreatic rhizobial populations are distinct.

Four soil cores (10 cm dia.) were collected in 0.3m increments to the depth of groundwater at 5m. Population sizes were estimated for surface (0–0.3m) and phreatic (3.9–4.5m) soil using a most probable number (MPN) technique (Garvin, Lindeman, 1983). Mesquite seedlings were used to trap rhizobia from each soil sample. Isolates were also obtained from root nodules of mesquite seedlings growing in a drainage at Harper's Well.

Surface soil rhizobial populations ranged from 0 to <10 cells g soil^{-1}, densities comparable to soils of ecosystems without legumes. Phreatic soil population densities ranged from 6×10^3 to 3×10^4 cells g soil^{-1} and are comparable to R. meliloti populations in established alfalfa fields. These data suggest that effective nodulation occurs in the phreatic rooting zone. Both Rhizobium and Bradyrhizobium spp. were isolated from mesquite root-nodules. Sixty-five percent of the total surface soil rhizobial nodule occupants were Rhizobium spp., whereas 83% of the nodule occupants from phreatic soil were Bradyrhizobium spp. Four colony morphologies were observed within each rhizobial genus. The distribution of these colony types suggest that the Rhizobium and Bradyrhizobium populations of each soil zone are distinct. Results of effectiveness tests and relative efficiency estimates show that nearly all isolates tested were effective; 45% of the surface Rhizobium and 17% of phreatic Bradyrhizobium isolates had relative efficiencies >0.90 and were assumed to be Hup$^+$. Since Hup$^+$ Bradyrhizobium spp. have been shown to live and grow chemoautotrophically, we hypothesize that this characteristic may be a factor in survival outside of the nodule in the phreatic rooting zone. In conclusion: 1. Nodulation and fixation occur in the deep phreatic zone. 2. Prosopis is nodulated by both Rhizobium and Bradyrhizobium. 3. Bradyrhizobium spp. dominate in the phreatic zone soil; Rhizobium spp. dominate in the surface soil. 4. The deep phreatic zone population is distinct from the surface soil population. 5. Both Rhizobium and Bradyrhizobium populations, based on colony types, effectiveness and relative efficiency, are diverse.

Garvin A and Lindemann WC (1983) Soil Sci. Soc. Am. J. 47, 604–606.
Shearer G et al (1983) Oecologia 56, 365–373.

MET⁻ AND ORN⁻ MUTANTS OF RHIZOBIUM LEGUMINOSARUM

R.BELTRA, G.DEL SOLAR, J.J.SANCHEZ-SERRANO and E.ALONSO. INSTITUTO DE MI-
CROBIOLOGIA "JAIME FERRAN". CONSEJO SUPERIOR DE INVESTIGACIONES CIENTIFI-
CAS. VELAZQUEZ, 144. 28006-MADRID, SPAIN.

1. INTRODUCTION

Mutation to auxotrophy is sometimes associated with the loss of symbiotic effectiveness in strains of Rhizobium leguminosarum (Swinghamer,-
1969). The degree of association between auxotrophy and effectiveness varies with the type of mutation and with the strain. Some of these mutants
also showed back mutation to prototrophy and was often accompanied by par
tial or full restoration of effectiveness.

2. MATERIALS AND METHODS

The mutagenesis of the 113 strain of R.leguminosarum was carried out
by means of NTG, following the technique of Beringer (1974). The auxotrophy of 583 colonies was studied using the test of Holliday (1956). We carried out biochemical assays, determining its plasmid pattern by the tech
nique of Eckhardt (1978). Its reversion frequency to prototrophy was as
well calculated. With all the mutants obtained we have done infectivity
tests in hydroponic cultures of Vicia faba L, var:Talo, using solution -
of Lie (1964). From the obtained nodules the bacteria were isolated in -
order to check its auxotrophy.

3. RESULTS AND DISCUSSION

Three mutants auxotrophs to methionine and one to ornithine were i-
solated. The results obtained while studying its reversion to prototrophy were in the same order. The plants inoculated with the mutants showed that: 1º In all cases was a delay in the aparition of the nodules in
relation to the wild type. 2º Excepting the controls, all plants showed
nodules after 11 days. The bacteria from the nodules were revertients -
prototrophs. In the nodulation experiments with the mutants to which methionine or ornithine was added (80 μg/ml), it was observed that: 1º The
controls did not nodulate. 2º Plants inoculated with the parent strain,
nodulate equally. 3º The mutants Orn⁻ added to ornithine not show delay
in nodulation. 4º Plants inoculated with the mutants Met⁻ plus methionine did not nodulate. 5º Plants inoculated with the mutants Met⁻ and without the requirement showed a delay in nodulation. From these nodules
revertients prototrophs were isolated. The biochemical behaviour of the
wild strain and its mutants, was very similar, finding for both three -
plasmids of 350, 230 and 140 Md.

We suggest the existence of a dynamic process of natural selection,
with advantage to prototrophic revertants. We also suggest that methio-
nine-requiring mutants change their symbiotic properties, independt of
the methionine presence in the hydroponic culture. This change is not -
showed in their biochemical behaviour and plasmid pattern.

4. REFERENCES

Beringer JE (1974) J. gen. Microbiol. 84, 188-198.
Eckhardt T (1978) Plasmid 1, 584-588.
Holliday R (1956) Nature (London) 178, 987.
Lie T (1964) In H Veenman and NV Zonem, eds. Wageningen. Nodulation of -
leguminous plants affected by roots secretions and red light, pp 1-89.
Schwinghamer EA (1968) Can. J. Microbiol. 14, 355-367.
Schwinghamer EA (1969) Can. J. Microbiol. 15, 611-622.

AMMONIUM ASSIMILATION IN THE FREE-LIVING STATE OF <u>RHIZOBIUM</u> <u>PHASEOLI</u>

A. BRAVO, G. ESPIN, D. ENREQUEZ, J. MORA. APDO. POSTAL 565-A,
CUERNAVACA-MOR., MEXICO

The enzymes involved in ammonium assimilation by <u>Rhizobium</u> <u>phaseoli</u>
CFN42 and their regulation on minimal medium containing different nitrogen
sources were studied.

<u>Rhizobium</u> <u>phaseoli</u> CF42 assimilates inorganic nitrogen through the
GS-GOGAT pathway.

Physiological and biochemical analysis have shown that GSI activity
is present in cells grown on complex medium and during ammonium limitation.
This activity is adenilated when cells are transferred to a minimal medium
containing only one nitrogen source. In contrast GSII activity is not
present in cells grown on complex medium. This activity increases in cells
growing in minimal medium. The levels of induction of GSII activity depends on the nitrogen source used.

We have found that GSII enzyme is inactivated after ammonium shock,
at the end of the exponential phase and in a mutant obtained as sensitive
to methionine sulfoximine. The inactivation process leads to a loss of
transferase activity, a high Km for glutamate and in a lower half life at
50°C. The inactivation of GSII activity correlates with a decrease in a
ratio α-ketoglutarate/glutamine.

We have found that cells that reach the stationary phase at growth,
in a minimal medium containing one carbon and one nitrogen source, inactivate the ammonium assimilation enzymes, excrete ammonium to the
medium and are unable to grow again in the same medium.

We have also studied the effects of the expression of the GDH gene
from <u>E. coli</u> in <u>R. phaseoli</u> in relation to growth, ammonium assimilation
and ammonium excretion.

STUDIES ON NH_4^+ UPTAKE AND EXPORT IN <u>RHIZOBIUM</u>

WILLIAM R. EVANS, SUSAN D. MARSH, JAMES L. CORBIN and ROBERT E. WYZA
Battelle-Kettering Laboratory, Yellow Springs, OH 45387

Introduction. Until recently, it was assumed that <u>Rhizobium</u> lacked an NH_4^+ transport system and NH_3 simply diffused across the cell membrane (Weigel, Kleiner 1982). However, NH_4^+ uptake was observed in micro-aerobically-grown, but not in aerobically-grown, <u>Rhizobium sp</u>. 32H1 (Gober, Kashket 1983) and continuous cultures of <u>R. leguminosarum</u> MNF3841 were utilized to demonstrate the presence of an NH_4^+ transporter (Glenn, Dilworth 1984). In this report aerobically-grown batch cultures of both slow and fast-growing <u>Rhizobium</u> are shown to possess active NH_4^+ transport systems. Preliminary results are also presented on the effect of NH_4^+ import and export on the intracellular level of gluta-mine.

Materials and Methods. The <u>Rhizobium</u> were grown aerobically in batch cultures in a mineral salts medium containing 0.15% l-malate and 0.1% casamino acids (pH6.8). Microaerobically-grown cultures were grown as previously described (Evans, Crist 1984). The uptake of $^{14}CH_3NH_3^+$ (10 µM, 46 µCi/µmol) was measured by a filtration procedure while the uptake of NH_4^+ was determined by following the decrease in extracellular NH_4^+ (Corbin 1984).

Results and Discussion. The existence of an NH_4^+ transport system can be adequately demonstrated if the NH_4^+ analogue, $CH_3NH_3^+$, is accumulated, the accumulation is inhibited by NH_4^+, and the accumulated $CH_3NH_3^+$ re-leased by the addition of NH_4^+ (Weigel, Kleiner 1982). Aerobically-grown <u>Rhizobium sp</u> 32H1 and <u>R. meliloti</u> AK631 were found to meet all these criteria. Similar results have been obtained with <u>R. japonicum</u> 61A76, <u>R. leguminosarum</u> 175G10, and <u>R. trifolii</u> T1. In contrast to the previous observations with <u>Rhizobium sp</u>. 32H1, aerobically-grown cells were found to transport NH_4^+ and $CH_3NH_3^+$ at 3 to 5 times the rate of microaerobically-grown bacteria. The uptake of $CH_3NH_3^+$ was not completely inhibited by NH_4^+ in the fast-growing species while in the slow-growing species uptake was completely inhibited. Growth of both the slow and fast-growing species in 10 mM NH_4^+ resulted in a greater inhibition of $CH_3NH_3^+$ uptake than of NH_4^+ uptake. Although the addition of enthetic NH_4^+ to aerobically-grown <u>Rhizobium sp</u>. 32H1 resulted in an immediate increase in the intracellular level of glutamine, a similar phenomenon was not observed when microaerobically-grown cells began to export endogenous NH_4^+.

References
Corbin, J. (1984) J. Appl. Environ. Microbiol 47: 1027-1030
Evans, WR and Crist, DK (1984) Arch. Microbiol 138: 26-30
Glenn, AR and Dilworth, MJ (1984) J. Gen Microbiol. 130: 1961-1968
Gober, JW and Kashket, ER (1983) J. Bacteriol. 153: 1196-1201
Weigel, I and Kleiner, D (1982) FEMS Microbiol. Lett. 15: 61-63

PLASMID PATTERN AND HUP PHENOTYPE OF EFFECTIVE RHIZOBIUM LEGUMINOSARUM
STRAINS ISOLATED OVER A PERIOD OF FIVE YEARS FROM TWO LOCATIONS.

HANS FEES, HANS-VOLKER TICHY, WOLFGANG LOTZ
INSTITUT FÜR MIKROBIOLOGIE UND BIOCHEMIE, UNIVERSITÄT ERLANGEN-NÜRNBERG,
D-8520 ERLANGEN, FRG.

A total of 265 Rhizobium leguminosarum strains has been isolated during
1980 (Tichy, Lotz, 1981), 1982, 1983 and 1984 from root nodules of Pisum
sativum, var. Poneka, grown at two locations near Erlangen. Location "B"
is a garden bed in which pea plants have been grown annually since 1975,
and "W" is a meadow where pea plants have not been grown for many years.
Only one bacterial strain was isolated per nodule; 135 strains were from
"B" and 130 strains from "W". All strains were Nod$^+$ Fix$^+$ with P. sativum.
Plasmids of the isolated strains were analyzed by agarose gel electrophoresis and Southern DNA hybridization. The nif-probes used for hybridization were pRmR2 (Ruvkun et al., 1982) and pGB5 (Schetgens et al.,
1984). As hup-probes pHU1 (Cantrell et al., 1983) and pHVT109 (H.V.
Tichy, unpublished) were used. Plasmid pHF5 was used as nod-probe (H.
Fees, unpublished). The Hup phenotype was tested by methylene blue reduction (Tichy, Lotz, 1985).

The following results were obtained: (1) About 70 % of the B-strains were
Hup$^+$ and hybridized with the hup-probes. In contrast, only two of the
130 W-strains have been identified has Hup$^+$. (2) In Hup$^+$ strains only the
Sym-plasmid carried hup-specific sequences. Except for two strains, the
largest plasmid (> 500 Mdal) per strain was not a Sym-plasmid. (3) Two to
six plasmids of different size were found per strain. A wide variety of
plasmid patterns was obtained per location and year of isolation. Plasmid
patterns most common per year and location occurred with frequencies of
about 25 to 50 %. For B- and W-strains, respectively, different plasmid
patterns dominated in each year of isolation.

The high frequency of Hup$^+$ strains from "B" may be linked to the yearly
cultivation of pea plants at this location. Alternatively, the hup
determinants may be located on highly "competitive" transferable plasmids.
The wide variety of plasmid patterns occurring in strains from both
locations may indicate a relatively high frequency of plasmid transfer
within the rhizobial population and possibly between rhizobial and
non-rhizobial populations in the soil.

REFERENCES

Cantrell MA et al (1983) Proc. Natl. Acad. Sci. 80, 181.
Ruvkun GB et al (1982) Cell 29, 551.
Schetgens TMP et al (1984) J. Mol. Appl. Genet. 2, 406.
Tichy HV and Lotz W (1981) FEMS Microbiol. Lett. 10, 203.
Tichy HV and Lotz W (1985) FEMS Microbiol. Lett. 27, 107.

VARIABILITY IN PROTEIN COMPOSITION OF
BRADYRHIZOBIUM JAPONICUM SEROGROUP 123

W. J. HICKEY, T. E. LOYNACHAN, A. AYANABA, and R. M. ZABLOTOWICZ, Iowa
State University, Ames, IA 50011, and Allied Corporation, Solvay, NY
13209.

This study examined the variability in protein composition within
Bradyrhizobium japonicum serogroup 123, the dominant serogroup found in
Iowa soils. Isolates were collected systematically from a field of bulk
soybeans (Corsoy variety) growing on a Webster soil. Two rows (3 m apart)
were sampled at 3 m intervals along a 15 m transect. Five plants were
taken at each sampling location and seven nodules were randomly chosen
from each plant for serological identification via immune diffusion.
Isolates from nodules reacting positively with USDA 123 antiserum were
then further purified for use in sodium dodecyl sulfate polyacrylamide
gel electrophoresis (SDS-PAGE). Protein profiles of several USDA strains
(obtained from the Beltsville collection) classified as serogroup 123
(USDA 2, 10, 105, and 123) were compared to those of field isolates.
Physiological characterization of selected isolates was performed on
solid media using a Cathra Replicator System (Cathra, St. Paul, MN).
Approximately 0.1 ml of 10-day broth cultures of selected isolates and
USDA 123 were added to each of two wells on the Cathra master plate.
Duplicate plates of each physiological media were inoculated from this
master and incubated for 10 days. Response of the cultures to each test
was then ranked as "+" for visible colony growth or "-" for no visible
colony growth.
Examination of protein profiles of 32 isolates indicated at least
three gel groups based on the presence or absence of two particularly
distinct bands (MW $\cong$ 62 Kd and 58 Kd respectively). Isolates in group
A contained both bands while those in group C lacked the 58 Kd band.
Group B isolates were intermediate between A and C, possessing the 58 Kd
band but at lower concentrations. Isolates in group A also possessed a
doublet (MW $\cong$ 130 Kd), which groups B and C lacked. Fifty-seven percent
of the isolates examined fell into group A, 31% into group B, and 12%
into group C. The 62 Kd and 58 Kd bands associated with gel group A
isolates were also observed in the profiles of USDA 2, 10, 105, and 123.
USDA 2, 10, and 105 possessed an additional pair of bands between the
62 Kd and 58 Kd bands not present in USDA 123 or gel group A isolates.
The USDA strains also possessed the doublet (MW $\cong$ 130 Kd) found in gel
group A. Further similarities and differences between USDA strains and
gel group A isolates were observed with several lower molecular weight
proteins ($\geq$ 50 Kd). Four representative isolates from each gel group
were selected for physiological characterization. The three groups
differed in their ability to: utilize arginine as a sole carbon and
nitrogen source, utilize succinate as a sole carbon source, and tolerate
osmotic stress. Several other physiological tests (utilization of a
variety of carbohydrates and tolerance to: NaCl, high temperature, and
high pH) were ineffective in yielding phenotypic differences between
isolates.

ACKNOWLEDGEMENTS

The authors wish to extend their gratitude to Joan M. Rettig for invaluable
assistance and preparation of materials for this presentation.

DENITRIFICATION BY RHIZOBIUM FREDII

HYNES, R.K., A.-L. DING[**], L.M. NELSON
Plant Biotechnology Institute
National Research Council of Canada
Saskatoon, Saskatchewan, Canada

[**] Institute of Crop Breeding and Cultivation
Chinese Academy of Agriculture Sciences
Beijing, People's Republic of China

Denitrification can provide energy for growth or maintenance of microorganisms under anaerobic conditions. Denitrification by Rhizobium spp was recently reviewed (O'Hara, Daniel, 1985). Van Berkum and Keyser (1985) reported that 270 of 321 isolates of soybean rhizobia were capable of denitrification and anaerobic growth. Since the initial report of the fast-growing Chinese isolates of R. japonicum (Keyser et al., 1982), currently named R. fredii, comparative studies with the slow-growing Bradyrhizobium japonicum have been reported. We report here denitrification and anaerobic growth by isolates of R. fredii and compare their growth rate to that of B. japonicum.

The isolates of R. fredii and B. japonicum supplied by the USDA, Beltsville, Md were 191, 192, 193, 194 and 110, 113, respectively. Isolates of B. japonicum supplied by the Nitragin Co. were 61A101, 61A118, 61A124, 61A148 and from the People's Republic of China, 005 and 113-2. One mL aliquots of 3 and 5 day-old cultures of the fast- and slow-growing isolates, respectively, grown in YEM salts medium were transferred to 10 mL of the same medium plus NO_3^- (about 5.5 mM) in 66.5 mL side-arm Erlenmeyer flasks. Controls contained no cells. The flasks were sealed, evacuated and backfilled with He. Liquid samples were taken at specific times, filtered and the filtrate was frozen for analysis of NO_2^- and NO_3^-. Analysis of N_2O was carried out using a EC or TC detector.

Anaerobic growth and reduction of NO_3^- to N_2O in the absence and presence of C_2H_2 was observed in all isolates of R. fredii examined. When the isolates were incubated for 35h under anaerobic conditions in the absence of NO_3^-, N_2O was produced in very low concentration or was undetectable. Denitrification and anaerobic growth were observed in B. japonicum isolates 005, 113-2, 110, 113, 61A118 and 61A148. Nitrous oxide was the terminal product of denitrification in isolates 113 and 61A118 and the remaining isolates capable of denitrification reduced NO_3^- to N_2.

Denitrification and anaerobic growth are common to most isolates of R. fredii and B. japonicum examined in this study. The end product of denitrification among R. fredii isolates was N_2O whereas N_2O or N_2 was the terminal product of denitrification by B. japonicum isolates. The difference in growth rate between these fast- and slow-growing microorganisms observed under aerobic conditions was also evident under anaerobic denitrifying conditions.

REFERENCES
Keyser HH et al (1985) Science 215, 1631-1632.
O'Hara GW and Daniel RM (1985) Soil Biol. Biochem. 17, 1-9.
Van Berkum P and Keyser HH (1985) Appl. Environ. Microbiol. 49, 772-777.

COMPARISON OF BIOCHEMICAL AND SYMBIOTIC PROPERTIES EXHIBITED BY TYPICAL FAST- AND SLOW-GROWING <u>LOTUS</u> RHIZOBIA AND A STRAIN WITH INTERMEDIATE GROWTH RATE ISOLATED FROM <u>LOTUS</u> <u>CORNICULATUS</u>

J.R. RAO, J.E. COOPER AND A.J. HOLDING

DEPARTMENT OF AGRICULTURAL AND FOOD BACTERIOLOGY, THE QUEEN'S UNIVERSITY OF BELFAST, NEWFORGE LANE, BELFAST BT9 5PX, N. IRELAND, U.K.

INTRODUCTION
Two groups of <u>Lotus</u> rhizobia are recognized. In addition to fast- and slow-growing rhizobia data has been accumulating on certain 'intermediate' types sharing properties of both groups.

In our present investigations, a strain (LC24) isolated from <u>Lotus</u> <u>corniculatus</u> showed that its growth rate, biochemical and symbiotic characteristics were intermediate between fast- and slow-growing <u>Lotus</u> strains.

METHODS
Comparisons were made between strain LC24, a slow-growing strain (CC814S) and a fast-growing <u>Lotus</u> strain (NZP2037). The local strain (LC24) was isolated from soil via <u>L. corniculatus</u> as host plant. Properties such as growth rates, cultural and morphological characteristics, symbiotic effectiveness on <u>L. corniculatus</u> and <u>L. pedunculatus</u>, carbohydrate and TCA intermediate utilization, 6-PGD activity and DNA homology studies were performed.

RESULTS
Results indicated that strain LC24 exhibited a growth rate intermediate between NZP2037 and CC814S and a similarity to NZP2037 in its acidic reaction on unbuffered YEMA. However, the biochemical and symbiotic properties of LC24 were similar to those of slow-growing strain CC814S and DNA homology tests showed 86% relatedness to CC814S. DNA homology between LC24 and NZP2037 was 24% while CC814S and NZP2037 showed only 6% homology. Data from 6-PGD activity indicated that NADP-linked activity was absent in strains LC24 and CC814S, but low levels of NAD-linked 6-PGD activity were present in LC24.

CONCLUSIONS
We suggest that the <u>Lotus</u> strain LC24 exhibits similarities to the fast-growing soybean nodulating rhizobia reported by Keyser <u>et al.</u> (1982) in showing properties which make it difficult to allocate to either the <u>Rhizobium</u> or <u>Bradyrhizobium</u> genus. In contrast to the soybean strains, however, this 'intermediate' strain shows more relatedness to <u>Bradyrhizobium</u> in DNA homology.

<u>Lotus</u> rhizobia are of interest to microbiologists due to their unusual specificity and nodulation patterns. Due to difficulties in manipulating <u>Bradyrhizobium</u> sp. for studies of genetic determinants of nodulation, "link strains" such as LC24 may be useful as substitutes for <u>Bradyrhizobium</u> in studies involving legumes which are normally nodulated by slow-growing strains.

REFERENCES - Keyser H H <u>et al.</u> (1982) <u>Science</u> <u>215</u>, 1631-1632.

CYANIDE ASSIMILATION BY NITROGENASE WITHOUT HYDROGEN PRODUCTION IN CHEMO-
STAT-CULTURES OF RHIZOBIUM ORS571 AND RHIZOBIUM LEGUMINOSARUM 128C30.

HEIN STAM, HENK W. VAN VERSEVELD, ADRIAAN H. STOUTHAMER
Biological Laboratory, Vrije Universiteit, P.O. Box 7161, 1007 MC
Amsterdam, The Netherlands

In succinate limited N_2-fixing chemostat cultures of Rhizobium ORS571
(specific growth rate (μ) is 0.1 h^{-1}) much ATP is used for hydrogen
production: The H_2/N_2 ratio (mol H_2 produced/mol N_2 fixed) is 7,5 and the
ATP/N_2 ratio (mol ATP used/mol N_2 fixed) is 42. An uptake hydrogenase is
induced, regaining a part of the lost energy (Stam et al., 1984).

Ammonia-assimilating cells of Rhizobium ORS571 grown at high oxygen
tension show a cyanide insensitive branch of the respiratory chain. With
cyanide in the medium, respiration via this branch is not favoured. Instead these cells show 3 adaptations: a higher cytochrome content, an
increased respiration rate and derepression of nitrogenase.

The respiration of N_2-fixing cells is much more sensitive to cyanide.
When cyanide is gradually added to a N_2-fixing culture the molar growth
yield on succinate (Y_{succ}) increases from 27 to 38. Hydrogenase activity
and acetylene reduction decrease gradually after the start of cyanide
addition. The respiratory chain of cells grown with 7 mM cyanide is still
very sensitive to cyanide. We conclude that cyanide is assimilated via
nitrogenase. Since cyanide is an inhibitor of the nitrogenase catalyzed
H_2 evolution (Li et al., 1982), induction of hydrogenase is halted.

TABLE 1. Theoretical growth yields in N_2-fixing chemostat cultures of a
hydrogenase positive and a hydrogenase negative strain of Rhizobium ORS571 at different H_2/N_2 ratios. For calculations see
Stam et al. (1984).

H_2/N_2	Y_{succ} (Hup$^+$)	Y_{succ} (Hup$^-$)	Y_{ATP}
0	35,7	35,7	4,82
1	34,3	33,9	4,57
7.5	27,2	25,8	3,14

The increase in growth yields is partly explained by this decreased H_2
production. This is confirmed by calculating the influence of H_2 production on Y_{succ} (Table 1). The products of cyanide assimilation are, besides
ammonia, methane and carbondioxide. During the oxidation of some intermediate products reduction equivalents are formed which can yield extra
ATP. Therefore the Y_{succ} (38) is higher than the highest calculated
Y_{succ} value (35,7). An ATP/2 KCN (mol ATP used/2 mol KCN assimilated) of
15,4 can be calculated.

Rhizobium leguminosarum 128C30 is only capable of N_2 fixation at a
low μ (0,02 h^{-1}, Stam et al., 1983). When cyanide is gradually added to
an ammonia-limited culture, grown at low oxygen tension, nitrogenase is
derepressed. In this way cyanide-assimilating cultures could be obtained
growing with a μ of 0,1 h^{-1} and an Y_{succ} of 37.

The results and calculations show that the H_2/N_2 ratio has much more
influence on Y_{succ} than the presence or absence of hydrogen reoxidation.

Li et al. (1982) Biochem. 21: 4393-4402.
Stam et al. (1983) Arch. Microbiol. 135: 199-204.
Stam et al. (1984) Arch. Microbiol. 139: 53-60.

POLYAMINE METABOLISM AND LEVELS IN RHIZOBIUM MELILOTI

PAULA A. TOWER AND A. J. FERRO
OREGON STATE UNIVERSITY, CORVALLIS, OR 97331 U.S.A.

Ornithine decarboxylase is a key enzyme in the biosynthesis of the diamine, putrescine, the precursor of the polyamine, spermidine. DL-α-difluoromethylornithine (DFMO), an enzyme-activated irreversible inhibitor of ornithine decarboxylase, was used to study the biological role of polyamines in the growth of Rhizobium meliloti. The generation time was increased (growth rate decreased) and the final cell density was reduced in the presence of DFMO. This growth inhibition was equally, but only partially, reversed with the addition of 1mM spermidine or 5 mM putrescine to the growth medium, whereas ornithine or sym-homospermidine were without effect. The in vitro activity of ornithine decarboxylase from R. meliloti was also inhibited by DFMO.

HPLC analyses of benzoylated cell extracts revealed that the intracellular concentration of sym-homospermidine, the uncommon but major polyamine in R. meliloti, decreased 32% in cultures grown in the presence of DFMO. In contrast, spermidine levels increased from 77 to 300 pmol/10^8 cells when treated with the inhibitor.

GENETIC DIVERSITY AND RELATEDNESS IN RHIZOBIUM

J P W YOUNG[1]   L DEMETRIOU[1]   S P HARRISON[1,2]   R G APTE[1,3]   D G JONES[2]

[1] John Innes Institute, Colney Lane, Norwich NR4 7UH, UK
[2] Dept. of Agricultural Botany, Univ. Coll. of Wales, Aberystwyth UK
[3] Indian Agricultural Research Institute, New Delhi, India

Rhizobium extracts were run on polyacrylamide gels, stained for specific enzyme activities (glucose-6-phosphate dehydrogenase, superoxide dismutase, beta-galactosidase). There were electrophoretic mobility variants for each enzyme, so each isolate could be assigned an enzyme type (ET) that reflected its phenotype for these three enzymes. The observed differences are probably encoded by chromosomal genes. Two methods were used to sample natural soil populations: (1) In situ: Rhizobium strains were isolated from root nodules on host plants growing at the site (one isolate from each nodule). (2) In vitro: Soil, suspended in water, was used to inoculate host plant seedlings growing individually on sterilized agar containing plant nutrients, and one Rhizobium strain was isolated from one nodule from each plant.

From our surveys of ET frequencies in a number of British populations of R. leguminosarum biovars viceae and trifolii, and also biovar phaseoli and R. meliloti at one site, we can draw a number of conclusions about the genetic structure of populations:
(1) **Rhizobium populations are mixtures of widespread lineages.** Enzyme electrophoresis reveals a limited number of distinct genotypes (ETs). Chromosomal recombination between them is probably infrequent, because there is very strong disequilibrium between enzyme loci. The same ETs have been found at widely separated sites in the UK.
(2) **Some lineages can carry alternative host-range determinants.** When several host species were nodulated in the same soil, the same ET was sometimes recovered from more than one species. Presumably such genotypes can carry any of several alternative Sym plasmids. Other ETs were restricted to a single host species. The three biovars of R. leguminosarum showed considerable overlap of ETs. R. meliloti was almost, but not entirely, distinct.
(3) **Diversity may be restricted in acid soils.** Acid soils at two different sites in Wales (20km apart) were dominated by the same single ET. By contrast, a neutral soil in between these sites yielded a diversity of ETs. The strains from acid sites have measurably greater acid tolerance.
(4) **A single plant is nodulated by many strains.** Up to 40 nodules were examined from each of 9 plants growing at 10cm intervals along a row within a pea crop. Diversity of ETs within plants was almost as great as for the population as a whole, and there was little correlation even between adjacent nodules. This implies that the strains are well mixed down to the millimetre scale. ET frequencies amongst strains nodulating the main root (early nodulation) were not very different from those on lateral roots (late).

REFERENCE
Young, J.P.W. (1985) J. Gen. Microbiol. 131, 2399-2408.

ACETYLENE REDUCTION BY ROOTS OF INTACT CORN PLANTS AFTER EXPOSURE TO REDUCED pO_2 AT SEVERAL STAGES OF GROWTH

D. B. ALEXANDER and D. A. ZUBERER
DEPT. OF SOIL AND CROP SCIENCES, TEXAS A&M UNIV.,
COLLEGE STATION, TX   77843   U.S.A.

To determine the effect of oxygen concentration on rates of N_2-fixation (C_2H_2) associated with roots of intact corn plants we conducted a series of experiments in which roots of intact plants were periodically exposed to reduced oxygen tensions (1-4%) then assayed for acetylene reduction activity (ARA). Corn plants (Zea mays, Mol7 x B73) were grown in plastic cylinders (50 cm x 10 cm diam.) containing approximately 3 kg of fritted (calcined) smectitic clay as a solid rooting medium providing a high degree of porosity and allowing gases to be flushed through the cylinders. Plants were inoculated with a soil suspension (Weswood silt loam) to provide indigenous diazotrophic bacteria. Plants were kept in a greenhouse and watered with 1/2-strength Hoagland's solution with nitrate. Plants could be grown to reproductive maturity under these conditions. Several days prior to conducting AR assays, the cylinders were leached with distilled H_2O to remove excess nitrate from the system. The oxygen concentration around the roots was controlled by flushing the cylinders with air diluted with N_2 by means of gas proportioners. The plants were sealed off several cm above the crown using a pourable silicone sealer (GE-RTV-11). Gases were flushed through the cylinders from the bottom to the top for 17-48 hrs prior to measurement of rates of AR.

AR associated with roots exposed to the low oxygen concentrations was immediate and linear for at least 3 hours. The activity of such plants was 5 to 15 fold greater than that of plants exposed to 8-14% oxygen concentrations. The highest rate of AR observed was 8 μmol C_2H_4/plant/hr for plants 53 days old. Prolonging the exposure to low oxygen concentrations led to greater rates of AR. Activity was frequently as much as two fold greater following a second day of maintenance of the roots at low pO_2. AR was variable among groups of plants sown at different times and within the same group of plants assayed at progressive stages of growth. Under our experimental conditions, AR associated with roots at low pO_2 was greatest during vegetative growth (32-55 days) and lower (2 to 3 fold) at the time plants began flowering. Plants treated with 8-14% O_2 exhibited greater AR at flowering than during the earlier vegetative growth period.

Our studies demonstrated conclusively that the concentration of oxygen in the root-zone is a major factor, among others, limiting the rates of nitrogen fixation in associative symbioses.

PHYSIOLOGICAL AND GENETIC CHARACTERIZATION OF A DIAZOTROPHIC PSEUDOMONAD

Yiu-Kwok Chan, Roger Wheatcroft and Robert Watson
C.B.R.I., Research Branch, Agriculture Canada
Ottawa, Ontario K1A 0C6 Canada

Introduction. The soil isolate, 4B, has been previously and tentatively identified to be a Pseudomonas species capable of fixing N_2 with simple phenolics as carbon source (Chan 1985). In view of the controversial status of diazotrophs in the Pseudomonas genus (Palleroni 1984; Postgate 1982), we carried out a more detailed physiological and genetic characterization of 4B including $^{15}N_2$ fixation and DNA homology studies with the nif structural genes of Klebsiella pneumoniae.

Materials and Methods. Two diazotrophic and one non-diazotrophic reference pseudomonads were used for comparative characterization of 4B. The former were Pseudomonas H8 (ATCC 35402) and Pseudomonas sp. (PD) isolated from grass rhizospheres (Barraquio et al. 1983; Haahtela et al. 1983). The non-diazotrophic reference was P. delafieldii (ATCC 17505). Maintenance and standard assays of the cultures were done as previously reported (Chan 1985; Stanier et al. 1966). $^{15}N_2$ incorporation was determined based on the technique outlined by Bergersen (1980). DNA sequence homology was tested by probing with pSA30 in which the nif HDK genes of K. pneumoniae M5al were cloned.

Results and Discussion. Maximum growth of 4B was found at 34°C and pH 6 when determined separately. At pO_2 0.05 in Burk's N-free medium, 4B reduced C_2H_2 at comparable rates with 28mM glucose or 5mM protocatechuate as carbon source, which was also confirmed by $^{15}N_2$ fixation. Under these conditions, the C_2H_4 formed/N_2 fixed ratio was 10. This apparently high value could be attributed to the sub-optimal assay conditions for nitrogenase activity. Pseudomonas species 4B, H8 and PD all contained DNA homologous to the nif HDK genes of K. pneumoniae. The DNA G+C mol% of 4B was determined by UV absorption to be 60.2 ± 1.2. Plasmid DNA was not detected in alkaline SDS lysates of any of these strains by agarose gel electrophoresis; only PD was found to accept the degradative (TOL) plasmid from P. putida (ATCC 33015). Biochemical results show that 4B is unequivocally a Pseudomonas species which has a similar carbon utilization pattern with P. delafieldii. Yet, Pseudomonas 4B was shown to be phenotypically and genetically distinct from H8 and PD. Isolate 4B, therefore, represents a new Pseudomonas species in which active nitrogenase genes have been identified.

References.
Chan YK (1985) In Skinner FA and Uomala P, eds, Nitrogen Fixation with Non-Legumes, Martinus Nijhoff/Junk, The Hague. In prep.
Barraquio WL et al. (1983) Can. J. Microbiol. 29, 867-873.
Bergersen FJ (1980) In Bergersen FJ, ed, Methods for Evaluating Biological Nitrogen Fixation, pp. 65-110, John Wiley & Sons, New York.
Haahtela et al. (1983) Can. J. Microbiol. 29, 874-880.
Palleroni NJ (1984) In Krieg NR and Holt JG, eds, Bergey's Manual of Systematic Bacteriology, Vol. 1, pp. 141-199, Williams & Wilkins, Baltimore.
Postgate JR (1982) Trans. R. Soc. London, Ser. B., 296, 343,-361.
Stanier RY et al. (1966) J. Gen. Microbiol. 43, 159-271.

BDELLOVIBRIO PARASITIC ON AZOSPIRILLUM BRASILENSE

IN SOILS OF NORTH EASTERN BRAZIL

J.J. GERMIDA
DEPARTMENT OF SOIL SCIENCE, UNIVERSITY OF SASKATCHEWAN,
SASKATOON, SASKATCHEWAN, S7N OWO, CANADA

Azospirillum brasilense is widely distributed in nature and associated with the roots of various grasses. Azospirilla and other nitrogen fixing bacteria are studied as inoculants for various crops based on reports of enhanced crop growth. Little is known about predators of azospirilla, yet predators are active in specialized habitats (such as those occupied by azospirilla) where large prey cell populations are found. This study was undertaken to assess soils from Brazil and Canada for bdellovibrios parasitic on A. brasilense.

Soils were assayed directly for populations of bdellovibrio, but none were recovered. After enrichment of soils with potential prey cells, however, bdellovibrio parasitic on A. brasilense were recovered from a Latosol and a Podzolic soil from Brazil. The addition of A. brasilense strain Cd or strain Sp 7 cells and nutrients to these soils stimulated growth of indigenous bdellovibrio. In the Podzolic soil strain Cd cells yielded a bdellovibrio population of 1340 per g soil, whereas strain Sp 7 cells yielded a population of only 50. Escherichia coli and Enterobacter aerogenes cells did not stimulate growth of bdellovibrio in this soil, but did stimulate growth of bdellovibrio in the Latosol as did strains Cd and Sp 7. The morphology of azospirilla-attacking bdellovibrio, isolated from the Podzolic soil, was typical of the genus; attack-phase cells were curved rods, 0.2-0.4 by 1.0-1.3 um, motile by means of a single polar flagellum. In broth culture, this bdellovibrio isolate parasitized several different gram negative bacteria, although the specificity towards host cells was A. brasilense strain Cd > strain Sp 7 > A. lipoferum strain Sp Br 17 = E. coli = En. aerogenes > A. brasilense strain Sp 35. Pseudomonas fluorescens, Ensifer adhaerens and 20 unidentified bacterial isolates from the Podzolic soil were not parasitized.

These results indicate that bdellovibrios are present in air-dry soils at undetectable levels but respond quickly to the presence of replicating prey cells. Bdellovibrios active against specific prey cells are found in soil and their detection in soil reflects the presence of specific prey cells or other prey cells able to support their growth in situ. The presence in soil of bdellovibrios that prefer A. brasilense strain Cd over strain Sp 7 cells and other azospirilla indicates potential problems when using strain Cd as a crop inoculant in certain soils.

BACTERIAL ASSOCIATIONS UTILISING STRAW TO FIX ATMOSPHERIC NITROGEN

ALAN H. GIBSON AND DOROTHY M. HALSALL
CSIRO DIVISION OF PLANT INDUSTRY, CANBERRA, A.C.T., 2601, AUSTRALIA.

Straw is a large energy resource which, under appropriate microbiological and physical conditions in the soil, can be utilised by N_2-fixing bacteria (Roper 1983). Azospirillum spp. are able to utilise xylan, a major constituent of straw (30%) as an energy source for nitrogen fixation (Halsall et al. 1985). Various combinations of cellulolytic fungi or bacteria and diazotrophic bacteria have been used to inoculate straw and other organic matter to promote nitrogen fixation (see Halsall and Gibson 1985). An obvious problem has been the separation of the two microbes in time or space, in order to achieve appropriate pO_2 conditions for both to function. We report here on the co-culture of Azospirillum spp. and Cellulomonas spp. on straw and cellulose substrates.

MATERIALS AND METHODS

Sterile straw or cellulose was mixed through sterile acid-washed sand in Universal bottles and inoculated with washed suspensions of the bacteria in inorganic nutrient solution with 10 mg $(NH_4)_2SO_4$/l. To examine pO_2 and pH effects on cellulose breakdown, shaken cultures (10 ml/125 ml flask) were used. Straw and cellulose utilisation was determined by gravimetric analysis and by CO_2 production, total N by Kjeldahl digestion and soluble carbohydrate by phenol/H_2SO_4. Cellulase activity in culture supernatants was determined by glucose production over 24 h in an 'Avicel'-Na azide mixture, pH 7.0.

RESULTS

Azospirillum brasilense Sp7, A. lipoferum 5A, and two Bacillus macerans cultures were unable to utilise cellulose, but in the presence of Cellulomonas gelida, nitrogenase activity was significant (1.3-2.8 µmoles/culture/24 h). C. gelida promoted nitrogenase activity by these diazotrophs with straw as the C-source, especially with the B. macerans strains. Even where the inoculum contained Azospirillum and Cellulomonas in the ratio 1:2000, the cultures reached an equilibrium of 1:3 in 7 days; nitrogenase activity was evident at 3 days.

A mutant strain of Cellulomonas sp. (Haggett et al 1979) was better able to break down cellulose at low pO_2 (1.0 and 0.1% in gas phase), and at pH 5.6, than C. gelida.

In co-cultures of the two Cellulomonas strains with A. brasilense Sp7 in straw-amended sand, 1.5 and 1.8 mg N_2 were fixed by C. gelida and the mutant strain over 38 days. Approx. 25% of the 100 mg substrate had been broken down. The estimated efficiency was 62.9 and 72.4 mg N_2 fixed/g straw, in contrast to figures of less than 20 mg N_2/g straw in previous work (see Halsall and Gibson 1985). This high level of efficiency was probably due to the close physical relationship of the two species and the low rate of supply of available carbon substrate to the Azospirillum. Significant promotion of nitrogenase activity was observed when Cellulomonas and Azospirillum were inoculated into straw-amended soil (both sterile and non-sterile).

Haggett, KD et al (1979) Eur. J. Appl. Microbiol. Technol. 9, 183-190.
Halsall DM et al (1985) Appl. Env. Microbiol. 49, 423-428.
Halsall DM and Gibson AH (1985) Appl Env. Microbiol. (submitted).
Roper MM (1983) Aust. J. Agric. Res. 34, 725-739.

REGULATION OF NITROGENASE ACTIVITY BY AMMONIUM CHLORIDE AND OXYGEN IN AZOSPIRILLUM spp.

ANTON HARTMANN, HAIAN FU, AND ROBERT H. BURRIS
Department of Biochemistry, Univ. of Wisconsin-Madison,
College of Agricultural & Life Sciences, 420 Henry Mall,
Madison, WI   53706   U.S.A.

We investigated the mechanism of nitrogenase regulation in $\underline{A}$. brasilense Sp7, $\underline{A}$. lipoferum SpRG20a & · SpBr17, and $\underline{A}$. amazonense Y1.

Ammonium chloride inhibited the nitrogenase activity of all three Azospirillum spp. reversibly at 50-100 µM concentration (NH_4-switch off/on). When the "switch off" with 1 mM NH_4Cl was performed under oxygen-controlled conditions, the inhibition of nitrogenase was complete after a transition period in $\underline{A}$. brasilense and $\underline{A}$. lipoferum, but it was not complete in $\underline{A}$. amazonense. Even 10 mM ammonium chloride did not completely inhibit its nitrogenase activity. In $\underline{A}$. brasilense and $\underline{A}$. lipoferum the in vitro-nitrogenase activity was reduced after the addition of ammonium chloride to the cells. "Quick extracts" were prepared according to Kanemoto and Ludden from cultures before and after NH_4-treatment and the status of Fe protein was examined with immuno-blotting. An additional Fe protein subunit appeared on SDS-PAGE during the "NH_4-switch off", and it migrated like the modified subunit of the inactive Fe protein of $\underline{R}$. rubrum. An incorporation of ^{32}P into this subunit was observed in a ^{32}P-labelled culture of $\underline{A}$. brasilense Sp7. In $\underline{A}$. amazonense the in vitro-nitrogenase activity was not affected by ammonium chloride. A closely migrating double band of Fe protein was present before and after "NH_4-switch off", but no modified Fe protein subunit appeared.

The influence of dissolved oxygen on the nitrogenase activity was measured in a specially designed oxygen chamber that provided a quick equilibration between the gas and liquid phase and allowed the simultaneous measurement of dissolved oxygen and of acetylene reduction activity. The optimal range for nitrogenase activity by Azospirillum spp. was 0.001-0.002 atm O_2 (0.1-0.2 kPa). The nitrogenase activity of $\underline{A}$. brasilense Sp7 was inhibited more effectively by increased O_2-levels than it was in $\underline{A}$. lipoferum SpRG20a. The highest oxygen tolerance was shown by $\underline{A}$. amazonense Y1. The inhibition by oxygen was partially reversible in all Azospirillum spp. The mechanism of the "oxygen switch off" was examined in "quick extracts" of oxygen-inhibited cultures with the immunoblotting method. No modification of the Fe protein was found in cells with oxygen-inhibited nitrogen fixation. The oxidation of reducing equivalents for nitrogenase function may have caused the inhibition. Under anaerobic conditions, which inhibit the nitrogenase activity of Azospirillum spp. reversibly, the Fe protein of $\underline{A}$. brasilense and $\underline{A}$. lipoferum were modified. This probably resembles the "dark-switch off" of nitrogenase in $\underline{R}$. rubrum.

We found clear differences in the regulation of nitrogenase activity by ammonium chloride and oxygen among the Azospirillum spp. Because of the weak inhibition by ammonium and relatively high oxygen tolerance of its nitrogenase, $\underline{A}$. amazonense Y1 isolated by J. Döbereiner's group, appeared a favorable organism for N_2 fixation in the field.

SPECIFICITY OF POLYCLONAL ANTISERA AGAINST
DINITROGEN FIXING AZOSPIRILLUM SPP.

PETRA G. LACHMANN[1] and MICHAEL KLOSS[2]
1) Abteilung für Immunologie, Medizinische Hochschule Hannover,
   Konstanty - Gutschow - Str. 8, D 3000 Hannover 71, FRG
2) Institut für Biophysik / Isotopenlaboratorium, Universität Hannover,
   Herrenhäuser Str. 2, D 3000 Hannover 21, FRG

Introduction

Serological techniques have been widely used to identify Rhizobia grown
in culture or in the bacteroid form from crushed root nodules.
Compared with this, relativly little work has been done to employ
these techniques for identification of organisms involved in
associative nitrogen fixation. This is due to the lack of sufficient
knowledge of serological identification methods for Azospirillum and
the limited availibility of monospecific antisera. For this reason the
antigenicity of Azospirillum was investigated.

Results

All experiments were carried out with two Antisera raised against the
type strains Azospirillum lipoferum Sp 59b and A. brasilense Sp 7
respectively. Dott - Blot revealed that both antisera are species
specific. Ouchterlonys revealed antigenic differences between
different strains. Crossreactivity between different species of the
same genus was not observed. No antigenic differents were observed
between nitrogen fixing and non - fixing cultures. SDS PAGE showed
clear differences between A. brasilense and A. lipoferum. Differences
of protein patterns between strains of A. lipoferum were only observed
for the root isolates from Kallar grass (RP 5, RP 16).
Immunoblots confirmed the serological differences of strains observed
for A. brasilense by immunodiffussion. The Isolate ER 8, which is
supposed to be a different genus, as determined by physiological tests
(see also,Reinhold, et al., this meeting) and confirmed by the
SDS PAGE protein pattern, showed significant crossreactivity with both
antisera.

The experiments show clearly the possibility for a serological strain
specific identification system of the genus Azospirillum.

RHIZOBIUM AND ENDOMYCORRHIZAL INTERACTIONS IN SOYBEAN

R. S. PACOVSKY AND G. FULLER
USDA, ARS, WESTERN REGIONAL RESEARCH CENTER
ALBANY, CA 94710 U.S.A.

1. INTRODUCTION

In most soils, N_2-fixing legumes exist in a tripartite association. The host plant provides C to the Rhizobium and the vesicular-arbuscular mycorrhizal (VAM) fungus. The P input due to VAM fungi influences host growth as well as nodulation and N_2 fixation (Asimi et al., 1980). In P-fixing soils, VAM fungi may be very effective at promoting growth, whereas in sand/perlite VAM infection can be deleterious. The question remains whether VAM-induced growth inhibition will occur in extremely N-deficient environments, and what effect will it have on N_2 fixation? The purpose of this study was to examine the effects of dual inoculation with VAM fungi and Rhizobium (RHIZO) on the growth, mineral content and N_2 fixation of soybeans grown in a P-fixing soil and in sand/perlite.

2. MATERIALS AND METHODS

Soybean (Glycine max L. Merr. cv. Amsoy 71) plants were grown in either a P-fixing soil (mesic Typic Haploxerult) or sand/perlite that received 19 mg P. Plants were inoculated with one or both of the microsymbionts (R. japonicum strain 61A118 or Glomus fasciculatum) or were left uninoculated (control). All combinations received an N- and P-free nutrient solution. Each treatment was replicated six times and all plants were harvested at week 9. Nitrogenase activity of excised roots (H_2 evolution and C_2H_2 reduction) was determined as described previously (Pacovsky et al, 1984). Plant and fungal characteristics were measured and statistical analyses were evaluated as before (Pacovsky et al, 1985).

3. RESULTS AND DISCUSSION

Plants in both soils showed the greatest response to inoculation with Rhizobium. Increases in plant dry weight, N content and C_2H_2 reduction for dually-infected plants was due to a positive VAM x RHIZO interaction. Nitrogen fixation, calculated from C_2H_4 and H_2 data, was significantly higher in the tripartite symbiosis due to nodule weight increases. The relative efficiency for N_2 fixation was much lower in sand/perlite due to greater H_2 evolution by the nodules. This trend towards greater H_2 production in sand/perlite was reversed following VAM colonization possibly due to decreased C allocation to the nodules. Root colonization, by VAM increased significantly following nodulation. The data indicated that N_2 fixation was essential to soybean growth and was dependent on the supply of P from the root. The efficient allocation of P to the nodules indicated a difference in the critical P concentration between host root and nodules. Enhancement of N_2 fixation in VAM plants decreased H_2 production, which in turn may be due to carbohydrate utilization following fungal proliferation. The synergistic nature of the VAM x Rhizobium interaction suggests that the influence these endophytes have on the host is not linked to N and P nutrition alone, but depends on a balance between the three members of the symbioses.

4. REFERENCES

Asimi S et al (1980) Can. J. Bot. 58, 2200-2205.
Pacovsky R S et al (1984) Crop Sci. 24, 101-105.
Pacovsky R S et al (1985) Soil Biol. Biochem. 17, 525-531.

OXYGEN RECOGNITION IN AEROTACTIC BEHAVIOUR OF AZOSPIRILLUM BRASILENSE

ORLY REINER*, YAACOV OKON* and MICHAEL EISENBACH**
* DEPARTMENT OF PLANT PATHOLOGY AND MICROBIOLOGY, FACULTY OF AGRICULTURE, THE HEBREW UNIVERSITY OF JERUSALEM, REHOVOT 76100, ISRAEL
** DEPARTMENT OF MEMBRANE RESEARCH, THE WEIZMANN INSTITUTE OF SCIENCE, REHOVOT 76100, ISRAEL.

Nitrogen fixing bacteria of the genus Azospirillum colonize the roots of grasses, promoting growth as a result of the association (Okon, 1985).

A. brasilense are microaerophilic bacteria which actively move towards self-created gradients of low oxygen concentrations. The purpose of this work was to identify receptor(s) involved in oxygen recognition by A. brasilense strain Cd ATCC 29729. Spectral studies of membrane preparations of A. brasilense revealed three terminal oxidases: cytochromes aa_3, O and d. We tried to examine the possible involvement of each of these cytochromes in oxygen reception.

It appears that cytochrome aa_3 is not the oxygen receptor for the following reasons: a) reducing the content of cytochrome aa_3 in bacteria (by growing the bacteria under low [0.01 mM] dissolved oxygen) did not affect their aerotactic behaviour, as measured by capillary assay (Barak et al., 1982); b) Inhibiting cytochrome aa_3 by low concentrations of cyanide did not affect aerotaxis.

High concentrations of cyanide that inhibited cytochrome O and d inhibited aerotaxis, as measured with tethered cells (Silverman and Simon, 1974). Between these two, cytochrome O seems to be the preferred candidate for the oxygen receptor since Antimycin A which inhibits the reduction of this cytochrome (but not that of cytochrome d) completely inhibited aerotaxis.

REFERENCES

Barak R et al (1982) J. Bacteriol. 152, 643-649
Okon Y (1985) In Ludden FW and Burris JE, eds.
Nitrogen Fixation and CO_2 Metabolism, pp. 165-174, Elsevier
Silverman M and Simon M (1974) Nature 249, 73-74.

AZOSPIRILLUM HALOPRAEFERANS SP.NOV., A DIAZOTROPH
ASSOCIATED WITH ROOTS OF LEPTOCHLOA FUSCA (LINN.) KUNTH

BARBARA REINHOLD, THOMAS HUREK AND ISTVAN FENDRIK
INSTITUT FÜR BIOPHYSIK/ISOTOPENLABORATORIUM
HERRENHÄUSER STR.2, D-3000 HANNOVER 21, FED.REP.GERMANY

Bacteria of the genus Azospirillum are known to be present in
the rhizosphere of various tropical grasses. Four species
have been described: A.lipoferum, A.brasilense, A.amazonense
and A. seropedicae. During screening for occurrence of the
most abundant diazotrophs in the rhizosphere of Kallar grass,
a salt tolerant pioneer plant on saline-sodic soils in Pakistan, a new Azospirillum species could be isolated from the
rhizoplane.

Eight isolates were tested. Like Azospirillum, it fixes N_2
by forming a thin subsurface pellicle in semisolid media, it
is vibroid to S-shaped and has a corkscrew-like motion by a
polar flagellum, and its DNA base composition is within the
range of Azospirillum. From the known Azospirillum species
it can mainly be distinguished by (I) Biotin requirement
without ability of glucose utilisation (II) higher optimum
temperature for growth at 41°C (III) stimulation of growth
by 0.25% NaCl.

Formal Description of the Organism:

Azospirillum halopraeferans sp.nov. (Gr.n. halo, halos,salt,
the sea; L.v. praeferre, to prefer. M.L. part. adj. salt preferring). Cells vibroid to S-shaped, only few helical cells
in alkaline media. Cell width 0.7-1.0 μm, can increase to
1.2 μm on alkaline media. Motile by one polar flagellum.Fixes
N_2 under microaerobic conditions at a concentration of 1 μM
dissolved O_2. Under these conditions efficiency 10 mgN/g
L-malate and generation time 4,2h. Oxidase, urease, nitrate
reductase, denitrification positive. Requires biotin. No
acidification of peptone based glucose broth or anerobically
of glucose or fructose broth. N_2-dependent growth on α-ketoglutarate, on mannitol weak and not on glucose. Acidification
of fructose. Sole carbon sources for growth pyruvate, malate
ß-hydroxybutyrate, α-ketoglutarate, succinate, fumarate,
glycerol, mannitol, fructose, arabinose, gluconate. Good
growth on tryptic soy agar, colonies cream coloured, flat,
with concentrical rings. No growth on potato agar. Optimum
temperature for growth 41°C. pH-range for good growth 6.8-8.0.
The mol% G+C of DNA is 68.1. Type strain Azospirillum halopraeferans Au 4.

ENUMERATION AND LOCALISATION OF DIAZOTROPHS IN THE
RHIZOSPHERE OF KALLAR GRASS

BARBARA REINHOLD, THOMAS HUREK, ISTVAN FENDRIK
AND ERNST-GEORG NIEMANN
INSTITUT FÜR BIOPHYSIK/ISOTOPENLABORATORIUM
HERRENHÄUSER STR.2, D-3000 HANNOVER 21, FED.REP. GERMANY

Leptochloa fusca (Linn.) Kunth, commonly called Kallar grass,
is a salt tolerant grass used as a pioneer plant on
saline-sodic soils, having a low nitrogen content, in
Pakistan. In this study, we estimated the population of
microaerophilic N_2-fixing microorganisms associated with
Kallar grass roots.
Root samples were collected from Shakot area, Punjab.
Enumeration of diazotrophs was carried out by the MPN-method
on two semisolid malate media, one of them adapted to the
alcaline and saline conditions found at the Kallar grass
site. Bacterial numbers from root-free soil, from rhizoplane
and endorhizosphere were determined in several independent
experiments. "Surface sterilisation" of the roots was
carried out with NaOCl and checked by determining the
reduction of the rhizoplane population. Evaluation was done
by acetylene reduction assay, microscopic examination of
positive cultures and isolation of the dominating diazotrophs followed by serological and physiological characterisation. Additionally, total numbers of non-diazotrophic
microorganisms were estimated by plating on nutrient agar.

The diazotrophic population of the rhizoplane was entirely
different from that of the endorhizosphere. The dominating
diazotrophs on the rhizoplane were Azospirillum spp.
($2 \times 10^7 \pm 1$ per g root dry weight), which were identified
as A. halopraeferans and A. lipoferum. They were not
colonizing the endorhizosphere. There, we found motile
straight rods in high numbers ($7 \times 10^7 \pm 7$ per g rdw), which
still have to be characterized. Compared with non diazotrophs, N_2-fixing bacteria were preferentially enriched in
the Kallar grass rhizosphere. They made up 0.2% $\pm$ 0.15 of
the total aerobic microflora in the root-free soil,
7% $\pm$ 5 on the rhizoplane and 85% $\pm$ 20 in the endorhizosphere.

High numbers of diazotrophs in and on the roots as well as
their preferential enrichment indicate a close association
between Kallar grass and diazotrophs.

Acknowledgements
This research was funded in part by KFK (project no. 044 2a).

COMPARATIVE CHARACTERIZATION OF SOLUBLE AND MEMBRANE-BOUND
HYDROGENASES FROM ANABAENA AZOTICA LEY HB696

CHENG SHUANGQI, CHEN ZHAOPING, MO XIMU, BIOLOGY, SOUTH CHINA NORMAL
UNIVERSITY, CANTON, CHINA

Two forms of hydrogenases,soluble and membrane-bound hydrogenases, had been isolated from A. azotica HB686 and partially purified. The membrane-bound hydrogenase, which accounted for 67% of the total activity of H_2 uptake, was the main form of H_2-uptake hydrogenase.

Some properties of both hydrogenases were similar.The two forms of hydrogenases could carry on bidirectional catalysis, in vitro. The most efficient electron acceptor was methylene blue (MB) for H_2-uptake reaction, and the donor was methyl viologen (MV) reduced for H_2-evolution reaction. The pH optimum for H_2 uptake was 7.4 with MB as the electron acceptor. The both hydrogenases were heat-stable and not inactivated when heated at 50 C for 10 min. They were inhibited by CO and C_2H_2, high ionic strength, sulphydryl reagents, and transition metals, in their degrees.

Differences existed in properties of the two forms of hydrogenases. Under optimum condition, the Km (MB) and Km (H_2) for the soluble hydrogenase were 0.410 mM and 0.082 mM, and for the membrane-bound one were 0.190 mM and 0.044 mM, respectively. In comparison to the soluble hydrogenase, the membrane-bound one had a higher activity of H_2 uptake whereas a lower activity of H_2 evolution. The ratio of H_2 uptake to H_2 evolution was about 19 for the membrane-bound hydrogenase, whereas about 2 for the soluble one. It appears that the membrane-bound hydrogenase mainly carries out the function of H_2 uptake, in vivo. The two forms of partially purified hydrogenase did not react with O_2 directly, but the isolated membrane, in which the uptake hydrogenase is bound, could catalyze an oxyhydrogen reaction. In addition,the activity of soluble hydrogenase was relatively equal in the extracts of both vegetative cells and heterocysts,whereas the membrane-bound one was mainly detected in extracts of heterocysts. Although both forms of hydrogenases showed cool instability, and the activity of H_2 uptake decreased quickly when stored at -20 C, the membrane-bound enzyme was more sensitive to low temperature, being more stable at 23 C, whereas the soluble one was more stable at 4 C.

GLYCOGEN SYNTHESIS AND THE INDUCTION OF NITROGENASE SYNTHESIS IN
HETEROCYSTOUS CYANOBACTERIA.

ANNELIESE ERNST and PETER BöGER, Lehrstuhl für Physiologie und Biochemie d.
Pflanzen, Universität Konstanz, D-7750 Konstanz, Germany.

The cyanobacterium Anabaena variabilis performs oxygenic photosynthesis in
the vegetative cells and at the same time is able to express oxygen
sensitive nitrogenase in specialized cells, the heterocysts. As in other
diazotrophic organisms nitrogenase synthesis results from an imbalance in
the C:N supply of the culture. In this organism the C is supplied via
photosynthesis and glycogen, a photosynthetic storage product was shown to
influence both, nitrogenase acitivity and stability (1). Prior to
heterocyst differentiation and nitrogenase induction an increase in the
cellular C:N ratio and in glycogen were observed (2,3). This process is
therefore thought to be also involved in the regulation of nitrogenase
synthesis. To further study this hypothesis, heterocyst-free cultures of
A. variabilis were grown on a limited amount of NH_3. 32 h after
inoculation, NH_3 was completely consumed by the cells. The dry weight
continued to increase, while chlorophyll synthesis was temporarily
inhibited until nitrogenase had been induced. During this experiment two
different types of glycogen accumulation were observed:
  a. after inocculation of a new batch culture;
  b. after depletion of NH_3 in the medium.
Only the b-type accumulation was accompanied by heterocyst differentiation
and the induction of nitrogenase.
A decrease in endogenous NH_3-pool was observed during growth, resulting
in a lower NH_3-level during b-type glycogen accumulation than during
a-type accumulation. However a similar decrease was observed in
diazotrophic cultures, not leading to enhanced nitrogenase synthesis.
From these results we conclude:
  1. Glycogen accumulation is forced by an increase in average incident
     irradiation or after a dark-light transition.
  2. Glycogen accumulation is stimulated by nitrogen limitation (lack of
     combined and molecular N).
  3. Nitrogenase synthesis is supressed, when N-supply is abundant, even
     when photosynthetic activity leads to a build-up of glycogen.
  4. A metabolic C:N ratio is unlikely to influence expression of
     nitrogenase.
  5. No conclusive evidence for NH_3 being a specific inhibitor of
     differentiation was observed.

References:
1) A. Ernst, H. Kirschenlohr, J. Diez and P. Böger (1984) Arch. Microbiol.
140: 120-125
2) S.A. Kulasooriya, N.J.Lang and P. Fay (1972) Proc. R. Soc. (B), 181:
199-209
3) R. Rippka and R.Y. Stanier (1978) J. Gen. Microbiol. 105: 83-94

LONG TERM EFFECTS OF EXPOSURE TO O_2 ON N_2 FIXATION IN GLOEOTHECE

JOHN R. GALLON/ALAN E. CHAPLIN/PAUL S. MARYAN*/ROBERT R. EADY*
DEPARTMENT OF BIOCHEMISTRY, UNIVERSITY COLLEGE OF SWANSEA, SINGLETON PARK, SWANSEA SA2 8PP, U.K.
*AGRICULTURAL AND FOOD RESEARCH COUNCIL, UNIT OF NITROGEN FIXATION, UNIVERSITY OF SUSSEX, BRIGHTON BN1 9RQ, U.K.

Cultures of the unicellular cyanobacterium Gloeothece do not fix N_2 (reduce acetylene) when exposed to O_2 concentrations greater than 0.7 atm. Although exposure to 1 atm of O_2 for as little as 2 min completely inhibited acetylene reduction, the ability to fix N_2 reappeared following return to air (Gallon, Hamadi, 1984).

The pattern of reappearance of N_2 fixation following transfer to air depended upon the length of previous exposure to O_2. After 1 h exposure to O_2, there was a lag of 1 - 2 h before nitrogenase activity reappeared in air but after 6 - 24 h exposure to O_2, nitrogenase activity reappeared without any lag. Furthermore, as the length of exposure to O_2 was increased within these limits, the initial rate of recovery in air became more rapid. However, after 48 h exposure to O_2, the initial recovery of nitrogenase activity in air was greatly decreased and, after longer exposure to O_2, no activity reappeared for at least 18 h after transfer to air.

Addition of 0.1 mg/ml of chloramphenicol or 2 mM NH_4Cl, simultaneously with transfer to air, abolished all recovery of nitrogenase activity, suggesting that recovery depends upon protein (probably nitrogenase) synthesis. In addition, radioimmunoassay of $^{35}SO_4^{2-}$ incorporation into the MoFe protein of Gloeothece nitrogenase demonstrated that, following transfer of cultures to O_2, nitrogenase synthesis initially ceased but subsequently occurred at a rapid rate. Furthermore, these data supported the idea that the initial rate of recovery of N_2 fixation, following transfer of cultures to air, reflects the rate of nitrogenase synthesis under O_2 immediately prior to this transfer.

It therefore appears that in cultures of Gloeothece incubated under 1 atm of O_2, nitrogenase synthesis is at first inhibited, perhaps repressed by O_2 itself. However, after 1 - 2 h of incubation, synthesis of nitrogenase gradually increases, reaching a maximum rate about 24 h after addition of O_2. This increase may be a consequence of derepression of nitrogenase synthesis caused by nitrogen starvation, since it may be delayed by addition of a non-repressive concentration of NH_4Cl. If control of nitrogenase synthesis under O_2 does involve repression/derepression, then it appears that derepression under conditions of nitrogen starvation can override any repression by O_2. Nevertheless, after more than 24 h under O_2, nitrogenase synthesis decreases, probably due to depletion of intracellular nitrogen reserves.

REFERENCES

Gallon J R and Hamadi A H (1984) J. Gen. Microbiol. 130, 495-503.

ACKNOWLEDGEMENTS

We thank S.E.R.C. for the award of a Research Studentship to P.S.M., and the Biochemical Society (London) for a travel grant in order to attend the symposium.

N$_2$-FIXATION ENHANCEMENT IN VIVO BY KREBS CYCLE INTERMEDIATES AND BY SUGARS IN THE CYANOBACTERIUM NOSTOC MUSCORUM

Lea Karni and Elisha Tel-Or
Department of Agricultural Botany, The Hebrew University, PO Box 12, Rehovot 76100, Israel

Recent reports on the enriched activity of isocitrate dehydrogenase in heterocysts of Nostoc muscorum (Karni, Tel-Or, 1983) suggested that isocitrate dehydrogenase is an important donor of reductant to nitrogenase in the heterocysts. Earlier observations of enriched activity of the oxidative pentose phosphate cycle in heterocysts and the absence of the Krebs cycle enzyme alpha-ketoglutarate dehydrogenase in cyanobacteria have led to the assumption that the oxidative pentose phosphate pathway is the major route for the provision of reductant for cyanobacterial nitrogenase (Smith, 1982).

Air grown filaments of Nostoc muscorum were incubated in different media in the presence of 0.5mM isocitrate and succinate in the light, and their consequent acetylene reduction was followed. Only the cultures suspended in HEPES buffer (25mM pH 7.8), but not in growth medium, or in HEPES + MgCl$_2$ showed isocitrate or succinate enhanced acetylene reduction. NADP at low concentration was found to further enhance the nitrogenase stimulated activity in the presence of isocitrate and succinate. A comparison of the organic substrates supported nitrogense activity showed that sucrose was the most effective sugar, better than fructose, glucose, and glucose-6-P, while the organic acids tested, pyruvate isocitrate, oxaloacetate, and malate, were similarly active. The observed stimulation by the organic substrates should involve the uptake and metabolism of the compounds during 22 hours preincubation, prior to the acetylene reduction assay. Sucrose was found to be the most effective substrate taken up in the dark, while isocitrate and malate were the most effective substrate taken up in the light.

This study suggests that Nostoc muscorum cells are provided with a variety of transport devices for a large number of substrates and that few metabolites may be translocated from the vegetative cell to the heterocyst.

References

Karni L and Tel-Or E (1983) In Papageorgiou GC and Packer L, eds, Photosynthetic Prokaryotes: Cell Differentiation and Function, pp 303-314, Elsevier, New York.
Smith AJ (1982) In Carr NG and Whitton BA, eds, The Biology of Cyanobacteria, pp 47-85, Blackwell, Oxford.

ELECTRON TRANSPORT TO NITROGENASE IN HETEROCYSTS
OF CYANOBACTERIA: RECENT ASPECTS

G. Neuer, H. Papen, H. Bothe

Botanical Institute, The University of Cologne, Gyrhofstr. 15, D-5000
Köln 41, F.R.G.

Introduction

The transfer of reducing equivalents to nitrogenase in heterocysts of
cyanobacteria has been a matter of dispute for years. Ferredoxin is the
immediate electron carrier transferring electrons to the nitrogenase
complex. Ferredoxin is reduced by several electron donors (H_2, NADPH,
NADH or pyruvate). The reaction between electron donors and ferredoxin is
enzyme mediated. Pyruvate is split by the pyruvate clastic reaction
catalyzed by the pyruvate:ferredoxin oxidoreductase in the cyanobacterium
Anabaena. The occurrence of this enzyme in heterocysts was reported
(Neuer and Bothe, BBA 716, 359, 1982; Neuer and Bothe, Arch. Microbiol.,
in press and this communication).The following reactions have been demonstrated:
a) decarboxylation of pyruvate
b) synthesis of citrate from oxaloacetate and acetylcoenzyme A generated
   by the ferredoxin and coenzyme A dependent cleavage of pyruvate
c) synthesis of pyruvate from acetylcoenzyme A and reduced ferredoxin
d) reduction of methyl viologen
e) reduction of ferredoxin
f) pyruvate-dependent C_2H_2-reduction in heterocysts of Anabaena variabilis and Anabaena cylindrica
g) inhibition of pyruvate-dependent C_2H_2-reduction by glyoxylate

Pyruvate is, however, not the only electron donor in heterocysts. These
cells utilize a variety of electron donors which are apparently not
related to each other. The transfer of electrons from some electron
donors to nitrogenase is photosystem I dependent (H_2, NADH, erythrose),
whereas that one from others proceeds in the dark (NADPH, pyruvate). The
regulation of the electron flow is not yet understood. Nitrogenase activity in heterocysts appears to be controlled by the availability of
electron donors (carbon compounds). Vegetative cells appear to supply
heterocysts with all sorts of carbon compounds. In exchange, heterocysts
transfer glutamine to vegetative cells.

PHOTOSYSTEM II AND AMMONIA LIBERATION BY CYANOBACTERIA

JACK W. NEWTON, NORTHERN REGIONAL RESEARCH CENTER, ARS, USDA, PEORIA, IL 61604

Cyanobacteria isolated from the Azolla-Anabaena symbiosis can be grown autotrophically in light or heterotrophically on carbohydrates in the dark. Heterotrophically grown cyanobacteria have a specific activity for acetylene reduction several-fold greater than autotrophs, and, like many nitrogen fixers, liberate ammonia into the medium on treatment with methionine sulfoximine (MSX). When incubated with MSX under N_2 in the light, the heterotrophs liberate large amounts of ammonia and also accumulate ammonia intracellularly. Addition of the photosystem II inhibitors DCMU or atrazine to these cells further increases the amount of ammonia liberated and depletes the ammonia pool without affecting acetylene reduction activity, hence the efficiency of ammonia formation is increased. Heterotrophs incubated with MSX under argon also liberate ammonia and the liberation is enhanced by DCMU, indicating that recently fixed nitrogen was not the sole source of ammonia production enhanced by inhibition of photosystem II. Furthermore cells grown on fructose and ammonia, which have no acetylene reduction activity, release ammonia when incubated with MSX under argon in a nitrogen free medium, and this liberation of ammonia is also enhanced by DCMU. DCMU and atrazine at very low concentration inhibit sustained ^{14}C methylamine uptake by the cells under conditions identical to that which were used to study ammonia liberation. The data support a role for photosystem II in ammonia liberation by the cyanobacteria, and suggest that it functions in a directional manner.

UPTAKE AND REVERSIBLE HYDROGENASE ACTIVITY IN UNICELLULAR AEROBIC NITROGEN
FIXING CYANOBACTERIA.

ROSMARIE RIPPKA, GERMAINE COHEN-BAZIRE
INSTITUT PASTEUR, PARIS, FRANCE.

We have examined six axenic strains of the genus <u>Gloeothece</u> and two of
<u>Cyanothece</u> for their potential to take up hydrogen <u>in vivo</u> in the presence
of C_2H_2 and under conditions where a) nitrogenase synthesis was repressed
by inclusion of 10 mM NH_4Cl in the medium and b) under nitrogenase dere-
pressive conditions, in the absence of a combined nitrogen source. These
experiments were performed in the light (10 μEinstein m^{-2} sec^{-1}) either
under a gas phase containing Ar/1% CO_2/10% C_2H_2 or with the same gas phase
supplemented with 20% O_2. In all cases 1% H_2 was added as overpressure.
All strains with one exception, exhibited relatively high H_2 consumption
rates (10-30 nmoles/min/mg protein) in the absence of combined nitrogen.
The rates under low oxygen concentrations (only photosynthetically produced O_2 being present) were at least 50% higher than those in the presence of 20% O_2. The kinetics of appearance of nitrogenase activity and
of H_2 consumption were very similar. This observation, together with repression of H_2 consumption in the presence of NH_4Cl, suggests that both
enzymes are under the same metabolic control. <u>Cyanothece</u> PCC 7822 exhibited very low H_2 consumption (1 nmole/min mg protein) even under low O_2
concentrations. On the contrary, this strain exhibits high activities of
a reversible hydrogenase both in the presence and absence of NH_4Cl after
cultivation under N_2/CO_2. Inhibitor studies <u>in vivo</u> on the activity of
this hydrogenase suggest the involvement of ferredoxin in the pathway of
H_2 evolution.

PROPERTIES OF A SPECIAL FERREDOXIN FOR NITROGEN FIXATION IN HETEROCYSTS ISOLATED FROM ANABAENA VARIABILIS

Berhard Schrautemeier and Herbert Böhme, Lehrstuhl für Physiologie und Biochemie der Pflanzen, Universität Konstanz, D-7750 Konstanz, Germany

Biological activities of ferredoxin isolated from heterocysts and of ferredoxin from vegetative cells of Anabaena variabilis were compared. Both plant-type ferredoxins catalyzed NADP-photoreduction equally well by reconstituted heterocyst thylakoids, with electrons from H_2 or NADH feeding into photosystem I. When photoreduced, heterocyst ferredoxin was twice as active as ferredoxin from vegetative cells in transferring electrons to Anabaena nitrogenase. In the dark, only the heterocyst ferredoxin was able to link reducing power to nitrogenase, as generated by soluble systems such as H_2/hydrogenase (from Clostridium pasteurianum) and NADPH/ferredoxin:oxidoreductase (from A. variabilis). Using heterocyst homogenates with glucose-6-phosphate as electron donor, only heterocyst ferredoxin was able to further stimulate nitrogenase activity (comp. B. Schrautemeier, H. Böhme (1985) FEBS Lett. 184, 304-308).

Biochemical and immunological studies support the view that the ferredoxin in vegetative cells is distinct from the major component of the heterocyst ferredoxin (Fd I, ~ 90%), suggesting a specific role of the latter in nitrogen fixation. Heterocysts contain an additional, minor component of ferredoxin (Fd II, ~ 10%) more tightly bound to DEAE sepharose, with properties very similar to the vegetative-cell ferredoxin. The molecular weights as determined by SDS-polyacrylamide gelelectrophoresis in the presence of 4 M urea were 23.5 kDa for heterocyst ferredoxin I, and 27.3 kDa for heterocyst ferredoxin II and vegetative-cell ferredoxin. The isoelectric points were 3.0 for ferredoxin I and >2.7 for ferredoxin II and vegetative-cell ferredoxin. Optical absorbance maxima were 465, 420, 329, 275 nm (Fd I) and 460, 420, 329, 276 nm (Fd II, veg. cell Fd). Preliminary results from EPR spectroscopy revealed again differences in g-values at 12 K: $g_x = 1.898$, $g_y = 1.955$, $g_z = 2.046$ (heterocyst ferredoxin), and $g_x = 1.886$, $g_y = 1.960$, $g_z = 2.051$ (vegetative-cell ferredoxin). A rhombic EPR signal in the reduced state was observed, characteristic of a 2 iron/2 sulfur plant-type ferredoxin in both cases. With an antiserum against ferredoxin from vegetative cells, cross-reaction was obtained with ferredoxin II, but not with ferredoxin I from heterocysts.

Acknowledgment: This study was supported by the Deutsche Forschungsgemeinschaft.

USE OF ELECTRON MICROSCOPY IN IDENTIFYING FRANKIA ISOLATES.

R. HOWARD BERG and MARY P. LECHEVALIER, University of
Florida, Gainesville, FL 32611, and Waksman Institute of
Microbiology, Rutgers University, Piscataway, NJ 08854, USA.

Sometimes the screening of isolates as frankiae can
require several weeks. Described here is a much faster
method that merely requires examination of spore walls, using
transmission electron microscopy (TEM). Mature Frankia
spores are known to have a wall surface lamination (van Dijk,
Merkus 1976). Seven genera of actinomycetes have been
examined here to determine the uniqueness of this spore wall
structure.

Broth and plate cultures (processed at 25°C) were fixed
1 h in 2% glutaraldehyde buffered in 0.1 M Na cacodylate, pH
7.2, postfixed in buffered osmium tetroxide, embedded in
Spurr's plastic, and sections poststained in uranyl and lead
salts (UPb) or permanganate followed by uranyl and lead salts
(PUPb) as described by Hoch 1977.

FIGURES 1,2. TEM of the surface lamination in
the frankia spore cell wall, the spore surface
in both figures is on the left. Shown is UFG
026008, isolated from spore (+) Casuarina
equisetifolia. Fig. 1 section poststained in
UPb, fig. 2 is of a different spore and is
poststained in PUPb, which is more reactive
with the inner and outer surfaces of the layer.
Both micrographs are at the same magnification,
bar= 10nm.

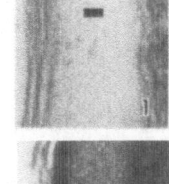
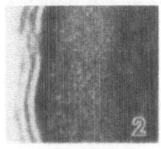

The surface lamination is most commonly comprised of two
or three 3-4 nm thick undulating layers, the exact number
varying even with spores from the same sporangia.
Staining, thickness and other properties of the individual
layers is similar to the laminations described in vesicles
(Torrey, Callaham 1982) and they may have a common chemical
composition. However, the vesicle lamination is comprised of
many more layers, and they are discontinuous. On spores the
lamination is continuous and reminiscent of a membrane.

A diverse group of actinomycetes was examined for the
presence of any similar spore wall lamination: Actinomadura
madurae, Microellobosporia sp., Micromonospora chalcea,
Microtetraspora glauca, Planomonospora parontospora,
Spirillospora albida, and Streptosporangium roseum. Among
the Frankia examined were: CpI1, MpI1, R43, and CcI3. When
cells are prepared as described, the surface lamination is
not found on spores of actinomycetes other than Frankia,
making its presence a useful criterion in identifying
actinomycetous cultures as Frankia.

References
Hoch HC (1977) Mycologia 69, 1209-1213.
Torrey JG and Callaham D (1982) Can. J. Micro. 28, 749-757.
van Dijk C and Merkus E (1976) New Phytol. 77, 73-91.

GROWTH KINETICS OF FRANKIAE IN BATCH CULTURE

MARK A. COLLINS, MICHAEL E. BUSHELL AND GORDON LETHBRIDGE, DEPARTMENT OF MICROBIOLOGY, UNIVERSITY OF SURREY, GUILDFORD, SURREY, GU25YH, UK./ICI, JEALOTTS HILL RESEARCH STATION, BRACKNELL, HERTS. RG121EY, UK

The nitrogen fixing actinomycete Frankia has great potential for use in forestry agriculture and land reclamation. This potential is not being realised because growth is slow and yields poor. The aim of this study is to obtain an understanding of the growth kinetics and physiology of Frankia spp in batch culture in order to optimise its growth in vitro.

Surfactants increased static culture yield and substrate utilisation when Frankia strains were grown on different C + N sources. Agitation increased yield in the presence of yeast extract. The addition of surfactants to agitated cultures depressed yield and substrate utilisation.

Detailed analysis of static culture kinetics was more consistent with linear than exponential growth on propionate. Ammonium and nitrate disappeared in parallel with subsequent nitrite accumulation, possibly due to nitrate respiration.

Data obtained from stirred aerated batch cultures grow on propionate indicated a multiphasic mode of growth, characterised by discontinuities in the rate of biomass accretion, which are reflected in growth limiting substrate utilisation.

The following kinetic parameters were calculated, u, $0.012h^{-1}$, Y propionate, 0.282; K_s, $0.01mmolm^{-3}$; umax occurred at 76h.

These phases of growth were further defined by their q (g substrate g biomass $^{-1}h^{-1}$) values, qmax NH_3 corresponding to qmax propionate in the second phase of growth. A qmax propionate in the initial growth phase pointed to carbon assimilation not used directly for biomass. Increases in the protein:RNA ratio suggested alterations in ribosomal activity prior to each phase.

By analysis of macro-molecular and molecular biomass composition the growth rate potential of Frankia was shown to be low, evident by a high protein:RNA ratio. A reductance value of 4.93 was consistent with low stored carbohydrate.

A molecular formula of $CH_{1.76}N_{0.146}O_{0.537}$ was calculated from elemental analysis. Using this a partial mass balance was constructed which indicated a considerable proportion of the assimilated N was unaccounted for in the final biomass.

Further work is in progress to elucidate this and the nature of other growth limiting factors of Frankia.

ENZYMATIC MECHANISMS FOR THE PROTECTION OF
NITROGENASE FROM OXYGEN IN FRANKIA

D.B. Steele and M.D. Stowers
NPI, University Research Park, Salt Lake City, Utah 84108 USA

INTRODUCTION
Superoxide dismutase (SOD) and catalase (CAT) are important enzymes in
the protection of cellular components from O_2 toxicity (Fridovich, 1974).
Buchanan (1977) has shown that the superoxide anion is capable of inhibiting nitrogen fixation and additional data by Tozum and Gallon (1979)
showed a correlation between nitrogen fixation and the level of enzyme
activity for SOD in the non-heterocystous cyanobacterium Gloeothece.

METHODS
SOD was visualized on polyacrylamide gels by the method described by
Steinman (1985). SOD assays were performed by the method of McCord and
Fridovich (1969) as described by Asada (1974). Catalase activity was
determined by the method of Beers and Sizer (1952).

RESULTS AND DISCUSSION
An iron-containing SOD (FeSOD) was induced under nitrogen-fixing conditions, whereas a manganese SOD (MnSOD) appeared to be constituitive.
Methyl viologen induced a four-fold increase in MnSOD activity and a
three-fold increase in CAT activity in NH_4^+-grown cells. Derepression of
nitrogen fixation induced an eight-fold increase in SOD activity and a
two-fold increase in CAT activity.

The coordinate induction of SOD and CAT activity with derepression of
nitrogen fixation suggests that these enzymes are involved in the protection of cellular components that are crucial to the overall process of
dinitrogen fixation.

LITERATURE CITED
Asada K et al. (1974) Agr. Biol. Chem. 38, 471-473.
Beers RF and Sizer IW (1952) J. Biol. Chem. 195, 133-140.
Buchanan AG (1977) Can. J. Microbiol. 23, 1548-1553.
Fridovich I (1974) Adv. Enzymol. 41, 35097.
McCord JM and Fridovich I (1969) J. Biol. Chem. 244, 6049-6055.
Steinman HM (1985) J. Bacteriol. 162, 1255-1260.
Tozum SRD and Gallon RJ (1979) J. Gen. Microbiol. 111, 313-326.

ACKNOWLEDGEMENTS
This research was partially supported by the National Science Foundation
grant PCM-8313855.

CHARACTERIZATION OF VESICLES ISOLATED FROM FRANKIA ISOLATE EAN1$_{pec}$.

LOUIS S. TISA and JERALD C. ENSIGN
UNIVERSITY OF WISCONSIN, MADISON, WI 53706 U.S.A.

Frankia isolates fix N_2 when incubated aerobically in defined growth medium. Appearance of N_2 fixation following transfer of Frankia from a NH_4 to a N_2 medium coincides with formation of a specialized cell structure, the vesicle. Evidence suggests that the function of the vesicle is to protect the nitrogen fixation system from inactivation by O_2. In order to determine the role of vesicles during nitrogen fixation, we developed procedures for their isolation from strain EAN1. Vesicles are purified by passing a culture through a french pressure cell under an argon atmosphere. This completely disrupts the mycelia while vesicles remain intact. Vesicles are separated and purified by differential centrifugation or by isopynic gradient centrifugation in Renografin. The purified vesicles require a low potential reductant and an ATP regenerating system for acetylene reduction under anaerobic conditions. They do not reduce acetylene when incubated in air. Whole cells reduce acetylene anaerobically when supplied ATP and a low potential reductant. However, the rate of acetylene reduction is at least 5 fold lower than the rate obtained with isolated vesicles. Atractyloside and carboxyatractyloside, specific inhibitors of mitochondrial ADP/ATP translocase, decrease anaerobic acetylene reduction activity by whole cells and isolated vesicles by as much as 70%. These inhibitors are more effective with whole cells and vesicles isolated from fresh cultures than with permeabilized vesicles. These data suggest the possible presence of an ADP/ATP translocase in Frankia.

A PRELIMINARY MECHANISTIC SIMULATION OF N_2 FIXATION BY A LEGUME

D. Bachelet, University of California, Riverside, CA 92521 USA.

Increased awareness of the benefits of woody legumes as agricultural (Brewbaker et al., 1984; Okigbo, 1984; Wijewardene, 1984) and ornamental crops (Nair et al., 1984), due to N_2 fixation and, consequently, low management costs, reinforces the need for models as tools to be applied for legume production and management. Most of the existing simulation models which include N_2 fixation as a major process, estimate fixation rate by decreasing a maximum potential rate of fixation with various functions of (1) exogenous variables such as temperature and humidity (Balandreau et al., 1982) (2), state variables such as NO3 concentration in the soil or N_2-fixing organism population size (Brugge, Thornley, 1984), or (3) material flows such as photosynthate translocation (Pate et al., 1979) or NO3 uptake. It has become necessary for modelers to more mechanistically represent N_2 fixation by synthesizing the current knowledge of microbiologists and agronomists to evaluate its variation in space (related to the absence or presence of N_2 fixing organisms (Dommergues, 1982) and structures) and in time (related to the phenology of the host plant; Rao et al., 1984) and quantify its importance in the system studied.

A predictive model (COLSIM) of N_2 fixation in a woody legume (_Prosopis glandulosa_) - Rhizobium system was developed to simulate growth of mesquite trees in a phreatophytic regime. Data from a greenhouse experiment were used to determine rates of carbon, nitrogen and water distribution between leaves, branches (and stem), roots and root nodules. Labile available pools of material were separated from structural immobile pools. Major physiological processes such as photosynthesis, translocation, respiration, NO_3 uptake, N_2 fixation, N distribution, water uptake, transpiration, were mathematically described. The aim of such a study was to create a theoretical framework to summarize, understand and predict (Brugge, Thornley, 1984) nutrient cycling both at the whole-plant level and at the lower level of mechanism and process. This model is currently being modified to represent the field situation (DESCYC) and the spatial variation in the distribution of the limiting resources.

REFERENCES

Balandreau J, Guckert A and Weinhard P (1982) Acta Oecologica, Oecologia generalis 3, 91-110.
Brewbaker J L, Van den Beldt R and Mac Dicken K (1984) Pesq. Agropec. bras., Brasilia, 19, 193-204.
Brugge R and Thornley J H M (1984) Annals of Botany, 54, 653-671.
Dommergues Y (1981) In Graham P H and Harns S C, eds, Biological N_2 fixation technology for tropical agriculture, pp 395-411.
Nair P K R, Fernandes E C M and Wangubu P N (1984) Pesq. agropec. bros., Brasilia, 19, 295-213.
Okigbo B N (1984) Pesq. agropec. bras., Brasilia, 19, 325-330.
Pate J S, Layzell D B and McNeil D L (1979) Plant Physiol., 63, 730-737.
Rao R S, Luthra Y P, Sheoran I S and Singh R (1984) J. of experimental Botany, 35, 774-784.
Wijewardene R, Pesq. agropec. bras., Brasilia, 19, 315-324.

EFFECT OF OXYGEN ON NITROGENASE OF
RHODOPSEUDOMONAS CAPSULATA AND KLEBSIELLA PNEUMONIAE

Ayala Hochman, Iris Reich and Varda Nadler, Department of Biochemistry,
The George S. Wise Center for Life Sciences, Tel Aviv University, Ramat
Aviv, Tel Aviv 69978, Israel.

Introduction: Both protein components of nitrogenase are irreversibly
inactivated by oxygen, with a half life in air of 3/4-10 min. Oxygen also
represses nitrogenase synthesis; and causes inhibition of the expression
of nitrogenase activity, which is reversible and, in some cases, is not a
direct effect on the enzyme itself. O_2 and O^{-2} are not normally reactive
enough to be toxic, but the intermediate reduction products (O_2^{-1},O_2^{-2} and
OH radical) are undoubtedly so. However, despite the O_2 toxicity, various
organisms can fix nitrogen in its presence. This is because the cells
possess protection mechanisms, among which are, respiration, which scav-
enges oxygen from the vicinity of the enzyme, "Shethna" protein; and the
protection enzymes: catalase, peroxidase and superoxide dismutase which
eliminate the inhibiting reduced forms of oxygen. We studied the effect of
molecular oxygen and peroxide on nitrogenase activity in Rhodopseudomonas
capsulata and Klebsiella pneumonia and the role of the protection enzymes.

Results and discussion: In whole cells of R. capsulata 50% of the reversible inhibition of C_2H_2 reduction is at 0.73uM O_2, but full activity is
restored under anaerobic conditions. Incubation of the cells with O_2 concentrations higher than 4uM cause a complete "switch-off" of the enzyme,
but even after 20 min with 222uM, 50% of the activity is restored. 0.09mM
H_2O_2 cause 50% inhibition of C_2H_2 reduction. In the mutant strain AH4
which has 70 times higher catalase activity and normal levels of SOD and
peroxidase, nitrogenase is more resistant to peroxide (50% inhibition at
1.4mM) but not to O_2, in both the reversible and the irreversible mode of
inhibition. A green mutant of R. capsulata·VN1 is sensitive to O_2 similar
to the wild type. C_2H_2 reduction in whole cells of K. pneumoniae also
shows the reversible "switch off" phenomenon by O_2 with 50% inhibition at
0.36uM, and it is inhibited by H_2O_2 (50% at 0.18mM). 5.3mM O_2 cause complete inhibition of the activity but it is partially reversible even after
incubation in air saturated medium. In a mutant strain AH10 which has
higher cellular levels of catalase peroxidase and SOD and also higher
respiration rates, nitrogenase activity is less sensitive to H_2O_2 (50%
inhibition of C_2H_2 reduction at 1.06mM) but more sensitive to O_2 (50%
inhibition at 0.18uM). It is suggested that the amount of catalase
normally present in R. capsulata and K. pneumoniae is sufficient to
eliminate the H_2O_2 produced in vivo and therefore higher levels of the
enzyme do not provide more protection against oxygen damage. AS we have
suggested for R. capsulata (Hochman and Burris, 1981, J. Bact. 147, 492-
499) the reversible inhibition of nitrogen fixation in K. pneumoniae is
also caused by the shortage of electrons which results from the
competition for them with oxygen.

NITROGEN ASSIMILATORY ENZYMES OF DESULFOVIBRIO

PATRICK L. BRUNO, JAMES M. ODOM, AND JUDY D. WALL
BIOCHEMISTRY DEPARTMENT, UNIV. OF MISSOURI, COLUMBIA, MO
65211   U.S.A.

Introduction. Nitrogen assimilation was examined in five strains of sulphate-reducing bacteria, Desulfovibrio gigas, D. vulgaris ATCC 29579, D. desulfuricans-Norway, D. desulfuricans ATCC 13541, and D. desulfuricans ATCC 27774. In addition, the nitrogen fixation ability of strains 13541 and 27774 is addressed.

Materials and Methods. Enzyme assays were conducted similarly to reported methods for nitrogenase (5), glutamine synthetase (6), glutamate dehydrogenase (3), and alanine dehydrogenase (1). Crude extracts of cells grown on standard lactate-sulphate medium (4), or standard medium lacking NH_4^+ and cysteine but containing 0.01% (w/v) yeast extract (2), were assayed after disruption by a French press. Nitrogenase was assayed on whole cells.

Results and Discussion. After three days growth under low ammonium conditions, strain 13541 exhibited a maximum C_2H_2 reduction of 1.8 nmol per min·mg protein. Strain 27774 did not demonstrate any detectable activity, even when assayed over a six day period.

Glutamine synthetase (GS) activity in D. desulfuricans-Norway was stimulated 3-fold when grown on limiting nitrogen, while all other strains studied increased at least 30-fold. Derepressed levels of GS varied between strains from 0.9 to 7.2 umol per min·mg protein. No GS activity was detectable in strain 27774 grown on high ammonium.

Glutamate dehydrogenase (GDH) did not appear to be regulated by nitrogen availability in any of the strains investigated here. GDH levels were highest in strains 27774 and 13541 (0.6 umol per min·mg protein), two strains which exhibit low GS activities when grown on high NH_4^+. D. desulfuricans-Norway exhibited 10-fold less GDH activity than strains 27774 and 13541, in contrast to D. gigas and D. vulgaris which had no detectable activity. When present, this enzyme preferred NADPH to NADH as substrate.

Alanine dehydrogenase did not appear to play an assimilatory role in the Desulfovibrio strains studied here. Alanine dehydrogenase activity was NADH dependent and was observed in low levels (0.01-0.08 umol per min·mg protein) in D. gigas, D. desulfuricans ATCC 13541, and D. desulfuricans-Norway.

1.   Johansson B and Gest H (1976) J. Bacteriol. 128, 683-688.
2.   Lespinat PA et al (1984) In Veeger C and Newton WE, eds, Advances in Nitrogen Fixation Research, p 233, Nijhoff, The Hague.
3.   Lewis AJ and Miller JDA (1975) J. Gen. Micro. 90, 286-292.
4.   Odom JM and Wall JD (1985) J. Environ. Microbiol. submitted.
5.   Postgate JR et al (1985) In Ludden PW and Burris JE, eds, Nitrogen Fixation and CO_2 Metabolism, pp 225-234, Elsevier, New York.
6.   Shapiro BM and Stadtman ER (1970) Meth. Enzymol. 17, 910-922.

ENRICHMENT OF MICROAEROBIC DIAZOTROPHS UNDER GLUCOSE LIMITATION

MICHAEL KLOSS
Institut für Biophysik / Isotopenlaboratorium, Universität Hannover,
Herrenhäuser Str. 2, D 3000 Hannover 21, FRG

Korean rice soil was used as starting material for the enrichment
cultures. The soil was suspended in nitrogen free mineral medium to
give a final concentration of 0.001 (g/ml). This suspension was given
into the culture vessel of a continuous culture apparatus. The
dilution rate was set in the range from D=0.003 to D=0.05 (h^{-1}). For
enrichment of copeotrophic organisms 5 g of glucose were added to the
culture vessel and the dilution rate was set to D=0 (h^{-1}). The
isolation of the diazotrophs was carried out with the double layer
technique. At higher substrate concentrations (s=5mg/l)
Beijerinckia indica was found to be the most competetive organism. At
low substrate concentrations (s=1mg/l) an unidentified rod was always
isolated as the predominant nitrogen fixing organism. No other
diazotroph could be isolated at all. The frequency of nitrogen fixers
never exeeded 30%. The efficiency of nitrogen fixation rose from
0.2 (mg N_2/ g Glucose) at D=0.003 (h^{-1}) up to 19.3 (mg N_2 / g Glucose)
at D=0.05 (h^{-1}). Free ammonia was observed only in traces in the
culture supernatant. All parameters were found to decrease with
the dilution rate, which may be interpreted as an effect of the
increasing maintenance requirements.

PURIFICATION TO HOMOGENEITY OF <u>AZOTOBACTER</u> <u>VINELANDII</u> HYDROGENASE: A
NICKEL AND IRON CONTAINING αβ DIMER.

Lance C. Seefeldt and Daniel J. Arp, Department of Biochemistry, University of California, Riverside, CA 92521 USA.

The hydrogenase enzymes from aerobic, N_2-fixing microorganisms are
involved in recycling H_2 produced by nitrogenase. The hydrogenases
from the free-living, N_2-fixing microorganisms <u>Azotobacter</u> <u>vinelandii</u>
(Y.W. Kow, R.H. Burris, 1984) and <u>Azotobacter</u> <u>chroococcum</u> (A.N. Van der
Werf, M.G. Yates, 1978) had previously been partially purified. It was
concluded from these studies that the <u>Azotobacter</u> hydrogenase was a
60,000 dalton monomer. We report here on the purification to homogeneity of the hydrogenase from <u>A. vinelandii</u> membranes and demonstrate
that the enzyme is an αβ dimer. In addition, we provide direct evidence
that the <u>A. vinelandii</u> hydrogenase contains nickel and iron.

The hydrogenase was solubilized from <u>A. vinelandii</u> membranes with the
nonionic detergent Triton X-100. This was followed by extraction with
hexane to remove lipids and detergent. This extraction procedure
allowed the hydrogenase to be purified by carboxymethyl-Sepharose and
octyl-Sepharose column chromatography. All purification steps were
carried out under anaerobic conditions in the presence of dithionite and
dithiothreitol. The enzyme was purified by this procedure 143-fold
from membranes to a specific activity of 124 μmoles of H_2 uptake·min^{-1}·
mg^{-1} (units·mg^{-1}). Nondenaturing polyacrylamide gel electrophoresis of
the purified hydrogenase revealed a single band which stained for both
activity and protein. Sodium dodecyl sulfate-polyacrylamide gel electrophoresis revealed two bands corresponding to peptides of 67,000 and
31,000 daltons. Densitometric scans of the SDS-gel indicated a molar
ratio of the two bands of 1.07 ± 0.05. While the native molecular
weight of the enzyme by gel permeation was 53,000, sucrose density
gradient centrifugation and native polyacrylamide gel electrophoresis
gave molecular weights of 98,600 ± 10,000 and 98,600 ± 2,000, respectively. We conclude that <u>A. vinelandii</u> hydrogenase is an αβ dimer
(98,000 daltons) with subunits of 67,000 and 31,000 daltons. Analyses
for metals by atomic absorption spectrophotometry indicated 0.68 ±
0.01 mol Ni/mol hydrogenase and 6.6 ± 0.5 mol Fe/mol hydrogenase. Thus,
for the first time, the <u>A. vinelandii</u> hydrogenase has been directly
shown to contain Ni. Several catalytic properties have also been
examined. The K_m for H_2 was 0.86 μM, and H_2 evolution was observed in
the presence of reduced methyl viologen. K_m and V_{max} values were determined for the electron acceptors methylene blue (17.2 μM; 122 units·
mg^{-1}), phenazine methosulfate (7.6 μM; 122 units·mg^{-1}) and benzyl viologen (45 mM; 255 units·mg^{-1}).

Kow YW and Burris RH (1984) J. Bacteriol. 159, 564-569.
Van der Werf AN and Yates MG (1978) In Schlegel HG and Schneider K,
eds, Hydrogenases: Their Catalytic Activity, Structure and Function,
pp 307-326, Erich Goltze, K.G., Göttingen, FRG.

INFLUENCE OF PALM OIL MILL CAKE ON GROWTH OF TROPICAL LEGUMES

Z.H. SHAMSUDDIN AND J. SABANG, UNIVERSITI PERTANIAN MALAYSIA, SERDANG, SELANGOR, MALAYSIA

Arachis hypogaea L., a food legume, and Calopogonium caeruleum (Benth.) Hensl., a leguminous cover, were planted on limed peat under glasshouse conditions for 6 weeks. Two high rates of Palm Oil Mill (POM) Cake ($\equiv$ 370 and 740 t/ha) were applied and compared to 4 levels of potassium ($\equiv$ 0, 50, 100 and 200 kg K_2O/ha). Liming ($\equiv$ 12.5 t GML/ha), and an allowance for 4 weeks of equilibration, raised the pH from 3.7 to 4.3 for the K treatments and 5.2 and 4.9 for the lower and higher rates of POM Cake, respectively.

N and P ($\equiv$ 20 kg N/ha and 100 kg P_2O_5/ha) were applied only to the K treatments. However, all the K and POM Cake treatments were supplemented with Cu and Zn ($\equiv$ 2.5 kg Cu/ha and 5 kg Zn/ha).

Results showed that the application of POM Cake (chemical analysis : 0.2% N, 0.3% P_2O_5, 4.0% K_2O) at 740 t/ha lowered P and K concentrations of tops, depressed growth and nodulation of groundnut but produced no significant effect on the relatively slow-growing Calopogonium. However, at 370 t/ha, POM Cake increased the P and K concentrations of tops, growth and nodulation of groundnut above all other treatments, but not for Calopogonium. All treatments had no effect on N concentration of both crops.

The sole application of POM Cake as an organic fertilizer ($\equiv$ 370 t/ha) promoted growth and nodulation of groundnut but not Calopogonium.

GROWTH AND NITROGENASE FORMATION IN *BACILLUS POLYMYXA*

YOSHIO SHINTANI
FACULTY OF TECHNOLOGY, KINKI UNIVERSITY, HIROMACHI, KURE, JAPAN

1. INTRODUCTION

Reports have demonstrated that cells of some nitrogen-fixing bacteria grown on a limiting amount of ammonia, glutamate or yeast extracts had severalfold higher nitrogenase activity than cells grown on molecular nitrogen (1, 2, 3). In this study, we examined growth, nitrogenase formation and denitrification of *Bacillus polymyxa* cultured with various nitrogen sources under nitrogen or argon gas. We also studied the effect of molybdenum (Mo) or tungsten (W) on nitrogenase synthesis in the bacteria.

2. MATERIAL AND METHODS

Bacillus polymyxa strain Hino (this strain was recently classified as a strain of a new species, *Bacillus azotofixans*, by Seldin *et al* (4)) was grown in the medium described by Hino (5). In some cultures containing NO_2^- or NH_4^+, argon replaced N_2 as the gas phase. The growth was monitored by measuring optical density at 660 nm. The nitrogenase activities of cell-free extracts were determined by the C_2H_2 reduction technique. The Gas in a flask before and after culture was analyzed by gas chromatography or mass spectrography.

3. RESULTS AND DISCUSSION

Bacillus polymyxa was grown anaerobically with NO_2^-, NH_4^+ or N_2 as the nitrogen source. Nitrogenase activity diminished rapidly, when the cells were subcultured in the nitrite medium (NO_2^-: 7.5 mM), from which Mo had been eliminated. When the nitrogenase depressed cells were cultured under argon in the medium containing NO_2^- and Mo, the cells showed almost the same level of nitrogenase activity as that of the cells grown by nitrogen fixation. A possibility was examined that the denitrification from NO_2^- produces N_2, which in turn stimulates the derepression of nitrogenase. N_2 could not be detected by gas chromatography in the gas phase of the flasks before and after culturing cells in the nitrite medium. Mass spectrographic analysis did not show the production of $^{15}N_2$ from $Na^{15}NO_2$ by *B. polymyxa*.

Nitrogenase activity was low in the cells grown under argon with 3.75 mM NH_4^+, and the synthesis of the enzyme in the nitrite medium was inhibited by the addition of NH_4^+. Tungsten, in the concentration ten times that of Mo, did not inhibit nitrogenase synthesis during nitrogen fixation or NO_2^- utilization. The amounts of Mo consumed by the cells were between 10 to 15 µg when cultured for 24 hours in one liter of the nitrite or nitrogen free medium containing 25 to 200 µg Mo.

4. REFERENCES

(1) Upchurch RG, Mortenson LE (1980) J.Bacteriol. 143, 274-284
(2) Sweet WJ, Burris RH (1981) J.Bacteriol. 145, 824-831
(3) Arp DJ, Zumft WG (1983) J.Bacteriol. 153, 1322-1330
(4) Seldin L, et al (1984) Int.J.Syst.Bacteriol. 34, 451-456
(5) Hino S (1960) J.Biochem. 47, 482-494

5. ACKNOWLEDGMENT

We thank Dr. S. Hino for confirming some of our results and for his helpful comments.

PHYSIOLOGY AND ECOLOGY OF NITROGEN-FIXING MICRO-ORGANISMS

M.G. YATES
AFRC Unit of Nitrogen Fixation, University of Sussex, Brighton BN1 9RQ, UK.

Several research areas fell within the broad title of this session and
the numerous posters precluded individual consideration during discussion.

Competition and survival of Rhizobia

The most important agronomic requirement for biological nitrogen fixation
is to achieve the right association between symbiont and cultivar to
maximise food yield. Continued detailed investigations are necessary to
lay the groundwork for future selection or genetic manipulation of plants
and microbes. Soil acidity, salinity, aridity, depth, nutrients, aspect
and texture as well as light conditions all affect rhizobial
competitiveness or survival. The role of the cultivar (Demezus and
Bottomley; Busse et al.) and phage susceptibility of rhizobia (Hashen
and Angle) were also emphasised. Strain identification continues to be
a major concern: methods include colony size, plasmid profiles, phage
typing, antibiotic resistance, effectiveness in symbiosis, substrate
utilisation, cytochrome and total protein profiles, fluorescent antibodies and ELISA; a rapid method for the latter was reported (Olsen and
Rice). Denitrification products (N_2 or N_2O) may also help to identify
R. japonicum strains (Hynes et al.). Care should be exercised when
interpreting cytochrome and protein patterns: these may vary with culture
age and nutrient status. Obviously a rapid and reliable screen for
strain identification remains a prime requirement in survival and
competition studies. Lepo drew attention to the role of succinate in
effectiveness and the need for more knowledge about carbon metabolism in
both rhizobia and nodules.

Associative bacteria

Azospirillum species are the most investigated organisms in this area and
a fifth species, Azospirillum halopraeferens has been isolated (Reinhold
et al.). Reports of a diazotrophic pseudomonad (Chan et al.) and an
unidentified motile rod (not Azospirillum) associated with surface
sterilised roots of the salt-tolerant Kaller grass (Reinhold et al.)
should stimulate interest in other microbes. A tripartite symbiosis
between rhizobia mycorrhiza and legumes has also been investigated
(Pacvosky). Whether the effect of such organisms is by fixing nitrogen,
mineral solubilisation, plant growth substances, disease combating or
combinations of these remains unknown. In the latter context the report
by Chahal and Chahal on the biological control of the tomato pest
Meloidogyne incognita by Azotobacter chroococcum breaks new ground. The
problem of identifying Azospirillum strains, particularly the more
effective ones has to be recognised and overcome if comparisons between
results in different laboratories are to be meaningful. Klos and Lachman
used a combination polyclonal antibody and protein patterns to attempt to
distinguish strains. Finally, the role of the cultivar on nitrogenase
activity must again be emphasised (Shelley et al.). Gibson and Halsall
inoculated straw with mixtures of Azospirillum or Bacillus species and

Cellulamonas gelida and observed substantial levels of nitrogen fixation. These systems are potentially a useful soil N source.

Frankia

Two out of five posters on this topic described faster methods of identifying Frankia: by the consistent appearance of a super oxide dismutase protein band in all isolates tested (Puppo et al.) and by the characteristic spore wall lamination seen under the transmission electron microscope (Berg and Lechevalier). Obviously an improvement in identification techniques would advantage field and forest studies. Other topics which remain unsolved include the role of vesicles in protection against O_2 (Tisa and Ensign; Torrey) and the prevalence of spores (Holman et al.).

Blue-green algae

A new high molecular weight ferredoxin with specificity for nitrogenase in heterocysts (Schrautemeier and Hohme) is welcome information for those interested in the difficult problem of electron transfer to nitrogenase in aerobic systems, including heterocysts where several electron donor systems may operate (Neuer et al.). Cyanobacteria isolated from the Azolla symbiosis liberate NH_4^+ in response to small levels of photosystem II inhibitors without affecting nitrogenase activity suggesting a role for photosystem II in NH_4^+ excretion in these systems (Newton). The mechanism of this effect is presumably complex and 'at a distance' since ammonia production 'nitrogen fixation' is in the heterocysts and photosystem II is in the vegetative cells.

H_2 metabolism and hydrogenase

The question whether H_2 recycling increases yield in legumes is not resolved: a survey of 256 R. leguminosarum strains showed 75% Hup^+ (Fees et al.). Hup activities are usually low in R. leguminosarum bacteroids and indiscernible in the free-living rhizobium. Drevon et al. again observes the inoculation with R. japonicum PH17 (Hup^-) gave better yields than with its Hup^+ revertant PH117-1. It was clearly felt that the nutritional status of the nodule (carbon or O_2-limited) was a deciding factor as to whether recycling H_2 improved yields. Stam & Stouthamer stress that the ratio of H_2 produced to N_2 reduced (the $H_2:N_2$ ratio) by nitrogenase affects cyanide-supported growth of R. sesbania ORS571 more than the presence of Hup activity. Controlled experiments to measure $H_2:N_2$ ratios with Hup^- mutants should reveal useful data for both metabolic and mechanistic reasons (Bishop and Eady).

O_2 and nitrogenase

O_2 protection mechanisms for nitrogenase vary according to the organism. Steel and Stowers suggest that catalase and/or superoxide dismutase protect Frankia nitrogenase against damage by O_2: an Mn SOD is constitutive, but both an Fe SOD and catalase activities are induced under N_2-fixing conditions. Catalase protects nitrogenase in Rhodospirillum capsulata and in Azospirillum brasilense against H_2O_2 but not against O_2 (Nadler et al.; Higuti and Pedrosa). If such enzymes are truly protective against O_2 it should be useful to include their genes when inserting nif in foreign hosts.

GENETICS OF NITROGEN FIXATION
IN FREE-LIVING BACTERIA

INVITED PAPERS

POSTER SUMMARIES:

(i)   Identification, cloning and mapping of nif.
(ii)   Analysis of nif sequences.
(iii)   Rearrangement of nif DNA.
(iv)   nif regulation.
(v)   Biochemistry and physiology of mutants.

DISCUSSION GROUP SUMMARY

PROGRESS IN UNDERSTANDING ORGANIZATION AND EXPRESSION OF NIF GENES
IN KLEBSIELLA

Frank Cannon, Jim Beynon, Vicky Buchanan-Wollaston, Robert Burghoff,
Maura Cannon, Robert Kwiatkowski, Gail Lauer and Robert Rubin
BioTechnica International, Inc./85 Bolton Street/Cambridge,
Massachusetts 02140/USA

INTRODUCTION

Our current understanding of nitrogen fixation (nif) genes in Klebsiella
pneumoniae is uniquely comprehensive. This has emerged from a decade
and a half of concerted studies in the biochemistry and physiology of
nitrogen fixation and in the molecular genetics of both nitrogen fixation and nitrogen regulation (ntr). A brief summary of the main findings
in molecular genetics and a discussion of recent progress in gene function, organization and expression will be presented in this paper.

THE nif GENES

Seventeen nif genes have been identified in K. pneumoniae and mapped as
a contiguous cluster between hisG and shiA (Fig. 1). It is now reasonable to assume that all genes specific to nitrogenase biosynthesis are
contained in this cluster since none has been found elsewhere in the
genome and the intergeneric transfer of the cluster confers nitrogen
fixation capability on other enteric bacteria which are naturally non-
nitrogen-fixing. The identification of genes and their order, hisG...
nifQBALFMVSUXNEYKDHJ...shiA, were determined by a combination of genetic
analysis and physical characterization of Nif⁻ mutants and of cloned DNA
restriction fragments. The physical locations of all the nif genes have
been mapped, and detailed restriction maps of the 23 kb of DNA which they
occupy have been generated. Nucleotide sequences for several nif genes
have been reported; these include nifH, nifD (Scott et al 1981;
Sundaresan, Ausubel 1981), nifA (Buikema et al, in press) and nifF
(Drummond, in press). Comparisons with sequences of corresponding genes
from other bacteria have highlighted extensive regions of sequence conservation which presumably are important for protein function.

PRODUCTS AND FUNCTIONS OF THE nif GENES

The molecular masses and probable functions of most nif gene products
are listed in Table 1. Fifteen of these products fall into four groups
of related function (Fig. 1): (1) synthesis of nitrogenase component 1
(nifD, K, E, N, Q, B, S, U); (2) synthesis of nitrogenase component 2
(nifH, M); (3) electron transport to nitrogenase (nifF, J); and (4)
nif transcriptional regulation (nifL, A). The functions of nifX and
nifY genes are still unknown.

So far, only four nif-specific proteins have been purified in an active
form. The two component proteins (Kp1 and Kp2) of nitrogenase which have
subunits specified by nifH, D and K (Fig. 1) have been studied extensively. Two nif-specific (nifF and nifJ) components of the electron

454

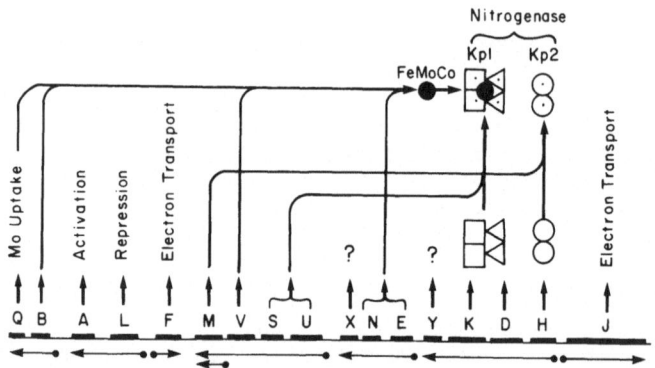

FIG. 1.   Map of K. pneumoniae nif genes. Vertical arrows indicate gene
functions and horizontal arrows indicate gene transcripts.

TABLE 1. Nif gene products

Gene	$Mr(10^{-3})$ of product		Function
Q	?		Mo uptake (1)
B	49	(2)	FeMoco synthesis (7)
A	57	(3)	Transcription activation (8)
L	45	(3)	Transcription repression (8)
F	19	(4)	Flavodoxin subunit (4)
M	28	(5)	Kp2 processing (6)
V	42	(5)	FeMoco synthesis (7)
S	45	(6)	Kp1 processing (6)
U	25	(5)	Kp1 processing (6)
X	18	(5)	Unknown
N	50	(6)	FeMoco synthesis (7)
E	40	(6)	FeMoco synthesis (7)
Y	24	(5)	Unknown
K	60	(6)	Kp1 β-subunit (6)
D	56	(6)	Kp2 λ-subunit (6)
H	35	(6)	Kp2 subunit (6)
J	120	(6)	Pyruvate-flavodoxin oxido-reductase subunit (9)

References for product Mr and functions are: (1) Imperial et al 1984;
(2) Pühler et al 1984;   (3) Cannon 1981;   (4) Deistung et al, in press;
(5) Pühler, Klipp 1981;   (6) Roberts et al 1978;   (7) V. Shah, private
communication;   (8) Buchanan-Wollaston et al 1981a;   (9) Shah et al 1983.

transfer pathway to nitrogenase have now been isolated and used with a physiological electron donor, pyruvate, and the component proteins of nitrogenase to obtain substrate reduction in vitro (Shah et al 1983)(see Fig. 2). A flavodoxin purported to be the nifF gene product has been shown to mediate electron transfer to the Fe protein of nitrogenase (Kp2) (Nieva-Gomez et al 1980; Shah et al 1983). More recently Deistung et al (in press) cloned a putative nifF gene into a multicopy plasmid to obtain an elevated concentration of the gene product in vivo. Characterization of the isolated flavodoxin which served as an electron donor to nitrogenase showed unambiguously that it was the product of the nifF gene described by Beynon et al (1983). A nif-specific pyruvate-flavodoxin oxidoreductase purported to be specified by nifJ has been purified and characterized by Shah et al (1983). This protein is probably the same as that purified earlier by Bogusz et al (1981). Our present knowledge of the electron transfer sequence from pyruvate to nitrogenase substrates in vivo is depicted in Fig. 2. Studies with crude extracts suggest that the activity of nitrogenase in vivo is limited by the concentration of the nif-specified flavodoxin (Hill, Kavanagh 1980; Bogusz et al 1981). This observation is consistent with the need for stoichiometric concentrations of the flavodoxin and Kp2 whereas the oxidoreductase (nifJ) may only be required at the catalytic concentration.

Kpl contains an iron-molybdenum co-factor known as FeMoco (Shah, Brill 1977) and it has recently been confirmed that FeMoco provides at least part of the site for nitrogenase substrate reduction (Hawkes et al 1984). The products of five nif genes (nifE, N, V, B, Q) are now known to be required for the synthesis of wild-type levels of intact FeMoco (Hawkes et al 1984; Imperial et al 1984; V. Shah, private communication).

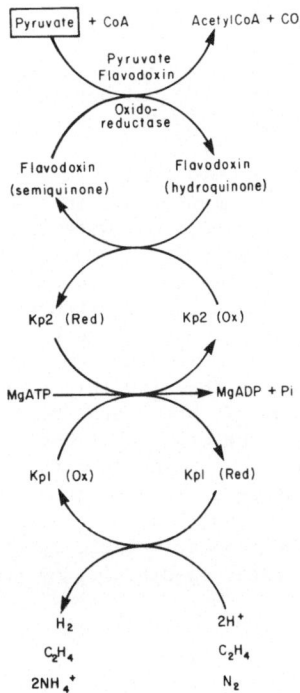

Kp2 is a dimeric iron-sulphur protein which is produced in an inactive form in nifM mutants (Roberts et al 1978). There is now evidence to show that nifH and nifM are the only nif gene products required to make an active Kp2 (K. Howard, private communication).

OPERON STRUCTURE AND TRANSCRIPTION OF THE nif GENES

The nif genes are organized in eight operons (Fig. 1). The transcription initiation points for all the nif operons except that of nifpJ have either been determined or confirmed by S1 mapping (see Beynon et al 1983 and relevant references cited therein). The direction of nifJ transcription was found by S1 mapping to be the same as that of nifF and opposite to that of all the other nif operons (Cannon et al 1984). These are also unique in being the only monocistronic operons in the nif cluster.

FIG. 2. Electron transport chain to nitrogenase in K. pneumoniae.

A noteworthy observation is that the nifM gene is expressed from transcripts initiated at two promoters (Fig. 1). It has been suggested that read-through transcription of nifM from the preceding nifpUSV operon is required to give a wild-type level of nitrogenase activity (Beynon et al 1983).

The maximum transcription rates of four nif operons have been reported by Cannon et al (1985). These results showed that of the four operons examined nifpHDKY was expressed at the highest level and the relative transcription rates of nifpLA, nifpBQ and nifpJ were 77%, 58% and 39%, respectively.

Polarity in the nifpHDKY operon was observed by Buchanan-Wollaston et al (1981a). The transcription rate of nifDKY was 20% less than that of nifH which is consistent with the observation that the cellular concentration of Kp2 is higher than that of Kp1 (R. Eady, private communication).

THE STRUCTURE OF THE nif PROMOTERS

The sequences of all the K. pneumoniae nif promoters are shown in Fig. 3. Beynon et al (1983) found that the primary structures of these promoters share characteristics which clearly distinguish them from the consensus promoter of enteric bacteria (Fig. 4)(Rosenberg, Court 1979; Hawley, McClure 1983). A 26 bp structure in the nif promoters situated between -26 and -1 bp relative to transcription start contains two regions of conserved sequence separated by 8 bp: CTGG at -26 to -23 bp, TTGCA at -14 to -10 bp (Fig. 4). This structure has also been found in nif promoters of several Rhizobium species (Alvarez-Morales, Hennecke 1985) and in promoters of enteric bacteria which are controlled by the nitrogen regulation gene products of ntrBC (Kustu et al, in press).

Previous results showed that the product of ntrA, which appears to be synthesized constitutively, is required for transcriptional activation of ntr and nif promoters (Merrick 1983; Ow, Ausubel 1983). Recent evidence from studies with Salmonella typhimurium strongly suggests that this product acts as the sigma subunit of the RNA polymerase required for transcription initiation from ntr promoters (Hirschman et al, in press). The unique structure of ntr and nif promoters is consistent with their utilization of an altered form of RNA polymerase which has a modified transcriptional specificity. This is a highly significant development in our understanding of gene expression under conditions of nitrogen starvation in two groups of taxonomically distant gram negative bacteria.

Deletion and point mutations have been isolated in the nifH (Bitoun et al 1983; Brown, Ausubel 1984; Ow et al 1985) and nifL promoters (Drummond et al 1983; Dixon et al 1984). These results show that mutations extending upstream as far as -184 in the nifH and -150 in the nifL promoters and within the conserved sequences cause a decrease in promoter activity. Of particular interest is the C to T transition at -15 in the nifH promoter which resulted in this promoter being activatable by ntrC (Ow et al 1985). This suggests a role in activator specificity by the conserved -10 sequence.

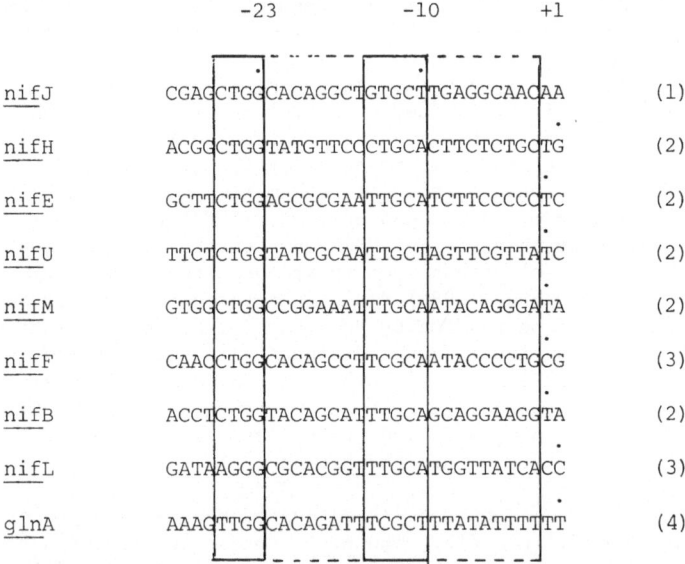

FIG. 3.   Comparison of nif and ntr promoter sequences from K. pneumoniae. References for sequences are (1) Shen et al 1983;   (2)
Beynon et al 1983;   (3) Drummond et al 1983;   (4) Dixon 1984.

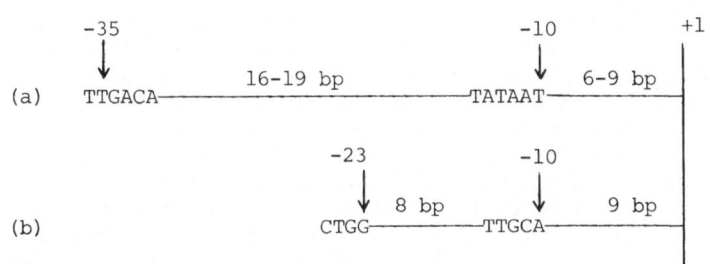

FIG. 4.   Comparison of ntr (b) with the "consensus" (a) promoter
for enteric bacteria (Beynon et al 1983).

REGULATION OF THE nif GENES

The expression of nif is inhibited by temperatures above 35°C, by oxygen
and by fixed nitrogen including ammonia, nitrate and amino acids. There
is no evidence however that nitrogenase activity is regulated by oxygen,
fixed nitrogen sources or temperature, though the enzyme is irreversibly
damaged by oxygen.

Quantitative studies on the rate of nitrogenase synthesis under maximum
O_2 or NH_4^+ repression have shown that it decreased with a half-life of
approximately 12 minutes in cultures obtained from different growth

conditions (Eady et al 1978; Kaluza, Hennecke 1981). Since the physical and functional half-life of HDKY mRNA during repression by O_2 or NH_4^+ was also found by Kaluza and Hennecke (1981) to be approximately 12 minutes, this strongly suggests that inhibition of nifHDKY expression by O_2 or NH_4^+ is entirely at the level of transcription. Kahn et al (1982) showed that the stability of K. pneumoniae mRNA was dependent on the nitrogen status of the culture and that the functional decay rate of nifHDKY mRNA was the same as that of bulk mRNA. These results together with the observation by Kaluza and Hennecke (1981) that NH_4^+ and O_2 inhibited de novo synthesis of nifHDKY mRNA further suggest that transcription inhibition of nifHDKY occurs at the site of transcription initiation. Cannon et al (1985) found in a recent study that during nif derepression and repression by NH_4^+ and O_2, nif transcription and translation remained tightly coupled, thus ruling out a translational control of nif expression. Therefore, all available evidence strongly suggests that genetic regulation of derepression and nif-specific repression by NH_4^+ and O_2 are directed exclusively at the sites of transcription initiation within the nif gene cluster.

Transcription of the nif operons is activated by the nifA protein and repressed by the nifL protein. The nifL protein is present during nif derepression and only becomes a repressor in the presence of fixed nitrogen or oxygen (see Buchanan-Wollaston et al 1981a and references cited therein). The nifA protein is temperature sensitive, thus explaining the absence of nif expression at temperatures above $35^{\circ}C$ (Buchanan-Wollaston et al 1981b; Zhu, Brill 1981).

The products of four genes play a direct role in the regulation of nif expression at the transcriptional level (Fig. 3). These are the nitrogen regulation genes, ntrBC (see Magasanik 1982 for a review of gln/ntr expression) and nifLA (MacNeil, Brill 1980; Buchanan-Wollaston et al 1981a, b; Sibold, Elmerich 1982; Espin et al 1982; Ow, Ausubel 1983). The ntrBC genes along with glnA constitute an operon which has three promoters (Dixon 1984). Two of these are ntr promoters, one of which precedes glnA and the other ntrB. The third promoter which precedes glnA upstream from the ntr promoter is probably the site at which gln transcription is initiated under conditions of nitrogen sufficiency. At low levels of NH_4^+ (< 4 mM) under aerobic or anaerobic conditions, expression of glnA/ntrBC is greatly enhanced, due most likely to autogenous activation of transcription at the ntr promoter preceding glnA by the ntrC protein in conjunction with an RNA polymerase containing the ntrA gene product as a sigma subunit (Kustu et al, in press). Transcription of nifLA and other operons whose products are involved in the assimilation of nitrogen sources is similarly activated (Magasanik 1982). The product of nifA then activates transcription at the remaining nif promoters while the nifL protein is maintained in an inactive form (Buchanan-Wollaston et al 1981a).

Repression of nifA-mediated transcription occurs at intermediate levels of NH_4^+ (> 4 mM) and at low levels of dissolved O_2 (>0.1 μM)(Bergersen et al 1982) by a mechanism which involves the nifL product (Hill et al 1981; Merrick et al 1982; Filser et al 1983). Under these conditions the repressor form of the nifL protein predominates and most likely inactivates the nifA protein (Buchanan-Wollaston, Cannon 1984). At high levels of NH_4^+ (> 20 mM), ntrC-mediated transcription of nifLA and other

ntr operons is repressed (Ow, Ausubel 1983). This repression involves
the product of ntrB and though not affected by oxygen (Cannon et al 1985),
the repressor mechanism may be analogous to that of nifL. There is
however conflicting evidence about the precise role of the ntrB protein
(Alvarez-Morales et al 1984 and references cited therein). Repression
of nif in nitrogen-rich cultures is therefore maintained by the repression of both ntrC and autogenous nifA-mediated transcription of the
nifLA operon.

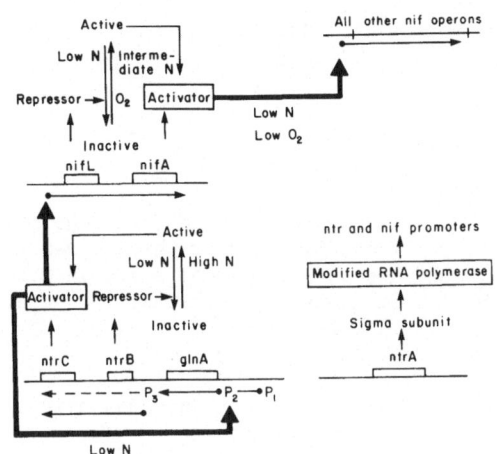

FIG. 5.

A model for nif
regulation in K. pneu-
moniae. Heavy arrows
indicate activation
sequence. Horizontal
arrows indicate gene
transcripts.

ACKNOWLEDGMENT

We thank Judith Grisham for generous help in preparing the manuscript.

REFERENCES

Alvarez-Morales A et al (1984) EMBO J. 3, 501-507.
Alvarez-Morales A and Hennecke H (1985) Mol. Gen. Genet. 199, 306-314.
Bergersen FJ et al (1982) J. Gen. Microbiol. 128, 909-915.
Beynon, J et al (1983) Cell 34, 665-671.
Bitoun R et al (1983) Proc. Natl. Acad. Sci. USA 80, 5812-5816.
Bogusz D et al (1981) Eur. J. Biochem. 120, 421-426.
Brown SE and Ausubel FM (1984) J. Bacteriol. 157, 143-147.
Buchanan-Wollaston V et al (1981a) Mol. Gen. Genet. 184, 102-106.
Buchanan-Wollaston V et al (1981b) Nature 294, 776-778.
Buchanan-Wollaston V and Cannon FC (1984) In Veeger C and Newton WE, eds,
    Advances in Nitrogen Fixation Research, p. 732, Martinus Nijhoff/Junk,
    The Hague.
Cannon FC (1981) In Gibson AH and Newton WE, eds, Current Perspectives
    in Nitrogen Fixation, p. 151, Australian Academy of Science, Canberra.
Cannon MC et al (1984) In Veeger C and Newton WE, eds, Advances in
    Nitrogen Fixation Research, p. 734, Martinus Nijhoff/Junk, The Hague.
Cannon MC et al (1985) Mol. Gen. Genet. 198, 198-206.
Dixon RA (1984) Nucl. Acids Res. 12, 7811-7830.
Dixon R et al (1984) In Veeger C and Newton WE, eds, Advances in Nitro-
    gen Fixation Research, pp. 635-642, Martinus Nijhoff/Junk, The Hague.

Drummond M et al (1983) Nature 301, 302-307.

Eady RR et al (1978) J. Gen. Microbiol. 104, 277-285.

Espin G et al (1982) Mol. Gen. Genet. 186, 518-524.

Filser M et al (1983) Mol. Gen. Genet. 191, 485-491.

Hawkes T et al (1984) Biochem. J. 217, 317-321.

Hawley DK and McClure WK (1983) Nucl. Acids Res. 11, 2237-2255.

Hill S and Kavanagh E (1980) J. Bacteriol. 141, 470-475.

Hill S et al (1981) Nature 290, 424-426.

Imperial J et al (1984) J. Bacteriol. 158, 187-194.

Kahn D et al (1982) J. Gen. Microbiol. 128, 779-787.

Kaluza K and Hennecke H (1981) Arch. Microbiol. 130, 38-43.

MacNeil D and Brill WJ (1980) J. Bacteriol. 144, 744-751.

Magasanik B (1982) Ann. Rev. Genet. 16, 135-168.

Merrick M et al (1982) Mol. Gen. Genet. 185, 75-81.

Merrick M (1983) EMBO J. 2, 39-44.

Nieva-Gomez D et al (1980) Proc. Natl. Acad. Sci. USA 77, 3346-3350.

Ow DW and Ausubel FM (1983) Nature 301, 307-313.

Ow DW et al (1985) J. Bacteriol. 161, 868-874.

Pühler A and Klipp W (1981) In Boethe H and Trebst A, eds, Biology of
  Inorganic Nitrogen and Sulphur, pp. 276-286, Springer-Verlag, Berlin.

Pühler A et al (1984) In Veeger C and Newton WE, eds, Advances in Nitrogen Fixation Research, pp. 609-619, Martinus Nijhoff/Junk, The Hague.

Roberts GP et al (1978) J. Bacteriol. 136, 267-279.

Rosenberg M and Court D (1979) Ann. Rev. Genet. 13, 319-353.

Scott KE et al (1981) J. Mol. Appl. Genet. 1, 71-81.

Shah V and Brill WJ (1977) Proc. Natl. Acad. Sci. USA 74, 3249-3253.

Shah VK et al (1983) J. Biol. Chem. 258, 12064-12068.

Shen S-C et al (1983) Nucl. Acids Res. 11, 4241-4250.

Sibold L and Elmerich C (1982) EMBO J. 1, 1551-1558.

Sundaresan V and Ausubel FM (1981) J. Biol. Chem. 256, 2808-2812.

Zhu J and Brill WJ (1981) J. Bacteriol. 145, 1116-1118.

REGULATION OF KLEBSIELLA PNEUMONIAE NITROGEN FIXATION GENE PROMOTERS BY REGULATORY PROTEINS ntrC, nifA, and nifL

DAVID W. OW*, QING GU, YUE XIONG, JIA-BI ZHU, AND SAN-CHIUN SHEN
SHANGHAI INSTITUTE OF PLANT PHYSIOLOGY, ACADEMIA SINICA, SHANGHAI, CHINA
PRESENT ADDRESS: *DEPARTMENT OF BIOLOGY, UNIVERSITY OF CALIFORNIA AT
SAN DIEGO, LA JOLLA, CA 92093

Klebsiella pneumoniae, a free living enteric bacterium, reduces N_2 to NH_4^+ only under conditions of nitrogen starvation and low O_2 tension. Its cluster of 17 nitrogen fixation (nif) genes are organized in 8 transcription units near the his operon (Beynon et al, 1983). One operon, nifLA, codes for a repressor and an activator. The nifL product mediates repression under conditions of N-excess and high O_2 tension, and the nifA product is required for transcription of all other nif genes (for review, see Roberts, Brill, 1981). Recent findings have shown that the nifA product has the capacity to activate promoters of the central nitrogen regulatory (NTR) system (Ow, Ausubel, 1983; Merrick, 1983). Whether nifA-mediated activation of NTR pathways is physiologically and evolutionarily significant is not clear. But in light of a possible contribution by nifA in activating NTR genes, the following model was proposed (Ow, Ausubel, 1983).

A working model on nif regulation.

This model states the following (see Fig. 1): The global NTR system (for review, see Magasanik, 1983) is repressed when cells are growing in nitrogen-rich media; and glutamine synthetase (GS), the glnA product, is expressed at the basal level. This low level of GS is solely for glutamine biosynthesis, as the major pathway of N-assimilation during NH_4^+ excess is carried out by glutamate dehydrogenase. Two products, ntrB (also known as glnL) and ntrC (also known as glnG), are necessary for repression of the NTR system (McFarland et al, 1981, Pahel et al, 1982).

Under N-deprivation, however, the NTR system activates a large set of genes that are involved in the transport, degradation, and assimilation of certain N-containing compounds (Kustu et al, 1979b; Magasanik, 1982), as well as genes involved in N_2 reduction. The current belief is that proteins involved in the GS adenylation system (glnB and glnD products) sense the α-ketogluturate/glutamine ratio in the cell as a measure of the intracellular level of fixed nitrogen, and this information is transmitted to the ntrB product (Reitzer, Magasanik, 1985). The ntrB product would then become inactive as a repressor, thereby allowing the ntrC product to assume an activator conformation. ntrC-mediated activation (Pahel, Tyler, 1979; Kustu et al, 1979a), however, requires another regulatory protein, the ntrA (also known as glnF) product (Garcia et al, 1977; Gaillardin, Magasanik, 1978), which was recently shown to be a sigma factor (Kustu, Magasanik, personal communication).

The short term remedy to N-starvation is to assimilate the low levels of NH_4^+. This ATP-driven reaction is catalyzed by GS. Hence, the first response to N-starvation is autoactivation of the glnA-ntrBC operon by ntrC and ntrA products. In addition to raising the level of GS, this also raises the levels of ntrB and ntrC proteins. While the ntrB product is inactive as a repressor, but plays a monitoring role for the level of fixed nitrogen, the higher level of ntrC product is essential (along with the ntrA product) for the activation of other NTR genes (Pahel et al, 1982), such as those involved in the transport and utilization of histidine*. This higher level of ntrC product also activates the nifLA operon.

Although N_2 reduction is an option to K. pneumoniae, the sensitivity of nitrogenase to O_2 and its high demand for energy make it likely to be the pathway of last resort. This assurance is attributed to both the nifL and nifA products. Due to the repressive effect of the nifL product, the nif regulon cannot be expressed under conditions of either high O_2 tension, or intermediate levels of fixed nitrogen (Hill et al, 1981; Merrick, 1982). Only when O_2 tension becomes low, and fixed nitrogen is still limiting, will nifL become inactive as a repressor and assume a monitoring role. This

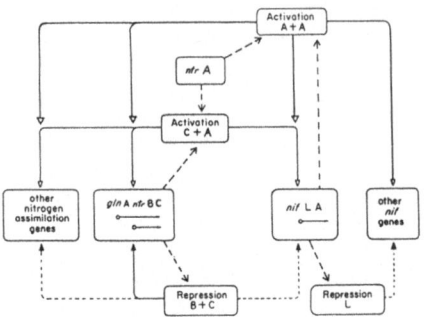

would then allow the nifA product to activate the NIF system. As with ntrC-mediated activation, nifA-mediated activation also requires the ntrA sigma factor as co-activator (Ow, Ausubel, 1983; Merrick, 1983).

The first step in activating the NIF system may be autoactivation of the nifLA operon, similar to that described for the glnA-ntrBC transcription unit (Ow, Ausubel, 1983; Drummond, 1983). A higher level of nifA product would efficiently activate other nif genes. In addition, it is possible that nifA may contribute to ntrC-mediated activation of NTR genes, such as hut and glnA. Consequently, this would insure that (i) all available sources of fixed nitrogen are exhausted before resorting to the most ATP-expensive pathway, and (ii) that optimal amounts of GS are produced for maximum efficiency assimilation of the expensively produced NH_4^+.

---

*Two points need clarification. (i) Many of these NTR pathways, such as hut (histidine utilization), also have their own specific regulators and high level activation can only be attained via substrate induction. This insures that high levels of catabolic enzymes would not be made in the absence of substrate. (ii) Since many N-containing compounds can act as donors of carbon as well as nitrogen, many of these pathways are also subject to global carbon regulation, i.e. cAMP-CAP (Magasanik, 1978).

The NIF system is turned off when either fixed nitrogen is no longer limiting, or when O_2 tension rises to an inhibitory level for nitrogenase. This is mediated by the nifL product, which blocks nifA-mediated activation at presumably the protein-protein level (see below). Alleviating nitrogen starvation would also shut down the NTR system. As the message for nitrogen sufficiency is relayed to the ntrB product, ntrB and ntrC products would assume repressor conformations. Direct repression at the glnA and ntrBC promoters would lower the level of glnA, ntrB and ntrC proteins (Dixon, 1984; Reitzer, Magasanik, 1985). Lowering the level of ntrC protein would further de-activate the NTR genes, including nifLA.

Ancestral relationship of NTR and NIF systems.

NIF and NTR regulatory systems share many similar features: (i) Both ntrB and nifL products mediate NH_4^+ repression. (ii) Both ntrC and nifA products require the ntrA sigma factor as co-activator. (iii) ntrC-activated promoters can also be activated via nifA. These facts have led to the proposal that the NIF regulatory system evolved from the central NTR system (Ow, Ausubel, 1983). Because of the suspected ancestral relationship between the two systems, it is not entirely surprising that primary structures of NTR and NIF promoters are strikingly similar to each other (Ow et al, 1983, Ausubel, 1984). The recent proposal that these genes require a specific sigma factor, the ntrA product, is also consistent with the fact that NIF/NTR consensus sequences bear little resemblance to canonic E. coli -35 and -10 promoter sequences.

What is astonishing, however, is that the NIF/NTR sequences are observed in nif gene promoters of nearly all nitrogen fixing bacteria examined thus far (the exception is Anabaena) (for review, see Ausubel, 1984). The implication here is that activation of these promoters is mediated by protein combinations similar to ntrA + nifA or ntrA + ntrC. Since there is no compelling reason to propose convergent evolution of a K. pneumoniae-like NIF regulatory system for these taxonomically unrelated species, one is left with the alternative hypothesis: that the nif regulatory elements, along with nif genes, radiated from free living diazotrophs such as K. pneumoniae. Conservation of this regulatory system lends further support for the use of K. pneumoniae as a model system.

Promoters of NIF and NTR systems.

Promoters of the NTR system are normally activated via ntrC, but can also be activated via nifA. These can be referred to as ntrC-activated promoters. In contrast, promoters of the NIF system such as the nitrogenase (nifH) promoter can only be activated via nifA, but not via ntrC. These can be referred to as nifA-activated promoters. Structurally, there is but slight difference between these two classes of promoters. This is best exemplified by the nitrogenase promoters from K. pneumoniae (Kp) and Rhizobium meliloti (Rm) (Sundaresan et al, 1983a, b). When examined in E. coli, the Kp nifH promoter is a typical nifA-activated promoter, while the Rm nifH promoter behaves as a typical ntrC-activated promoter. Both contain the CTGG sequence at the -25 region. In the -14 region, however, the Kp nifH promoter has the sequence CCCTGCA, while the corresponding sequence in Rm nifH is

TTTTGCA. Tentative conclusions drawn from sequence comparison studies have suggested that CTGG-6bp-TTGCA is required for nifA activation, while additional T bases before TTGCA (such as TTTTGCA) are required for ntrC-mediated activation (Ow et al, 1983; Beynon et al, 1983).

In the sections that follow, we summarize recent results obtained in this laboratory aimed at pinpointing what specific bases are involved in promoter activation mediated by nifA and ntrC products (Ow et al, 1985). In addition, our attempt to isolate nifL independent mutants of the nifH promoter leads to the conclusion that the nifL product is unlike the classical DNA binding repressor. The alternative is that it mediates negative regulation by interaction with the nifA product.

Directed point mutagenesis of the K. pneumoniae nifH promoter.

Our goal was to uncover point mutations affecting regulation of the nitrogenase promoter. This required efficient methods for (i) promoter-directed point mutagenesis, (ii) detection of promoter mutations, (iii) determination of promoter strength in vivo, and (iv) structural characterization of mutant promoters. The strategy we took is diagrammed in Fig. 2. A nifH promoter-lacZ fusion plasmid was used in deletion loop sodium bisulfite mutagenesis (Peden, Nathans, 1982) after hetero-duplex formation with the promoter-less lacZ plasmid. Mutagenized plasmid DNA was transformed into strains of the appropriate geno- type, and putative promoter mutants were visually detected on chromogenic indicator plates. Genetic analysis of promoter strength was determined by assaying for beta-galactosidase. Finally, base changes were determined by the chain-termination method of 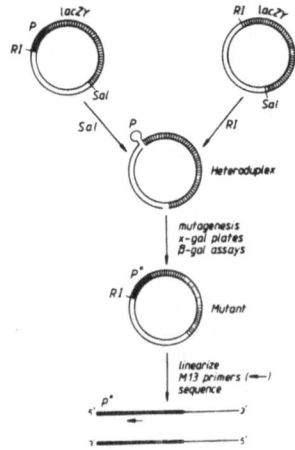 DNA sequencing, using linearized double stranded plasmid DNA as template (Smith et al, 1979). Due to the proximity of promoter to the 12th codon of lacZ, it was convenient for us to use commercially available primers designed for M13 sequencing. The novelty of this scheme is that a single plasmid can be used for mutagenesis, determination of promoter strength in vivo, and dideoxy DNA sequencing.

Class A: Mutations that affect nifA-mediated activation.

Mutagenized plasmid DNA was introduced into an E. coli host harboring a compatible nifA producing plasmid. Whereas wild type promoter-lacZ fusions gave rise to dark blue colonies on indicator plates, mutants insensitive to nifA-mediated activation conferred a pale blue phenotype. From this screen, a total of 19 independent mutants were found. Fifteen of these mutants have base changes clustered between the region from −25 to −12. A correlation between genotype and relative promoter strength (in parentheses) is shown in Fig. 3. The remaining four mutants do not affect nifA-specific activation. Two of these mutants contain base changes at the ribosome binding site that affect translation of lacZ. The remaining two mutants have base changes outside the promoter region. Since this last class of mutant was not found in abundance, it seems unlikely that another regulatory site lies outside the promoter region. Our current belief is that these two mutants represent rare spontaneous occurrences, such as with additional alterations within the lacZ coding region.

				Activation	
				nifA	ntrC (glnG)
			-14		
Kp nifH wt	CTGG	-6bp-	CCCTGCA	+ (100)	-
H A1	----		-----T-	- (16)	-
H A2	----		----A--	- (1)	-
H A3	---A		-------	- (1)	-
H A4	—A-		-------	- (16)	-
H B1	----		T------	+	+ (22)
H B2	----		--T----	+	+ (46)
H B3	----		T-T----	+	+ (100)
Rm nifH wt	----		TTT----	+	+
Kp nifL wt	AG--		GTT----	+	+
Kp nifB wt	----		ATT----	+	-
Kp nifM wt	----		ATT----	+	-
Kp nifE wt	----		AAT----	+	-
Kp nifU wt	----		AAT---T	+	-
Kp nifF wt	----		-TTC---	+	-

Class B: Mutations that affect ntrC-mediated activation.

The Kp nifH promoter can only be activated via nifA but not via ntrC. In contrast, the Rm nifH promoter can be activated with either proteins. Similarity in primary structures between the two promoters compelled us to isolate Kp nifH promoter mutants that respond to ntrC-mediated activation. Mutagenized plasmid DNA was introduced into a ntrC⁺ host and plated on N-poor medium to derepress the NTR system. Mutants that exhibited beta-galactosidase activity were further tested in an ntrC⁻ host to screen out promoters with constitutive phenotypes. Using this scheme, nine independent clones were found. In only one mutant, which gave a very low level of ntrC-activated expression, did we fail to detect base alterations at the promoter region. Although base changes were found near the protein initiation codon, it is not clear whether these changes are related to ntrC-specific regulation.

The other eight mutants, however, all acquired base changes at the consensus sequence. Three of these mutants have in common the change from CCCTGCA (found in wild type) to TCCTGCA, and they gave low levels of ntrC-activated expression (see Fig. 3, promoter strength shown in parentheses). Another class, represented by four mutants with the sequence CCTTGCA, gave an intermediate level of ntrC-activated expression. A high level of ntrC-mediated activation was observed in one mutant with the sequence TCTTGCA. These results suggest that the sequence TTTTGCA is important to ntrC-specific activation. It is surprising that a single base change is sufficient to alter an nifA-activated promoter to an ntrC-activated one. Even more surprising is the class of mutant promoters with the sequence CCTTGCA. This is because TTGCA is common to many nifA-activated promoters that do not respond to ntrC-mediated activation (such as nifB, M, E, and U, see Fig. 3). To account for this anomaly, one possible explanation may be that the A base at the first position of the heptameric sequence prevents ntrC-specific activation.

Class C: Pseudo mutants that escape nifL-mediated repression.

It is not known whether repression by the nifL protein occurs by interaction with (i) the nifA protein, (ii) the transcription complex, or (iii) the promoter itself. If the last case is true that the nifL product is a classical DNA binding repressor, then there should exist operator mutations that would render the promoter insensitive to nifL-mediated repression. We sought to test this possibility by attempting to isolate promoter mutants with an nifL insensitive phenotype. Our strategy was to introduce mutagenized nifH-lacZ plasmid into a wild type host harboring the nifLA operon on a compatible plasmid. Under aerobic growth, the nifH promoter is only expressed at a low level due to oxygen repression mediated by the nifL product. Clones with high levels of promoter activity, however, would be either (i) those with pseudo-nifL independent promoters such as constitutive or ntrC-activated promoters, or (ii) those with true nifL-independent mutations.

After extensive screening, we failed to detect mutants with true nifL-independent mutations. We did find, however, many with a constitutive or ntrC-activated phenotype. These pseudo nifL-independent promoters were distinguished in an $ntrC^+$ host grown in either rich or poor nitrogen source. We did not examine the base alterations of our collection of constitutive promoter mutants because they could be due to new canonic promoter-like sequences with transcriptional start points at new locations. Since these new constitutive promoters could be far from the hypothetical operator site, then even if nifL were a classical DNA binding repressor, binding at its operator site would not necessarily prevent transcription. Hence, we reasoned that constitutive promoter mutants would yield no useful information.

In contrast, this argument would be different with ntrC converted promoters. Sequence analysis of a number of these mutants showed that they contain the same base alterations in the consensus sequence as our ntrC-activated (class B) mutants. This suggests that the transcriptional complex recognizes the same region of DNA as the wild type nifA-activated promoter. Therefore, if an nifL operator site exists that affects nifA activation, then repressor binding at this site should have affected ntrC activation as well. The fact that it did not is indicative of the lack of an operator site. Along with our inability to recover true nifL-insensitive mutants, the tentative conclusion we draw from this study is that the nifL protein is not likely to bind to DNA as classical repressors do. This favors the alternative hypothesis: that the nifL product interacts with the nifA protein or with the nifA-specific (but not the ntrC-specific) transcriptional complex (Gu et al, manuscript in preparation).

References.

Ausubel FM (1984) Cell 37, 5-6
Beynon J. et al (1983) Cell 34, 665-671
Dixon R (1984) Nuc Acids Res 20, 7811-7830
Drummond M et al (1983) Nature 301, 302-307
Gaillardin CM, Magasanik B (1978) J Bacteriol 133, 1329-1388
Garcia F et al (1977) Proc Natl Acad Sci USA 74, 1662-1666
Hill S et al (1981) Nature 290, 424-426
Kustu S et al (1979a) Proc Natl Acad Sci USA 76, 4576-4580

Kustu S et al (1979b) J Bacteriol 138, 218-234
Magasanik B (1978) In Miller J, Reznikoff W, eds, The Operon, pp 373-
    387, Cold Spring Harbor Laboratory, New York
Magasanik B (1982) Annu Rev Genet 16, 135-168
McFarland N et al (1981) Proc Natl Acad Sci USA 78, 2135-2159
Merrick M (1983) EMBO J 2, 39-44
Merrick M et al (1982) Mol Gen Genet 185, 75-81
Ow DW, Ausubel FM (1983) Nature 301, 307-313
Ow DW et al (1983) Proc Natl Acad Sci USA 80, 2524-2528
Ow DW et al (1985) J Bacteriol 161, 868-874
Pahel G, Tyler B (1979) Proc Natl Acad Sci USA 76, 4544-4548
Pahel G et al (1982) J Bacteriol 150, 202-213
Peden K, Nathan D (1982) Proc Natl Acad Sci USA 79, 7214-7219
Reitzer LJ, Magasanik B (1985) Proc Natl Acad Sci 82, 1979-1983
Roberts GP, Brill WJ (1981) Annu Rev Microbiol 35, 207-235
Smith et al (1979) Cell 16, 753-761
Sundaresan V et al (1983a) Nature 301, 728-731
Sundaresan V et al (1983b) Proc Natl Acad Sci USA 80, 4030-4034

GENETIC AND PHYSICAL CHARACTERISATION OF nif AND ntr GENES
IN Azotobacter chroococcum AND A. vinelandii

C. KENNEDY, R. ROBSON, R. JONES, P. WOODLEY, D. EVANS, P. BISHOP, R. EADY,
R. GAMAL, R. HUMPHREY, J. RAMOS, D. DEAN*, K. BRIGLE*, A. TOUKDARIAN and
J. POSTGATE
AFRC Unit of Nitrogen Fixation, University of Sussex, Brighton, U.K.
*Battelle-Kettering Research Labs., Yellow Springs, Ohio, U.S.A.

Characterisation of genes affecting nitrogen fixation in
Azotobacters has progressed rapidly using molecular methods: nif genes
similar to those in K. pneumoniae have been revealed and also other genes
whose function is related to the distinctive ability of Azotobacters to
fix nitrogen under aerobic conditions or in the absence of detectable Mo.
Recent advances in understanding the genetic basis for nitrogen fixation
and its regulation in Azotobacters are described in this paper.

1.   IDENTIFICATION OF nif GENES IN A. chroococcum BY HYBRIDISATION AND
COMPLEMENTATION STUDIES WITH K. pneumoniae nif

A 70 Kb region of the A. chroococcum genome, identified on cosmids
that hybridised to a nifHD gene probe from A. vinelandii, has been
analysed for nif genes corresponding to those of K. pneumoniae by DNA-DNA
hybridisation and by complementation of nif⁻ mutants. Previous work
delineated regions that hybridised to nifHDK and nif(M)V(S) probes (Fig.
1)(Jones et al., 1984). Also, subclones with nifHDK, pJRW1 and pJRW11,
restored appreciable acetylene reducing activity in K. pneumoniae nifH,
D and K mutants and in nifD and K mutants of A. vinelandii.

Further analysis using K. pneumoniae nif probes revealed sequences
of A. chroococcum homologous to fragments carrying 1) nif(L)F(M),
2) nifUX and 3) nifNE. Strong hybridisation was observed to regions of
pACB1 (Fig. 1). Thus a major nif cluster of A. chroococcum has
sequences homologous to most or all nif genes of K. pneumoniae from nifF
through nifH. No hybridisation to nifQBAL nor to nifJ has been found
in the 70 Kb cluster carried together by pACB1 and pACD30 (Jones et al.,
1984). Of interest are the 8 and 9 Kb gaps between the nifF, nifMVSU,
and nifNEKDH regions: do they contain sequences involved in nitrogen
fixation which are unique to Azotobacters or has evolution resulted in
clustering but not contiguity of nitrogen fixation genes in these
organisms with non-nif genes interspersed among nif genes?

DNA sequencing of nifHDK has shown, as expected, 3 open reading
frames (ORF) of 867, 1476, and 1575 bp homologous to nifH, D and K genes
from other diazotrophs, separated by 124 and 100 bp. Features of this
region include 1) a sequence preceding nifH similar to the consensus
sequence for ntrA-dependent promoters (see Ausubel, 1984 for review),
2) inverted repeat (IR) sequences preceding the translational start sites
of all 3 coding regions and 3) tandem 'stop' codons terminating nifH and
nifD (Jones, Woodley, Robson, this symposium). (Similar features have
been found in A. vinelandii nifHDK sequences; Brigle, Newton, Dean, 1985).

470

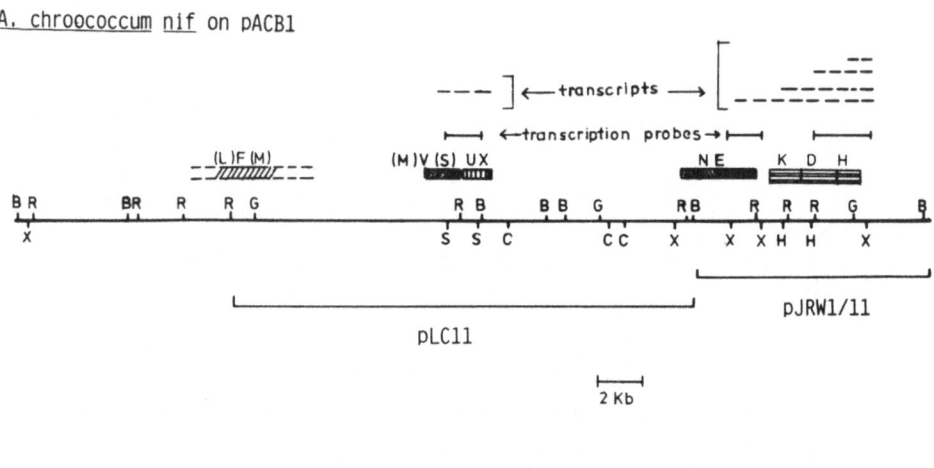

A. chroococcum nif on pACB1

K. pneumoniae nif

FIGURE 1. A. chroococcum nif genes. Regions of K. pneumoniae nif that hybridise to pACB1 are indicated by similar cross-hatching.

Previously, transcripts of 4.3, 1.1 and 2.6 Kb which hybrisided to a nifH probe were detected in A. chroococcum during nitrogenase derepression (Jones et al., 1984 and Fig. 1). Transcription mapping has been advanced using probes from the "nifNE" region (a 1.4 Kb XhoI fragment) and another from the "nifMVSU" region (a 1.4 Kb SalI fragment). mRNA from cultures of A. chroococcum was used in hybridisation experiments, and under conditions of N-limitation transcripts of 6.4 and 2.5 Kb were found which hybridised to each probe respectively (Fig. 1). Additional experiments with the nifH probe revealed a 6.4 Kb transcript present in small amounts relative to the 3 mentioned previously, and it is likely that 4 transcript species are made from nifHDK and downstream genes. Regulatory features revealed by DNA sequences that are consistent with this proposal are the IR's in front of nifH, D and K mentioned previously and also that the nifK ORF is not followed by tandem translation terminators nor by obvious transcription stop or attenuator sequences. Also the sizes of the 3 smaller transcripts are exactly that expected if they end shortly after the stop codons of nifH, nifD and nifK within the HDK intergenic regions. Thus, expression of genes in the 35 Kb insert of pACB1 and apparently corresponding to nif genes found in K. pneumoniae, is controlled by cellular N-status, a condition expected for genes participating in nitrogen fixation.

Another cosmid carrying A. chroococcum nif genes was isolated from a pLAFR1 gene bank after transfer to a nif::Tn5 mutant of A. vinelandii, MV21 (described below). The complementing cosmid, pLC11, had EcoRI fragments of 11 and 10.5 Kb identical to two found in the mid-portion of pACB1 (Fig. 1). pLC11 was found to complement K. pneumoniae nif mutants: nifM⁻, V⁻, S⁻ and U⁻ strains with pLC11 were able to reduce acetylene at rates similar to wild-type and also could grow on NH₄⁺-free medium. The nifB, F, E, N and J mutants tested were not complemented by pLC11 nor

by pJRW1 or pJRW11. Expression of the A. chroococcum nifMVSU genes in
K. pneumoniae was not enhanced by the nifA gene expressed constitutively
from pCK1.

TABLE 1. Complementation of K. pneumoniae nif mutants by pLC11

Strain	nif mutation	% of wild-type acetylene reduction (growth on N-free medium)	
		no plasmid	+pLC11
UNF50231	Nif$^+$	100 (+)	62.5(+)
UNF812	nifV2585::Tn5	35 (−)	68 (+)
UNF828	nifM2568::Tn5	6.7(−)	100+ (+)
UNF2050	nifM2104	8 (−)	nt (+)
UNF866	nifS2554::Tn5	2.4(−)	100+ (+)
UNF2142	nifS2442	3.2(−)	27.5(+)
UNF811	nifU2573::Tn5	1 (−)	62 (+)

Cultures were derepressed in NFDM + 20 µg histidine ml^{-1} + 100 µg aspartate
ml^{-1} for 18 hrs (UNF811 for 5.5 hrs) then assayed for acetylene reduction.
nt, not tested.

Thus, genes corresponding in both structure and function to nifMVSU
and nifKDH are clustered in A. chroococcum. Genes similar in DNA
sequence to nifN and nifE are also present and located near to nifKDH as
in K. pneumoniae, and a nifF-like gene marks, at least for the moment,
the "left-hand" side of the cluster.

2.   IDENTIFICATION AND DNA SEQUENCING OF A SECOND nifH GENE IN
      A. chroococcum CONTIGUOUS WITH A FERREDOXIN-LIKE GENE

The nifH gene of K. pneumoniae hybridised equally well to two
regions of the A. chroococcum genome: the first is within the nif cluster
described above while the other lies at least 15 Kb distant from the ends
of pACB1 (Jones et al., 1984; Robson, unpublished results). Two
recombinant clones of the latter were isolated from colonies which
hybridised to a nifH probe. DNA sequencing of 2 Kb of DNA spanning the
cloned 2.5 and 3.5 Kb XhoI fragments revealed a complete nifH-like ORF,
called nifH*. The nifH and nifH* genes are 88% homologous in DNA
sequence; comparison of the amino acid sequences of their encoded
proteins is shown in Fig. 2.

Separated by 124 bp and downstream from nifH* is an ORF, preceded by
a sequence analogous to the E. coli ribosome binding site and followed by
an IR typical of rho-independent terminators. The coding sequence of
this ORF determines a protein of 63 amino acids; 9 are cysteinyl residues
spaced in a manner characteristic of 2[4Fe-4S] ferredoxins [Fd].
Cotranscription of this gene and nifH* is indicated by finding no
obvious termination site between them and also by gene fusion
experiments (Robson et al., unpublished). Also the consensus sequence
for ntrA-dependent promoters was found 51 bp before the nifH* coding
sequences. A possible role for these genes in Azotobacter nitrogen
fixation is suggested by the observation that a DNA fragment from the
A. chroococcum Fd-like gene hybridised to mRNA prepared from A. vinelandii
grown under Mo-deficient conditions (Jacobson, Bishop, this symposium).

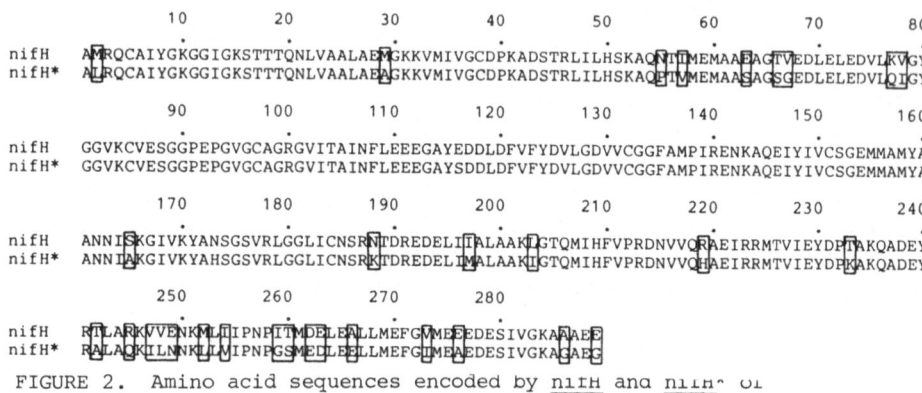

FIGURE 2. Amino acid sequences encoded by nifH and nifH* of A. chroococcum

3. NITROGEN FIXATION IN A. vinelandii UNDER MOLYBDENUM DEPRIVATION: CHEMOSTAT STUDIES OF A nifHDK DELETION MUTANT

The detection of nitrogen fixation in both wild-type and mutant strains of A. vinelandii grown in media depleted of detectable Mo led Bishop et al. (1980) to propose that an 'alternative' nitrogenase was synthesised under these conditions. In order to further characterise the system, these workers have constructed a mutant strain, CA11, deleted for a 5.25 Kb length of DNA spanning nifHDK.

Chemostat cultures of CA11 were established to define the N_2 dependence of growth and sensitivity to MoO_4 of the 'alternative' system. In steady-state culture supplied with Mo-depleted Burks sucrose medium flowing to give dilution rates of 0.174 h^{-1}, CA11 attained a density of 1.14 x 10^8 cells ml^{-1} and produced ethylene from acetylene at 1.36 nmol min^{-1} mg $protein^{-1}$ with concomitant H_2 evolution of 75.3 nmol min^{-1}mg $protein^{-1}$. Under air, the N_2 fixation rate calculated from the N content of the culture was 17.8 nmol min^{-1}mg $protein^{-1}$ with H_2 evolution of 2.65. Clearly the N_2 fixation rate calculated from the C_2H_2 reduction rate is underestimated 10-fold. Two important experiments then showed that 1) when the gas supply was changed from air to Ar/O_2 (80:20 v/v), CA11 and also wild-type stopped growing and washed out of the chemostat (though it did not if NH_4Cl was added) and 2) when 10 nM Na MoO_4 was added to the medium reservoir, the density of wild-type UW136 increased from OD540 0.94 to 1.24 over the next 47 hrs while that of CA11 decreased from 0.65 to 0.115. Addition of NH_4Cl to CA11 'poisoned' by MoO_4 stimulated its growth. Thus the ability of A. vinelandii to grow on Mo-depleted medium is dependent on N_2 and the nitrogenase activity that supports nitrogen fixation under these conditions is repressed by added MoO_4. These experiments are discussed in more detail by Bishop, Eady, this symposium. The possible involvement of nifM in the 'alternative' system is suggested by experiments described in the following section.

4. CHARACTERISATION OF Tn5 MUTANTS OF A. vinelandii

Tn5 mutagenesis was achieved in A. vinelandii UW136 by transfer from E. coli of suicide vector pGS9, a pACYC184 derivative carrying Tn5 and the IncN Tra genes (Selveraj, Iyer, 1983). Km^r (and Sm^r) colonies arose at a frequency of 10^{-4} to 10^{-5}. About 3% of these were apparently auxotrophic mutants, growing on enriched but not minimal medium. A number of

Nif⁻ mutants were isolated though most were unstable, segregating Nif⁺ colonies at high frequency even after several subcultures. The stable mutants fell into 3 classes, represented by MV6, MV21 and MV22.

MV6 was complemented by pJRW11 and hybridisation experiments showed that Tn5 was inserted in nifH. MV6 was 'corrected' by a pLAFR1 gene bank cosmid, pLV10, isolated by its ability to complement UW91, a nifH point mutant. Since pLV10 carries nifH and D but not nifK then its failure to complement MV6 was probably due to a polarity effect of Tn5 which prevented expression of nifD and nifK. Since pLV10 has nifH, it could correct MV6 by reciprocal recombination with the chromosome.

MV22 contains Tn5 in a 5.7 Kb EcoRI fragment of unknown distance from nifHDK but presumably within the major nif cluster of A. vinelandii. This inference was drawn from the fact that it was complemented by pLC11 described above which carries A. chroococcum nif genes. A cosmid pLV1 from the A. vinelandii pLAFR1 bank also complemented this mutant (see Kennedy et al., poster abstract).

Of interest was the finding that MV22 was complemented for growth on N-free medium by pRD1 carrying the K. pneumoniae nif genes. This was unexpected in light of the failure of pRD1 to complement nifHDK mutants of A. vinelandii, including UW6 and UW10 (Cannon, Postgate, 1984 and confirmed in these experiments). pRD1 also failed to complement MV6, the nifH mutant described above, but did complement MV21, described below. The failure of pRD1 to complement mutations in nifHDK is consistent with the observed failure of the K. pneumoniae nifH promoter to be expressed in A. vinelandii (Kennedy, Drummond, 1985). Preliminary experiments with mutant derivatives of pRD1 suggest that nifE and nifN mutants of pRD1 failed to complement MV22.

MV21 grew on neither Mo-sufficient nor -deficient agar media. Other mutants such as UW6, UW10, UW91 and UW1, do grow on Mo-deficient media (Bishop et al., 1980) as do MV6 and MV22, and apparently express an "alternative" system for nitrogen fixation under these conditions (see above). The site of Tn5 insertion in MV21 is within a 5.7 Kb EcoRI fragment different from that described for MV22. This mutant, like MV22, was complemented by pLV1, pLC11 and pRD1. Preliminary results suggest that nifM mutants of pRD1 failed to complement MV21. Therefore the nifM product may be necessary for nitrogen fixation in A. vinelandii under both +Mo and -Mo conditions. In K. pneumoniae nifM is required for activation of nifH polypeptide (Howard et al., 1985) so the requirement for nifM is consistent with the suggestion that a second Fe protein is involved in "alternative" nitrogen fixation (Premakumar, Lemos, Bishop, 1984).

5.   REGULATION OF nif GENE EXPRESSION IN AZOTOBACTERS

Recent studies of nif regulation in Azotobacters have mainly sought features similar to the Klebsiella control system where activation at the nifLA promoter by the ntrA and ntrC gene products is followed by activation at the 7 other nif promoters by the ntrA and nifA gene products. Earlier work established that both NH_4^+ and O_2 repress nitrogenase synthesis in A. chroococcum and A. vinelandii (Robson, 1979; Krol et al., 1982; Kennedy, Robson, 1983); that regulatory mutants at least superficially like K. pneumoniae nifA⁻ strains could be isolated (Shah et al., 1973); and the K. pneumoniae nifA gene corrected such mutants and resulted in constitutive nif expression (Kennedy, Robson, 1983). A search for genes similar to nifA, ntrA and ntrC began with the transfer of cosmids from pLAFR1 gene banks carrying genomic fragments of A. chroococcum and A. vinelandii to various K. pneumoniae, E. coli and A. vinelandii mutants. Selection was for restoration of the mutant phenotypes. No cosmids were found that complemented nifA⁻ mutants of K. pneumoniae or Nif⁻ regulatory mutants of A. vinelandii. Cosmids were, however, isolated from the A. vinelandii gene bank by complementation of E. coli ntrC and ntrA mutants for growth on arginine as N-source.

The complementing genes were subcloned from pLV72, the ntrA cosmid, and from pLV50, which carried glnA in addition to ntrC. Hybridisation of cosmids and A. vinelandii genomic DNA to probes carrying K. pneumoniae ntrA, ntrC or glnA genes revealed structural homology. The cosmids and their sub-clones (Fig. 3) also complemented ntrC or ntrA mutants of K. pneumoniae for nitrogen fixation. A promoter arrangment of the A. vinelandii glnA-ntrC genes similar to that of the P1-glnA-P2-ntrBC operon of enteric organisms (Magasanik, 1984) was suggested by the complementation patterns of pAT512, pAT523, and pAT524. All three complemented E. coli ntrC mutants for growth on arginine but only the former two restored GlnA⁺, apparently because the 'natural' P1 promoter for this operon is located between the EcoRI and ClaI sites. pAT512 has this region which is expressed in E. coli, while in pAT523 a strong promoter is provided from the vector Cmʳ gene. These results also show that the entire coding region for glnA is contained within the 6 Kb EcoRI fragment. Tn5 insertion mutants in the A. vinelandii ntrA and ntrC plasmids were constructed and introduced into the A. vinelandii genome by marker exchange. Hybridisation experiments using Tn5 and the ntrC and ntrA subclones (pAT523 and pAT705) to probe DNA from the putative mutants showed that Tn5 was in a chromosomal position identical to that of the original insertion; 'wild-type' fragments were not detected.

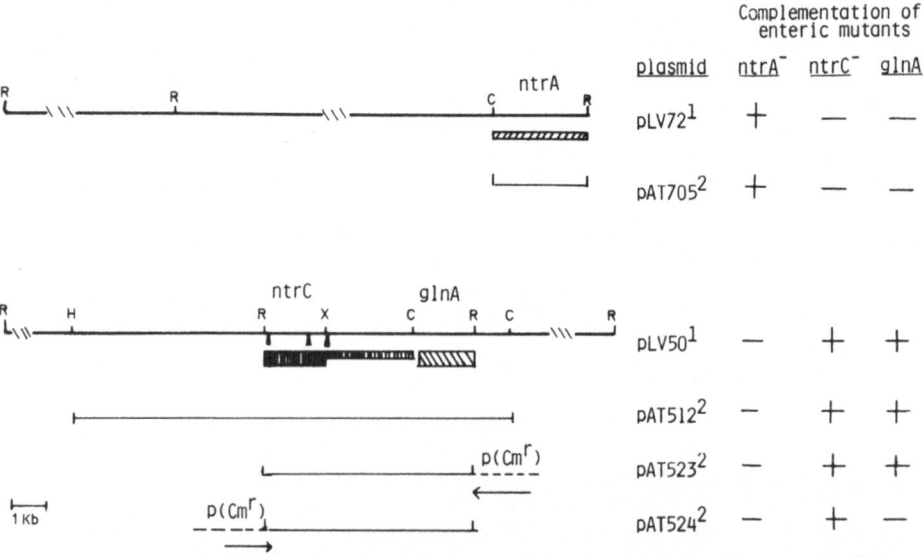

FIGURE 3. The ntrA and glnA-ntrC genes of A. vinelandii. Regions hybridising to K. pneumoniae ntrA, ⬚⬚⬚ ; glnA, ⬚⬚⬚ ; and ntrC, ⬚⬚⬚. Plasmid vectors were 1, pLAFR1; 2, pBR325. ▲ are sites of Tn5.

The ntrA mutant MV700 failed to grow on N_2 or to reduce acetylene under any conditions tested. The only other phenotype detected was the failure of MV700 to grow on NO_3^- as N source. Therefore, in A. vinelandii as in K. pneumoniae, ntrA is necessary for nitrogen fixation.

The ntrC mutants MV511, MV512, and MV516 unexpectedly grew on N_2 and had nitrogenase activity similar to that of the Nif$^+$ strain UW. The mutants were however chlorate resistant and had little nitrate reductase activity. No other phenotype was detected. Thus ntrC is apparently not necessary for nif expression in A. vinelandii, a surprising result because it can restore nitrogen fixation to ntrC mutants of K. pneumoniae.

If ntrC has no role in activation of nif genes, what does? Although K. pneumoniae nifA works well in Azotobacters, attempts to find a homologous gene by hybridisation or by complementation of regulatory mutants has failed. That activation of nif genes in Azotobacter does occur is suggested by experiments where expression of β-galactosidase was measured from Azotobacter nifH promoters fused to lacZ. In E. coli, expression of β-galactosidase from the fusion plasmids only occurred if the K. pneumoniae nifA gene was also present (although the structure of pDB10, with Av nifH-lac, permitted significant read-through from vector sequences)(Table 2). Sequences found upstream from the nifH ORF of both A. chroococcum and A. vinelandii are nearly identical and contain the ntrA-dependent promoter consensus sequence GCTGGCACAGACGCTGCA at positions -78 and -76, respectively (Robson et al., this symposium and Brigle, Dean, 1985).

TABLE 2. Expression of β-galactosidase from A. chroococcum (Ac) or A. vinelandii (Av) nifH-lac fusions in E. coli

Strain	Plasmid				-nifA	+nifA[1]
ET8000 Ntr$^+$	pFAC11: 2.4 Kb BamHI-BglII insert with Ac nifH promoter in IncP fusion vector pGD500[2]				0	878
ET8556 NtrC$^-$	"	"	"	"	0	1302
ET8000 Ntr$^+$	pDB10: 2.7Kb XhoI-HindIII insert with Av nifH promoter in IncP fusion vector pLC101[3]			558[4]	2066
ET8556 NtrC$^-$	"	"	"	"	705[4]	2302

Activity was measured in 18 hr anaerobic cultures in NFDM + 200 µg serine ml^{-1}. [1]provided by pCK1 (Kennedy, Robson, 1983) with pFAC11 or by pCK3 (Kennedy, Drummond, 1985) with pDB10. [2]Ditta et al. (1985) [3]Dean (unpublished). [4]Read-through from a Km^r promoter through nifH sequences is responsible for high unactivated levels of lac expression.

The recent identification of a nifA-like gene (also called fixY) in R. leguminosarum (Downie, personal communication) prompted further hybridisation experiments. A cloned 1.4 Kb fragment which carries the 3' portion of fixY and only 15 bp of the intergenic fixY-Z region was tested against a number of genomic digests of Azotobacters, and at conditions of low stringency, hybridised to 2 to 5 restriction fragments, depending on the enzyme used. Importantly, one of the hybridising bands in each of two A. vinelandii digests corresponded in size to that carrying ntrC. Thus a gene of the ntrC-nifA class showed some homology to this regulatory gene of Rhizobium. In neither digest was the ntrC gene the most intensely labelled; one or two other bands were darker. Perhaps these fragments carry an activator gene necessary for nif gene expression in Azotobacter. Experiments are underway to identify clones of such a gene by hybridisation to R. leguminosarum nifA.

6. REFERENCES

Ausubel FM (1984) Cell, 29, 1-2.
Bishop PE et al. (1980) Proc. Natl. Acad. Sci. USA, 77, 7342-7346.
Brigle K, Newton W, Dean D (1985) Gene (in press).
Cannon FC, Postgate JR (1984) Nature, 306, 290.
Ditta G et al. (1985) Plasmid 13, 149-153.
Jones R, et al. (1984) Mol. Gen. Genetics 197, 318-327.
Kennedy C, Robson, R. (1983) Nature 301, 626-628.
Kennedy C, Drummond M (1985) J. Gen. Microbiol. 131, 1787-1795.
Krol AJM et al. (1982) Nucl. Acids Res. 10, 4147-4157.
Howard KS et al. (1985) Submitted to J. Biol. Chem.
Pahel G et al. (1982) J. Bacteriol. 150, 202-213.
Premakumar R. et al. (1984) Biochim. Biophys. Acta, 797, 64-71
Robson R (1979) FEMS Microbiol. Lett. 5, 259-262.
Selveraj G, Iyer VN (1983) J. Bacteriol. 156, 1292-1300.
Shah VK et al. (1973) Biochim. Biophys. Acta, 292, 246-255.

7. ACKNOWLEDGEMENTS

We thank Mike Merrick, Martin Drummond, Ray Dixon and Eduardo Santero for strains, plasmids and useful discussions, also Beryl Scutt for typing the manuscript.

ADVANCES IN THE GENETICS OF AZOSPIRILLUM.

CLAUDINE ELMERICH, CORRADO FOGHER, HERVE BOZOUKLIAN, BERTRAND PERROUD, ILONA DUSHA.
UNITE DE PHYSIOLOGIE CELLULAIRE, INSTITUT PASTEUR, 75724 PARIS CEDEX 15, FRANCE.

1. INTRODUCTION

Bacteria of the genus Azospirillum are diazotrophs associated with the root of grasses, without formation of differentiated structure. Very little is known on the genetics of these bacteria and on the molecular biology of their association with plants. The bacteria, first described in 1922 by Beijerinck and rediscovered in 1963 by Becking, were called under the name of Spirillum lipoferum. Their potential agronomic importance was raised when Döbereiner and Day (1976) described their association with plants from various geographical origin. Taxonomic studies led to the creation of a new genus: Azospirillum. This genus, defined by Tarrand et al. in 1978, comprises two species: A. brasilense and A. lipoferum. A third species A. amazonense was recently discovered (Magalhaes et al. 1984).

The bacteria are gram-negative aerobes, curved rod shaped, with a polar flagellum and containing globules of poly-beta-hydroxybutyrate. They have a DNA base composition of 67-70 moles per cent G+C. The bacteria can grow on organic acids such as malate. Strains of A. lipoferum can utilize a large number of carbohydrates including glucose, which is not used by A. brasilense. A. amazonense can utilize saccharose. Ability to fix nitrogen in pure culture was established by the ^{15}N isotopic method. Nitrogen fixation occurs only under microaerobic conditions (Okon 1985).

2. TOOLS FOR GENETIC ANALYSIS

2.1. Plasmids and phages. All Azospirillum strains contain at least one plasmid and some contain up to six molecular species (Franche, Elmerich 1981; Plazinsky et al. 1983). The largest plasmids detected had MW over 300×10^6 Md, and thus they are comparable in size to the megaplasmids found in rhizobia. The two taxonomic groups, A. brasilense and A. lipoferum, cannot be differentiated on the basis of their plasmid content, and no phenotypic property was demonstrated as plasmid-borne, except for the prophage Al-1 which is maintained as a plasmid in A. lipoferum Br17 (Elmerich et al. 1982). Plasmid location for nif genes was not observed in Azospirillum (Elmerich 1983, Plazinski et al. 1983). However, it is tempting to speculate that functions related to bacteria-plant interaction might be coded by Azospirillum plasmids.

Most Azospirillum spp. are lysogenic for defective prophages (Franche, Elmerich 1981; Elmerich et al. 1982). A temperate phage forming plaques on A. lipoferum Br17, named Al-1, was studied. Its morphology and its size were similar to that of the coliphage lambda (Elmerich et al. 1982). No transducing property of Al-1 has yet been detected.

2.2. Transposon mutagenesis. Classical techniques of mutagenesis previously described in E. coli using chemical mutagens can be successfully applied to Azospirillum. Transposon mutagenesis with Tn5, which codes for kanamycin resistance (Km^R), was investigated. Suicide plasmids containing Mu and Tn5 could not be used for transposon mutagenesis (see Elmerich 1983). An alternative procedure was developed by using the mobilizable plamid pSUP2021, which contains Tn5 and cannot replicate outside enteric bacteria (Simon et al. 1983). This plasmid was introduced into Sp7 to perform random mutagenesis. The frequency of Tn5 transposition was 10^{-7}, and two per cent auxotrophs were found among the Km^R mutants (this laboratory, unpublished; J. Vanderleyden, personal communication).

2.3. Genetic exchange and construction of partial diploids. Phages and plasmids of Azospirillum cannot be used, as yet, for genetic exchange. Mishra et al. (1979) reported DNA mediated transformation in A. brasilense strain Sp7. Plasmid R68-45 can promote chromosome mobilization (Franche et al. 1981; Bazzicalupo et al. 1983). IncP-1 plasmids such as RK2, RP4, or their derivatives, with broad-host-range are very stable in Azospirillum. In particular, pRK290 (Ditta et al. 1980) was used to construct nif partial diploids of strain Sp7 (Jara et al. 1983).

3. GENETICS OF NITROGEN FIXATION

3.1. Homology with Klebsiella pneumoniae nif genes and with Rhizobium japonicum fix genes . In K. pneumoniae, 17 nitrogen fixation (nif) genes organized in seven or eight transcriptional units have been identified (Dixon 1984; Elmerich 1984). Hybridization was performed between K. pneumoniae nif probes covering the entire cluster and total DNA of several Azospirillum strains. Homology was detected with nifHDK, the structural genes for the nitrogenase complex, and with nifA, which codes for an activator of nif transcription, (Quiviger et al. 1982; Nair et al 1983). The same type of experiment was performed with probes containing fixA, or fixBC from R. japonicum (Fuhrmann et al. 1985). These genes are involved in symbiotic nitrogen fixation but their function is not identified. Hybridization was performed with total DNA of two A. brasilense strains, including Sp7 and R07 and two A. lipoferum strains, including Br17 and S28. Homology to the fixA gene was detected, with the four DNAs, using as a probe a 2.1 kb PstI fragment purified from pRJ7101 (Fuhrmann et al. 1985). In the case of strain Sp7, hybridization was found with a 15 kb EcoRI fragment and two SalI fragments of 2.4 and 12 kb (this laboratory, unpublished).

3.2. Cloning of a nif cluster of A. brasilense Sp7. Using the nifHDK cluster of K. pneumoniae as a probe, a 6.7 kb EcoRI fragment was cloned from total DNA of A. brasilense Sp7 (Quiviger et al. 1982). Heteroduplex analysis performed with the K. pneumoniae genes established the approximate location of the corresponding nifH, -D, -K genes. A more precise location of nifH was obtained by probing with an internal part of the K. pneumoniae nifH gene containing codons 71 to 206. Recently, 6 kb of DNA adjacent to the nifHDK cluster was cloned (see Figure 1). In A. lipoferum the nifHDK cluster was recently cloned from two different strains using the probe fom Sp7 (M. Singh, personal communication).

3.3. Isolation of Nif⁻ mutants. Attempts to isolate Nif⁻ mutants were carried out in several laboratories including this one. As reported by Bani et al. (1980), most of the mutants isolated as non- or poorly-growing on nitrogen free medium had an Asm⁻ phenotype. The other mutants isolated after chemical mutagenesis fall into two classes : regulatory and nitrogenase mutants (Pedrosa, Yates 1984; Jara et al. 1983).

3.4. Tn5 site-directed mutagenesis of the Sp7 nif cluster. Isolation of Nif⁻ mutants was performed by Tn5 site-directed mutagenesis according to Simon et al. (1983). The localization of Tn5 insertions obtained is shown in Figure 1. Tn5 insertions were recombined in the Sp7 genome and the resulting Nif phenotype was determined. As expected, insertions in nifH, nifD, and nifK led to a Nif⁺ phenotype. Two insertions located between nifH and nifD led to a Nif⁺ phenotype, suggesting that the nifHDK cluster might be composed of two transcription units. However, it was established that in these two particular mutants transcription proceeds from a Tn5 promoter or from the creation of a promoter-like sequence due to the Tn5 insertions. Insertions in nifH were polar on nifD and nifK, as determined by genetic complementation and by gene-product analysis. Thus the nifHDK genes are trancribed as a single operon in Azospirillum (Perroud et al. 1985). Some of the insertions in the 5.8 kb fragment led also to a Nif⁻ phenotype suggesting the presence of nif genes in the region adjacent to nifHDK, as in K. pneumoniae. Hybridization with K. pneumoniae and ORS571 Rhizobium (Norel et al. 1985) nifE probes suggests that the newly identified nif region contains the equivallent of the nifE gene (this laboratory, unpublished).

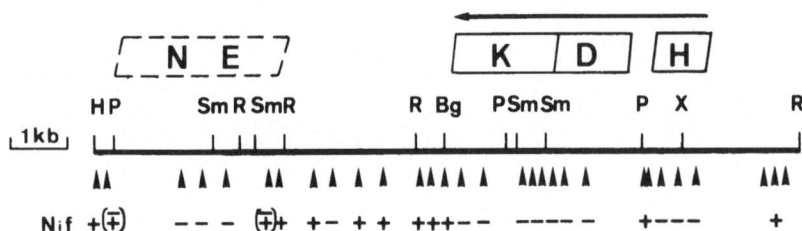

FIGURE 1. The nif cluster of A. brasilense Sp7. Restriction sites: Bg: BglII; P: PstI; R: EcoRI; Sm: SmaI; X: XhoI; vertical arrows: location of the Tn5 insertions; horizontal arrow: direction of transcription; Nif phenotype: -: <0.3 %; +: >50 %; (+): between 10 and 20 % of the wild type.

4. REGULATION OF NITROGEN FIXATION

Azospirillum mutants impaired both in glutamine synthetase activity and nitrogen fixation were isolated (Gauthier, Elmerich 1977). In particular strain 7029 of Sp7 has a Gln⁻ Nif⁻ phenotype, and thus resembles to glnA or ntrBC mutants of K. pneumoniae. Pedrosa and Yates (1984) reported the isolation of Nif⁻ mutants of Sp7 with nifA and ntrC phenotypes. The ntrC like mutant was complemented by plasmids pGE10 and pCK3, which carry the glnAntrBC regulon and the nifA-gene of K. pneumoniae respectively. The nifA like mutant was complemented by pCK3 only. This is in favor of regulatory mechanisms related to those existing in K. pneumoniae (Merrick 1983; Dixon 1984).

In order to clone the glnA gene of strain Sp7, a gene library of total DNA of A. brasilense Sp7 was constructed in the broad host range cosmid vector pVK100 (Knauf, Nester 1983). A plasmid, designated pAB44, that restored by complementation a Gln⁺ Nif⁺ phenotype to the Gln⁻ Nif⁻ mutant 7029, was isolated. The glnA gene was localized on subclones of pAB44 by Tn5 insertion and DNA hybridization with a K. pneumoniae glnA specific probe (see Figure 2). The glnA product was identified as a 50 kd polypeptide after labeling of maxicells. The glutamine synthetase was identified on non-denaturing polyacrylamide gels by in situ revelation of the enzyme activity.

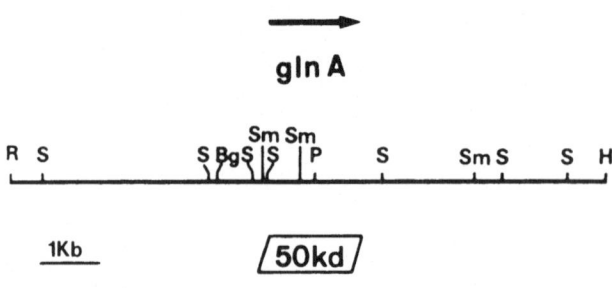

FIGURE 2. Physical map of the glnA region of A. brasilense Sp7. Restriction sites: as Figure 1, S: SalI; arrow: direction of transcription.

No complementation of E. coli or K. pneumoniae Gln⁻ mutants by pAB44 was observed. However, plasmid mutants which restored glutamine independant growth were obtained in E. coli strain ET8051, that carries a glnAntrBC deletion. The plasmid containing the glnA gene restored a wild type phenotype to different Gln⁻ Nif⁻ and Gln⁻ Nif^c mutants of K. pneumoniae and A. brasilense (see Fogher et al., this volume). It is thus tempting to speculate that plasmid pAB44 carries the equivalent of the ntrBC genes. However, pAB44 did not complement the mutant strains isolated by Pedrosa and Yates. This raises the question of the localization of the ntrBC genes in Azospirillum. Further experiments are in progress to answer this question.

5. HOMOLOGY TO RHIZOBIUM NOD GENES

5.1.Process of the association with grasses. Effect of Azospirillum
inoculation was recently reviewed (Patriquin et al. 1983; Elmerich
1984; Okon 1985). In most cases, Azospirillum spp. were isolated from
the rhizosphere of grasses after surface sterilization of the roots.
No differentiated structures were formed, but pictures of root hair
deformation were reported (Patriquin et al. 1983). Lectin-like
substances produced in root exudates and bacterial pectinase activity
that could play a role in root colonization were described. After
Azospirillum inoculation, a large enhancement of the number of lateral
roots and of root hairs was observed. This proliferation was
concommittant to an increase in mineral uptake, and was attributed to
phytohormone production rather than to nitrogen fixation (Okon 1985).
Baldani and Döbereiner (1980) showed that most of the strains isolated
from maize were A. lipoferum and that most of the strains isolated
from wheat or rice were A. brasilense Nir⁻, suggesting a difference of
specificity between the two species towards C_4 and C_3 plants
(Döbereiner, De Polli 1980). Recent reports of Plazinsky and Rolfe
(1985) showed that Azospirillum could inhibit Rhizobium trifolii
nodulation of clover.

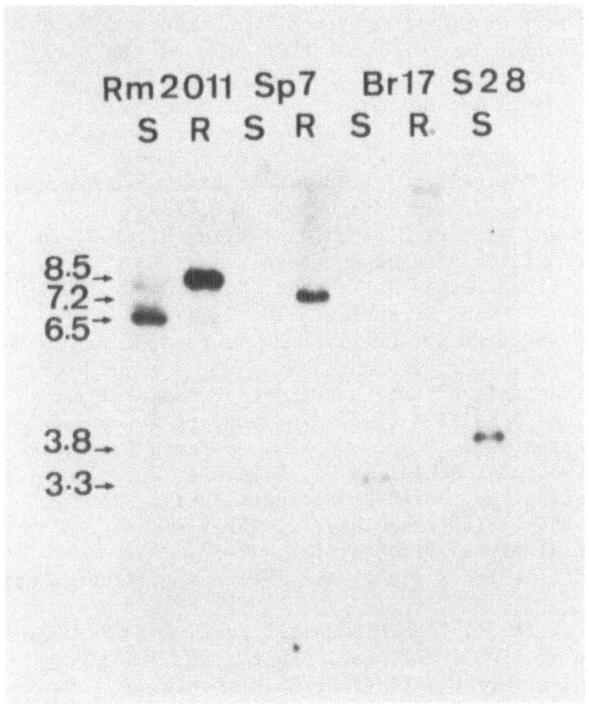

FIGURE 3. Hybridization between the nod genes of R. meliloti and
restriction digests of total DNA of Azospirillum strains. S: SalI
digest; R: EcoRI digest; 2011: DNA of R. meliloti used as control;
Br17 and S28: DNA of A. lipoferum; Sp7: DNA of A. brasilense.

5.2. Hybridization with nod and hsn probes from Rhizobium meliloti. In R. meliloti, two DNA regions involved in early stages of the bacteria-plant interaction, located on plasmid near the nifHDK and the fixABC cluster, were identified. One of them is referred to as the "common nod genes" and the other is referred to as the host specificity region (hsn) (Kondorosi et al. 1984). Some of the Tn5 insertions in the two regions led to a Nod⁻ phenotype and in both cases the mutants did not produce root hair curling.

Using R. meliloti Rm41 probes carrying either the common nod genes, (8.5 EcoRI fragment purified from pSK5) or the hsn region (6.5 kb EcoRI fragment purified from pEK10) (Kondorosi et al 1984), hybridization was performed with total DNA of 4 Azospirillum strains (same strains as for the fixA probe). Homology was detected in each case, and Figure 3 shows an example of the results obtained with the probe containing the common nod gene. EcoRI and SalI fragments of similar size were detected with strains Sp7 and RO7 (not shown).

Though it is too early to conclude on the significance of the homology detected, a few hypotheses can be raised. As Azospirillum contains large plasmids, it is possible that some of them are related to those found in Rhizobia and that the sequences homologous to the nod and hsn regions originate from common ancestors. This would not necessarily imply that the corresponding genes are functional in Azospirillum. Moreover, it cannot be ruled out that some of the early steps of the recognition between plant and bacteria, e.g. root hair deformation, proceed from common mechanisms.

REFERENCES
Baldani VLD and Döbereiner J (1980) Soil Biol. Biochem. 12, 434-44.
Bani D et al (1979) J. Gen. Microbiol. 119, 239-244.
Bazzicalupo M and Gallori E (1983) In: Klingmüller W, ed, Azospirillum II, pp. 24-28, EXS48, Birkhaüser, Basel.
Ditta G et al (1980) Proc. Natl. Acad. Sci. USA 77, 7347-7351.
Dixon RA (1984) J. Gen. Microbiol. 130, 2745-2755.
Döbereiner J and Day JM (1976) In: Newton WE and Nyman CJ, eds, Proceedings of the 1st International Symposium on Nitrogen fixation, pp. 518-536, Washington State University Press, Pullman.
Döbereiner J and DePolli H (1980) In: Stewart WDP and Gallon JR , eds, Nitrogen fixation, vol 18 pp. 301-333, Academic Press, London.
Elmerich C (1983) In: Pühler A, ed, Molecular Genetics of the Bacteria Plant Interaction, pp. 367-372, Springer Verlag, Berlin.
Elmerich C (1984) Bio/Technology, 2, 967-978.
Elmerich C et al (1982) Virology 122, 29-37.
Franche C and Elmerich C (1981) Ann. Microbiol. (Inst. Pasteur), 132A, 3-17.
Franche C et al (1981) FEMS Microbiol. Lett. 10, 199-202.
Fuhrmann M et al (1985) Mol. Gen. Genet. 195, 315-322.
Gauthier D and Elmerich C (1977) FEMS Microbiology Letters 2, 101-104.
Jara P et al (1983) Can. J. Microbiol. 29, 968-972.
Kondorosi E et al (1984) Mol. Gen. Genet. 193, 445-452.
Knauf VC and Nester EW (1982) Plasmid 8, 45-54.
Maghalhaes FM et al (1983) An. Acad. brasil. Cién. 55, 417-430.
Merrick MJ (1983) EMBO J. 3, 501-507.
Mishra AK et al (1979) J. Bacteriol. 137, 1425-1427.

Nair SK et al (1983) In: Klingmüller W, ed, *Azospirillum* II, pp. 29-38, EXS48, Birkhaüser, Basel.

Norel F et al (1985) Mol. Gen. Genet. 199, 352-356.

Okon Y (1985) In: Ludden PW and Burris JE, eds, Nitrogen Fixation and CO_2 Metabolism, pp. 165-174, Elsevier, New-York.

Patriquin DG et al(1983) Can. J. Microbiol. 29, 900-915.

Pedrosa FO and Yates MG (1984) FEMS Microbiol. Lett. 29, 95-101.

Perroud B et al (1985) in press.

Plazinsky J. and Rolfe B. (1985) Appl. Environm. Microbiol. 49, 984-989 and 990-993.

Plazinsky J et al (1983) J. Bacteriol. 155, 1429-1433.

Quiviger B et al (1982) Biochimie 64, 495-502.

Simon R et al (1984) BIO/TECHNOLOGY 1, 784-791.

Tarrand JJ et al (1978) Can. J. Microbiol. 24, 967-980.

ACKNOWLEDGEMENTS
C.F. was the recipient of a postdoctoral fellowship from research contract funds. This work was supported by funds from the University Paris VII, and by a Research Contract from Elf Bio-Recherche, Entreprise Minière et Chimique, Rhône-Poulenc Recherche, and CDF Chimie.

ORGANIZATION OF THE GENES FOR NITROGEN FIXATION IN THE CYANOBACTERIUM ANABAENA

ROBERT HASELKORN, JAMES W. GOLDEN, PETER J. LAMMERS, AND MARTIN E. MULLIGAN
DEPT. OF MOLECULAR GENETICS & CELL BIOLOGY, THE UNIVERSITY OF CHICAGO, CHICAGO, IL. 60637 USA

1. INTRODUCTION

The nifHDK genes encoding the nitrogenase complex polypeptides of Klebsiella pneumoniae were cloned on a single 6 kbp DNA fragment (Cannon et al 1979) and found to hybridize with DNA from every nitrogen-fixing microorganism tested (Ruvkun, Ausubel 1980; Mazur et al 1980). The cloned Klebsiella nifHDK DNA was subsequently used both to clone and to physically map the homologous genes from other organisms, including various Rhizobia, photosynthetic bacteria, and the cyanobacterium Anabaena. In most cases, the nifHDK gene organization as well as the gene sequences were found to be conserved. In general, the three genes are contiguous and transribed as a single unit. The exceptions are the slow-growing Rhizobia, described elsewhere in this volume, and the heterocyst-forming cyanobacteria, of which Anabaena is the best studied example.

The nifH gene encodes dinitrogenase reductase; the nifD and nifK genes encode the α and β subunits of dinitrogenase. Thus it is not unreasonable to find the Rhizobium arrangement of nifH separated from nifDK by many kbp of DNA (Scott et al 1983; Kaluza, Hennecke 1984). It was, however, quite unexpected to find nifHD separated from nifK by 11 kbp in the vegetative cell DNA of Anabaena (Rice et al 1982). Moreover, our early work showed that most of the DNA in the region between nifD and nifK of Anabaena is not transcribed into messenger RNA during derepression of nitrogenase, and therefore is not likely to contain nif genes. Resolution of the paradox posed by these results was provided by the discovery of Golden et al. (1985) that the 11 kbp is excised from Anabaena DNA during heterocyst differentiation, restoring the nifHDK operon structure. In the course of that work it was observed that a second rearrangement of vegetative cell DNA occurred to the right of the nifHDK region. All of the joint fragments involved in that rearrangement have now been cloned and sequenced (J. W. Golden and M. E. Mulligan, unpublished). Both rearrangements result from site-specific recombination within short repeated sequences, although the sequences are different in the two cases. Some progress has also been made in our understanding of developmental aspects of these rearrangements, based on the observation that the 11 kbp element can excise from cloned Anabaena DNA during propagation in E. coli (P. J. Lammers and J. W. Golden, unpublished).

Heterocysts differentiate at regular intervals, roughly every tenth cell, along filaments of Anabaena deprived of ammonia or nitrate (Haselkorn 1978). Heterocysts, which do not divide, provide an anaerobic micro-environment for nitrogen fixation. The ultimate product of nitrogen fixation, glutamine, is exported to neighboring vegetative cells. At the same time, carbohydrate is imported from vegetative cells to provide reductant for nitrogen fixation. Complete differentiation of a vegetative cell into a mature, nitrogen-fixing heterocyst requires about 30 hours. Early in the differentiation, unique glycolipid and polysaccharide components of the heterocyst envelope are synthesized. Proteases are induced which break down the phycobiliprotein light-harvesting apparatus as well as enzymes that fix CO_2. The photosynthetic membranes are

486

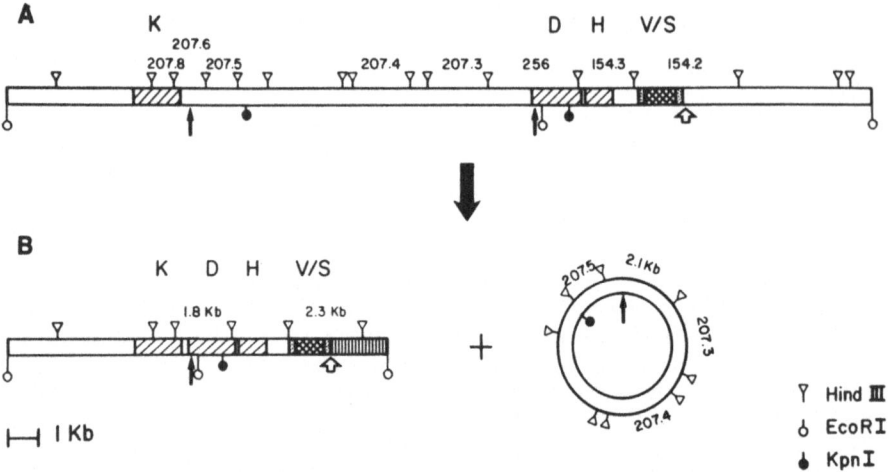

FIGURE 1. Organization of nitrogen-fixation genes in Anabaena vegetative cell and heterocyst DNA. The upper figure (A) shows two contiguous EcoRI fragments in vegetative cell DNA containing the genes nifK, nifD, nifH, and nifS (labeled V/S in the figure). In heterocysts (B), the DNA is rearranged such that nifK and nifD are adjacent on the chromosome. The intervening DNA is excised from the chromosome and is found as a circle in heterocysts. A second rearrangement places new sequences near the nifS gene. The sequence originally to the right of the nifS gene has moved to an unknown location in heterocyst DNA. The nifK, nifD, and nifH genes have been sequenced and their coding regions are shown. The approximate location of the open reading frame corresponding to the nifS gene is shown by crosshatching. HindIII restriction fragments described in the text are labeled. The position of 11 base pair direct repeats involved in the excision of 11 kbp from nifD are indicated by solid arrows. An open arrow indicates the approximate position of one of the breakpoints involved in the rearrangement near the nifS gene.

reorganized to contain photosystem I units only, thus producing ATP by cyclic photophosphorylation without the generation of O_2 associated with photosystem II. Nitrogenase, hydrogenase, and enzymes of the oxidative pentose pathway are induced, along with glutamine synthetase and elements of the transport systems for glutamine and carbohydrate.

Complete differentiation of mature heterocysts is not essential for induction of nitrogenase in Anabaena if anaerobiosis is established in another way. This can be accomplished by removal of combined nitrogen, addition of a herbicide to inhibit O_2 evolution by photosystem II, and bubbling the culture with argon. Under these conditions, Anabaena differentiation is arrested at the proheterocyst stage (Rippka, Stanier 1978; Haselkorn et al 1983; Tumer et al 1983). Admission of O_2 to such an induced culture results in rapid inactivation of nitrogenase and disappearance of nif gene transcripts (Haselkorn et al 1983). We present below the results of our studies in differentiating Anabaena and in E. coli of developmentally regulated nif gene rearrangements.

2. RESULTS AND DISCUSSION

The key to these experiments is the development of a method for the preparation of high molecular weight DNA from heterocysts (Golden et al 1985). With this material it was possible to compare the restriction patterns of vegetative cell and heterocyst DNA following digestion with several enzymes and probing with nifK, nifD and nifH-specific DNA fragments. We will consider the nifKD rearrangement first, summarized in Figure 1.

Vegetative cell DNA digested with EcoRI and probed with the nifK internal fragment 207.8 shows a single band at 17 kbp. This band is missing from heterocyst DNA, replaced by one at 6 kbp. If the DNAs are digested instead with HindIII and probed with 256, a HindIII fragment that contains nifD, we find the 2.9 kbp fragment of vegetative cell DNA replaced with two new ones at 2.1 and 1.8 kbp in heterocyst DNA. The HindIII fragment 207.6, which contains 5'-flanking sequences of the nifK gene, identifies the same 2.1 and 1.8 kbp fragments in heterocyst DNA. These and other data in Golden et al (1985) lead to the conclusion shown in Figure 1: recombination between sequences in fragments 207.6 and 256 results in the excision of an 11 kbp circular DNA molecule and fusion of the nifD and nifK genes in the chromosome.

A

				1350					
CCC	TTC	CGT	CAA	ATG	CAC	TCT	TGG	<u>GAT</u>	<u>TAC</u>
Pro	Phe	Arg	Gln	Met	His	Ser	Trp	Asp	Tyr
						450			

												1400		
<u>TCC</u>	GGC	CCT	TAT	CAC	GGT	TAC	GAC	GGA	TTC	GCT	ATC	TTC	GCC	CGT
Ser	Gly	Pro	Tyr	His	Gly	Tyr	Asp	Gly	Phe	Ala	Ile	Phe	Ala	Arg
★						460								

GAC	ATG	GAT	TTA	GCC	CTC	AAC	AGC	CCA	ACT	TGG	AGC	TTG	ATT	GGC
Asp	Met	Asp	Leu	Ser	Leu	Asn	Ser	Pro	Thr	Trp	Ser	Leu	Ile	Gly
	470										480			

1450													1491	
GCT	CCT	TGG	AAG	AAA	GCG	GCT	GCA	AAG	GCT	AAG	GCT	GCG	TCC	TAA
Ala	Pro	Trp	Lys	Lys	Ala	Ala	Ala	Lys	Ala	Lys	Ala	Ala	Ser	End
						490							497	

B

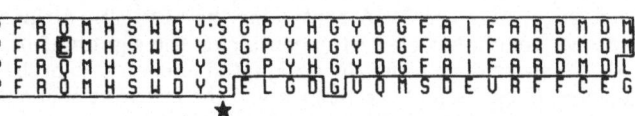

FIGURE 2. (A) Carboxy-terminal sequence of the rearranged Anabaena nifD gene present in heterocysts. The 11 base pair repeated sequence is shown underlined. Nucleotide and amino acid numbering is from Lammers and Haselkorn (1983). The rearrangement has altered the amino acid sequence distal to the serine at position 454. (B) Comparison of the carboxy-terminal amino acid sequence of the ∝-subunit of dinitrogenase from two Rhizobium species (Weinman et al 1984; Yun, Szalay 1984) and the Anabaena sequence present in heterocysts (rearranged) and in vegetative cells. Serine 454 is marked with a star.

Cloning and sequencing the new 2.1 and 1.8 kbp HindIII fragments from heterocyst DNA showed that the recombination event resulting in excision of the 11 kbp DNA circle occurs within an 11 base pair directly repeated sequence found at each end of the 11 kbp element. Originally we thought that the 11 kbp element separated the nifD and nifK coding regions. After the 2.1 and 1.8 kbp fragments from heterocysts were sequenced, we realized that the 11 kbp element in fact interrupted the nifD open reading frame; the C-terminal 43 amino acids of the Anabaena nitrogenase α-subunit are encoded beyond the 11 kbp element, underlined in Figure 2A, terminating just 199 base pairs from the start of the nifK open reading frame.

The predicted C-terminal amino acid sequence of the rearranged nifD gene found in heterocyst DNA is highly homologous with that of two Rhizobium species, whereas the predicted C-terminal sequence of the Anabaena vegetative cell gene shows no homology to these other two genes distal to the site of the rearrangement (Figure 2B).

The nucleotide sequence of the repeated region of the nifD gene, within which recombination occurs, is shown in Figure 3A. The actual recombination must occur within the 11 base pair conserved central region, outside of which there are a few short stretches of homology. These sequences are not related to the repeated sequences found at the ends of other bacterial elements that invert, occasionally, as a result of recombination between inverted repeat sequences (Silverman, Simon 1983).

Less is known about the rearrangement to the right of nifH in Figure 1. This reorganization was first detected as a reduction in size of the EcoRI fragment containing the nifH gene, from 10.5 kbp in vegetative cell DNA to 6 kbp in heterocyst DNA (Golden et al 1985). Subsequently it was found that HindIII fragment 154.2 in vegetative cell DNA was replaced by two new fragments of 2.3 and 4.7 kbp in heterocyst DNA. The smaller of these new fragments is part of the nifH-containing 6 kbp EcoRI fragment, but the larger one is elsewhere in heterocyst DNA and, to date, imprecisely located.

In order to determine where the recombination event leading to this rearrangement occurs, it was necessary to clone two more DNA fragments: one from vegetative cell DNA containing the vertically striped segment in Figure 1B and one from heterocyst DNA containing the right half of segment 154.2 (Figure 1A). This was done using the designated elements to probe recombinant DNA libraries in bacteriophage lambda and the relevant portions of the cloned DNA were sequenced (J. W. Golden and M. E. Mulligan, unpublished). As in the case of the nifD element, this rearrangement is characterized by a repeated 11 base pair core outside which there are short stretches of lower homology. However, here the 11 base pair repeat is not perfect. In one vegetative cell clone and one heterocyst clone, the fifth and sixth residues are GA, while in the other two clones they are CT. These differences allow us to conclude that the recombination event takes place somewhere within the sequence GAATA (Figure 3B).

The 11 base pair sequence just described is found to the right of an open reading frame labeled nifV/S in Figure 1. More detailed Southern hybridization with well-characterized Klebsiella nif gene DNA allowed us to conclude that the Anabaena open reading frame is homologous to the nifS gene (M. E. Mulligan, unpublished), whose product is believed to function in the assembly of nitrogenase. The Anabaena open reading frame could encode a protein the size of Klebsiella nifS. More sequencing will have to be done, along with S1 nuclease protection

(A) **nifD** GGCA——T-C—GCCTCATTAGG——CAC—AA—

(B) **nifS** - - -T—G-AAA-TTCT$^{GA}_{CT}$ GAATA—A-T——C-A-

FIGURE 3. Conserved nucleotide sequences within which recombination is observed during <u>Anabaena</u> heterocyst differentiation. (A) Sequences directly repeated at the ends of the 11 kbp element interrupting the <u>nifD</u> gene. (B) Sequences common to all four breakpoints of the rearrangement next to the <u>nifS</u> gene. Dashes indicate nucleotides that are not conserved.

---

experiments, to determine whether this rearrangement results in operon fusion or brings a new promoter into proximity of the <u>nifS</u> gene. The <u>nifS</u> gene reads from right to left, i.e. in the direction of <u>nifH</u>, away from the site of recombination.

Although the sequences of the recombining fragments are now known, their relative orientation is not. They appear to be quite far apart, at least 25 kbp and perhaps more than 50 kbp. Chromosome walking has not yet linked the repeated sequences in either vegetative cell or heterocyst DNA. Until that is accomplished, we cannot distinguish among several possibilities: deletion of a circle, inversion, or insertion of a circle during heterocyst differentiation.

The two rearrangements occur at roughly the same (late) stage of heterocyst development, beginning around the time that anaerobiosis is established (Golden et al 1985). Thus the environmental cues required for the rearrangements appear to be both nitrogen starvation and the absence of oxygen. Since these are exactly the requirements for transcription of the <u>nif</u> genes in <u>Anabaena</u> (as well as in <u>Klebsiella</u>) it is not a simple matter to determine whether the rearrangements are also required for <u>nif</u> gene transcription or vice versa. To pursue this question we attempted to construct plasmids, in <u>E. coli</u>, that contained the <u>nifD</u> 11 kbp element together with flanking sequences modified in such a way that the excision process could be monitored <u>in vitro</u> or <u>in vivo</u> after being returned to <u>Anabaena</u> by conjugation from <u>E. coli</u>. In the course of that work, it was observed that the 11 kbp element excised, occasionally, during propagation of the plasmid containing it, in <u>E. coli</u>. Excision was detected by noticing, in a preparation of the 21 kbp plasmid consisting of pBR322 containing fragment 207 (Figure 1), a contaminating plasmid of only 10 kbp. This plasmid was found to contain exactly the restriction fragments predicted for precise excision of the 11 kbp element from <u>nifD</u>.

To study the excision process in <u>E. coli</u> quantitatively, it was useful to develop an assay less laborious than the preparation of plasmid DNA from hundreds of colonies. For this purpose, the procedure of Castilho et al (1984) was used to introduce miniMu-<u>lac</u> into the plasmid An207. The miniMu-<u>lac</u> element contains a selectable kanamycin-resistance marker and a promoter-less β-galactosidase gene. Plasmids containing this element generally provide enough read-through transcription to make some β-galactosidase, which turns the chromogenic substrate X-gal blue. If the miniMu-<u>lac</u> is inserted within the 11 kbp <u>nifD</u> "excison", <u>E. coli</u> cells containing that construction produce blue colonies on X-gal plates. However, if the excison recombines out of the plasmid, the miniMu-<u>lac</u> comes out too. Failing to replicate, it is diluted by continued growth of the <u>E. coli</u>, giving rise to a white colony on the X-gal plate. Thus, excision of an <u>Anabaena</u> element in <u>E. coli</u> is measured by determining the frequency of white colonies on X-gal plates (P. Lammers and J. Golden, unpublished).

The white colony assay has not yielded useful information on the environmental regulation of the excision. However, it has made possible the assignment of a region of the excison as being essential for the excision process. It should be recalled that the insertion of miniMu-lac, like the insertion of any transposon, inactivates the gene (or operon) it enters. It was observed that introduction of miniMu-lac into fragment 207.6 (Figure 1A) to the right of the 11 base pair repeat completely prevented the appearance of white colonies. The transposon interrupts a gene, rather than interfering with the site of recombination, because it was possible to clone a fragment containing 207.6 and 207.5 on a compatible plasmid and to complement in trans the original mutated excison with respect to white colony formation. The gene required for excision has been sequenced; it encodes a 39 kD protein oriented from left to right in Figure 1A (P. J. Lammers and J. W. Golden, unpublished).

The finding that the nifD excison codes for a protein needed to excise itself raises interesting possibilities for the regulation of the excision in Anabaena. Regulation might be at the level of transcription of the excisase gene. If that is so, the promoter for that gene need only be a typical nif gene promoter, turned on in response to nitrogen starvation and anaerobiosis. Alternatively, or in addition, the excisase might be regulated at the level of function. In this case it would be necessary to postulate the existence of another gene, whose product inactivates the excisase unless both nitrogen and oxygen are absent.

3. CONCLUSIONS

Two genome rearrangements occur as late events during Anabaena heterocyst differentiation. One is the excision of an 11 kbp element that interrupts the nifD gene in vegetative cell DNA. Excision results from recombination within an 11 base pair directly repeated sequence at the ends of the element, catalyzed by an enzyme encoded within the element. The products of the excision process are a fused chromosome, in which the nifD open reading frame and the nifHDK operon structure are restored, and an 11 kbp circular molecule whose functions, if any, are unknown. The second rearrangement involves a different 11 base pair repeated sequence at the ends of an element of unknown size. DNA to the right of the nifS gene in vegetative cells is replaced, in heterocysts, by material originating at least 25 kbp away. This rearrangement could result from the excision of a circle, the insertion of a circle, or an inversion.

4. REFERENCES

Cannon FC et al (1979) Mol. Gen. Genet. 174, 59-67.
Castilho BA et al (1984) J. Bacteriol. 158, 488-495.
Golden JW et al (1985) Nature 314, 419-423.
Haselkorn R (1978) Ann. Rev. Plant Physiol. 29, 319-344.
Haselkorn R et al (1983) Ann. Microbiol. Inst. Pasteur 134B, 181-193.
Kaluza K and Hennecke H (1984) Mol. Gen. Genet. 196, 35-42.
Lammers PJ and Haselkorn R (1983) Proc. Natl. Acad. Sci. USA 80, 4723-4727.
Mazur BJ et al (1980) Proc. Natl. Acad. Sci. USA 77, 186-190.
Rice D et al (1982) J. Biol. chem. 257, 13157-13163.
Rippka R and Stanier RY (1978) J. Gen. Microbiol. 105, 83-94.
Ruvkun G and Ausubel FM (1980) Proc. Natl. Acad. Sci. USA 77, 191-195.
Scott KF et al (1983) DNA 2, 141-148.
Silverman M and Simon M (1983) In Shapiro JA ed, Mobile Genetic Elements, pp. 537-557, Academic Press, New York.
Tumer NE et al (1983) Nature 306, 337-342.
Weinman JJ et al (1984) Nucleic Acids Res. 12, 8329-8344.
Yun AC and Szalay A (1984) Proc. Natl. Acad. Sci. USA 81, 7358-7362.

DEVELOPMENT OF THE GENETICS OF HETEROCYST-FORMING CYANOBACTERIA.

C. PETER WOLK, ENRIQUE FLORES, GEORG SCHMETTERER, ANTONIA HERRERO AND JEFFREY ELHAI. MSU-DOE PLANT RESEARCH LABORATORY, MICHIGAN STATE UNIVERSITY, E. LANSING, MI 48824, U.S.A.

Heterocyst-forming cyanobacteria are of scientific interest for a number of reasons (Wolk, 1973; Stanier, Cohen-Bazire, 1977). These bacteria are among the relatively small number of organisms that are capable of nitrogen fixation under aerobic conditions. They can express this capacity even while producing oxygen by photosynthesis; their ability to fix nitrogen aerobically is dependent upon cellular differentiation, and upon intercellular interactions between their vegetative cells and the heterocysts in which fixation actually takes place; and they enter into nitrogen-fixing symbioses with a variety of eukaryotes, including fungi, liverworts, cycads, the fern, Azolla, and the angiosperm, Gunnera. The cyanobacteria, in general, have a photosynthetic apparatus that is very similar to that of eukaryotic plants. The heterocyst-forming cyanobacteria include many of the facultatively heterotrophic strains that, because they can grow non-photosynthetically, would be particularly well suited for genetic analysis of photosynthesis. Light-dependent modification of the array of photosynthetic pigments, so-called complementary chromatic adaptation, is observed often within this group of microorganisms. These same bacteria are among the few prokaryotes, the cells of which are capable of alternative modes of morphological and physiological differentiation, so that they may be used to inquire into possible relationships between such alternative pathways. Moreover, their cellular differentiation is, as in many cases of eukaryotic morphogenesis, under the control of intercellular interactions, both inductive and inhibitory: they therefore provide prokaryotic models of major embryogenetic mechanisms. Finally, it may be mentioned that these ecologically cosmopolitan microorganisms have various additional characteristics, including (for example) gliding motility and production of interesting cellular organelles (gas vesicles, carboxysomes, and cyanophycin granules), that invite investigation.

Whereas many of the phenomena just alluded to have been the subject of physiological investigation for decades, such investigation was less fruitful than it might have been because few of the powerful techniques of prokaryotic genetics were found to be applicable to their analysis.

The work to be described had, as its initial goal, development of techniques for genetic analysis of heterocyst-forming cyanobacteria. We hope then to apply those techniques to study the physiology of the cyanobacteria.

The elements of classical genetic analysis are the generation and selection of mutants; identification of the mutated genes, e.g., by complementation with cloned DNA; the mapping of genes on the chromosome; and (in recent years) transcriptional reporting, i.e., quantitation of the rate of transcription of particular genes. In the following, we will consider each of these elements in turn.

Mutagenesis. Nitrosoguanidine (NTG) has been found to be an effective mutagen for strains of Anabaena (Currier et al., 1977), giving rise to revertible mutants, and therefore presumably not to clustered, multiple mutations. Nonetheless, because the number of kinds of mutations observed was much less than might have been found; because NTG probably generated additional, cryptic mutations; and because of the potential health hazards of working with so potent a mutagen, we have sought alternative means by which to mutagenize. One easily controllable mutagenic agent that has been used successfully with unicellular cyanobacteria (Astier et al., 1979; Lambert et al., 1980) is ultraviolet light. Normally, UV-irradiated microorganisms are grown in the dark to permit replication of mutated DNA before they are exposed to the photoreactivating effects of white light. Such a procedure is inapplicable to those cyanobacteria, such as Anabaena PCC 7120, that are obligately photoautotrophic. However, under non-photoreactivating conditions of illumination with yellow "Bug-lites" (General Electric) (Anderson et al., 1984), cyanobacteria can grow. By following UV irradiation with 36h of yellow illumination, a period of growth in white light to provide further segregation of mutated genomes, and then penicillin enrichment, we have isolated strains of Anabaena 7120 that are phenotypically nif-minus under aerobic conditions (E. Flores et al., unpublished). Nostoc ATCC 29150 has proven to be remarkably resistant to -- and not effectively mutated by -- prolonged, intense UV-irradiation. However, we have isolated spontaneous and NTG-induced mutants of this strain that are defective in aerobic fixation of dinitrogen, and others that, unlike the wild type, do not grow heterotrophically (G. Schmetterer et al., unpublished).

Transposon mutagenesis, used successfully with unicellular cyanobacteria (Tandeau de Marsac et al., 1982), has yet to be achieved with filamentous cyanobacteria. Restriction appears to pose a significant barrier to the transfer of DNA to many filamentous cyanobacteria (see below); but with the possible exception of Nostoc sp. ATCC 29133, which is said to lack type-II restriction endonucleases (Lambert, Carr, 1984), there are as yet no reports of the isolation of restriction-minus strains of filamentous cyanobacteria. Transposon Tn5 has 8 AvaI sites and 6 AvaII sites (Mazodier et al., 1985, and references therein). We have constructed derivatives of Tn5 that have only one (in the neomycin phosphotransferase (NPT) gene) or no site (if a gene for chloramphenicol acetyl transferase is substituted for the NPT gene) for AvaII and only a single site for AvaI (in the transposase gene of IS50R), and these derivatives have been shown to transpose in E. coli (G. Schmetterer, C.P. Wolk, unpublished). To date, however, they have not been observed to transpose stably in cyanobacteria. A final method of mutagenesis that has proven practicable in other prokaryotes is insertional: an operon conferring antibiotic resistance is inserted within a cloned gene, and the construct returned to the organism of origin, permitting (upon selection) recombinational replacement of the wild-type gene by the interrupted gene. Our efforts to isolate nif$^-$ and glnA$^-$ derivatives of N$_2$-fixing cyanobacteria by this approach have produced antibiotic-resistant chromosomal recombinants, but not yet any having the predicted chromosomal structure (E. Flores et al., unpublished).

Gene transfer. A number of unicellular cyanobacteria, but not yet any filamentous strain, have been shown clearly to be transformable (Herdman, 1982). Moreover, although a number of viruses, including at least one temperate virus (Hudyakov, Gromov, 1973), are available for heterocyst-forming cyanobacteria (Hu et al., 1981), transduction has not been achieved. Neither has mobilization of the chromosome been observed. However, it was found that broad-host-range conjugative plasmid RP-4, although it itself has yet to be recovered from a heterocyst-forming cyanobacterium, is able to effect conjugation between E. coli and various strains of cyanobacteria, both unicellular and filamentous (Wolk et al., 1984), including facultative heterotrophs and organisms capable of chromatic adaptation (Flores, Wolk, 1985). In the presence of helper plasmids pDS4101 or pGJ28, derived from colicinogenic plasmids ColK and ColD, RP-4 can mobilize pBR322-based plasmids to the cyanobacteria. If these pBR322-based plasmids include an appropriate cyanobacterial replicon, they can replicate in the cyanobacteria and serve as shuttle vectors. By omission of the cyanobacterial portion, the plasmids are converted to suicide vectors, capable of entering the cyanobacteria and conferring unstable antibiotic resistance upon them. Such plasmids are incapable of replicating in the cyanobacteria unless they integrate into the cyanobacterial DNA, for example, by homologous recombination between a cloned fragment and the chromosome.

In order to observe gene transfer to strains of Anabaena, it was necessary to identify markers suitable for selection in these cyanobacteria, and it appeared necessary to remove from the vectors most of the sites recognized by Anabaena restriction endonucleases. Our most generally useful marker is the Tn5 gene that confers resistance to neomycin and kanamycin, despite the presence of an AvaII site in this gene. A streptomycin-resistance gene, from R300B, with one AvaI and two AvaII sites, can also be used with some organisms, as can genes conferring resistance to chloramphenicol and erythromycin. The ampicillin resistance gene from pBR322 is expressed by Anabaena 7120 (J. Elhai, C.P. Wolk, unpublished), but ampicillin-based selection for exconjugants would be complicated by the fact that β-lactamase secreted by donor bacteria would protect recipient cyanobacteria on a mating plate. It is quite possible that restriction will prove not to be a serious impediment to reintroduction of cloned Anabaena DNA into Anabaena: cloned chromosomal DNA from Anabaena variabilis ATCC 29413 has been found to be highly deficient in sites for the type-II restriction endonucleases present in that strain, and deficient also in certain sites recognized by such enzymes from other heterocyst-forming cyanobacteria (Herrero et al., 1984). A deficiency of sites for BamHI and BglII seemed anomalous when that paper was written; subsequently, isoschizomers of those enzymes have been found in closely related cyanobacteria (Lau et al., 1985; A. de Waard, personal communication).

To date, vectors based on 6.3-kb plasmid pDU1 from Nostoc PCC 7524 have been transferred to, and have been shown to replicate in, Anabaena strains PCC 7120, U. Leningrad 458 (PCC 7118), and U. Tokyo M-131, and Nostoc strains ATCC 27896, 29133 and 29150, and Nostoc ellipsosporum, U. Göttingen B1453-7. We have now observed that a 2.3-kb NotI-ScaI fragment of pDU1 suffices to confer replication in cyanobacteria (G. Schmetterer, C.P. Wolk, unpublished).

Mapping. We are making use of recombinant DNA techniques to map the chromosome of A. variabilis ATCC 29413. A cosmid library of the DNA from this strain was generated (Herrero et al., 1984), and approximately 1000 cosmids (6^+ genome equivalents) isolated and purified. The insert in one of these cosmids contains, on the average, about 2/3 of 1% of the Anabaena genome. By nick-translating an individual cosmid and hybridizing it to the library, we could identify other overlapping cosmids. We have in this way organized our library into approximately forty linkage groups. The serial order of the cosmids within each linkage group is known. Currently, we are trying to link the linkage groups, so as to construct a physical map of the chromosome as a whole; and by use of hybridization with heterologous probes, we are trying to convert the physical map into a genetic map. In one 60-kb interval of the chromosome, represented by 10 overlapping cosmids in our library, we find a sequence rbcB rbcA-9 kb-nifH nifD-11 kb-nifK (A. Herrero, C.P. Wolk, unpublished). The fact that the ribulose bisphosphate carboxylase (rbc) and nitrogenase (nif) genes, subjected respectively to negative and positive regulation within heterocysts, are closely linked suggests that the two groups of genes may be in some way coordinately regulated.

Transcriptional reporting. Hydrolysis of a chromogenic substrate by β-lactamase (see above) is potentially useful for transcriptional reporting from Anabaena 7120. Much of our effort, however, is directed to reporter genes chosen because they may have the potential to permit quantitation of the transcription from single genes in single cells. If such sensitivity were observed, it would be possible to identify, in vivo, transcriptional differences between vegetative cells and heterocysts, or gradients of transcription of specific genes along filaments. In particular, we are making use of the luciferase genes from strains of Vibrio fischeri and Vibrio harveyi (Belas et al., 1982; Engebrecht et al., 1983). The lux AB regions from these two strains, incorporated into our shuttle vectors, have been transferred to Anabaena M-131 and Anabaena 7120, respectively. Upon addition of n-decanal as substrate, both cyanobacteria "glow" (G. Schmetterer et al., unpublished). Strong cyanobacterial promoters (rbc, nif) have now been inserted into the constructions; the resulting plasmids are being transferred to Anabaena. We hope to be able (quite literally) to see whether these promoters enhance light production, and do so in a cell-type specific manner.

Summary. Plasmids were constructed containing Nostoc replicon pDU1 or parts thereof; determinants for resistance to chloramphenicol, streptomycin, neomycin and erythromycin; and portions of pBR322 lacking sites for AvaI and AvaII. These plasmids can be mobilized from Escherichia coli to a variety of cyanobacteria by conjugative plasmid RP-4 in the presence of helper plasmids. Nitrogen-fixing strains in which the shuttle vectors replicate include Anabaena PCC 7120, as well as Nostoc isolates capable of heterotrophic growth and (in two cases) chromatic adaptation. Derivatives of Anabaena 7120 and Nostoc 29150 that are phenotypically nif$^-$ under aerobic conditions have been isolated after mutagenesis with UV or NTG. Analysis of the mutants is in progress. Initial mapping of the chromosomal DNA of Anabaena ATCC 29413 by use of overlapping cosmid clones has established that in this strain, nif and rbc (ribulose bisphosphate carboxylase) genes, presumptively identified by hybridization with heterologous probes, are closely linked. Upon transfer to strains of Anabaena of lux genes from Vibrio fischeri and Vibrio harveyi, the cyanobacteria become capable of light production; these genes therefore have potential for transcriptional reporting from cyanobacteria.

Acknowledgments. G.S. was supported, in part, by a Max Kade Fellowship, and E.F. by Fundacion Juan March and the Fulbright Foundation. Other support was provided by N.S.F. grants PCM-8202665 and PCM-8402500, and by the U.S. Department of Energy under Contract DE-ACO2-76ERO-1338.

References

Anderson LK et al (1984) Arch. Microbiol. 138, 237-243.
Astier C et al (1979) Arch. Microbiol. 120, 93-96.
Belas R et al (1982) Science 218, 791-793.
Currier TC et al (1977) J. Bacteriol. 129, 1556-1562.
Engebrecht J et al (1983) Cell 32, 773-781.
Flores E, Wolk CP (1985) J. Bacteriol. 162, 1339-1341.
Herdman M (1982) In Carr NG and Whitton BA, eds, The Biology of Cyanobacteria, pp. 263-305, Blackwell, Oxford.
Herrero A et al (1984) J. Bacteriol. 160, 781-784; corrected in J. Bacteriol. 162, 858 (1985).
Hu N-T et al (1981) Virology 114, 236-246.
Hudyakov I Ya, Gromov BV (1973) Mikrobiologiya 42, 904-907.
Lambert GR, Carr NG (1984) Biochim. Biophys. Acta 781, 45-55.
Lambert JAM et al (1980) J. Gen. Microbiol. 121, 213-219.
Lau RH et al (1985) FEBS Lett. 179, 129-132.
Mazodier P et al (1985) Nucleic Acids Res. 13, 195-205.
Stanier RY, Cohen-Bazire G (1977) Annu. Rev. Microbiol. 31, 225-274.
Tandeau de Marsac N et al (1982) Gene 20, 111-119.
Wolk CP (1973) Bacteriol. Rev. 37, 32-101.
Wolk CP et al (1984) Proc. Natl. Acad. Sci. U.S.A. 81, 1561-1565.

GENETICS OF NITROGEN FIXATION IN PHOTOSYNTHETIC BACTERIA

J. D. WALL, A. GOLDENBERG, A. FIGUEREDO, B. J. RAPP AND D. C. LANDRUM
BIOCHEMISTRY DEPARTMENT, UNIVERSITY OF MISSOURI, COLUMBIA, MO 65211,
U.S.A.

1. INTRODUCTION

Most members of the anoxygenic phototrophic bacteria are capable of
fixing molecular nitrogen (Johansson et al 1983). As a result several
interesting metabolic properties may be displayed simultaneously in
these bacteria and their interactions analyzed. In addition, the
purple bacteria were the first diazotrophs in which the nitrogenase
was demonstrated to be subject to inhibition by reversible covalent
modification in response to ammonium levels (Ludden, Burris 1976;
Nordlund et al 1977). In Rhodospirillum rubrum, this inhibition has
now been shown to result from inactivation of the Fe protein by the
addition of an ADP ribosyl group to an arginine residue (Pope et al
1985). The generality of this regulatory mechanism among diazotrophs
is currently being evaluated. As a consequence of these properties of
the nitrogen fixation system of the purple phototrophs and their
relative genetic accessibility (Marrs 1983), efforts to elucidate the
genetics of this system have intensified. Here we shall summarize
some of the recent results from several laboratories concerning the
overall arrangement and regulation of nif genes in one species,
Rhodopseudomonas capsulata.

2. NIF⁻ MUTANTS OF RHODOPSEUDOMONAS CAPSULATA

2.1 Isolation and Characterization of Mutants

The first reports of the isolation of Nif⁻ mutants in the phototroph
Rhodopseudomonas capsulata was made ten years ago (Wall et al 1975).
Both the successful transduction of these mutations and the link
between hydrogen production and nitrogenase activity were
demonstrated. Among the mutants isolated was one (W15) subsequently
reported to contain a regulatory lesion as judged from patterns of
ammonium-repressible protein after two-dimensional gel electrophoresis
(Hallenbeck et al 1982). Because the lesion carried in this mutant
has been mapped within the nitrogenase structural gene region (Wall,
Braddock 1984), a regulatory role for one or more of the structural
proteins may be inferred, as has been reported in Klebsiella
pneumoniae (Roberts et al 1978).

A second type of mutant from that same study was subsequently shown to
be pleiotropic, being defective in the utilization of nitrogen sources
other than free ammonium (Wall et al 1977). Analogous mutants have
been isolated for Salmonella typhimurium (Broach et al 1976) and K.
pneumoniae (Close, Shanmugam 1980). Although the pleiotropic effect
of these mutations is unclear, a generalized regulatory element
involved in sensing ammonium levels or an ammonium transport protein
(Castorph, Kleiner 1984) may be altered.

Willison and Vignais (1982) reported the isolation of a large number

of Nif⁻ mutants and their classification into one of five different
groups based on residual in vivo and/or in vitro nitrogenase activity.
Their study demonstrated the usefulness of the antibiotic metronidazole for the enrichment of Nif⁻ mutants of R. capsulata. Two
mutants from among those isolated appeared to have regulatory
properties; RC5, perhaps unable to inactivate nitrogenase by covalent
modification, and RC34, pleiotropic for utilization of a number of
nitrogen sources and believed to be similar to strains with ntr C
mutations (Willison et al 1983).

More recently, a battery of Nif⁻ mutants generated by Tn5 insertional
inactivation has been obtained (Klipp, Pühler 1984; W Klipp, personal
communication) and used to identify clones of nif containing R.
capsulata DNA. Biochemical studies have allowed the mutants to be
catalogued into groups for which tentative functional assignments
could be made (W Klipp personal communication).

2.2 Mapping of nif Mutations

A fine structure analysis of a small number of nif mutations has been
carried out by transduction in this lab (Wall, Braddock 1984; Wall et
al 1984; Goldenberg, Wall, unpublished). The transducing vector used
was the defective phage-like particle unique to this species called
the Gene Transfer Agent (GTA, Marrs 1974) which carries 4600 bp of

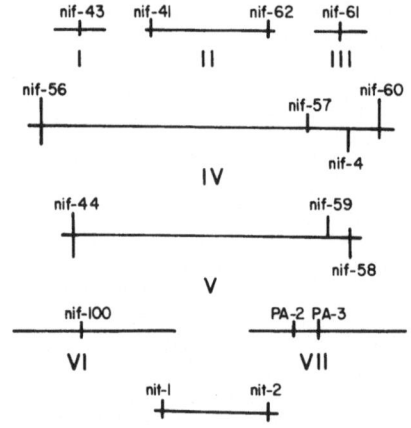

Figure 1. Transductional linkage
groups of mutations conferring a
Nif⁻ phenotype on R. capsulata.
I, II, V and VII code for unknown
functions; III and VI, regulation;
IV, nitrogenase structural genes,
and the nit mutations are
pleiotropic for nitrogen
metabolism. Lengths of lines are
drawn proportional to the map
distances determined by genetic
analysis and a reference scale
(Taylor et al 1983) is included
at bottom of figure.

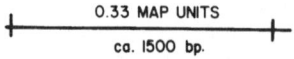

randomly packaged chromosomal DNA (Yen et al 1979). Presently seven
transductional linkage groups have been identified (Figure 1;
excluding nit mutations). Two, III and VI, contain genes believed to
be regulatory in nature. Even though GTA carries a small fragment of
DNA, the large number of linkage groups was unexpected and was
tentatively interpreted to mean that the genes for nitrogen fixation
in R. capsulata might be scattered on the chromosome.

Mapping studies have also been initiated by Willison et al (1983) both by transduction and by conjugation mediated by pTH10, a broad host-range P1-incompatibility plasmid temperature-sensitive for replication. Linkage of some nif mutations was suggested and linkage to drug resistance markers was demonstrated.

To examine the overall arrangement of the nif linkage groups, a conjugational mapping study was carried out (Goldenberg, Wall, unpublished). The P1-incompatibility plasmid, pBLM2 (Marrs 1981), which has increased chromosome mobilizing ability in R. capsulata, was introduced into a representative mutant from linkage groups I-VI. These Nif⁻ donors were used in conjugations with each of sixteen auxotrophs representing ten transductionally unlinked regions of the chromosome (Table 1). The interpretation of co-inheritance frequencies is limited by the lack of knowledge of the size of fragments transferred and the possibility of preferred origins of chromosomal transfer. Because of the occurrence of reciprocal transfer of single markers via GTA during conjugation, double selection against both the donor and recipient was required.

Table 1. Co-inheritance frequencies of nif mutations and auxotrophic markers.[a]

Transductional Linkage Group Representatives

Selected auxotrophic marker	I nif-43	V nif-58	II nif-62	IV nif-60	VI nif-100	III nif-61
leu-201[b]	7	3	0.5	0.4	0.4	0
leu-203[c]	7	1	0	0.9	0	0.7
ilv-211[d]	12	15	0	0	1	0
ade-241	0	0	53	36	37	0
ade-243	0	0	0	0	0	3
arg-231	2	1	0	0	0	0
trp-11	0	0	2	1	3	0

[a] Donors, His⁻ Str^r derivatives of the Nif⁻ mutants containing pBLM2, were conjugated to Nif⁺ Rif^r auxotrophic recipients. Kan^r Rif^r prototrophic exconjugants were scored for their Nif phenotype. Usually 100 CFU were tested.

[b] Five leu markers in the same transductional linkage group were used: leu-201, 202, 205, 208 and 209.

[c] Three leu markers in the same transductional linkage group were tested: leu-203, 206 and 207.

[d] Two ilv markers in the same transductional linkage group were tested: ilv-211 and 212.

Therefore, failure to observe co-transfer in a given cross might be a reflection of the gene order of the selected markers rather than an indication of map distance. We do not consider lack of co-inheritance in these crosses as meaningful and only those in which some linkage was observed are shown in Table 1.

Although percentages of co-inheritance were quite small in some cases, the patterns of linkage were clear. Groups I and V were linked to isoleucine-valine markers, a cluster of leucine markers and an arginine mutation. In contrast Groups II, IV and VI showed the strongest linkage to an adenine auxotrophic marker, a small co-inheritance with a tryptophan marker and just detectable linkage to the leucine cluster. Group III did not show significant linkage with the other nif mutations. Thus from this data alone Groups I and V and Groups II, IV and VI appear to form two larger nif clusters with at least one additional gene mapping elsewhere. After consideration of the physical mapping data in addition, we conclude that a large loose cluster of the five groups occurs with at least Group III mapping apart.

2.3 Physical Mapping of nif Genes

R. capsulata was among the diazotrophs screened by Ruvkun and Ausubel (1980) and shown to have homology to K. pneumoniae nif structural genes. Using the K. pneumoniae probe pSA30, Avtges et al (1983) were able to identify nif gene-containing R. capsulata clones which complemented Nif⁻ mutants of linkage group IV. Also, the order of the structural genes was elegantly shown to be conserved as nifHDK.

The hybridization experiments with nif structural gene probes have revealed the presence of multiple copies of nif genes in the purple phototrophs (Fornari, Kaplan 1983; Scolnik, Haselkorn 1984). The significance of these extra copies is unknown, but one has apparently been made to produce active nitrogenase in a background deleted for a normally functional gene (Scolnik, Haselkorn 1984). One of the reiterated copies of nifH has been located within the photosynthetic gene cluster of R. capsulata (Hearst et al 1985). As we have been unable to complement mutations in linkage group IV with a plasmid containing the photosynthetic gene cluster of R. capsulata, pRPS404 (Marrs 1981), this nifH gene may be nonfunctional.

Additional nif gene clones have now been identified by Haselkorn and coworkers (1985) by complementation of representatives of the transductional linkage groups and mutants isolated in that lab as well as those supplied by A. Hochman. A large cluster has been found which contains sequences complementing both mutations from groups I and II; however, clones complementing mutations in other linkage groups were apparently not contiguous (R. Haselkorn, personal communication). These results are in agreement with those obtained by Klipp and Pühler (1984) who have identified a similar large cluster of nif genes in R. capsulata following transposon mutagenesis. Transposons inactivating the structural genes were found outside of that cluster as was another group of mutations leading to loss of all nitrogenase related activities (W Klipp, personal communication).

2.4   Regulation of Nif Genes

Nitrogenase in K. pneumoniae is regulated both at a nif- specific
level through the actions of nifA and nifL proteins and through a
generalized nitrogen control system by the products of the ntrA and
ntrBC genes (Ausubel 1984).   The ntrBC genes are a part of the glnA
operon in this organism, and mutations in glnA, the structural gene
for glutamine synthetase (GS), are often polar onto these genes
resulting in a Nif⁻ phenotype (Leonardo, Goldberg 1980).   In contrast,
all mutations so far isolated in glnA of R. capsulata (ca. 40) result
in derepression of nitrogenase biosynthesis in the presence of
ammonium (Wall, Gest 1979).   This finding suggests that positive-
acting regulatory elements required for nitrogenase expression do not
reside in the same operon.   The restoration of nitrogen control to
glnA mutants by reintroduction of the GS structural gene alone
(Scolnik et al 1983), provides evidence that GS or a product of its
activity is required for wild-type nitrogen regulation.

Although unlinked to glnA (unpublished), at least two of the
transductionally determined Nif linkage groups (III and VI) contain
genes which were reported to have regulatory functions (Wall,
Braddock, 1984; Wall et al 1984).   Their regulatory nature has been
confirmed by an examination of the effects of mutations in these
regions on the expression of nifH::lac fusions and the DNA
complementing Group III mutations also has some homology with the ntrC
gene of E. coli (R. Haselkorn, personal communication).

Clear physiological evidence for the existence of a generalized
nitrogen control system in R. capsulata has not been reported.   In
contrast to ntr mutations in K. pneumoniae or E. coli (Magasanik
1982), R. capsulata strains with mutations in either Nif Groups III or
VI were not impaired in the utilization of various organic nitrogen
sources.   However during an examination of the properties of ammonium
uptake systems in R. capsulata, we have obtained evidence supporting
the operation of an Ntr system.

Many microorganisms have been shown to possess an energy-linked
ammonium uptake system (Kleiner 1981) which, in K. pneumoniae (Kleiner
1982) and E. coli (Servín-González, Bastarrachea 1984), appears to be
regulated by ntr genes.   To help differentiate between transport and
assimilation, the ammonium analog, methylammonium has frequently been
used (Kleiner 1981).   When uptake into R. capsulata cells of this
radioactively labelled analog was tested, only cells grown under N-
limiting conditions were found to be competent (Table 2).   These
results suggest that synthesis of the methylammonium transport system
is regulated by nitrogen availability in a manner similar to that of
nitrogenase.   A kinetic characterization of this system was carried
out (See Rapp et al, this volume) for comparison to an earlier study of
ammonium uptake (Genthner, Wall 1985).

Methylammonium uptake assays were then performed on Nif⁻ mutants of
Groups III, IV and VI grown with glutamate as nitrogen source.   With
mutants containing lesions mapping to Group VI (regulatory for
nitrogenase) or to IV (nitrogenase structural genes), the rates

obtained were not significantly different from wild type. By
comparison, the mutant J61 representing Group III was unable to
transport methylammonium into cells grown under any nitrogen condition
(Table 2), while glutamine synthetase activities in extracts of this

Table 2. Ammonium and Methylammonium Uptake Rates of
Rhodopseudomonos capsulata Strains

Strain	Phenotype	Nitrogen Source	Substrate	
			$CH_3NH_3^{+b}$	NH_4^{+c}
B100	wild type	NH_4^+	<0.1	80
		Glu	1.1	58
		N-strv	1.5	80
J61	Nif⁻	NH_4^+	<0.1	25
		Glu	<0.1	35
		N-strv	<0.1	60

[a] Nitrogen source for aerobic growth of cells was: NH_4^+ = 7.5 mM
$(NH_4)_2 SO_4$; Glu = 15 mM Na glutamate; N-strv = nitrogen
starved by overnight growth with an initial ammonium
concentration of 2 mM.

[b] nmol per min·mg protein; uptake of $[^{14}C]$ – $CH_3NH_3^+$ was assayed
as counts retained by whole cells after filtration and washing.

[c] nmol per min·mg; uptake of ammonium was determined from
residual ammonium in the assay medium by the indophenol method.

mutant were not greatly different from the wild type. Mutations
resulting from transposon insertion in this region (a generous gift of
R. Haselkorn) also resulted in an inability to express the
methylammonium transport system, as did additional Nif⁻ transposon
mutations mapping 1.3 and 1.6 kbp away. Because the latter two
inserts were not polar onto the gene(s) of Group III (R. Haselkorn,
personal communication), they identify an additional regulatory element
or elements necessary for expression of nitrogen controlled genes.

3. REFERENCES

Ausubel FM (1984) Cell 37, 5-6.
Avtges et al (1983) J. Bacteriol. 156, 251-256.
Broach et al (1976) J. Bacteriol. 128, 86-98.
Castorph H and Kleiner D (1984) Arch. Microbiol. 139, 245-247.
Close TJ and Shanmugam KT (1980) J. Gen. Microbiol. 116, 501-510.
Fornari CS and Kaplan S (1983) Gene 25, 291-299.
Hallenbeck et al (1982) J. Bacteriol. 151, 1612-1616.

Haselkorn et al (1985) In Ludden P and Burris JE, eds, Nitrogen Fixation and CO_2 Metabolism, pp 83-90, Elsevier, Amsterdam.
Hearst et al (1985) Cell 40, 219-220.
Johansson BC et al (1983) In Ormerod JG, ed, The Phototrophic Bacteria, pp 120-145, Univ. of Calif. Press, Berkeley.
Kleiner D (1981) Biochim. Biophys. Acta 639, 41-52.
Kleiner D (1982) Biochim. Biophys. Acta 688, 702-708.
Klipp W and Pühler A (1984) In Veeger C and Newton WE, eds, Advances in Nitrogen Fixation Research, p 738, Martinus Nijhoff/Junk, The Hague.
Leonardo JM and Goldberg RB (1980) J. Bacteriol. 142, 99-110.
Ludden PW and Burris RH (1976) Science 194, 424-426.
Magasanik B (1982) Ann. Rev. Genet. 16, 135-168.
Marrs B (1974) Proc. Natl. Acad. Sci. USA 71, 971-973.
Marrs B (1981) J. Bacteriol. 146, 1003-1012.
Marrs BL (1983) In Ormerod JG, ed, The Phototrophic Bacteria, pp 186-214, Univ. of Calif. Press, Berkeley.
Nordlund S et al (1977) Biochim. Biophys. Acta 462, 187-195.
Pope MR et al (1985) Proc. Natl. Acad. Sci. USA 82, 3173-3177.
Roberts et al (1978) J. Bacteriol. 136, 267-279.
Ruvkun GB and Ausubel FM (1980) Proc. Natl. Acad. Sci. USA 77, 191-195.
Scolnik PA and Haselkorn R (1984) Nature 307, 289-292.
Scolnik PA et al (1983) J. Bacteriol. 155, 180-185.
Servín-González L and Bastarrachea F (1984) J. Gen. Microbiol. 130, 3071-3077.
Taylor et al (1983) J. Bacteriol. 154, 580-590.
Wall JD and Braddock K (1984) J. Bacteriol. 158, 404-410.
Wall JD and Gest H (1979) J. Bacteriol. 137, 1459-1463.
Wall et al (1975) Nature 258, 630-631.
Wall JD et al (1977) Arch. Microbiol. 115, 259-263.
Wall JD et al (1984) J. Bacteriol. 159, 652-657.
Willison JC and Vignais PM (1982) J. Gen. Microbiol. 128, 3001-3010.
Willison JC et al (1983) In Papageorgiou GC and Packer L, ed, Photosynthetic Prokaryotes: Cell Differentiation and Function, pp 333-352, Elsevier Biomedical, New York.
Yen et al (1979) J. Mol. Biol. 131, 157-168.

4. ACKNOWLEDGEMEMTS

We thank Dr. R. Haselkorn for mutants and Dr. Haselkorn and Dr. W. Klipp for communication of results before publication. Also we thank Ms. Jackie Pruneau for typing the manuscript. The work was supported by NSF Grant PCM-8302942, U. S. Department of Agriculture CRGO Grant 84-CRCR-1-1468 and the Missouri Agricultural Experiment Station from which this is paper No. 9933. Ann Goldenberg was supported by NIH pre-doctoral training grant DHHS 2-T-32-GMO7494, and Antonio Figueredo was a LASPAU Scholar.

THE COMPLETE NUCLEOTIDE SEQUENCE OF THE NITROGENASE STRUCTURAL GENES FROM AZOTOBACTER VINELANDII

KEVIN E. BRIGLE, R. PREMAKUMAR, MARTY JACOBSON and DENNIS R. DEAN.
Virginia Polytechnic Institute & State University, Blacksburg, Virginia
24061 and North Carolina State University, Raleigh, NC 27650.

In vitro biochemical data have indicated that Azotobacter vinelandii nitrogenase has a higher turnover rate and is more efficient when the Fe protein (nifH gene product) - MoFe protein (nifD/nifK gene products) molar ratio is greater than one. These measurements suggest that physiological mechanisms could exist which ensure the excess accumulation of Fe protein relative to MoFe protein. Because the differential accumulation of the individual nif structural gene mRNAs represent a simple mechanism for regulating the balance of these component proteins, we have examined this possibility.

The nif structural genes coding A. vinelandii nitrogenase have been isolated and sequenced. These genes are arranged in the order promoter:nifH:nifD:nifK. There are 129 nucleotides separating nifH and nifD and 101 nucleotides separating nifD and nifK. Secondary structures can be inferred from the sequences in these intercistronic regions.

Northern and S1 analyses have revealed differential accumulation of nif structural gene mRNAs. In A. vinelandii derepressed for nitrogenase, two major nif structural gene mRNAs accumulate, one coding for nifH and one coding for the entire nifHDK cluster. A third, minor transcript coding for nifHD is also apparent. Because of their abundance and rate of appearance, it is unlikely that the accumulation of the individual transcripts represents degradation intermediates. Assuming equal translation of the mRNAs, these results demonstrate a mechanism for the differential accumulation of the component proteins. It is unclear why there is slightly more nifD mRNA than nifK mRNA as the products are present in the MoFe protein in equimolar amounts. Perhaps the nifD gene product has some other unknown structural or regulatory role.

The A. vinelandii nif structural genes are clustered and their transcription is directed by single promoter. Consequently, the accumulation of distinct transcripts must be due to post-transcriptional processing or to an attenuation mechanism. Of these possibilities, we favor an attenuation model where transcription is partially terminated between nifH and nifD and between nifD and nifK. This model is supported by the absence of individual nifD and nifK transcripts and by the presence of potential structural features within the intercistronic regions of the nifHDK cluster.

NIF GENE ORGANISATION IN AZOTOBACTER CHROOCOCCUM

DAVID EVANS, ROBERT JONES*, PAUL WOODLEY, CHRISTINA KENNEDY and
ROBERT ROBSON
AFRC Unit of Nitrogen Fixation, University of Sussex, Brighton, East
Sussex, BN1 9RQ, U.K.
*Department of Molecular Genetics and Cell Biology, The University of
Chicago, Chicago, Illinois 60637, U.S.A.

Fragments, of the Klebsiella pneumoniae nif gene cluster, were used
as probes to identify and map homologous regions on the recombinant
cosmids pACB1, pACD30, and pACD37 which contain about 40 Kb of DNA
spanning 70 Kb of the A. chroococcum chromosome. Homology to nitrogenase
structural genes nifHDK and a fragment spanning nifV has been reported
previously (Jones, Woodley, Robson, 1984). We have also demonstrated
homology between the cosmids and K. pneumoniae nif fragments bearing
nifEN, nifUX and nifF. The homologous sequences are ordered as in K.
pneumoniae but span a total of 31 Kb compared to 14 Kb, with gaps of 9
and 7 Kb respectively separating the nifEN- and nifUX-like regions and
the nifV- and nifF-like sequences. Homology was not observed to nifJ nor
to a region from nifQ to nifL. A second copy of nifH (nifH*) has also
been identified at least 15 Kb from the main nif gene cluster.

Complementation studies using defined nif mutants of K. pneumoniae
demonstrated functional copies of nifU, nifS, nifV and nifM on the
cosmid pLC11, which contains a 21 Kb fragment in pLAFR1 which maps
internally to the cosmid pACB1 spanning the regions homologous to nifUX
and nif(M)V(S). pLC11 restored the ability both to reduce acetylene and
grow on N-free media under anaerobic conditions to these mutants, but did
not complement mutations in nifJ, E, N or B.

Subcloned DNA fragments from the A. chroococcum nif gene cluster
were used to map transcripts synthesised along this region under
conditions of N-starvation, by probing Northern blots of total cell RNA.
For the region containing the nifHDK operon four transcripts of 1.1, 2.6,
4.3 and 6.4 Kb were detected. Each probably originates from the nifH
promoter with the longest species being present at relatively low levels.
It is likely that this operon extends a further 2.1 Kb beyond the end of
nifK. Two DNA fragments from the nifV-like region both hybridised with
a transcript of 2.5 Kb. A further two fragments from the gap between
the nifEN- and nifUX-like regions hybridised to a transcript of 4.7 Kb,
suggesting that the region immediately adjacent to nifUX-like may also
be involved in nitrogen fixation in A. chroococcum.

REFERENCE

Jones R et al. (1984) Mol. Gen. Genet. 197, 318-327.

IDENTIFICATION AND ORGANIZATION OF NIF GENES
OF AZOTOBACTER VINELANDII

ROSALIND J. HELFRICH, JAMES M. LIGON, and ROBERT G. UPCHURCH, Corp
Science Laboratory, Allied Corporation, Solvay, NY   13209, U.S.A.

From a clone library of Azotobater vinelandii (AV) constructed
in cosmid vector pHC79, over 90 clones were isolated by in situ
colony hybridization with nif H, D, and K gene probes from
Klebsiella pneumoniae (Kp). Two overlapping clones, pAv20 and
pAv80, that combined, cover 30 Kb to the left and to the right of
the common nifHDK cluster, respectively, were identified. The Nif⁻
phenotypes of Av mutants UW102 and UW4, both of which fail to
synthesize the MoFe and Fe proteins of nitrogenase, were corrected
by genetic transformation using these two plasmids. A detailed
restriction map of this 66 Kb region of the Azotobacter genome
surrounding the nif HDK cluster has been generated and analyzed.
Hybridization analysis with several Kp nif probes suggests that the
organization of the Nif operon in Av is similar to that of the Kp
Nif operon. A region of approximately 3Kb was identified 2Kb
downstream from nif K which strongly hybridized to Kp nif E and nif
N, the protein products of which are essential for active FeMo
cofactor centers in the MoFe protein of nitrogenase. A region
covering approximately 7Kb adjacent and downstream to Av nif N was
identified by its hybridization to both Kp nif USVM and nif MLF
gene probes. Analysis of the restriction map and hybridization
data of this region suggest that Av nif M, which in Kp encodes a
protein necessary for the activation of the Fe protein of nitrogenase resides approximately 6Kb downstream of nif N.

ISOLATION OF Tn5 Nif⁻ MUTANTS OF A. VINELANDII AND THEIR COMPLEMENTATION
BY pLAFR1 GENE BANKS OF A. VINELANDII AND A. CHROOCOCCUM.

C. KENNEDY, R. GAMAL, J. RAMOS, R. HUMPHREY, D. DEAN* and K. BRIGLE*.
AFRC Unit of Nitrogen Fixation, University of Sussex, Brighton, UK
and *Battelle-Kettering Labs, Yellow Springs, Ohio, USA

Nif⁻ mutants were isolated after transfer of an IncN-Tn5 plasmid,
pGS9 (Selveraj, Iyer, 1983), to A. vinelandii Nif⁺ Rif^r strain UW136.
Km^r recipients were also Sm^r indicating that the gene encoding the
latter on Tn5 was also expressed. Screening of Km^r colonies for
growth on N_2 yielded 24 mutants, most of which were unstable even
after subculturing several times on medium with Km. Three classes
of stable Nif⁻ mutants were isolated, represented by strains MV6,
MV21 and MV22.

MV6 had Tn5 inserted in nifH carried on an 8.5Kb EcoR1 fragment.
It failed to synthesize nitrogenase polypeptides, could grow on Mo-
deficient medium, and was 'corrected' by recombination with pLV10,
a cosmid isolated from the pLAFR1 gene bank that complemented UW91,
a nifH point mutant.

MV21 had Tn5 inserted in a 5.7Kb EcoR1 fragment and failed to
grow on either Mo-sufficient or Mo-deficient medium. However,
acetylene reduction levels were high (20-40% of wild type) in this
mutant. Since this phenotype was reminiscent of nifV mutants, the
EcoR1 fragment carrying the insert was cloned into pACYC184 and used
in hybridization experiments against a nifV probe from K. pneumoniae.
No hybridization was observed. A cosmid was isolated from each of
the A. chroococcum and A. vinelandii pLAFR1 gene banks: pLC11 and
pLV1, respectively. pLC11 had two EcoR1 fragments of 9.5 and 10.5
KB which corresponded to the 'middle' part of the nif gene cluster
of A. chroococcum (see Kennedy et al., this Symposium Proceedings).
pLV1 had EcoR1 fragments of 7, 5.7, 5.7 and 3.6 Kb. The latter two
correspond in size to those known to carry the nifE and N genes.
pRD1 carrying the nif gene cluster of K. pneumoniae fully complemented
MV21 but nifM⁻ plasmid mutants did not, suggesting that the mutation
is in the nifM gene which is apparently required for nitrogen fixation
under both Mo-sufficient and -deficient growth conditions.

MV22 had Tn5 inserted in a 5.7 KB EcoR1 fragment different from
that mutated in MV21; the former has a BamH1 site. MV22 grows on
Mo-deficient medium and is complemented for growth on Mo-sufficient
medium and for acetylene reduction by pLC11, pLV1 and pRD1. Neither
nifE⁻ nor nifN⁻ mutants of pRD1 could complement MV22. The mutation
must therefore be in nifN since the nifE gene is carried on a 3.6Kb
EcoR1 fragment.

REFERENCES

Selveraj G, Iyer VN (1983) J. Bacteriol. 156, 1292-1300.

IDENTIFICATION AND CLONING OF GENES FOR NITROGEN FIXATION FROM A SOIL PSEUDOMONAS SPECIES ISOLATED FROM THE RHIZOSPHERE OF RICE

William T. Tucker and Peter J. Dart.
Department of Genetics, Research School of Biological Sciences, Australian National University, Canberra, A.C.T. 2601, Australia.

In the Philippines, Barraquio et al.(1983) have identified a Pseudomonas cepacia-like organism which accounts for 29-50% of the heterotrophic nitrogen-fixing bacteria isolated from the rhizosphere of rice. In an attempt to identify and study the genes involved in this process in this novel nitrogen-fixing bacterium, mutagenisis with transposon Tn5 has been performed. Strains of this organism (H8 and KLH76) were tested as recipients for the broad host range plasmid RP4. Transfer to spontaneous rif^r mutants occurred at between $1x10^{-1}$ and $1x10^{-2}$ transconjugants per donor cell. Conjugation experiments using two "suicide" vectors, pUW964 (Weiss et al., 1983) and pSUP1011 (Simon et al., 1983) gave transposition frequencies of $1x10^{-4}$ to $1x10^{-5}$ per donor. Kanamycin resistant colonies were screened for differential growth on M9 mininal medium with and without inorganic nitrogen. Wild-type strains grow, albeit poorly, on this N-free medium and from several thousand Km^r, approximately 40 were selected and screened for acetylene reduction ability in M9 semi-solid medium with 10% acetylene in the gas phase. Six mutants showed consistently lower levels of ethylene production.

Total genomic DNA from these 6 mutants was digested with EcoRl and the fragments cloned into pBR322. Km^r transformants of E. coli strain RRl were selected and the cloned genomic fragment carrying the Tn5 insertion mapped using restirction enzymes. All six fragments showed different restriction enzyme digestion patterns. The cloned Tn5 insertion fragments were tested for homology to the Klebsiella pneumoniae nifHDK region in plasmid pSA30 (Cannon et al., 1979). One clone showed homology, apparently to the nifD or nifK genes. A partial Sau3A genomic library of strain H8 was prepared in the vector EMBL3 and screened by hybridization to this Tn5 insertion. From two overlapping clones, a restriction map of a 24kb region of the chromosome was deduced. Hybridization to digests of these clones by the nifHDK region showed that a number of non-contiguous fragments showed homology to the probe.

References:
Barraquio WL et al. (1983) Can. J. Microbiol. 29, 867-873
Cannon FC et al. (1979) Molec. Gen. Genetics 174, 59-66
Simon R et al. (1983) In Puhler A, ed, Molecular Genetics of the Bacteria-Plant Interaction, pp 98-105, Springer-Verlag, Berlin Heidelberg.
Weiss AA et al. (1983) Infect. Immun. 42, 33-41

CHROMOSOMAL ORGANIZATION OF NIF GENES IN RHODOPSEUDOMONAS CAPSULATA

J.C. WILLISON, G. AHOMBO, J.P. MAGNIN, P.M. VIGNAIS
LABORATOIRE DE BIOCHIMIE MICROBIENNE, DRF, CEN-G,
85X, 38041 GRENOBLE CEDEX, FRANCE

A circular linkage map of the Rhodopseudomonas capsulata chromosome has been established by conjugation with the mutant R plasmid pTH10 (Willison et al., 1985). This plasmid has been shown to transfer the chromosome of R. capsulata strain B10, probably from several points of origin (Willison et al., 1983, 1985).
Nitrogen fixation (nif) mutations were shown by conjugation to be dispersed in at least five distinct regions of the chromosome. One region was shown by analysis with gene transfer agent to contain at least two genes separated by more than 2.7 kb of DNA. The organization of nif genes is the photosynthetic bacterium R. capsulata is therefore very different from that in Klebsiella pneumoniae, where the 15 to 17 genes required for nitrogen fixation are clustered in a 24 kb region of the chromosome (Roberts, Brill, 1981 ; Dixon, 1984). Our results are in agreement with those of Wall et al., (1984), who have identified six linkage groups of nif mutations using gene transfer agent alone.
Biochemical characterization of nif mutants was carried out by using rocket immunoelectrophoresis to determine the concentration of individual nitrogenase components, and by titrating crude extracts with purified MoFe protein or Fe protein to determine the catalytic activity of the components present. The structural gene for the nitrogenase Fe protein (nifH) was identified, and was found to map is the same region of the chromosome as a nifA-like regulatory gene. Mutations in two other regions of the chromosome were found to decrase synthesis and/or stabi- lity of the MoFe protein. One such group of mutations appeared to affect the assimilation of Mo, since the level of MoFe protein and the nitrogenase activity of mutants was stimulated 5- to 6- fold by the addition of 1 mM molybdate to the growth medium.
In addition to the nif mutant of R. capsulata, ntr-like mutants were isolated that were defective is growth on all N sources tested other than NH_4^+. These ntr-like mutations mapped in the same region of the chromosome as, but were not closely linked to, the structural gene for glutamine synthetase (glnA).

REFERENCES

Dixon RA (1984) J. Gen. Microbiol. 130, 2745-2755.
Roberts GP and Brill WJ (1981) Annu. Rev. Microbiol. 35, 207-235.
Wall JD et al (1984) J. Bacteriol 159, 652-657.
Willison JC et al (1983) In Papageorgiou GC and Packer L, eds, Photosynthetic Prokaryotes: Cell Differentiation and Function, pp.335- 352, Elsevier Biomedical, New-York/Amsterdam/Oxford.
Willison JC et al (1985) J. Gen. Microbiol. (in press).

THE STRUCTURAL GENES FOR NITROGENASE (NIFKDH) ARE SEPARATED FROM THE MAJORITY OF NIF GENES IN RHODOPSEUDOMONAS CAPSULATA

Werner Klipp, Bernd Masepohl and Alfred Pühler
Universität Bielefeld, D-4800 Bielefeld 1, FRG

Using a broad host range transposon mutagenesis system, based on mobilizable E.coli vector plasmids (Simon et al, 1983), 60 Tn5 induced nif mutants of Rhodopseudomonas capsulata were isolated. By cloning and restriction analysis of nif::Tn5 containing DNA fragments it could be shown that these mutants map within three regions separated from each other. The extent of region A is about 18 kb, of region B about 4 kb and of region C about 2 kb. The corresponding wildtype nif DNA fragments were isolated by two different methods:
- nif::Tn5 containing DNA fragments were used to identify the appropriate clones from a cosmid gene bank of R.capsulata by colony hybridization.
- Mobilizable, in R.capsulata nonreplicative plasmids, carrying the nif::Tn5 DNA fragment were integrated into the R.capsulata chromosome by single cross-over recombination. Spontaneous excision of the plasmid followed by mobilization back to E.coli results in plasmids containing the original nif::Tn5 fragment plasmids with the unmutated nif-DNA fragment or plasmids with the unmutated nif-DNA fragment.

Using single gene-probes of all 17 Klebsiella pneumoniae nif genes for DNA hybridization experiments with these cloned R.capsulata nif regions the following homolgies were detected: region A: nifA, nifE and nifS; region B: nifD and nifH; region C: no significant homology. By immunological and biochemical analysis it could be confirmed that region B contains the structural genes for nitrogenase (nifKDH). A second region with homology to nifH and nifA was localized about 10 kb from nifKDH.

At least 8 different transcription units were identified by complementation analysis. Their direction of transcription was determined by gene fusions. For this purpose the promotorless kanamycin resistance gene derived from Tn5 was linked to a constitutive tetracycline resistance gene. This composite DNA fragment was cloned in both orientations at various positions into R.capsulata nif DNA. Due to the selectable tetracycline resistance these gene fusions could be inserted into the R.capsulata chromosome by homologous recombination. Compairing the kanamycin resistance of the resulting R.capsulata strains under repressed and derepressed growth conditions regulated nif promotors were identified. With these gene fusions it could be shown that mutants in region C inhibit transcription of all other nif operons so far tested.

Simon R et al (1983) Biotechnol. 1, 784-791.

NITROGENASE STRUCTURAL GENES AND MULTIPLE nifH-LIKE SEQUENCES IN CLOSTRIDIUM PASTEURIANUM W5

KATHERINE C. CHEN, J.-S. CHEN, AND J. L. JOHNSON
DEPARTMENT OF ANAEROBIC MICROBIOLOGY, VIRGINIA POLYTECHNIC INSTITUTE
AND STATE UNIVERISTY, BLACKSBURG, VIRGINIA 24061, U.S.A.

We are interested in $\underline{C}$. pasteurianum (Cp) nitrogenase structural genes because 1) the amino acid sequence of the mature Fe and MoFe proteins is either completely or partially determined, 2) the Cp nitrogenase is distinctive in its lower H_2 sensitivity, higher nucleotide specificity, and incompatibility with heterologous components, and 3) $\underline{C}$. pasteurianum is a Gram-positive anaerobe with a low G + C content of 28%. The Cp genes can thus provide information concerning posttranslational modification, codon usage, and regulatory genetic elements in this type of organism; they also allow studies to locate the protein regions that determine the enzyme's distinctive properties.

MATERIAL AND METHODS. Klebsiella pneumoniae nifKDH genes, contained in fragment A (KDH) or subfragments A1 (K), A2 (KD) and A3 (DH), were isolated from plasmid pSA30. EcoRI-digested $\underline{C}$. pasteurianum DNA or isolated insert DNA was probed with A or A1-A3 for nif genes using the Southern procedure; the hybridizations were carried out at 42°C in the presence of 20-50% formamide. The dideoxy method and the M13 mp18 and mp19 phages were used in DNA sequencing.

RESULTS AND DISCUSSION. The G + C content of the $\underline{C}$. pasteurianum and $\underline{K}$. pneumoniae DNA differs by about 30%. One would not expect extensive nucleotide sequence similarity between genes from these two organisms, if the G + C content of the genes reflects that of the genome. Indeed, we were unable to detect homology between Cp DNA and Kp fragments A, A1 or A2. Using fragment A3 (Kp nifDH) and 25% formamide, we detected six weakly hybridizing bands (~10, ~7, 3.8, 2.6, 2.1, and 1.7 kb) in EcoRI-digested Cp DNA. The 3.8 kb band was the strongest, and the fragment was cloned into pBR325 as pCP114. DNA sequence analysis of the 3.8 kb insert revealed two nifH-related sequences (nifH1 and H2) and nifD in the order of nifH2 → nifH1 → nifD. nifH1 could code for a protein (273 amino acid residues) identical to the purified Fe protein. nifH2 was separated from nifH1 by 412 bp and could code for a protein (272 residues) very similar to the Fe protein (23/273 or 8% different residues). nifH1 and H2 were also separated by a potential stem (15 bp) and loop structure. The nifD region in the insert covered the first 166 amino acids of the larger (α) subunit of the purified MoFe protein. From the nucleotide sequence, GUG (f-Met) was identified as the initiation codon for nifD which preceded the N-terminus (Ser) of the isolated protein, suggesting posttranslational processing of nifD but not nifH1 polypeptides. Another nifH-like sequence (nifH3) was located on the 2.6 kb EcoRI fragment (cloned in pCP3). The putative nifH3 product would be more different than the putative nifH2 product from the Fe protein. The codon usage in Cp nifH1, H2, and D was very biased, with the third base predominantly in A/U. For arginine, only one (AGA) of the six synonymous codons was used in Cp nifH1. For leucine, CUU and UUA, instead of CUG or UUG were the major codons. Overall, 38 of the 61 codons was used in Cp nifH1. The longest stretch of homologous nucleotide sequence between Cp nifH1 and Kp nifH was for 11 nucleotides. At the level of triplet codons, the homology was 20% between Cp nifH1 and Kp nifH. The different codon usage patterns may also affect the efficiency of nif translation in new host cells.

THE AZOTOBACTER VINELANDII NIFD AND NIFK ENCODED POLYPEPTIDES SHARE STRIKING STRUCTURAL HOMOLOGY WITH THE NIFE AND NIFN ENCODED POLYPEPTIDES

Dennis R. Dean and Kevin E. Brigle
The Virginia Polytechnic Institute and State University
Blacksburg, Virginia 24061 USA

The structural comparison of functionally related biomolecules provides a basis for identifying features important for their activity. In this study we have isolated and analyzed the A.vinelandii genomic regions flanking the nitrogenase structural genes.

1) Three A.vinelandii nif specific promoters have been identified. One promoter directs the synthesis of the nifHDK cluster, another promoter directs the synthesis of the nifEN cluster and a third divergent promoter directs the synthesis of a gene(s) whose function is not known. 2) Three open reading frames have been identified in the region between the end of nifK and the start of nifE. Whether these genes are actually expressed and the potential function of their products in nitrogen fixation is not known. 3) Comparisons of A.vinelandii and K.pneumoniae nifH and nifE promoters reveal homologies which occur not only within the consensus nif promoter sequence but also within regions upstream. These latter homologies are promoter specific and possibly represent sequences important for the differential expression of the individual nif clusters. 4) The complete sequence of the nifE gene and a portion of the nifN has been determined. The ribosome binding site for the nifN gene product overlaps with the termination signal for the nifE gene product. Such overlapping start and stop signals often indicates translational coupling (also called sequential translation). 5) The nifD and nifK gene products share striking homology with the nifE and nifN gene products. These homologies imply an evolutionary as well as a functional relationship between the nifDK and nifEN genes and their products. The nifDK-nifEN gene product similarities should be considered in site directed mutagenesis studies aimed at identifying amino acids wich participate in binding FeMo cofactor to the MoFe protein. We suggest that the FeMo cofactor is assembled upon a nifEN product complex prior to its donation to the MoFe protein.

This work was supported by Grant 84-CRCR-1-1519 from the United States Department of Agriculture.

THE DNA SEQUENCE OF NITROGENASE GENES FROM AZOTOBACTER CHROOCOCCUM

ROBERT ROBSON(1), ROBERT JONES(2), PAUL WOODLEY(1) and DAVID EVANS(1)

(1) AFRC UNIT OF NITROGEN FIXATION, UNIVERSITY OF SUSSEX, BRIGHTON, U.K.
(2) UNIVERSITY OF CHICAGO, CHICAGO, ILLINOIS 60637, U.S.A.

The cloning of the nifHDK genes from Azotobacter chroococcum has been described (1) and the sequence of about 5 kbp of DNA encoding these genes has now been determined. The amino acid sequences of the gene products are very similar to those of other species. 76 bp upstream of the start of the nifH gene is a sequence that closely resembles the consensus nif promoter. The nifHDK genes are expressed in K.pneumoniae, presumeably from this promoter. Four discrete lengths of transcript that hybridize to nifH are found in nitrogen-fixing A.chroococcum. Assuming that transcription is intiated adjacent to the proposed promoter, the 1.1 kb transcript would terminate within the nifH-nifD intergenic region, the 2.6 kb transcript would terminate between nifD and nifK, and the 4.3 kb transcript would terminate after the end of nifK. Each of these regions has the potential to form secondary structures within mRNA transcripts and we propose that these features influence the transcript stability, thereby reducing probably by the level of expression of downstream genes within the operon. The fourth transcript that hybridizes to nifH, of 6.4 kbp, is present at a relatively low level. No obvious transcription terminator is found in the 150 bp of sequence downstream of nifK and the start of an open reading frame has been identified 130 bp from the end of the gene. This would indicate that the nifHDK operon extends beyond three genes.

A second copy of the nifH gene has been cloned from A.chroococcum and has been given the interim designation nifH*. The sequence of 1.7 kbp of DNA in this region has been determined. The predicted amino acid sequence of the nifH* gene product is very similar to other iron proteins. 124 bp downstream from nifH* is an open reading frame that encodes a ferredoxin of the bacterial 4Fe-4S type. This is most similar to ferredoxin I from the green sulphur bacterium Chlorobium limicola. 50 bp upstream of the nifH* gene start there is a nif promoter-like sequence, although lac fusions to nifH* indicate that expression is not significantly activated by nifA from K.pneumoniae. The nifH*-ferredoxin intergenic region shows no recognisable sequences that would influence transcript initiation or termination. Sequence 70 bp downstream from the end of the ferredoxin gene strongly resembles a rho-independent transcription terminator. It is probable, therefore, that the operon comprises only these two genes.

We believe that the proteins encoded by this operon are involved in the form of nitrogenase that is produced under molybdenum limitation. Evidence for this involvement is the isolation from molybdenum-starved, nitrogen-starved A.vinelandii of a 1.4 kb mRNA species that hybridizes both to nifH and to the ferredoxin gene. This species is not found in molybdenum-sufficient, nitrogen-starved cells (Jacobsen and Bishop, pers.comm.).

1. Jones R et al (1984) Mol. Gen. Genet. 197, 318-327.

ANALYSIS OF A NIFH-LIKE GENE FROM THE ARCHAEBACTERIUM METHANOCOCCUS VOLTAE

L. SIBOLD, N. SOUILLARD and M. HENRY. Unité de Physiologie Cellulaire, Institut Pasteur, 28 rue du Dr Roux, F-75724 Paris Cedex 15, FRANCE.

Recently, nitrogen fixation was reported in 2 methane producing archaebacteria (Belay et al. 1984; Murray, Zinder 1984). We have used the Southern hybridization technique to investigate the presence of nif genes in four methanogenic strains and found sequences homologous to Klebsiella pneumoniae and Anabaena nifHDK in the four strains tested (Sibold et al. 1984, 1985).

In Methanococcus voltae, a strain not known as Nif$^+$, homology to nifH was found on a 3.0 kbp HindIII fragment and homology to nifD and nifK on a 3.8 kbp HindIII fragment. The 3.0 kbp fragment was cloned and the region homologous to nifH localized more precisely. In Escherichia coli maxi--cells, the fragment directed the synthesis of a 30 kd polypeptide and the coding region corresponded to the nifH homologous region. Plasmids carrying the fragment did not complement K. pneumoniae nifH mutants and did not inhibit nitrogen fixation of a Nif$^+$ strain.

The complete nucleotide sequence of the nifH homologous region was determined. It contained an open reading frame (ORF) of 834 bp encoding 278 amino acid residues (MW 30,362). This ORF is surrounded by regions of very high A+T content as it is the case for other Mc. voltae genes (Hamilton, Reeve 1985; Cue et al. 1985). The region upstream of the ORF contains prokaryotic-like promoters and a sequence sharing striking similarities with the -24 consensus region of the eubacterial nifA/ntrC-regulated promoters (Dixon 1984). A ribosome binding site is located 5 bp preceding the translation initiation sequence. Though the usage codon is characteristic of Mc. voltae, the nifH-like gene is very similar to eubacterial nifH genes, in particular the position of the cysteine residues is very conserved. These data confirm the high conservation of nifH sequences. S_{AB} values (binary matching coefficients) of 0.5 were found with eubacterial nifH genes at the nucleotide or amino acid level suggesting that the Mc. voltae nifH sequence is distantly related to eubacterial nifH sequences, among which the lowest S_{AB} value is 0.6. It thus appears that nifH genes have evolved as the bacteria that harbor them as it was proposed in the case of eubacterial nifH genes (Hennecke et al. 1985).

Belay N et al. (1984) Nature 312, 286-288.
Cue D et al. (1985) Proc. Natl. Acad. Sci. USA 82, 4207-4211.
Dixon R (1984) J. Gen. Microbiol. 130, 2745-2755.
Hamilton PT and Reeve JN (1985) Mol. Gen. Genet. 200, 47-59.
Hennecke H et al. (1985) Arch. Microbiol., in press.
Murray PA and Zinder S (1984) Nature 312, 284-286.
Sibold L et al. (1984) In Antonopoulos A, ed, Proc. of the 1st Symp. on Biotechnol. Advances in processing Municipal Wastes for Fuels and Chemicals, Academic Press, New York, in press.
Sibold L et al. (1985) Mol. Gen. Genet. 200, 40-46.

ORGANIZATION OF NIFH, D, K IN THE NONHETEROCYSTOUS, FILAMENTOUS CYANOBACTERIUM, PLECTONEMA BORYANUM.

Susan Ruttenberg Barnum and Steven M. Gendel, Department of Genetics, Iowa State University, Ames, Iowa 50011 U.S.A.

Cyanobacteria are a highly diverse group of oxygenic photosynthetic prokaryotes with some species capable of fixing atmospheric nitrogen. Studies on the organization of nitrogen fixation genes in cyanobacteria have been limited to the heterocystous filamentous species (e.g. Golden et al., 1985) and one strain of Gleothece, a unicellular fixer (Kallas et al., 1983). Our goal was to determine the arrangement of nifH, D, K genes in the nonheterocystous, filamentous cyanobacterium, Plectonema boryanum PCC 6306, that fixes nitrogen only when grown under anaerobic conditions. The genes for nitrogenase (nifK and nifD) and nitrogenase reductase (nifH) from Anabaena were used to probe Plectonema DNA, and the resulting patterns of hybridization were used to construct a physical map of the Plectonema nif region. Whole cell DNA was prepared from 2L cultures grown either in the presence of air and combined nitrogen (fix⁻ cells) or starving for nitrogen in the absence of air (fix⁺). Cells were lysed using lysozyme followed by Tris-EDTA-sarkosyl and DNA were recovered using cesium chloride-ethidium bromide equilibrium ultracentrifugation. Gene Screen Plus (New England Nuclear) blots were prepared from agarose gels of Plectonema DNA digested with single and double digests of Hind III, Cla I, Xba I, and Bgl II restriction enzymes. The map of the Plectonema nif region from fix⁻ cells deduced from the hybridization patterns obtained is shown:

All three nif probes hybridized to a single 12 kbp Cla I fragement. When Cla I digested Plectonema DNA from fix⁺ cells were hybridized to Anabaena nifH, nifD, and nifK, a 4.5 kbp Cla I fragment was seen in addition to the 12 kbp band. The appearance of the 4.5 kbp fragment may be due to either demethylation of the DNA or a DNA rearrangement. Experiments are currently underway to distinguish between these two types of DNA modifications.

REFERENCES

Golden JW et al (1985) Nature 314, 419-423.
Kallas T et al (1983) J. Bacteriol. 155, 427-431.

REARRANGEMENT OF NIF STRUCTURAL GENES IN NOSTOC PCC 7906.

THIERRY DAMERVAL, CLAUDINE FRANCHE, ROSMARIE RIPPKA, GERMAINE COHEN-BAZIRE
INSTITUT PASTEUR, PARIS, FRANCE.

We examined the nif structural gene arrangement by hybridization with labelled probes carrying the nif K, D and H genes of Anabaena PCC 7120 to
Southern blots of DNA from Nostoc PCC 7906 (Het$^+$) and from a Het$^-$ mutant
of the same strain. The restriction fragments hybridizing to the probes
were identical for both strains, and the nif organization of DNA from cells
grown in the presence of combined nitrogen was similar to that reported
previously for Anabaena PCC 7120 and Calothrix PCC 7601, nif K being located distantly from nif DH. Examination of DNA isolated from heterocysts of
Nostoc PCC 7906 revealed that a nif rearrangement must have occurred in the
process of heterocyst differentiation since in this DNA both nif K and nif
D were located on the same PvuI fragment of about 2 kb. Furthermore, total
DNA from cells of the Het$^-$ mutant exposed to a period of nitrogen starvation under air exhibited the same nif rearrangement as DNA of the heterocysts and the nif arrangement typical of vegetative cells (i.e. nif K and
nif D being located on the 14 kb PvuI fragment) was completely lacking.
This observation implies that the nif rearrangement occurred in all cells
of the mutant after nitrogen starvation in air and not just in a small
fraction of the population giving rise to the nonfunctional proheterocysts.

AN ANABAENA NIF GENE REARRANGEMENT OCCURS IN E. COLI

PETER J. LAMMERS/JAMES W. GOLDEN/ROBERT HASELKORN
DEPARTMENT OF MOLECULAR GENETICS AND CELL BIOLOGY, THE UNIVERSITY OF
CHICAGO, CHICAGO, ILLINOIS 60637, USA

Anabaena 7120 is a filamentous cyanobacterium with the capacity to differentiate specialized cells, called heterocysts, in the absence of combined nitrogen. These cells become anaerobic and specialize in nitrogen fixation, exporting glutamine to surrounding vegetative cells. During the development of heterocysts, an 11 kilobase DNA element is removed from the 3' coding region of the nifD gene by site-specific recombination within eleven base-pair directly repeated sequences at the ends of the element (1). We have discovered that this same deletion event, which is under developmental control in Anabaena, can also occur at low frequency from an Anabaena DNA fragment cloned in pBR322 and maintained in E. coli. A visual assay for the deletion event was developed by Mini-Mulac transposition into the eleven kbp element (2). Cells that retain the excision element are blue on XGal agar, while excision events result in plasmids without the lac genes and thus yield white colonies on XGal agar. Excision of this element occurs in 0.05-0.3% of cells grown in batch culture. This frequency is not greatly affected by nitrogen limitation, anaerobiosis, or heat shock. Mini-Mulac insertion in the left end of the element (near nifK) leads to undetectable levels of rearrangement. Complementation with subclones of DNA from the left end allow such mutant plasmids to undergo excision at nearly normal frequencies, suggesting that a trans-acting factor required for excision is encoded in this region of the element. DNA sequence analysis has revealed an open reading frame which encodes a 39 kilodalton protein. The reading frame is on the opposite DNA strand relative to the nif genes. Both transposon insertion and BAL-31 deletions into the reading frame completely abolish excision of the element, proving that the gene is absolutely required for the rearrangement process, although not necessarily sufficient. Comparison of the amino acid sequence predicted from this gene with the Protein Sequence Database (NIH) failed to reveal significant homologies with other known proteins.

1) Golden, J., Robinson, S. & Haselkorn, R. (1985) Nature 314: 419-423.

2) Castilho, B., Olfson, P. & Casadaban, M. (1984) J. Bacteriol. 158: 488-495.

CHARACTERIZATION OF A SECOND DNA REARRANGEMENT
NEAR THE NIF GENES IN ANABAENA

MARTIN E. MULLIGAN, JAMES W. GOLDEN AND ROBERT HASELKORN
DEPT. OF MOLECULAR GENETICS & CELL BIOLOGY, THE UNIVERSITY OF CHICAGO
CHICAGO, IL. 60637, USA

The cyanobacterium Anabaena 7120 undergoes at least two developmentally regulated DNA rearrangements during heterocyst differentiation. The first rearrangement is the excision of an 11 kbp element that interrupts the gene for the α subunit of dinitrogenase, nifD, near the 3'-end of the gene. The same authors also observed that a second rearrangement occurred upstream of the gene for dinitrogenase reductase nifH. We report here the further characterization of this latter rearrangement.

The breakpoint of the rearrangement was observed to be on a 3.3 kbp HindIII fragment (called 154.2) adjacent to the HindIII fragment containing the nifH gene. When HindIII-digested heterocyst DNA was probed with 154.2, the 3.3 kbp fragment was absent and was replaced by two fragments of length 2.3 kbp and 4.7 kbp (called H20.1 and H35 respectively). H20.1 is adjacent to nifH in heterocyst DNA while H35 is elsewhere on the heterocyst chromosome. Both of these fragments have been cloned from a λgt7 library of heterocyst DNA. In addition, we have cloned a 3.6 kbp fragment (called 264) from a λL47 library of vegetative cell DNA, which contains DNA that is brought adjacent to nifH as a result of rearrangement.

The DNA sequence of the breakpoints has been determined for all four clones. The sequence shows that the rearrangement is a conservative, site-specific recombination. The 11 bp sequence TTC$^{GA}_{CT}$GAATA is conserved at the center of each breakpoint and recombination cleavage must occur within the sequence GAATA. This conserved sequence has no similarity with the sequences found at the breakpoints of the nifD rearrangement suggesting that the two rearrangements are independent of one another despite the fact that they both occur at a similar time during heterocyst differentiation. There are three possible mechanisms for the rearrangement. These are the inversion of a segment of the chromosome, deletion of a segment of the chromosome (analogous to the nifD rearrangement) or insertion of a circular piece of DNA. We do not yet know which of these possibilities is correct. However, the rearrangement must involve a minimum of 25 kbp in the case of the latter two mechanisms, or in the case of an inversion, 50 kbp.

The region of the Anabaena chromosome between nifH and the breakpoint located in 154.2 has been shown previously to contain a gene homologous either to nifV or to nifS of Klebsiella. We have probed digests of Klebsiella DNA, containing the nifMVSU genes, and determined that the Anabaena gene is homologous to the Klebsiella nifS gene. The region containing the Anabaena nifS gene has been sequenced. We have found an open reading frame that could code for a protein the size of Klebsiella nifS. A second reading frame in this region and in the same orientation as nifH and nifS could code for a small (174 amino acid) protein. This region of the Anabaena chromosome is transcribed during heterocyst differentiation, but we have not yet determined the size of the transcript that is made nor where it is initiated. In view of the proximity of the 5'-end of the putative nifS gene to the breakpoint of the rearrangement, it is possible that the rearrangement is required for the expression of nifS.

NIF PROMOTER STRUCTURE IN KLEBSIELLA PNEUMONIAE

MARTIN BUCK, STEPHEN MILLER, HASEENA KHAN, MARTIN DRUMMOND and RAY DIXON
AFRC Unit of Nitrogen Fixation, University of Sussex, Brighton BN1 9RQ UK.

The promoters of nitrogen fixation genes in Klebsiella pneumoniae are characterised by a lack of the canonical -35 and -10 sequences typical of many prokaryotic promoters. Instead, transcriptional activation requires nucleotide sequences located around -24 and -12 which are conserved amongst nif promoters. Our analysis of point mutations in the nifL and nifH promoters has demonstrated that the conserved residues play an important role in transcriptional activation. We have found that the nifH promoter can be activated weakly by the ntrC gene product. A comparison of mutations in both promoters reveals that transitions at -12 are strong down mutations with respect to ntrC activation but such mutations are silent with respect to nifA activation of the nifH promoter, whereas transitions at -13 are strong down mutations with respect to ntrC and nifA activation of both promoters.

In addition to the -24, -12 elements we have identified a sequence: TGT N_{10} ACA which is located more than 100 bp upstream of the -24, -12 consensus sequence in the nifH, nifU, nifB and ORF promoters. This sequence is essential for both efficient nifA activation (but not ntrC activation) and multicopy inhibition. The upstream element is also conserved amongst many Rhizobium nif promoters. Analysis of the upstream element from the nifH promoter indicates:

(1) It acts in cis with the downstream elements to produce a fully active promoter,
(2) it is itself transcriptionally inactive,
(3) it can be displaced at least 20 bp further upstream and still retains activity.

It seems likely that the nifA protein interacts with the upstream element to facilitate transcriptional activation. The properties of the upstream element are analogous to those of eukaryotic enhancers and demonstrate that an extended promoter structure is involved in nif promoter activation and nif gene expression.

CLONING AND CHARACTERIZATION OF THE GLNA GENE OF AZOSPIRILLUM BRASILENSE Sp7.

CORRADO FOGHER, HERVE BOZOUKLIAN, SUKHPAL BHANDARI AND CLAUDINE ELMERICH. UNITE DE PHYSIOLOGIE CELLULAIRE, INSTITUT PASTEUR, 75724 PARIS CEDEX 15, FRANCE.

A plasmid, designated pAB44, complementing the glutamine auxotrophy of the A. brasilense mutant 7029 was isolated from a gene library of Sp7. Plasmid pAB44 did not complement E. coli or K. pneumoniae Gln⁻ mutants, but plasmid mutants which restored prototrophy were isolated. In pAB450 and -452, the glnA gene was under the control of the Tc promoter, and selection of mutants was not necessary. Tn5 insertions and hybridization with a K. pneumoniae glnA probe established that the glnA gene was located within a 1.9 kb SalI fragment (Figure 1).

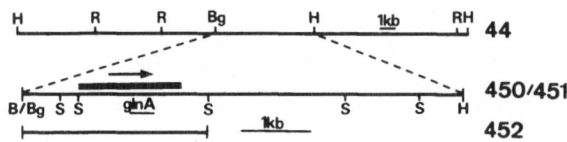

FIGURE 1. Physical map of plasmids containing the glnA gene. Restriction sites: B: BamHI; Bg: BglII; H: HindIII; R: EcoRI; S: SalI; arrow: direction of transcription; pAB44 and -451 vector is pVK100; pAB450 and -452 vector is pSUP202.

Introduction of Azospirillum glnA in K. pneumoniae glnA mutants (e.g. UNF1787) restored a wild type control of nitrogen fixation. In most of the ntrC mutants (e.g. UNF1816) no complementation occurred, suggesting that the glnA plasmids did not contain ntr functions. However, complementation of UNF1812 was observed. Introduction in Azospirillum mutants of plasmids containing Azospirillum glnA (pAB451) or K. pneumoniae glnA (pGE10, pPC940) restored a wild type control of nitrogen fixation. Strains 7028 and 7029 are likely to be glnA mutants. Thus, a direct involvement of Azospirillum glutamine synthetase in the regulation of nitrogen fixation cannot be ruled out.

	K. PNEUMONIAE[*]			A. BRASILENSE	
	UNF1787	UNF1812	UNF1816	7029	7028
	glnA	ntrC	ntrC	Gln⁻ Nif⁻	Gln⁻Nif^c
NH₄⁺	− +	− +	−	−	− +
no plasmid	94 42	0.5	−	2.7	18 200
pGE100/10	200 <0.3	69	nd	83	132 13
pGE102/PC940	53 <0.3	3	nd	30	30 0.8
pAB450/1	120 3.4	29	<0.1	122	68 <0.5
pAB452	295 0.5	42	<0.1	nd	nd nd

Nitrogenase activity is expressed in % of the wild type; pGE10 and pGE100 contain the glnAntrBC regulon of K. pneumoniae; pGE102 and pPC940 (this work) contain only glnA; * : Espin et al (1981) (1982) Mol. Gen. Genet. 184, 213-217 and 186, 518-524.

Nif A⁻ MUTANTS OF AZOSPIRILLUM BRASILENSE STRAIN Sp. 245

S.FUNAYAMA, L.U. RIGO and F.O. PEDROSA
Dept. of Biochemistry, Universidade Federal do Paraná, Caixa Postal 939,
80.000 - Curitiba, Pr, Brazil.

A. brasilense strain Sp 245 has been shown to associate specifically to
the roots of wheat plants and to give rise to high levels of N_2 fixing
efficiency in this association (1). Mutants of Sp 245 deficient in
nitrogen fixation are needed to determine its actual contribution in
fixed N to the associated plants. We have isolated and characterized
five mutants of Sp 245 Nal^{20}. Sm^{200} (= SFo) following NTG mutagenesis.
Four of these mutants were complemented by nif A from Klebsiella
pneumoniae. These are regulatory nif A⁻ mutants as those described for
A.brasilense FP2 (2).

MATERIALS AND METHODS

SFo was grown in NFbHP.NH$_4$Cl medium, mutagenised (NTG 20 μg/ml) and
screened for Nif⁻ mutants as described before (2). The final screening
was in liquid NFbHP medium containing glutamate. Growth on N source was
carried out on solid media. Nitrogenase activity was measured by C_2H_2
reduction.

RESULTS

Nif mutants: Small colonies comprised 3% of the mutagenized SFo; 12
isolate showed less than 1% of SFo acetylene-reducing activity in N-
free semi-solid medium and only 5 failed to derepress nitrogenase in
liquid medium containing glutamate at 30ºC. These were called SF2, SF3,
SF4, SF5 and SF9 and were characterized further.

Growth on N sources: All strains grew well with NH$_4$Cl (1 or 20 mM), and
in the other N sources (1 mg/ml Arg, Pro, Ala; 5 mM Glu, and 10 mM NO$_3$).
SF2 and SF9 grew poorly on His (1 mg/ml).

Genetic complementation: Plasmid pCK$_3$ (K. pneumoniae nif A^c, Tc^r, Km^r)
kindly supplied by Dr. C. Kennedy (UNF, Brighton, England) was
transfered into SFo and all Nif mutants, using plasmid pRK2013 as
helper. Transconjugants of SF2, SF3, SF5 and SF9 showed 70 to 100% of
the wild-type nitrogenase activity in liquid medium containing glutamate
at 30ºC but failed to derepress at 37ºC. pCK$_3$ was present in all
transconjugants after growth in both temperatures, as confirmed by
agarose gel electrophoresis. These are possibly nif A⁻ mutants. SF4
failed to be complemented by pCK$_3$. SFo, SF2, SF3, SF5 and SF9
transconjugants derepressed nitrogenase (up to 17% of the maximum
activity in glutamate medium) in the presence of 20 mM NH$_4$Cl at 30ºC.
NH$_4^+$ "Switched off" nitrogenase activity in glutamate grown SFo,SF3,
SF5 and SF9 transconjugants, but not in SF2. Elimination of pCK$_3$ by
pR68.45 abolished nif expression.

Conclusions: SF2, 3, 5 and 9 were complemented by K. pneumoniae nif A⁻
and are therefore regulatory mutants. SF3, 5 and 9 are probably nif A⁻
mutants, while SF2, by analogy with mutants FP8 and FP9 (2) is probably
a ntrC mutant. These results support previous studies which indicated
that nif expression in A.brasilense is regulated by a nif A type gene
analogous to that of K. pneumoniae (3).

References and Acknowledgments.
1. Baldani, V.L.D.; Baldani,J.I. and Dobereiner,J. 1983 Can.J.Microbiol
   29, 924 - 929.
2. Pedrosa, F.O. and Yates, M.G. 1984 FEMS Microbiol. Lett.23, 95-101
3. Merrick,M. 1983 EMBO J. 2, 39-44. We thank CNPq - Brasil

INTERACTION OF PURIFIED NtrC PROTEIN WITH NITROGEN REGULATED PROMOTERS FROM Klebsiella pneumoniae

TIMOTHY HAWKES, MIKE MERRICK and RAY DIXON
AFRC Unit of Nitrogen Fixation, University of Sussex, Brighton BN1 9RQ UK.

The product of the ntrC gene (which we designate as NtrC) is a bifunctional regulatory protein which can positively activate transcription from a number of nitrogen regulated promoters in Klebsiella pneumoniae including the nitrogen fixation, nifLA and the glutamine synthetase, glnA (RNA1) promoters as well as acting negatively to repress transcription from the ntrBC and glnA (RNA2) promoters.

We have purified NtrC protein from K. pneumoniae and analysed its interaction with promoters which are subject to regulation by the ntr system. Our results and conclusions are summarised below:

1)   K. pneumoniae NtrC is a dimeric site-specific DNA binding protein with M_R ∿108 kd.
2)   NtrC inhibits expression from both the glnA and ntrBC promoters in vitro and binds to both promoters with equal affinity.
3)   On the basis of nucleotide sequence homology among ntr repressible promoters in enteric bacteria, we propose a consensus repressor binding site for NtrC:

$$5' \text{ TGCACTA } {}^{\text{AAA}}_{\text{TTT}} \text{ TGGTGCAA } 3'$$

This sequence has dyad symmetry and can accommodate a dimer of identical sub-units in either a direct or an inverted repeat arrangement.
4)   DNaseI and methylation protection experiments with the K. pneumoniae glnA (RNA2) promoter show that NtrC makes close contact with guanine residues which flank the -10 region.
5)   Two-dimensional modelling of the protected region indicates that NtrC binds as a dimer on one face of the DNA molecule, each subunit contacting successive major grooves in the helix, as has been shown for other sequence-specific DNA binding proteins.
6)   We did not detect binding of NtrC to the ntr activated nifLA, nifF or glnA (RNA1) promoters. Although it is possible that we have not yet determined the appropriate conditions for binding to these promoters, it is also feasible that NtrC does not interact with these sequences in the absence of other regulatory factors (e.g. ntrA and RNA polymerase).

TRANSCRIPTIONAL REGULATION OF NITROGEN FIXATION BY MOLYBDENUM IN AZOTOBACTER VINELANDII

MARTY R. JACOBSON*, R. PREMAKUMAR AND PAUL E. BISHOP. North Carolina State University and USDA/ARS, Raleigh, NC. 27695-7615.

Since a Mo-containing dinitrogenase has been observed in all known diazotrophs, one might speculate that Mo might also regulate nitrogenase expression. This has led to a number of studies on the effect of Mo-deprivation on nitrogenase in a variety of organisms, revealing no uniform response. Bishop et. al. (Proc. Natl. Acad. Sci. USA.. 1980. 77: 7342-7346) have suggested the existence of an alternative nitrogen fixation system in Azotobacter vinelandii, based on the observation that Mo-deprivation (omitting Mo from the growth medium) caused Nif$^-$ mutants to undergo phenotypic reversal to Nif$^+$. Since then, the existence of two dinitrogenase reductases under regulatory control by Mo in A. vinelandii has been reported (Premakumar et. al.. 1984. Biochim. Biophys. Acta.. 797: 64-70). Evidence for a second dinitrogenase reductase-like enzyme (alternative reductase) was based on the observation that a dinitrogenase reductase-like activity is present in Mo-deprived cell-free extracts of Nif$^-$ mutant strains of A. vinelandii which either lack a conventional dinitrogenase reductase (UW1 or UW3) or contain a defective enzyme (UW91) under conditions of Mo-sufficiency.

In this study we demonstrate that multiple genomic regions homologous to nifH are present in the diazotroph Azotobacter vinelandii and are transcriptionally regulated by molybdenum. The nifHDK gene cluster is located on a 12.8-kb XhoI fragment. Two additional XhoI fragments (7.4 kb and 8.4 kb) are also found to hybridize to nifH but not to nifD or nifK. In vivo transcription of the nifHDK gene cluster occurs from a common promoter, is ammonia repressible and requires the presence of at least 50 nM Mo in the derepression medium. Three mRNA species are transcribed from the nifHDK gene cluster, a 4.2-kb transcript homologous to nifH-, nifD- and nifK-specific DNA templates, a 2.6-kb transcript homologous to nifH- and nifD-specific DNA templates and a 1.2-kb transcript homologous to the nifH-specific DNA template.

Strain CA11, which is a nifHDK deletion mutant (unpublished data), will not grow using N$_2$ as a sole source of nitrogen in batch cultures containing Mo concentrations of 50 nM or greater. The nifHDK-specific transcripts are not produced when this strain is derepressed for N$_2$ fixation in medium containing 50 nM Mo or greater. However, at Mo concentrations of 25 nM or less strain CA11 is capable of growth using N$_2$ as a sole source of nitrogen and two nifH (but not nifD or nifK) homologous transcripts are observed (1.2 kb and 1.8 kb). Presumably, these transcripts result from transcription of the additional ("alternative") nifH homologous sequences present in the genome of A. vinelandii.

Sequences homologous to a ferredoxin-like gene of Azotobacter chrococcum (obtained from R. Robson) have also been identified in Azotobacter vinelandii. These Fd-like sequences appear to co-migrate on the same XhoI and SmaI restriction fragments with one of the additional nifH homologous regions. In A. vinelandii a 1.8-kb transcript which is homologous to the Fd-like gene of A. chrococcum, is repressed by both ammonia and molybdenum. This 1.8-kb transcript is the same size as one of the nifH homologous transcripts which is expressed only under N$_2$ fixation and Mo-deprivation conditions suggestion that these two regions might be in the same transcriptional unit.

TN5-INDUCED NIF⁻ MUTANTS OF AZOTOBACTER VINELANDII

ROLF D. JOERGER, JENNIFER B. KOPCZYNSKI, AND PAUL E. BISHOP.
Dept. of Microbiology and U. S. Dept. of Agriculture, Agricultural
Research Service, NC State University, Raleigh, NC 27695-7615, USA

Based on genetic and biochemical evidence, an "alternative" N_2 fixation system has been proposed to operate in A. vinelandii under Mo-deficient conditions (Bishop et al., 1980 PNAS, 77:7342-7346). The genetic basis of N_2 fixation under Mo deficiency is still largely unknown. Therefore we isolated Tn5-induced mutants affected in N_2 dependent growth under Mo-deficient conditions. Utilizing the suicide plasmid pSUP1011 (Simon et al., 1983, Biotechnology 1:784-791), we obtained 107 mutants, which can be classified into at least three phenotypic classes.

The 48 mutants of the first class did not grow on N-free medium regardless of the Mo content. One of these mutants, probably carrying Tn5 in a gene involved in FeMoco synthesis, did not reduce acetylene in vivo after 3 h of derepression. A second mutant in this class still reduced acetylene at 6% of the wild-type level in the presence of $1 \mu M$ Mo. Both mutants synthesized the nifHDK gene products under Mo-sufficient conditions and the NH_4^+-repressible proteins, previously observed under Mo deficiency.

The second class consists of 58 mutants which failed to grow on N-free, Mo-deficient medium, but grew slower than the wild type in the presence of $1 \mu M$ Mo. Under Mo deficiency acetylene reduction was less than 5% of the wild type activity. No changes in the pattern of NH_4^+-repressible proteins on two-dimensional gels could be detected. Two-factor crosses involving previously described A. vindelandii Nif⁻ mutants (Shah et al., 1973, Biochim. Biophys. Acta 292:246-255) with one of the Tn5 mutants indicate that Tn5 (Kan^r) was located near nif-1 and nif-45 (linkage 42% and 53% respectively). In another mutant the Kan^r phenotype was linked to nif-6 and nif-3 by about 20%, but by less than 1% to nif-1 and nif-45. Strains constructed by transformational transfer of the Kan^r phenotype of these two mutants in to a nifHDK deletion mutant (strain CA11) were unable to reduce acetylene in the presence of $1 \mu M$ Mo, but still reduced acetylene under Mo deficiency. This suggests that the low levels of acetylene reduction activity of these mutants under Mo deficiency can be attributed to an acetylene-reducing agent operating in the absence of the nifHDK gene products.

The third class consists of a mutant that grows as fast as the wild type under Mo-sufficient conditions. Under Mo deficiency growth and acetylene reduction activity were reduced and could not be stimulated by Vanadium. NH_4^+-repressible "alternative" proteins could not be detected and growth in Mo-deficient medium was probably due to leakiness in the expression of the conventional nitrogenase, since the nifHDK gene products could be seen on 2-D gels of extracts from cells derepressed under Mo deficiency. Transfer of the Kan^r phenotype of this mutant into a nifHDK deletion strain which is capable of fixing N_2 under Mo deficient conditions, resulted in a mutant unable to grow or reduce acetylene under Mo deficiency.

In conclusion, our preliminary studies with Tn5 induced mutants show that several loci exist in A. vinelandii that contribute to the N_2 fixation activity observed under Mo deficient conditions. The genetic information of some of these loci appears to be shared by the "conventional" and "alternative" N_2 fixation systems.

TRANSCRIPTIONAL REGULATION OF NITROGEN FIXATION GENES IN RHODOPSEUDOMONAS CAPSULATA

ROBERT G. KRANZ, ROBERT JONES AND ROBERT HASELKORN, DEPARTMENT OF MOLECULAR GENETICS AND CELL BIOLOGY, THE UNIVERSITY OF CHICAGO, CHICAGO IL. 60637 USA

Translational fusions of the E. coli lacZYA operon to Rhodopseudomonas capsulata nif genes were obtained by using mini-mu lac transposons (Castilho et al 1984) inserted into cloned fragments of R. capsulata DNA. A lac fusion to the nifH gene, which encodes dinitrogenase reductase, was used to identify nif regulatory mutants. A total of 25 nif mutants were screened for ability to derepress the nifH::lacZ fusion. Nine mutations were unable to activate nifHDK transcription. Four different genes present on two different cosmids complementing these mutations have been mapped. Three of these regulatory genes are adjacent on the same EcoRI DNA fragment and possess independent promoters; one is homologous with the ntrC gene from E. coli, based on Southern hybridization. These four genes are presently being sequenced. The fourth regulatory gene is present on a different EcoRI fragment and mini-mu transposons mapped it to a 1.5 to 1.9 kbp region. Mutations in these four regulatory genes do not result in Ntr phenotypes (i.e. all grow on glutamine, arginine or proline as the only nitrogen source). Therefore, R. capsulata, unlike Klebsiella, may not require two levels of nif transcriptional control. E. coli lacZ gene fusions to all four regulatory genes were constructed. The direction of transcription and expression of each gene was determined. Each regulatory gene is weakly expressed compared to derepressed nifH and only partially repressed in the presence of ammonia.

Thirteen mutants unable to repress the nifH::lac fusion in the presence of ammonia were isolated. None of these mutations is located in the nifH promoter; all made the nifH, D, and K gene products in the presence of ammonia, glutamine or other fixed nitrogen sources. All the mutants were still able to repress nifH directed lac expression when oxygen was present. This result suggests that different mechanisms exist for oxygen and ammonia repression of nif genes in R. capsulata.

REFERENCES

Castilho, B. A. et al (1984)   J. Bacteriol. 158, 488-495.

ACKNOWLEDGEMENTS

We thank Peter Avtges, Judy Wall, Ayala Hochman and Malcolm Casadaban for strains.

TAC-PROMOTED NIFA IN BINARY PLASMID SYSTEMS PRODUCES K. PNEUMONIAE NIF GENE PRODUCTS FROM NATIVE PROMOTERS IN E. COLI

PAUL V. LEMLEY/GEORGIANNA M. HARRIS/KAREN S. HOWARD/KENNETH S. KOBLAN/COLLEEN FARRIS/JANICE FLORY/W. H. ORME-JOHNSON. MASSACHUSETTS INSTITUTE OF TECHNOLOGY, CAMBRIDGE, MASSACHU- SETTS 02139 U.S.A.

The K. pneumoniae nifA gene product, in combination with the ntrA gene product, activates native nif promoters in E. coli. We present the construction of plasmids which produce nifA gene product from the very strong and inducible tac promoter (deBoer, et al.). Although initially constructed to aid in the biochemical characterization of nifA gene product, these plasmids have found usefulness as activators of native nif promoters coharbored on a secondary plasmid (Howard et al.,a). The tac promotion of the nif positive activator circumvents the physiological regulation of the nifLA operon, precludes classical derepression, and allows cultures to be grown in media with ammonia. This results in high cell and gene pro- duct yields. It also permits radio-label and β-galactosidase assay experiments to be done aerobically. We can grow cul- tures at 37°C to begin production of nif proteins.

The application of these tac-promoted nifA plasmids is pre- sented. In a binary plasmid system in E. coli, we have de- fined the minimum number of genes required to produce active iron protein (Howard et al., b). Efforts are underway to similarly define gene packages which will produce molybdenum- iron protein and FeMoco-activatable apo-molybdenum-iron pro- tein (Harris et al.). The binary system has been employed to produce single nif gene products, isolated nif operons and various gene packages or combinations. We have noted that there is not a linear response (by all nif promoters) to the production of nifA. We have apparently non-limiting nifA protein, however, production of gene products on the second- ary plasmid is limited by other factors. We present strate- gies for engineering fragments of nif DNA in vitro to examine individual gene functions and the effects of specific nif gene deletions.

This work is supported by NIH, grant #GM30943.

deBoer, H. A. et al. (1983) Proc. Natl. Acad. Sci. USA, 80, 21-25.
Harris. G. et al. (1985) these proceedings.
Howard, K. S. et al.,a (1985) these proceedings.
Howard, K. S. et al.,b (1985) manuscript submitted.

THE ROLE OF THE NTRA GENE PRODUCT IN POSITIVE CONTROL OF THE NITROGEN FIXATION GENES IN KLEBSIELLA PNEUMONIAE

M.J. MERRICK, J.R. GIBBINS, R.A. DIXON and W.D.P. STEWART*
AFRC Unit of Nitrogen Fixation, University of Sussex, BN1 9RQ, U.K.
*present address - Dept. of Biology, University of Dundee, Dundee DD1 4HN
U.K.

In K. pneumoniae the nitrogen fixation (nif) genes are subject to positive control and transcriptional activation requires the product of the ntrA gene together with the ntrC or the nifA product. The nif promoters are atypical having a consensus sequence CTGGCAC N_5 TTGCA between positions -26 and -10 instead of the normal -35, -10 consensus. The absolute requirement for ntrA for transcription initiation together with the atypical promoter sequence has led to suggestions that ntrA could encode an alternative RNA polymerase sigma factor.

The K. pneumoniae ntrA gene has been cloned and the gene product identified by 2-D SDS PAGE as a 75 kDal acidic polypeptide. NtrA-lacZ fusions were used to demonstrate that the gene is not transcriptionally regulated in response to the N-status of the cell and is not subject to control by ntrB or ntrC (de Bruijn, Ausubel, 1983; Merrick, Stewart, 1985). A 1.9 kb DNA fragment carrying ntrA has now been sequenced and an open-reading frame (ORF) of 1431 bp has been identified which we believe to be ntrA and which would encode a polypeptide (NtrA) of 54 kDal. This ORF has two potential ATG initiation codons and the amino acid sequence of the longest ORF is shown in Table 1.

MKQGLQLRLSQQLAMTPQLQQAIRLLQLSTLELQQELQQALDSNFLLEQTDLHDEVETKE
AEDRESLDTVDALEQKEMPEELFLDASWDEIYTAGTPSGNGVDYQDDELPVYQGETTQSL
QDYLMWQVELTPFTDTDRAIATSIVDAVDDTGYLTISVEDIVESIGDDEIGLEEVEAVLK
RIQRFDPVGVAAKDLRDCLLVQLSQFAKETPWIEEARLIISDHLDLLANHDFRSLMRVTR
LKEEVLKEAVNLIQSLDFRFGQSIQTGEPEYVIPDVLVRKVNDRWVVELNSDSLPRLKIN
QQYAAMGNSTRNDADGQFIRSNLQEARWLIKSLESRNDTLLRVSRCIVEQQQAFFEQGEE
FMKPMVLADIAQAVEMHESTISRVTTQKYLHSPRGIFELKYFFSSHVNTEGGGEASSTAI
RALVKKLIAAENPAKPLSDSKLTTMLSDQGIMVARRTVAKYRESLSIPPSNQRKQLV

As predicted from 2D PAGE (Merrick, Stewart, 1985), NtrA is markedly acidic which may account for its aberrant mobility on SDS gels. NtrA has an overall amino acid composition similar to that of other known sigma factors but the amino acid sequence shows no striking homology to other sigmas. Overproduction of NtrA has been achieved by placing the gene under control of the inducible tac promoter such that induction by IPTG results in synthesis of NtrA to 1.5% of total cell protein. Such strains are currently being used to provide material for purification of NtrA in order to analyse further the role of this protein in nif regulation.

REFERENCES

de Bruijn FJ and Ausubel FM (1983) Mol.Gen.Genet. 192, 342-353.
Merrick MJ and Stewart WDP (1985) Gene 35, 297-303.

REGULATION OF ANABAENA VARIABILIS NITROGENASE GENE EXPRESSION

J.S. THRUTCHLEY[1], J.A. MYERS[1], J.E. MILLER[1], L.R. YARBROUGH[2], AND R. HIRSCHBERG[1]. UNIVERSITY OF MISSOURI- KANSAS CITY, KANSAS CITY, MO 64110, USA and [2] UNIVERSITY OF KANSAS MEDICAL CENTER, KANSAS CITY, KS 66103, USA.

Like other nitrogen-fixing organisms, the filamentous, heterocyst-forming cyanobacterium Anabaena variabilis 29413 fixes nitrogen only when other nitrogen sources are not available. Thus, both heterocyst formation and synthesis of nitrogenase are repressed by ammonia and certain other nitrogen compounds. Although this has been assumed to represent control of gene expression at the transcription level, only recently has this been shown directly. We have been investigating this question by measuring nitrogenase mRNA levels using recombinant DNA techniques. Under aerobic conditions, short filament fragments (of 3-5 cells) form heterocysts in 17-18 hr when starved for nitrogen; proheterocysts appear at 7-8 hr. The appearance of nitrogeanse activity and nitrogenase mRNA during development were monitored by acetylene reduction and by RNA dot blot hybridization analysis. Previously cloned Anabaena nif genes were used as probes. Ammonia grown cells have very low levels of mRNA and no nitrogenase activity. Nitrogenase mRNA levels increase 50-100 fold during derepression. Although only heterocysts have enzyme activity under aerobic conditions, low but significant levels of nitrogenase mRNA were found in nitrogen-starved vegetative cells. When starved for nitrogen under anaerobic conditions, A. variabilis fixes nitrogen without forming heterocysts. Nitrogenase activity appeared after 2.0 to 2.5 hr under such conditions, and mRNA levels increased within 1.5 to 2 hr. NH_4Cl prevented the appearance of nitrogease activity and mRNA. Nitrate repressed nitrogenase gene transcription, whereas glutamine and glutamate had little effect. Methionine sulfoximine, but not aza-tryptophan, reversed the repressive effect of NH_4Cl on nitrogenase gene transcription. Our results support the conclusion that nitrogenase gene expression is regulated primarily at the transcription level.

REGULATION OF NITROGEN METABOLISM IN AZOTOBACTER: ISOLATION OF NTR GENES

Aresa Toukdarian and Christina Kennedy
AFRC Unit of Nitrogen Fixation,
University of Sussex, Brighton, BN1 9RQ, England.

Nitrogen regulatory genes analogous to ntrA and ntrC of enteric bacteria have been isolated from A. vinelandii. The roles of these genes in regulating N-metabolism were established by using Tn5 insertions in the cloned genes to construct A. vinelandii ntrA and ntrC mutants. These mutants show that ntrA but not ntrC is required for nif expression while both genes are required for induction of nitrate reductase.

Complementation of growth phenotypes of E. coli ntrA, ntrC and Δ(glnA ntrBC) mutants by an A. vinelandii pLAFR1 gene bank yielded cosmid pLV72, which corrected the ntrA mutant, and cosmid pLV50, which provided both glnA and ntrC functions. These cosmids also complemented K. pneumoniae mutants for growth phenotypes including nif expression. A combination of subcloning and Tn5 mutagenesis identified a 2.4 kb EcoR1-Cla1 fragment of pLV72 coding for the ntrA gene of A. vinelandii and a 7.3 kb EcoR1-Cla1 fragment of pLV50 coding for glnA and ntrC. Hybridization of K. pneumoniae ntrA, ntrC, and glnA genes to these cosmids and to genomic DNA showed that the A. vinelandii genes had physical as well as functional homology to the enteric sequences.

A. vinelandii ntr mutants were constructed by transformation of competent wild type with various plasmids containing Tn5 insertions. Hybridization analysis of putative mutants confirmed that the site of Tn5 insertion was as expected for marker exchange and wild-type sequences without Tn5 inserts were not detected. One Km^r isolate representing each of the plasmids used was characterized for nitrogenase and nitrate reductase activities:

| Strain | Genotype | Acetylene Reduction[a] | | Nitrate Reductase[a] | |
		$+NH_4^+$	$-N$	$-NO_3^-$	$+NO_3^-$
UW136	ntr$^+$	O	100%	6.6%	100%
MV700	ntrA$^-$	O	O	2.0	3.5
MV511	ntrC$^-$	O	95	3.8	5.3
MV512	ntrC$^-$	O	78	6.4	4.4
MV513	ntrC$^-$	O	80	4.2	8.5
MV516	ntrC$^-$	O	120	3.8	11

a Acetylene reduction and nitrate reductase activities reported as the % of wild type derepressed or induced activities.

Transfer of pLV72 into MV700 or pLV50 into strains MV511, MV512, MV513, and MV516 restored enzyme activities.

THE CHARACTERIZATION OF THE PURIFIED PRODUCT FROM THE CLONED <u>NIFF</u> GENE FROM <u>KLEBSIELLA</u> <u>PNEUMONIAE</u> UNAMBIGUOUSLY CONFIRMS A FLAVODOXIN

JANET DEISTUNG, MAURA C. CANNON*, SUSAN HILL, FRANK C. CANNON*, ROGER N.F. THORNELEY. AFRC Unit of Nitrogen Fixation, University of Sussex, Brighton, BN1 9RQ, UK. *BioTechnica International Inc., 85 Bolton Street, Cambridge, Massachusetts 02140, USA.

The <u>nifF</u> gene product is purported to be a flavodoxin that is involved in the electron transfer sequence shown in Eqn. (1) (Shah et al., 1983).

$$\text{Pyruvate} \rightarrow \text{Pyruvate-Flavodoxin} \rightarrow \text{Flavodoxin} \rightarrow \text{Kp2} \rightarrow \text{Kp1} \rightarrow \text{substrate}$$
$$\text{oxido-reductase} \qquad\qquad\qquad (N_2, H^+)$$

<u>nifJ</u>   <u>nifF</u>   Eqn. (1)

An elevated synthesis of the <u>nifF</u> gene product <u>in vivo</u> was achieved by cloning the <u>nifF</u> gene on a multicopy plasmid (Deistung et al., 1985). The transformant, UNF5112, a <u>K. pneumoniae</u> strain deleted for chromosomal <u>nif</u> genes, contained the constructed plasmid pBCC13, a derivative of pBR328 carrying the <u>nifF</u> gene. Because maximal transcription of <u>nifF</u> requires the <u>nifA</u> gene product, the transformant also contained a plasmid (pMC71a) carrying the <u>nifA</u> gene under the control of a constitutive promoter.

The <u>nifF</u> gene product was purified using a method based on that described by Shah et al. (1983) with the assay procedure developed by Hill, Kavanagh (1980). Cells were ruptured by French pressing and a crude extract supernatant obtained by centrifugation. Chromatography with DEAE cellulose, gel filtration with Sephadex G-50 and finally preparative gel electrophoresis yielded a yellow protein (M_r=19K) with $\varepsilon_{450nm} = 10.1$ mM cm^{-1} which was homogeneous as judged by SDS-PAGE. The amino acid composition and N- and C-terminal amino acid residues of the purified protein are the same as from the DNA sequence of the <u>nifF</u> gene (M. Drummond, personal communication). This confirms that the purified protein is the <u>nifF</u> gene product.

The E_m for the semiquinone/hydroquinone couple of Kp flavodoxin is surprisingly positive (e.g. -421 mV at pH 8.0) compared to Ac flavodoxin (e.g. -526 mV at pH 8.0)(Deistung et al., 1985).

References

Deistung J et al (1985) Biochem.J. in press.
Hill S and Kavanagh JP (1980) J. Bact. 141, 470-475.
Shah VK et al (1983) J. Biol. Chem. 258, 12064-12068.

MUTANT OF AZOSPIRILLUM BRASILENSE Sp7 THAT OVERPRODUCES CATALASE

ILMA H.HIGUTI and F.O.PEDROSA
DEPARTMENT OF BIOCHEMISTRY, UNIVERSIDADE FEDERAL DO PARANA
C.POSTAL 939, 80.000 - CURITIBA,PR , BRAZIL.

We have isolated a mutant of Azospirillum brasilense Sp7 (AT CC29145) resistant to high concentrations of H_2O_2. This mutant synthesises 500 to 1000 times more catalase than the parent strain Sp 7 .Comparative studies on nitrogenase derepression and activity showed that the excess catalase activity afforded no additional protection against oxygen.

MATERIALS AND METHODS

Sp 7($Nal^5 Sm^{50}$) was grown in $NFbHP-NH_4Cl$ medium(Pedrosa,Yates,1984). Catalase was determined according to Del Rio (1977).Nitrogenase activity was measured by C_2H_2 reduction . Oxygen consumption was determined polaro graphically.

RESULTS

Catalase overproducing mutant: Sp 7 does not grow in the presence of H_2O_2, however,after 20 subculturings in media containing increasing amounts of H_2O_2 , a mutant capable of growing in 50 mM H_2O_2 was isolated.This mutant IH1 showed 500 to 1000 times more catalase activity than the parent strain Characterization of IH1:mutant IH1 grew in the presence of Nal(5ug/ml), Sm(50 ug/ml) or both.It did not grow on D-glucose nor required biotin. It gave rise to the same plasmid pattern as Sp 7 after alkaline lysis followed by agarose gel electrophoresys. IH1 had similar levels of nitrogenase activity as those of the wild type.

Effect of oxygen on nitrogenase derepression:cell suspensions of Sp 7 and IH1 in N-free medium were adjusted to the same respiratory activity (0.05 umol O_2 consumed per min per ml), and exposed to different oxygen tensions for 3 h at 37ºC and 160 rpm .Nitrogenase activity was determined as a function of O_2 tensions. Maximum derepression occurred at 8% O_2 for Sp 7 and 7% O_2 for IH1.

Effect of oxygen on nitrogenase activity:cell suspensions of Sp 7 and IH1 showing similar nitrogenase and respiratory activities were assayed for nitrogenase activity under various oxygen tensions. Both strains had the same activity pattern with a maximum at 3% O_2.

CONCLUSIONS : IH1 is a mutant of Sp 7 resistant to high H_2O_2 concentrations due to the presence of high levels of catalase. IH1 was equally prone to oxygen inactivation and repression of nitrogenase as Sp 7 , in spite of its high content of catalase . Catalase "per se" therefore does not afford any additional protection of nitrogenase against oxygen in Azospirillum brasilense.

REFERENCES AND ACKNOWLEDGEMENTS

Del Rio et al (1977) Anal.Biochem.80,409-415.
Pedrosa,F.O. and Yates,M.G. (1984) FEMS Microbiol.Letts 23,95-101.

We thank CNPq - Brazil .

CHARACTERIZATION OF A MOLYBDENUM-BINDING PROTEIN IN CLOSTRIDIUM PASTEURIANUM

STEVE HINTON[+], CLIVE SLAUGHTER[+], GREG FREYER[+], BILL EISNER[+], AND BRENDA MERRITT
[+]COLD SPRING HARBOR LABORATORY, COLD SPRING HARBOR, N.Y. 11724 U.S.A.
EXXON CORPORATE RESEARCH LABS, CLINTON, N.J. 08801 U.S.A.

We have identified four molybdoproteins in crude extracts of Clostridium pasteurianum that have characteristics suggesting they may play a role in molybdenum (Mo) metabolism (Hinton, 1985). The major Mo-binding protein (Mop) in C. pasteurianum, suspected to be involved in Mo metabolism, was purified and characterized. Biochemical studies have shown that the purified Mop contains Mo and a pterin-derivative, the components of the molybdenum-cofactor (Mo-co). It has been proposed that the Mo atom in Mo-co is bound to a pterin-derivative via sulfur ligands. The structure of the Mo environment in Mop was shown by X-ray absorbance spectroscopy (XAS) to be unusual in that it has three terminal oxo-groups and three oxygen or nitrogen ligands (Cramer, Hinton, unpublished data). Recently, clostridial formate dehydrogenase (containing active Mo-co) has been shown to have a Mo site (Lui, 1984) almost identical to Mop except the molybdopterin species dissociated from Mop was not able to reconstitute the activity of Mo-co deficient nitrate reductase in extracts of Neurospora crassa mutant, nit-1. We identified an expression clone of Mop by immunoscreening genomic libraries of C. pasteurianum. DNA sequence analysis of the expression clone (mop gene) revealed an open reading frame coding for 68 amino acids (- 7,000 MW). Amino acid and DNA sequence analysis showed that Mop does not contain aromatic nor cysteine residues, therefore, the UV absorbancy (peak 293nm) of native Mop can be attributed to the pterin chromophore and the protein has no potential sulfur ligands for the Mo atom. The amino acid sequence analysis also showed that there are two or more related proteins in the purified Mop preparations. Southern blot analysis of C. pasteurianum chromosomal DNA showed that the mop gene has been duplicated and DNA sequence analysis of one duplicate gene showed 93% homology with the original clone. Presently, we are sequencing a third copy of the mop gene (family), and determining the structure of the molybdopterin and how it binds to the protein. Evidence suggests that the molybdopterin species associated with Mop might be a precursor to active Mo-co which is eventually inserted in formate dehydrogenase, the end product of the molybdenum processing pathway.

Hinton SM, Mortenson LE (1985) J. Bacteriol. 162, 485-493.
Lui C-L, Mortenson LE (1984) J. Bacteriol. 159, 375-380.

PRODUCTION OF KLEBSIELLA PNEUMONIAE NIFH PROTEIN IN THE ABSENCE OF NIFM BY ESCHERICIA COLI

KAREN S. HOWARD/KENNETH S. KOBLAN/PAUL V. LEMLEY/FINN B. HANSEN/W. H. ORME-JOHNSON. MASSACHUSETTS INSTITUTE OF TECHNOLOGY, CAMBRIDGE, MASSACHUSETTS 02139 U.S.A.

Earlier work defined the nitrogen fixation (nif) genes from K. pneumoniae required to produce active nitrogenase Fe protein in E. coli as nifA, H, and M (K.S. Howard, et al.). nifA protein from a tac-promoted plasmid, pVL4 (P.V. Lemley, et al.), in combination with the ntrA gene product provided by the E. coli host, promoted the nifH and nifM genes from their native promoters on a secondary plasmid, pKH733. Our current work utilized this binary plasmid design to overproduce the nifH gene product in the absence of nifM in order to isolate and biochemically characterize the nifH protein and to provide substrate for nifM activity assays.

Initial attempts to overproduce the nifH gene product in the absence of nifM involved a single plasmid, pAHDKY#2, or a binary system, pMC71A (V. Buchanon-Wollaston, et al.) + pBS1ΔBam. Both designs used the nifA gene product to provide activator for nifH promotion. In each case, nifH protein was formed, as evidenced by ^{35}S-methionine labeling, however, the steady state level, observed by Coomassie Blue staining of similar samples, was marginally detectable. This indicated that nifH protein made in the absence of nifM was less stable in vivo than mature Fe protein (see also J. Berman, et al.).

Fe protein made by pKH733 constituted 10% of the whole cell protein. Therefore, a direct modification of pKH733 was performed to delete nifM. The new plasmid, pKK8, in combination with a pVL4 derivative, pVL15 (P.V. Lemley, et al.), produced nifH protein equivalent to 4.7% of the whole cell protein. Crude extracts showed no C_2H_2-reducing activity in standard assays, confirming the nifM requirement for Fe protein activity. The steady state levels of nifH protein produced in the absence of nifM never reach those levels produced in the presence of nifM in parallel genetic systems. It is not yet clear whether the nifM protein is involved in maximizing the production of nifH protein, increasing the half life of nifH protein in whole cells, or both, as it assists in the formation of active Fe protein.

This project was supported by the National Institutes of Health, grant #GM30943 and fellowship #GM09380.

Berman, J., et al. (1985) J. Biol. Chem. 260, 5240-5243.
Buchanon-Wollaston, V. et al. (1981) Nature 294, 776-778.
Howard, K.S., et al. (1985) manuscript submitted.
Lemley, P.V., et al. (1985) these proceedings.

AMMONIUM AND METHYLAMMONIUM TRANSPORT BY RHODOPSEUDOMONAS CAPSULATA

BARBARA J. RAPP, DEBORAH C. LANDRUM AND JUDY D. WALL.
University of Missouri, Columbia, Mo. 65211 USA

Introduction Transport of ammonium across cell membranes by a specific transport system has been established in many bacteria (2). Our studies investigated the ammonium transport system of aerobically grown cultures of the photosynthetic, N_2-fixing bacterium Rhodopseudomonas capsulata through the use of the analog methylammonium for comparison to earlier work on ammonium uptake (1).

Materials and Methods The minimal medium contained 15 mM $(NH_4)_2SO_4$ which, for nitrogen limitation, was modified to contain only 15 mM glutamate or 2mM $(NH_4)_2SO_4$. $[^{14}C]CH_3NH_3^+$ uptake was measured in whole cells grown to early stationary phase aerobically in darkness. The assay consisted of 2 ml of cell suspension (equivalent to 0.3 mg protein/ml) in medium lacking NH_4^+. After a 30 min preincubation at 30°C, the reaction was started with substrate. Sampling took less than 7s.

Results When the ammonium analog $[^{14}C]$ methylammonium was used in transport assays, uptake was observed only when cells were grown on glutamate or were N-starved. In contrast, ammonium was taken up by cells grown on any nitrogen source tested. Within 20s with an initial concentration of 75 uM methylammonium, there was greater than a 3 fold accumulation of methylammonium within the cells as measured by the comigration of radioactivity with standard methylammonium during thin layer chromatography of cell extracts. Uptake of methylammonium exhibited typical saturation kinetics and the K_m for methylammonium was determined to be about 22 uM with a V_{max} of about 10 nmol/min·mg. An analysis of the inhibition of methylammonium transport by ammonium yielded a K_i for ammonium of around 1 uM suggesting that ammonium may be the natural substrate of the system. In addition, the inhibition of methylammonium uptake by various energy poisons, including CCCP, cyanide and arsenate indicated an essential role for energy. Studies of Nif⁻ mutants, W11 (3) and J61 (4) suggest that two distinct regulatory elements are involved in the nitrogen control of transport.

Conclusions These studies demonstrate the existence of nitrogen regulated transport system for R. capsulata which is capable of concentrating the ammonium analog against a gradient.

References
1. Genthner BRS and Wall JD (1985) Arch Microbiol 141, 219-224
2. Kleiner D (1981) Biochim et Biophys Acta 639, 41-52
3. Wall JD et al (1977) Arch Microbiol 115, 259-263
4. Wall JD and Braddock K (1984) J Bacteriol 158, 404-410

ISOLATION AND CHARACTERISATION OF HYDROGEN UPTAKE GENES FROM AZOTOBACTER CHROOCOCCUM

KARL H. TIBELIUS and M.G. YATES
AFRC Unit of Nitrogen Fixation, Sussex University, Brighton, BN1 9RQ, U.K.

Recombinant cosmids carrying genes involved in hydrogenase activity were isolated from a gene library of Azotobacter chroococcum (Jones et al. 1984) by colony hybridisation using as a probe the 6 Kb HindIII fragment from pHU1 (a gift from Professor H.J. Evans). The cosmid pHU1 contains hup-specific sequences from Rhizobium japonicum (Cantrell et al. 1983) and complements Hup⁻ mutants of A. chroococcum (Yates, Robson, 1985). Further hybridisation studies using the 6 Kb HindIII, 6.4 Kb BglII, and 5 Kb EcoRI fragments from pHU1 confirmed that the isolated cosmids (pACD101 and pACD102) contained sequences homologous to R. japonicum DNA on pHU1. Some homology to the structural genes of hydrogenase from Desulfovibrio vulgaris (Hildenborough) (Voordouw et al. 1985) carried on the plasmid pHV150 (a gift from Dr. Gerrit Voordouw) was also observed.

The cosmids pACD101 and pACD102 contain about 41 and 42 Kb of insert DNA respectively with about 27 Kb of overlapping DNA. Restriction fragments of pACD101 and pACD102 were subcloned into the broad host-range vector pKT210 (Bagdasarian et al. 1981) and used to complement Hup⁻ mutants of A. chroococcum, thus indicating the presence of functional copies of hup genes on these cosmids.

Hup⁻ mutants of A. chroococcum were also complemented for hydrogenase activity by the cosmids pHU2 and pHU53, as well as pHU1, which together contain all the R. japonicum hup genes (Haugland et al. 1984; H.J. Evans, personal communication). Plasmids pHU1 and pHU53 both complemented the same 12 out of 17 Hup⁻ mutants to give specific hydrogenase activities of 72 to 554 and 11 to 191 nmol H_2 (mg protein)$^{-1}$min^{-1} respectively. Every pHU53 transconjugant was less active than the corresponding pHU1 transconjugant. Four of the remaining mutants were weakly complemented by plasmid pHU2 (specific activities 5.4 to 9.1). Two mutants were complemented and one was unaffected by all three plasmids. The presence of the plasmid in all transconjugants was confirmed by agarose gel electrophoresis.

Conclusion: DNA hybridisation and complementation studies suggest similarities in organisation and function of the hup-specific genes of R. japonicum and A. chroococcum. This similarity has allowed us to isolate hup-specific determinants from A. chroococcum.

REFERENCES

Bagdasarian M et al. (1981) Gene 16, 237-247.
Cantrell MA et al. (1983) Proc. Natl. Acad. Sci. USA 80, 181-185.
Haugland RA et al. (1984) J. Bacteriol. 159, 1006-1012.
Jones R et al. (1984) Mol. Gen. Genet. 197, 318-327.
Voordouw G et al. (1985) Eur. J. Biochem. 148, 509-514.
Yates MG and Robson RL (1985) J. Gen. Microbiol. 131, 1459-1466.

MUTANTS OF *AZOSPIRILLUM BRASILENSE* RESISTANT TO METHYLAMMONIUM

Luca Turbanti, Marco Bazzicalupo, Renato Fani, Enzo Gallori and Mario Polsinelli.
Dipartimento di Biologia Animale e Genetica, University of Florence, Florence, Italy.

The availability of strains of nitrogen fixing microorganisms with improved nitrogen fixing ability, could be very useful both for soil fertilization and for studies of the N_2-fixation regulation. For this pourpose we have isolated mutants of the bacterium *Azospirillum brasilense* resistant to the toxic ammonium analog methylammonium (MEA).

Spontaneous mutants resistant to MEA were isolated from strain Sp6 (1) on nitrogen free medium containing 300mM MEA. In this condition the analog reduces the growth of the parental strain, while resistant mutants appeared as larger colonies on a background of microcolonies. Six of these mutants were further characterized in respect to their ability to reduce acetylene (ARA) and to grow in the presence of MEA and different concentrations of NH_4Cl. Results obtained are reported in the Table.

STRAIN	ARA	ADDITIONS				
		MEA(120mM)		NH_4Cl (4mM)		NH_4Cl(300mM)
		Growth	ARA	Growth	ARA	Growth
Sp6(w.t.)	+	−	−	+	−	−
SPF400	−	+	−	+	−	−
SPF401	−	+	−	+	−	−
SPF402	−	+	−	+	−	−
SPF403	+	+	+	+	+	+
SPF404	+	+	+	+	+	+
SPF405	+	+	+	+	+	+

Data obtained indicated that mutants belong to two different groups. SPF400, SPF401 and SPF402 showed a Nif⁻ phenotype; they had a very low ARA (less then 2% of the parental strain), both in the presence or absence of MEA. Mutants SPF403, SPF404 and SPF405 showed ability to reduce acetylene also in the presence of 4mM NH_4Cl and to grow up to a concentration of 300mM NH_4Cl, inhibitory for the parental strain.

1) Bani D., Barberio C., Bazzicalupo M., Favilli F., Gallori E. and Polsinelli M. (1980) J.Gen.Microbiol. 119: 239-244.

POSTER DISCUSSION 8.   GENETICS OF NITROGEN FIXATION IN FREE-LIVING
BACTERIA

C. ELMERICH, Unité de Physiologie Cellulaire, Institut Pasteur, 75724
Paris Cedex, France.

In 1972, Dixon and Postgate reported the transfer of the nif region of
Klebsiella pneumoniae to Escherichia coli, creating a new nitrogen-
fixing species. At that time, none of the nif genes was identified and
no one might be aware of the complexity of the mechanisms that
controlled nif expression. For long, K. pneumoniae was the only
organism in which nif genetics could be studied. During the past five
years, due to the development of DNA recombinant techniques and to the
remarkable conservation of the nif genes, a tremendous amount of
information has been accumulated in other free-living diazotrophs.
This aspect was particularly well illustrated in the Poster Session.

With regard to the general methodology, it was very interesting to
note that genetic studies was performed at two levels of analysis. One
of them based on classical genetics and hybridization techniques was
used for mapping and cloning structural and regulatory genes and to
compare their organization to the Klebsiella model system. The other
employed more sophisticated techniques such as in vitro site directed
mutagenesis or overproduction of nif polypeptides in expression
vectors. In addition, a considerable amount of nucleotide sequences
was presented.

1. Nif gene organization.
Many Posters dealt with the organization of the nif genes. Two systems
have been particularly studied: Azotobacter and Rhodopseudomonas. For
example, it was found that in Azotobacter the organization of the nif
genes in a main cluster was similar to K. pneumoniae, whereas the nif
genes are rather scattered on the genome of Rhodopseudomonas. Among
the interesting features revealed by nucleotide sequence analysis, it
is worth noting that, in A. vinelandii, homology was found between
nifDK and nifEN suggesting functional relationship between the
MoFe-protein and the nifEN product complex. I was pleased to find a
poster on the cloning of nif genes of a Pseudomonas sp. isolated from
the rice rhizosphere, since no data on the genetics of newly
identified nitrogen-fixing pseudomonas has been reported yet.

Very spectacular data on nif DNA rearrangement in cyanobacteria were
reported. This was the subject of the lecture of Pr. Haselkorn, but
additional information was displayed in the Poster Session. Two DNA
rearrangements have now been discovered in Anabaena and it was
established that one of them could occur in E. coli. This indeed
raises exciting questions on the significance of this phenomenon. From
another point of view, the discovery of nitrogenase genes in
archaebacteria raises interesting questions on the origin and on the
evolution of the nif genes.

2. Regulation.
Though regulation of nif expression in K. pneumoniae was extensively
studied in the past years, the precise mechanism of action of the
regulatory proteins is still unknown. However, the most sophisticated

techniques of the molecular biology are now used to elucidate the remaining questions. Several Posters dealt with the purification and the role of the ntr products, and the DNA target of the regulatory proteins was explored using in vitro site directed mutagenesis. It was previously reported that the nif promoters differ from the canonical prokaryotic promoter. In addition, a large homology in the upstream sequences was also found. A common six bases DNA sequence, located in position around - 130 and which is likely involved in the binding of the nifA product, was discovered. This upstream element was found in front of most K. pneumoniae nif promoters, several Rhizobium promoters and in Azotobacter.

3. Alternative system for nitrogen fixation.
Evidence for a second system of nitrogen fixation in Azotobacter, which did not involved molybdenum and which was induced under Mo deficiency was initially reported by Bishop's group some years ago. Several posters dealt with this exciting new system. The alternative system does not involve the nifHDK or the nifNE genes since insertion or deletion mutants could still display nitrogen fixation under Mo deprivation. Transcript analysis suggested that the components of the second system involve an extra copy of the nifH gene, which is only expressed under Mo deprivation and which is not regulated through a nifA-like product. A second gene encoding a ferredoxin protein is likely to be also involved.

4. Other topics.
A few Posters dealt with the purification and the role of other nif proteins and a Conference Session should be devoted to this topic in the next Symposium. There were also some Posters dealing with related aspects of nitrogen fixation, in particular cloning of the Azotobacter structural gene for hydrogenase was reported.

ENZYMOLOGY AND CHEMISTRY OF
NITROGEN FIXATION

A. NITROGENASE

INVITED PAPERS

POSTER SUMMARIES:

(i)   Mechanism.
(ii)   Proteins and cofactors.
(iii)   Purification and properties.
(iv)   Regulation.
(v)   ATP effects.

DISCUSSION GROUP SUMMARY

B. CHEMICAL APPROACHES

INVITED PAPERS

POSTER SUMMARIES

DISCUSSION GROUP SUMMARY

NITROGENASE MECHANISM --- AN OVERVIEW

Barbara K. Burgess, Department of Molecular Biology and Biochemistry, University of California, Irvine, CA 92717

1. INTRODUCTION

Nitrogenase is composed of two separately purified proteins called the molybdenum-iron protein (MoFe protein) and the iron protein (Fe protein). N_2 fixation and all other reductions catalyzed by N_2ase require both component proteins, a source of reducing equivalents, MgATP, protons and an anaerobic environment. This overview covers what has been learned recently about the events that occur during N_2ase turnover and the chemical mechanism of substrate reduction.

2. NITROGENASE TURNOVER

2.1 <u>Reduction of the Fe protein</u>. The physical and redox properties of the Fe protein have been reviewed recently (Burgess, 1984; Orme-Johnson, 1985; Thorneley, Lowe, 1985; Stephens, 1985). The consensus view is that the Fe protein is a 60,000 dalton dimer which contains a single $[Fe_4S_4(Cys)_4]^{n+}$ cluster which functions between the n=1 and 2 oxidation states during turnover. The following three pieces of information have appeared inconsistent with this view: a) one piece of analytical data (Braaksma et al, 1983); b) the observation that the S=1/2 EPR signal exhibited by the reduced Fe protein integrates to less than one spin per molecule (Orme-Johnson, 1985); and c) that n=2 type behavior is sometimes observed for the purified Fe protein (Braaksma et. al. 1982) or for the Fe protein during turnover (Hallenbeck, 1983). The EPR problem has been resolved. The reduced Fe protein appears to exist in a mixture of spin states (S=1/2 and S=3/2) and only one EPR signal was being integrated (Orme-Johnson, 1985; Watt, McDonald, 1985). The n=2 behavior has been suggested to be an artifact of the method used to set the redox potential in those experiments (Thorneley, Lowe 1985).

The physiological electron donation system to the oxidized Fe protein has recently been established for <u>Klebsiella pneumoniae</u> (Shah et al., 1983). The <u>nifJ</u> gene product catalyzed the two electron oxidation of pyruvate to acetyl-SCoA + CO_2 with the two electrons being used to reduce two molecules of the <u>nifF</u> gene product, a flavodoxin (Nieva-Gomez et al., 1980). The flavodoxin functions between the semi-quinone and hydroquinone states (Shah et al., 1983, Klugkist et al., 1985) in transferring one electron to the oxidized Fe protein. The electron donation system in other organisms is likely to be different with <u>Clostridium pasteurianum</u> preferring ferredoxin to flavodoxin (Mortenson, Thorneley, 1979) and aerobes like <u>Azotobacter</u> probably using quite different systems altogether (Haaker, Veeger, 1977). Unfortunately, mechanistic information has only come from studies with the artificial electron donor dithionite where SO_2^- is the actual one electron donor (Thorneley, Lowe, 1985). The oxidized Fe protein can exist uncomplexed, complexed with two MgATPs or complexed with two MgADPs (Burgess, 1984; Cordewener et al., 1985). The rates of reduction of all three states by SO_2^- have been studied and the rate of reduction of the MgADP form is comparable to the rate of reduction of the Fe protein during turnover suggesting the first step in N_2ase turnover to be (Burgess, 1985; Thorneley, Lowe, 1985):

$$Fe_{ox}(MgADP)_2 + SO_2^{\cdot-} \rightleftharpoons Fe_{red}(MgADP)_2 + HSO_3^-$$

However, for the Fe protein to participate further in turnover, MgATP must exchange with MgADP so that the first step is usually given as:

$$Fe_{ox}(MgADP)_2 + 2MgATP + SO_2^{\cdot-} \rightleftharpoons Fe_{red}(MgATP)_2 + HSO_3^- + 2MgADP$$

2.2 Complex formation. The physical properties of the MoFe protein have have recently been reviewed (Burgess, 1984; Orme-Johnson, 1985; Stephens, 1985). There is general agreement from spectroscopic and kinetic data that the MoFe protein can be viewed as two functional dimers which do not interact during turnover. In the second step a complex is formed, reversibly, which contains $Fe_{red}(MgATP)_2$ and the MoFe protein in its dithionite reduced state ($MoFe^0$). The MoFe protein can bind a maximum of two Fe proteins presumably with one Fe protein on each half of the MoFe protein. However, one:one complexes are also active (Burgess, 1984) so that complex formation is given as:

$$[Fe_{red}(MgATP)_2]_x + MoFe^0 \rightleftharpoons [Fe_{red}(MgATP)_2]_x MoFe^0$$

where x usually equals 2 but can be 1. It had been suggested that the second MgATP site on the reduced Fe protein is generated only after complex formation (Cordewener et al., 1983) but this view has recently been revised (Cordewener et al., 1985).

2.3 Electron transfer coupled to MgATP hydrolysis. The third step in nitrogenase turnover is:

$$[Fe_{red}(MgATP)_2]_x MoFe^0 \longrightarrow [Fe_{ox}(MgADP + Pi)_2]_x MoFe^X$$

where x=1 or 2. If two Fe proteins are present in the complex then each half of the MoFe protein is believed to be reduced by one electron in this step. There is general agreement (Thorneley, Lowe, 1985) that this step is: a) rapid; b) irreversible (if you give the enzyme ADP + Pi it will not make ATP); and c) coupled (MgATP hydrolysis and electron transfer appear to occur at the same time). Unfortunately, there is currently no published information to suggest what the chemical mechanism of ATP hydrolysis might be. Most available data indicate that the minimum stoichiometry of this reaction is two MgATPs hydrolyzed per electron transfered (Thorneley, Lowe, 1985; Burgess 1985) but there is one report suggesting that the enzyme might be more efficient, hydrolyzing only one ATP per electron (Mortenson, Upchurch, 1980). Under some circumstances (e.g. low temperature, low Fe protein, CN^-, CH_3NC) the enzyme can be very inefficient, hydrolyzing many more than two MgATP per electron transferred (Orme-Johnson, 1985; Li et al., 1982; Rubinson et al., 1983). It has been proposed that this inefficiency can be explained by a futile cycling of electrons between the two component proteins (Orme-Johnson et al., 1977). In this futile cycle it is suggested that if the MoFe protein is somehow prevented from using its electron rapidly then, the electron will fall back into the Fe protein. This proposal has yet to be tested experimentally.

2.4 Complex dissociation. The last step in a single catalytic cycle of electron transfer is now given as:

$$[Fe_{ox}(MgADP + Pi)_2]_xMoFe^X \rightleftharpoons Fe_{ox}(MgADP)_2 + 2Pi + MoFe^X$$

That the two proteins dissociate following each electron transfer was first shown by Hageman and Burris (1978) and has been confirmed recently by Thorneley and Lowe (1983). The idea that free reduced MoFe protein exists and might later transfer its electrons to substrate, without the help of the Fe protein, forms the basis of the nomenclature proposed by Hageman and Burris (1978).

2.5 The rate of N_2ase Turnover. N_2ase is an extremely slow enzyme and as a consequence of this N_2-fixing organism must expend a great deal of energy and fixed nitrogen making large quantities of N_2ase just to get enough fixed nitrogen to grow (Thorneley, Lowe, 1985). The rate at which N_2ase turns over will depend on the rates of all of the forward and reverse reactions described above. If there are high concentrations of reductant, Fe protein and MgATP then the enzyme will move rapidly through the first three steps while if there is insufficient reductant, a low concentration of Fe protein or high concentration of MgADP then the enzyme will be slowed. Thorneley and Lowe (1983) have recently obtained data showing that if steps one through three are optimized the fourth, complex dissociation step, becomes rate determining. If the rate determining step is the dissociation, then the turnover rate of the enzyme is expected to be independent of the substrate being reduced. This is generally true for N_2ase but there are a few examples, reviewed elsewhere (Burgess, 1985) suggesting turnover is not strictly independent of substrate. Since the dissociation step involves protein-protein interactions its being the rate determining step offers an explanation for why N_2ases from different organisms or heterologous crosses (Emerich, Burris, 1978) turn over at different rates. That N_2ase turnover occurs via the sequence of reactions indicated above, with the complex dissociation step rate limiting, forms a working hypothesis around which further experiments can be designed. For example, it predicts that substituting reduced flavodoxin for SO_2^- should not affect the turnover rate as long as both are present at saturating concentrations. It also predicts a maximum turnover rate for the enzyme. One report, that the enzyme turns over more rapidly than this maximum rate has yet to be rationalized (Kluskist, 1985).

3. SUBSTRATE REDUCTION

3.1 Redox States of the MoFe protein. N_2ase catalyzes the six electron reduction of N_2 to $2NH_3$, the two electron reduction of $2H^+$ to H_2 and the reduction of a wide variety of non-physiological substrates involving two to 14 electrons (Burgess, 1985). Because no N_2ase substrate is completely reduced by only one electron the enzyme must go through more that one catalytic cycle before products can appear. This leads to the concept that there are different redox "states" of the MoFe protein. For the sake of this discussion we will use the nomenclature (Thorneley, Lowe, 1985) E_0 = 1/2 MoFe in its dithionite reduced state. Thus,

$$E_0 \xrightarrow{\ +e^-\ } E_1 \xrightarrow{\ +e^-\ } E_2 \xrightarrow{\ +e^-\ } E_3 \xrightarrow{\ +e^-\ } \text{etc.}$$

where each arrow represents a single catalytic cycle, involving a rate determining step, as described above.

3.2 <u>Competition between H_2 evolution and substrate reduction</u>. Protons (from H_3O^+) are required for all N_2ase catalyzed substrate reductions. We do not have any information about when the MoFe protein picks up those protons. It is possible that the Fe protein or MgATP are somehow involved or that the reduced MoFe protein picks up protons from solvent after complex dissociation. What we do know is that if the Fe protein and the MoFe protein are combined together with reductant and MgATP, in the absence of an added reducible substrate, electrons will go into the enzyme from SO_2^- and come out of the enzyme as H_2. If a reducible substrate is added the enzyme has a choice. It can use its electrons productively to reduce that substrate or it can, in a sense, waste them by evolving H_2. A number of experiments have been performed where the distribution of electrons between substrate reduction and H_2 evolution was varied either by increasing the substrate concentration or by changing the rate of N_2ase turnover (primarily by varying the Fe protein to MoFe protein ratio). These experiments have been performed for a number of substrates and the substrates seem to fall into two catagories. 1. At saturating concentrations of C_2H_2, HCN, CH_3NC and HN_3 hydrogen evolution is eliminated and high ratios of Fe protein to MoFe protein favor H_2 evolution over substrate reduction (Burgess, 1985; Rubinson et al, 1985). 2. At saturating N_2 and N_3^- some hydrogen evolution continues and high turnover rates now favor substrate reduction over hydrogen evolution (Burgess, 1985; Rubinson et al., 1985). These data have been rationalized (Burgess, 1985; Thorneley, Lowe, 1985) by suggesting that substrates in the first catagory bind to E_0 and that once they are bound the enzyme is committed to their reduction and cannot evolve H_2 (the H_2 evolution reaction is believed to occur at E_2, E_3 or more reduced states) (Thorneley, Lowe, 1984; Lowe, Thorneley, 1984). It is further suggested that the substrates in the second catagory, including N_2, bind only to a more reduced state of the enzyme (E_3) than that which can evolve H_2 (E_2) (Thorneley, Lowe, 1984; Lowe, Thorneley, 1984). It should be noted that there is some biophysical evidence to support the idea that C_2H_2 and CH_3NC bind to E_0 (see below). However, there is no biophysical evidence to suggest to which state of the enzyme N_2 or N_3^- might bind.

If this proposal (which is consistent with all currently available data) is correct then N_2ase has a very serious problem. It must somehow prevent H_2 evolution at E_2 and force the enzyme to go onto state E_3 so that N_2 can bind and be reduced. Thorneley and Lowe (1985) have proposed a mechanism to explain how this might be accomplished. They suggest that E_2 can be prevented from evolving H_2 by binding $Fe_{red}(MgATP)_2$ or by rebinding $Fe_{ox}(MgADP)_2$. Their conclusions are; a) when Fe protein is bound the active site is "covered" such that H_2 cannot be evolved; b) a second function of the Fe protein is to prevent H_2 evolution at E_2; and 3) N_2ase must be present in high concentrations in the cell to maximize the chances of E_2 binding to Fe protein in order to minimize H_2 evolution at that state.

The data demonstrate that even though N_2ase may be able to minimize H_2 evolution during N_2 reduction it cannot eliminate it. Thus, even when N_2 is present at 50 atm, and under conditions which should favor the binding of $Fe_{red}(MgATP)_2$ to E_2, hydrogen evolution continues and there is an apparent minimum stoichiometry of one H_2 evolved per N_2 reduced (Simpson, Burris, 1984). This stoichiometry could be coincidental as it is also observed for N_3^- (Rubinson et al, 1985). All H_2 evolution could represent a leakage of electrons at a more oxidized state of N_2ase with 75% efficiency being the best the enzyme can do. It is also possible that some of this H_2 represents a leakage of electrons and some of it represents H_2 inhibition of N_2 reduction (see below). It may seem more probable, however, that the stoichiometry is mechanistically significant, and this has been suggested by a number of investigators (e.g. Thorneley, Lowe, 1985; Guth, Burris 1983). The general proposal is that N_2 only binds to a reduced state of the enzyme which contains metal bound hydrides (E $_H^H$) and that when N_2 binds it displaces a dihydride to form H_2. At present the formation of metal hydrides at the active site is the most chemically plausible proposed mechanism for N_2ase catalyzed H_2 evolution (Burgess, 1985). However, attempts to demonstrate the presence of hydrides using D_2 or DCL traps have so far met with negative results (Burgess, 1985; Thorneley, Lowe, 1985).

3.3 Chemical Mechanism of N_2 Reduction. N_2 is reduced by six electrons to $2NH_3$. Even though no less highly reduced products are released during this reaction it has been established that enzyme bound intermediates are formed. Thorneley et al. (1978) have shown that an intermediate is formed, which, when quenched with acid or base, gives rise to free N_2H_4. Based on reactivity comparisons, with low valent Mo- and W-phosphine dinitrogen and dinitrogen hydride complexes, the authors propose the intermediate to be a hydrazido (2-) ligand (=N-NH). These data do not eliminate the possibility, however, that the intermediate could be some unknown structure which has not yet been chemically synthesized. Pre-steady state kinetic experiments have shown that the intermediate is formed following four electron transfers. However, the kinetics can best be fit to an eight electron reduction scheme for N_2, involving concomitant H_2 evolution, consistent with the hydrazido (2-) proposal (Thorneley, Lowe 1984b).

A very different mechanism for N_2 reduction has been proposed involving the formation of a two electron reduced diazine (HN=NH) intermediate. This proposal is supported, in part, by studies of alternative substrates, HCN and CH_3NC. These substrates have higher K_ms than N_2 and their six electron reductions appear to occur via enzyme bond intermediates which also do not bind as tightly to N_2ase as the N_2 reduction intermediates. Thus, both HCN and CH_3NC give rise to less highly reduced, two and four electron products which have been identified and would be analogous to diazine and hydrazine level intermediates in N_2 reduction (Li et al., 1982; Rubinson et al., 1983). The diazine intermediate was proposed based on the extensively investigated H_2-inhibition and HD-formation reactions (Burgess, 1985; Jensen, Burris, 1985). The stoichiometry of these reactions is:

$$E + N_2 + 2e^- + 2H^+ + D_2(H_2) \longrightarrow E + N_2 + 2HD(2H_2)$$

N_2 is the only substrate which is sensitive to H_2 inhibition and which gives rise to HD under D_2. Thus, any mechanism which tries to explain these reactions should rely on some property which is unique to N_2. The structure of a chemical intermediate in N_2 reduction would be unique to that substrate which leads to the proposal that HD is formed via the interception by D_2 of a diazine intermediate (Burgess, 1985). The major argument against this proposal has been that H_2 is a competitive inhibitor of N_2 reduction, however, Guth and Burris (1983) have presented a scheme which explains the kinetics and allows D_2 and N_2 to be bound simultaneously.

Another possible explanation for the H_2 inhibition and HD formation reactions has been proposed which does not rely on the interception of a N_2 reduction intermediate by D_2 (Thorneley, Lowe, 1985; Guth, Burris, 1983):

$$E + 2e^- + 2H^+ \longrightarrow E{<}^H_H \xrightarrow{\ N_2\ H_2\ } E-N_2 \xrightarrow{\ D_2\ N_2\ } E{<}^D_D \xrightarrow{\ +2H^+\ } E + 2HD$$

However, this scheme requires a maximum stoichiometry of 2HD formed per H_2 evolved and a stoichiometry of 6HD per H_2 has been determined experimentally (Simpson, Burris personal communication and this volume).

3.4 <u>Where do Substrates Bind</u>? The controversy described above concerning the chemical mechanism of N_2 reduction would probably be resolved if we knew exactly how N_2 binds to N_2ase. For example, end on binding to a single metal atom would support the hydrazido (2-) proposal while side on bound or bridged N_2 would support the diazine proposal. There is now direct evidence that C_2H_2 (Smith, 1983), CH_3NC (Orme-Johnson, 1985), N_2 and CO (Hawkes et al., 1984) bind to the FeMoco center of the MoFe protein and it is likely that all substrates bind to this site. Very little information is available concerning how substrates bind except in the case of C_2H_2 where end on binding is eliminated and side on binding is supported by kinetic studies of C_2H_2 reduction in D_2O (McKenna et al., 1979). Kinetic studies also indicate that all sustrates do not bind identically because HCN and N_2O appear to bind simultaneously (Li et al., 1982) while CH_3NC has both inhibitor and substrate binding modes (Rubinson et al., 1983). The FeMoco center contains Mo, Fe and sulfide and presumable any of these atoms might be involved in substrate binding. There is also some possibility that an amino acid R group could also participitate, for example, by nucleophilic attack on a metal bound CO or CH_3NC. For CH_3NC there is now some evidence for the involvement of both Mo and Fe in binding (Orme-Johnson, 1985) but for the physiological substrate N_2 there is not yet any direct evidence for how binding occurs. Hopefully, structural and spectroscopic data will soon provided this much needed information.

4. REFERENCES

Braaksma A, Haaker H, Grande H and Veeger C (1982) Eur. J. Biochem.
121, 483-91.
Braaksma A, Haaker H and Veeger C (1983) Eur. J. Biochem. 133, 71-76.
Burgess BK (1984) In Veeger C and Newton WE, eds,
Advances in Nitrogen Fixation Research, pp 103-114,
Martinus Nijhoff/Junk, The Hague.
Burgess BK (1985) In Spiro T. ed. Metal Ions in Biology Vol. 7, in press.
Cordewener J, Haaker H and Veeger C (1983) Eur. J. Biochem. 132, 47-54.
Cordewener J, Haaker H, Van Ewijk P and Veeger C (1985) Eur. J. Biochem.
148, 499-508.
Emerich DW and Burris RH (1978) J. Bacteriol 134, 936 -942.
Guth JH and Burris RH (1983) Biochem. J. 213, 741-749.
Haaker H and Veeger C (1977) Eur. J. Biochem. 77, 1-10.
Hageman RV and Burris RH (1978) Proc. Natl. Acad. Sci. USA 75, 2699-2702.
Hallenbeck PC (1983) Arch. Biochem. Biophys. 220, 657-660.
Hawkes TE, McLean PA and Smith BE (1984) Biochem. J. 217, 317-321.
Jensen BB and Burris RH (1985) Biochemistry 24, 1141-1147.
Kluskist J, Haaker H, Wassink H and Veeger C (1985) Eur. J. Biochem.
146, 509-515 (1985).
Li J-G, Burgess BK and Corbin JL (1982) Biochemistry 21, 4393-4402.
Lowe DJ and Thorneley RNF (1984) Biochem. J. 224, 895-901.
McKenna CE, McKenna M-C and Huang CW (1979) Proc. Natl. Acad. Sci. 76,
4773-4777.
Mortenson LE and Thorneley RNF (1979) Ann. Rev. Biochem. 48, 387-418.
Nieva-Gomez D, Roberts GP, Klevickis S, and Brill WJ (1980) Proc. Natl.
Acad. Sci. USA 77, 2555-2558.
Orme-Johnson WH, Davis LC, Henzl MT, Averill BA, Orme-Johnson NR, Munck
E and Zimmerman R (1977) in Newton WE, Postgate JR and Rodriquez-
Barrueco C, eds., Recent Developments in Nitrogen Fixation, pp 131-178,
Academic Press, London.
Orme-Johnson WH (1985) Ann. Rev. Biophys. Biophys. Chem. 14, 419-459.
Rubinson JF, Corbin JL and Burgess BK (1983) Biochemistry 22, 6260-6268.
Rubinson JF, Burgess BK, Corbin JL and Dilworth MJ (1985) Biochemistry
24, 273-283.
Shah VK, Stacey G and Brill WJ (1983) J. Biol. Chem. 258, 12064-12068.
Simpson FB and Burris RH (1984) Science 224, 1095-1097.
Smith BE (1983) In Muller A and Newton WE eds, Nitrogen Fixation: The
Chemical - Biochemical - Genetic Interface pp 23-62, Plenum Press,
New York.
Stephens PJ (1985) In Spiro T, ed. Metal Ions in Biology vol. 7 in press.
Thorneley RNF, Eddy RR and Lowe DJ (1979) Nature 272, 557-558.
Thorneley RNF and Lowe DJ (1983) Biochem. J. 215, 393-403.
Thorneley RNF and Lowe DJ (1984a) Biochem. J. 224, 903-909.
Thorneley RNF, and Lowe DJ (1984b) Biochem. J. 224, 887-894.
Thorneley RNF, Lowe DJ (1985) In Spiro T. ed. Metal Ions in Biology vol.
7 in press.
Upchurch RG and Mortenson LE (1980) J. Bacteriol 143, 274-279.
Watt GD and McDonald JW (1985) Biochemistry in press.

REGULATION OF NITROGENASE ACTIVITY IN THE ANOXYGENIC PHOTOTROPHIC BACTERIA

WALTER G. ZUMFT
LEHRSTUHL FÜR MIKROBIOLOGIE DER UNIVERSITÄT, KAISERSTR. 12
D-7500 KARLSRUHE 1, FEDERAL REPUBLIC OF GERMANY

1. INTRODUCTION

At a time when the problems of processing the nitrogenase components of heterotrophic bacteria in vitro had been mastered and an understanding of their properties was well under way, it remained difficult to consistently achieve cell-free activity of nitrogenase from Rhodospirillum rubrum (Munson and Burris, 1969). This situation lasted until about 10 years ago, despite the fact that the experimentally essential detail of direct coupling of nitrogenase to dithionite had been discovered in both Azotobacter and Rhodospirillum (Bulen et al., 1965). Persistency in studying the nitrogen-fixing system in photosynthetic diazotrophs was amply rewarded, when an activating enzyme (AE) was found, whose incorporation into a cell-free assay overcame the problem of low or nil nitrogenase activity in vitro (Ludden and Burris, 1976; Nordlund et al., 1977).

2. ACTIVATING ENZYME

AE is membrane-bound and can be solubilized by 0.5 M NaCl (Ludden and Burris, 1976). Its synthesis is constitutive and exempt from ammonia control (Triplett et al., 1982). AE acts catalytically on the Fe protein (Guth and Burris, 1983a) and cleaves from it a modifying group (Ludden and Burris, 1979). The M_r, determined by gel filtration, is about 20 000 (Zumft and Nordlund, 1981; Guth and Burris, 1983a; Saari et al., 1984); however, in SDS-PAGE AE migrates as a 33 kDa protein (Saari et al., 1984). AE has a half life in air of ~2 min (Zumft and Nordlund, 1981). A 12 000-fold purification was necessary to obtain about 80 µg homogeneous protein from 1 kg cell paste (Saari et al., 1984). AE requires a M^{2+}-ATP complex for activity, where M^{2+} can be Mn, Fe, or Mg (Gotto and Yoch, 1982; Guth and Burris, 1983a; Nordlund and Norén, 1984). Binding of Mn occurs at a site on AE with an apparent K_d approx. 20 µM (Guth and Burris, 1983a); Fe is about equally effective (Nordlund and Norén, 1984), but the affinity of Mg^{2+} is considerably less; K_d ~20 mM.

3. PROPERTIES OF MoFe AND Fe PROTEINS

The nitrogenase components have been purified and partially characterized from R. rubrum (Ludden and Burris, 1978; Nordlund et al., 1978), Rhodopseudomonas capsulata (Hallenbeck et al., 1982) and Rhodopseudomonas palustris (Zumft et al., 1985). Although these proteins have not been characterized as detailed as their homologous components of heterotrophic bacteria, they exhibit the same basic characteristics and catalytic properties. No structural data are yet available. From an angular dependence of the low-field EPR signal of the MoFe protein in oriented cells of R. rubrum, association with the membrane was suggested (Howard et al., 1985). Both active and inactive forms of the Fe protein were isolated and shown to be immuno- and proteinchemically identical (Ludden et al., 1982a). The two forms were distinguishable by PAGE because modification of the protein lowers the electrophoretic mobility of the modified subunit (Ludden et al., 1982a; Gotto and Yoch, 1982b; Jouanneau et al., 1983). The chemical

552

composition of the two protein forms differs with respect to the non-proteinaceous component (see below).

A general view is the provision of reductant to nitrogenase via a ferredoxin (Fd). Of two soluble ferredoxins from R. rubrum, the 2[4Fe-4S] protein (Fd I) was more effective for nitrogenase in a chloroplast-coupled assay (Yoch and Arnon, 1975). Two ferredoxins were found in R. capsulata, proving there Fd I again more active than Fd II (Yakunin and Gogotov, 1983). It has been claimed that with Fd I no AE is required for nitrogenase activity (Yakunin and Gogotov, 1984). Since the synthesis of Fd I is repressed by ammonia and absence of light, an in situ function as electron carrier to nitrogenase is suggested. Inhibition of nitrogenase activity by the presumed ferredoxin antagonist metronidazole was also shown (Hallenbeck and Vignais, 1981; Kelley and Nicholas, 1981). Of two ferredoxins found in N_2-fixing cells of R. palustris, Fd I was sequenced (Minami et al., 1984). While the R. rubrum Fd closely resembles that of Clostridium pasteurianum, the R. palustris protein is structurally related to ferredoxins from the green bacteria and other purple nonsulfur bacteria. In the light of a specific electron-donating flavoprotein which is part of the commonly regulated nif operons in Klebsiella, it is of interest to see, whether a specific Fd is part of a nif gene cluster in the phototrophic bacteria.

4. ADP-RIBOSYLATION OF THE Fe PROTEIN

The long enigmatic, modifying group of the Fe protein has now been identified as adenosine diphosphoribose (Fig. 1; Pope et al., 1985a). The result has been surprising; once, for consisting in a known mechanism of protein modification, on the other hand, for establishing a prominent physiological example for ADP-ribosylation in prokaryotes.

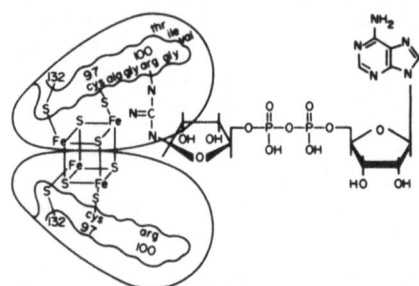

Fig. 1. Model for the ADP-ribosylated Fe protein from R. rubrum.

A recent monograph on the subject of ADP-ribosylation reactions is available (Hayaishi and Ueda, 1982). The reaction is relatively common in eukaryotes. Only a few reports document the reaction in bacteria, e. g., modification of E. coli polymerase (Goff, 1974), Vibrio cholerae NAD-glycohydrolase (Fernandes et al., 1979), and several E. coli proteins of unknown function (Skórko and Kur, 1981). The definitive function within the bacterial physiology is an open problem. A conspicuous case of ADP-ribosylation, however, is the modification of the eukaryotic (Van Ness et al., 1980) and archaebacterial (Kessel and Klink, 1980) elongation factor EF2 and the GTP-binding protein of the adenylate cyclase reaction (Holmgren, 1981) by bacterial toxins. The mode of action of these ADP-ribosyl transferases might serve as model to understand the process of modifying the Fe protein, for which no physical component has yet been detected.

A key observation in the successful isolation and purification of the modifying group was the finding of heat activation of the Fe protein (Dowling et al., 1982). The entire group is being split from the protein upon treatment at 60°C for 3 h and pH 8.5 (Pope et al., 1985b). The removal of the group shifts the two-band pattern of inactive Fe protein towards a one-band pattern observed for the active form (Dowling et al., 1982). The heat-released group was purified on a boronate column and by HPLC-IEC (Pope et al., 1985b). The former procedure was suggested from the inhibition of the activation process of the Fe protein by borate, which resulted most likely from the interaction of borate with the cis-hydroxyl groups of the sugar (Ludden, 1981). The isolated modifying group was shown to be composed of adenine, ribose, and phosphate in a 1:2:1 stoichiometry (Pope et al., 1985b), correcting previous determinations of 1:1:1 (Ludden et al., 1982a). Apparently, the phosphomolybdate assay of Chen et al. (1956) underestimates the amount of bound P_i on the Fe protein (Pope et al., 1985b). Values of one P_i, possibly of two, were also reported from independently obtained preparations (Gotto and Yoch, 1982b).

The linkage of the components of the modifying group was established by fast atom bombardment (FAB) mass spectrometry (MS) and FAB/collisionally-activated decomposition MS (Pope et al., 1985a). Comparison of the heat released, purified modifying group from Fe protein with authentic ADP-ribose gave a virtually identical pattern of breakdown products in MS. The site of attachment of ADP-ribose on the Fe protein was deduced from a hexapeptide carrying the modifying group, which was isolated by boronate and reversed phase HPLC from a subtilisin digest of the Fe protein (Pope et al., 1985a). The bonding amino acid is arginine within the peptide sequence Gly-Arg-Gly-Val-Ile-Tyr. The peptide can be aligned to amino acids 99 to 104 of the known primary structure of Fe proteins from other bacteria. This region is within a highly conserved part of the protein harboring a cystein residue with a presumed function of FeS cluster ligation (Hausinger and Howard, 1983). This region has also been discussed as possible nucleotide-binding domain, based on a structural comparison with other nucleotide-binding proteins (Robson, 1984). The possibility of introducing a heterologous nifH gene into a Fe protein-less mutant of R. rubrum and modifying its gene product in situ has been indicated (Falk and Johansson, 1983).

The ADP-ribosylated Fe protein was investigated with respect to nucleotide-binding and the associated conformational shift as measured by the accessibility of chelators to the [4Fe-4S] cluster (Ludden et al., 1982b; Guth and Burris, 1983b). Both properties are unaffected in the ribosylated protein which still can be reversibly reduced and oxidized. R. rubrum Fe protein has two binding sites for MgATP, the affinity of which is not greatly changed by ADP-ribosylation (Guth and Burris, 1983b). This indicates that the sites of ADP-ribosylation and ATP-binding are not in close proximity. The modified Fe protein cannot transfer electrons rapidly to the MoFe protein, and thus does not support acetylene reduction.

Several observations are consistent with an ADP-ribosyl Fe protein. As of putative precursors [^{32}P]phosphate (Ludden et al., 1982a; Gotto and Yoch, 1982b; Jouanneau et al., 1983), [^{14}C]ATP (Michalski et al., 1983), and [^{3}H]adenine (Nordlund and Ludden, 1983) were incorporated into the

Fe protein. The fact that the Fe protein is rapidly labelled by ATP indicates that the phototrophic cell must have a very active pyridine nucleotide cycle (Foster and Moat, 1980). Within it, nicotinic acid mononucleotide adenylyltransferase (EC 2.7.7.18) and NAD synthetase (EC 6.3.5.1) require ATP for the synthesis of NAD as the presumptive donor molecule for ADP-ribosylation. It is of additional interest to note that NAD synthetase also requires glutamine, a likely, regulatory molecule for the inactivation process.

5. ORGANIZATION OF nif GENES

Existence of this so far unique mechanism of nitrogenase regulation provides the rationale to explore the organization of nif genes in the anoxygenic phototrophic bacteria. Homology between the structural genes of Klebsiella and R. capsulata had been shown (Ruvkun and Ausubel, 1980) and was used to identify these genes in a cosmid library of the latter organism (Avtges et al., 1983). The order of the structural genes nifKDH which reside also in one transcriptional unit is identical to that of Klebsiella. In addition, extra, usually silent copies of the structural genes were found (Scolnik and Haselkorn, 1984).

Insertional mutagenesis with Tn5 was used as a general strategy to localize nif-coding genes (Klipp and Pühler, 1984). Three nif-coding regions were identified (Fig. 2; Klipp and Pühler, 1984; Zumft et al., 1985). A 17 kb ClaI fragment harbors the structural genes for nitrogenase as shown by activity analyses, in vitro protein hybridization assays, immunochemical evidence, and DNA-DNA hybridization with cloned Klebsiella structural genes. Within this fragment, DNA hybridization with nifA and nifH also was found. The main nif-coding region consists of a 30 kb fragment which, in part, hybridizes with genes nifE, nifS, and nifA (Klipp and Pühler, this volume). Biochemical tests of mutants with insertions within a third region indicate a putative regulatory phenotype. Mutants with clear assignments to components of the covalent modification system have not yet been reported. However, a spontaneous R. capsulata mutant, RC5,

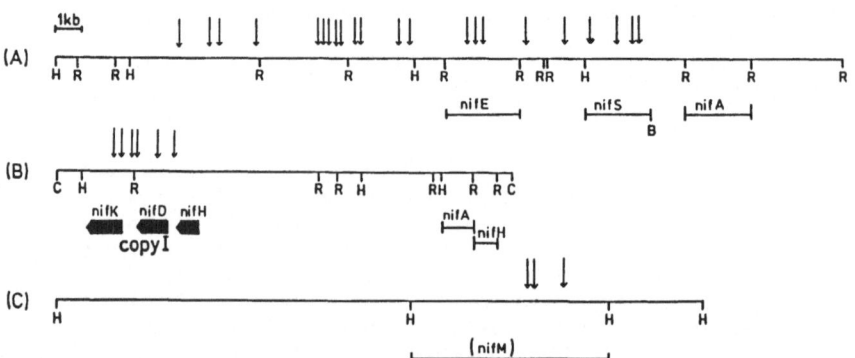

Fig. 2. Localization of nif::Tn5 mutations of R. capsulata within restriction maps of the three regions coding for nif genes (E, EcoRI; H, HindIII; C, ClaI). Arrows denote the sites of Tn5 insertions. In the lower part, regions are indicated where hybridization with nif genes (K, D, H, A, M, E, and S) from Klebsiella pneumoniae was found. Within fragment B, the structural genes, denoted as copy I, are located (W. Klipp, unpublished).

selected on metronidazole, had lost ammonia sensitivity and may be affected in a yet to be defined component of the regulatory system of nitrogenase activity (Willison and Vignais, 1982).

6. NITROGENASE SWITCH-OFF

Cessation of nitrogen uptake on addition of ammonia was shown by mass
spectrometry (Pratt and Frenkel, 1959) and by manometric techniques
(Schick, 1971), constituting the two cases of early experimental evidence
for short-term nitrogenase inhibition. At a time when N_2 fixation and
H_2 evolution were not yet recognized as activities of the same enzyme,
fast inhibition of H_2 production was documented (Gest et al., 1950),
shortly after the discovery of N_2 fixation and its repression by ammonia
in R. rubrum (Kamen and Gest, 1949). This short-term inhibition of
nitrogenase and the modification of the Fe protein, is now viewed as one
mechanism. Occurrence of nitrogenase switch-off (Zumft and Castillo, 1978)
is widely distributed among the Rhodospirillaceae (Neilson and Nordlund,
1975; Hillmer and Gest, 1977; Meyer et al., 1978; Jones and Monty, 1979;
Yoch, 1979; Zumft et al., 1980; Sweet and Burris, 1981; Haaker et al.,
1982; Howard et al., 1983; Kanemoto and Ludden, 1984). The effect has
recently also been shown to occur in Ectothiorhodospira (Bognar et al.,
1982) and in Chromatium (Gotto and Yoch, 1985).

Examples outside the phototrophic bacteria require still more experimental evidence, such as Azospirillum (Ludden et al., 1978) and Azotobacter (Laane et al., 1980). Whether nitrogenase of the latter exhibits a
short-term inhibition has long been a subject of controversy. The dissolved O_2 concentration, the pH of the medium, and the respiratory rate
of the cells were found to influence nitrogenase regulation (Klugkist and
Haaker, 1984). Recently, it was shown for Azotobacter chroococcum that an
ammonia pulse inhibits acetylene reduction. The inhibition persists until
ammonia is assimilated but can be reversed by methionine sulfoximine
(Cejudo et al., 1984). Other inhibitors of nitrogen assimilaton indicate
that a product of ammonia is required to exert nitrogenase switch-off.

General consensus views not ammonia but a product of its assimilation
as trigger in nitrogenase switch-off. Thus, the glutamine pool was proposed as regulator for nitrogenase activity (Jones and Monty, 1979) —
although there is no evidence that nitrogenase would behave allosterically
towards glutamine. Several reports have involved glutamine synthetase (GS)
in nitrogenase regulation to the extend that its role consists in providing glutamine as the effector molecule for Fe protein modification (Arp
and Zumft, 1983; Michalski and Nicholas, 1984; Jouanneau et al., 1984).
This conclusion is also supported by work with a mutant lacking GS (Scolnik et al., 1983); and by mimicking the ammonia effect with methylamine
which is taken up by the ammonia transport system (Alef and Kleiner, 1982)
and converted to γ-glutamyl-methylamide (Yoch et al., 1983). However, the
view of a principal role of glutamine is not shared unanimously (Falk et
al., 1982). The adenylylation state of GS is without effect on nitrogenase
switch-off (Yoch, 1980; Alef et al., 1981; Jouanneau et al., 1984). Similarly, changes in the adenine nucleotide pool are not closely correlated
with activity changes of nitrogenase (Paul and Ludden, 1984).

It is attractive to postulate an ADP-ribosyl transferase for inactivation of the Fe protein. This transferase should be regulated and respond
to glutamine as one of several metabolic signals. The physiology behind

nitrogenase switch-off appears complex as judged from several studies aimed at this aspect. Darkness inactivates the Fe protein of R. rubrum (Kanemoto and Ludden, 1984), but the same organism grown fermentatively on fructose is ammonia-sensitive (Schultz et al., 1985). Ammonia sensitivity depends on the nitrogen nutrional status (Sweet and Burris, 1982; Alef et al., 1981) with nitrogen-starved cells being insensitive. Loss and gain of ammonia sensitivity are not readily reversible (Zumft et al., 1985). Nitrogenase activity is also sensitive to uncouplers, electron acceptors, oxygen, and light intensity (Yoch and Gotto, 1982; Kanemoto and Ludden, 1984). With the establishment of the chemical nature of Fe protein modification and the advancements in the function and organization of nif genes in the phototrophic bacteria, the groundwork has now been laid for a better understanding of the physiology of nitrogenase regulation.

7. REFERENCES

Alef K et al (1981) Arch Microbiol 130, 138-142.
Alef K and Kleiner D (1982) Arch Microbiol 132, 79-81.
Arp DJ and Zumft WG (1983) J Bacteriol 153, 1322-1330.
Avtges P et al (1983) J Bacteriol 156, 251-256.
Bognar A et al (1982) J Bacteriol 152, 706-713
Bulen WA et al (1965) Proc Natl Acad Sci USA 53, 532-539.
Cejudo FJ et al (1984) Biochem Biophys Res Commun 123, 431-437.
Chen PS et al (1956) Anal Chem 28, 1756-1758.
Cordewener J et al (1985) Eur J Biochem 148, 499-508.
Dowling TE et al (1982) J Biol Chem 257, 13987-13992.
Falk G et al (1982) Arch Microbiol 132, 251-253.
Falk G and Johansson BC (1983) FEMS Microbiol Lett 19, 145-149.
Fernandes PB et al (1979) J Biol Chem 254, 9254-9261.
Foster JW and Moat AG (1980) Microbiol Rev 44, 83-105.
Gest H et al (1950) J Biol Chem 182, 153-170.
Goff CG (1974) J Biol Chem 249, 6181-6190.
Gotto JW and Yoch DC (1982a) J Bacteriol 152, 714-721.
Gotto JW and Yoch DC (1982b) J Biol Chem 257, 2868-2873.
Gotto JW and Yoch DC (1985) Arch Microbiol 141, 40-43.
Guth JH and Burris RH (1983a) Biochem J 213, 741-749.
Guth JH and Burris RH (1983b) Biochim Biophys Acta 749, 91-100.
Haaker H et al (1982) Eur J Biochem 127, 639-645.
Hallenbeck PC and Vignais, PM (1981) FEMS Microbiol Lett 12, 15-18.
Hallenbeck PC et al (1982) J Bacteriol 149, 708-717.
Hausinger RP and Howard JB (1983) J Biol Chem 258, 13486-13492.
Hayaishi O and Ueda K, eds, (1982) ADP-Ribosylation Reactions, Academic Press, New York
Hillmer P and Gest H (1977) J Bacteriol 129, 732-739.
Holmgren J (1981) Nature 292, 413-417.
Howard KS et al (1983) J Bacteriol 155, 107-112.
Howard KS et al (1985) Biochim Biophys Acta 812, 575-585.
Jones BL and Monty KJ (1979) J Bacteriol 139, 1007-1013.
Jouanneau Y et al (1983) Biochim Biophys Acta 749, 318-328.
Jouanneau Y et al (1984) Arch Microbiol 139, 326-331.
Kamen MD and Gest H (1949) Science 109, 560.
Kanemoto RH and Ludden PW (1984) J Bacteriol 158, 713-720.
Kelley BC and Nicholas DJD (1981) Arch Microbiol 129, 344-348.
Kessel M and Klink F (1980) Nature 287, 250-251.
Klipp W and Pühler A (1984) In Veeger C and Newton WE, eds, Advances in Nitrogen Fixation Research, p 738, Martinus Nijhoff/Junk, The Hague.

Klugkist J and Haaker H (1984) J Bacteriol 157, 148–151.
Laane C et al (1980) Eur J Biochem 103, 39–46.
Ludden PW (1981) Biochem J 197, 503–505.
Ludden PW et al (1978) Biochem J 173, 1001–1003.
Ludden PW et al (1982a) Biochem J 203, 663–668.
Ludden PW et al (1982b) Biochim Biophys Acta 700, 213–216.
Ludden PW and Burris RH (1976) Science 194, 424–426.
Ludden PW and Burris RH (1978) Biochem J 175, 251–259.
Ludden PW and Burris RH (1979) Proc Natl Acad Sci USA 76, 6201–6205.
Meyer J et al (1978) Biochimie 60, 245–260.
Michalski WP et al (1983) Biochim Biophys Acta 743, 136–148.
Michalski WP and Nicholas DJD (1984) J Gen Microbiol 130, 1069–1077.
Minami Y et al (1984) J Biochem 96, 585–592.
Munson TO and Burris RH (1969) J Bacteriol 97, 1093–1098.
Neilson AH and Nordlund S (1975) J Gen Microbiol 91, 53–62.
Nordlund S et al (1977) Biochim Biophys Acta 462, 187–195.
Nordlund S et al (1978) Biochim Biophys Acta 504, 248–254.
Nordlund S and Ludden PW (1983) Biochem J 209, 881–884.
Nordlund S and Norén A (1984) Biochim Biophys Acta 791, 21–27.
Paul TD and Ludden PW (1984) Biochem J 224, 961–969.
Pope MR et al (1985a) Proc Natl Acad Sci USA 82, 3173–3177.
Pope MR et al (1985b) Biochemistry 24, 2374–2380.
Pratt DC and Frenkel AW (1959) Plant Physiol 34, 333–337.
Robson RL (1984) FEBS Lett 173, 394–398.
Ruvkun GB and Ausubel FM (1980) Proc Natl Acad Sci USA 77, 191–195.
Saari LL et al (1984) J Biol Chem 259, 15502–15508.
Schick HJ (1971) Arch Mikrobiol 75, 89–101.
Schultz JE et al (1985) J Bacteriol 162, 1322–1324.
Scolnik PA et al (1983) J Bacteriol 155, 180–185.
Scolnik PA and Haselkorn R (1984) Nature 307, 289–292.
Skórko R and Kur J (1981) Eur J Biochem 116, 317–322.
Sweet WJ and Burris RH (1981) J Bacteriol 145, 824–831.
Sweet WJ and Burris RH (1982) Biochim Biophys Acta 680, 17–21.
Triplett EW et al (1982) J Bacteriol 152, 786–791.
Van Ness BG et al (1980) J Biol Chem 255, 10717–10720.
Willison JC and Vignais PM (1982) J Gen Microbiol 128, 3001–3010.
Yakunin AF and Gogotov IN (1983) Biochim Biophys Acta 725, 298–308.
Yakunin AF and Gogotov IN (1984) FEMS Microbiol Lett 23, 217–220.
Yoch DC (1979) J Bacteriol 140, 987–995.
Yoch DC (1980) Biochem J 181, 273–276.
Yoch DC et al (1983) Arch Microbiol 134, 45–48.
Yoch DC and Arnon DI (1975) J Bacteriol 121, 743–745.
Yoch DC and Gotto JW (1982) J Bacteriol 151, 800–806.
Zumft WG et al (1980) In Gibson AH and Newton WE, eds, Current Perspectives in Nitrogen Fixation, pp 190–193, Austral Acad Sci, Canberra
Zumft WG et al (1985) In Ovchinnikov YA, ed, Proceedings 16th FEBS Congress, Part A, pp 411–424, VNU Science Press, Utrecht
Zumft WG and Castillo F (1978) Arch Microbiol 117, 53–60.
Zumft WG and Nordlund S (1981) FEBS Lett 127, 79–82.

8. ACKNOWLEDGEMENTS

I wish to thank Dr. W. Klipp for sharing unpublished data with me and Drs. P. M. Vignais and P. W. Ludden for providing preprints. Work cited from this laboratory was supported by the Deutsche Forschungsgemeinschaft.

STRUCTURE AND MECHANISM OF THE Azotobacter vinelandii Fe-PROTEIN

JAMES BRYANT HOWARD, GRETCHEN L. ANDERSON, THOMAS L. DEITS
DEPARTMENT OF BIOCHEMISTRY, UNIVERSITY OF MINNESOTA, 4-225 MILLARD
HALL, 435 DELAWARE ST., MINNEAPOLIS, MN 55455.

INTRODUCTION

Transfer of electrons between the Fe-protein and the MoFe-protein of the nitrogenase complex is a highly specific reaction requiring that a number of criteria be satisfied. For example, the Fe-protein is the unique donor for the reduction of substrates. Low redox potential alone does not appear to be sufficient for electron transfer as other low potential electron carrier proteins are inactive. Futhermore, Fe-proteins having similar amino acid sequences will not necessarily be equally active in heterocomplexes with the MoFe-protein. Because electron transfer occurs in the binary complex of the MoFe-protein and the MgAXP:Fe-protein, the correct orientation of the proteins to align the electron donor and acceptor sites, as encoded in the primary sequences, must be maintained. After each cycle of electron transfer, the protein complex and hydrolyzed nucleotide must dissocate. (Where in the cycle the ATP is hydrolyzed has not been determined.)

The Fe-protein has two unusual properties compared to other Fe:S proteins. First, the Fe-protein contains a single 4Fe:4S center which appears to bridge the identical subunits. Using iodoacetic acid (IAA) to label exposed cysteinyl residues, we have identified the putative ligands for the center as residues 97 and 132 (numbering based upon the A. vinelandii Fe-protein, Av2, sequence)(Hausinger,Howard 1983). In addition, recently, Munck, Orme-Johnson, and coworkers have shown that the center has two interconverable spin states in the reduced (-3) form, viz. spin 1/2 and 3/2 states (see Orme-Johnson, et al., This proceedings.).

Because of the unique role of the Fe-protein and the unusual physiochemical properties of the center, we have begun a systematic comparison of the oxidized and reduced states of Av2. As the Fe-protein undergoes a redox cycle associated with the nitogenase turnover, both protein states will be involved. By comparing the properties of the two states, we hope to learn more about the electron transfer process. In this report, we discuss our recent results on Fe release by nucleotide dependent chelation, modification and inactivation by IAA, and characterization of a 2Fe form of Av2.

RESULTS AND DISCUSSION

Fe Release Properties of Reduced Av2. Elegant studies by Walker and Mortenson (1974) and Ljones and Burris (1978) demonstrated the relationship between ATP binding and protein conformational change in the Fe-protein. Namely, they observed the rapid release of Fe to chelators in the presence of ATP but not in buffer alone or ADP. The fraction of ATP-dependent Fe release has become a measure of the amount of active enzyme. The initial rate of Fe release appears to be ATP concentration dependent and has been used to ascertain the binding

onstant for ATP. We have investigated the Fe release reaction for Av2
over the entire time course using aa-dipyridyl (aadp) as the chelator.
The results for one ATP concentration are shown in Fig 1A. When the
reaction was analyzed as a pseudo-first order process (see Fig 1B),
the plot had significant curvature for the early time points. The
data were successively fit to rate equations for one, two, and

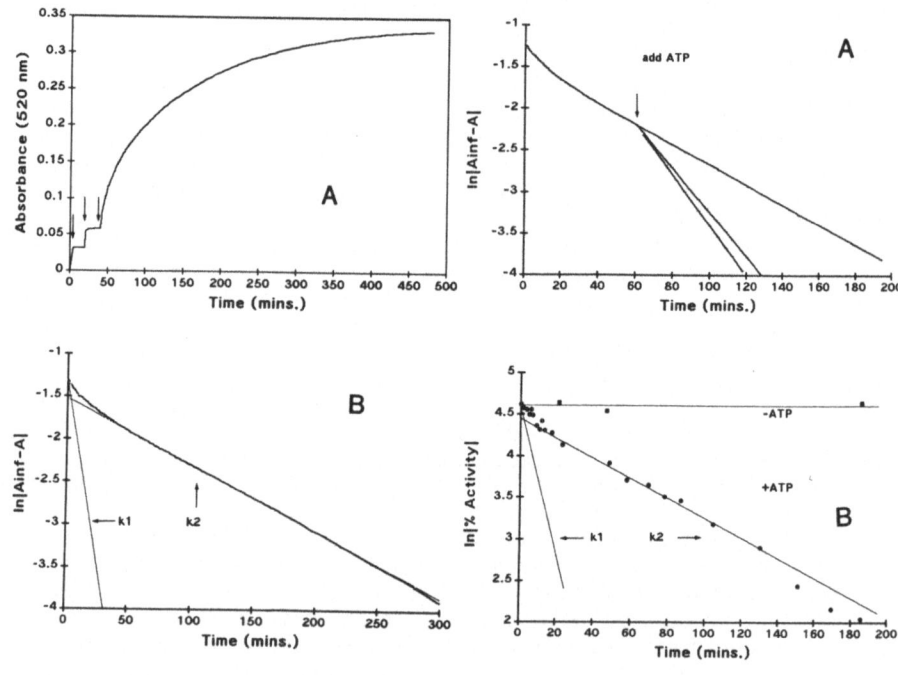

Fig 1. A. Fe release from Av2
at pH 8 with MgATP and aadp.
Order of addition was protein
aadp, and ATP. B. Replot of A.

Fig 2. A. Same as Fig 1B.
ATP at higher concentra-
tions added. B. activity
loss for conditions in
Fig 1.

three exponential terms using a nonlinear least squares analysis.
Equations with one term poorly fit the data while the three term
equation did not improve the fit over a two term equation. Previously,
we reported the preferential fit to the three term equation (Howard,
et al. 1985). However, in the earlier studies, the third phase was
generated by the slow turnover of ATP due to a very low level of
contaminating MoFe-protein (less than 0.1%). Because ADP is an
inhibitor of Fe release, an ATP regenerating system has been included.
From the the pre-exponential constants for the two term equation, the
relative magnitude for the two phases was calculated to be ca. 3 to 1
with the slow phase larger.
    Two simple models as outlined below could be considered to ex-
plain the biphasic Fe release. Both models predict that after the

$$\text{(a)} \quad \text{Av2(4Fe)} \xrightarrow{k_1} \text{Av2(3Fe)} \xrightarrow{k_2} \text{Av2(apo)}$$

(b)
$$\text{Av2(4Fe)} \xrightarrow{k_1} \quad\quad \text{Av2(apo)}$$
$$\text{Av2'(4Fe)} \xrightarrow{k_2}$$

fast phase is complete, there should be only a single component in the slow phase and the rate should respond to the ATP concentration. As shown in Fig 2A, the slow phase remains a single phase even when the rate is increased several fold. The results in Fig 2A also indicate the rate of release saturates with ATP. For the sequential model (a), an intermediate containing 3Fe should be isolated from the slow phase. Samples taken throughout the time course had identical EPR spectra, albeit the intensity decreased with the overall Fe release. This is inconsistent with the quantitative conversion of 4Fe to 3Fe protein in the initial phase. Furthermore, the enzyme activity loss during the Fe release (see Fig 2B) cannot be explained by a sequential model unless the the 3Fe form is equally active as the 4Fe protein. The results to this point seem more consistent with the parallel release of Fe from two forms of the 4Fe Av2, model (b).

Attempts to isolate Av2 from the slow phase provided an unexpected result. The protein was separated from the reaction mixture by gel filtration and the Fe release resumed. As for the original protein, two phases were observed. This suggested that the two forms interconverted during the gel filtration step but were prevented from interconverting during the conditions of the Fe release. The latter can be excluded by the observed biphasic nature of the release. Thus, some component of the reaction mixture must prevent the equilibration of the two forms. The isolation was repeated with either ATP or aadp in the column buffer. The protein isolated in the presence of aadp was biphasic in the subsequent Fe release experiment. In contrast, Av2 isolated in the presence of saturating concentrations of ATP had only a single phase comparable in rate to the slow phase of the mixture from which it was isolated.

To account for these results, model (b) was modified as shown.

$$
\begin{array}{ccccccc}
\text{Av2:ATP} & \leftrightarrow & \text{Av2} & \leftrightarrow & \text{Av2'} & \leftrightarrow & \text{Av2':ATP} \\
\updownarrow & & \updownarrow & & \updownarrow & & \updownarrow \\
\text{Av2:2ATP} & \leftrightarrow & \text{Av2:ATP} & & \text{Av2':ATP} & \leftrightarrow & \text{Av2':2ATP} \\
\downarrow k(1) & & & & & & \downarrow k(2) \\
\text{apoAv2} & & & & & & \text{apoAv2}
\end{array}
$$

For this model, the two forms of Av2 are in equilibrium in the ratio of 3:1. At saturating ATP the two states cannot equilibrate because the only forms which can isomerize, viz. Av2 and Av2', are in vanishingly low concentration. At less than saturating concentration of ATP, the two forms of Av2 that release Fe, viz. Av2 and Av2' with 2 ATP bound, are in low concentration which slows the rate of Fe release. Under these conditions, re-equilibration of the two states can occur. The latter suggests that at low ATP, the Fe release ought to become monophasic. Indeed, as the concentration of ATP is decreased below 250 micromolar, the fraction of the reaction in the first phase becomes less until it is no longer detectable at less than 25 micromolar ATP. This model is consistent with the observations that the Fe release is "all or none"-- that there are no intermediates. The mechanistic implications for the two states of Av2 are not established. How the two Fe release states are related to the the recently observed multiple spin states is under investigation. Finally it is tempting to compare the two states of Av2 to the "tight" and "relaxed" states in other enzymes. However, it should be noted that Av2 is not behaving like a typical allosteric enzyme in that the substrate, ATP, does not appear to be altering the ratio of the states, rather it is "freezing" a pre-existing equilibrium.

Fe Release From Oxidized Av2. In contrast to reduced Av2, oxidized Av2 will release Fe in buffer alone, albeit slower that with ATP. As for the reduced protein, ADP prevented Fe release. The most striking difference is the nature of the Fe release. In the section above we have provided evidence for two states of reduced Av2 which release their Fe by an "all or none" mechanism. For oxidized Av2 we also found biphasic Fe release (Anderson, Howard 1984). However, in oxidized Av2, the two phases are equal in magnitude and appear to follow a sequential mechanism.

The putative 2Fe Av2 was isolated by gel chromatography at the early stages of the second phase. Several different approaches have been used to characterize the nature of the remaining center. Most of these have been described previously (Anderson, Howard 1984). In summary, the Fe center has the properties of a plant-type ferredoxin. The visible spectrum has maxima at 325, 416, and 460 with extinction coefficents identical to putidaredoxin. Furthermore, the new EPR specta had the temperature dependence of 2Fe centers rather than 4Fe center. To substantiate that a single 2Fe center accounted for the total Fe in the protein, we have undertaken Mossbauer studies with E. Munck and coworkers. Av2 was prepared from A. vinelandii grown on Fe-57. The Mossbauer specta of oxidized Av2 had a single quadrupole doublet with an isomer shift of 0.45 mm/sec and splitting of 1.1 mm/sec, both of which are typical for 4Fe:4S centers in the +2 state. The Fe associated with this center accounted for 98+% of the Fe in the sample. The 2Fe Av2 was isolated from the same sample and the Mossbauer spectra measured. The spectra was distinctly different from the oxidized protein and showed a single quadrupole doublet with an isomer shift of 0.26 mm/sec and a splitting of 0.52 mm/sec. The latter are typical for high spin ferrous in a tetrahedral environment and are the values reported for 2Fe feredoxins in the oxidized state. The Fe in the quadrupole doublet accounted for 97+% of that in the sample and indicates that a single Fe:S center is the intermediate in the Fe release from oxidized Av2.

A puzzling feature of the 2Fe center generated from oxidized Av2 has been the quantitative EPR specta. The spin quantitation in the g=2 region of the reduced 4Fe Av2 EPR spectra is usually well less than 0.5 (our preparations are usually 0.2-0.3 spins). The origin of this enigma has been elucidated by Orme-Johnson, Munck and coworkers. Likewise, the newly generated 2Fe protein had substantially less than 1 spin/2Fe when the center was reduced with dithionite. Because the Mossbauer spectra indicated a high yield of a simple 2Fe center, we have reinvestigated the state of the Fe in the reduced 2Fe protein using EPR. The results are shown in Fig 3. 4Fe centers can be characterized, in part, by having an EPR spectra which is so line broadened above 20-25 K that it is not observed. For 2Fe proteins the EPR spectra can be observed to 100 K. Consistent with these general propeties, the native, reduced 4Fe Av2 EPR spectra is observed at 15 K but not at 35 K. We reported previously that the partially reduced 2Fe protein had an EPR spectra unline-broadened at 90 K. However, when the spectra was measured at 20 K and below, an additional signal was observed, a signal similar to the native 4Fe protein. When excess dithionite was added, the EPR spectra at 30 K disappeared. This obser- vation appeared to contradict the usual titration behavior for 2Fe proteins which are EPR active in the reduced state. At lower tempera- tures, an EPR spectra identical to the native 4Fe protein was ob- tained. Spin quantitation was 0.4-0.5 per 4 Fe. The Mossbauer spectra on a sample excess reduced 2Fe protein was consistent with the presence of 4Fe centers. Thus, the 2Fe center appears to be unstable in the reduced form and undergoes a rearrangement to a 4Fe center.

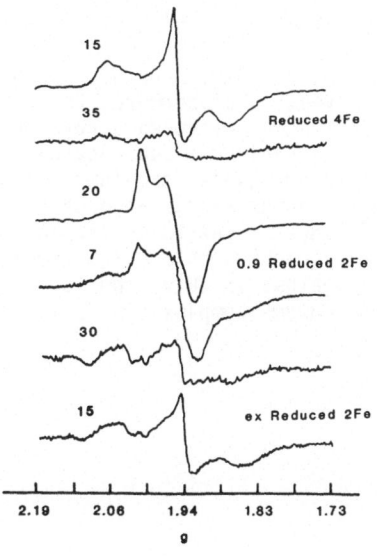

Fig 3. EPR spectra of Av2. Spectra are normalized to 100 micromolar protein. All spectra are at 200 micro- watts except reduced 4Fe Av2 which is at 40 microwatts.

Because the excessively reduced 2Fe protein had an EPR spectrum similar to native 4Fe Av2, we have investigated the reconstitution of 2Fe Av2. Upon incubating the 2Fe protein with 1 millimolar ferrous ammonium sulfate, 2-mercaptoethanol, and sodium sulfide, a protein with 3.5 Fe / protein can be obtained. This material has as high as 30% regain of activity. For unknown reasons the reconstitution procedure does not give reproducible activity regain. When inactive protein is isolated the Fe release with aapd is immediate and ATP independent. Similar results have been obtained in attempts to reconstitute apo-Av2.

Modification and Inactivation of Av2 with Iodoacetic Acid. We have used the reactivity of the Av2 cysteines with IAA as a probe of differences in the protein structure for the reduced and oxidized states. When the reduced protein was incubated with 10 millimolar IAA and 5 millimolar MgATP, the Fe:S center was destroyed as measured by the decrease in absorbance at 410 nm. By 30 min less than 50% of the original center remained. The extent of cysteine modification was 2.2 residues. The activity was more rapidly lost with no activity detected after 20 min. Indeed, the activity loss appeared to parallel the modification of the first cysteine rather than absorbance loss. The specific cysteines preferentially modified by the reaction were residues 97 and 132 which we have postulated are the center ligands (Hausinger, Howard 1983). Thus, reaction with any cysteine ligand is sufficient to inactivate the protein whereas more extensive modification is required to disrupt the center. A similar rapid modification and inactivation occured for the reduced protein with MgADP. This is in striking contrast to the Fe release studies where ADP inhibited the reaction. This suggests that IAA may be participating as a allosteric modulator as well as a chemical modifier.

When oxidized Av2 was incubated with IAA with or without nucleotides, no absorbance change was observed. Furthermore, the activity was unchanged and no carboxymethylation was detected after 30 min. This major difference in reactivity must reflect a substantial difference in the exposure of the cycteinyl ligands or their inherent chemical reactivity when part of a +2 center versus when part of the more reduced +1 center. This difference in IAA inactivation for the two oxidation states may explain the protective effect of sub-stoichiometric amounts of MoFe-protein against inactivation of the Fe protein by IAA during turnover of the enzyme complex.

CONCLUSIONS AND SUMMARY

The chemical reactivity of several states of Av2 have been investigated using Fe chelators and cycteinyl labeling. The relationship of the states are outlined below.

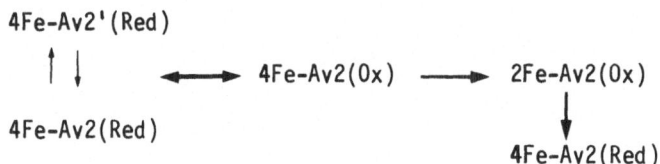

4Fe-Av2'(Red)

4Fe-Av2(Red)   4Fe-Av2(0x)   2Fe-Av2(0x)

4Fe-Av2(Red)

1. Two forms of 4Fe-Av2(Reduced)
   Equilibrium between forms frozen by ATP

2. Reduced and oxidized 4Fe-Av2 have different
   Inactivation and modification by IAA
   Fe chelation properties

3. 2Fe-Av2 can be generated from 4Fe-Av2
   Reduction converts 2Fe-Av2 to inactive 4Fe-Av2

REFERENCES

Anderson GL and Howard JB (1984) Biochem 23, 2118-2122
Hausinger RP and Howard JB (1983) J Biol Chem 258 13486-13492
Howard JB et al (1985) In Ludden P and Burris J ed
 Nitrogen Fixation and Carbon Dioxide Metabolism Elsevier 153-162
Ljones T and Burris R (1978) Biochem 17 1866-1872
Walker G and Mortenson L (1974) Biochem 13 2382-2388

THE ROLE OF Fe PROTEIN IN NITROGENASE CATALYSIS

HUUB HAAKER, JAN CORDEWENER, ANNELOOR TEN ASBROEK, HANS WASSINK,
ROBERT EADY* and CEES VEEGER
Department of Biochemistry, Agricultural University, De Dreijen 11,
6703 BC Wageningen, The Netherlands
*AFRC Unit of Nitrogen Fixation, University of Sussex, Brighton, BN1
9RQ, U.K.

1.   INTRODUCTION

The role of Fe protein in nitrogenase catalysis seems to be understood
(Mortenson, Thorneley, 1979; Hageman, Burris, 1980). Fe protein contains a [4Fe-4S] cluster that functions as a one electron donor.
Reduced Fe protein, MoFe protein and two MgATP rapidly form a ternary
complex. In the complex, ATP hydrolysis is coupled to electron
transfer from Fe protein to MoFe protein (Eady et al., 1978) and it
has been calculated that two MgATP molecules are hydrolyzed for each
electron transferred (Hageman et al., 1980). The rate of electron
transfer is relatively fast compared to the flux of electrons through
nitrogenase during turnover (Thorneley, 1975). After electron
transfer, the slowest step in the cycle of nitrogenase catalysis
occurs: that of dissociation of oxidized Fe protein from MoFe protein
(Thorneley, Lowe, 1983). Dissociation is necessary because it has been
suggested that Fe protein cannot be reduced when bound to MoFe protein
(Hageman, Burris, 1978).
In this paper factors that determine the specific activity of Fe protein will be discussed, and experiments that give new information
about the role of MgATP and Fe protein in nitrogenase catalysis will
be presented.

2.   MATERIALS AND METHODS

Azotobacter vinelandii ATCC 478 was grown on a Burk's nitrogen-free
basic salts medium with 20 g/l sucrose as the sole carbon source in a
Bioengineering fermentor of 20 l or 200 l. Isolation of the proteins
and standard nitrogenase assays were run as described previously
(Braaksma et al., 1983). Av1 used had a specific activity from
1800-2500 nmoles C_2H_4 formed.min^{-1}.mgAv$_1$$^{-1}$. Av1 was oxidized with
solid thionine and Av2 by PMS as described earlier (Braaksma et al.,
1983). Protein blotting was performed as described earlier (Klugkist
et al., 1985). Fe and S^{2-} were estimated as described by Braaksma et
al. (1983). The rapid-mix apparatus was from Update Instruments.
Approximately 200 μl of a concentrated nitrogenase containing mixture
(20-80 μM Av1, 50-200 μM Av2 in 50 mM TES-NaOH, 5 mM MgCl$_2$, 5 mM
Na$_2$S$_2$O$_4$) was rapidly mixed with 200 μl 10 mM MgATP in 50 mM TES-NaOH,
5 mM Na$_2$S$_2$O$_4$, final pH 7.4 and passed into a delay line were the reaction continues. Flavodoxin when used, was added to the nitrogenase
containing solution. The temperature was 22°C. The length of the delay
line determined the time of the reaction. After passing the delay line
the reaction was quenched in 400 μl 10% TCA contained in a stoppered
bottle and stored on ice. The exact volume was determined by weighing

After estimation of the H_2 concentration in the gas phase, the mixture was centrifugated thoroughly to precipitate the protein. Pi was estimated according to Ottelenghi (1975). Fe protein used in the rapid quench experiments had a Fe/S content between 3-4 atoms/molecule. Protein was estimated by the microbiuret method after precipitation of the protein with TCA and deoxycholate.

3.   RESULTS AND DISCUSSION

3.1.   Influence of the growth conditions on Fe/S content and on the specific activity of Fe protein

Braaksma et al. (1983) reported that fully active Fe protein contained up to 8.8 iron and 8.6 sulfide atoms/molecule. The cells used were grown in a batch culture of 2500 l. Since harvesting took 24 h, the growth conditions of the different collected cells were not the same. To investigate the effect of oxygen during growth on the Fe/S content of Fe protein, cells were grown on a smaller scale.
Figure 1 shows how Azotobacter can be grown logarithmically without exposing the cells to high oxygen concentrations. To maintain logarithmic growth the air input rate (aeration and/or stirring speed) must be increased. It is then possible to harvest logarithmically grown cells in large quantities. In these cells nitrogenase has not been exposed to high oxygen concentrations for long periods, which is not the case with cells grown as described in fig. 2.

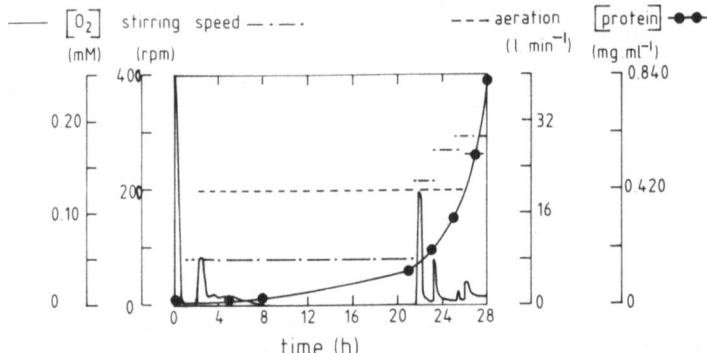

Figure 1. Growth curve of Azotobacter vinelandii in a 20 l fermentor at 30°C. The fermentation was started by inoculating Burk's medium with 500 ml of an end log phase culture. At the time indicated, the stirring speed (-•-) and/or the aeration (---) was increased. The free oxygen concentration was measured continuously (___). Whole cell activity during logarithmic growth (23-28 h) was 80-88 nmoles C_2H_4 formed.min^{-1}.mg protein^{-1}.

Cells growing under oxygen limitation were exposed to a sudden increase in the oxygen input rate. The free oxygen concentration increased and due to inhibition of nitrogenase, growth stopped. Cells responded to the high oxygen concentration by increasing their rate of respiration, and consequently the free oxygen concentration decreased. When cells were removed from the fermentor and incubated at different oxygen input rates, it was found that nitrogenase of the whole cells

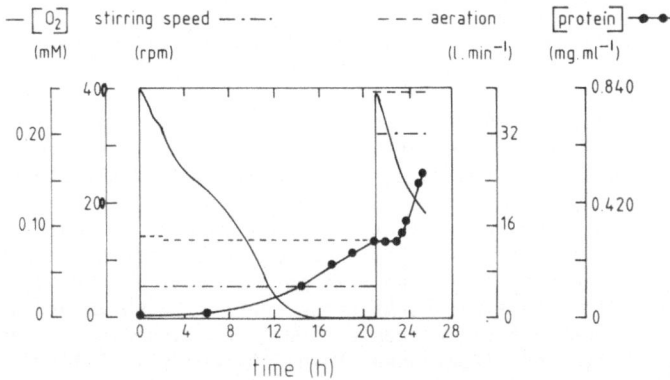

Figure 2. Growth curve of <u>Azotobacter</u> <u>vinelandii</u> in a 20 l fermentor at 30°C. The fermentation was started by inoculating Burk's medium with 200 ml of an endlog phase culture. At the time indicated, the stirring speed (-•-) and the aeration (---) was increased. The free oxygen concentration (___) was measured continuously. Whole cell activity before, at the end of the oxygen shock, and at harvest time was 70, 65 and 100 nmoles C_2H_4 formed.min^{-1}.mg protein^{-1} respectively.

had not been inactivated. Approximately 2 hours after the oxygen input was increased, growth resumed. As long as the oxygen concentration was above 50 μM, growth was linear not logarithmic. During the adaptation period, the respiration rate of the cells increased, as did whole cell nitrogenase activity. This is a general phenomenon, since nitrogenase activity of Azotobacter <u>in</u> <u>vivo</u> is strongly dependent upon the respiration rate of the cells (J. Klugkist, unpublished). In table 1

Table 1. The effect of the growth conditions on the activity of the nitrogenase components and on the Fe/S content of Fe protein. Cells were grown as described in figures 1 and 2.

Growth Conditions	Whole Cell Nitrogenase Activity	In Vivo Catalytic Activity		In Vitro Catalytic Activity		Isolated Av2 Specific Activity	Av2 Fe/S Content
		Av1	Av2	Av1	Av2		
Low free oxygen	70	1200	3000	1700	2100	2650	3.6
Oxygen shock	120	1800	4300	1900	2800	2580	5.1

the properties of nitrogenase from the two different types of cells are compared. Protein blotting was used to determine the amount of the nitrogenase proteins in the cells (Klugkist et al., 1985). The whole cell and crude extract activities were measured. With these data the catalytic activity of both nitrogenase components in whole cells and crude extracts was calculated.

As can be seen in table 1, the increase in whole cell nitrogenase activity of oxygen shocked cells almost correlates with a similar increase in the activity of the nitrogenase components in vivo but not in vitro. It is also clear that the Av2 activity in vivo can be significantly higher than the activity in vitro. Klugkist et al. (1985) reported even higher values for the catalytic activity of Av2 in vivo (6200 nmoles C_2H_4 formed.min^{-1}.mg Fe protein^{-1}). These differences are explained by the fact that chemostat grown cells were used, which have higher oxygen uptake rates than batch grown cells. These high in vivo activities are even more remarkable because of the high intracellular protein concentrations (Av1 = 52-60 µM and Av2 = 74-90 µM). At these protein concentrations, the dissociation model predicts inhibition of nitrogenase (Lowe, Thorneley; 1984). To minimize inhibition due to association of oxidized Fe protein with MoFe protein, it is necessary to effectively decrease the concentration of oxidized Fe protein during catalysis. This must mean that oxidized protein is reduced very efficiently in vivo or that the physiological electron donors can reduce oxidized Fe protein still bound to MoFe protein bypassing the slow dissociation step. Another explanation is offered by Thorneley and Lowe (1984). They suggested that the protein association and dissociation rates are altered in vivo.

The activity of Av1 in cell-free extracts when measured with excess Av2, indicates that the specific activity of Av1 in both types of cells is similar and that the higher catalytic activity of Av1 observed in oxygen shocked cells is probably caused by the higher catalytic activity of Av2. In contrast to Av1, the catalytic activity of Av2 in crude extracts from oxygen-shocked cells is somewhat higher than that of regularly grown cells. However Av2 with the same specific activity can be isolated from both types of cultures. The main difference is the Fe/S content of the protein. There is no indication that the polypeptides are different since their behaviour on a two dimensional gel is the same. We have isolated Fe protein from 10 different batches of oxygen-shocked cells and from 6 batches of regularly grown cells. The properties of Av2 from these isolations are presented in figure 3. From 5 isolations from oxygen-shocked cells, Fe protein was obtained with an Fe/S content significantly above 4. We also observed that from some batches Fe protein readily lost Fe and S^{2-}, the loss of Fe occurring faster than S^{2-}. This loss of Fe and S^{2-} plateaus around 3-4 Fe and S^{-2} atoms per Av2. This was not observed with Fe protein isolated from regularly grown cells. Although no systematic study has been undertaken, there are indications that the intensity of the oxygen shock (a high stirring rate) damages Fe protein in the cell and that during isolation this protein loses Fe, S^{2-} and sometimes activity. Additional evidence that Fe protein can contain more than the 4Fe and S^{2-} atoms per molecule is provided from reconstitution experiments. Fe protein was incubated with $FeSO_4$, Na_2S and DTT. The Fe content of the protein was estimated by the reaction of bathophenantroline disulphonate with Fe in the presence of MgATP. In a typical experiment, Fe protein initially containing 3.6 Fe atoms was reconstituted and separated from the incubation mixture on a molecular sieve and found to contain 8.3 Fe/Av2. From these experiments and those presented in table 1 and figure 3, it is clear that Fe protein can contain more than 4Fe and $4S^{2-}$ atoms per molecule, but there is no linear correlation between Fe/S content and the specific acti-

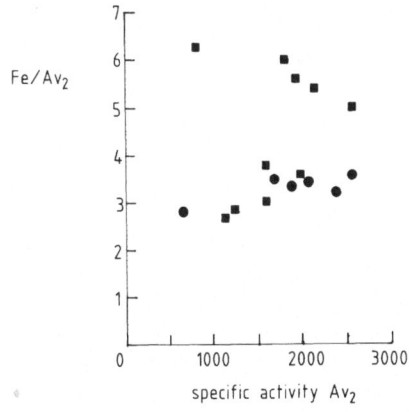

Fe/Av$_2$

specific activity Av$_2$

Figure 3. The effect of the growth conditions on the specific activity and Fe/S content of Fe protein. Cells are grown as described in figures 1 and 2. Isolations, activity measurements and Fe and S^{2-} determinations as described in materials and methods. •-• Protein isolated from regular grown cells (see figure 1). ■-■ Protein isolated from oxygen shocked cells (see figure 2).

vity of Fe protein (Fig. 3). Moessbauer and EPR studies on several samples of different specific activities and Fe/S content do not support the idea that a significant proportion of the Fe atoms present in preparations of low activity and/or in preparations with a high Fe/S content are in an environment different from Fe in active preparations with 3-4 Fe and S^{2-} atoms per molecule. See Dunham et al. (this volume). It must therefore be concluded that the Fe/S cluster in both types of preparation is similar, and that regular Fe protein preparations might contain a significant amount of apo-protein or in case of 2 [4Fe-4S] clusters per Av2 half apo-protein. What the nature is of the Fe/S cluster in Fe protein is not clear. Although the Fe/S cluster looks homogeneous in Moessbauer, it is not homogeneous in EPR. We have recently detected a novel s=3/2 EPR signal in Av$_2$, Ac$_2$ and Kp$_2$. The intensity of the s=3/2 signal decreased by the addition of MgADP or MgATP with no concomittant changes in the intensity of the regular s=½ signal (W. Hagen, unpublished). We think that further work is necessary to clarify the nature of the Fe/S clusters of Fe protein.

3.2. <u>Electron transfer and MgATP hydrolysis by nitrogenase</u>.

Now it has been shown that there is no linear relation between specific activity and Fe/S content of Fe protein, the precise role of Fe protein in the general accepted dissociation model is not so clear anymore. To recapitulate, the only role of Fe protein is donation of electrons to MoFe protein in a reaction that is coupled to MgATP hydrolysis (Hageman, et al., 1980). The rapid burst of ATP hydrolysis in a pre-steady-state reaction is occurring on the same time scale as electron transfer from Fe protein to MoFe protein (Eady et al., 1978). This reaction can only be studied with rapid-quench techniques. We have studied the ATPase activity of nitrogenase by rapid mixing a concentrated nitrogenase solution with MgATP followed by an acid(chemical) quench of the reaction mixture. The P$_i$ and H$_2$ concentration of the quenched reaction mixture were analyzed. An example of such an experiment is shown in figure 4.

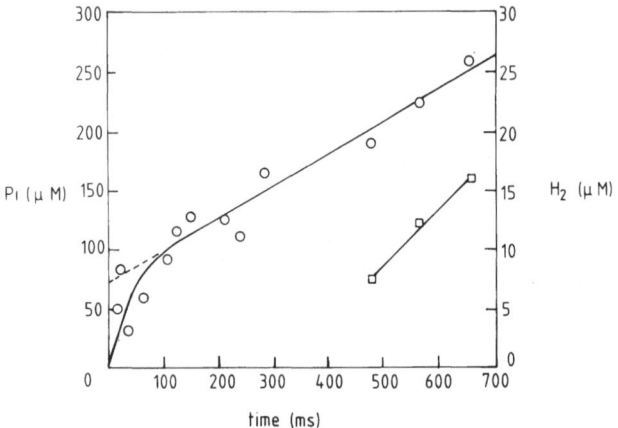

Figure 4. Time course of P_i and H_2 production by nitrogenase. Rapid-quench was performed as described in Materials and Methods. Nitrogenase concentrations after mixing were: Av_1 10 μM, Av_2 100 μM; temperature was 22°C. Specific activities of proteins used under standard conditions were: Av_1 1800, Av_2 1200. o-o P_i; □-□ H_2.

An initial burst of MgATP hydrolysis is followed by a slower linear rate of MgATP hydrolysis. Under the conditions used the reaction is too fast to determine the rate content. According to the literature, the burst of MgATP hydrolysis is associated with a fast electron transfer from Fe protein to MoFe protein (Eady et al., 1978; Hageman et al., 1980). The slow rate of MgATP hydrolysis is coupled to the steady-state rate of electron transfer through nitrogenase. Besides P_i production, H_2 production was also measured at longer reacton times. Beyond 400 ms linear rates of H_2 production were observed. In this experiment the Av_1 concentration was 10 μM and Av_2 was in excess (100 μM). The specific activity of Av_1 used in this experiment and measured under standard conditions at 30°C was 1800 nmoles C_2H_4 formed. $min^{-1}.mgAv_1^{-1}$. An activity of 932 nmoles C_2H_4 formed.$min^{-1}.mgAv_1^{-1}$ was measured at 22°C. The activity calculated from the experiment shown in figure 4 is 1164 nmoles H_2 formed. $min^{-1}.mgAv_1^{-1}$. This indicates that under these conditions the flow of electrons through nitrogenase is not inhibited. The steady-state rate of MgATP hydrolysis was higher than that measured at low protein concentrations, resulting in an ATP/2e ratio of 6.4 which is high for dilute proteins at this temperature (ATP/2e at 22°C = 4). The nitrogenase and ATPase activity were also measured at other (high) protein concentrations. At an Av_1 concentration of 25 μM and varying concentrations of Av_2 from 20-170 μM, the flux of electrons is inhibited from 70-50% respectivley but the ATPase activity is less inhibited (35-25% respectively). This results in ATP/2e values between 7 and 9. These experiments also indicate that ATP hydrolysis and electron transfer are not tightly coupled reactions and that the ATP/2e ratio in vivo might be significantly above 4. To test the hypothesis that the high in vivo activities are due to the physiological electron donor, flavodoxin, experiments were performed with flavodoxin as electron donor. In figure 5 an experiment is shown with flavodoxin and dithionite as electron donors. The time course of P_i production is similar to that in the presence of dithionite alone. Flavodoxin concentrations up to 670 µM were tested but no effects of flavodoxin on the burst or on the steady-state rate of P_i production up to 660 ms could be detected.

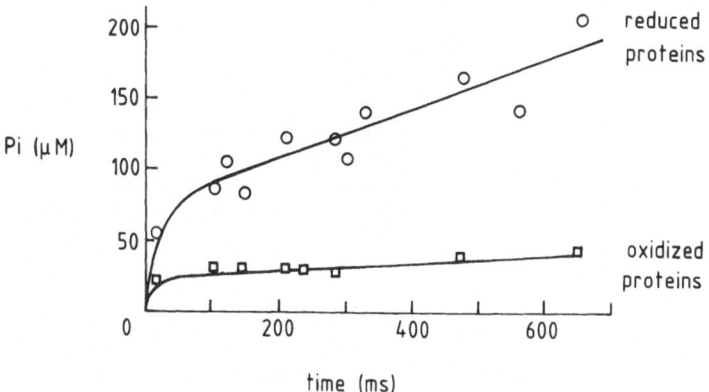

Figure 5. Time course of P_i production by nitrogenase. Rapid-quench was performed as described in Materials and Methods. Nitrogenase concentrations after mixing were: o-o Av_1 15 µM, Av_2 46 µM, flavodoxin 68 µM; temperature was 22°C. Specific activities of proteins used under standard conditions were: Av_1 1800, Av_2 1400. ▫-▫ experiment without $Na_2S_2O_4$ in the buffers and dye-oxidized proteins. Av_1 8 µM, Av_2 22 µM. Specific activities of proteins used under standard conditions were: Av_1 1800, Av_2 1200.

Besides the ATPase activity and the H_2 production, the stoichiometry of the burst of MgATP hydrolysis with protein was calculated by extrapolating the steady-state rate of MgATP hydrolysis back to t=0. There are only two reports in the literature concerning the stoichiometry of the burst reaction. Eady et al. (1978) reported a ratio of 2.3 mol P_i/mol Kp_1 at a ratio of Kp_2/Kp_1=5. Hageman et al. (1980) gave a value of 3.2 mol P_i/Av_1 at Av_2/Av_1 = 2.5. These values were based upon total protein concentrations and not corrected for the presence of inactive protein. Hageman et al. (1980) analysed their data further by comparison with the active site concentration, arguing that only then a meaningful relationship of the stoichiometry is obtained. Mechanistic interpretations of these experiments are based on the idea that the burst of MgATP hydrolysis is tightly coupled to electron transfer from Fe protein to MoFe protein. The other experiment shown in figure 5 demonstrates that this is not the case, since when dye-oxidized proteins were used, the burst of MgATP hydrolysis was still present. The steady-state rate of MgATP hydrolysis was much lower (842 nmoles P_i.min^{-1}.mgAv$_1$$^{-1}$) as compared with the rate of MgATP hydrolysis in the presence of reductant (3191 nmoles P_i.min^{-1}.mgAv$_1$$^{-1}$) but was higher than the reductant independent ATPase activity at low protein concentrations (488 nmoles P_i.min^{-1}.mgAv$_1$$^{-1}$). The proteins separately did not have a significant ATPase activity. (<1 nmol P_i.min^{-1}.mg

protein^{-1}). There might be a similarity between the two iron containing acid phosphatase, which activity is dependent upon its Fe/S
cluster. (Chichitek et al., 1985).
To gain more insight to the ATPase reaction the effect of different
ratios of Fe protein/MoFe protein on the burst of MgATP hydrolysis was
studied. The results are presented in figure 6.

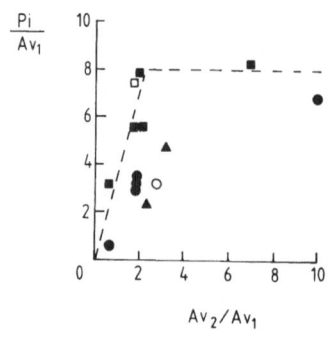

Figure 6. P_i production in the
burst. Rapid-quench experiments were
performed as described in figure 5.
The burst was determined as
described in the text. Nitrogenase
concentrations after mixing were:
Av_1 10-40 μM, Av_2 27-100 μM and flavodoxin 68-670 μM. Specific activity
of proteins used were measured under
standard conditions as described in
Materials and Methods. Av_1
1800-2400. ■ Av_2 2100-2300, ▲ Av_2
1700, ● Av_2 1200 ; □ and ○ dye-
oxidized proteins.

Two important conclusions can be drawn from these experiments. Firstly
when Av_2 of specific activity >2000 was used the amount of MgATP
hydrolyzed in the burst increases with increasing ratio of Av_2/Av_1 up
to a ratio of 2-3. Above this ratio additional Av_2 has no effect on
the amount of P_i produced in the burst. The maximum amount of MgATP
hydrolyzed in the burst phase is 8 per Av_1. Secondly, there is a relation between the specific activity of Av_2 used and the amount of P_i
produced in the burst. When Fe protein with a specific activity above
2000 was used at the same ratio of Av_2/Av_1 a higher burst was observed
compared with protein of lower activity while the effect on the
steady-state rate of P_i production was minimal. With high or low
activity protein, there was no difference in the magnitude of the
burst between oxidized or reduced protein. This shows that the burst
of MgATP hydrolysis is independent of electron transfer between Fe
protein and MoFe protein and is not a consequence of futile cycling of
electrons between the two proteins. Since it is known that there is
only electron transfer in the presence of MgATP (Thorneley, 1975) and
that this occurs with the same rate constant as P_i production in the
burst (Eady et al., 1978), this must mean that MgATP hydrolysis must
preceed electron transfer.

3.3. Model for the mechanism of action of Fe protein in nitrogenase catalysis

We have shown that the growth conditions of Azotobacter can determine
the Fe/S content of Fe protein, that the specific activity of isolated
Fe protein is not determined by the amount of Fe/S clusters present in
the protein but by its capacity to hydrolyse MgATP in a complex with
MoFe protein. MgATP hydrolysis is independent of electron transfer and
is stoichiometric with MoFe protein. We explain the burst of MgATP
hydrolysis as being the formation of an "activated" (possibly
phosphorylated intermediate) complex and the rate of relaxation of

this "activated" complex determines the ATPase activity of nitrogenase (see scheme 1). The presence of electrons on the "activated" complex stimulates the rate of relaxation of this complex. Since Av_2 has only 2 MgATP binding sites (Cordewener et al., 1985) this means that in a $Av_1/Av_2 = \frac{1}{2}$ complex each MgATP binding site hydrolyzes 2 MgATP molecules, and this reaction is independent of the presence of electrons. At the moment it is not known if covalent intermediates of ADP or phosphate with the protein are formed. One can think of a phosphorylation of FeMoCo or of the P-clusters in MoFe protein. When the clusters are reduced, dephosphorylation is faster and this might lower the redox potential of the cluster(s) and make substrate reduction possible.

SCHEME 1

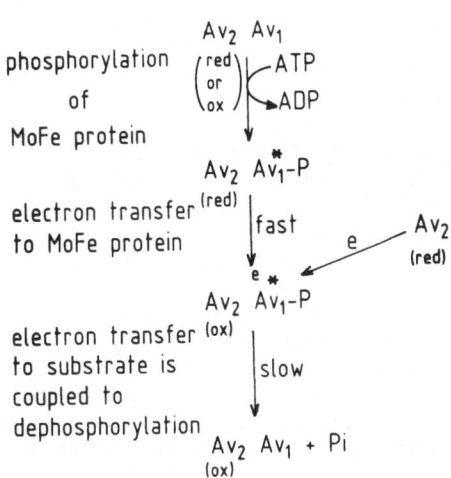

phosphorylation of MoFe protein

electron transfer to MoFe protein

electron transfer to substrate is coupled to dephosphorylation

Are the results of our experiments conflicting with the literature? We think not and are convinced that several experiments can be explained better now. To understand the role of Fe protein in nitrogenase catalysis two situations must be distinguished. 1) The situation where a significant proportion of Fe protein is not bound to MoFe protein. This occurs at low protein concentrations (MoFe protein=Fe protein = 1 μM) or MoFe protein is 1 μM and Fe protein is in excess. 2) The other situation is that where almost all Fe protein is bound to MoFe protein. This is at protein concentrations above 15 μM and at a ratio of Fe protein/MoFe protein <3 or at all protein concentrations with excess MoFe protein.

Under the conditions of situation 1 the specific activity of Fe protein is normally measured. We postulate that the ATPase properties of Fe protein (the formation of an "activated" complex) are independent of (a) functional Fe/S cluster(s), since electron transfer is not necessary for ATP hydrolysis (see figure 4 and 5). Electron transfer to this "activated" complex is fast and can arise from Fe protein already bound to MoFe protein or by free Fe protein with a functional Fe/S cluster. The concentration of the "activated" complex determines the rate of catalysis. When Fe protein with a Fe/S cluster binds but the protein cannot "activate" the complex by MgATP hydrolysis, no electron transfer occurs and this lowers the specific activity of Fe protein. Dissociation of Fe protein from MoFe protein can still be the rate limiting step.

Under the conditions of situation 2, two properties of Fe protein are important. The ATPase activity in the complex and the presence of a Fe/S cluster because all Fe protein is bound to MoFe protein or all binding sites on MoFe protein for Fe protein are occupied. When Fe protein has activated the complex but it has no electron to transfer, the "activated" complex will relax without transferring an electron to substrates, and when a Fe protein is bound that has a Fe/S cluster but it cannot activate the complex by MgATP⁻hydrolysis no electron transfer can occur. This explains why protein with a good specific activity (and a Fe/S content of 4) is inhibited at high protein concentrations. Our hypothesis predicts that only protein with a high specific activity and a high Fe/S content will not be inhibited at high protein concentrations. Our experiments have also shown that a physiological electron donor for nitrogenase, flavodoxin, does not stimulate the burst of MgATP hydrolysis or the flux of electrons through nitrogenase at higher protein concentrations. Since nitrogenase activity in vivo is not inhibited despite the high protein concentrations, it might be possible that in vivo there is a Fe/S cluster synthesizing system operating (extra activated by dioxygen) that continuously adds Fe/S to Fe protein to maintain 2[4Fe-4S] clusters in Fe protein. In vitro experiments with this type of protein need to be done to understand how Fe protein can function in vivo.

4. REFERENCES

Braaksma A, Haaker H and Veeger C. (1983) Eur.J.Biochem. 133, 71-76.
Cichutek K, Witzel H and Parak F (1985) Rev.Port.Quim. 27, 157.
Cordewener J, Haaker H, Van Ewijk P and Veeger C (1985) Eur.J.Biochem. 148, 499-508.
Eady RR, Lowe DJ and Thorneley RNF (1978) FEBS Lett. 95, 211-213.
Hageman RV and Burris RH (1978) Proc.Natl.Acad.Sci.USA 75, 2699-2702.
Hageman RV and Burris RH (1980) Curr.Top.Bioeng. 10, 279-291.
Hageman RV, Orme-Johnson and Burris RH (1980) Biochemistry 19, 2333-2342.
Klugkist J, Haaker H, Wassink, H and Veeger C (1985) Eur.J.Biochem. 146, 509-515.
Lowe DJ and Thorneley RNF (1984) Biochem.J. 244, 877-886.
Mortenson LE and Thorneley RNF (1979) Annu.Rev.Biochem. 48, 387-418.
Ottelenghi P (1975) Biochem.J. 151, 61-66.
Thorneley, RNF (1975) Biochem.J. 145, 391-396.
Thorneley, RNF and Lowe DJ (1983) Biochem.J. 215, 393-403.
Thorneley, RNF and Lowe DJ (1984) Biochem.J. 244, 903-909.

5. ACKNOWLEDGEMENTS

We thank Mrs. J.C. Toppenberg-Fang for typing the manuscript and Mr. M.M. Bouwmans for drawing the figures. This investigation was supported by the Netherlands Foundation for Chemical Research (SON) with financial aid from the Netherlands Organization for Advancement of Pure Research (ZWO).

STUDIES ON THE STRUCTURE AND FUNCTION OF THE MOFE PROTEIN OF NITROGENASE

L. E. MORTENSON, M. WEBB*, R. BARE, S. P. CRAMER AND T. V. MORGAN
EXXON RESEARCH AND ENGINEERING COMPANY,
ANNANDALE, NEW JERSEY 08801   U.S.A.

*NATIONAL INSTITUTE FOR MEDICAL RESEARCH
MILL HILL, LONDON

1.   INTRODUCTION

Analysis of the iron molybdenum center of nitrogenase, called FeMo-co, shows it to contain Mo, Fe and sulfur. EXAFS analysis has shown that this Mo center has ligands of Fe and S and a structure with four or five sulfur ligands and two or three iron atoms was deduced in which the iron atoms were at a somewhat longer distance than the sulfurs (S.P. Cramer, et al., 1978). Two models were proposed, one where Mo substituted for an iron in a 4Fe-4S cluster and another more linear structure. Subsequently, several other models were proposed to account for the measured ratio of Mo/Fe/S in the isolated center (S.P. Cramer et al., 1985). Although crystals of the MoFe protein are presently being analyzed for structure and results from their study will rule out most, if not all, of the proposed structures for FeMo-co, it was felt that examination of such crystals by polarized X-ray absorption spectroscopy would give us some immediate information on what types of clusters are possible. These results will be summarized in this paper.

Another problem that has remained unanswered for some time concerns what role ATP plays in the functioning of nitrogenase. Although it has been established (L.E. Mortenson, R.N.F. Thorneley, 1979) that electron transfer from the Fe protein to the MoFe protein only occurs with concommitant ATP hydrolysis, the type of ATP hydrolysis and the role it plays have not been shown. Recently, Knowles (J.R. Knowles, 1980) has developed techniques to study the stereochemistry of phosphoryl group transfer and results showed that when a phosphoryl group is transferred enzymatically, each step involves an inversion of configuration.   For this technique to determine the stereochemistry, each of the four ligands of the inorganic phosphate released on hydrolysis must be different. (Fig. 1)

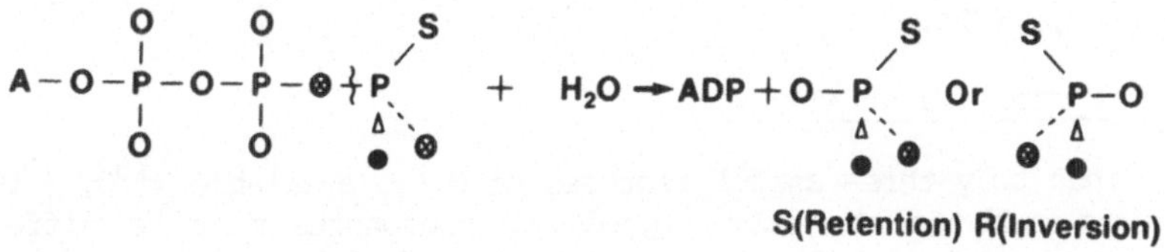

S(Retention) R(Inversion)

FIGURE 1. Stereochemistry of ATP hydrolysis. O, $\mathbb{N}$ and $\bullet$ represent ^{16}O, ^{17}O and ^{18}O respectively.

If there is a single direct transfer between substrates, the overall reaction shows inversion, but if a phosphorylation followed by hydrolysis occurs, then retention of configuration results (two inversions). In this talk, we wish to present the results of the use of this technique to determine the stereochemistry of ATP hydrolysis by nitrogenase.

2. PROCEDURES

2.1 EXAFS Studies

2.1.1 Sample Preparation.

Nitrogenase crystals were prepared by our previously described method (M.S. Weininger, L.E. Mortenson, 1982). The largest crystals were selected for easier and more reliable data collection. They were mounted in capillary tubes in an anaerobic chamber and were oriented by the use of precession photographs so that they could be systematically aligned with respect to the X-ray beam.

2.1.2 Data collection.

Spectra were recorded at the Stanford Synchrotron Radiation Laboratory using a wiggler beam line for greater flux. Both fluorescence and transmission detection were used and, in all cases, a molybdenum foil was used for an internal energy calibration. The details of this experimentation are presented elsewhere (S.P. Cramer et al., 1985).

2.1.3 Data Analysis.

For calibration, processing and fitting of the data previously described programs and functions were used (S.P. Cramer et al., 1985).

2.2 Stereochemical Studies of ATP Hydrolysis

2.2.1 Preparations Used.

The growing of cells, preparation of crude extracts and purification of the Fe and MoFe proteins of nitrogenase are described elsewhere (L.E. Mortenson, 1972; M.R. Webb et al., 1986). Briefly, frozen cells were lysed and, after removing insoluble components by centrifugation, the Fe and MoFe proteins were separated anaerobically by a series of columns of DE-52 and DEAE-sephacel using linear gradients. The specific activities of the Fe and MoFe proteins used were about 2000 each when measured by the standard assay (L.E. Mortenson, 1972).

2.2.2 Hydrolysis of ATPγS.

Since only three useful isotopes of O are available (^{16}O, ^{17}O and ^{18}O) and each of the four ligands of phosphorus must be different, sulfur was substituted for oxygen in the fourth position. To check that the resulting ATPγS was hydrolyzed, an obvious requirement, ATPγS was mixed with the Fe and MoFe proteins as in an acetylene reduction assay (M.R. Webb et al., 1986). The hydrolysis of ATPγS was monitored by HPLC at 45 and 90 min as described in the above reference.

2.2.3 Preparation of (γ-S [$\beta\gamma$-^{17}O; γ-^{17}O; ^{18}O] ATPγS (O-labelled ATPγS).

This reagent was synthesized by the method of Webb (M.R. Webb, 1982) using [^{17}O] and [^{18}O] water.

2.2.4 Reaction Mixtures.

The ^{17}O, ^{18}O-labelled ATPγS was incubated at pH 7.5 with MgCl$_2$, adenylate kinase (to remove ADP), sodium dithionite, Fe protein and MoFe protein in the presence of acetylene/argon (20/80, V/V). The resulting thiophosphate released after a 90% hydrolysis requiring 30-45 min., was incorporated into the β position of ATP by a series of enzymatic steps (M.R. Webb et al., 1986). The resulting ATPβS was examined for its isotope distribution (^{18}O) by ^{31}P NMR.

Potential exchange of [$\gamma^{18}O_3$]ATP with H$_2$O (M.R. Webb et al., 1986) was performed using a similar reaction mixture as above. Samples were removed periodically and the inorganic phosphate derivatized as triethylphosphate, was examined for ^{18}O content by mass spectrometry.

3. RESULTS AND DISCUSSION

3.1 EXAFS Experiments.

For background information and to test the accuracy of the single crystal EXAFS experiments, model compounds were examined first (S.P. Cramer et al., 1985). Figure 2 shows two orientations of (Ph$_4$P)$_2$[Cl$_2$FeS$_2$MoS$_2$FeCl)$_2$].

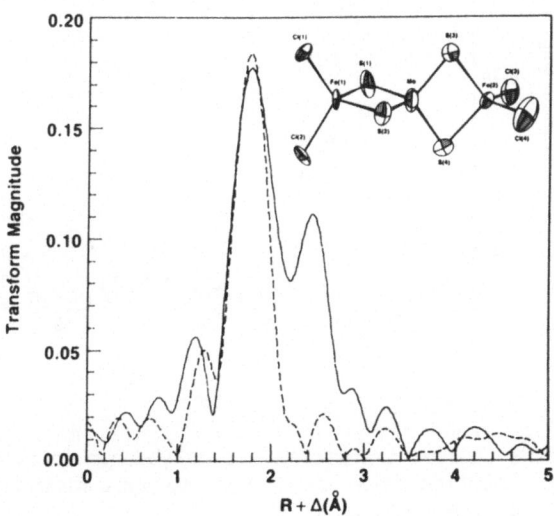

FIGURE 2. EXAFS fourier transforms for molybdenum in (Ph$_4$P)$_2$[Cl$_2$FeS$_2$MoS$_2$FeCl$_2$].

580

The Mo-Fe component of this crystal can be strongly enhanced or almost nullified depending on the orientation of the crystal. With another crystal $(Et_4N)_3[Fe_6Mo_2S_8(SEt)_9]$ the Mo-Fe component was more intense when the polarization was along its molecular 3-fold axis, than when perpendicular to it. The curve-fitting analysis showed a factor of 3.2 for the change in Mo-Fe intensity where a ratio of 4.0 would be expected for a perfect tetrahedron under optimal polarizing and alignment conditions.

For the MoFe protein of nitrogenase there are 8-32 possible MoFe vectors to define because of symmetry of the iron neighbors to Mo and the relative orientation of the four molybdenum sites in the unit cell. Because certain types of symmetry can be assumed and because the crystals have both a crystallographic and a molecular two-fold axis, the data analysis is simplified by assuming a particular cluster symmetry and postulating the orientation of one cluster (S.P. Cramer et al., 1985). When assumptions about the cluster 2-fold or 3-fold axes were made and rotations about the a and c*-axes were examined, the results as expected, clearly showed that EXAFS orientation dependence was sensitive to both symmetry and the orientation of the clusters assumed to be present.

The EXAFS Fourier transforms for nitrogenase in solution was compared with the most extreme single crystal spectra and it is easily seen that whereas the Mo-S feature is relatively invariant,

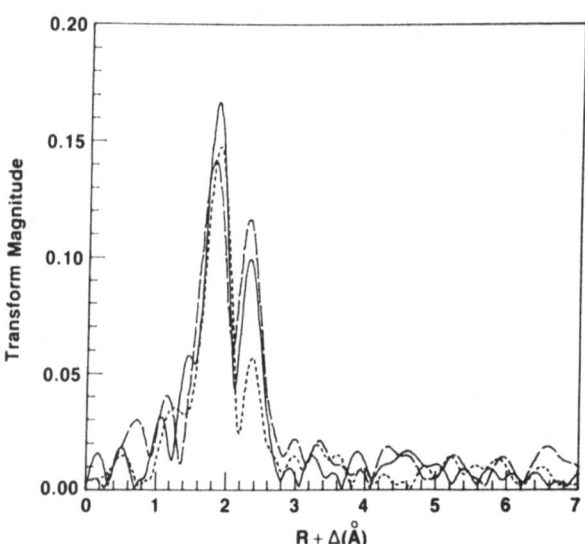

Figure 3. EXAFS fourier transforms for molybdenum in nitrogenase. (-) represents solution data, (-.-.-) represents maximum amplitude and (----) represents minimum amplitude.

there are significant changes in the intensity of the Mo-Fe component (Fig. 3). Curve fitting analysis shows that the Mo-Fe amplitude

changes by a factor of 2.5 from one orientation to another but that
the Mo–S amplitude varied only by ±15%.

From this data, and with knowledge of the limitations of the
methodology used, one can draw some tentative and reasonable
conclusions. The first is that the orientation of the iron neighbors
around the molybdenum is clearly not cubic nor is the molybdenum at
the center of a tetrahedron of iron atoms. This is because such
models would yield isotropic Mo–Fe EXAFS. Secondly, the orientations
of the Mo–Fe cluster axis lie close to the bc* plane. Finally, a nonlinear Mo–Fe cluster geometry better fits the observed anisotropy and
a linear Mo–Fe model can be ruled out.

3.2 Stereochemical Analysis of ATP Hydrolysis. ATPγS is hydrolyzed
by nitrogenase (Fe + MoFe proteins) at a rate 1% of that of ATP
whereas its Km is close to that of ATP. Under these conditions of
slow ATPγS hydrolysis, no ethylene or H_2 was detected in the
atmosphere above the reaction nor did the reaction require the high
concentrations of dithionite needed for the optimal reaction with
ATP. This suggests that under conditions of greatly decreased
nitrogenase turnover, electrons from the MoFe protein are transferred
back to the Fe protein (the so-called futile cycle) similar to
"reductant independent" ATP'ase. The thiophosphate released during
hydrolysis of O labelled ATPγS and incorporated without stereochemical
change into the β position of ATP can be analyzed for its ^{18}O position
by ^{31}P NMR. However, during incorporation into ATP, one of the
oxygens of the thiophosphate is lost, i.e. one third of the molecules
lose ^{16}O, one third ^{17}O and one third ^{18}O. One does not see ^{17}O in
molecules that retain ^{17}O in the β phosphorous since the NMR signal is
very broad as a result of the quadrupole moment of the ^{17}O. Thus only
the ^{18}O that is bridging or non-bridging will effect the NMR signal
(Fig. 4), the ^{31}P NMR is sharp and the chemical shift depends on the
location of the ^{18}O (M.R. Webb, D.R. Trentham, 1981).

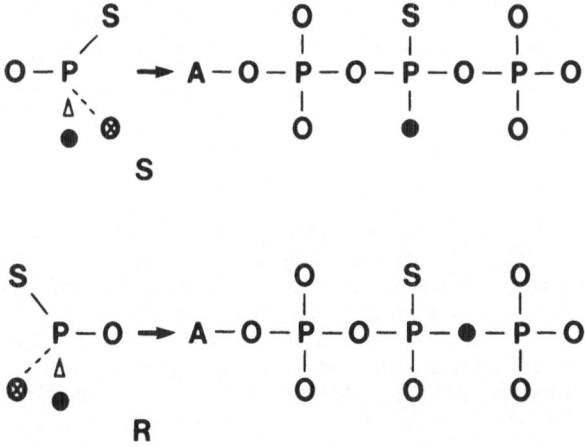

Figure 4. Species of ^{18}O labelled ATP detected by ^{31}P NMR from the
two stereoisomers of inorganic phosphate released during ATP
hydrolysis and in-corporated into the β position of ATP.

The results of the hydrolysis of (^{18}O) labelled ATPγS showed from the configuration of the thiophosphate released that the ^{18}O label (after correction for impure oxygen isotopes in the original O enriched ATPγS) was in the bridging ^{18}O (Fig. 5). The results, therefore, strongly indicate that ATP hydrolysis by nitrogenase involves a single step phosphotransfer to water yielding ADP and inorganic phosphate. Furthermore, it was found that ^{18}O of [γ-^{18}O] ATP did not exchange with water oxygen. Since ^{18}O of [γ-^{18}O] ATP does exchange with water oxygen in myosin ATPase where the conversion of the bound ADP·Pi complex to ADP + Pi is slow, one explanation for the lack of exchange is that the ADP·Pi complex is converted to ADP + Pi at a rate an order of magnitude faster than the reformation of ATP from ADP·Pi and H$_2$O. The kinetics established (D.J. Lowe, R.N.F. Thorneley, 1985) showed that the cleavage step was 200 S^{-1} and the breakdown of the ADP·Pi complex was 6.4 S^{-1} and rate limiting. To not have ^{18}O of [γ-^{18}O] ATP exchange with water oxygen, the reverse of the hydrolysis step would need to occur at less than 0.3 S^{-1}, just within

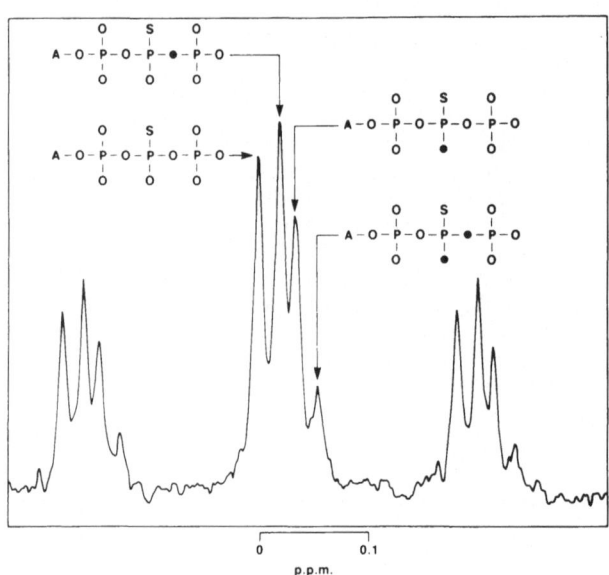

FIGURE 5. ^{31}P NMR of ^{18}O labeled thiophosphate released during hydrolysis of ^{17}O, ^{18}O, labeled ATPγS by nitrogenase. The ^{18}O labeled thiophosphate was incorporated into the β position of ATP and the ^{31}P NMR of this species was recorded. The peaks for no ^{18}O, ^{18}O in the non-bridging position and the double label were eliminated when the results were corrected for percent contaminating isotopes in the original O labeled ATPγS. After correction only the bridging ^{18}O remained.

the sensitivity of this assay. Another explanation, and a potentially intriguing one, is that the hydrolysis of ATP to ADP·Pi + H^{+} occurs in a "pocket" created when the Fe and Mo-Fe proteins associate and that the hydrolysis is the source of the H^{+} needed for substrate reduction. If so, the water needed for ATP hydrolysis would (a) have to be associated either with the Fe protein·MgATP complex before complexing with the MoFe protein, or with the MoFe protein before the association step, or (b) have to have limited access to the pocket. For example Fe protein·MgATP(H_2O) + MoFe protein complex would associate, the H_2O complexed with the Fe protein·MgATP could be the source of the water for the ATP hydrolysis and the H^+ released could be used in substrate reduction. Recent results (V.L. Tsuprun et al., 1985) of an analysis of the MoFe protein of Azotobacter vinelandii by electron microscopy show that the particle size of the MoFe protein does not change when the Fe protein is compelxed with it suggesting that the Fe-protein could be located in a central cavity of the MoFe protein. This is consistent with the creation of an active site "pocket" where ATP hydrolysis occurs, H^+ are produced and electrons are transferred. In fact it may be that this H^+ release at the hydrolysis site creates a charge separation that drives electron transfer from the 4Fe·4S center of the Fe protein to the substrate reduction site of the MoFe protein where H^+ and electrons are combined during substrate reduction.

4. REFERENCES

Cramer SP et al (1978) J. Am. Chem. Soc. 100, 3814-3819.
Cramer SP et al (1985) J. Am. Chem. Soc. In press.
Knowles JR (1980) Ann. Rev. Biochem. 49, 877-919.
Lowe DJ, Thorneley RNF (1985) Biochem. J. 224, 877-886.
Mortenson LE (1972) Methods Enzymol. 24, 446-456.
Mortenson LE, Thorneley RNF (1979) Ann. Rev. Biochem. 48, 387-418.
Tsuprun VL et al (1985) Eur. J. Biochem. 149, 389-392.
Webb MR (1982) Methods Enzymol. 87, 301-316.
Webb MR et al (1986) J. Biol. Chem. In press.
Webb MR, Trentham DR (1981) J. Biol. Chem. 256, 10910-10916.
Weininger MS, Mortenson LE (1982), Proc. of Natl. Acad. Sci. USA 79, 378-380.

REDOX PROPERTIES OF THE NITROGENASE COMPONENT PROTEINS
FROM AZOTOBACTER VINELANDII

GERALD D. WATT
BATTELLE-KETTERING RESEARCH LABORATORY, YELLOW SPRINGS, OH 45387, USA

Nitrogenase is a two component enzyme system that is responsible for the six-electron reduction of N_2 into ammonia known as biological nitrogen fixation. Both components comprising the nitrogenase system are redox active, iron-sulfur containing proteins with similar characteristics independent of their biological source. The smaller of the two proteins, the iron protein (Av_2) contains one Fe_4S_4 center while the larger protein, the MoFe protein (Av_1) contains a fairly complex array of metal sulfur clusters including one which uniquely contains both Fe and Mo. An understanding of the functioning of these proteins in the catalytic cycle of N_2 reduction is the objective of much research carried out in various laboratories throughout the world (Burgess, 1984). Although much has been learned about the nitrogenase proteins, there are still many gaps in our knowledge and many incomplete and contradictory results reported which require extension and reevaluation.

The chemical mechanism of biological nitrogen fixation catalyzed by the nitrogenase enzyme encompasses many varied and important aspects of the nitrogenase proteins and their interaction with each other as well as with the known substrates such as N_2, MgATP, H^+ and low potential reductants. A very prominent feature of nitrogenase catalysis is the six-electron requirement for N_2 reduction which implies the need for a rich and varied redox capability provided by the enzyme system to facilitate this process. This feature of nitrogenase has been recognized for some time now with both component proteins known to be redox active and involved in electron transfer during the catalytic cycle. In this paper we describe our recent results directed toward determining the redox properties of the component proteins from Azotobacter vinelandii and the influence that nucleotides have in altering the redox capability.

Redox properties of Av_2. Av_2 has been reported to be both a two and (Thorneley et al., 1976; Braaksma et al., 1982) a one (Ljonnes, Burris) electron transfer protein. The redox capability of this protein is important information in gaining an understanding of the mechanism of biological nitrogen fixation. Fig. 1 is an EPR titration of Av_2red with standardized indigodisulfonate which clearly shows that the EPR signal height and signal integration reach a minimum after a one electron removal. Similar results were obtained by titrating Av_2ox with standardized $S_2O_4^{2-}$. We have used a wide range of oxidants for preparing Av_2ox all of which give the same result. This same result has been obtained also by oxidative EPR titrations of Av_2red as well as with both oxidative and reductive titrations of Av_2 as monitored by CD optical spectroscopy.

A more detailed view of the Av_2ox reduction process is shown in Fig. 2 where both the reduction stoichiometry and the reduction potential were both accurately measured by microcoulometry. The reduction stoichiometry is seen to be one electron per Av_2 in complete agreement with the titration results discussed above. In addition, a well behaved Nernstian reduction curve is obtained with an $E_{1/2}$ value of -310mv and an n value of one. In both the titration experiments and in the microcoulometric

586

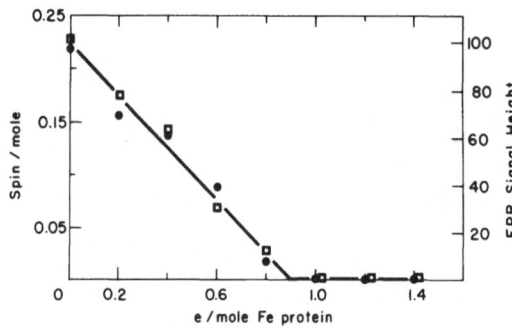

FIGURE 1. EPR titration of Av_2ox with standardized $S_2O_4{}^{2-}$ showing the signal, height at $g = 1.94$ and signal double integration as a function of added reductant.

experiments, we have shown that Av_2ox obtained from the fixing system in which $S_2O_4{}^{2-}$ is limiting also undergoes only a single electron reduction step. Thus, both naturally occurring oxidation of Av_2 and artificially induced oxidation by chemical reagents produce only one-electron oxidized Av_2.

Fig. 2 further shows the effect that the presence of nucleotides has on the reduction properties of Av_2ox. MgATP causes the reduction potential to be shifted to -430 mv, a value 140 mv more negative than for the reduction of Av_2ox in the absence of MgATP. MgADP causes an even larger negative shift in the reduction potential of Av_2ox, giving a value of -480 mv for reduction. Finally, the ATP analog in which the γ oxygen atom in the phosphate chain has been replaced by a CH_2 group, shifts the reduction potential of Av_2ox to -385 mv.

The shift of -140 mv for the reduction of Av_2ox in the presence and absence of MgATP is a result of the change in binding of MgATP to the two redox states of Av_2. If 2 MgATP molecules are assumed to bind to each redox state of Av_2, a binding of 2 MgATP molecules to Av_2ox 233-times stronger than to Av_2red can be calculated from the shift of -140 mv in the reduction potential. More details of the binding of MgATP to Av_2 are discussed below.

Nucleotide Binding to Av_2. The binding differences of 233 for MgATP and 1648 for MgADP binding to the two redox states of Av_2 determined from the shifts in reduction potentials, demonstrate a significant effect of redox state of Av_2 on its ability to bind nucleotides. The binding of nucleotides to Av_2ox has been extensively studied in our laboratory by a

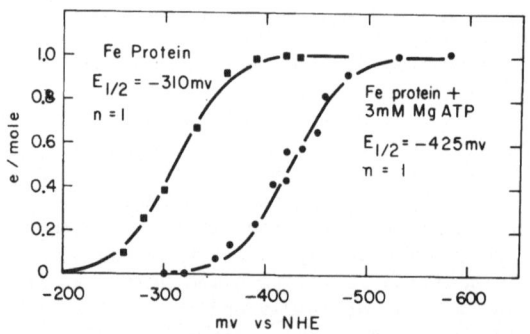

FIGURE 2. Microcoulometric reduction of Av_2ox, Av_2ox + ATP analog, Av_2ox + MgATP and Av_2ox + MgATP (from left to right).

variety of techniques. However, at present, due to technical and chemical problems, only limited studies of nucleotide binding to Av_{red} have been completed.

Fig. 3 presents a binding curve for MgATP binding to Av_2ox as determined by equilibrium dialysis of six and twelve hours duration. Also shown are several representative points where binding was detetermined by the rapid-flow dialysis method of Colwick, Womack (1969). The figure shows a sigmoidal binding curve and demonstrates the binding stoichiometry to be two MgATP molecules bound per Av_2ox. Analysis of this curve yields a binding constant (K_1K_2) of 5×10^7 M^{-2}. Similar binding curves have been determined by monitoring the CD and optical spectroscopic changes in Av_2ox as MgATP is added and binding constant consistent with that from equilibrium dialysis has been observed. These spectroscopic methods were also used to measure the binding of MgADP to Av_2ox. In this case, the binding is much tighter (consistent with the redox measurements discussed above) and essentially stoichiometric and only the ratio of $MgADP/Av_2ox$ of 1.83 ± 0.20 was obtained.

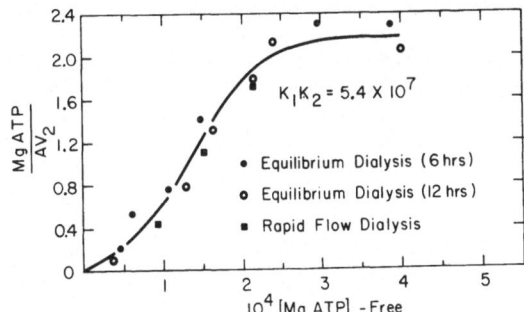

FIGURE 3. Equilibrium dialysis of Av_2ox + MgATP.

Redox Properties of Av_1. As with Av_2, we begin evaluation of Av_1 by attempting to define its redox stoichiometry which can then be related to the numbers of redox centers present. However, due to the complexity of this system in terms of the number of redox centers present and their diverse nature, this is not an easy task. We have used various forms of spectroscopy to monitor spectral changes corresponding to defined redox conditions as well as electrochemical methods for defining redox conditions for measuring electrons transferred under these conditions. In all cases reported here, we have dealt with Av_1 at or more oxidized than the $S_2O_4{}^{2-}$ reduced level.

Table I contains information regarding Av_1 and the effect that varying the oxidation potential has on the properties of Av_1. For these experiments, the oxidant to Av_1 ratio was maintained at 10-20. As the redox potential of the oxidant becomes more positive, oxidation of Av_1 increases until three electrons have been removed. The EPR signal is uneffected by this oxidation but well defined optical and CD changes occur demonstrating protein oxidation. At potentials more positive than -125 mv, further oxidation of Av_1 occurs during which the intensity of the EPR signal decreases to near zero at potentials more positive than 0 mv. At these potentials a total of six electrons have been removed from Av_1 with complete retention of activity. Increasing the potential of the oxidant to values of +200-300 mv causes further oxidation of Av_1 from which species

TABLE I. Properties of Av_1 as a Function of Oxidation Potential.

		PROTEIN REDUCTION AT		
OXIDANT	E^O_{pH7}	-400 mv	-600 mv	EPR
None	–	0	0	Present
Methyl viologen	-440	0	0	Present
Neutral red	-325	0	$\sim$ 1	Present
Phenosafranine	-250	0	3	Present
FMN	-220	0	3	Present
Brilliant alizarin blue	-173	0	3	Present
Indigo disulfonate	-125	0	3	Present
Methyl capri blue	- 61	–	–	–
Methylene blue	+ 11	3	3	Absent
Thionine	+ 64	3	3	Absent
DCPIP	+217	3	3	Absent

oxidized by nine and twelve electrons can be isolated. In these cases, a careful control of oxidant concentration is required in order to prevent Av_1 activity loss.

Oxidation stoichiometry of the isolated nitrogenase complex from Azotobacter vinelandii was determined by exhaustively oxidizing the Av complex with excess oxidants whose potentials were more positive than 0 mv. This treatment produced Av complex oxidized by about 15 electrons. This corresponded to the presence of 2 Shethna proteins oxidized by 1 electron each, $1AV_2$ protein oxidized by 1 electron and 1 AV_1 oxidized by twelve electrons. Contrary to free Av_2 discussed above, the Av_2 component of the Av complex does not undergo any shift in reduction potential in the presence of MgATP when complexed with the other proteins in the Av nitrogenase complex.

The oxidation behavior of reduced Av_1 under mild oxidation conditions is shown in Fig. 4a while Fig. 4b displays the reduction behavior of three-electron oxidized Av_1. A three electron process is observed for each process but different $E_{\frac{1}{2}}$ values are observed depending on the redox direction followed. We have previously reported the reduction properties of six-electron-oxidized Av_1 (Watt et al 1980).

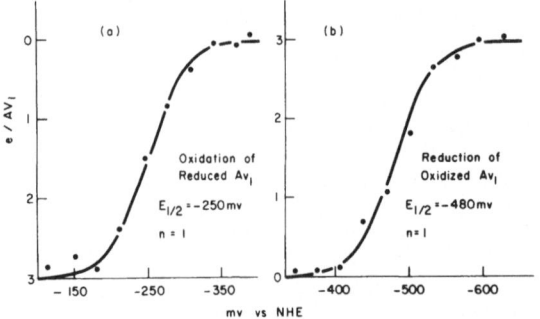

FIGURE 4. Microcoulometric reduction of: a) three-electron oxidized Av_1 b) six electron oxidized Av_1 c) nine electron oxidized Av_1 and d) twelve-electron oxidized Av_1.

Discussion. The redox stoichiometry of the iron protein has been reported to be both a one and a two electron process. However, our results for Av_2 reported here demonstrate that only a one electron process is observed whether Av_2 is naturally oxidized by the nitrogenase system or oxidized by various artificial oxidants. The Av_2 used in these studies had activities of 2500-3200 nmH_2/min-mg and iron analysis which ranged from 3.3-4.5 Fe per 63,000 daltons molecular weight. The redox stoichiometry and the analytical data are all consistent with the presence of a single Fe_4S_4 center in Av_2.

Previously reported redox measurements of the iron protein have established a negative shift of the reduction potential of Av_2 in the presence of MgATP or MgADP (Burgess, 1984). We have confirmed this result with several separately prepared Av_2 samples which indicate a shift of -140 mv for MgATP and -190 mv for MgADP. The MgATP results are summarized in (1) and (2) below.

$$Av_2ox + e = Av_2red, \quad E_{\frac{1}{2}} = -290 \text{ mv} \tag{1}$$

$$Av_2(MgATP)_2ox + e = Av_2(MgATP)_2red, \quad E_{\frac{1}{2}} = -430 \text{ mv} \tag{2}$$

The MgATP binding stoichiometry for $Av_2(MgATP)_2ox$ shown in (2) was obtained from direct binding studies. That for $Av_2(MgATP)_2red$ is suggested by EPR and rapid flow dialysis measurements but firm data for this reaction have not been obtained at this time due to technical problems (self oxidation and other Av_2 reactions). However, the reduction potentials can be used to support (2) by including the binding constant discussed below for 2 MgATP binding to Av_2ox. This approach is outlined by (3)-(5).

$$Av_2ox + e = Av_2red \qquad \Delta G = + 6.69 \text{ kcal/mole} \tag{3}$$

$$Av_2(MgATP)_2ox + e = Av_2ox + 2 \text{ MgATP} \quad \Delta G = + 10.5 \text{ kcal/mole} \tag{4}$$

$$Av_2(MgATP)_2ox + e = Av_2red + 2 \text{ MgATP} \quad \Delta G = + 17.2 \text{ kcal/mole} \tag{5}$$

where the ΔG values in (3) and (4) were calculated from $E_{\frac{1}{2}}$ value of -290 mv for (1) and the MgATP dissociation constant of 2.0×10^{-8} M^{-2} for (4). From reaction (5) and the calculated ΔG value of + 17.2 kcal/mole, a reduction potential of -720 mv would be expected, if complete dissociation of MgATP from Av_2red occurred when $Av_2(MgATP)ox$ undergoes reduction. However, a reduction potential of -430 mv is actually observed, so that the difference between the -720 mv predicted for no MgATP binding to Av_2red and that actually measured demonstrates an MgATP interaction with Av_2red. The magnitude of this binding interaction is determined from $E_{\frac{1}{2}}$ value of -430 mv

$$Av_2(MgATP)_2ox + e = Av_2red + 2 \text{ MgATP} \quad \Delta G = + 17.2 \text{ kcal/mole} \tag{5}$$

$$Av_2(MgATP)_2red = e + Av_2(MgATP)_2ox \qquad \Delta G = - 9.92 \text{ kcal/mole} \tag{6}$$

$$Av_2(MgATP)_2red = Av_2red + 2MgATP \qquad \Delta G = + 7.28 \text{ kcal/mole} \tag{7}$$

for (6) and the value of + 17.2 kcal/mole from (5) and is given by (7). A value for the dissociation constant of 4.5×10^{-6} M^{-2} for (7) results.

The dissociation constant of 2.0×10^{-8} M^{-2} for (5) was measured directly by equilibrium and rapid flow dialysis measurements as well as by spectroscopic methods. Av_2ox is stable for at least 12 hrs under strictly anaerobic conditions. However, we have not directly measured MgATP binding to Av_2red with much reliability as yet. Long term dialysis is an unusable method because of Av_2 self oxidation and as yet we have had only limited success using rapid flow dialysis. The change in EPR spectra of Av_2red ($S_2O_4^{2-}$) produced by varying the concentration of MgATP is consistent with a K_d of $\sim 10^{-6}$ M^{-2} but the changes are too small to determine K_d with much precision. The value quoted above for (7) of 4.5 x 10^{-6} M^{-2} obtained from redox potential variation seems to be the most reliable value at present. The combined results indicate that binding of MgATP to oxidized Av_2 is much stronger than that to reduced Av_2.

Av$_1$ has a rich redox chemistry. We report that Av$_1$ can be oxidized by 3,6,9 and 12 electrons depending upon the oxidation potential of the artificial oxidants used. The nine- and twelve-electron oxidized states have reduction potentials nearly coincident with that for the six-electron oxidized state except that the n value for reduction is 2 instead of 1. We have not as yet demonstrated that any of these oxidized states function during enzyme turnover. At present their utility is simply to demonstrate that a number of redox clusters are present in the MoFe protein that can undergo facile electron transfer reactions at interesting reduction potentials. The proposal by Thorneley,Lowe that some of these are involved in enzyme turnover is without substantiation but continues to be examined.

References

Braaksma A et al (1982) Eur. J. Biochem. 121, 483.
Burgess BK (1984) in Veeger C and Newton WE, eds. Proceedings of the 5th International Symposium on Nitrogen Fixation, p. 104, Martinus Nijhoff/ Dr. W. Junk, The Hague.
Colwick SP and Womack FC (1969) J. Biol. Chem. 244, 774.
Ljonnes T and Burris R (1978) Biochemistry 17, 1866.
Thorneley RNF (1981) in Gibson AH and Newton WE, eds, Current Perspectives in Nitrogen Fixation, p. 53, Australian Academy of Science, Canberra.
Watt GD et al (1980) Biochemistry 21, 4926.

MOSSBAUER AND EPR STUDIES ON NITROGENASE

W.R. DUNHAM[1]/W.R. HAGEN[2]/A. BRAAKSMA[2]/H. HAAKER[2]/S. GHELLER[3]/W.E. NEWTON[3] AND B. SMITH[4]. [1]BIOPHYSICS RESEARCH DIVISION, INSTITUTE OF SCIENCE AND TECHNOLOGY AND DEPARTMENT OF RADIOLOGY, MEDICAL CENTER, THE UNIVERSITY OF MICHIGAN, ANN ARBOR, MI 48109, USA, [2]BIOCHEMISTRY DEPARTMENT, AGRICULTURAL UNIVERSITY, WAGENINGEN, THE NETHERLANDS, [3]USDA-ARS, WESTERN REGIONAL RE-SEARCH CENTER, BERKELEY, CA 94710, USA, [4]AFRC UNIT OF NITROGEN FIXATION, BRIGHTON, BN19RQ, U.K.

1. INTRODUCTION

Magnetic resonance spectroscopy has played an important role in the studies of the nitrogenase active sites. The electron transport center(s) in the Fe-protein can be studied by EPR in the reduced state and by Mossbauer spectroscopy in both oxidation states. Likewise, several of the oxidation states of the MoFe-protein are amenable to investigation by EPR and Mossbauer spectroscopy. In the following, we will update some recent results on the Fe-protein and the high temperature Mossbauer spectra of the MoFe-protein as isolated. Our focus will be to re-examine the structural information obtainable in these spectra with special interest given to identifying the number of irons in these centers.

2. Fe-PROTEIN

It is generally thought that this protein contains a 4Fe, cubane center, whose oxidized state accepts one electron from flavodoxin and then passes this electron to the MoFe-protein with the aid of MgATP. There are, however, several pieces of spectroscopic information that indicate that this picture of the protein active site is oversimplified. Implicit in the preceding picture is the requirement that the reduced Fe-protein have a low temperature EPR spectrum that integrates to one electron per molecule. In fact, the EPR signal, which is characteristic of a [4Fe-4S] center, usually integrates to approximately 0.2 electrons per molecule. Low-lying excited EPR states (Orme-Johnson et al., 1977) and nearby, fast relaxing paramagnets (Lowe, 1978) have been proposed to explain this lack of EPR intensity. Either of these explanations should show itself in a non-standard feature of the EPR signal. Searching for these features we have re-examined the EPR spectra of the Fe-protein. Our first effort was to computer simulate the EPR spectrum from Av2 at 9 and 15 GHz.

In Fig. 1 we show the fit to the 9GHz spectrum using our g-strain program (Hagen et al., 1985a,b). The high quality of the fit is expected for an Fe/S center as the computer program was designed to account for the g-strained spin systems in these centers. We conclude that there is no evidence for spin-spin interaction in this spectrum. Furthermore, the temperature dependence of the signal is also characteristic of a magnetically-isolated [4Fe-4S] cluster (Hagen et al., 1985c). Since the signal quantitates to 0.2 spins per molecule, we conclude that it is impossible that the Fe-protein contains a single type of [4Fe-4S] cluster.

592

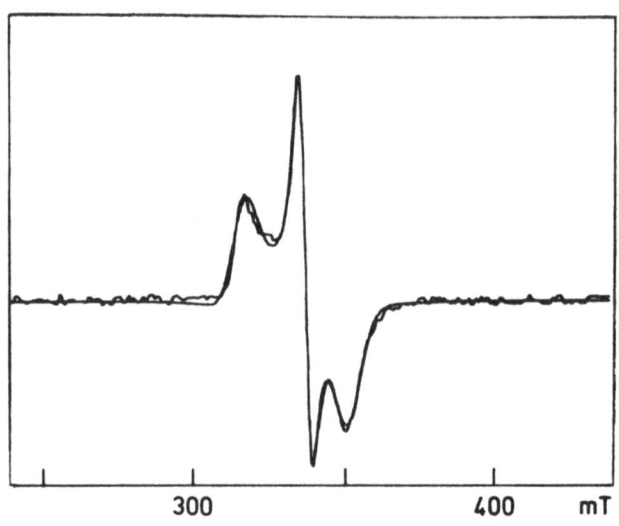

Fig. 1. Computer simulation of the EPR signal from Av2 at 9°K, 9.13GHz, 13µW power, 0.63 mT modulation amplitude. The simulation parameters are: $g_{1,2,3}$ = 1.854,1.936,2.063; $\Delta g_{11,22,33,12,13,23}$ = 0.024,0.010, -0.030, 0.004,-0.005,0-.001, residual broadening, 0.006.

The Mossbauer spectra of reduced Av2 add to the mystery of the missing EPR signal. At high sample temperatures (Fig. 2a), reduced Av2 shows a narrow quadrupole pair. The resolution in this spectrum is high enough to exclude the possibility of there being two lines in the spectrum that amount to 15% of the total intensity of the spectrum and are not in the main two lines of the spectrum. As 0.2 equiv.(EPR) x 4Fe equals 0.8 Fe equiv., and there are 5.6 Fe per molecule in this sample, the specie giving rise to the EPR signal must have a Mossbauer resonance that is part of the two lines in Fig. 2a. In other words, the 4.8 Fe atoms per molecule not accounted for by EPR have the same Mossbauer spectrum as the 0.8 Fe atoms per molecule that are attributable to the S=1/2 EPR signal of Av2. Since we have previously identified the S=1/2 EPR signal with a [4Fe-4S] cluster, we are led to identify the 4.8 Fe per molecule with another [4Fe-4S] cluster. There are two main problems with drawing this conclusion: the first is that the iron assay gives (indirectly) 4.8 Fe/molecule as a minimum for the iron content of the cluster. The second problem is that the chloro- and thiophenoxy-prismanes (Kanatzidis et al., 1985) are virtually indistinguishable at high temperature from their cubane analogs. This Mossbauer result is consistent with the fact that the prismanes and cubanes are isoelectronic and, in fact, two prismanes decompose on heating to yield three cubanes (Kanatzidis et al., 1984). We are not concluding that there is a prismane in Av2, but we do point out that Mossbauer spectroscopy does not compel the conclusion that there is a cubane in Av2 that accounts for the major component although the minor component is probably a cubane.

At low temperatures, reduced Av2 shows the Mossbauer spectrum of a paramagnetic system. In Fig. 2b we show that there is at most a very

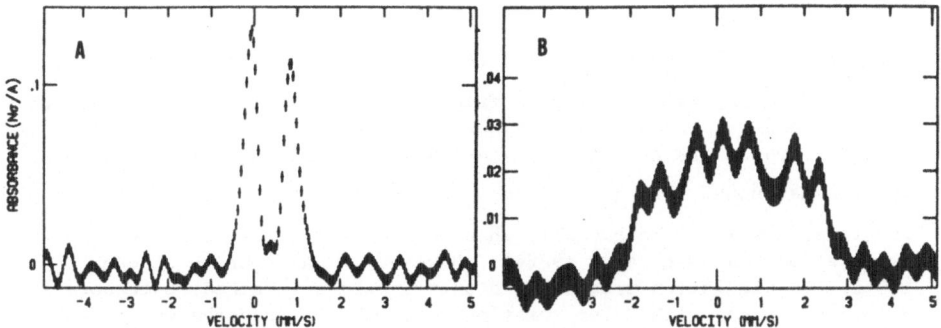

Fig. 2. Mossbauer data for reduced Av2, free of nucleotide (A) 126°K, zero applied field. (B) 4.4°K, 0.05T applied field.

small component of diamagnetic iron in this sample. We conclude that all of the iron in reduced Av2 is in a spin-system(s) that is half-integer (S = 1/2 or 3/2...) because all integer spin systems are diamagnetic at some orientations at low applied magnetic fields (see Hagen et al. , 1985d, for example). Any half-integer system that has a T_1 long enough to magnetically broaden a Mossbauer spectrum must have an EPR signal. Therefore, we have looked for this mysterious missing signal.

Recently, Watt and MacDonald (J. Biol. Chem., in press) have found an S=3/2 signal in Av2 and Cp2, while we have also found a similar signal in Av2, Kp2 and Ac2 preparations (Hagen et al., 1985e). This EPR signal has visible resonances at g5.9 ($m_s=\pm3/2$) and g4.8 and 3.4 ($m_s=\pm1/2$) with zero-field splitting parameters D=-5 cm^{-1} and E=.6 cm^{-1} (Hagen et al., 1985e). EPR Quantitations based on Cu^{2+} calibrations gave the following results:

Table 1. EPR Spin Equivalents per Molecule of Component 2

	[S=3/2]	[S=1/2]
Av2	0.8	0.17
Kp2	0.6	0.40
Ac2	~1	0.18

We are told that Watt and MacDonald have results similar to these as well as the MIT group of Orme-Johnson. The signal intensities depend on the addition of urea, ethylene glycol, MgADP, MgATP, removal of MgATP, phenazine methosulfate, subsequent reduction by dithionite (Hagen et al., 1985e), but these data are not germane to the present discussion. The point under discussion here is quantitation of the S=3/2 as shown in the above table.

As mentioned previously, these signals have been quantitated against a Cu^{2+} standard (standard practice). However, quantitation of an S=3/2 with an S=1/2 standard is a very dangerous procedure. To begin with, the EPR transition rate, by Fermi's Golden Rule, is proportional to $g_\perp^2 |<S_+>|^2 \delta(h\nu-E)$. Whereas, the δ-function and $g_\perp^2$ are the same for S=1/2 and S=3/2, $|<S_+>|^2$ changes from 1 to 4 for the $m_s=\pm1/2$ transition when one changes from S=1/2 to S=3/2. In addition, the real g-values of an S=3/2 system are near to g2, whereas the apparent g-values are near g5. We had hoped that $g_\perp^2$ would compensate for $|<S_+>|^2$ so that the Cu^{2+} quantitations would be accurate when we submitted this work for publication (Hagen et al., 1985e). However, we also wrote a general S=3/2, EPR simulation program with anisotropic g-strained, g-tensors and zero-field terms with Euler rotation to check the previous work. At the time of this writing, curve fitting is in its preliminary stages and the program is unproven in its youth. We make the tentative conclusion that the S=3/2 EPR intensity is around 25% underestimated by the Cu^{2+} calibration. Thus, the aesthetic property of the EPR quantitation table, i.e., several of the total intensity numbers are near value one, could be fortuitous. We are working on this problem and have discovered the need for multi-frequency EPR data to enable its solution. The question to be answered here is whether component 2 of nitrogenase has a [4Fe-4S] cluster that is capable of generating an S=1/2 or S=3/2 signal <u>or</u> whether the S=3/2 in the active site of the protein with the S=1/2 system being an "accidental" decomposition of the active site. In either case, it now seems that the S=1/2 system may not be from the main biochemical system so that all previous work based on EPR quantitations should be re-evaluated in this light.

3. MoFe-PROTEIN

In a recent paper (Dunham et al., 1985), we demonstrated a weakness in the previous analyses of the high temperature Mossbauer spectra from MoFe-protein from <u>Klebsiella pneumoniae</u> (Smith and Lang, 1974) and <u>Azotobacter vinelandii</u> (Munck et al., 1975). In our work, the conventional percent transmission data were converted to absorption data by a Fourier deconvolution algorithm. These processed data were then computer simulated as four quadrupole pairs. Our best fits were obtained by assuming that there are four irons in the "Fe^{2+}" doublet, two iron atoms in the "S" doublet, twelve iron atoms in the "D" doublet and sixteen iron atoms in the "M" doublet. The implication of this work is that the MoFe cofactor contains one molybdenum and eight iron atoms, rather than the six iron atoms that had been previously proposed.

The main point of our work was that the estimate of the total iron content of the Kp1 and Av1 preparations depend critically on the intensity ratios of the quadrupole pairs "M" and "D", with respect to the "Fe^{2+}" quadrupole pair. These ratios had been underestimated by the previous workers due to their having ignored the importance of Beer's law to these spectra. In Fig. 3, we show the "raw" data and the Fourier deconvolved data from Av1 at 125°K. The effect of Beer's law on these spectra is subtle, but it is enough to force the spectroscopist to add four iron atoms to his model for Av1 based on the intensity of the "Fe^{2+}" doublet.

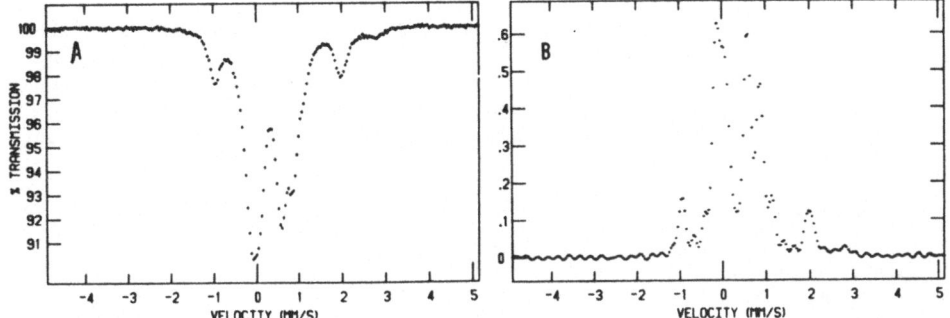

Fig. 3.   Mossbauer data for Av1 as isolated at 125K and zero applied
field. (A) Raw data, corrected for pulse pile-up and solid angle effects.
(B) Background-corrected and deconvolved data, plotted as logarithmic
values.

        In all the analyses of these spectra, including ours, it is assumed
that the "Fe^{2+}" doublet intensity is a valid yardstick by which to
measure the other intensities. While there is some room for doubt in
this assumption, there remains the need for the Mossbauer spectroscopist
to give his best estimate for the iron content of these centers, and this
assumption seems to be the best alternative at present. It remains for
us to prove that these Mossbauer spectra represent the in vivo,
fully-iron-containing form of component 1 in nitrogenase. Along these
lines, we are currently engaged in studies where different component 1
preparations are compared. At present, we have not been able to show
that the component 1 intensities can be affected by any other means than
be chemical oxidation-reduction reactions, but small differences in the
"M" to "D" ratios can be seen by comparing samples. We are presently
trying to extend our data set to clarify these differences.

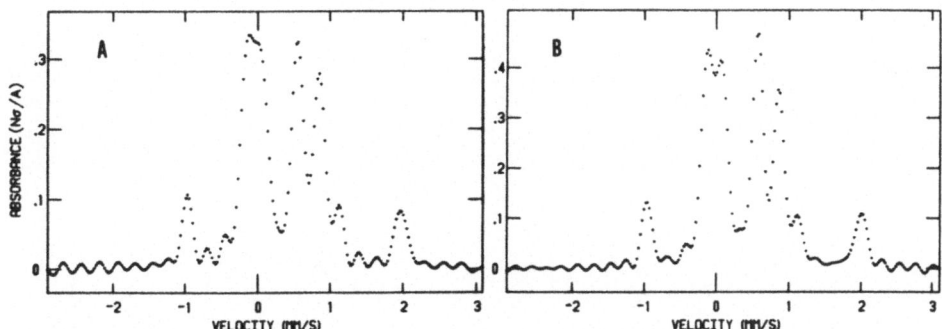

Fig. 4.   Mossbauer data for the MoFe-protein at 125°K and zero applied
field. (A) Kp1, as isolated. (B) Av1, as isolated.

In Fig. 4 we show the spectra of Kp1 and Av1. The similarity of these spectra suggests that our previous conclusions concerning the iron content of component 1 are valid, but we are still very concerned by the fact that the iron analyses of component 1 preparations invariably fall short of the iron requirements of our best estimate of the iron content of these preparations from Mossbauer spectroscopy. The main point here is that the chemists trying to synthesize the MoFe-cofactor should consider seven and eight iron strucures as possibilities in addition to six iron structures.

4. REFERENCES

Dunham WR et al (1985) Eur J Biochem 146, 497-501.
Hagen WR et al (1985a) J Magn Reson 61, 220-232.
Hagen WR et al (1985b) J Magn Reson 61, 233-244.
Hagen WR et al (1985c) Eur J Biochem (in press).
Hagen WR et al (1985d) Biochim Biophys Acta 828, 369-374.
Hagen WR et al (1985e) FEBS letters (submitted)
Kanatzidis MG et al (1984) J Chem Soc, Chem Commun, 356-358.
Kanatzidis MG et al (1985) J Am Chem Soc 107, 953-961.
Lowe DJ (1978) Biochem J 175, 955-957.
Munck E et al (1975) Biochim Biophys Acta 400, 32-53.
Orme-Johnson WR et al (1977) in Recent Developments in Nitrogen Fixation (Newton W et al, eds) pp 131-178, Acad Press, London.
Smith BE, Lang G (1974) Biochem J 137, 169-180.

5. ACKNOWLEDGEMENTS

The authors wish to thank Ms. Shirley Mieras for her help in preparing the manuscript.

THE IRON-MOLYBDENUM COFACTOR OF NITROGENASE

B.E. SMITH, P.E. BISHOP, R.A. DIXON, R.R. EADY, W.A. FILLER, D.J. LOWE,
A.J.M. RICHARDS*, A.J. THOMSON*, R.N.F. THORNELEY and J.R. POSTGATE
AFRC Unit of Nitrogen Fixation, University of Sussex, Brighton, BN1 9RQ
*Department of Chemistry, University of East Anglia, Norwich, U.K.

INTRODUCTION

It is now eight years since Shah, Brill (1977) published their
method for isolating the iron-molybdenum cofactor (FeMoco) from the MoFe
protein. Their results indicated that the Mo and Fe were combined with
inorganic sulphide in an unique extractable cluster that could activate
extracts of certain mutants lacking MoFe protein activity. There had
long been a prejudice, based on circumstantial evidence (Smith, 1977),
that Mo was at the substrate binding site of nitrogenase and it was
assumed that FeMoco was this site. Since that seminal paper a number of
groups have studied FeMoco and there is now better (but not absolutely
conclusive) evidence on its function. However its structure is not
known, its composition is not well defined and its biosynthesis has
proved extremely complex. In this paper these topics are reviewed and
recent experiments on them described.

THE FUNCTION OF FEMOCO

The best preparations of FeMoco can have an activity (after
combination with the inactive MoFe protein from various mutants) of
275 ± 27 units/ng atom Mo (where 1 unit = 1 nmol C_2H_2 reduced/min). A
similar activity/Mo has been found in partially and highly purified
preparations of Kpl (specific activities from 600-1770 units/mg) from
which other Mo-containing enzymes have been removed (Hawkes et al. 1984).
This activity/Mo extrapolates to a specific activity for the MoFe
protein with $M_r = 220,000$ containing 2 Mo atoms/molecule of $2,500\pm250$
units/mg in good agreement with the observed maximum specific activities
of MoFe protein preparations and confirming a linear relationship between
their FeMoco contents and specific activities (Lowe et al. 1985).

Further confirmation of this relationship comes from an analysis of
the kinetics of nitrogenase action. The rate-limiting step in the
catalytic cycle is the dissociation of the oxidised Fe protein from the
MoFe protein after the MgATP-activated protein-protein electron transfer.
The rate constant for this process (6.4 ± 0.8 s^{-1} at 23°) can be used to
calculate the specific activity of a MoFe protein preparation from its
Mo content on the basis of one active site/Mo atom. Such calculations
give a maximum specific activity (at 30°) of 2400 units/mg MoFe protein
containing 2 Mo atoms/molecule, consistent with the values quoted above
(Thorneley, Lowe, 1983). In addition computer simulation of the steady-
state concentration of the N_2H_4-releasing intermediate formed during N_2
reduction requires that one N_2 molecule is bound and reduced at each
FeMoco site (Thorneley, Lowe, 1984).

The best evidence linking the FeMoco sites (rather than another species present in stoichiometric equivalence) to the substrate-binding site has come from a study of nifV mutants (Hawkes et al., 1984). The MoFe protein (NifV⁻Kpl) from these mutants is associated with substrate-reduction properties that differ from the wild-type (see below). When the FeMoco from the NifV⁻Kpl was extracted and bound to the MoFe protein (NifB⁻Kpl) from nifB mutants unable to synthesise FeMoco, the active protein formed had the substrate-reduction properties of NifV⁻Kpl. If wild-type Kpl was used as the source of the FeMoco then MoFe protein with wild-type substrate-reduction properties was formed.

These experiments clearly linked the FeMoco sites with substrate binding and reduction but direct evidence of substrate binding to these sites is difficult to obtain. The pH dependence of the $S = 3/2$ epr spectrum of the FeMoco sites in Kpl is affected by C_2H_2 (Smith et al., 1973) and their spin state can be changed to $S = 1/2$, in the presence of the product C_2H_4, during turnover (Hawkes et al., 1983). However there was no evidence for a direct interaction between the FeMoco sites in the protein and H^+ or C_2H_4. Recently we have observed an interaction between isolated FeMoco and the substrate cyanide (Richards, 1985). Provided FeMoco of good activity was used its epr spectrum changed on binding cyanide from the usual rhombic form of g values near 4.6, 3.3 and 2.0 to a near axial form with g values near 4.3, 3.8 and 2.0. These spectra were unaffected by addition of alkali suggesting that it was the cyanide ion CN^- that was bound. The spectral change was complete after addition of <2.5 CN^- ions/Mo atom. It has not been possible to simulate a titration curve obtained in such experiments with a single binding constant nor with two equivalent binding constants for the cyanide-FeMoco interaction. The data therefore indicate the binding of two CN^- ions, with different binding constants, to each FeMoco molecule. In the enzymatic reduction of cyanide, HCN is reported as a substrate with CN^- ion acting as an inhibitor but binding at a different site (Li et al., 1982), suggesting that there may be two binding sites for cyanide on the enzyme as well.

Taken together the above data provide strong evidence that the FeMoco sites are the sites of substrate binding to nitrogenase. However there is evidence, from chemostat studies with a nifKDH deletion mutant of Azotobacter vinelandii (Kennedy et al., and Bishop, Eady, this volume) that FeMoco may not always be essential for N_2 reduction. These experiments show that the alternative nitrogenase system proposed by Bishop et al (1980) is regulated by low levels of Mo and does not require a functional MoFe protein. Although the MoFe protein is absent from the deletion strain it is conceivable that FeMoco is present, bound to another nif product, acting in place of the MoFe protein. This seems unlikely since preliminary experiments (discussed by Kennedy et al., this volume) with another mutant of A. vinelandii, carrying Tn5 apparently inserted in nifN (or nifE), also grew on N_2 under Mo-deficient conditions. Assuming that nifN and nifE are involved in FeMoco formation in A. vinelandii it is doubtful that FeMoco (and therefore Mo) is a component of the alternative nitrogenase.

The above data on the alternative nitrogenase do not negate the earlier evidence on FeMoco as the active site of conventional nitrogenase. However they do suggest that in the conventional system Fe rather than Mo may be the site of N_2 binding.

THE COMPOSITION OF FEMOCO

Any attempt to model FeMoco requires as initial parameters a good estimate of its chemical composition. There is general agreement that each molecule of FeMoco contains 1Mo and 7±1 Fe atoms (Shah et al., 1984; Lowe et al., 1985), however its sulphur content is much more controversial.

Shah, Brill (1977) found 6 acid-labile sulphur atoms per Mo in their preparations. They used the normal chemical method for determination of acid-labile sulphur in Fe-S proteins in which the sulphide is released by acid and then reacted with para-methylene-diamine to form methylene blue. Two later reports (Eady et al., 1980; Yang et al., 1982) using the same method found 4 S^{2-}:Mo however the method underestimated the S^{2-} content of metal complexes of known structure containing Mo as well as Fe. Therefore alternative analytical methods were sought.

Analysis of Mo EXAFS spectra of the MoFe protein indicated that the Mo atom had at least 4 S atoms in its first coordination sphere (Cramer et al., 1978). Thus 4 S:Mo is probably the minimum S:Mo ratio likely to be found.

In principle it should be possible to determine the S content of FeMoco using cells grown on radioactive $^{35}SO_4^{2-}$. This method requires an accurate knowledge of the quantity of ^{35}S, relative to total S available for growth, together with a method for estimating or removing S-containing impurities from FeMoco preparations. The technique will measure total S and not just acid-labile S^{2-} in FeMoco. Preliminary data from this laboratory (Eady et al., 1980) using this radioactive method indicated that there were 4 S atoms/Mo in FeMoco. However subsequent experiments gave results ranging between 3.7 and 12.5 S/Mo. It was provisionally concluded that this method suffered from the difficulty of removing S-containing impurities since the disparate results had been obtained even after all the preparations had been chromatographed on Sephadex G-100 in N-methylformamide. However Nelson et al. (1983) did manage to obtain self consistent results with this method for FeMoco isolated from Cp1 and found 8.7±1.0 nmol of non-amino acid sulphur/ng atom Mo. They calibrated the ratio of ^{35}S to total S in the medium by using purified 8Fe-8S ferredoxin and Cp2 from the same organism. Their preparations were contaminated with about 2 cysteine and 1 methionine residue per Mo atom. In addition the reported activity/ Mo for their FeMoco preparations was only 175±25 units/ng atom Mo. This activity is only about two-thirds of that associated with the best FeMoco preparations (see above), and indicates that their preparations probably contained Mo from inactivated FeMoco. The S content of such inactivated species is unknown.

Because of these uncertainties an alternative method of identifying those S atoms associated with active FeMoco was sought. In this method ^{35}S-labelled FeMoco was isolated as above and then combined with NifB⁻Kp1 to form Kp1. The total activities of the FeMoco and NifB⁻Kp1 were matched so that only active FeMoco bound to NifB⁻Kp1 to form Kp1. The latter was then purified. Radioactivity measurements were standardised against the ^{35}S-Kp2 isolated from the same cells as the ^{35}S-Kp1 from which the ^{35}S-FeMoco was prepared. One molecule of Kp2 (M_r = 66,000) contains 18 cysteine and 30 methionine residues and 4 acid-labile S atoms.

Unfortunately direct measurement of the Mo content of the purified Kp1 was not useful since all NifB⁻Kp1 preparations contain some Mo that apparently copurifies with the active protein (Hawkes, Smith, 1984). However the Mo content of the active protein, generated from the active FeMoco could be estimated from the known relationship between the Mo content and activity of the best FeMoco and Kp1 preparations i.e. 275±25 units/ng atom Mo. The Kp1 assembled from ^{35}S-FeMoco and NifB⁻Kp1 was purified to a constant relationship between C_2H_2-reducing activity and radioactivity. The S:Mo ratios determined for the last two purification stages of four preparations were averaged to yield S:Mo of 8.82±0.55.

We therefore consider the stoichiometry of the FeMoco centre to be Mo Fe$_{6-8}$S$_{8-10}$.

There have been no definitive studies on the presence or absence of an essential organic component in extracted FeMoco. It is generally agreed that it contains no essential amino acids (Smith, 1980; Yang et al. 1982) and suggestions (Levchenko et al., 1980) of coenzyme A or lipoic acid as endogenous components have been discounted in later work which also ruled out the presence of common sugars (Yang et al., 1982). Nevertheless FeMoco, unlike other FeS complexes, is remarkably resistant to attack from iron-chelating reagents. This stability would be more easily understood if an endogenous organic component were present and bound to the Fe atoms.

THE DIFFERENCE BETWEEN WILD-TYPE AND NIFV⁻FEMOCO

The substrate reduction properties associated with the MoFe protein from NifV mutants can be ascribed to an incompletely processed FeMoco. Briefly, NifV⁻ nitrogenase reduces N_2 poorly, its hydrogen evolution (unlike that of the wild-type enzyme) is partly inhibited by CO and it is more effective at reducing cyanide than the wild-type enzyme (Smith et al. 1984). These properties, as we have argued elsewhere (Smith et al., 1984) are consistent with the mutant enzyme operating at a higher redox potential than the wild-type.

It is clearly of great interest to discover what differences between wild-type and NifV⁻FeMoco lead to these altered substrate reduction properties. Modification of FeMoco by the nifV gene product leads to the formation of a more effective nitrogenase and thus definition of this modification could considerably aid our understanding of the specific requirements for the activation of N_2.

Within experimental error the Fe:Mo ratio in NifV⁻FeMoco is indistinguishable from that in wild-type (Hawkes et al., 1984). These measurements do not rule out the possibility that there could be one Fe more or less in the mutant species.

Mo EXAFS spectroscopy has provided the most information on the immediate coordination sphere of the Mo in the MoFe protein and FeMoco. We have therefore compared (Eidsness et al., this volume) the Mo EXAFS of wild-type and NifV⁻Kp1 in three oxidation levels; (a) as prepared in $Na_2S_2O_4$, (b) super-reduced during enzyme turnover and (c) after oxidation with methylene blue. In no case was any significant difference in the Mo environment observed. We have also compared the low-temperature magnetic circular dichroism of the two types of MoFe protein and find no significant difference.

One well-characterised reaction of FeMoco is its binding of thio-phenol which results in a sharpening of the epr spectrum in N-methyl-formamide. Quantitation of this effect with the wild-type enzyme indicated that one thiophenol molecule bound to each FeMoco molecule with a relatively tight binding constant (Burgess et al., 1980). We have compared the binding of the thiophenol to wild-type and NifV⁻FeMoco and once again no significant differences were found. Thus the differences between the two types of FeMoco remain to be elucidated but must be rather subtle. If an organic component is present in FeMoco it is a probable candidate for the site of modification by the nifV gene product.

BIOSYNTHESIS OF FEMOCO IN KLEBSIELLA PNEUMONIAE

The biosynthesis of FeMoco is clearly very complex. The nifB, nifN and nifE genes are essential under all conditions and the nifQ gene is required for growth in media with low levels of Mo. Furthermore the nifV gene is necessary for complete processing of FeMoco (Dixon, 1984).

We have found (Filler et al., this volume) that strains carrying polar insertions in nifD, which are unable to synthesise the MoFe protein polypeptides are nevertheless able to synthesise FeMoco. Thus FeMoco is synthesised before insertion into the MoFe protein. However no FeMoco activity was found in extracts of a strain with a nifKDHJ deletion. These observations indicated that the nifH product and/or the nifJ product were involved in FeMoco biosynthesis.

Further investigations of this phenomenon revealed that a mutant with a promoter-distal insert in nifH could also synthesise FeMoco, but one with a promoter proximal insertion did not. These data indicated that at least part of the nifH gene or its product were essential for FeMoco synthesis. To clarify the matter a low copy number nifH plasmid and a multicopy number nifJ plasmid were transformed, separately and together, into the nifKDHJ deletion strain. A low copy number nifH plasmid was used since multiple copies of the nifH promoter have been shown to prevent expression from other nif promoters. Crude extracts of strains containing the nifH plasmid exhibited FeMoco activity whereas those with the nifJ plasmid alone did not. When both plasmids were present the FeMoco activity of the extract was approximately the same as that observed in extracts of strains containing just the nifH plasmid, indicating that there was no multicopy inhibition by the nifJ promoter.

The above data demonstrate that the nifH gene or its product (or part of it) is essential for observable FeMoco activity in the absence of the MoFe protein polypeptides. This was a very surprising result since the nifH gene product had previously been thought of solely as the Fe protein polypeptide, an essential part of nitrogenase. Its additional role could be an involvement in (a) the expression of the genes known to be essential for FeMoco synthesis or (b) the stability of their products or (c) some other stage of FeMoco biosynthesis.

We have constructed lac fusions within the nifENX and nifUSVM operons in strains carrying polar mutations in nifH. These strains express normal levels of β-galactosidase on derepression demonstrating that these operons can all be transcribed in the absence of nifHDK expression. Thus it is unlikely that nifH is involved in the control of the transcription of genes involved in FeMoco biosynthesis. It may however play a role in stabilising FeMoco as it is formed. Either by forming a complex with proteins binding FeMoco (and thus e.g. preventing proteolytic attack) or by maintaining a low redox potential in the cell thus reducing and stabilising FeMoco.

A further comment on the role of nifJ is apposite. The nifJ product is a pyruvate flavodoxin oxidoreductase which metabolises pyruvate to provide the reducing power for nitrogen fixation (Dixon, 1984). We have confirmed earlier reports (Hill, Kavanagh, 1980) that extracts of nifJ mutants have low Kpl activities and have found that such extracts can be further activated by addition of FeMoco. However our results above demonstrate clearly that nifJ is not required for FeMoco biosynthesis. It may have a role in stabilising the MoFe protein polypeptides before their activation with FeMoco or in helping to provide a low potential to stabilise FeMoco.

In summary, our results have demonstrated a definite requirement for the nifH gene or its product in the synthesis of FeMoco in the absence of the MoFe protein polypeptides. NifH is probably not involved in the control of the transcription of genes required for FeMoco biosynthesis.

CONCLUSIONS

1. There is now strong evidence that FeMoco forms the substrate-binding site in conventional nitrogenase. Isolated FeMoco binds two CN^- ions per Mo.
2. The stoichiometry of isolated FeMoco is Mo $Fe_{6-8} S_{8-10}$. There is no definitive evidence on the presence or absence of an endogenous organic component.
3. No physicochemical differences have yet been found between wild-type and $NifV^-$FeMoco.
4. FeMoco can be biosynthesised in the absence of the MoFe protein polypeptides. NifH but not nifJ is required for this process.

REFERENCES

Bishop PE et al. (1980) Proc.Natl.Acad.Sci.U.S.A. 77, 7342-7346.
Burgess BK et al. (1980) J.Biol.Chem. 255, 353-356.
Cramer SP et al. (1978) J.Amer.Chem.Soc. 100, 3814-3819.
Dixon RA (1984) J.Gen.Microbiol. 130, 2745-2755.
Eady RR et al. (1980) In Stewart WDP and Gallon JR, eds, Nitrogen Fixation, pp.19-35, Academic, London.
Hawkes TR and Smith BE (1984) Biochem.J. 223, 783-792.
Hawkes TR et al. (1983) Biochem.J. 211, 495-497.
Hawkes TR et al. (1984) Biochem.J. 217, 317-321.
Hill S and Kavanagh EP (1980) J.Bacteriol. 141, 470-475.
Levchenko LA et al. (1980) Biochem.Biophys.Res.Comm. 96, 1384-1392.
Li J-g et al. (1982) Biochemistry 21, 4393-4402.
Lowe DJ et al. (1985) In Harrison PM ed, Metalloproteins, Part I, pp.207-249, Verlag Chemie, Weinheim.
Nelson MJ et al. (1983) Proc.Natl.Acad.Sci.USA, 80, 147-150.

Richards AJM (1985) Ph.D. Thesis, University of East Anglia, U.K.
Shah VK and Brill WJ (1977) Proc.Natl.Acad.Sci.U.S.A. 74, 3249-3253.
Shah VK et al. (1984) Ann.Rev.Biochem. 53, 231-257.
Smith BE (1977) J.Less Common Met. 54, 465-475.
Smith BE (1980) In Newton WE and Otsuka S, eds., Molybdenum Chemistry of Biological Significance pp.171-190, Plenum, New York and London.
Smith BE et al. (1973) Biochem.J. 135, 331-341.
Smith BE et al. (1984) In Veeger C and Newton WE, eds., Advances in Nitrogen Fixation Research, pp.139-142, Martinus Nijhoff/Junk, The Hague.
Thorneley RNF and Lowe DJ (1983) Biochem.J. 215, 393-403.
Thorneley RNF and Lowe DJ (1984) Biochem.J. 224, 887-894.
Yang S-S et al. (1982) J.Biol.Chem. 257, 8042-8048.

REDOX AND COMPOSITIONAL INSIGHTS INTO THE IRON-MOLYBDENUM COFACTOR OF AZOTOBACTER VINELANDII NITROGENASE AS A GUIDE TO SYNTHESIS OF NEW Mo-Fe-S CLUSTERS

WILLIAM E. NEWTON[1]/STEPHEN GHELLER[1]/FRANKLIN A. SCHULTZ[2]/BARBARA K. BURGESS[3]/STEVEN D. CONRADSON[4]/JOHN W. MCDONALD[5]/BRIT HEDMAN[6]/KEITH O. HODGSON[6]. [1]WESTERN REGIONAL RESEARCH CENTER, USDA-ARS, BERKELEY, CA 94710, USA AND DEPARTMENT OF BIOCHEMISTRY AND BIOPHYSICS, UNIVERSITY OF CALIFORNIA, DAVIS, CA 95616, USA. [2]DEPARTMENT OF CHEMISTRY, FLORIDA ATLANTIC UNIVERSITY, BOCA RATON, FL 33432, USA. [3]DEPARTMENT OF MOLECULAR BIOLOGY AND BIOCHEMISTRY, UNIVERSITY OF CALIFORNIA, IRVINE, CA 92717, USA. [4]DEPARTMENT OF CHEMISTRY, HARVARD UNIVERSITY, CAMBRIDGE, MA 02136, USA. [5]BATTELLE-KETTERING LABORATORY, YELLOW SPRINGS, OH 45387, USA. [6]DEPARTMENT OF CHEMISTRY, STANFORD UNIVERSITY, STANFORD, CA 94305, USA.

1. INTRODUCTION

The larger of the two component proteins of nitrogenase, the molybdenum-iron protein (MoFe protein), contains the iron-molybdenum cofactor (FeMoco), which is generally recognized as the enzymic N_2-reducing site (Hawkes et al., 1984). Two such cofactors are contained within the MoFe protein, each of which accounts for one Mo, ca. 50% (6-8 atoms) of the Fe and probably 8-9 S atoms (Shah, Brill, 1977; Yang et al., 1982; Nelson et al., 1983), arranged in a cluster so far unrecognized elsewhere in biology. Many on-going syntheses of Mo-Fe-S cluster compounds to simulate the structure or duplicate the chemical properties of FeMoco (Holm, 1981; Coucouvanis, 1981; Averill, 1983), although interesting in their own right, have yet to produce a close chemical model. Additional redox and compositional probes of FeMoco, extruded from Azotobacter vinelandii nitrogenase, have been undertaken recently to gain added insight into this entity and to produce guidelines for the bioinorganic chemist in his quest for the successful synthesis of FeMoco. These probes cover the application of physical techniques as well as successful attempts to perform chemistry on the natural cluster.

On-going XAS studies of nitrogenase (Newton et al., 1984) continue to give the most definitive information about the environment of the Mo atoms in both bound and extruded FeMoco. Recently acquired, good quality Mo K edge x-ray absorption data have confirmed that the Mo in the dithionite-reduced state of both the MoFe protein and free FeMoco has light atom (O or N), as well as S and Fe, neighbors. Information gleaned from Se EXAFS on phenylselenol-treated FeMoco are reported and identify the binding site. Independent compositional insights have been obtained from oxidative decomposition of FeMoco in the presence of the thiomolybdate trapping agent, $(PPh_3)_2CuNO_3$ (Muller et al., 1983; Gheller et al., 1984).

As the putative active site, FeMoco must function in electron transfer to substrate. Thus, the characterization of its electrochemical properties, both to compare with those of protein-bound FeMoco and to establish criteria for chemical models, is a high priority endeavor. The redox properties of isolated FeMoco have been characterized by both cyclic voltammetry (CV) and chemical redox titrations with voltammetric, potentiometric and EPR spectroscopic monitoring (Schultz et al., 1985) and are compared here with those of both the protein-bound system and some cluster compounds.

2. PROCEDURES

2.1 Materials and Methods

2.1.1. MoFe protein of A. vinelandii. Purification to homogeneity was as described previously (Burgess et al., 1980a); specific activities were > 2000 nmole C2H2 reduced (min)$^{-1}$(mg protein)$^{-1}$.

2.1.2. FeMoco. The HCl/NaOH modification (Yang et al., 1982) of the original isolation method (Shah, Brill, 1977) was used. The NMF extracts were concentrated by either vacuum distillation and centrifugation (Burgess et al., 1980a) or chromatography on DE-52 DEAE-cellulose in NMF (Schultz et al., 1985). Sample solutions had [Mo] = 0.6-1.7 mM, Fe:Mo = 6.05-7.10, and activities in the UW45 reconstitution assay (Shah, Brill, 1977) of 200-275 nmol C2H2 reduced (min)$^{-1}$ (ng-atom Mo)$^{-1}$. All sample solutions were 1-2 mM in Na2S2O4. Activity and EPR measurements determined that sample integrity decreased by <10% after XAS and electrochemical measurements.

2.1.3 Synthetic Compounds. The synthetic Mo compounds used were prepared by methods in the indicated literature references.

2.1.4. XAS Measurements. All data were acquired at the Stanford Synchrotron Radiation Laboratory under dedicated synchrotron radiation production conditions. Synthetic Mo and Se compounds were measured as solids at ambient temperature and -150° respectively. Mo spectra of MoFe protein and FeMoco samples were obtained in aqueous buffer at 2° to 4°C and NMF solutions at -15° to -25°C, respectively. Se data on phenylselenol-treated FeMoco were collected at ca. -150°C. All measurements were performed under anaerobic conditions.

2.1.5. Electrochemistry and Chemical Redox Titrations. Cyclic voltammetry was conducted anaerobically in a Vacuum Atmosphere glove box using a small-volume (100 ul) Teflon cell with an inverted glassy carbon (working) electrode for a base. A Pt wire auxilliary electrode was inserted through the Teflon wall. A miniature Ag/AgCl(1M to KCl) electrode was used as reference by immersion directly in the sample solution. About 60 ul of FeMoco solution was used for electrochemical measurements. Cyclic voltammograms were generated with a Bioanalytical Systems CV-1B potentiostat. All electrochemical and recording equipment was connected to the cell by shielded cable passing through an air-tight seal in the glove box. Except as noted, all potentials are corrected to NHE. For chemical titrations, calculated quantities of aqueous 0.02 M Na2S2O4 were added to 350 ul aliquots of FeMoco solution. 60 ul of each aliquot were measured into the electrochemical cell and the remainder frozen in a matched, quartz EPR tube. Equilibrium potentials were the potential differences between the glassy carbon and Ag/AgCl electrodes and the CV trace was recorded.

2.1.6. Oxidative Decompositions. Treatment of chromatographically purified FeMoco (300-350 ul of a >1 mM in Mo anaerobic NMF solution) with an excess of (Ph3P)2CuNO3 (10 mg) in acetonitrile in air resulted in the formation of red, but poorly formed, crystals after several days at 5°C.

Visible (λ max, 380 nm in CHCl3) and infrared (νMo=O, 910 cm^{-1}; νMo-S-Cu, 445/452 cm^{-1}) spectroscopy and low precision x-ray data are all consistent with the cubane (Muller et al., 1983) formulation, [OMo(CuPPh3)3S3Br].

3. RESULTS AND DISCUSSION

3.1 Compositional Insights

3.1.1. X-Ray Absorption Spectroscopy (XAS) at the Mo Edge. The most incisive structural information concerning FeMoco continues to come from XAS studies (Newton et al., 1984; Conradson et al., 1985). Very recent studies of the Mo XANES of FeMoco, both protein-bound and free, plus various Mo compounds show a distinct correlation between the shapes of the absorption edge features and the environment of nearest neighbor atoms around the Mo atom. The Mo compounds examined cover mononuclear and cluster types, some of which also contain ligated thiolates and/or alkoxides. Their XANES divide these compounds into three subsets, which correlate closely with structural aspects.

All members of one subset contain **tetrathiomolybdate** as the common unit and include [NH4]2[MoS4], [X2FeMoS4]$^{2-}$ (X = PhS, Cl) (Coucouvanis et al., 1983; Tieckelmann et al., 1980), [(Cl2Fe)2MoS4]$^{2-}$(Coucouvanis et al., 1984), and [Fe(MoS4)2]$^{3-}$ (McDonald et al., 1980; Coucouvanis et al., 1980). Spectral features include inflection points near 20004 and 20011 eV, corresponding to a lower energy shoulder and the principal absorption edge discontinuity, respectively. The double-cubanes [Mo2Fe6S8(SEt)9]$^{3-}$ (Acott et al., 1979; Wolff et al., 1979) and [Mo2Fe7S8(SEt)12]$^{3-/4-}$ (Wolff et al., 1980) plus Mo(S2C6H4)3 (Stiefel et al., 1966) have the Mo-S6 coordination unit as a common feature. This group of spectra reveal a single inflection point at ca. 20011 eV, which corresponds to the principal absorption discontinuity. The third group encompasses the single-cubane, [MoFe3S4(SEt)3Fe(cat)3]$^{3-}$ (Wolff et al., 1981) and the methoxybridged dicubane [Mo2Fe6S8(SEt)6(OMe)3]$^{3-}$ (Acott et al., 1979), both of which have MoS3O3 coordinated units. The inflection point of the principal absorption discontinuity occurs at 20013 eV, which is slightly, but consistently, higher than those for the Mo-S6 subset. In contrast to other subsets, these spectra exhibit a shoulder at higher energies.

Because the XANES correlates strictly with the structure of the first shell of the Mo atom's nearest neighbors, insight should be gained about the Mo site of FeMoco by simply comparing XANES spectra. Close correspondence would imply structural and chemical similarity more than identity, while a difference is an unequivocal indication of a structural difference. The XANES of both the bound and free FeMoco samples are very similar to one another with their inflection points at 20,012-20,013 eV with a shoulder at 20,018-20,020 eV. Empirical comparisons produce the following unambiguous conclusions about the environment around Mo: it is most closely approached by compounds with a MoS3O3 coordination unit; the nearest neighbors of Mo in nitrogenase and in free FeMoco consist of a significant number of both singly bound hard and soft ligands; and any changes in the structure resulting from the extraction of FeMoco from the protein must be sufficiently small so that the basic structure is conserved. These conclusions based on XANES studies are completely consistent with and complementary to results from Mo EXAFS studies (Newton et al., 1984). Thus, a structural criterion is available against which the bioinorganic chemist can judge his progress in cofactor synthesis.

3.1.2 <u>XAS at the Se Edge</u>. Phenylthiol addition to dilute (ca. 0.1 mM in Mo) FeMoco solutions sharpens the EPR signal (Rawlings et al., 1978; Burgess et al., 1980b) to more closely resemble that of the intact protein. We have found that phenylselenol has an identical effect. This similarity has been used to probe the binding site on FeMoco for selenol (and by analogy thiol) by XAS on the phenylselenol-treated cofactor. Mo EXAFS results have indicated that Se is not a nearest neighbor of Mo (S. Conradson, K. Hodgson, B. Burgess, W. Newton, unpublished) and 19-F NMR studies of p-trifluoromethylphenylthiol-treated FeMoco have suggested Fe as the probable thiol-binding site (Mascharak et al., 1982). EXAFS data collected at the Se edge are consistent with these other spectroscopic results and clearly show Fe and C as nearest neighbors of Se. Thus, a thiol-displaceable ligand is an additional requirement for an exact chemical model of FeMoco.

3.1.3. <u>Oxidative Elicitation of Thiomolybdates</u>. Support of the conclusions from the Mo K-edge XAS studies of a mixed O (or N) and S coordination sphere for FeMoco was sought through controlled oxidative degradation of FeMoco. Because thiomolybdates can be released from the intact MoFe protein by appropriate treatment (Zumft, 1978) and with the knowledge that FeMoco is decomposed with respect to its iron atoms by dioxygen (Yang et al., 1982), we attempted the release of Mo by simple oxgenation of FeMoco solutions. Spectral monitoring over 350–800 nm showed some striking changes, consistent with the release of thiomolybdates. A steeply rising background starting at ca. 400 nm in the final solution, however, made quantitative spectral simulation difficult. This problem was overcome when $(PPh_3)_2CuNO_3$ was employed as a trap. The red crystals, formed when FeMoco was mixed with an aerobic $(PPh_3)_2CuNO_3$ solution, were shown to be $[OMo(CuPPh_3)_3S_3Br]$ by a combination of visible, infrared and Raman spectroscopies, plus x-ray data. Even though these last data were severely limited by poor crystal quality, they clearly showed the basic cubane structure and bromide.

These data may reasonably be interpreted as a MoS_3 core for FeMoco, consistent with our EXAFS and XANES results. In fact, these data constitute a refinement in the number of S neighbors to Mo in extruded FeMoco, the softest parameter from the EXAFS investigation, by clearly distinguishing between the possibilities of 3 and 4. Further because a mixture of tetra- and tri-thiomolybdates is observed with the intact MoFe protein (Zumft, 1978), a real difference is indicated in line with the EXAFS results in the first coordination sphere of Mo within and outside the protein matrix.

3.2. <u>Oxidation-Reduction Properties</u>

3.2.1 <u>FeMoco</u>. Three FeMoco oxidation states exist within the protein: (i) FeMoco(s-r), the EPR-active, S=3/2, semi-reduced form, maintained in excess $Na_2S_2O_4$; (ii) FeMoco(ox), with S=0 and EPR silent, prepared by dye oxidation of FeMoco(s-r); and (iii) FeMoco(red), the integer spin, EPR-silent, substrate-reducing state, achieved only in the full nitrogenase system (Munck et al., 1975; Zimmerman et al., 1978; O'Donnell, Smith, 1978; Lough et al., 1983). Even though the MoFe protein does not exchange electrons directly with electrodes, redox mediators, such as viologens (Lough et al., 1983), have allowed the measurement of its redox couples. Both observed couples exhibit considerable hysteresis and the average potentials for Av1 are −0.17 V and −0.465 V vs. NHE. Cyclic voltammetry of oxidized, EPR-silent FeMoco also shows two redox processes,

one beginning at ca. −0.4 V vs. NHE and a second at ca. −1.1 V with the related oxidation waves at −0.21 V and −0.85 V, respectively. The first
couple likely involves interconversion of oxidized and semi-reduced forms,
while the second quasi-reversible redox process likely corresponds to reduction to the substrate-reducing state. Peak potential separation in
both couples of ca. 120-240 mV (at $v = 0.1$ V s^{-1}) indicates some charge
transfer irreversibility accompanying redox, in line with the hysteresis
in the protein. The values of the current parameter for reduction or oxi-
dation of FeMoco(s-r) are 331 and 284 uA s$^{\frac{1}{2}}$ v$^{-\frac{1}{2}}$ cm^{-2} mM^{-1}, respectively.
These two values are not unreasonable for one-electron transfers in NMF
(cf. 375 for methyl viologen) and support a one-electron stoichiometry
for both redox processes.

The one-electron stoichiometry of the −0.3 mV couple was confirmed by
titration of EPR-silent FeMoco with sodium dithionite while the CV,
potentiometric and EPR responses were monitored. As dithionite was added: (i) the solution rest potential shifted to values more negative
than −0.3 V; (ii) primary reduction current decreased and primary oxidation current increased; and (iii) the S=3/2 EPR signal developed to
maximum intensity after addition of 0.78 electron per g-atom Mo. A
Nernst plot of these data had a potential at half-intensity of −0.30 V
and a reciprocal slope of 65 mV (n = 0.91 at 25°C). These results confirm that FeMoco is converted from its oxidized to semi-reduced state by
a one-electron transfer. Because the potentials involved are very similar
and because we could monitor oxidized and reduced species during the
course of these titrations, the chemical and electrochemical processes
are the same.

3.2.2. Synthetic Clusters. Attempts were made to determine the electrochemical behavior of the following compounds in 0.1 M [Et$_4$N]Br/NMF:
[Fe$_4$S$_4$(SPh)$_4$]$^{2-}$ (Hill et al., 1977); [Fe(MoS$_4$)$_2$]$^{3-}$ (McDonald et al., 1980);
[Fe(WS$_4$)$_2$]$^{3-}$ (Friesen et al., 1983); and [Mo$_2$Fe$_6$S$_8$(SPh)$_9$]$^{3-}$ (Christou,
Garner, 1980), but each either decomposes in NMF or exhibits behavior so
different from its reported electrochemistry in other solvents that a
significant chemical change must have taken place. This instability is an
obvious indication of considerable differences between these models and
FeMoco. Thus, a new criterion for cofactor models is two quasi-reversible
redox couples, separated by 0.7 V, at ca. −0.3 V and −1.0 V vs. NHE in a
polar aprotic solvent (or at ca. + 0.1 V and −0.6 V vs. NHE in water).
Currently available electrochemical data for Mo-Fe-S-O cluster complexes
show that none fully satisfy this criterion, although of course, no data
set was collected in NMF.

4. ACKNOWLEDGEMENTS

We gratefully acknowledge support through grants No. PCM-81-10355 (to
WEN and BKB), PCM-82-08115 (to KOH) and CHE-82-11694 (to FAS) from NSF
and 5-RO1-AM-30812 (to WEN and BKB) from NIH. Stanford Synchrotron
Radiation Laboratory is supported by DOE, NIH and NSF.

5. REFERENCES

Acott SR et al. (1979) Inorg. Chim. Acta 35, L337.
Averill BA (1983) Struct. Bonding (Berlin) 53,59.
Burgess BK et al. (1980a) Biochim. Biophys. Acta 614, 196.
Burgess BK et al. (1980b) J. Biol. Chem. 255, 353.

Christou G and Garner CD (1980) J. Chem. Soc., Dalton Trans., 2354.

Conradson et al (1985) J. Am. Chem. Soc., in press.

Coucouvanis D et al. (1980) J. Am. Chem. Soc. 102, 6644.

Coucouvanis D (1981) Acc. Chem. Res. 14, 201.

Coucouvanis D et al. (1983) Inorg. Chem. 22, 293.

Coucouvanis D et al. (1984) Inorg. Chem. 23, 741.

Friesen GD et al. (1983) Inorg. Chem. 22, 2202.

Gheller SF et al. (1984) Inorg. Chem. 23, 2519.

Hawkes TR et al. (1984) Biochem. J. 217, 317.

Hill CL et al. (1977) J. Am. Chem. Soc. 99, 2549.

Holm RH (1981) Chem. Soc. Rev. 10, 455.

Lough S et al. (1983) Biochemistry 22, 4062 and refs. therein.

Mascharak PK et al. (1982) Proc. Nat. Acad. Sci. USA 79, 7056.

McDonald JW et al. (1980) Inorg. Chim. Acta 46, L79.

Muller A et al. (1983) Inorg. Chim. Acta 69, 5.

Munck E et al. (1975) Biochim. Biophys. Acta 400, 32.

Nelson MJ et al. (1983) Proc. Nat. Acad. Sci. USA 80, 147.

Newton WE et al. (1984) In Veeger C and Newton WE, eds. Advances in Nitrogen Fixation Research, p 160, Martinus Nijoff/Junk, The Hague.

O'Donnell MJ and Smith BE (1978) Biochem. J. 173, 831.

Rawlings J et al. (1978) J. Biol. Chem. 253, 1001.

Schultz FA et al. (1985) J. Am. Chem. Soc., in press.

Shah VK and Brill WJ (1977) Proc. Nat. Acad. Sci. USA 74, 3249.

Stiefel EI et al. (1966) J. Am. Chem. Soc. 88, 2956.

Tieckelmann RH et al. (1980) J. Am. Chem. Soc. 102, 5550.

Wolff TE et al. (1979) J. Am. Chem. Soc. 101, 4140.

Wolff TE et al. (1980) J. Am. Chem. Soc. 102, 4694.

Wolff TE et al. (1981) Inorg. Chem. 20, 174.

Yang S-S et al. (1982) J. Biol. Chem. 257, 8042.

Zimmerman R et al. (1978) Biochim. Biophys. Acta 537, 185.

Zumft WG (1978) Eur. J. Biochem. 91, 345.

RAPID, SENSITIVE HPLC ANALYSIS OF AMMONIA AND METHYLAMINE IN NITROGENASE REDUCTIONS

H. ERAN, F.X. ZHANG, M. BRAVO AND C.E. McKENNA, DEPARTMENT OF CHEMISTRY, UNIVERSITY OF SOUTHERN CALIFORNIA, LOS ANGELES, CA 90089-1062, U.S.A.

A dansyl chloride (DNSCl) pre-column derivatization method has been developed for High Performance Liquid Chromatography (HPLC) analysis of NH_3 and/or CH_3NH_2 produced by nitrogenase-catalyzed reduction of substrates such as N_2 or diazirine (McKenna et al, 1984). The DNSCl reagent can be used immediately after preparation and is stable at 4^0 for 1 mo. The derivatization products from NH_3 and CH_3NH_2 were prepared by direct treatment of the assay mixture (terminated by 0.1 mL 1 N HCl in sat. KIO_3 and centrifuged briefly at 2000 rpm) with 1 eq. 0.045% DNSCL (acetone)and 1 eq. 0.164 M Na borate. After 30 min incubation, the amine chromophores are stable in air and ambient light at room temperature. They were separated isocratically within several min on a 150 x 4.6 mm Ultrasphere-ODS 5 µm C-18 reversed-phase (RP) column (Altex) at 2000-2500 psi (1.5 mL/min) using an eluent of 7:7:3 H_2O:methanol:acetonitrile and were detected fluorometrically (360 nm excitation; 500 nm emission). Enhanced performance was obtained by use of an anion-exchange resin guard column. The NH_3 sensitivity is limited by the background NH_3 in the reagents and under practical conditions is 0.01-0.02 nmol (20 µl injection volume). CH_3NH_2 sensitivity is higher. The fluorescence intensity for either product is linear over an extended concentration range. NaN_3, KCN, N_2H_4 and $(CH_3)_2NH$ did not interfere with NH_3 and CH_3NH_2 response.

The DNSCl HPLC method described here offers the advantages of simultaneous as well as individual determination of NH_3 and CH_3NH_2 with high sensitivity, convenience, flexibility and speed. The reagents and derivatives are stable and the procedure is simple to use. We have also found it useful in other applications, such as measurement of NH_3 in culture media.

REFERENCES

McKenna CE et al (1984) In Veeger C and Newton WE, eds., Advances in Nitrogen Fixation Research, pp 115-122, Martinus Nijhoff/Junk, The Hague.

ACKNOWLEDGEMENTS

We gratefully acknowledge the support of The National Institutes of Health, USDA and the H. Frasch Foundation. F. X. Zhang was a UNESCO Fellow.

NITROGEN FIXATION IN <u>NIF</u> <u>HDK</u> DELETION STRAINS OF <u>AZOTOBACTER</u> <u>VINELANDII</u>

Brian J. Hales, Ellen E. Case and Dieter Langosch
Department of Chemistry
Louisiana State University
Baton Rouge, Louisiana 70803

A nif HDK deletion of Azotobacter vinelandii was constructed by transformation and recombination with plasmid pDB 12. The resultant strain (LS10), lacking the structural genes for nitrogenase, is Nif⁻ on norman N-free Burk's medium (i.e., containing 10μ M Mo). LS10 was spread onto agar plates containing normal N-free Burk's medium. After 10-14 days several colonies appeared and were isolated. These isolates, although mutants of LS10, grew on N-free medium in either the presence or the absence of added Mo as well as the presence of 10^{-2}M tungsten.

One of these isolates (LS15) was grown and lysed by osmotic rupture, heat treated at 51°C for 5 min., centrifuged and the supernatent loaded onto a DEAE column previously equilibrated with 50mM Tris pH 8.0 containing 2mM dithionite. Using a linear gradient (.1 - .5M NaCl) component 1 activity eluted at .15 M NaCl similar to the conventional component 1 while component 2 activity eluted at .20M NaCl compared to .27M NaCl for the wild-type component.

Both components were concentrated on DEAE and fractionated on a Sephacryl S-200 column. Component 2 migrated with an apparent molecular weight of ca. 60k making it slightly smaller than Av_2 (Mol. Wt. 63k). On the other hand, component 1 migrated in the 110 - 150k range making it's molecular weight approximately half that of the wild-type component. The esr spectrum of the partialy purified component 1 exhibited no signals with g = 3.6, a region normally characteristic of the S=3/2 state of FeMo-co, nor did analysis of the component yield any detectable amounts of Mo.

The LS15 component 2 was further purified using Sephadex G-100. The polypeptides for this component on SDS-PAGE showed a dominant band at ca. 29k compared to 31.4k for Av_2. The esr spectrum was broader than that of the wild-type in the g = 2 region and contained a higher relative contribution from the S = 3/2 state in the g = 5 region in Tris buffer at pH 7.4. Using C_2H_2-reduction to monitor activity, it was found that the LS15 component 1 coupled more efficiently to its own component 2 than to Av_2. Both LS15 and wild-type component 2, on the other hand, coupled equally well to Av_1. Finally Ouchterlong plates showed crossreaction of polyclonal Anti-Av_2 with LS15 nitrogenase while no crossreaction was observed for Anti-Av_1.

All of these results are consistent with the existence of a alternate nitrogen-fixation system in azotabacter which is completely different from the wild-type enzyme, as originally proposed by Bishop et al. (1980).

References
Bishop, P. E., Jarlenski, D. M. L., and Hetherington, D. R. L. <u>Proc.</u> <u>Natl.</u> <u>Acad.</u> <u>Sci</u> <u>USA</u> (1980) <u>77</u> 7342.

INHIBITION BY NITRITE AND NITRIC OXIDE OF NITROGENASE IN
Rhodopseudomonas sphaeroides f.sp. *denitrificans*

W. P. Michalski, D. F. Mathew and D. J. D. Nicholas
Department of Agricultural Biochemistry, Waite Agricultural Research
Institute, University of Adelaide, 5064, S.A., Australia.

Phototrophic bacterium *R.sphaeroides* f.sp. *denitrificans* can grow
either in light/anaerobic or dark/aerobic conditions.   It also generates
ATP by nitrate respiration in the dark.   (1,2)   Cells grown photoheterotrophically in the presence of nitrate have high activities of
nitrate, nitrite and nitrous oxide reductases, and accumulate a large
amount of nitrite (5 µmol/mg dry wt) in the culture medium (3).   The incubation of denitrifying cells for 30 min with 10% (v/v) acetylene
completely inhibits nitrous oxide (N_2O) reductase activity.   In these
cells the reduction of nitrite (10 µmol/mg dry wt) results in a rapid
production of N_2O (3 umol/mg dry wt/h) followed by an accumulation of
nitric oxide (NO; 0.1 µmol/mg dry wt/h), as analysed by head space gas
chromatography.   Cells grown in the presence of nitrate in light require
ammonia or glutamate as a nitrogen source, since under these conditions
dinitrogen fixation is markedly reduced.   Nitrite ($\geq$1µmol/mg dry wt) or
nitric oxide ($\geq$50 nmol/mg dry wt) completely inhibited nitrogenase (N_2ase)
activity either in washed cells or in those treated with toluene and, in
addition, the rate of $^{14}CO_2$ fixation in washed cells was reduced by a
half.   In the presence of nitrite (1mM) or nitric oxide (1µM) the EPR
signal characteristic of Fe-S centre (g=1.94 at -100°C) was drastically
reduced in washed cells and a signal associated with the iron-nitrosyl
complexes (g=2.035) appeared.   The subsequent addition of $Na_2S_2S_2O_4$ or
sodium ascorbate enhanced these effects six-fold.   Neither nitrite or
nitric oxide affected the protonmotive force or its components ($\Delta\Psi,\Delta$pH)
in N_2-fixing cells.   Thus, it is likely that the inhibition of N_2ase
activity as well as CO_2 fixation by nitrite or NO is associated with a
chemical modification (iron-nitrosyl complex) of Fe-S proteins, e.g.
ferredoxins which transfer electrons to component II (Fe protein) of
nitrogenase and to NAD for CO_2 fixation.

1.   Satoh, T. *et al.* (1976) Arch. Microbiol. **108**, 265-269.

2.   Satoh, T. (1981) Plant & Cell Physiol. **22**, 443-452.

3.   Michalski, W.P. & Nicholas, D.J.D. (1984) J. Gen. Microbiol. **130**,
     155-165.

I. QUANTITATION OF THE NUMBER OF ELECTRONS STORED BY DINITROGENASE DURING THE LAG PERIODS THAT PRECEDE FORMATION OF H_2, HD, AND UPTAKE OF N_2. II. THE EFFECT OF HIGH PRESSURES OF H_2 AND D_2 ON HD FORMATION BY NITROGENASE.

FRANK B. SIMPSON*/JURGEN THIELE**/JIHONG LIANG*/ROBERT H. BURRIS*
*DEPARTMENT OF BIOCHEMISTRY, UNIVERSITY OF WISCONSIN, MADISON, WI
53706, USA. **DEPARTMENT OF BIOCHEMISTRY, MICHIGAN STATE UNIVERSITY,
EAST LANSING, MICHIGAN 48824, USA.

I. We have utilized a membrane leak mass spectrometer to measure the lag periods that precede steady-state formation of H_2, HD, and uptake of N_2. Lags were generated by limiting dinitrogenase reductase. We found the following: H_2 lag, 3.72 sec; HD lag, 10.20 sec; N_2 lag, 10.28 sec. The lag period for N_2 was not affected by inclusion of D_2, and the lag periods for HD and N_2 were equal within experimental error. With the assumption that the electron flux was constant from time zero and equal to the steady state rate, and from knowledge of the [Mo], we calculated the number of electrons (e^-) that entered dinitrogenase per Mo during the lag periods for H_2 and N_2/HD. 7.83 e^- entered dinitrogenase per Mo during the N_2/HD lag, 2.12 of which were evolved as H_2 and 5.71 of which were stored. Evolution of H_2 preceded uptake of N_2 and began after dinitrogenase accumulated 2.82 e^-/Mo. Under conditions of saturating N_2, zero D_2, and high e^- flux, a burst of H_2 evolution of 1.14 H_2/Mo preceded uptake of N_2, and this burst period was equal to the lag period for uptake of N_2. N_2 uptake began after dinitrogenase accumulated 5.66 e^-/Mo. By the time point at which 1.06 N_2/Mo had been taken up, 15.87 e^-/Mo had entered dinitrogenase, 2.50 H_2/Mo had been evolved, and 11.32 e^-/Mo had accumulated within dinitrogenase and/or as reduced forms of N_2. During the lag period that preceded uptake of N_2, 1.14 H_2/Mo were evolved per zero N_2 taken up, and subsequent to the N_2 lag period, i.e., during steady state, 1.56 H_2/Mo were evolved per 1 N_2/Mo taken up. Under argon or in the presence of CO, the lag period for evolution of H_2 was longer than in the presence of N_2, and 5.73 e^-/Mo accumulated during this lag. This result suggests that H_2 can be evolved by two different forms of dinitrogenase. H_2 evolution begins from the 3 e^-/Mo form, and in the absence of N_2 or in the presence of CO, it apparently occurs also from the 6 e^-/Mo form. Thus when dinitrogenase was under argon or in the presence of CO, an average lag for H_2 of between 3 e^-/Mo and 8 e^-/Mo was observed.

II. We have placed dinitrogenase at saturating electron flux in D_2O under 50 atm H_2 plus 2 atm N_2, and also in H_2O under 50 atm D_2 plus 2 atm N_2, and we have measured the H_2, D_2, HD, NH_3, and ND_3 formed. These conditions approximate saturating pN_2 and infinite pH_2 or pD_2. The model for formation of HD proposed by Cleland predicts $1D_2:2HD:0ND_3$ or $1H_2:2HD:0NH_3$ under these conditions, while Guth and Burris predicted $1D_2:6HD:0ND_3$ or $1H_2:6HD:0NH_3$. In 2 experiments under 50 atm H_2, dinitrogenase formed $1D_2:6.55HD:0.077ND_3$ and $1D_2:5.95HD:0.11ND_3$. In 2 experiments under 50 atm D_2, dinitrogenase formed $1H_2:6.78HD:0.14NH_3$ and $1H_2:6.36HD:0.27NH_3$. These results suggest that the prediction of Guth and Burris for formation of HD by nitrogenase is more nearly correct.

KINETICS AND MECHANISM OF THE NITROGENASE ENZYME SYSTEM

Roger N.F. Thorneley, David J. Lowe, Janet Deistung
AFRC Unit of Nitrogen Fixation, University of Sussex, Brighton, BN1 9RQ,
U.K.

Klebsiella pneumoniae nitrogenase is usually present in vivo at high
concentrations (ca. 200 μM Kp1, 500 μM Kp2). It functions very slowly
with a rate dissociation of $Kp2_{ox}(MgADP)_2$ from Kp1 (the rate limiting
step) $k_{-3} = 6.4 \pm 0.8$ s^{-1} (Thorneley, Lowe 1984 and references therein).
Computer simulations have been used to determine the consequences of
increasing the rate of the limiting step and of varying the relative and
absolute concentrations of Kp1 and Kp2. They show that as the Kp2:Kp1
ratio increases above 2:1, the total NH_3 production and percentage of
electron flux used for NH_3 production increase rapidly. As the value of
k_{-3} increases above 6.4 s^{-1} both total NH_3 and efficiency (NH_3:H_2) fall.
Thus any attempt to modify nitrogenase by increasing the rate of the
limiting step would give less NH_3 and more H_2. However, when [Kp1] is
very low (induced by Mo deficiency), the efficiency (NH_3:H_2) is not
impaired by increasing the value of k_{-3} and the total NH_3 formation rate
increases. The synthesis of an alternative Fe protein which dissociates
more quickly from Kp1 would therefore be an advantage under these
conditions. If this involved a different structural gene, this would
have the additional advantage of uncoupling the transcription of the
MoFe and Fe protein structural genes.

Kinetics of electron transfer from the nifF gene product (Kp
flavodoxin hydroquinone form) to $Kp2_{ox}(MgADP)_2$ are rapid $k > 2 \times 10^6$
M^{-1}s^{-1}. The reaction is inhibited by Kp1 protein consistent with the
obligate dissociation of the $Kp2_{ox}(MgADP)_2Kp1$ complex prior to electron
transfer. Dissociation constants for the complexes formed between
$KpFld_{SQ}$ and Kp2, Kp2(MgADP)$_2$ and Kp2(MgATP)$_2$ are >730 μM, 47 μM and
13 μM respectively. This suggests that $KpFld_{SQ}$ can distinguish between
the three different protein conformations associated with Kp2, Kp2(MgATP)$_2$
and Kp2(MgADP)$_2$. Electron transfer to $KpFld_{SQ}$ within these complexes is
relatively slow, being >6.7 s^{-1} for $KpFld_{SQ}$-Kp2, 1.2 s^{-1} for $KpFld_{SQ}$-
Kp2(MgADP)$_2$ and 0.49 s^{-1} for $KpFld_{SQ}$-Kp2(MgATP)$_2$ (Deistung, 1985).

REFERENCES

Thorneley R and Lowe DJ (1984) Biochem. J. 224, 903-909.
Deistung J (1985) D. Phil. Thesis, Submitted, University of Sussex.

KLEBSIELLA PNEUMONIA NITROGENASE ELECTRON TRANSFER COMPONENTS, FLAVODOXIN (NIFF PROTEIN) AND PYRUVATE: FLAVODOXIN OXIDOREDUCTASE (NIFJ PROTEIN)

ROBERT C. WAHL/W. H. ORME-JOHNSON. MASSACHUSETTS INSTITUTE OF TECHNOLOGY, CAMBRIDGE, MASSACHUSETTS 02139 U.S.A.

The nifF protein was isolated from an overproducing strain of Kp which harbored a nif chimeric plasmid in which the nifF gene was under the transcriptional control of both its natural promoter and the nifH promoter, both of which require the nifA gene product for expression. The nifF protein was overproduced at least 250-fold the above level found in wild type, and allowed the purification of 2 umol of flavodoxin in 60% yield from 50 gm of cells grown under N-limiting conditions. The flavodoxin preparation appeared to contain apoprotein, as the molecular weight based on flavin content was 35,000 gm/mol, while SDS-PAGE indicated a subunit molecular weight of 15,000 gm/mol. The flavodoxin supported up to 550 nmol acetylene reduced/min/mg flavodoxin in a nifF crude extract, and was also active in a reconstituted system of purified nifJ protein, Kp1, and Kp2. The midpoint redox potential of the semiquinone to hydroquinone reduction was determined to be -405 mV versus the hydrogen electrode. Under identical conditions, Azotobacter vinelandii flavodoxin midpoint potential was -470 mV, and methyl viologen gave a value of -422 mV.

The catalytic activity of the nifJ protein was assayed by spectrophotometrically monitoring the reduction of methyl viologen, or by acetylene reduction in the presence of nifF protein, Kp2 and Kp1. The specific activity of methyl viologen reduction in crude extracts of derepressed wild type Kp was 0.1-0.3 umol/min/mg, and the specific activity of pure nifJ was 6.8 umol/min/mg. The major EPR signal (at 12K) from dithionite-reduced enzyme had g-values of 2.044, 1.94, 1.88.

This was supported by NIH, grant #GM30943 and fellowship #GM09633.

MOLYBDENUM X-RAY ABSORPTION STUDIES OF ALTERNATE FORMS OF NITROGENASE Fe-Mo PROTEIN

MARLY K. EIDSNESS,[a] STEPHEN P. CRAMER,[b] BARRY E. SMITH,[c] ROGER N. F. THORNELEY,[c] DAVID J. LOWE,[c] C. DAVID GARNER,[d] and ANNETTE FLOOD,[d] [a]Stanford University, Stanford, CA 94305, [b]Exxon Research and Engineering Co., Annandale, NJ 08801, [c]AFRC Unit of Nitrogen Fixation, University of Sussex, Brighton BN1 9RQ, U.K., and [d]University of Manchester, Manchester M13 9PL, U.K.

Comparison of Kp nifV⁻ and Kp wild-type. The Klebsiella pneumoniae nitrogenase mutant Kp nifV⁻ is defective in the reduction of N_2, and the mutation appears to involve a structural change in the Fe-Mo cofactor. To elucidate possible structural changes at the Mo site, Mo edge x-ray absorption spectra were recorded on the 1) semi-reduced (as-isolated), 2) dye-oxidized, and 3) "fixing" enzyme system.

The nifV⁻and wild-type Mo edge structure is nearly identical (see Fig. 1). Small edge shifts are observed in the alternate redox states, but a shift of 1 eV or less is expected for the oxidation of Mo with S ligands. Thus the dye-oxidized forms of nifV⁻ and wild-type probably contain oxidized Mo. Surprisingly, the fixing state edge positions are also at higher energies. Conceivably, Mo is oxidized in the fixing state.

Mo EXAFS of nifV⁻and wild-type were collected at 4 K and the data were analyzed over a k space range of 4-16 Å^{-1}. Fig. 2 shows the Fourier transformed data of the semi-reduced species. Curve-fitting indicates that both forms are best described as containing 4-5 S atoms at 2.36 Å, 2-4 Fe atoms at 2.69 Å and 0-2 O or N atom(s) at 2.19 Å. The spectral similarity suggests that nifV⁻and wild-type have the same local Mo environment.

Rapid Freeze EXAFS. The rapid freeze technique has been used to trap transient intermediates in the N_2 reduction cycle. Mo EXAFS have been measured on Kp wild-type intermediates trapped at 300 ms in the catalytic cycle. Analysis of these preliminary data suggest the presence of 1 N bound to Mo at 1.98 Å.

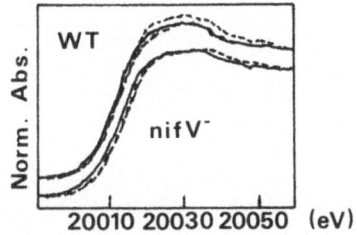

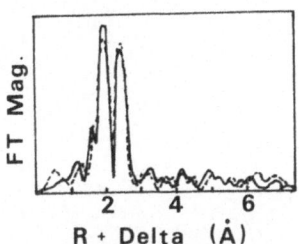

Fig. 1. Mo K edges of Kpl wild-type and Kp nifV. (—) semi-reduced; (- -) dye-oxidized (— —) fixing state.

Fig. 2. Fourier transformed data of Kpl wild-type (—) and Kpl nifV (- -).

618

FeMoco ACTIVITY IN Klebsiella pneumoniae REQUIRES THE nifH GENE

WENDY A. FILLER, RAY A. DIXON and BARRY E. SMITH
AFRC Unit of Nitrogen Fixation, University of Sussex, Brighton, BN1 9RQ
UK.

The MoFe protein of nitrogenase from Klebsiella pneumoniae contains
an iron-molybdenum cofactor, FeMoco, which has been detected in extracts
of strains (NifDK⁻) which synthesise neither of the MoFe protein subunits. However a strain carrying a transposon insertion in nifH (polar
on nifD and nifK), which codes for the Fe protein, Kp2, had no
detectable FeMoco activity. The "FeMoco-less" phenotype of a NifJHDK⁻
deletion strain could be complemented by a low-copy-number plasmid
carrying nifH, confirming that the nifH gene is essential for observable
FeMoco activity in strains lacking the MoFe protein subunits. However
active Kp2 is not required since nifM mutant strains[1] and strains lacking
the C-terminal end of nifH synthesise FeMoco but have no Kp2 activity.
Results from strains carrying nifH insertion mutations and transcriptional
lac fusions in each of the three polycistronic operons (nifENX, nifUSVM
and nifBQ) containing genes known to be essential for FeMoco synthesis,
indicated that nifH is not required for transcription of these genes.

A strain carrying an in-frame deletion in nifH and expressing nifD
and nifK did not exhibit Kp1 activity thus supporting the hypothesis that
nifH is absolutely required for FeMoco synthesis or activity. The MoFe
protein polypeptides present in this strain could not be activated by the
addition of NMF-extracted FeMoco. It is possible that, in the absence of
nifH, an intermediate of the FeMoco biosynthetic pathway or a breakdown
product of active FeMoco accumulates and blocks the FeMoco-binding site
of the Kp1 protein inhibiting its activation by FeMoco added in vitro.

1.   Roberts GP et al. (1978)   J.Bacteriol. 136, 267-279.

CONSTRUCTION OF A BINARY PLASMID SYSTEM TO PRODUCE KLEBSIELLA PNEUMONIAE MOFE PROTEIN AND APO-MOFE PROTEIN IN ESCHERICHIA COLI

GEORGIANNA M. HARRIS/ PAUL V. LEMLEY/ W. H. ORME-JOHNSON. MASSACHUSETTS INSTITUTE OF TECHNOLOGY, CAMBRIDGE, MASSACHU-SETTS 02139 U.S.A.

The aim of this project is to define the minimum set of nif genes required to synthesize FeMoCo-reactivatable molybdenum iron protein (apo-MoFe protein) and catalytically active molybdenum iron protein (MoFe protein). In addition, it should be possible to define processing pathways and to determine the functions of the individual nif gene products in the formation of FeMoCo and apo-MoFe protein. We report the construction and biochemical characterization of a series of plasmids containing various portions of the nif region from the Klebsiella pneumoniae chromosome. These constructions were designed to produce active nitrogenase components. Our strategy will be to delete individual nif genes in order to determine their effect on the apo-MoFe protein and MoFe protein. Utilizing a binary plasmid system to express the nif gene products (P.V. Lemley et al.), Escherichia coli has been cotransformed with a tac-promoted nifA plasmid, pVL15 or pVL16a, and a second plasmid containing the nif genes of interest, pGH1, pGH2, or pGH3. Each of these plasmid combinations was then examined to determine the amounts of MoFe protein and apo-MoFe protein produced.

Results were presented that pGH1 (nif MVSUXNEYKDH) in combination with pVL16a (nif BQA) enabled E. coli to produce Fe protein, MoFe protein and apo-MoFe protein. Active nitrogenase was measured by acetylene reduction, both in vivo and in vitro. Although some catalytically active MoFe protein was produced, addition of FeMoCo to an extract resulted in a 4-40 fold increase in the MoFe protein activity. Deletions of nifM from pGH1 appeared to have little effect on the synthesis of MoFe or apo-MoFe protein in E. coli, while in this system no Fe protein was produced, a result consistent with Howard et al. Although deletion of nifN and E had no effect on the activity of the Fe protein, it did not result in an apo-MoFe protein which could be reactivated by the addition of FeMoCo. Subsequent plasmid constructions are in progress to further investigate these results.

This work was supported by NIH, grant #GM30943.

Lemley, P.V. et al. (1985) these proceedings.
Howard, K. S. et al. (1985) manuscript submitted.

Acknowledgement: The authors wish to thank Dr. S. K. Chapman for his generous gift of FeMoCo.

Mo-EXAFS, MOSSBAUER SPECTROSCOPY, AND IRON, PROTON AND Mo ENDOR OF THE
MoFe PROTEIN FROM KLEBSIELLA PNEUMONIAE nifV MUTANTS.

PAUL MCLEAN(1), ANNE TRUE(2), STEPHEN CHAPMAN(1), MARK NELSON(3), BOON-K
TEO(4), ECKARD MUNCK(5), BRIAN HOFFMAN(2) AND WILLIAM H.
ORME-JOHNSON(1).
1)MASS. INST. TECH., CAMBRIBGE, MA 02139; 2) NORTHWESTERN UNIV., ILL; 3)
E.I.DUPONT DE NEMOURS, DE; 4) BELL LABS., NJ; 5) UNIV. OF MINN., MN.

The nitrogenase MoFe protein (Kp1) from Klebsiella pneumoniae nifV
mutants contains an altered molybdenum-iron cofactor (1,2). We have
studied the altered Kp1 protein by Mössbauer, Mo-EXAFS (Extended X-ray
Absorption fine structure), and 57-Fe, 1-H and 95-Mo ENDOR (Electron
Nuclear Double Resonance) spectroscopy. Protein enriched in the
appropriate isotopes was purified from the mutant and wild type strains
of K. pneumoniae for these experiments.

Mössbauer spectra were collected at 4.2K on native (dithionite-
reduced) and on thionine oxidized (6 electrons removed / Kp1 molecule).
No detectable difference was observed between the mutant and wild type
in either oxidation state.

Mo EXAFS data were collected between 100-150°K and 21 spectra summed
for analysis. Data was truncated from 3-13 k in k space (1/Angstroms)
and fitted using theoretical sine wave functions. Fits were refined
using model compounds. Mo EXAFS indicated that wild type and mutant had
Mo-S distances of 2.36$\overset{\circ}{A}$ (+/-0.012) and 2.35$\overset{\circ}{A}$ (+/-0.011) respectively;
and coordination numbers of 4.56 (+/-0.92) and 4.4 (+/-1.6). Mo-Fe
distances were 2.702 (+/-0.022) for wild type, and 2.702 (+/-0.018) for
nifV Kp1. Coordination numbers for Mo-Fe were 2.7 (+/-1.2) and 2.8
(+/-2.3) for WT and mutant. The data show that there is no difference
in Mo-S or Mo-Fe bond lengths between mutant or wild type. Coordination
numbers are not accurately given by this technique, but the Mo-S
coordination number is apparently unchanged.

57-Fe ENDOR indicates that the Fe coupling constants are essentially
unchanged in the mutant. Four classes of Fe are seen at Gx, and are
unchanged in the mutant. Five classes of Fe are seen at Gz, and one of
these is changed by 1%. Proton ENDOR spectra are very similar.
However, the difference spectrum for proton ENDOR (wild type minus
mutant) indicates that the mutant may have an extra class of protons not
present in wild type. These extra protons are exchangeable with
deuterated water. ^{95}Mo ENDOR indicates a significant change (10-15%) in
coupling constant both at g_x and g_z.

REFERENCES

McLean PA (1981) D.Phil Thesis, University of Sussex, Sussex, England.
Hawkes TR et al (1984) J. Biochem (UK) 217, 317-321.

This work was supported by NIH (GM30943) and NSF (PCM 8205764).

CONSTRUCTION AND IN VITRO RECONSTITUTION OF AZOTOBACTER
VINELANDII MUTANTS DELETED FOR NITROGENASE STRUCTURAL GENES

Amy C. Robinson,[a] Dennis R. Dean[b] and Barbara K. Burgess[a]
a) University of California, Irvine, CA 92717
b) Virginia Polytechnic Institute and State University, Blacksburg, VA
   24061

There are three structural proteins which comprise the enzyme nitrogenase. The Fe protein polypeptide is coded for by the nifH gene while
the alpha and beta subunits of the MoFe protein are coded for by the
nifD and nifK genes, respectively. In order to better understand the
structure and reactivity of the MoFe protein mutant strains of Azotobacter
vinelandii have been constructed by deleting specific portions of the
nifHDK genomic region. One mutant, CA12, is missing the entire nifHDK
region. Another mutant, DJ33, is missing the nifDK region. A third
mutant, DJ100, is specifically missing the nifD region. A fourth mutant,
CA13 (Bishop et al., 1985), is deleted for the nifK gene. All deletions
are in phase, and the exact end points of the deletions have been confirmed by sequence analysis (Brigle et al., 1985). All of these mutants
contain all other nif genes.

Two-dimensional gel electrophoresis has been performed on derepressed cells from wild type Av and each of the deletion mutants. The data
demonstrate that: 1) the Fe protein polypeptide is synthesized in 75 -
100% wild type levels in DJ100, CA13 and DJ33 but is not synthesized in
CA12; and 2) the alpha and beta subunits of the MoFe protein are synthesized in 75-100% wild type levels in CA13 and DJ100, respectively.
Thus, the alpha and beta subunits are not required individually or together for wild type levels of expression of the nifHDK genes in Av.

Activity assays have also been performed and demonstrate that:
1) all mutants which contain the Fe protein gene have active Fe protein;
2) all mutants which are missing one or more of the nifD or nifK genes
do not have MoFe protein activity; and 3) FeMoco is synthesized in the
absence of the nifD and/or nifK gene products but does not appear to be
synthesized in the absence of the Fe protein. These data show that the
nif gene products involved in Fe protein processing and FeMoco biosynthesis are functioning in these mutants, and that neither MoFe protein
subunit by itself can function in the acetylene reduction assay.

Reconstitution activity is obtained when extracts of CA13 (beta
deletion) and DJ100 (alpha deletion) are mixed together without the addition of FeMoco or [Fe-S] clusters. Although we have not yet optimized
this assay (currently we see about 5% of the expected maximum activity)
we do now have an assay which should allow for the purification of the
individual, nondenatured, alpha and beta subunits of the MoFe protein.

Acknowledgements: We wish to thank Deloria Jacobs for performing the
two-dimensional gel electrophoresis and Paul Bishop for providing the
mutant CA13 and constructing CA12 using our deletion plasmid. This work
was supported by the USDA/CGO (to DRD) and by UCI set-up funds (to BKB)

References

Bishop PE, Rizzo TM and Bott KF (1985) J. Bacteriol. 163, 21-28.
Brigle KE, Newton WE and Dean DR (1985) GENE, in press.

NITROGEN FIXATION BY A nifHDK DELETION STRAIN OF Azotobacter vinelandii

PAUL E. BISHOP[+] and ROBERT R. EADY[*]
[+]ARS-USDA North Carolina State University, Raleigh, North Carolina 27695-7615 U.S.A., [*]AFRC Unit of Nitrogen Fixation, University of Sussex, Brighton, BN1 9RQ, U.K.

In addition to conventional nitrogenase, A. vinelandii has an alternative N_2 fixation system which is expressed under conditions of Mo deficiency (1-3). To study this system in isolation, a mutant strain (CAll) was constructed which has the structural genes for nitrogenase (nifHDK) deleted. Steady-state chemostat cultures of strain CAll were established in a simple defined medium chemically purified to minimise Mo and that contained no utilisable combined N source. Growth was dependent on N_2 fixation (1.1×10^8 viable cells ml^{-1} at D = 0.175 h^{-1}) and was inhibited by added Mo (20 nM). DNA hybridisation showed the deletion to be stable during prolonged (55 days) growth in the chemostat. Batch cultures in unsupplemented 'spent' chemostat medium showed good growth (1.9×10^8 cells ml^{-1}) so no requirement for sub nanomolar concentrations of Mo was found.

The biomass yield as the dilution rate (D) was varied showed that the N content of the culture, protein and dry.wt. increased as D was decreased, indicating that neither N_2 or O_2 were limiting growth. The limiting nutrient was not identified.

Nitrilotriacetic acid present in the medium inhibited O_2-dependent hydrogenase activity and substantial amounts of H_2 evolution, which was not inhibited by C_2H_2, were observed. C_2H_2 was a poor substrate, underestimating the rate of N_2 fixation, calculated from the N content of the cells, by 7 to 10 fold. Over a range of D approximately 50% of the electron flux through the alternative system was allocated to H^+ reduction. In unperturbed chemostat cultures the product distribution had the stoichiometry

$$N_2 + 12H^+ + 12e^- \longrightarrow 2NH_3 + 3H_2$$

SO_4^{2-}-limited steady state continuous cultures of the strain UW136 wild-type for nifHDK had a 2-fold greater biomass in the presence of MoO_4^{2-} (1 μM). A previous study (4) attributed the stimulation in yield when MoO_4^{2-} was added to Mo-deficient steady-state as indication of Mo-limitation. Since their cultures were almost certainly SO_4^{2-}-limited the stimulation can be attributed to a more effective utilisation of SO_4^{2-}.

REFERENCES

1. Bishop PE et al. (1980) Proc. Natl. Acad. Sci. USA 77, 7342-7346.
2. Bishop PE et al. (1982) J. Bacteriol. 150, 1244-1251.
3. Premakumar et al. (1984) Biochim. Biophys. Acta 797, 64-71.
4. Eady RR, Robson RL (1984) Biochem. J. 224, 853-862.

PARTIAL PURIFICATION OF AN ALTERNATIVE NITROGENASE FROM A nifHDK DELETION STRAIN OF AZOTOBACTER VINELANDII.

John R. Chisnell and Paul E. Bishop, Dept. of Microbiology and USDA/ARS, North Carolina State University, Raleigh, NC 27695-7615, U.S.A.

The A. vinelandii mutant strain CA11.6 (ΔnifHDK, W-tolerant) (R. Joerger, personal communication) was grown in N-free Burk's medium with no added Mo. Maximum nitrogenase activity in cell-free extracts was obtained when cells were harvested from cultures at a density of 8×10^7 $CFU \cdot ml^{-1}$. Specific activities for nitrogenase in these cell-free extracts have been measured as high as 3.5 nmol NH_3, 2.0 nmol C_2H_4, and 58.3 nmol H_2 (in the presence of C_2H_2) formed $min^{-1} \cdot mg$ protein^{-1}. The ratios of these three nitrogenase activities (1:0.6:17) differ significantly from those found for cell-free extracts of wild type A. vinelandii grown in N-free, +Mo medium (1:3.8:0.9). The nitrogenase activity of the cell-free extract was enhanced when it was mixed with A.v. dinitrogenase (N$_2$-ase) but not when mixed with A.v. N$_2$-ase reductase. This alternative N$_2$-ase reductase has previously been described (Premakumar et al., B.B.A. (1984) 797:64). When the cell-free extract was chromatographed on a column of DEAE-cellulose, the bulk of the nitrogenase activity eluted with 0.25 M NaCl. No increase in nitrogenase activity was found when this fraction was mixed with either a 0.5 M NaCl DEAE fraction or A.v. N$_2$-ase or A.v.N$_2$-ase reductase. A pH of 7.1 and a sodium dithionite concentration of 10mM in the assay mixture were optimal for the nitrogenase activity of the 0.25 M NaCl DEAE fraction. When the 0.25 M NaCl DEAE fraction was subjected to chromatography on either Sephadex G-100 or Sephacryl S-300 it was resolved into at least two brown colored fractions which had nitrogenase activity only when mixed together. When a high mw (130,000 to 300,000 molecular weight) fraction and a low mw (40,000 to 70,000 molecular weight) fraction eluted from the Sephacryl S-300 column were combined they had maximum specific activities of 8.40 and 4.39 nmol C_2H_4 formed $min^{-1} \cdot mg$ protein^{-1}, respectively. When the high mw S-300 fraction (185 µg) was mixed with the low mw S-300 fraction (210 µg) specific activities of 2.0 nmol NH_3, 1.8 nmol C_2H_4, and 44.7 nmol H_2 (in the presence of C_2H_2) formed $min^{-1} \cdot mg$ protein^{-1} were obtained. The ratios of these activities (1:0.9:22) are similar to those found for these activities in the cell-free extract. The nitrogenase activity of the high mw S-300 fraction was not complemented by A.v. N$_2$-ase and only low activity was obtained when the fraction was mixed with either A.v. N$_2$-ase reductase or the alternative N$_2$-ase reductase in the 0.5 M NaCl DEAE fraction. The nitrogenase activity of the low mw S-300 fraction was slightly increased by the addition of A.v. N$_2$-ase and essentially not affected by A.v. N$_2$-ase reductase. We propose that the low mw S-300 fraction contains a third N$_2$-ase reductase of A. vinelandii which primarily complements a putative alternative N$_2$-ase found in the high mw S-300 fraction and is only slightly active with conventional A.v. N$_2$-ase. It differs from both the alternative N$_2$-ase reductase found in the 0.5 M NaCl DEAE fraction and the conventional A.v. N$_2$-ase reductase both of which efficiently complement the nitrogenase activity of conventional A.v. N$_2$-ase but are less effective in stimulating the nitrogenase activity of the putative alternative N$_2$-ase. Anaerobic gel electrophoresis revealed no major protein bands corresponding to A.v. N$_2$-ase in any of the CA11.6 nitrogenase-containing fractions. Also, two-dimensional SDS-PAGE of the high mw and the low mw S-300 fractions lacked protein spots corresponding to the subunits of A.v. N$_2$-ase or A.v. N$_2$-ase reductase.

PURIFICATION AND CHARACTERISTICS OF THE NITROGENASE OF AZOSPIRILLUM AMAZONENSE

SEUNG-DAL SONG*, ANTON HARTMANN and ROBERT H. BURRIS
* Kyungpook National University, Taegu 635 KOREA,
University of Wisconsin, Madison, WI 53706 U.S.A.

It was deemed important to isolate and characterize the nitrogenase system of the free-living, microaerobic N_2 fixing bacterium Azospirillum amazonense, because the organism differs from the other azospirilla in acid tolerance, optimal pO_2 and ready growth and nitrogen fixation with sucrose. The nitrogenase of A. amazonense(strain Y1) was purified by chromatography on DEAE-52 cellulose, heat treatment and by preparative polyacrylamide gel electrophoresis. The specific nitrogenase activities were 2,400 nmoles C_2H_4 formed min^{-1}mg protein^{-1} for dinitrogenase(Aa1, MoFe-protein) and 1,800 nmoles C_2H_4 formed min^{-1} mg protein^{-1} for dinitrogenase reductase(Aa2, Fe-protein). The optimal pH for nitrogenase activity was in the range 6.0 - 7.5. During short exposure time to low pH, about 50% activity was observed at pH 5.0. The optimal temperature appeared at 35C, and purified MoFe-protein was stable for more than 30 days at 4C when precautions are taken to keep the preparation anaerobic, whereas at room temperature the activity lasted only about 4 days. The Fe protein lost its activity in 10 days at 4C and in only 5 hours at room temperature.

The MoFe-protein was composed of a minimum 1,852 amino acid residues, had an isoelectric point of 5.2 and contained 2 atoms of Mo, 24 atoms of Fe and 28 atoms of acid-labile sulfide per molecule. The Fe-protein had 624 amino acid residues, an isoelectric point of 4.6 and contained 4 atoms of Fe and 6 atoms of acid-labile sulfide per molecule. The difference index of amino acid composition of the nitrogenase system showed closest relatedness between A. amazonense Y1 and Rhizobium lupini and R. japonicum. The least related Fe-protein was that of Rhodospirillum rubrum.

The purified MoFe-protein showed two types of subunits with molecular weights of 55,000 and 50,000. Also the Fe-protein revealed two nonidentical subunits on SDS-PAGE with apparent molecular weights of 35,000 and 31,000. The two types of Fe-protein subunits were demonstrated in the purified, highly active enzyme as well as in extracts with immunological techniques. In the high resolving SDS-PAGE system used, Azotobacter vinelandii Fe-protein also showed two closely migrating subunits that migrated differently from the Fe-protein subunits of A. brasilense or R. rubrum. The nitrogenase activity of A. amazonense Y1 was independent of Mn^{+2} and the addition of activating enzyme had no effect.

THE K. PNEUMONIAE NIFHDKY OPERON: STRUCTURAL AND REGULATORY ASPECTS; ASSEMBLY OF NITROGENASE POLYPEPTIDES IN FOREIGN HOSTS

A. Zamir, D. Holland, A. Lers, R. Bitoun, J. Berman, D. Salomon and A. Zilberstein. Biochem. Dept. Weizmann Institute, Rehovot, Israel.

1. nifK sequence and comparative analysis of MoFe protein α and β subunits

The nucleotide sequence has been determined for the 3' end of nifD, the entire nifK and the putative start of nifY. The deduced amino acid sequences of nifK from K. pneumoniae, anabaena 7120 and Parasponia Rhizobium were simultaneously aligned (1), according to the mutation data matrix of Dayhoff. Approximately 40% of all amino acid residues including 4 cysteines are conserved in the 3 sequences, and most of the remaining residues are similar. However, the alignment is clearly broken in two short regions, where extensive sequence divergence had occurred. When the 3 β subunit sequences were compared to α subunit sequences, two blocks of amino acids, containing conserved cysteines were found to be highly similar between the two subunits.

2. Transcriptional analysis of nifHDKY promoter mutations

Previously isolated promoter mutations that allowed expression of the nifHDKY operon in the absence of nifA (2) were characterized according to 2 criteria: 1, the requirement for ntrA and 2, the sites of transcription initiation. Plasmids pRB1 and pRB5 containing, respectively, point and duplication mutations in the nifHDKY regulatory region were transformed into E. coli and K. pneumoniae hosts with different nifA and ntrA backgrounds, and nif transcription start sites were determined by nuclease S1 mapping. It was found that the point mutation apparently increased the activity of a pre-existing promoter located upstream to the nif promoter element, while expression from pRB5 is presumably due to a canonical promoter sequence generated by the sequence duplication.

3. Assembly of nitrogenase polypeptides in foreign hosts

While genes required for the synthesis and activity of nitrogenase in K. pneumoniae have been identified, the minimal genetic requirements, the pathway of assembly and the role, if any, of the host organism in this process are not fully understood. The synthesis (3) and assembly of nitrogenase polypeptides was examined in E. coli and in the yeast Saccharomyces cerevisiae. A sensitive assay was developed to enable the detection and characterization of small amounts of native nitrogenase components in crude cell lysates (4). Nitrogenase components were identified as O_2-sensitive antigens, or ^{55}Fe-labeled proteins.

Fe Protein: as shown (4) in both E. coli and yeast, the nifH product was the only nif product required for the assembly of a structure resembling Kp2 in electrophoretic mobility, apparent MW and high sensitivity to oxygen. Evidence was also provided for the specific binding of MgATP to the Kp2-like structures. MoFe protein: in E. coli assays of assembly did not reveal any products co-migrating with native Kp1, or its presumably oxidized forms. However, some nif-specific antigens were observed, and may represent incomplete or intermediate forms in the assembly of Kp1. Assembly assays in yeast are currently underway.

References

1. Murata, M., Richardson, J.S. and Sussuman, J.L. (1985) PNAS 82, 3073.
2. Bitoun, R. et al. (1983) PNAS 80, 5812.
3. Berman, J. et al. (1985) Gene 35, 1.
4. Berman, J. et al. (1985) J. Biol. Chem. 260, 5240.

REGULATION OF NITROGENASE ACTIVITY BY COVALENT MODIFICATION
OF NITROGENASE Fe PROTEIN IN RHODOPSEUDOMONAS CAPSULATA.

Y. JOUANNEAU, C. MEYER AND P.M. VIGNAIS
DRF/Biochimie Microbienne, (CNRS UA 1130), CEN-G, 85X
F-38041 Grenoble Cédex, France

In photosynthetic bacteria, nitrogenase activity is regulated both at
the transcriptional and the enzymatic levels in response to several
environmental factors including oxygen tension, concentration of nitrogen
source and light intensity (cf. Vignais et al 1985). In addition, ammonia
brings about short-term total and reversible inhibition of nitrogenase.
The mechanism of this regulation involves a reversible inactivation of
the iron (Fe) protein by attachment to or release from one of its
subunits of a specific modifying group recently identified in
Rhodospirillum rubrum as ADP-ribose (Pope et al 1985a).
The binding site of the adenine and phosphate-containing modifying group
on the Fe protein of Rhodopseudomonas capsulata was investigated using
^{32}P-labeled Fe protein. Cells were harvested in the log phase,
resuspended in fresh growth medium containing 5 mCi ^{32}P-Pi, 0.4 mM
phosphate and 10 mM MOPS buffer pH 7.0. After incubation in the light for
6-8h, NH_4Cl (15 mM final) was added to inactivate nitrogenase and cells
were harvested. ^{32}P-labeled Fe protein was purified as described for non
labeled protein (Jouanneau et al 1983). The apoprotein obtained by tri-
chloracetic acid precipitation was treated by 8 M urea at pH 8.6,
reduced wich β-mercaptoethanol then carboxymethylated with iodoacetamide.
The carboxymethylated Fe protein was first digested by trypsin for 1h at
37°C then by pronase, over a 27h incubation period at 37°C. ^{32}P-labeled
polypeptides highly negatively charged were isolated by horizontal
electrophoresis on cellulose plates at pH 1.9 (75% acetic acid, 2.5%
formic acid, 75 to 90 min at 1600 V, under cooling). Amino acid analysis
was carried out at 62°C by HPLC in a Waters apparatus. Comparison of the
peptide amino acid composition (expressed as mol AA/mol Adenine) with
known sequences of Fe proteins from other sources indicates that the
binding site for the modifying group is located in the region "C", defined by Robson (1984) as a possible nucleotide binding domain; this
region includes Cys 97 (in the sequence of Fe protein from Klebsiella
pneumoniae or from Azotobacter vinelandii (Hausinger, Howard, 1982)), a
probable ligand for the Fe-S center (Hausinger, Howard, 1983). These
results are in ageement with those found with R. rubrum Fe protein by Pope
et al (1985b).

REFERENCES
Hausinger RP, Howard JB (1982) J. Biol. Chem. 258, 2483-2490.
Hausinger RP, Howard JB (1983) J. Biol. Chem. 258, 13486-13492.
Jouanneau Y et al (1983) Biochim. Biophys. Acta 749, 318-328.
Pope MR et al (1985a) Biochemistry 24, 2374-2380.
Pope MR et al (1985b) Proc. Natl. Acad. Sci. USA (1985) 82, 3173-3177.
Robson RL (1984) FEBS-Let. 394-398.
Vignais PM et al (1985) Adv Microbial. Physiology 26, 155-234.

AMINO ACID LEVELS DURING SWITCH-OFF OF NITROGENASE FROM RHODOSPIRILLUM RUBRUM

ROY H. KANEMOTO and PAUL W. LUDDEN
UNIVERSITY OF WISCONSIN-MADISON, MADISON, WI. 53706 U.S.A.

Ammonia can be assimilated to glutamate by the glutamine synthetase-glutamate synthase (GS-GOGAT) pathway, or by glutamate dehydrogenase (GDH) utilizing α-ketoglutarate. Subsequent transamination reactions transfer the glutamate N to other carbon skeletons to yield a variety of amino acids. Glutamate thus serves as a nitrogen reservoir for cellular N-metabolism.

Glutamine formed by GS is also involved in a number of biosynthetic reactions. In addition, glutamine (or a related metabolite) and/or GS have been proposed to mediate the expression of the nif genes, and in the photosynthetic bacteria, also mediate nitrogenase activity (switch-off) by the inactivation of the Fe protein component through ADP-ribosylation. This covalent regulatory process can be initiated by NH_4Cl, darkness, oxygen, CCCP, and various oxidants. If glutamine is responsible for this regulation then changes in glutamine should be observed during switch-off. We examined the amino acid levels in R. rubrum cultured under repressed and derepressed conditions, and followed the change in glutamine and other amino acids during NH_4Cl switch-off of nitrogenase.

Perchloric acid extracts were prepared from repressed (NH_4Cl as N) and derepressed (glutamate as N) cultures of R. rubrum. The amino acids were derivatized with o-phthalaldehyde and separated by reverse phase HPLC. Glutamine concentrations were greater in the derepressed culture reflecting the growth on glutamate. However, the Gln:Glu ratio was similar under the two conditions implying that absolute glutamine levels alone do not regulate nif expression. Some other factor not measured by this analysis, such as carbon skeleton, may also be involved.

When R. rubrum is derepressed on glutamate medium, the nitrogenase is regulated in vivo by Fe protein modification. NH_4Cl addition to these cultures inhibited nitrogenase and caused a rapid though transient increase in Gln (10X). Transient fluctuations in Glu and Asp were observed as the culture compensated for the Gln synthesis and corresponding Glu decrease. Ser, Ala, and Val gradually increased 2X. Treatment of a culture with methionine sulfoximine (MSX) prior to NH_4Cl addition blocked the increase in Gln. Asp, Ala, and Val gradually increased while Ser did not change. In spite of MSX, the added NH_4Cl was assimilated, probably by GDH.

R. rubrum cultured under limited N conditions do not switch-off their nitrogenase. NH_4Cl addition to a limited N culture also showed a transient increase in Gln (10X). Asp and Glu were found to increase while Ser and Ala were found to transiently decrease. Val did not change. The response of limited N cultures to NH_4Cl was not identical to that of glutamate cultures, but similar increases in Gln were observed. This observation suggests that the inactivating system is absent in limited N cultures, or glutamine alone does not regulate nitrogenase activity.

628

REVERSIBLE REGULATION OF THE IRON PROTEIN OF NITROGENASE FROM
RHODOSPIRILLUM RUBRUM BY ADP-RIBOSYLATION IN VITRO

ROBERT G. LOWERY AND PAUL W. LUDDEN
Department of Biochemistry
University of Wisconsin-Madison
Madison, WI 53706

Nitrogenase activity in the photosynthetic bacterium Rhodospirillum rubrum is regulated by reversible ADP-ribosylation of a specific arginine residue in the Fe protein (Pope et al. 1985). The Fe protein regulatory cycle has been studied extensively in vivo, and is known to respond to fixed N, dark/light,O_2, and various other agents (Kanemoto and Ludden 1985). The activating enzyme (AE), which catalyzes the activation of Fe protein by removal of the ADP-ribose moiety has been purified to homogeneity (Saari et al. 1984). Further characterization of the Fe protein regulatory cycle has been prevented by the inability to demonstrate the modification/inactivation of the Fe protein in vitro. We report the reversible inactivation of Fe protein in R. rubrum cell extracts. In addition, we report the partial purification of the inactivating enzyme (IE), determination of its requirements and cross reactivity with FE proteins from other species.

The Fe protein is composed of two identical subunits and ADP-ribosylation of one subunit of the dimer results in inactivation. The modified subunit can be separated from the unmodified subunit by SDS-PAGE (Preston and Ludden 1982). Incubation of active (unmodified) Fe protein with R. rubrum cell extract in the presence of ^{32}P-NAD resulted in: 1) modification of Fe protein, 2) incorporation of radioactivity specifically into the modified subunit, and 3) inhibition of nitrogenase activity (as measured by the acetylene reduction assay in the presence of MoFe protein) at similar rates. Incubation of the in vitro-modified Fe protein with purified AE resulted in removal of the ADP-ribose moiety and recovery of nitrogenase activity. Subtilisin digestion of the in vitro-modified Fe protein yielded a ^{32}P-labelled product which comigrated with an ADP-ribosylated hexapeptide purified after identical treatment of the in vivo-modified Fe protein (Pope et al. 1985). These results provide direct evidence that the physiologically relevant ADP-ribosylation of Fe protein occurs under our assay conditions.

We have obtained an 18x purification of the IE with >100% yield from R. rubrum cells using DEAE chromatography. NAD and DTT stabilize the enzyme. ATP or ADP and a divalent metal ion (MgII and MnII are the most effective of those tested) are required for IE activity. IE is O_2-stable but requires anaerobic Fe protein for optimal activity. Dithionite inhibits IE completely (probably by reacting with NAD). The Fe proteins from Azotobacter vinelandii and Klebsiella pneumoniae are substrates for the IE, but FE protein from Azospirillum amazonense is not a substrate under the same conditions.

REFERENCES
Kanemoto RH and Ludden PW (1984) J. Bacteriol. 158, 713-720.
Pope MP et al. (1985) Proc. Natl. Acad. Sci. 82, 3173-3177.
Preston GG and Ludden PW (1982) Biochem. J. 205, 489-494.
Saari LL et al. (1984) J. Biol. Chem. 259, 15502-15508.

STUDIES ON INACTIVATION OF NITROGENASE Fe-PROTEIN IN RHODOSPIRILLUM RUBRUM

Stefan Nordlund, Erica Brostedt and Abdo Soliman, Department of Bio-chemistry, University of Stockholm, Sweden

Nitrogenase activity in the photosynthetic bacterium Rhodospirillum rubrum is regulated by the availability of fixed nitrogen in the culture. As shown by a number of investigators a decrease in nitrogenase activity occurrs upon the addition of ammonium ions to a nitrogen fixing culture ("switch-off"). The molecular basis for this loss of activity is the reversible inactivation of the Fe-protein by covalent modification. As shown recently by Ludden and coworkes the Fe-protein is modified by addition of ADP-ribose to a arginine residue in the protein.
Very little is known about the molecular events leading to the inactivation of the Fe-protein. It has been shown that changes in adenine or pyridine nucleotide pools do not seem to be correlated to the inactivation. The aim of this investigation was to study changes in the glutamate and glutamine pools upon "switch-off" by the addition of ammonium ions and also to study the inactivation in vitro.

MATERIALS & METHODS

Rhodospirillum rubrum was grown photoheterotrophically under N_2. Whole cell nitrogenase activity was determined. Determinations of glutamate and glutamine were done on cell-free extracts from 10 ml samples of the culture. Glutamate concentrations were determined by using glutamate dehydrogenase and bacterial luciferase. Glutamine was determined as glutamate after preincubation with glutaminase. Glutamine synthetase (GS) activity was estimated by the transferase assay. Studies on inactivation in vitro were performed on cell-free extracts from N-starved cells produced according to our standard methods. Nitrogenase activity was determined as acetylene reduction.

RESULTS AND DISCUSSION

In "switch-off" experiments the activities of nitrogenase and GS decrease in an essentially parallell manner. The decrease in GS activity is however initially somewhat slower, which could indicate that this enzyme is regulated by a different enzyme. Alternatively the inactivating enzyme(s) acting on the Fe-protein is less efficient with GS as substrate.
The glutamate and glutamine concentrations increase during "switch-off" and decrease when the enzyme activities start to recover. In a number of experiments we have however observed that the glutamine pool increase again and that nitrogenase activity is not fully recovered. We are still investigating this phenomenon.
In vitro investigations on inactivation of nitrogenase show that NAD^+ is needed for inactivation to occurr in cell-free extract. The inactivation is strongly stimulated by the addition of Mg^{2+} and ATP. No effect of glutamine could be demonstrated, which would indicate that glutamine is not an activator of the inactivation system.

REGULATORY ADP-RIBOSYLATION OF THE FE PROTEIN OF NITROGENASE FROM
RHODOSPIRILLUM RUBRUM AND ACTIVATING ENZYME CATALYSIS.

MARK R. POPE, SCOTT A. MURRELL AND PAUL W. LUDDEN, Dept. of Biochemistry,
Univ. of Wisconsin, Madison, WI 53706 USA.

In response to a number of external stimuli, Rhodospirillum rubrum
has been shown to regulate its nitrogen fixing activity through a rapid
and reversible covalent modification of the Fe protein of the nitrogenase
complex (Kanemoto, Ludden 1984). Recently, we have shown the modification
to be adenosinediphospho ribosylation of a specific arginine residue
(Pope et al. 1985). ADP-ribose (ADPR) is linked through the terminal ribose to a guanidine nitrogen of arginine 100 (Azotobacter vinelandii sequence, Hausinger, Howard 1982) which is only 3 amino acids removed from
one of the proposed cysteinyl ligands for the iron sulfur cluster.
Inactive Fe protein may be activated in vitro by an activating enzyme
which removes at least a portion of the modifying group. The study of
activating enzyme catalysis is complicated by the variables introduced
by a protein substrate, particularly when catalysis is monitored by
following acetyline reduction in the coupled reaction with the Mo Fe
protein. In order to simplify the study of activating enzyme catalysis,
we have developed HPLC-radio label assays utilizing small molecule
substrates for the activating enzyme. An ADP-ribosylated hexapeptide
has been isolated and purified from proteolytic digests of inactive Fe
protein. The modified hexapeptide and the products of activating enzyme
catalysis may be separated by reverse-phase high pressure liquid chromatography. This allows both the analysis of the products and a means of
assaying the reaction. Similarly, a synthetic substrate, ADP-ribosyl-α-
dansyl-L-arginine methyl ester (ADPR-DAME), has been used to probe the
activating enzyme reaction.

The products of the activating enzyme reaction with ADPR-hexapeptide
have been analyzed by fast atom bombardment mass spectrometry and
chromatography. The results show that activating enzyme acts as an N-
glycohydrolase yielding ADP-ribose and a free arginyl guanidinium group.

From HPLC assays using ^{3}H-Adenine labeled ADPR-hexapeptide as a
probe, several conclusions have been drawn for the activating enzyme
reaction. The K_m for ADPR hexapeptide is 10µM, with an apparent V_{max} of
630 pmol·min^{-1}µgAE^{-1}. The reaction requires a divalent cation, displaying
optimal activity at 1mM Mn^{+2}. From its influence on the reaction, a
K_{diss} of 50 to 60 µM was estimated for Mn^{+2}. ADPR-DAME has been shown
to be a suitable substrate for the activating enzyme, and may be prepared
using the cholera toxin method outlined by Oppenheimer (1978). An
estimated K_m of 25µM was made for ADPR-DAME. These assay techniques
should provide further information in the study of the activating enzyme
reaction.

REFERENCES
Hausinger, R. P. & Howard, J. B. (1982) J. Biol. Chem. 257, 2483-2490.
Kanemoto, R. H. & Ludden, P. W. (1984) J. Bacteriol. 158, 713-720.
Oppenheimer, N. J. (1978) J. Biol. Chem. 253, 4907-4910.
Pope, M. R., Murrell, S. A. and Ludden, P. W. (1985) Proc. Natl. Acad.
Sci. USA 82, 3173-3177.

STUDIES ON THE ACTIVATING ENZYME OF IRON
PROTEIN FROM RHODOSPIRILLUM RUBRUM

Leonard L. Saari and Paul W. Ludden, Department of Biochemistry,
University of Wisconsin-Madison, 420 Henry Mall, Madison, WI 53706

Nitrogenase activity in Rhodospirillum rubrum is regulated in a
non-genetic manner via the ADP-ribosylation of the Fe protein (Ludden,
Burris, 1978; Pope et al., 1985). An activating enzyme (AE) isolated
from the chromatophore-containing membranes of R. rubrum can activate
the inactive (ADP-ribosylated) Fe protein in vitro (Ludden, Burris,
1976; Nordlund et al., 1977). A number of studies have attempted to
determine which components of the ADP-ribose moiety are removed from
the inactive Fe protein upon activation (Ludden, Burris, 1979; Preston,
Ludden, 1982; Gotto, Yoch, 1982; Nordlund, Ludden, 1983); the results
have been inconclusive possibly because the protein components used
were impure. AE purified to homogeneity (Saari et al., 1984) was used
to activate purified, inactive Fe protein (Ludden, Burris, 1978). The
moiety removed by AE was determined by mass spectrometry to be ADP-
ribose.

Previous work utilized an indirect acetylene reduction assay to
quantitate the AE activity (Ludden, Burris, 1976). In addition to
being indirect, the assay suffered from being dependent on varying
nitrogenase component ratios and possibly not measuring the initial
velocity of the activation reaction. A radioassay that directly
measures the initial velocity of Fe protein activation by AE was
developed using Fe protein radiolabelled with either ^{3}H- or ^{32}P-ADP-
ribose. The release of labelled ADP-ribose was linearly correlated
with the increase in the specific activity of the Fe protein as
measured by C_2H_2 reduction. The optimal ratio of $[MnCl_2]/[ATP]$ in the
radioassay was 2.0 and the optimal total concentrations were 4 mM and
2 mM for $[MnCl_2]$ and $[ATP]$, respectively. Both ATP and Mn^{2+} were
required for the activation of native, inactive Fe protein by AE.
Oxygen exposure of the Fe protein resulted in the loss of the ATP
requirement for activation. Guanosine, cytidine, or uridine mono-,
di, or triphosphate nucleotides were not substitutes for ATP in
the activation reaction. ADP-ribose but not ADP or AMP appeared
to inhibit activation. The K_m for the inactive Fe protein was 74
μM and the V_m was 628 pmol ^{32}P-ADP-ribose released min^{-1} μg AE^{-1}.
Polyclonal antibodies specific for AE were prepared.

REFERENCES

Gotto JW and Yoch DC (1982) J. Bacteriol. 152, 714-721.
Ludden PW and Burris RH (1976) Science 194, 424-426.
Ludden PW and Burris RH (1978) Biochem. J. 175, 251-259.
Ludden PW and Burris RH (1979) Proc. Natl. Acad. Sci. USA 76, 6201-6205.
Nordlund S et al (1977) Biochim. Biophys. Acta 462, 178-195.
Nordlund S and Ludden PW (1983) Biochem. J. 209, 881-884.
Pope MR et al (1985) Proc. Natl. Acad. Sci. USA 82, 3173-3177.
Preston GG and Ludden PW Biochem. J. 205, 489-494.
Saari LL et al (1984) J. Biol. Chem. 259, 15502-15508.

KINETICS OF ATP-INDUCED FE CHELATION FROM AV2

THOMAS LLOYD DEITS and JAMES BRYANT HOWARD
University of Minnesota, Minneapolis, MN 55455 USA

In the presence of ATP, an ATP regenerating system and a,adipyridyl (a,adp) or orthobathophenanthroline (obp), chelation of Fe from the Fe protein of Azotobacter vinlandii (Av2) releases 4 Fe per mol Av2. Stoichiometries and binding constants for ATP have been determined from initial rates studies of this process (Walker,Mortenson,1973;Ljones,Burris,1978). The time course of Fe release, when followed to completion, occurs via a two exponential process, with rate constants differing by a factor of ca. 10, and with 23 % of the Fe associated with the more rapid rate constant. Neither the products of the Fe release reaction nor the products of nitrogenase turnover influence the time course of reaction. AdPPNP is competent to effect release with identical biphasic kinetics. When samples of Av2 are withdrawn from an Fe release reaction mixture and asayed for their ability to support nitrogenase activity, catalytic activity is found to be strictly proportional to the amount of Fe remaining unchelated. EPR signal intensity in the g = 2.05 to 1.94 region decays in concert with the slower rate constant; any changes associated with the more rapid rate constant have not been detected. The rate of Fe release is at a minimum ca. pH 8.2 and ca. 35 degrees C. Neither variation of pH from 7.2 to 9.0 nor variation in temperature from 20 to 40 degrees C significantly alters the relative amplitudes of the two phases. The rate of release depends upon the square of a,adp concentration, while the relative amplitude of the two phases is independent of the concentration of a,adp. At ATP concentrations above ca. 0.1 mM, the rate of Fe release is sensitive to [ATP] while the relative amplitudes of the two phases are invariant. Below 0.1 mM ATP, the rate of Fe release slows while the more rapid phase diminishes in amplitude, becoming indetectable below 0.03 mM ATP; Fe release is a single exponential process at these ATP concentrations. Acceleration of such a monoexponential Fe release process by increasing the concentration of ATP yields biphasic kinetics for the release of the remaining Fe, with rate constants and phase amplitudes identical to those of native Av2. Reisolation of Av2 from Fe release reaction mixtures yields material homogeneous towards further Fe release experiments only if reisolation is carried out in the presence of high concentrations of ATP. This data can be quantitatively accounted for by a model in which two conformers of Av2, differing in their reactivity towards chelators, can equilibrate only when they have no ATP bound.

Ljones T and Burris RH (1978) BBRC 53, 904-909
Walker GA and Mortenson LE (1973) Biochemistry 17, 1866-1872

ATPase PROPERTIES OF NITROGENASE

HUUB HAAKER, JAN CORDEWENER, ANNELOOR TEN ASBROEK and HANS WASSINK
Department of Biochemistry, Agricultural University, De Dreijen 11,
6703 BC Wageningen, The Netherlands

RESULTS

The induction of the ATPase activity of nitrogenase has been studied. In
the absence of reductant, with oxidized proteins, 4 molecules of Av_2 can
be maximally activated by one molecule Av_1. The specific activity is 600
nmoles $P_i.min^{-1}.mg\ Av_2^{-1}$. In the presence of reductant 3 molecules of Av_1
are necessary to obtain maximal ATPase activity per molecule Av_2. The
specific activity is about 8000 nmoles $P_i.min^{-1}.mg\ Av_2^{-1}$. The ATPase reaction of nitrogenase has been studied further with rapid-quench techniques. Figure 1 shows that the pre-steady-state burst of MgATP hydrolysis, which is observed when nitrogenase is mixed with MgATP, is not
coupled to electron transfer. The slower steady-state rate of MgATP hydrolysis is stimulated by the presence of electrons.
As shown in figure 2, the P_i production in the burst reaction is depending
upon the ratio of the proteins and on the specific activity of Fe protein.
With active Av_2, 2 molecules of Av_2 per Av_1 give the maximal P_i production in the burst (8 $P_i./Av_1$). With 2 binding sites, this means that each
binding site has two turn-overs. Since there is no electron transfer
without MgATP and the pre-steady-state ATPase activity is dependent on
the presence of electrons, this must mean that the ATPase activity of
nitrogenase preceeds electron transfer.

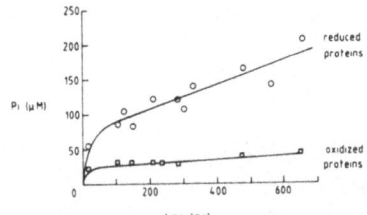
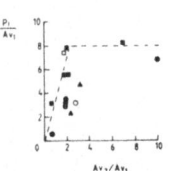

Figure 1: Time course of P_i production by nitrogenase. Rapid-quench
was performed as described earlier. Nitrogenase concentrations
after mixing were: o-o Av_1 15 μM, Av_2 46 μM, flavodoxin 68 μM;
temperature was 22°. Specific activities of proteins used under
standard conditions were: Av_1 1800, Av_2 1400. □-□ experiment without
$Na_2S_2O_4$ in the buffers and dye-oxidized proteins. Av_1 8 μM, Av_2
22 μM. Specific activies of proteins used under standard conditions
were: Av_1 1800, Av_2 1200.

Figure 2: P_i production in the burst. The burst was determined by
extrapolation to t=0. Nitrogenase concentrations after mixing were:
Av_1 10-40 μM, Av_2 27-100 μM and flavodoxin 68-670 μM. Specific activity
of proteins used were measured under standard conditions as described
in Materials and Methods. Av_1 1800-2400. ● Av2 2100-2300, ▲ Av_2 1700, ●
Av_2 1200; □ and 0 dye-oxidized proteins.

CONCLUSIONS

Our results suggest the hypothesis that the ATP binding sites are located
on Fe protein. Complexed to MoFe protein ATP rapidly phosphorylates MoFe
protein (8 $P_i./Av_1$) and ATP is slowly hydrolyzed at its binding sites on
the Fe protein. In the presence of reductant, dephosphorylation of MoFe
protein is stimulated which results in an increased ATPase activity
(phosphorylation/dephosphorylation) coupled to electron transfer to substrates.

634

A MODEL FOR THE MgATP BINDING SITE OF NITROGENASE IRON PROTEIN

ROBERT JONES (1), WILLIAM TAYLOR (2) and ROBERT ROBSON (3)

(1) UNIVERSITY OF CHICAGO, CHICAGO, ILLINOIS 60637, U.S.A.
(2) BIRKBECK COLLEGE, LONDON, U.K.
(3) AFRC UNIT OF NITROGEN FIXATION, UNIVERSITY OF SUSSEX, BRIGHTON, U.K.

We are attempting to model the ATP binding site of the iron protein
subunit (1). The model presented here is based on a nucleotide binding
domain that is highly conserved within the known structures of several
dehydrogenases, consisting of a beta strand - alpha helix - beta strand
unit that contains the conserved sequence Gly X Gly X X Gly (where X is
any residue) at the amino terminal end of the alpha helix (2). A
partial positive charge exists at the amino terminal end of the helix
which stabilizes the negative charge on the terminal phosphate of the
nucleotide. The glycines clustered around the mouth of the helix allow
close approach of the phosphate. The conserved sequences of iron
protein contain the pattern of glycine residues close to the amino
terminus of the polypeptide in a region of the chain which is predicted
to fold into the required strand-helix-strand unit (Taylor,
unpublished). In A.vinelandii the sequence that matches this pattern is

1 AMRQCAIYGKGGIGKSTTTQNLVAALAEMGKKVMIVGCDP 40

Computer graphics have been employed to model this site using the domain
from lactate dehydrogenase as a basis, replacing important side chains
with those of iron protein and examining their potential interactions
with an ATP molecule.

The figure shows a stereo view of an ATP molecule at the mouth of the
helix in the current model.

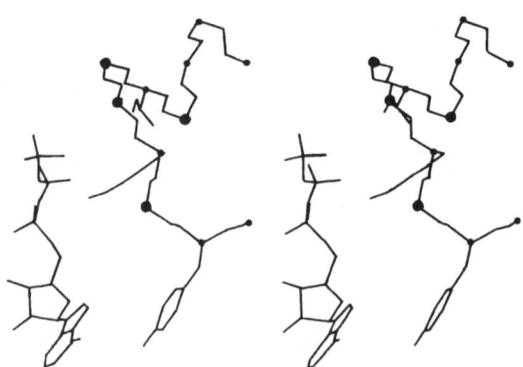

1. Robson R (1984) FEBS letts. 173, 394-398
2. Sternberg MJ and Taylor W (1984) FEBS letts. 175, 387-392

ATP HYDROLYSIS BY NITROGENASE: POSITION OF P-O BOND CLEAVAGE AND PIX ANALYSIS

CHARLES E. McKENNA AND WILLIAM G. GUTHEIL, DEPARTMENT OF CHEMISTRY, UNIVERSITY OF SOUTHERN CALIFORNIA, LOS ANGELES, CA 90089-1062, U.S.A.; GEORGE L. KENYON AND TERRY O. MATSUNAGA, DEPARTMENT OF PHARMACEUTICAL CHEMISTRY, SCHOOL OF PHARMACY, UNIVERSITY OF CALIFORNIA SAN FRANCISCO, SAN FRANCISCO, CA 94143, U.S.A.

Nitrogenase catalyzes reduction-dependent hydrolysis of ATP to ADP and phosphate. In principle, cleavage of the ATP P_β-O-P_γ linkage by nitrogenase could occur at either the P_β-O or P_γ-O bond. To determine which bond is actually broken by the enzyme, we have prepared (P_β-^{18}O-P_γ) ATP (75.4% enrichment) and incubated it with purified nitrogenase FeMo and Fe proteins from A. vinelandii (Av1, Av2) using dithionite as electron donor. The phosphate produced was isolated chromato-graphically, converted to trimethyl phosphate, and analyzed by high-resolution mass spectrometry. The ^{16}O/^{18}O ratio found in the product phosphate was 13.8 (σ 2.8); calculated values for P_β-O cleavage and P_γ-O cleavage were 0.23 and 15.6, respectively. Av nitrogenase thus cleaves the P_γ-O bond of ATP, in this respect behaving like a normal ATPase. Our result is consistent with a hydrolysis mechanism involving reaction at P_γ, e.g. by direct attack of H_2O or via an enzyme-derived nucleophile.

We have also used ^{31}P NMR of recovered substrate (P_β-^{18}O-P_γ) ATP from fixing systems of Av1 and Av2 to investigate the possibility that nitrogenase might catalyze positional isotope exchange (PIX) under standard turnover conditions. Relative amounts of ^{18}O-bridge, ^{18}O-nonbridge, and ^{16}O species can be determined from ^{18}O effects on the P_β ^{31}P NMR chemical shift (Reynolds et al, 1983). Using this technique we find no evidence for PIX in the fixing system relative to a non-fixing control (-Av1) to within the limit of spectral peak area determination (< 5%). The absence of PIX indicates that once the P_γ-O bond is cleaved at the active site of nitrogenase, forward reaction must be considerably more rapid than the reverse steps to reform and release ATP, unless P_α-O-P_β bond rotation in the bound ADP fragment is unusually restricted.

REFERENCES

Reynolds MA et al (1983) J. Am. Chem. Soc. 105, 6663-6667.

ACKNOWLEDGEMENTS

We gratefully acknowledge the support of the National Institutes of Health, USDA and the H. Frasch Foundation.

636

MEASUREMENT OF ATP BINDING TO NITROGENASE PROTEINS BY A MICRO MODIFICATION OF THE FLOW DIALYSIS METHOD.

RICHARD MILLER, CHEMISTRY AND BIOLOGY RESEARCH INSTITUTE, OTTAWA, CANADA/
ROBERT EADY, AFRC UNIT OF NITROGEN FIXATION, BRIGHTON, SUSSEX, UK

Although ATP hydrolysis is required for the reduction of all substrates of nitrogenase, the mechanism of energy transduction is not clear. A micro modification of the flow dialysis technique has been applied to this problem allowing smaller amounts of the enzyme to be used (1,2).
Methods - Nitrogenase proteins were purified from K. pneumoniae by anaerobic ion exchange chromatography on DEAE-Sephacel. The proteins were completely resolved by 2 passages through Sephadex G-200 (Kp1) and G-100 (Kp2) and equilibrated with 20 mM Tris buffer containing 9 mM magnesium chloride and 1 mM dithionite. Protein concentration was less than 35 mg/ml to avoid aggregation. Flow dialysis was carried out with a modified Feldman cell (1) having a spiral dialysate chamber, volume = 50 μL. Nitrogenase proteins (0.45 ml) were placed in the cell with an initial level of α^{32}P-ATP. Aliquots of unlabelled ATP were added and dialysis rates determined.
Results - Data (Fig. 1, 2) indicate stronger binding of ATP to Kp2 than to Kp1. The binding of ATP and ADP to Kp1 is not due to artifacts of dilution or other disturbances in the dialysis rates as shown by the lack of change in controls to which buffer alone was added or in controls containing catalase at protein concentrations similar to that of Kp1. Scatchard plots for Kp2 data conformed within experimental error to a linear least squares fit. The calculated dissociation constant for ATP was 125 μM for 2 equivalent sites. Data for ATP binding to Kp1 indicated 2 types of binding sites with low (K=4000 μM) and higher affinity. The most likely interpretation is a combination of non-specific binding and specific binding with K=500 μM and N=2 (3).
Conclusions - Although the data obtained in flow dialysis experiments with Kp1 and Kp2 indicate ATP and ADP binding to both nitrogenase proteins, Scatchard plots for Kp1 were curved. In both cases, multiple binding sites were indicated but the observed variable time dependence of the ligand dialysis rate, especially at low ligand concentrations, suggests that the proteins may be undergoing ligand-induced conformational changes and that binding equilibria could be altered by this process.

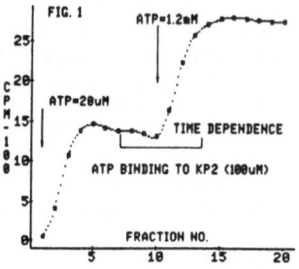

Implications for the mechanism of energy transduction by the functioning nitrogenase system are that ATP may be bound and hydrolysed at the Kp1 site during passage of electrons between the iron protein and the Mo-Fe protein.

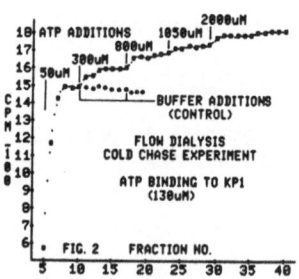

1. Feldman, K. (1978). Anal. Biochem. 88: 225-235.
2. Miller, R.W., Robson, R.L., Yates, M.G. and Eady, R.R. (1980). Can. J. Biochem. 58: 542-548.
3. Kimber, S.J., Bishop, E.O. and Smith, B.E. (1982). Biochem. Biophys. Acta. 705: 385-395.

DISCUSSION - STRUCTURE AND ENZYMOLOGY OF NITROGENASE

Robert H. Burris

Alternative Nitrogenase. Paul Bishop and colleagues in 1980 reported that an alternative form of nitrogenase, apparently devoid of Mo, could be demonstrated in a mutant of Azotobacter vinelandii. Because of the importance attached to Mo in nitrogenase catalysis, the validity of the concept was questioned. Since 1980, Bishop's group has continued to accumulate supporting evidence for the alternative nitrogenase. During the past year, Bishop has worked in collaboration with the group at the ARC laboratory in Sussex to accumulate impressive support for the alternative nitrogenase. Continuous cultures were established and it was demonstrated that growth was dependent on N_2. C_2H_2, H^+ and N_2 were reduced. The culture was inhibited by added $MoO_4^=$. The alternative nitrogenase appears inefficient and generates more H_2 per N_2 reduced than normal nitrogenase. The stoichiometry of its MgATP requirements has not been established.

The case for an alternate nitrogenase also is supported by Hales, Case and Langosch who isolated a mutant of A. vinelandii lacking the structural genes for nitrogenase. The organism grew on N-free media plus or minus Mo. The nitrogenase components did not separate well by typical purification procedures, and the EPR signals differed from normal nitrogenase. Support for an alternative nitrogenase has been strengthened substantially during the past year.

FeMoco. Progress on the nature and role of FeMoco in nitrogenase was described by Newton in a general lecture, and work on this cofactor by the Sussex group has demonstrated that nif H is an absolute requirement for synthesis and activity of FeMoco, as well as the products from nif B, nif N, nif E and nif V.

Shah reported that he and his coworkers had achieved an in vitro synthesis of FeMoco and that this required proteins coded for by nif B, nif N and nif E. Mutants incapable of producing FeMoco can be complemented to restore the capability of producing the cofactor. The cofactor produced in vitro is capable of activating the deficient MoFe protein derived from mutants both of Klebsiella pneumoniae and A. vinelandii. The EPR signal from the MoFe protein supplemented with FeMoco synthesized in vitro is very similar to the MoFe protein from the wild type organism.

The consensus still is that the FeMoco center forms the binding site for N_2, although this concept now must be reconciled with the accumulating evidence that the alternative dinitrogenase lacks Mo.

ATP. Mortenson's general lecture covered progress in studies of the role of MgATP in nitrogenase action, and he described his recent investigations of ATP metabolism with ^{17}O, ^{18}O, and stable S as tracers, combined with NMR measurements. The studies revealed no exchange of the ^{18}O label with water, but the metabolism involved inversion in the phosphate group of ATP.

McKenna and associates have investigated the position of the P-O bond cleavage when ATP is hydrolyzed by nitrogenase. Again the study employed ^{18}O as a tracer to examine changes in the $P_\beta - ^{18}O - P_\gamma$ linkage. The mass analysis indicated cleavage of the P_γ-O bond as is typical of ATPases. Further studies indicated no terminal phosphate exchange catalyzed by nitrogenase action.

Robert Jones et al. described a highly conserved region of the nitrogenase Fe protein that could serve as a binding site for MgATP, and they constructed a binding model to accommodate the data.

Regulation. The area of regulation has progressed well. Ludden reported that his group had clarified the nature of the inactivating group of Rhodospirillum rubrum and had established that the inactivating ADP-ribose unit is attached to the Fe protein by a ribose-guanidinium bond to arginine 100 of the Azotobacter vinelandii sequence. Evidence was derived from proton NMR and fast atom bombardment mass spectrometry. The activating enzyme has been purified essentially to homogeneity. It also is interesting that attention now is turning to the neglected inactivating enzyme. It has been partially purified, and it incorporates ^{32}P from ^{32}P-labeled NAD at the same rate as inactivation occurs.

Yves Jouanneau, speaking for colleagues Meyer and Vignais, described their current work on inactivation of Rhodopseudomonas capsulata, also by covalent modification. The binding site for the modifying group is in a region of the Fe protein that is highly conserved.

Mechanisms. Thorneley and Lowe have developed a detailed kinetic model for nitrogenase activity with the Klebsiella pneumoniae system as their test nitrogenase. The enzyme is sluggish, and they find the rate-limiting reaction is the dissociation of the complex of the MoFe and Fe proteins. They have used their kinetic constants to examine the nature of the alternative nitrogenase system produced when Mo is limiting.

Simpson described studies in which he had used a membrane-leak mass spectrometer to measure the lag time between the mixing of nitrogenase components and the first appearance of H_2 and the products of N_2 reduction. It was interesting that the lag for N_2 reduction ($6e^-$ reduction) was 3 times as long as the lag for H_2 reduction ($2e^-$ reduction), and that the lag in HD production was almost identical to that for N_2 reduction. Calculating the electron flux to substrates indicated that 2 H^+ were not reduced to H_2 until 3 electrons had accumulated per catalytic unit, and N_2 was not reduced until 8 electrons had been transferred (2 of these used to produce H_2).

FROM PHOSPHORUS TOWARDS SULPHUR: THE DEVELOPMENT OF NEW DINITROGEN BINDING SITES

J.R. Dilworth[*], R.A. Henderson, D.L. Hughes, G.J. Leigh, C.J. Pickett and R.L. Richards
AFRC Unit of Nitrogen Fixation, University of Sussex, Brighton, BN1 9RQ.

The binding and activation of dinitrogen by molybdenum and tungsten tertiary phosphine complexes is now well established and has been reviewed extensively (Henderson et al. 1983). Recent work in this area at the Unit of Nitrogen Fixation has concentrated on detailed investigations of the mechanism of the protonation reaction and attempts to render it cyclic or catalytic via electrochemical methods.

In the mechanistic studies of the protonation of the complexes $[M(N_2)_2(PR_3)_4]$ (M=Mo, W; PR_3=PMe$_2$Ph) the details of how the intermediate NNH_2 ligand breaks down to give ammonia are very complex and remain obscure. Thus, in the case of molybdenum the NNH_2 complexes may undergo intramolecular electron and proton transfer to give N_2 and further reduction of the NNH_2 ligand.

There has been considerable speculation on the structure and nature of the species resuling from the addition of one proton to the NNH_2 intermediate. Based on analogues produced by protonation of NNR_2 (R=alkyl or aryl) species (Carroll J.A., Sutton D, 1980, Chatt J. et al., 1980) it was suggested that an η^2- or 'side-on' hydrazido(1-) species (I) could be involved.

I   III

---

[*]Present address: Department of Chemistry, University of Essex, Wivenhoe Park, Colchester CO4 3SQ.

Since there is no hydrazido(1-)-species is isolable from the phosphine systems, the kinetics of deprotonation and reprotonation of [Mo(NHNMePh)(NNMePh)(S$_2$CNMe$_2$)$_2$][BPh$_4$] (II) (Chatt et al., (1980) have been studied in detail. Stopped flow measurements of the reaction of (II) with Et$_3$N revealed the unequivocal initial formation of the η^2-hydrazido(2-)-species (III) followed by rapid ring opening to give the more conventianl η^1-hydrazido(2-) ligand. This is the first example of the detection of the η^2- geometry for the important hydrazido(2-)-intermediate. Reaction of [Mo(NNMePh)$_2$(S$_2$CNR$_2$)$_2$] with excess of acid HX gives the hydrazine, PhMeNNH$_2$, and [MoX$_2$(NNMePh)(S$_2$CNR$_2$)$_2$] whereas one equivalent of acid generates the hydrazido(1-) complex (II). Examination of the rates of the hydrazine-forming reaction shows that it is too fast for the η^2-hydrazido(1-) complex to be involved. Separate protonation of (II) shows that there is a relatively slow ring-opening process prior to hydrazine formation. It is concluded that in this instance the formation of the 'side-on' hydrazido(1-) species is an artefact of the addition of one equivalent of acid and it does not occur as an intermediate en route to hydrazine formation.

The protonation of the N$_2$ molecule to NH$_3$ in the complexes [M(N$_2$)$_2$(PR$_3$)$_4$] (M=Mo,W; PR$_3$=PMe$_2$Ph) is accompanied by oxidation of the metal to oxidation state (IV) M=Mo or VI (M=W) and consequent decomposition. There has been a continuing effort over the past few years to devise a system where the electrons necessary to reduce N$_2$ are provided at least in part by an electrode. The past few years have seen the development of an elegant cyclic system for the electrochemically mediated conversion of N$_2$ to organohydrazines (Hussain et al., 1982) using [Mo(N$_2$)$_2$(dppe)$_2$] (dppe=Ph$_2$PCH$_2$CH$_2$PPh$_2$, R=alkyl) alkyl halides and protons. The key step of this reaction was the 2e electron electrochemical reduction of the hydrazido(2-) complex [Mo(NNR$_2$)X(dppe)$_2$]X to the Mo(II) intermediate [Mo(NNR$_2$)(dppe)$_2$]. Further two electron reduction under N$_2$ in the presence of protons liberated the hydrazine and reformed the parent dinitrogen complex.

Attempted reduction of [MX(NNH$_2$)(dppe)$_2$]$^+$X$^-$ (X=halogen; M=Mo,W) invariably produced H$_2$ and N$_2$ and frustrated attempts to generate NH$_3$ from N$_2$ electrochemically. However, very recently it has been shown (Pickett, Talarmin, 1985) that use of the good leaving group 4-CH$_3$C$_6$H$_4$SO$_3^-$ ($^-$OTs) and a mercury pool electrode with its high overpotential for H$_2$ evolution dramatically changes the situation. The $^-$OTs group leaves the metal in preference to hydrogen evolution occurring. Thus under N$_2$ exhaustive electrolysis of [W(NNH$_2$)(OTs)(dppe)$_2$]$^+$ in thf under N$_2$ at -2.6v (vs ferrocene/ferrocenium) with [Bu$^n_4$N][BF$_4$] as support electrolyte gives 0.22-0.24 moles of NH$_3$ with quantitative formation of [W(N$_2$)$_2$(dppe)$_2$]. The source of the NH$_3$ was verified as being coordinated N$_2$ by ^{15}N labelling experiments. The reaction is complex, as part of the NNH$_2$ complex serves to protonate reduced NNH$_2$ complex. The overall stoichiometry of the reaction is represented by equation (1) where W=W(dppe)$_2$.

$$8TsOW(NNH_2) + 9N_2 + 16e \longrightarrow 2NH_3 + 8W(N_2)_2 + 8TsO^- + 5H_2 \quad (1)$$

Several successive electrolysis protonation sequences have been carried out on a single catholyte solution and after 3 cycles the yield of NH_3 was 0.73 moles/mole of tungsten. The sequence provides an excellent chemical model for the way in which the enzyme system might function by coupled protonation–reduction steps. In this respect it is in excellent agreement with the recently proposed computer model for the functioning of nitrogenase (Lowe, Thorneley, 1984).

There is considerable current interest in the chemistry of binuclear high oxidation state dinitrogen complexes. The first of these to be reported (Turner et al., 1980) was the dinuclear tantalum complex $[\{TaCl_3(THF)_2\}_2(N_2)]$ (THF=tetrahydrofuran) prepared by an indirect route from a carbene precursor. The niobium analogues could be synthesised only with great difficulty as the carbene precursors are much less stable. However, recently we have reported (Dilworth et al., 1984) a convenient one step synthesis of the bridged niobium dinitrogen complexes using tetrakis(trimethylsilyl) hydrazine. The stoichiometry of the synthesis is shown in equation (2).

$$2NbCl_5 + (Me_3Si)_2NN(SiMe_3)_2 \xrightarrow[CH_2Cl_2]{THF} [\{NbCl_3(THF)_2\}_2(N_2)] + 4Me_3SiCl \quad (2)$$

Subsequent reaction of the Nb μ–N_2 species with $Me_3SiS_2NEt_2$ gave the dithiocarbamato–N_2 complex $[\{Nb(S_2CNEt_2)_3\}_2(N_2)]$ in good yield. The presence of a bridging N_2 ligand has been verified by an X-ray crystal structure (Hughes, D.L., unpublished results). This to our knowledge represents the first fully characterised example of a dinitrogen complex where the metal ions are exclusively ligated by sulphur. However, the dinitrogen ligand in such complexes with its long N–N and short Nb–N bonds is very different in character from terminally bound dinitrogen. It remains one of the largest gaps in the inorganic chemistry of nitrogen fixation that despite the presence of sulphur–ligated metal ions at the active site of nitrogenase there are no examples of simple sulphur–containing dinitrogen complexes.

The strategy that we have adopted to prepare such species is to employ sterically hindered aryl thiolate ligands of the types (IV) and (V). Here we inferred that

the steric bulk would prevent loss of potential binding sites via bridge formation and the aryl C-S bond is comparatively resistant to cleavage to give metal sulphides.

Thus tri-isopropylthiophenol (TIPT, (IV)) reacts as a sodium salt with $[MoCl_4(THF)_2]$ in the presence of CO to give the stable five-coordinate carbonyl anion $[Mo(TIPT)_3(CO)_2]^-$. (VI) (Dilworth et al., 1983). The dicarbonyl complex can also be prepared in high yield by reaction of $[MoBr_2(CO)_4]$ with TIPT in methanol. If thiophenol itself is used instead of TIPT, brown unstable polymeric materials are obtained. An X-ray crystal structure showed a trigonal bipyramidal structure with apical CO ligands. Substitution occurred with a range of neutral ligands to give the purple monomeric species $[Mo(TIPT)_3(CO)L](L = MeCN, Bu^tCN, MeNC, Bu^tNC, PMe_2Ph)$. However, complex (VI) underwent complete decarbonylation on reaction with $[PhN_2]BF_4$ to give green $[Mo(N_2Ph)(TIPT)_3(MeCN)]$. An X-ray structure again showed a trigonal bipyramidal structure with apical NNPh and MeCN groups. The M-N-N system of the diazenido-system is linear, indicating that it is functioning formally as a three electron donor. The complex is isostructural with the closely related nitrosyl complex $[Mo(TIPT)_3(NO)(NH_3)]$ and both have a formal d^2 configuration with fourteen valence electrons. The NNPh ligand can be protonated to give the hydrazido(2-)-species $[Mo(TIPT)_3(NNHPh)(MeCN)]^+$.

The products obtained from the reactions of thiolates with molybdenum halides are highly solvent dependent. Thus reaction of $MoCl_4$ with NaTIPT in 1,2-dimethoxyethane gives a 20-25% yield of the red dimer $[Mo_2(TIPT)_6]$. This was shown by an X-ray crystal structure (Figure 1) to contain a molybdenum-molybdenum triple bond with an array of three thiolates on each molybdenum in a staggered conformation. In contrast to the formal reduction of Mo(IV) to Mo(III) observed here, $[MoCl_3(MeCN)_3]$ (Mo(III)) reacts with TIPT anion in MeCN to give the monomeric Mo(IV) species $[Mo(TIPT)_4(MeCN)](VII)$. Complex (VII) is related to $[Mo(SBu^t)_4]$ (Otsuka et al., 1981) but undergoes simple substitution reactions of the MeCN molecule without C-S bond cleavage. This provides access to a range of new Mo(IV) thiolato-species.

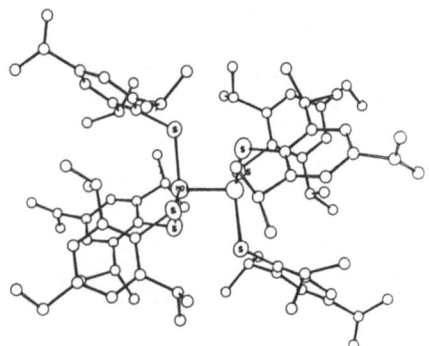

FIGURE 1.   Structure of $[Mo_2(TIPT)_6]$.

The presence of O–phenyl groups as in 2,6 diphenylthiophenol (DPT,(V)) causes a dramatic change in the observed co-ordination chemistry. Thus NaDPT reacts with [MoBr$_2$(CO)$_4$] in 1,2–dimethoxyethane at room temperature to give a green air–stable solid of empirical formula Mo(DPT)$_2$(CO) (VIII) (Dilworth et al., 1985). This shows GνN(CO) in the ir spectrum at 1915 cm^{-1} and the ^{1}H NMR spectrum shows resonances in the 4–7 ppm region characteristic of an η^6–bonded phenyl group. The presence of such co-ordination was verified by an X-ray crystal structure of complex (VIII). A simplified view of the co-ordination sphere is shown in Figure 2.

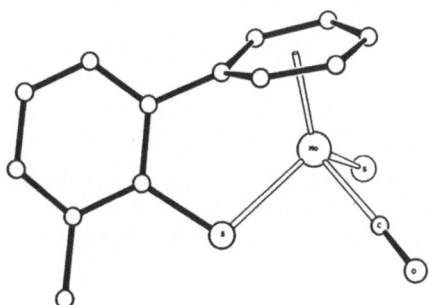

FIGURE 2.   Partial structure of [Mo(DPT)$_2$(CO)]

The structure is best described as distorted 'piano-stool' with one of the O–phenyl grups of the thiol η^6– bonded to the molybdenum. The most interesting feature of the chemistry of this complex is the reversible lability of the η^6– arene group. Thus complex (VIII) reacts with CO to give a polycarbonyl which on work-up loses CO to reform the parent arene complex. The removal of the arene provides access to three co-ordination sites at the metal and we are currently investigating their potential to bind small molecules. Reaction of complex (VIII) with bipyridyl is complex giving a small (<50%) yield of a purple di-carbonyl species[Mo(DPT)$_2$(bipy)(CO)$_2$] (bipy=bipyridyl). The mechanism of formation of a dicarbonyl product from a monocarbonyl precursor is obscure, but it may involve an intermediate with bridging CO groups. By contrast (VIII) reacts clearly with dppe to give a stable five co-ordinate pink complex [Mo(DPT)$_2$(dppe)(CO)] which is apparently reluctant to accept a sixth ligand.

The success in generating reactive monomeric species capable of interacting with small molecules by utilising bulky thiolates suggest that this is a valid approach. Some indications of transient N$_2$ interactions have been observed and indicate that anionic metal sulphur species with formal metal oxidation states of two or three should certainly be capable of binding dinitrogen. There will undoubtedly be a continuing effort to develop a sulphur–based chemistry to parallel that already so well established for the tertiary phosphine molybdenum and tungsten complexes.

REFERENCES

Bishop P.T., Dilworth J.R., Zubieta J., (1985), J. Chem. Comm., 257.
Carroll J.A., Sutton D., (1980), Inorg. Chem., 19, 3137.
Chatt J., Dilworth J.R., Dahlstrom P.L., Zubieta J., (1980),
 Chem. Comm. 786.
Dilworth J.R., Henderson R.A., Harrison S.A., Walton D.R.M., (1984),
 Chem. Comm. 196.
Dilworth J.R., Hutchinson J., Zubieta J., (1983). Chem. Comm. 1034,
Henderson R.A., Leigh G.J., Pickett G.J., (1983), Adv. Inorg.
 Radiochem., 27, 197,
Hussain W., Leigh G.J., Pickett C.J., (1982), Chem. Comm. 747.
Hughes D.L. (1985), unpublished results.
Otsuka S., Kamata M., Hirotsu K., Higuchi T., (1981),
 J. Amer. Chem. Soc., 103, 3011.
Pickett C.J., Talarmin J., (1985), unpublished results.

ACKNOWLEDGEMENTS

    We are indebted to Prof. J.A. Zubieta and co-workers of State
University of New York at Albany, Albany N.Y. for the determination
of X-ray crystal structures of the metal thiolato-complexes.

DETECTION OF INTERMEDIATES IN CHEMICAL N_2 FIXATION.

T. ADRIAN GEORGE AND ROBERT C. TISDALE
DEPARTMENT OF CHEMISTRY, UNIVERSITY OF NEBRASKA
LINCOLN, NEBRASKA 68588-0304, U.S.A.

1. INTRODUCTION

It has been suggested that the biological reduction of N_2 involves, after N_2 binding, a series of six electron-proton transfer steps (Thorneley, Lowe, 1984). No nitrogen-hydride intermediates have been identified in any biological system. However, indirect evidence has led to the following proposed intermediates: (1) diazene-level (N_2H_2) intermediate (HD formation; Burgess, Newton 1981), (ii) molybdenum-hydride-N_2 [Mo(H)(N$_2$)] complex (HD formation and pre-steady state kinetic data; Thorneley, Lowe, 1984), and (iii) hydrazido(2-) (NNH$_2$) species (quenching to form hydrazine; Thorneley, Lowe, 1984).

N_2 complexes of almost all the transition metals are known but in very few of these complexes can N_2 be reduced to ammonia. In fact the conversion of coordinated N_2 to ammonia is limited almost exclusively to the bis-N_2 complexes of molybdenum and tungsten (Henderson, et al., 1983). Considerable effort has been made to try and detect and identify the many intermediates that must be present in such a complex reaction as this. These intermediates may then be studied as possible models of intermediates in biological nitrogen fixation.

1.1. Detection of Nitrogen Hydride Intermediates.

A number of different approaches have been taken, some direct and some indirect, to obtain nitrogen hydrides species that could be true intermediates or analogues of actual intermediates in the reduction of N_2.

1.1.1. Isolation and Characterization.
This is the most direct method since intermediates are isolated either from a reaction leading to ammonia formation or from a modified reaction and shown subsequently to react further to form ammonia. Two such intermediates, the hydrazido(2-) and hydride-hydrazido, are shown in equations 1 (Chatt et al., 1978) and 2 (Takahashi, et al., 1980).

$$\underline{cis}\text{-Mo}(N_2)_2(PMe_2Ph)_4 \xrightarrow{HBr} [MoBr(NNH_2)(PMe_2Ph)_4]Br + N_2 \qquad (1)$$

$$[WBr_2(NNH_2)(PMe_2Ph)_3] + HCl \longrightarrow [WHClBr(NNH_2)(PMe_2Ph)_3]Br \qquad (2)$$

1.1.2. <u>Spectroscopic detection.</u> ^{15}N NMR spectroscopy has been particularly useful for the detection of nitrogen hydride intermediates in solution during ammonia-forming reactions. The hydrazido(2-) unit has been frequently identified (Anderson et al., 1981).

1.1.3. <u>Proposed from kinetic data.</u> Analysis of kinetic data from the protonation reactions of a number of bis-N_2 complexes of molybdenum has led to the following proposed intermediates: (a) dinitrogen hydride (HMoN$_2$). (b) diazenido (NNH), and (c) various hydrazido(2-) intermediates (Henderson, 1984).

1.1.4. <u>Indirect methods.</u> In a few cases it has been possible to reversibly mono-deprotonate the hydrazido(2-) ligand with base to produce the corresponding diazenido ligand, a proposed intermediate in the formation of the hydrazido(2-) ligand from coordinated N_2(Chatt et al., 1976). Other interesting indirect methods have involved using hydrazine as the dinitrogen hydride source. So far these reactions have led to bimetallic complexes with bridging N_2H_2 units. Examples are shown in equations 3 (Sellman et al., 1979) and 4 (Churchill et al., 1984).

$$[(C_5H_5)Mn(CO)_2NH_2NH_2Cr(CO)_5] \longrightarrow [(C_5H_5)Mn(CO)_2(NHNH)Cr(CO)_5] \quad (3)$$

$$2WCp(CCMe_3)X_2 + N_2H_4 \longrightarrow [WCp(CCMe_3)X]_2(\mu-N_2H_2) \quad (4)$$

$$Cp = C_5H_4Bu^t; \; X=Cl,I$$

Currently there is no data uniquely supporting N_2 reduction at a single metal rather than a multi-metal site in biological fixation.

2. RESULTS AND DISCUSSION

2.1. <u>Reactions of bis-dinitrogen complexes of molybdenum.</u>

Scheme I incorporates what is known and proposed for the general mechanism of the conversion of molybdenum-bound N_2 to ammonia with acid (George 1983; Baumann et al., 1985).

<u>Scheme I</u>

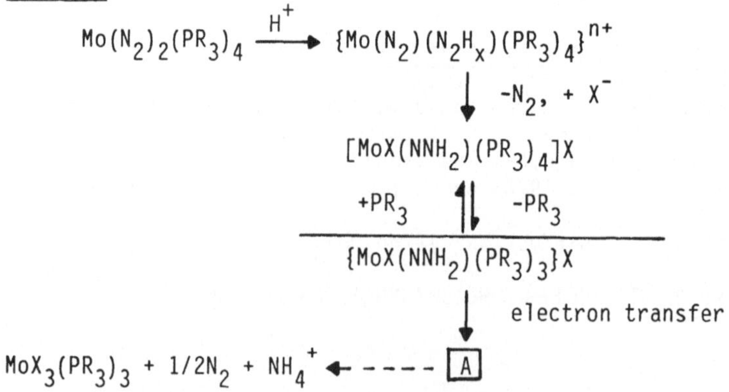

$$Mo(N_2)_2(PR_3)_4 \xrightarrow{\;H^+\;} \{Mo(N_2)(N_2H_x)(PR_3)_4\}^{n+}$$

$$\downarrow \; -N_2, \; + X^-$$

$$[MoX(NNH_2)(PR_3)_4]X$$

$$+PR_3 \; \updownarrow \; -PR_3$$

$$\{MoX(NNH_2)(PR_3)_3\}X$$

$$\downarrow \quad \text{electron transfer}$$

$$MoX_3(PR_3)_3 + 1/2N_2 + NH_4^+ \longleftarrow \; - \; - \; - \; \boxed{A}$$

Rapid protonation, loss of one N_2 and coordination of the conjugate base of the acid used leads to the formation of a stable hydrazido(2-) intermediate. The reaction proceeds no further if $4PR_3=2$(bidentate phosphines). However, if at least one monodentate phosphine is present the reaction proceeds. The next step is the reversible loss of one phosphine followed by electron transfer to produce a paramagnetic species A. We have isolated A where $3PR_3 =$ Triphos=PhP$(CH_2CH_2PPh_2)_2$ but have been unable to characterize the species. However, A has been shown to mimic one of the properties of nitrogenase (George, Koczon, 1983). At the present time a big question mark surrounds the chemistry occurring below the horizontal line in Scheme I, particularly, the electron transfer step(s).

2.2. Reactions of mono-N_2 complexes of molybdenum.

The only mono-N_2 complex of any transition metal element that reacts with acid to form an identifiable nitrogen hydride complex is $M(N_2)(RCN)(dppe)_2$ where M=Mo,W; R=n-Pr, dppe = $Ph_2PCH_2CH_2Ph_2$ (Chatt et al., 1980). However, it does not react further to give ammonia. Very recently, we have prepared an extensive new series of six-coordinate mono-N_2 complexes of molybdenum that not only afford dinitrogen hydride species when reacted with acid but also react further to give ammonia and hydrazine (George, Tisdale, 1985). The reduction of $MoCl_3$(Triphos) in the presence of either two equivalents of PMe_2Ph or one equivalent of a bidendant ligand and four equivalents of N_2 led to the formation of the corresponding new complexes in high yield (eq. 5).

$$MoCl_3(Triphos) + L_2(or2L) \xrightarrow{N_2, Na/Hg} Mo(N_2)(Triphos)(L_2) \qquad (5)$$

L = PMePh		1A
L_2 = $Me_2PCH_2PMe_2$, DMPM		1B
L_2 = 1,2-$(Me_2As)_2C_6H_4$		1C
L_2 = $Ph_2PCH_2PPh_2$, DPPM		1D
L_2 = $Ph_2PCH_2CH_2PPh_2$		1E
L_2 = 1,1'-$[Ph_2PC_5H_4]_2$Fe, DPPFe		1F

Both 1C and 1E exist as a pair of isomers while 1F is mixed with a non-N_2-containing complex Mo(Triphos)$(\eta^2$-DPPFe)$(\eta^2$-DPPFe)(32%). 1F is the first complex containing molybdenum, iron, and N_2 that gives ammonia upon treatment with strong acid: HBr(20 mol)/CH_2Cl_2/20°C/64h, yields 0.65 mol NH_3 and 0.09 mol of N_2H_4 per mol of 1F together with some H_2.

2.2.1. Reactions of 1A. 1A reacts slowly with one equivalent of HCl in benzene solution to yield the diazenido complex [Mo(NNH)(Triphos)$(PMe_2Ph)_2$]Cl (eq. 6) which has been principally

characterized by ^{15}N NMR spectroscopy (δ(ppm rel. to CD_3NO_3); -14.62

(N_α), -208.0 (N_β), $J(^{15}N^{15}N) = 10.0$ Hz). A second product was

$$\underline{1A} + 1HCl \longrightarrow [Mo(NNH)(Triphos)(PMe_2Ph)_2]Cl \qquad (6)$$

obtained from this reaction mixture that was not a hydrazido(2-) complex. This corresponds to the first direct isolation of a diazenido (NNH) complex.

$\underline{1A}$ reacts with two equivalents of HCl in benzene solution or with excess HCl in the absence of solvent to form a hydrazido(2-) complex $[Mo(NNH_2)(Triphos)(PMe_2Ph)_2]Cl_2$, without the loss of any phosphine ligand. Both the ^{31}P and ^{15}N NMR spectral data (δ(ppm relative to CD_3NO_2): -63.3 (N_α), $-212.3(N_\beta)$; $J(^{15}N_\beta H)=49.8$, $J(^{15}N^{15}N)=11.0$ Hz) and elemental analysis support this formulation. $\underline{1A}$ and the corresponding hydrazido(2-) complex react with excess HCl in toluene at 70°C and 50°C, respectively, for about 38 h to yield (mol/mol of Mo): $NH_3(0.36)$, N_2H_4 (0.48) and NH_3 (0.37) and N_2H_4 (0.45), respectively.

2.2.2. Reactions of 1B. The reaction of solid $\underline{1B}$ with 20 equivalents of HCl at room temperature led to the isolation of a hydride-hydrazido(2-) complex, $[MoH(NNH_2)(Triphos)(DMPM)]Cl_3$. The ^{31}P NMR spectrum shows that all five phosphorus atoms are still coordinated to molybdenum. However, this hydride-hydrazido complex affords little ammonia or hydrazine upon treatment with more acid in solution: e.g., HBr(80 mol)/CH_2Cl_2/23°C/16 h yielded (mol/mol of Mo) NH_3 (0.15) and N_2H_4 (0.06) and considerable H_2. By contrast $\underline{1B}$ reacted with HBr(23 mol)/CH_2Cl_2/20°C/48 h to yield NH_3 (0.36) and N_2H_4 (0.32) and considerable H_2.

2.2.3. Reactions of 1D.

The reaction of solid $\underline{1D}$ with excess HCl led to the isolation of $[Mo(NNH_2)(Triphos)(DPPM)]Cl_2$. This complex decomposes in solution at room temperature. However, at -78°C the ^{31}P NMR spectrum unambiguously confirms the coordination of the five phosphorus atoms to molybdenum. The yields of ammonia and hydrazine from this hydrazido(2-) complex and $\underline{1D}$ with HCl are poor. Yields are much higher using HBr and $\underline{1D}$ although there is some uncertainty as to the identity of the nitrogen hydride product formed when solid $\underline{1D}$ is reacted with excess HBr.

2.2.4. Reactions of 1C, 1E, and 1F.

These three complexes all produce ammonia and hydrazine upon treatment with acid. So far, no effort has been made to intercept or detect intermediates in these reactions.

2.3. Comparisons between the mono-N_2 and bis-N_2 systems.

The hydrazido(2-) stage of protonation of N_2 is common to both mono- and bis-N_2 series of molybdenum complexes. The major identifiable

difference that has been detected so far is the ready isolation of a
diazenido complex from the reaction of 1A with one equivalent of HCl.
This result may reflect a significant difference in the rates of the
second protonation step: e.g.   [MoCl(NNH)(PR$_3$)$_4$] >>
[Mo(NNH)(Triphos)(PMe$_2$Ph)$_2$]Cl. The former diazenido complex is
neutral with a π-electron-releasing halogen coordinated to molybdenum
whereas the latter diazenido complex is cationic with no halogen
coordinated.

1A contains two monodentate phosphines and would have been predicted
to give ammonia, provided initial protonation of N$_2$ occurred.
However, 1B-1F all contain a tridentate and a bidentate ligand and
might have been predicted not to react beyond the hydrazido(2-)
stage; cf. [MoX(NNH$_2$)(dppe)$_2$]X. Indeed, 1B reacts with excess HCl
to give a hydride-hydrazido complex without the loss of phosphine.
However, in the presence of excess acid in solution all these
complexes do give ammonia and hydrazine. It has been shown that
MoCl$_3$(Triphos) readily reacts with acid to cleave one of the Mo-P

bonds to give $\overline{\text{MoCl}_4(\text{Ph}_2\text{PCH}_2\text{CH}_2\text{PPhCH}_2\text{CH}_2\text{PHPh}_2)}$, with a pendant
phosphonium ion (George, Howell, 1984). It is similarly possible
that under acid conditions an arm of triphos swings out to allow
further chemistry to occur in nitrogen hydride derivatives of 1B-1F.

3. REFERENCES

Anderson S.N., Fakeley M.E., Richards, R.L. J. Chem. Soc., Dalton
Trans. 1981, 1973-1980.
Baumann J.A., Bossard G.E., George T.A., Howell D.B., Koczon L.M.,
Lester R.K., Noddings C.M. Inorg. Chem. 1985, in press.
Burgess B.K., Wherland S., Newton W.E., Stiefel E.I. Biochemistry
1981, 5140-5146.
Chatt J., Pearman A.J., Richards R.L. J. Chem. Soc. Dalton Trans.
1976, 1520-1524.
Chatt J., Pearman A.J., Richards R.L. J. Chem. Soc. Dalton Trans.
1978, 1766-1776.
Chatt J., Leigh G.J., Neukomm H., Pickett C.J., Stanley D.J. J. Chem.
Soc., Dalton Trans. 1980, 121-127.
Churchill M.R., Li Y.-L., Blum L., Schrock R.R. Organometallics,
1984, 3, 109-113.
George T.A. In Pignolet L.H., ed, Homogeneous Catalysis with Metal
Phosphine Complexes, 1983, pp405-441, Plenum Press, New York.
George T.A., Koczon L.M. J. Am. Chem. Soc. 1983, 105, 6334-6335.
George T.A., Howell D.B. Inorg. Chem. 1984, 23, 1502-1503.
George T.A., Tisdale R.C. J. Am. Chem. Soc. 1985, in press.
Henderson R.A. J. Chem. Soc., Dalton Trans. 1984, 2259-2263.
Henderson R.A., Leigh G.J., Pickett C.J. Adv. in Inorg. Chem. and
Radiochem. 1983, 27, 197-292.
Sellman D., Gerlach R., Jodden K. J. Organometal. Chem. 1979, 90,
309-318.
Takahashi T., Mizobe Y., Sato M., Uchida Y., Hidai M. J. Am. Chem.
Soc. 1980, 102, 7461-7467.
Thorneley R.N.F., Lowe D.J. Israel J. Bot. 1982, 31, 61-71.
Thorneley R.N.F., Lowe D.J. Biochem. J. 1984, 224, 877-909.

650

4. ACKNOWLEDGMENTS

The authors wish to thank the National Science Foundation for generous support of this work and the University of Nebraska Research Council.

^{95}Mo NMR AS A PROBE OF MOLYBDENUM CENTERS

JOHN H. ENEMARK, CHARLES G. YOUNG/UNIVERSITY OF ARIZONA
TUCSON, ARIZONA   85721   USA
JON R. DILWORTH/UNIVERSITY OF ESSEX
COLCHESTER, UNITED KINGDOM

The use of ^{95}Mo nuclear magnetic resonance as a direct probe of molybdenum centers in solution has expended rapidly since 1980. The utility of this low frequency (Ξ = 6.52 MHz), insensitive (15.7% abundant), quadrupolar (I = 5/2) nucleus as a direct probe of structure and reactivity has been clearly demonstrated (Minelli et al.).

The figure clearly demonstrates that ^{95}Mo NMR can be used to directly probe chemical features of potential relevance to nitrogen fixation. These include: interaction of thiomolybdates with other metal centers; binding of nitrogenous ligands to molybdenum centers; diamagnetic Fe-Mo-S clusters.

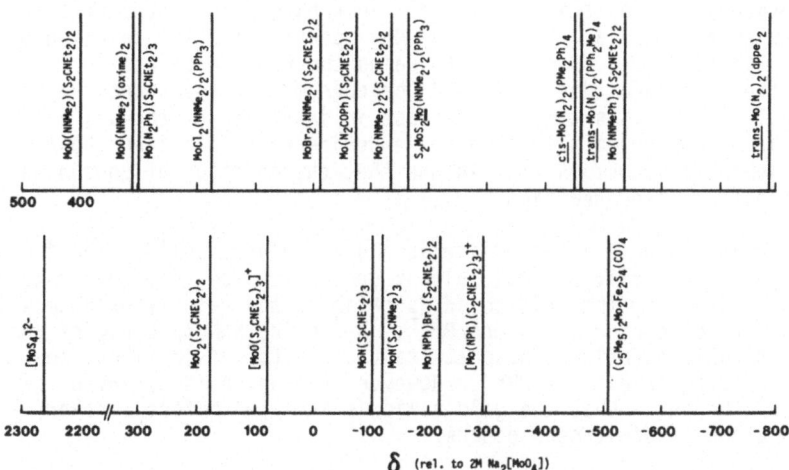

Potential complications for using ^{95}Mo NMR to directly probe the molybdenum centers of nitrogenases and FeMo-co include: availability of material; paramagnetic electronic states; extremely broad linewidths for certain molybdenum species such as mononuclear Mo(IV) and Mo-NNR complexes.

References: Minelli M et al   (1985) Coord. Chem. Rev. in press.

Acknowledgement: We thank the US Department of Agriculture for financial support of this work.

A BIS(DINITROGEN) COMPLEX MOLYBDENUM: A CHEMICAL RESEMBLANCE TO NITROGENASE.

T. ADRIAN GEORGE AND LENORE M.KOCZON
DEPARTMENT OF CHEMISTRY, UNIVERSITY OF NEBRASKA
LINCOLN, NEBRASKA 68588-0304, U.S.A.

The reaction of trans-$[Mo(N_2)_2(triphos)(PPh_3)]$, 1, where triphos = $PhP(CH_2CH_2PPh_2)$, with an excess of anhydrous HBr in tetrahydrofuran (THF) solution, over a period of $\geq$60h, produces ammonia in high yield, and occasionally a trace amount of hydrazine. However, if the reaction is analyzed at 1h, both ammonia and hydrazine are detected, the latter in greater yield.

The first step in the reaction involves the rapid evolution of 1 mol of dinitrogen and the formation of two isomeric hydrazido(2-) complexes, $[MoBr(NNH_2)(triphos)(PPh_3)]Br$, 1A and 1B. Structurally, the isomers differ in the position of the phenyl group on the middle phosphorus of the triphos ligand. Phosphine loss is the next step leading to ammonia. The $^{31}P\{^1H\}$ NMR spectra of a THF solution of the two hydrazido(2-) complexes, revealed the rapid decrease in intensity of the resonances due to 1B and the concomitant appearance of free PPh_3. During this time no decrease in intensity of the signals due to 1A was observed. In separate experiments it has been shown that phosphine loss occurs from 1B and that the addition of phosphine inhibits further reaction.

The next isolable intermediate is a gold-colored solid. The isolated solid is thermally unstable slowly decomposing at room temperature to lose nitrogen. The gold solid is cationic and readily exchanges its counter ion(s) for BPh_4^- and PF_6^-. A $^{31}P\{^1H\}$ NMR spectrum of the THF-soluble hexafluorophosphate salt exhibits no signal due to coordinated triphos or PPh_3. However, the anion is observed. It should be noted that the gold solid is isolated before further evolution of dinitrogen occurs.

The reaction mixture containing the gold solid proceeds to give ammonia as the reduced nitrogen-containing end-product. If large quantities of water are added while the gold solid is present, hydrazine is produced in significant yield. The production of hydrazine by this intermediate upon hydrolysis resembles the behavior of nitrogenase reported by Thorneley and coworkers. This suggests a chemical resemblance between these two systems at this stage in the reduction of dinitrogen. Work continues in this area to determine the structure of the gold solid. It is hoped that this work may provide a model for one of the intermediate stages in the reduction of dinitrogen by nitrogenase.

Thorneley, R.N.F.; Lowe, D.J., Biochem. J 1984, 224, 887-894 and references therein.

STRUCTURE AND PROPERTIES OF CYCLOPENTADIENYLTITANIUM IV COMPLEXES CONTAINING DERIVATIVES OF DINITROGEN

IAN A. LATHAM, G. JEFFERY LEIGH, DAVID L. HUGHES, AFRC Unit of Nitrogen Fixation, University of Sussex, Brighton U.K. GOTFREID HUTTNER, and IBRAHIM JIBRIL, Facultat der Chemie der Universitat Konstanz F.R.G.

Monocyclopentadienyltitanium trichloride, $[CPTiCl_3]$, reacts cleanly with a variety lithium or trimethylsilyl substituted aryl- and alkyl-hydrazines or diazenes to produce a series of Ti IV compounds containing diazenido (NNR), (Dilworth et al. 1978), hydrazido(2-) or (NNR_2), or hydrazido(1-) ($NRNR_2$)ligands in excellent yield, Scheme 1.

Scheme 1

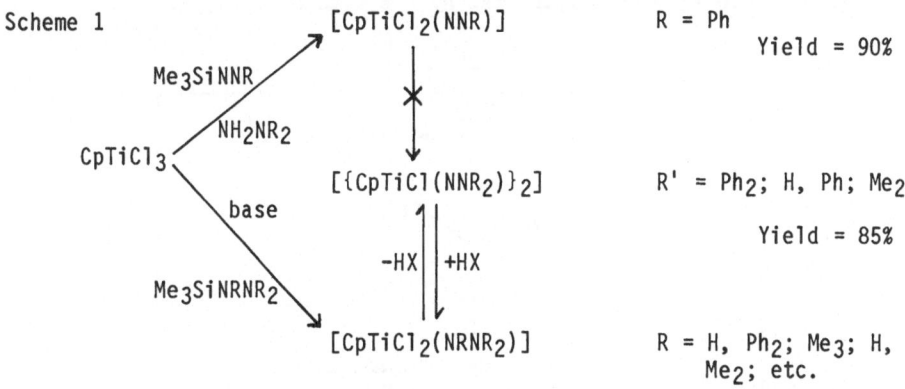

$[CpTiCl_2(NNR)]$	R = Ph Yield = 90%
$[\{CpTiCl(NNR_2)\}_2]$	R' = Ph_2; H, Ph; Me_2 Yield = 85%
$[CpTiCl_2(NRNR_2)]$	R = H, Ph_2; Me_3; H, Me_2; etc. Yield = 90%

X-ray structural characterization of these compounds revealed several unusual side on coordination modes in the organo-nitrogen fragments, a unique side on diazenido group, an unusual bridging side-on hydrazido(2-) group, and side-one hydrazido(1-) groups (Leigh et al 1983). The properties and interconversion reactions of these compounds have been investigated.

However reactions between $[CpTiCl_2]$ and $Me_3SiNHNHSiMe_3$ or $LiNHNH_2$ in tetrahydrofuran (thf), yield $[CpTiCl_2(thf)]$, N_2 and NH_3 in proportions dependent on stiochiometric ratio of the reactants. These results have been interpreted in terms of a reductive disproportionation of coordinated $NHNH^{2-}$ and $NHNH_2^-$ groups.

References

Dilworth J.R. et al. (1978) J. Organometallic Chem., 159, 47-51.
Leigh G.J. et al. (1983) J. Chem. Soc., Chem. Commun., 1368-1370.

ELECTROSYNTHESIS OF AMMONIA

C.J. PICKETT, K.S. RYDER and J. TALARMIN
AFRC Unit of Nitrogen Fixation, University of Sussex, Brighton BN1 9RQ UK.

We describe the electrosynthesis of ammonia from dinitrogen at room-temperature and atmospheric pressure.[1] Trans-[W(N$_2$)$_2$(Ph$_2$PCH$_2$CH$_2$PPh$_2$)$_2$] is converted to an hydrazido(2-)-complex, Scheme, (i). Reduction of this complex at a Hg-pool cathode affords NH$_3$ and re-generates the parent dinitrogen complex, Scheme, (ii). The system is cycled by re-protonation of the dinitrogen complex, this can be performed in situ. Free NH$_3$ is expelled from the catholyte during the course of the electrolysis into an aqueous acid-trap by purging the system with dinitrogen.

SCHEME

(W is trans-{W(Ph$_2$PCH$_2$CH$_2$PPh$_2$)$_2$})

$$TsO-\underline{W}^{IV}\equiv N-NH_2^{+} \xrightarrow{2e} TsO^{-} + \underline{W}^{II}\equiv N-NH_2 \qquad (1)$$

$$TsO-\underline{W}\equiv N-NH_2^{+} \rightleftharpoons H^{+} + TsO-\underline{W}-N\equiv NH \qquad (2)$$

(TsO = p-CH$_3$C$_6$H$_4$SO$_3$)

The success of the electrosynthesis depends upon two properties of the tosylate complex. First, it electronates according to reaction (1), secondly it behaves as an acid-source, reaction (2): protons necessary for the formation of NH$_3$ are supplied by sacrificial deprotonation of the hydrazido(2-)-complex.[1]

REFERENCE

1. C.J. Pickett and J. Talarmin, Nature, accepted for publication.

REPORT ON DISCUSSION OF POSTERS IN SESSION 2

G.J. LEIGH
AFRC UNIT OF NITROGEN FIXATION, UNIVERSITY OF SUSSEX, BRIGHTON, BN1 9RQ,
U.K.

1.   INTRODUCTION

Since only four posters were actually presented in this session, the
discussion was expanded to cover questions of general interest to chemists
and biochemists rather than being restricted to the poster material.   The
convenor posed a series of questions: what intermediates are involved in
the chemical conversion of N_2 to NH_3 on a single metal site?;   how does
this chemistry relate to nitrogenase chemistry?;   could this kind of
chemistry occur on clusters?;   what does chemistry tell us about the
iron-molybdenum cofactor (FeMoco)?   All these were touched on, though,
in the time available, generally superficially.

2.   DINITROGEN CHEMISTRY

Three of the four posters concerned dinitrogen chemistry.   Two
(George, Koczon, 2.04;   Latham et al., 2.05) relate to the breakdown of
hydrazide(2-), now generally accepted to be the first stage of
protonation of dinitrogen bound end-on to a single metal site.   The
former showed that protonation of $[Mo(Ph_2PCH_2CH_2PPhCH_2CH_2PPh_2)(PPh_3)(N_2)_2]$
produces a mixture of hydrazido(2-)-complexes, only one of which is
directly reactive in further reaction.   It was shown that hydrazine is
produced in this system if it is diluted, the yield of hydrazine depending
upon the amount of time which has elapsed since the initiation of the
reaction.   A parallel between this hydrazine production (time-scale of
the order of hours) and that occurring upon quenching of nitrogenase
(time-scale of the order of seconds) was suggested but questioned in
discussion.   It is not yet clear whether these observations really are
related.   Further reaction of the hydrazide(2-) complexes produces a
gold solid which is incompletely characterised.   Latham et al. presented
data which implicate disproportionation of hydrazide(1-) to produce NH_3
and N_2 and there is evidence that hydrazide(1-) can be produced by
protonation of hydrazide(2-).   Should this occur in nitrogenase, N_2
would be one product of its turnover.   In discussion it became apparent
that any evidence for disproportionation during nitrogenase turnover is,
as yet, unconvincing.

Pickett and Talarmin, 2.08, showed evidence for a second available
route for hydrazide(2-) degradation, involving coupled protonation and
reduction which leads to production of $2NH_3$ from each hydrazide(2-).
Further, this route can be made cyclic, with regeneration of the starting
dinitrogen complex.   Although this route may prove of great technical
interest, and although it shows that a cyclic system based on a single

metal site can be used to convert N_2 to NH_3 with high efficiency, there is as yet no evidence as to whether or not this route is exploited in nitrogenase.

3.   CHEMISTRY OF NITROGENASE

There was a lively discussion on the chemistry of nitrogenase in relation to dinitrogen chemistry.   It was agreed that quenching experiments with acids were unlikely to give definitive data on whether nitrogenase uses metallo-hydrides during its turnover since almost any conceivable result can be satisfactorily rationalised in terms of hydride chemistry.   It seems doubtful whether physical techniques such as Fourier-transform infra-red spectroscopy would be sensitive enough or appropriate to detect transient hydrido-complexes or dinitrogen complexes. Although ^{95}Mo n.m.r. spectroscopy (Enemark et al., 2.02) has given a vast amount of information concerning mononuclear complexes, it has not yet been successful in detecting molybdenum in molybdoenzymes, and is unlikely to be so.   E.s.r. techniques are more informative.   In fact, due to problems related to asymmetry, it has not yet proved possible to detect molybdenum in iron-molybdenum-sulphur clusters by ^{95}Mo n.m.r. spectroscopy.

The gap between what has been achieved in cluster chemistry in structural terms and the elucidation of the structure of FeMoco, as well as the gap between the established complex chemistry of N_2 and the chemistry of nitrogenase are still very wide.   Many still prefer to discuss nitrogenase chemistry in terms of side-on or bridging dinitrogen, although the precedents for either in model chemistry are far less convincing than those for mononuclear end-on binding.

4.   MECHANISM OF NITROGENASE

It was generally agreed that the mechanism of nitrogenase function presented elsewhere in this symposium provides the most complete description of the process of nitrogen fixation (Thorneley, Lowe, 3.33). This requires the presence of hydrido-species to explain the evolution of H_2 and the formation of $^1H^2H$ by nitrogenase fixing N_2 in the presence of $^2H_2$, and the identification of an intermediate hydronitrogen complex. This mechanism is consistent in many essentials with the established chemistry of N_2.   However, the data of Simpson et al., 3.29, seem greatly at variance with this scheme.   Questions were raised about the comparability of the two sets of data, because the former involves quenching the enzyme and the latter attempts to measure H_2 evolution from, and N_2 uptake by, the fixing enzyme as it starts to turn over.   It became clear that the question of instrumental response time needs to be very carefully investigated before the data in 3.29 can be taken at their apparent face value.

NITROGEN FIXATION IN
AGRICULTURE AND FORESTRY:
STRATEGIES, LIMITATIONS AND POTENTIAL

INVITED PAPERS

POSTER SUMMARIES:

(i)   Management practice.
(ii)   Rhizobia inoculant technology.
(iii)   <u>Frankia</u>.
(iv)   Free-living fixers.
(v)   Quantitation.

DISCUSSION GROUP SUMMARY

NITROGEN FIXATION IN AGRICULTURE IN TEMPERATE REGIONS

R.J. RENNIE, AGRICULTURE CANADA RESEARCH STATION, LETHBRIDGE, ALBERTA, CANADA T1J 4B1

INTRODUCTION

Research to improve the efficiency of N_2-fixing symbioses between plants and bacteria has been impeded by the inability to accurately quantify both the %Ndfa (% nitrogen (plant) derived from the atmosphere) and the actual amounts of N_2 fixed. Most emphasis has been placed on the latter which is an expression of phenotype but the %Ndfa may be an expression of genotype (Rennie, Kemp 1983a).

Some techniques used to quantify N_2 fixation are sufficiently precise and will at least rank treatments correctly. Others are inherently more variable and may even result in incorrect rankings. Some researchers can use destructive techniques; plant breeders require selection techniques that are rapid, inexpensive, partially or non-destructive of the plant and suitable for screening large plant populations for N_2 fixation. Several techniques do exist and although none of them is entirely satis-factory, they may be used successfully under specifically designed experimental conditions.

METHODS OF QUANTIFYING N_2 FIXATION

Indirect Methods

Acetylene Reduction Assay (ARA).--Detailed ARA procedures for greenhouse and field experiments have been described (Balandreau, Dommergues 1973; Dart et al 1972; Hardy et al 1968,1973). The limitations of ARA were also clearly described (Bergersen 1970; Hardy et al 1968,1973; Turner, Gibson 1980) but unfortunately have been largely ignored. ARA is a short-term kinetic measurement and the existence of diurnal and seasonal variations in N_2 fixation and plant cultivar-rhizobial strain speci-ficity in expression of H_2 evolution makes the ranking of treatments and extrapolation to total N_2 fixed over a growing season questionable. If the technique is used for qualitative evaluations the guidelines suggested by Lethbridge et al (1982) should be followed. ARA can be very useful, however, in ranking treatments imposed on single cultivar-single rhizobial strain interactions where H_2 evolution is not a consideration.

Direct Methods

Yield-Dependent Criteria.--A plant has two or three sources of N (atmosphere, soil, and perhaps fertilizer). Only two techniques are suitable to estimate N_2 fixation integrated over the growing season, the N balance (NB) method and methods based on the principle of ^{15}N

isotope dilution (ID). The NB method, used extensively to estimate N_2 fixation (LaRue, Patterson 1981), is based on the assumption that the fixing (fs) and non-fixing system (nfs) assimilate identical amounts of soil and fertilizer N. Thus,

$$\%Ndfa = \frac{N \text{ yield (fs)} - N \text{ yield (nfs)}}{N \text{ yield (fs)}} \times 100 \qquad (1)$$

If the N yield of the fs exceeds that of the nfs, the difference is attributed to N_2 fixation. Where sufficient soil N exists for the plant to attain its genetic yield potential, the plant can exhibit an ability to alter its three possible N sources to attain that yield potential (Table 1). Aurora and Great Northern 1140 had identical N yields but %Ndfa of 43.2 and 60.3%, respectively. Comtesse de Chambord and Limelight also had identical N yields but %Ndfa of 44.3 and 15.9%, respectively. When less N_2 was fixed, as in the case of Aurora and Limelight, more soil N was assimilated. Surprisingly, however, in the case of GN 1140 and Comtesse de Chambord, when more N_2 was fixed, the plants apparently ignored readily available soil N. This is a classical example of an instance where NB would have incorrectly ranked these cultivars for N_2 fixation.

Table 1. N yield, N_2 fixed and sources of N yield in four cultivars of <u>Phaseolus vulgaris</u> L. receiving 40 kg N ha^{-1} as $Ca(NO_3)_2$ (3.1250 atom %^{15}N) (Rennie, Kemp 1983b)

| Cultivar | Shoot N | N fixed | % N derived from | | |
			Atmosphere	Fertilizer	Soil
	-- kg ha^{-1} --		----------- % -----------		
Aurora	182	70	43.2	13.9	42.8
GN1140	184	112	60.3	9.7	29.9
Comtesse de Chambord	105	47	44.3	13.7	42.1
Limelight	102	16	15.9	20.6	63.4

Thus, positive or negative values for N_2 fixation in the absence of nodulation are sometimes encountered (particularly on high N soils) when N_2 fixation is estimated using the NB method (Broadbent et al 1982; Rennie et al 1978; Rennie 1979; Ruschel et al 1979).

Yield-Independent Criteria.--Several isotopes of N exist varying in mass number from 12 to 17. In agriculture, the two most important are the radioactive nuclide ^{13}N, whose half-life of 10.05 minutes severely limits its usefulness, and the stable nuclide ^{15}N.

$^{15}N_2$($^{13}N_2$) Reduction.--Traditionally, ^{15}N has been used primarily as a definitive test to prove that N_2 fixation has occurred. The second major use has been to establish the true $N_2:C_2H_2$ ratio. The use of $^{15}N_2$ (or $^{13}N_2$) is a short-term kinetic measurement like ARA and, as such, is no more applicable to field quantification of %Ndfa than ARA. It is a definitive but rather awkward means of proving N_2 fixation <u>in situ</u>. An excellent treatise on methodology of $^{15}N_2$ usage is now available (Bergersen 1980).

[15]N Isotope Dilution.--Recent improvements in ^{15}N tracer techniques have permitted quantification of N_2 fixation in legumes by labelling the soil (Kohl, Shearer 1981; Legg, Sloger 1975; Rennie 1982; Rennie et al 1982) or fertilizer N (Broeshart 1974; Fried, Broeshart 1975) with ^{15}N or by measuring changes in the abundance of naturally occurring [15]N (Rennie et al 1976; Virginia, Delwiche 1982) (for reviews see Rennie et al 1978; Rennie, Rennie 1983; Vose et al 1981).

$$\text{Thus,} \qquad \%Ndfa = 1 - \frac{\text{atom } \% \, ^{15}N \text{ ex. (fs)}}{\text{atom } \% \, ^{15}N \text{ ex. (nfs)}} \times 100 \qquad (2)$$

The calculation of %Ndfa is yield-independent, requiring only that the fs and nfs assimilate identical proportions (but not necessarily identical quantities) of N from the soil and/or fertilizer (Rennie, Rennie 1983).

ID is totally dependent on the comparison of a fixing and non-fixing system. The nfs has two possible N sources, soil and fertilizer, while the fs has a third source, atmosphere. Thus, N in the plant which is derived from the soil is slightly or highly enriched in ^{15}N and will be diluted by $^{14}N_2$ from the atmosphere via N_2 fixation.

Detailed studies on the choice and assumptions involved with various nfs are available (Rennie, Rennie 1983; Witty 1983; Witty, Ritz 1984).

The unique instance when NB and ID would yield similar estimates of %Ndfa and N_2 fixed occurs when the fs and nfs have identical fertilizer use efficiency (FUE) or, when no fertilizer N is applied, when the fs and nfs assimilate identical amounts of soil N. A comparison of %Ndfa and N_2 fixed in the bean cultivar Aurora inoculated with various single strains of Rhizobium phaseoli (Table 2) showed that NB and ID gave similar estimates only when the soil N uptakes of the fs and nfs were identical. In fact, NB would not even have ranked the rhizobium strain treatments correctly.

Table 2. N_2 fixation and %Ndfa in Phaseolus vulgaris cv 'Aurora' determined by [15]N (at levels of ^{15}N natural abundance) and by NB methods (Rennie 1984)

Rhizobium phaseoli strain	N fixation (kg ha⁻¹) ID	NB	%Ndfa ID	NB	Soil N (kg ha⁻¹)
57	0	-16	0	-27.1	59
127	56	55[1]	43.3	42.3	74a[2]
161	50	31	47.2	29.1	10
166	64	57	48.4	43.2	68a
255	72	65	51.4	46.4	68a
390	30	-6	43.6	-8.7	39
407	92	89	56.0	54.3	72a
727	38	26	42.8	25.7	63
904	69	42	58.9	35.9	48
952	59	46	48.7	38.0	62
Mix[3]	47	21	48.6	21.9	49

Uninoculated					
0 kg N[4]	0	0	0	0	75a
40 kg N	0	57	0	43.2	–

[1]Underlined values indicate significant agreement (P < 0.05) between methods as determined by t-test.
[2]Values with the same letters do not differ significantly (P < 0.05) according to Tukey's hsd.
[3]Mix = R. phaseoli strains 117K12, 127K80, 127K81.
[4]Non-fixing control.

In another study screening Phaseolus cultivars (Table 3), ID and NB gave identical estimates for %Ndfa and N_2 fixed only when the fs and nfs (Galt barley) had identical FUE, in accordance with the proposed theory (Rennie 1984). NB overestimated N_2 fixation in cultivars Aurora and Redkloud.

It is obvious that NB can sometimes result in an accurate estimate of N_2 fixation in field-grown legumes. Unfortunately, prior to experimentation it is not possible to predict when NB will accurately measure N_2 fixation even in soils of low N status; nor can NB be relied upon to accurately rank treatments. Thus, ^{15}N-ID results in the most precise and reproducible estimates of N_2 fixation and is the technique of choice to quantify N_2 fixation in field-grown annual legumes.

Table 3. N_2 fixation and %Ndfa in six cultivars of Phaseolus vulgaris receiving 10 kg N ha^{-1} as ^{15}N-labelled Ca(NO^3)$_2$ as determined by ^{15}N-ID and by NB methods (Rennie 1984)

Phaseolus cultivar	N fixed (kg ha^{-1})		%Ndfa		%FUE
	ID	NB	ID	NB	
Aurora	93	116	55.0	68.6	53.8
Comtesse	56	56[1]	53.9	51.4	35.6a[2]
Kentwood	75	75	55.4	57.1	40.6a
GN-1140	125	130	68.3	71.5	41.9a
Limelight	40	48	40.3	47.3	43.0a
Redkloud	45	64	38.2	54.7	51.0
Galt[3]	0	0	0	0	36.9a

[1]Underlined values indicate significant agreement (P < 0.05) between methods as determined by t-test.
[2]Values with the same letters do not differ significantly (P < 0.05) according to Tukey's hsd.
[3]Non-fixing control.

HOW MUCH N_2 IS FIXED?

N_2 fixation (by ^{15}N-ID) in soybeans grown in North America and Europe has been estimated to average 100 kg N ha^{-1}, which is approximately 50% of the plant's N yield (Table 4). N_2 fixation in field beans has been severely underestimated probably because of the acetylene reduction assay. Field beans receive an average of 75 kg N fixed ha^{-1} (a maximum of 125 kg N ha^{-1} in some cultivars) which is also approximately 50% of their N requirement.

Table 4. Maximum N₂ fixed for field-grown legumes as determined by
^{15}N isotope dilution

Crop and country	Cultivar	% Ndfa	N fixed (kg ha^{-1})	Reference
Glycine max				
Austria	Chippewa	23	10	Rennie 1982
Brazil	DT71-9330	44	–	Ruschel et al (1979)
Canada	X005	67	115	Rennie et al (1982)
France	Chippewa	58	–	Domenach et al (1979)
Hungary	Chippewa	38	92	Rennie et al (1978)
Sri Lanka	Chippewa	50	71	Rennie et al (1978)
U.S.A.	Chippewa	38	67	Rennie et al (1978)
U.S.A.	Clay	37	114	Ham & Caldwell (1978)
U.S.A.	Ford	60	108	Deibert et al (1979)
Phaseolus vulgaris				
Brazil	Goiano	37	25	Ruschel et al (1982)
	Carioca	68	65	Ruschel et al (1982)
Canada	Various	38–68	40–125	Rennie, Kemp (1983b)
U.S.A.	Various	30–50	89–90	Westerman et al (1981)
U.K.	Cascade	56	95	Witty (1983)
Cicer arietinum				
Canada	UC-5	47–60	24–84	Rennie, Dubetz (1985)
Lens culinaris				
Canada	Eston	79–86	188–192	Rennie, Dubetz (1985)
	Laird	72–86	129–162	
Pisum sativum				
U.K.	Kelvedon Wonder	62	83	Witty (1983)
Canada	Trapper	75–80	169–189	Rennie, Dubetz (1985)
	Century	77–82	166–178	
Vicia faba				
Austria	–	81	–	Fried et al (1983)
U.K.	Minden	64	174	Witty (1983)
Canada	Aladin	74–91	145–237	Rennie, Dubetz (1985)
	Ackerperle	73–91	138–236	
	Outlook	85–92	165–216	
Lotus corniculatus				
U.S.A.	Carroll	–	103	Heichel et al (1983)
Medicago sativa				
U.S.A.	MnPL-8-HB-A	43	138	Heichel et al (1981)
	Agate	–	224	Heichel et al (1983)
Trifolium repens				
U.S.A.	Ladino	85–100	200	Broadbent et al (1982)
Trifolium subterraneum				
U.S.A.	Woogenellup	88	180	Phillips, Bennett (1978)

Chickpea growth in southern Alberta fixed 84 kg N ha^{-1} with 60% of the
plant N being derived from N₂ fixation (Rennie 1985b) similar to

chickpea at four Syrian locations (ICARDA 1983) where N_2 fixed ranged from 14-120 kg N ha^{-1} with %Ndfa from 45-82%. Inoculated fababean in western Canada averaged 230 kg N fixed ha^{-1} in one location and 179 kg N ha^{-1} in another with %Ndfa of 79-91%. N_2 fixed was 174 kg N ha^{-1} with cv Minden in the United Kingdom (Witty 1983), 181 kg N ha^{-1} in Egyptian fababean (Fried, Broeshart 1975) and 144 kg N fixed ha^{-1} in cv Wieselburger in Austria (Wagner, Zapata 1982). %Ndfa varied from 61% (Wagner, Zapata 1982) to 81% (Fried et al 1983) in the temperate cultivar Wiselburger grown in Austria.

Field pea grown in Canada averaged 172 and 179 kg N fixed ha^{-1} over two years at two different locations when inoculated. %Ndfa was 83-76%. Lentil in western Canada fixed 162-190 kg N ha^{-1} with %Ndfa of 81-76%. In the only other report on ^{15}N-determined lentil fixation (ICARDA 1983), %Ndfa varied from 39-87% and N_2 fixed from 10-129 kg N fixed ha^{-1} in four Syrian locations.

Perennial legumes grown in temperate regions receive a high proportion of their plant N from N_2 fixation, over 200 kg N fixed each year from N_2 fixation accounting for over 90% of the plant's N yield. Other legumes have intermediate values for N_2 fixed.

Temperate legumes in N_2-fixing mode readily obtain over 50% of their plant N requirements from symbiotic N_2 fixation. The remaining N is from assimilation of mineral N from the soil and/or fertilizer N pools. The proportion of N from the three available plant N sources can be modified by selection of rhizobial strain-host plant cultivar combinations or by modifying agronomic conditions to favor N_2 fixation.

CONCLUSIONS

It is imperative to quantify %Ndfa under field conditions to establish the potential agronomic importance of N_2 fixation in various agricultural ecosystems. Further, %Ndfa must be determined to select for the ability of plants to support and benefit from symbiotic N_2 fixation.

Short-term kinetic measurements such as acetylene-reducing activity and reduction of $^{15}N_2$ or $^{13}N_2$ are not applicable to field quantification of %Ndfa. Nor are yield-dependent criteria such as dry matter and N yield or %N. Only techniques employing the principle of ^{15}N isotope dilution yield a reproducible and precise estimate of N_2 fixation integrated over time under field conditions. While ^{15}N techniques now permit the determination of %Ndfa in annual or biannual legumes, this technology can only be applied with difficulty and uncertainty to established stands of legumes such as alfalfa and clover (Rennie 1985a; Heichel et al 1984).

Isotope techniques are partially or totally destructive of the plant. While this presents no problem in investigations of varieties or advanced generations, it is a problem for plant breeders selecting early generation material (e.g., F2, F3) where each plant is genetically unique. Experiments must be undertaken to assess the accuracy of determining %Ndfa by sampling of a specific plant part at a certain early growth stage or by earlier samplings which would permit regeneration of the sampled plant.

REFERENCES

Balandreau J, Dommergues Y (1973) Bull. Ecol. Res. Comm. (Stockholm) 17, 146-154.
Bergersen FJ (1970) Aust. J. Biol. Sci. 23, 1015-1025.
Bergersen FJ (1980) In Bergersen FJ, ed, Methods for Evaluating Biological Nitrogen Fixation, pp 65-110, John Wiley & Sons, New York.
Broadbent FE et al (1982) Agron. J. 74, 625-628.
Broeshart H (1974) Neth. J. Agric. Sci. 22, 245-254.
Dart PJ et al (1972) In Uses of Isotopes for Study of Fertilizer Utilization by Legume Crops, pp 85-100, I.A.E.A., Vienna.
Diebert EJ et al (1979) Agron. J. 71, 717-723.
Domenach AM et al (1979) C.R. Hebd. Seances Acad. Sci. 289, 291-293.
Fried M, Broeshart H (1975) Plant Soil 43, 707-711.
Fried M et al (1983) Can. J. Microbiol. 29, 1053-1062.
Ham GE, Caldwell AC (1978) Agron. J. 70, 779-783.
Hardy RWF et al (1968) Plant Physiol. 43, 1185-1207.
Hardy RWF et al (1973) Soil Biol. Biochem. 5, 47-81.
Heichel GH et al (1981) Crop Sci. 21, 330-335
Heichel GH et al (1983) In Forage and Grassl. Conf., pp. 196-197, Am. For. Grassl. Counc, Eau Claire, Wis.
Heichel G et al (1984) Crop Sci. 24: 811-815.
ICARDA (1983) In ICARDA Research Highlights 1982, pp 22-23, ICARDA, Alleppo, Syria.
Kohl DH, Shearer GB (1981) Plant Soil 60, 487-489.
LaRue TA, Patterson TG (1981) Adv. Agron. 34, 15-38.
Legg JO, Sloger C (1975) In Klein ER, Klein PD, eds, Proc. 2nd Int'l. Symp. Stable Isotopes, pp 661-666.
Lethbridge G et al (1982) Soil Biol. Biochem. 14, 27-35.
Phillips DA, Bennett JP (1978) Agron. J. 70, 671-674.
Rennie RJ et al (1978) In International Atomic Energy Agency, Isotopes in Biological Dinitrogen Fixation, pp 107-134, I.A.E.A., Vienna.
Rennie RJ (1979) Rev. Ecol. Biol. Sol 16, 455-463.
Rennie RJ (1982) Can. J. Bot. 60, 856-861.
Rennie RJ (1984) Agron. J. 76, 785-790.
Rennie RJ (1985a) Agron. J. (in press).
Rennie RJ (1985b) ASA Spec. Pub. (in press).
Rennie RJ, Dubetz S (1985) Agron. J. (submitted).
Rennie RJ, Kemp GA (1983a) Agron. J. 75, 640-644.
Rennie RJ, Kemp GA (1983b) Agron. J. 75, 645-649.
Rennie RJ et al (1982) Agron. J. 74, 725-730.
Rennie RJ, Rennie DA (1983) Can. J. Microbiol. 29, 1022-1035.
Rennie DA et al (1976) Can. J. Soil Sci. 56, 43-50.
Ruschel AP et al (1979) Plant Soil 53, 513-525.
Ruschel AP et al (1982) Plant Soil 65, 397-407.
Turner GL, Gibson AH (1980) In Bergersen FJ, ed, Methods for Evaluating Biological Nitrogen Fixation, pp 111-138, John Wiley & Sons, New York.
Virgina RL, Delwiche CC (1982) Ocecologia (Berl) 54, 317-325.
Vose PB et al (1981) In Graham PH, Harris SE, eds, Biological Nitrogen Fixation for Tropical Agriculture, pp 575-592, CIAT, Cali.
Wagner GH, Zapata F (1982) Agron. J. 74, 607-612.
Westerman DT et al (1981) Agron. J. 73, 660-664.
Witty JF (1983) Soil Biol. Biochem. 15, 631-639.
Witty JF, Ritz K (1984) Soil Biol. Biochem. 16, 657-666.

CURRENT RESEARCH STRATEGIES FOR USE OF
ACTINORHIZAL SYMBIOSES IN FORESTRY

M. LALONDE and L. SIMON. CENTRE DE RECHERCHES EN BIOLOGIE FORESTIERE,
FACULTE DE FORESTERIE ET DE GEODESIE, UNIVERSITE LAVAL, STE-FOY, QUEBEC,
CANADA G1K 70R

1. INTRODUCTION

Since the first successful isolation of a <u>Frankia</u> strain by Callaham
et al in 1978, the field of research involving actinorhizal host plants
has undergone rapid expansion. With this availability of pure Frankia
strains, cultivatible in vitro, it is now feasible to apply the modern
techniques of microbiology, physiology, biochemistry, molecular biology
and genetics to this group of nitrogen-fixing actinomycetes. Furthermore, the ability to dispense large quantities of infective and effective
strains for various host plants of economic importance for forestry is
also increasing the interest both in evaluating and optimizing various
combinations of hosts and strains and in understanding the involvement of
both partners in order to manipulate the symbiotic system in the future.
In this short review, we will present the various strategies used in our
laboratory and by others for the optimization of the various actinorhizal
symbioses with an emphasis on the <u>Alnus-Frankia</u> symbiosis, although the
same strategies are currently used or applicable to others.

2. FRANKIA

2.1 <u>Isolation and in vitro cultivation.</u>
The isolation and cultivation in vitro of <u>Frankia</u> is now feasible for
species from numerous host plants, including <u>Alnus</u> (Baker et al, 1979;
Berry, Torrey, 1979; Lalonde, Calvert, 1979; Quispel, 1979; Benson, 1982;
Normand, Lalonde, 1982), <u>Casuarina</u> (Gauthier et al, 1981; Diem et al,
1983; Diem, Dommergues, 1985; Zhang et al, 1982; Zhang, Torrey, 1985),
<u>Elaeagnus</u> (Baker et al, 1980; Lalonde et al, 1981), <u>Shepherdia</u> (Lalonde
et al, 1981), <u>Colletia</u> (Gauthier et al, 1984) and other actinorhizal
systems. Various methods of isolation, complemented with different
nutrient media, can now be used with predictible success and for a good
number of host plants (Baker et al, 1979; Burggraaf et al, 1981; Lalonde
et al, 1981). At present, the strains associated with the <u>Alnus</u> species
are considered as being the easiest ones to isolate. It should be noted
that study and use of the important group of actinorhizal plants used in
the tropics, i.e., <u>Casuarina</u>, is now also developing at a fast pace with
the current availability of numerous <u>Frankia</u> species (Lie et al, 1984).

2.2 <u>Frankia taxonomy</u>
The recent availability of numerous strains (Baker, 1982) helps in the
clarification of relationships in this group of nitrogen-fixing actinomycetes known as the unique family, Frankiaceae. It should be noted that
only one genus and no species is recognized (Lechevalier, 1980).
Although bizarre for a taxonomist, this absence of recognition of species
within the genus <u>Frankia</u> is justified by the accumulating information
indicating a strong variability in the various morphological and physiological characters normally used for taxonomic purposes. Nevertheless,
the current trend is to use a chemotaxonomy approach (Lechevalier, 1980;
1983; Lechevalier et al, 1982), using various characteristics, such as

cell composition (Mort et al, 1983), serology (Baker et al, 1981),
protein pattern (Benson, Hanna, 1983; Benson et al, 1984), DNA GC content
and DNA homology (An et al, 1983) and plasmid mapping (Simonet et al,
1985). It should be noted that, although the available Frankia isolates
can belong to various host specificity groups, this character is not used
in the actual classification of the strains because some Frankia strains
can infect two host specificty groups, such as the Alnus and Elaegnus
groups (M. Lalonde, unpublished) and because some non-infective strains
of Frankia are known to exist. One stable character is the presence of a
unique sugar, identified as 2-0-methyl D-mannose, which is a unique trait
of the Frankiaceae family (Mort, Lalonde, 1981).

2.3 Physiology

The first research trends, following the availability of strains of
Frankia, were to study both the activity of its nitrogenase under in
vitro conditions (Tjepkema et al, 1980; 1981) and the various strategies
used by the microorganism or host cells to protect and control it (Murry
et al, 1984a; Tjepkema, 1984; Akkermans et al, 1984). It is not
surprising then that the possible implication of leghemoglobin or other
oxygen control mechanism was suggested (Murry et al, 1984b).
Furthermore, the possible mechanisms of nutrition and accumulation of
energy by Frankia either as a free-living organism or as a microsymbiont
are still being extensively evaluated (Akkermans et al, 1981; Blom,
Arking, 1981; Shipton, Burggraaf, 1982; Huss-Danell et al, 1982;
Akkermans et al, 1983; Tisa et al, 1983; Lopez et al, 1984; Lopez,
Torrey, 1985). The unique morphology of Frankia, its abiltity to form
hyphae, vesicles and spores inside sporangia (Figure 1), makes this
organism very complex to study. Even so, it still represents an
interesting nitrogen-fixing microorganism to study per se, if not just
from a developmental point of view.

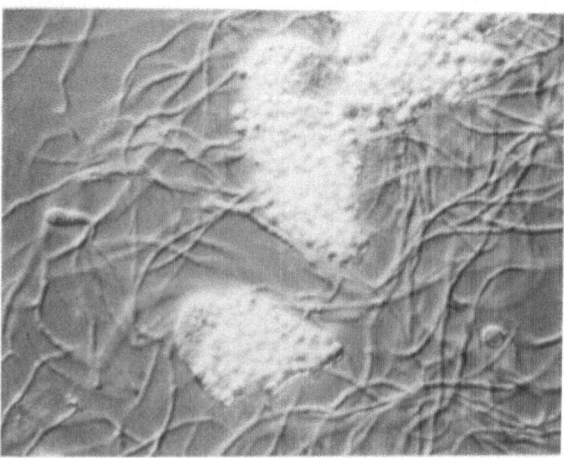

Figure 1. A Frankia strain isolated from Alnus crispa. This isolate
shows a typical morphology with hyphae, sporangia and spores, and
vesicles.

2.4 Genetics.

Although Frankia is characterized as a pleiomorphic, slow-growing actinomycete, with a doubling time of more than 24 hours, and with extreme phenotypic variability when grown in culture, genetic studies have begun with good success. For example, the size and GC content of Frankia DNA is known (An et al, 1983) and the presence of small molecular weight plasmids have been shown in some strains (Normand et al, 1983; Simonet et al, 1984). These plasmids are, however, too small to bear the symbiotic genes necessary for symbiosis. In this regard, the presence of a megaplasmid is still controversial (P. Simonet, personal communication) but should be sought. The cryptic plasmids found in Frankia and their specific restriction enzyme digestion patterns are presently used as specific markers to distinguish among Frankia strains showing very similar growth rate, morphology and infectiveness on a specific host. These genotypic characteristics complement quite well the specific phenotypic characteristics of protein patterns studied by Benson and co-workers and others (Parson et al, 1985; M. Gardes and M. Lalonde, unpublished). A current important step in the genetic manipulation of Frankia is the development of cloning vectors into E. coli (Normand et al, 1985). The results of Normand et al (1985) indicate that cloning vectors can be engineered, although no expression of Frankia genes occurs in Streptomyces. The use of Frankia protoplasms is another strategy under close scrutiny for the exchange of DNA between Frankia strains and other species, but is not yet fully developed (P. Simonet and P. Normand, personal communication). Finally, with the availability of Frankia DNA from easy to grow strains, from fragments of plasmids or genomic DNA cloned in E. coli, various workers are presently evaluating the presence of genes using probes developed from other microorganisms. This strategy is analogous to the one used in the evaluation of the symbiotic genes involved in the legume-Rhizobium symbiosis. In fact, the specific "symbiotic gene probes" of nodulins, developed from any legume-Rhizobium system, are prime probes to be used in the actinorhizal-Frankia symbioses. For further details, a full review on the genetics of actinorhizal Frankia has been recently published by Normand and Lalonde (1985).

2.5 Evaluation of infectiveness and effectiveness.

Most of the hundreds of Frankia strains currently available in institutional germplasm collections (Baker, 1982) have been evaluated for their infectiveness and effectiveness on specific host plants. With few exceptions, all strains were found to be infective and effective. In the case of positive infectiveness and non-effectiveness, the microscopic evaluation of the actinorhizae indicated the absence of vesicle formation by the endophytic actinomycete (Baker et al, 1980). This is in accordance with the close relationship observed in vitro between the presence of vesicles and induction and control of nitrogenase activity (Tjepkema et al, 1980). Although possible to evaluate under in vitro conditions in a selection program for detection of highly effective strains, it is required that the nitrogenase activity of a particular strain should be evaluated in association with the appropriate host plant.

2.6 Commercial use of effective strains.

Quite soon after the first modern isolation of a Frankia strain by Callaham et al (1978), Lalonde and Calvert (1979) were using pure cultures of the newly isolated strains of Frankia as infective inoculants for various Alnus species. Since that first demonstration of the ease of using Frankia cultures as inoculants for specific host plants, other researchers have studied the various factors involved in the successful nodulation of host plants by Frankia inoculants for commercial use (Berry, Torrey, 1985; Stowers, Smith, 1985; Perinet et al, 1985). The last authors review the first large scale inoculation of actinorhizal plants with Frankia, which involved more than seven million seedlings of Alnus crispa, A. glutinosa, Elaeagnus angustifolia, E. commutata, Hippophae rhamnoides, Myrica gale and Shepherdia canadensis. This large scale experiment over the last five years demonstrates clearly the feasibility of using Frankia inoculants, which can be easily prepared, preserved and used. This ease of use is mainly due to the advantageous propriety of Frankia in being able to form spores which are more resistant to the environmental conditions detrimental to vegetative cells, e.g., temperature, dessication, U.V. irradiation, etc. With this technology and the great number of strains available, it is now feasible to produce and use commercial inoculants of Frankia on demand. Nevertheless, there is still a lot to be done in optimizing the actinorhizal-Frankia symbioses. In this regard, as suggested by Hall and Maynard (1979), in any amelioration program of the actinorhizal symbiosis, it is imperative to select in parallel both the macro and the micro symbionts.

3. HOST PLANT

3.1 Participation of the host in the symbiosis.
As indicated in the preceding section, the host involvement in the actinorhizal symbiosis should be evaluated in parallel to any selection program applicable to Frankia. This is easier said than done with an organism with a doubling time of a few years, if not decades. Nevertheless, in the Alnus-Frankia symbiosis, it has been shown (Hall et al, 1979) that an important improvement in the rate of nitrogen fixation can be gained by the selection of a host plant which is more active in photosynthesis, the key element for nitrogen-fixation activity. Furthermore, the susceptibility of the host plant to a Frankia strain is variable and consequently selectible (Tremblay et al, 1984). The reverse is also true with strains which are more aggressive than others (Nesme et al, 1985) in infecting the root hairs of the host.

3.2 Host selection via micropropagation techniques.
In order to study the involvement of the host plant in the symbiosis, cuttings have been used (Dawson and Sun, 1981; Hennessey et al, 1985) in order to remove the genotypic variations observed between plants derived from seedlings. With the current developments in the in vitro micropropagation of plants, it is now feasible to produce in great number an interesting and uniform phenotype of particular plant material (Sommer et al, 1979; Mott, 1981). The alder genus is known as an easy material to clone (Garton et al, 1981). Consequently, we have used plantlets from various Alnus speices, obtained by in vitro micropropagation, in the study of the involvement of the host in the interactions with Frankia. The susceptibility of the host to be nodulated by various Frankiae was never lost by the micropropagated plantlets (Perinet, Lalonde, 1983). As expected, this type of material has shown great uniformity (Figure 2) in the plantlets obtained and their phenotypic stability has been useful in demonstrating significant variation both in the effectiveness of various strains of Frankia and in responses among clones of various origins (Simon et al, 1985) (Figure 3). Clonal material of Alnus, obtained by in

vitro micropropogation is presently used in studies of the cold and frost resistance of various species and populations of Alnus (F.M. Tremblay and M. Lalonde, unpublished). The same strategies are currently applied to other actinorhizal host plants, e.g., Elaeagnus (Bertrand, Lalonde, 1985), Shepherdia (L. Bertrand and M. Lalonde, unpublished) and Hippophae (D. Montpetit and M. Lalonde, unpublished).

Figure 2. Illustration of uniformity of Alnus glutinosa plantlets, clone AG-8, inoculated with Frankia sp. ArgN22d. The plant on the left is an uninoculated control. For more details, see Simon et al (1985).

3.3 Host variation via tissue culture.
The stability of the growth and nodulation characters observed in the micropropagated plantlets of seven Alnus species (Tremblay, Lalonde, 1984) is an interesting trait. However, in our improvement program of the Alnus-Frankia symbiosis, where variability of various characters are looked for (e.g., nodulation, photosynthesis and growth rate, frost and cold resistance, wood and fiber qualities, etc.), the variability that can be obtained due to the "somaclonal variation" phenomenon during a "protoplasm-regeneration cycle" was selected as a means to induce variability in Alnus. In this regard, although it is not yet possible to

672

Figure 3. Comparison of two A. glutinosa clones 74 days after
inoculation with Frankia sp. ARgN22d. The two plantlets on the left are
clone, AG-2, and two on the right, AG-4. In the center is an
uninoculated control.

obtain the regeneration of plantlets from Alnus protoplasms, it is at
least feasible to obtain protoplasm-derived calluses from Alnus spp.
(Hutinen et al, 1982; Tremblay, Lalonde, 1985). Similar strategies are
currently applied to other actinorhizal host plants, such as Casuarina
(H.R. Berg, J.G. Torrey, personal communications) and should provide, in
a relatively short period of time, major improvements in the knowledge
and the use of those economically important Frankia symbioses.

4. REFERENCES

Akkermans ADL et al (1981) Physiol. Plant. 53, 289-294.
Akkermans ADL et al (1983) Can. J. Bot. 61, 2793-2800.
Akkermans ADL et al (1984) In Veeger C. and Newton WE, eds. Advances in
Nitrogen Fixation Research, pp. 311-319, Martinus Nijhoff/Dr W. Junk, The
Hague.
Baker D et al (1979) Nature (London) 281, 76-78.
Baker D et al (1980) Can. J. Microbiol. 26, 1072-1089.
Baker D (1982) The Actinomycetes 17, 35-42.
Benson DR and Hanna (1982) Can. J. Bot. 61, 2919-2923.
Benson DR et al (1984) Appl. Environ. Microbiol. 44, 461-465.
Berry A and Torrey JG (1979) In Gordon JC, Wheeler CT and Perry DA, eds,
Symbiotic nitrogen fixation in the management of temperate forests, pp
69-83, Oregon State University, Corvallis.
Berry A and Torrey JG (1985) Plant and Soil, in press.
Bertrand L and Lalonde M (1985) Plant and Soil, in press.
Blom J and Harking R (1981) FEMS Microbiol. Lett. 11, 221-224.
Burggraaf AJP et al (1979) Plant and Soil 61, 157-168.

Callaham D et al (1978) Science (Washington, DC) 199, 899-902.

Dawson JO and Sun SH (1981) Can. J. For. Res. 11, 758-762.

Diem HG et al (1983) Can. J. Bot. 61, 2815-2821.

Garton S et al (1981) Hort. Science 16, 758-759.

Gauthier D et al (1981) Appl. Environ. Microbiol. 41, 306-308.

Gauthier D et al (1984) Acta Oecologica/Oecologia Plantarum 5 (19) 231-239.

Hall RB and Maynard CA (1979) In Gordon JC, Wheeler CT and Perry DA, eds. Symbiotic nitrogen fixation in the management of temperate forests, pp 322-344, Oregon State University, Corvallis.

Hall RB et al (1979) Bot. Gaz. (Chicago) 140S, S120-S126.

Hennessey TC, Bair LK and McNew RW (1985) Plant and Soil, in press.

Huss-Danell KW et al (1982) Physiol. Plant. 54, 461-466.

Hutinen O, Honkanen J and Simola LK (1982) Plant Sci. Lett. 28, 3-7.

Lalonde M and Calvert HE (1979) In Gordon JC, Wheeler Ct and Perry DA, eds. Symbiotic nitrogen fixation in the management of temperate forests, pp 95-110, Oregon State University, Corvallis.

Lalonde M et al (1981) In Gibson AH and Newton WE, eds., Current perspectives in nitrogen fixation, pp. 296-299 Australian Academy of Science, Canberra.

Lechevalier MP (1980) In Dietz A and Thayer DW, eds., Actinomycete taxonomy, Society for Industrial Microbiology Special Publication No. 6.

Lechevalier MP (1983) Can. J. Bot. 61, 2964-2967.

Lechevalier MP et al (1982) Dev. Ind. Microbiol. 23, 51-60.

Lie TA et al (1984) An. van Leeuwenhoek 50, 489-503.

Lopez M et al (1984) Can. J. Microbiol. 30, 746-752.

Lopez M and Torrey JB (1985) J. Bacteriol. 162, 110-116.

Normand P and Lalonde M (1982) Can. J. Microbiol. 28, 1133-1142.

Normand P et al (1983) J. Bacteriol. 155, 32-35.

Normand P and Lalonde M (1985) Plant and Soil, in press.

Normand P et al (1985) Gene 34, 367-370.

Mott RL (1981) I Conger BV ed., Cloning Agricultural Plants via in vitro Techniques, pp. 217-254, CRC Press, Boca Raton.

Mort A et al (1983) Can. J. Microbiol. 29, 993-1002.

Murry MA et al (1984a) Plant and Soil 78, 61-78.

Murry MA et al (1984b) Arch. Microbiol. 139, 162-166.

Nesme X et al (1985) Can. J. Bot. 63, in press.

Parson WL et al (1985) Plant and Soil, in press.

Perinet P and Lalonde M (1983a) Can. J. Bot. 61, 2883-2888.

Perinet P and Lalonde M (1983b) Plant Sci. Lett. 29, 9-17.

Perinet P et al (1985) Plant and Soil, in press,

Quispel A (1979) In Gordon JC, Wheeler CT and Perry DA, eds, Symbiotic nitrogen fixation in the management of temperate forests, pp 57-68, Oregon State University, Corvallis.

Shipton WA and Burggraaf AJP (1982) Plant and Soil 69, 1249-1261.

Simon L et al (1985) Plant and Soil, in press.

Simonet P et al (1984) Can. J. Microbiol. 30, 1292-1295.

Sommer HE et al (1979) In Sharp WR, Larsen, PO, Paddock EF and Raghavan, eds., Plant Cell and Tissue Culture, principles and applications, pp 461-491, Ohio State University, Columbus.

Stowers MD and Smith JE (1985) Plant and Soil, in press.

Tisa et al (1983) Can. J. Bot. 61, 2768-2773.

Tjepkema JD et al (1980) Nature (London) 287, 633-635.

Tjepkema JD et al (1980) Nature 287, 633-635.

Tjepkema JD et al (1981) Can. J. Microbiol. 27, 815-823.

Tjepkema JD (1984) In Veeger C and Newton WE, eds., Advances in Nitrogen Fixation Research, pp. 467-473, Martinus Nijhoff/Dr W. Junk, The Hague.
Tremblay FM and Lalonde M(1984) Plant Cell Tissue Organ Cult. 3, 189-199.
Tremblay, FM and Lalonde M (1985) Plant Sci. Lett., in press.
Vanden Bosch KA and Torrey JG (1983) Can. J. Bot. 61, 2989-2909.
Zhang Z et al (1984) Plant and Soil 78, 79-90.
Zhang Z and Torrey JG (1985) Plant and Soil, in press.

5. ACKNOWLEDGEMENTS
This paper was made possible by a NSERC of Canada grant NO. 7-209 to ML. We thank Louise Warner for her assistance with the manuscript.

BIOLOGICAL NITROGEN FIXATION IN TROPICAL AGRICULTURE.

JAKE HALLIDAY,
Battelle-Kettering Laboratory, Yellow Springs, OH 45387-0268 USA.

The most recent (1983) symposium in this series concluded thus: the pure sciences are telling us more about how the diazotroph works and therefore about how we can make it work for us more usefully...that, after all, is why we are all here (Postgate 1984). Proceeding on the assumption that this same premise still pertains, an overview is now attempted of the extent to which improved understanding is rendering biological nitrogen fixation more useful in tropical agriculture. Biological nitrogen fixation occurs by essentially the same process in the tropics as it does in the temperate zone. By confining the overview to the tropics, the emphasis is shifted from the process of nitrogen fixation and onto the microbial and plant and microbial systems which fix nitrogen. This overview also confines itself to biological nitrogen fixation systems of significance in tropical agriculture, and does not deal with those systems occurring more generally in natural ecosystems. By "agriculture" in this paper I am including managed production systems involving crops, or crops and trees.

I am choosing to draw an additional boundary on this overview that is not implicit in the title. I will concentrate on the significance of advances in biological nitrogen fixation in tropical agriculture in the decade following the first in this series of International Symposia on Biological Nitrogen Fixation held in Pullman in 1974. This period coincides the decade of renaissance in biological nitrogen fixation research inspired by the energy crises of 1973/74. This was also the era of major new program initiatives at the international level on biological nitrogen fixation. I was privileged to live and work in the tropics for that decade, initially at CIAT in Colombia (1975-79) and later with the USAID supported NifTAL Project at the University of Hawaii (1979-84). This is an overview, not an exhaustive review, tinged inevitably with my personal interests and perspectives.

A rather loose definition of the "tropics" is used in this overview, referring to more or less the area bounded by 30 degrees of latitude north and south, rather than by the tropics of Cancer and Capricorn. The tropics is a very diverse region, almost defying definition. Plant breeders have been looking for "super cultivars" to solve food production problems in the tropics, as if the word tropic was sufficient to define an environment...this is not the case (Cochrane 1984). The tropics is a region of alarming statistics: it is where more than half of the world's population lives, but where only about one quarter of the world's food is produced. In Africa, one third of the people depend on imported food grains.

All of those nations categorized as less developed countries (LDCs) are clustered in the tropics. This is not just a coincidence. The dominant sector of the economies of these countries is agriculture, which they must pursue in a region characterized by infertile soils, deficiencies, excesses, or disruptions in rainfall, and year round proliferation of pests and diseases affecting crops, livestock, and the people themselves.

In 1974, the world population was 3.6 billion and projections were that it would reach 7 billion and stabilize. A decade later the population is 4.76 billion and the projected ceiling is now 10 billion. Greatest growth is in the developing countries. The projected increase in Nigeria alone (or Bangladesh) exceeds that of North America, Europe and Russia combined. This is a sobering reminder that advances in food production continue to be negated by rampant population growth.

Despite the best efforts of all involved, more people are starving in 1985 than was the case 10 years ago. Per capita food production rose significantly between 1950 and 1975, but leveled off in the last decade. The famine in Ethiopia is now in the headlines, but it is not news:

> "The current situation in the sub-Sahara region of Africa is a grim reminder of Man's struggle for food," (Lovvorn, in the Keynote address of the First International Symposium on Nitrogen Fixation in 1974).

Desertification proceeds unabated, and topsoil is being depleted and lost from cropland at a rate of 25.4 billion tons per year (Brown 1985). Forest resources have dwindled from 7 million ha at the beginning of this century to 4.8 million ha in 1950, and will be reduced to 3.0 million ha by the year 2000. The energy crisis may have lessened, but the fuelwood crisis is certainly worsening. The outlook is alarming given that the forest clearance:replanting ratio is 29:1 in tropical Africa, 10:1 in Latin America, and 5:1 in tropical Asia. Nitrogen-fixing trees feature prominently among the fast-growing tree species targeted for development as fuelwood species and for reforestation of denuded natural forests. Research on these species and their biological nitrogen fixation will intensify in coming years.

Traditional and modern farming systems in the tropics almost invariably include legumes (Okigbo 1977, Rao et al. 1979) and have done for centuries. Yet, most farmers in the tropics remain unaware that legumes fix nitrogen. Biological nitrogen fixation is of significance to tropical agriculture only to the extent that it is an economically feasible alternative to nitrogenous fertilizers. Nitrogen is more limiting for more crops in more places than any other plant nutrient. Fertilizer application has been the most effective technology for alleviating this constraint.

Fertilizer technology has its roots in the chemistry research of Justus von Liebig in the mid-1800s, but became a common farming practice only after the World War II. The rate of increase in fertilizer use has decreased dramatically in the last decade compared with the decade before. Between 1964 and 1974 fertilizer use increased almost 400% (from 22 to 84 million metric tons), whereas from 1974 to 1984 it increased less than 50% (to 121 million metric tons). Half of all fertilizer produced is of nitrogenous compounds. The developing countries consume just over a third of the total fertilizer production, and their share is increasing at a greater rate than that of the industrialized nations. Fortunately, there is also a trend towards increased fertilizer

production in the developing countries, since some of these are endowed
with the natural gas resources that are the determinant of locations
where nitrogen fertilizer production will be ēconomically feasible.
Although only about 11% of the earth's land surface receives fertilizer
nitrogen (Hauk 1983), fertilizer supports one third of the world's food
production, in turn supporting about 1.5 billion people.

Countries of the tropics are the least able to derive the benefits of
fertilizer technology. Several countries, including Brazil restrict
import to save foreign exchange. Other countries, like Zaire, lack the
infrastructure for reliable distribution to production regions. Thus,
biological nitrogen fixation has an immediate appeal as an inexpensive
alternative to nitrogenous fertilizer for tropical agriculture.
Biological nitrogen fixation technology implies intential use of this
trait in ways, and to an extent, that would not have occurred without
deliberate management. Biological nitrogen fixation technology,
therefore, has two aspects: one relates to the deliberate inclusion of
legumes in cropping systems to derive benefits from their nitrogen fixing
abilities; and the other is the intentional use of practices designed to
maximize the quantity of nitrogen fixed by the system. Examples of the
use of legumes aspect are intercropping, crop rotation, green manures,
and legume-based pastures. These technologies have much greater
significance in the tropics than elsewhere, though green manures are used
extensively only in countries with centrally controlled economies (Norman
1982). About one billion ha in the tropics (as here defined) are
climatically capable of supporting pasture legumes (Mannetje 1978). The
only technology in use today in the tropics to enhance the quantity of
nitrogen fixed is inoculation with Rhizobium. In India, inoculation with
Azotobacter and cyanobacteria is also practiced, though the precise
nature of benefits to crop production have not been definitively linked
to biological nitrogen fixation. Inoculation remains a very imperfect
technology. In Brazil, less than 1% of farmers inoculate Phaseolus
vulgaris (Burton 1982). In the U.S.A., only 2% of the nodules on
soybeans are formed by inoculant strains, (Ham, Lawn, Brun 1976). The
problems with this technology in the tropics continue to be: inadequate
infrastructure for production and distribution of a quality inoculant
product; inappropriateness of the form of the technology for use by
subsistance farmers; and probable inability of the inoculant strain to
survive in sufficient numbers, or by other means, to compete effectively
with indigenous soil strains of Rhizobium.

Any suggestion of a substantial replacement of nitrogenous fertilizer use
for cereals and root crops by biologically fixed nitrogen is unrealistic
because these crops respond to levels of nitrogen fertilizer far greater
than those currently supplied by legumes. Thus, there is a need to
devise ways to increase the contribution that legumes can make to
cropping systems as a complement to nitrogen fertilizer based production,
rather than as an alternative to it.

Interestingly, the world's greatest concentrations of trained scientists
are in the temperate zone, whereas the tropics have the greatest plant
diversity. Appropriately, some of the most notable new findings in the
area of nitrogen fixation over the past decade have come from the
tropics: Spirillum (Dobereiner, Day 1976); Trema/Parasponia (Trinnick
1973); stem nodules in Sesbania (Dreyfus, Dommergues 1981) and

Aeschynomene (Yatazawa, Yoshida 1979); and others. The tropics is also a region about which there are many misconceptions based on inappropriate extrapolations and exagerations. Australian researchers have made a substantial contribution to the development of our understanding of agriculture in the tropics, though there has probably been a tendency to over extrapolate from Australian circumstances to other, less unique, environments. Australian Alfisols and Vertisols present more serious physical than chemical limitations, whereas in Latin America the dominant Oxisols and Ultisols present more serious limitations related to acid soil infertility. Tropical legume species adapted to Australian conditions are not likely to be successful in most Oxisols and Ultisols of Latin America, and vice verse (Sanchez, Isbell 1979).

The impression that tropical legumes are much more promiscuous than temperate legumes, that they nodulate freely with a wide range of tropical rhizobia, and that tropical soils are laden with so many such rhizobia that effective nodulation is virtually guaranteed without the need for inoculation is still widely held. This view is no longer well-founded. Some species and accessions from species previously considered to be promiscuous require specific strains of Rhizobium, or form highly effective associations with only a few out of the wide array of strains with which they nodulate. Intensification of interest in the tropical legumes and their rhizobia is revealing much greater variation in genetic compatability and nitrogen fixation effectiveness than is generally acknowledged. Allocation of tropical legumes to one of three categories has been recommended: Promiscuous Effective group, where nodulation occurs with a wide array of rhizobia isolated from many legume genera and where the resultant symbioses are predominantly effective in nitrogen fixation; Promiscuous Ineffective group, where nodulation occurs with an array of strains of rhizobia isolated from many legume genera, but where fully effective symbioses form with only a few of those strains; Specific group, where those strains from the same genus (or a restricted number of other genera) form effective symbioses.

Leucaena leucocephala is a fast growing tropical tree which has attracted a great deal of research attention in the past decade. Leucaena serves to illustrate the kind of misconceptions that can be perpetuated about tropical agricultural species in the face of data and evidence to the contrary. Leucaena has been attributed with some of the highest figures for annual nitrogen fixation (Vietmeyer, Cottom 1977). But interpretation of the high levels purported to occur must be tempered by an appreciation of the methods by which nitrogen fixation per se, as opposed to soil-nitrogen uptake, can be determined. For example, an exceptionally high figure for annual nitrogen fixation, 1 t N/ha (Dijkman 1950), was actually the accumulation of nitrogen in leafy matter of trees in a leucaena plantation harvested bimonthly for a one year period. The often quoted annual values of 500 kg N/ha given by Guevarra (1976) for K8 leucaena, and 600 kg N/ha for K341 leucaena are also total-nitrogen accumulation in the forage fraction. A more realistic figure for annual fixation of 110 kg N/ha was derived by Hogberg and Kvarnstrom (1982). This puts leucaena in much the same league as many other perennial tropical legumes with respect to its nitrogen fixation. Time will tell whether the 850 kg N/ha/yr reported for Parasponia holds up as a realistic value for fixation by that species in a field situation. Measuring nitrogen fixation in the field remains a difficult challenge.

Leucaena affords another example of how some of the notions that we have held about biological nitrogen fixation in the tropics do not hold up in practice. Despite being heralded as a miracle tree, Leucaena is notorious for its failure to grow in acidic soils. This has been attributed to many factors, including inability to nodulate and fix nitrogen under acid soil conditions. This, in turn, has been attributed to the unusual nature of its rhizobial symbiont, differing from "typical" tropical rhizobia in that it is moderate- to fast-growing, and that it is neutral or mildly acid-producing. Tropical rhizobia have been typified as slow-growing alkali producers. Alkali production has been inferred to be of ecological advantage to tropical rhizobia in the acid soils that predominate in the tropics. Conversely, rhizobia lacking this ability are considered disadvantaged. After calling into question the agricultural significance of acid and alkali production (Date, Halliday 1979) by rhizobia in laboratory media, selection programs were conducted in Colombia and Hawaii to identify strains of Rhizobium with high tolerance to soil acidity factors. Sites with widely differing soil pH were tested for the presence of rhizobia effective for leucaena, and the acid tolerance of these strains tested in a range of artificial media. Strains baited from acid soil grew well in acidic media, and strains from high pH soil did not (Halliday, Somasegaran 1983). Some strains with acid tolerance were also found to tolerate media at pH 8.5. Combining pre-screening of this type tests for nitrogen fixation effectiveness identified numerous strains that were acid tolerant and fully effective nitrogen fixers. When used as inoculants, these strains successfully nodulated leucaena abundantly and effectively in a highly acid (pH 4.1) field site (Llanos Orientales, Colombia). The best strain from that site, CIAT 1967 (now TAL 1145), was found to be highly competitive for nodulation of leucaena in an acid soil site in Hawaii when inoculated as a mixed inoculant in equal numbers with CB 81 and NGR 8. As TAL 1145 was neutral in its pH reaction on YMA and there was no correlation between ability to multiply in acidified media and production of alkali by any of the strains, the relevance of alkali production to success of an inoculant for acid soils is questionable. Successful initial nodulation of leucaena in acid soil did not render leucaena productive at the acid soil sites. The leucaena plant genotype is clearly limited by other factors, most likely calcium deficiency, under such conditions.

We should continue to be wary of over extrapolation from the knowledge base derived in one location to the circumstances elsewhere. We should be wary also of an overly simplistic understanding of, and reliance on, agrotechnology transfer concepts and trickle down (or across) theories. There is no substitute for well-trained scientists conducting on-site research and testing in the tropics. It is no accident that several program initiatives taken in the last decade have been designed to get North American, European, and Australian scientists more involved directly in tropical agriculture, to encourage accomplished scientists already resident in the tropics, and train cadres of new professionals to take up the challenge of original research on biological nitrogen fixation in their countries. Effective networks are now in place and the capacity for research on biological nitrogen fixation in the tropics is markedly improved over the situation a decade ago.

The composition of the list of systems which fix nitrogen biologically and are of significance in tropical agriculture has not changed in the past decade. There have been fluctuations in the relative research interest in these systems, however. Interest in the legume system continues steadily, in Azolla has increased dramatically and is now levelling out, in associative fixation has passed its peak, and in tree research is seeing a resurgence of interest. While research has provided a clearer understanding of biological nitrogen fixation in tropical systems, progress with translating that improved understanding into new technology that will impact in farmers' fields has been disappointing. Greatest near-term potential for putting biological nitrogen fixation to work to a greater extent in tropical agriculture lies in strategic management of the fate of nitrogen fixed biologically in the following contexts: multiple cropping systems including legumes for underutilized savannahs; agroforestry systems that combine fast-growing, nitrogen fixing trees legumes and other crops to meet the food and fuel requirements of the rural poor; fast-growing leguminous trees for reforesting water catchment areas following forest clearance; legume based cropping systems to give sustained productivity in tropical soils following jungle clearance; and selection of deep-rooted, drought tolerant leguminous trees that can serve as browse species in arid and semi-arid environments.

The resurgence in research into biological nitrogen fixation during the last decade was driven by: the world energy crisis of 1974; the presumed potential of biological nitrogen fixation to contribute significantly to food production by subsistence farmers in Third World countries; and an opportunity to contribute towards a more peaceful world by alleviating the dependence of poor nations on oil-rich states for something so basic as the fertilizer with which to produce food. Ten years have passed. There is a glut of oil and prices are down. The power of the OPEC cartel has been eroded. The dominant fuelstock for fertilizer production has shifted to natural gas, abundant in some developing countries. So, what is the driving force behind the continuing strong funding for research in biological nitrogen fixation? Several factors are at play. There is certainly a flywheel effect. Escalating attendance at this symposium series over the years is ample evidence of the momentum achieved. Current and future investments in biological nitrogen fixation research seem justified by: the intrinsically fascinating science of the process and the systems in nature that have evolved to accomplish it; and by the fortuitous coincidence that the tools of bacterial molecular genetics are being perfected on nitrogen-fixing microorganisms as subjects. Funding for research in biological nitrogen fixation as an international development strategy is likely to be more narrowly targeted in the future, concentrating on the management of nitrogen fixed in cropping systems to optimize its benefit to non-leguminous crops, primarily the cereals. Given the frustrations to date with strategies to improve the efficiency and the effectiveness of the legume-Rhizobium symbiosis, and the continuing anomaly that relatively few developing countries have been able to implement existing inoculation technology successfully, development support resources are being channelled into research to adapt the various components of inoculant technology to developing country circumstances.

I began this overview with an aspiration from the closing remarks at the most recent in this series of symposia. I will end it by referring, with the advantage of hindsight, to a speculative

statement made ten years ago in a paper at the First
International Symposium on Nitrogen Fixation:
>"I doubt that any alternate technology for improved provision
>of fixed N will be implemented in the next ten years, e.g.
>enhanced symbiotic nitrogen fixation, domestication of
>associative symbioses for cereals, abiological fixation under
>mild conditions in protic media)" (Hardy 1975).

Even though it is my assessment that in tropical agriculture this
prediction has been shown to be correct, I am sure the author and many of
the rest of us would be more content to have seen it proven wrong.
Despite the fact that no alternate technologies for improved provision of
fixed N emerged in the past decade, research over this period has
resulted directly in significant upgrades of existing (legume
inoculation) technology, and research milestones have been logged in the
steady progression towards emergence of technology that will
inevitably come about in the future.

REFERENCES

Brown LR (1985) State of the World 1985, Norton, New York.
Burton JC (1982) In Graham PH and Harris SC, eds, BNF Technology for
Tropical Agriculture, pp 105-114, CIAT, Colombia.
Cochrane TT (1984) In Uehara G, ed, A Multidisciplinary Approach to
Agrotechnology Transfer, pp 53-71, University of Hawaii.
Dobereiner J and Day JM (1976) In Newton WE and Nyman CJ, eds, 1st Int.
Symp. Nitrogen Fixation, pp 518-538, Washington State Univ, Pullman.
Date RA and Halliday J (1979) Nature 277, 62-64.
Dijkman MJ (1950) Economic Botany 4, 337-349.
Dreyfus BL and Dommergues YR (1981) In Gibson AH and Newton WE, eds,
Current Perspectives in Nitrogen Fixation, p 471, Australian Academy of
Science, Canberra.
Guevarra AB (1976) PhD Thesis University of Hawaii.
Halliday J and Somasegaran P (1983) In Leucaena Research in the Asia-
Pacific Region, pp 27-32, IDRC, Ottawa.
Ham GE Lawn RJ and Brun WA (1976) In Nutman PS, ed, Symbiotic Nitrogen
Fixation in Plants, pp 239-253, Cambridge Univ. Press, England.
Hauk RD (1983) in Freney JR and Simpson JR, eds, Gaseous Loss of Nitrogen
from Plant-Soil Systems, pp 285-312, Martinus Nijhoff/Junk, London.
Hogberg P and Kvarnstrom M (1982) Plant and Soil 66, 21-28.
Mannetje LT (1978) Tropical Grasslands 12, 1-9.
Norman MJT (1982) In Graham PH and Harris SC, eds, BNF Technology for
Tropical Agriculture, pp 9-26, CIAT, Colombia.
Okigbo BN (1977) In Dart PJ and Ayanaba A, eds, Biological Nitrogen
Fixation in Farming Systems in the Tropics, pp 61-72, IITA, Nigeria.
Postgate JR (1984) In Veeger C and Newton WE, eds, Advances in Nitrogen
Fixation Research, p X, Martinus Nijhoff/Junk, The Hague.
Rao MR (1979) in Ahmed S, ed., Final Inputs Review Meeting, pp 123-160,
East-West Center, Hawaii.
Sanchez PA and Isbell RF (1979) In Sanchez PA and Tergus LE, eds, Pasture
Production in Acid Soils of the Tropics, pp 25-54, CIAT, Colombia.
Trinnick MJ (1973) Nature 244, 459.
Vietmeyer N and Cottom B (1977) Leucaena - Promising Forage and Tree Crop
for the Tropics. National Academy of Sciences.
Yatazawa M and Yoshida S (1979) Physiol. Plant. 45, 293

APPLICATIONS OF NITROGEN FIXATION IN AGRICULTURE
AND FORESTRY

Ralph W. F. Hardy
BiotTechnica International Inc.
85 Bolton Street
Cambridge, MA 02140 USA
and
Cornell University
Ithaca, NY 14853 USA

INTRODUCTION

About 1970 an animated award-winning cartoon advertisement produced by a multinational company that had made seminal advances in biological nitrogen fixation knowledge was shown several times on U.S. national TV. This cartoon used a gleeful robust soybean plant to explain the agricultural facts and expectations for nitrogen fixation to an envious less-than-robust wheat plant. It said that soybeans can obtain essential nitrogen from the air while wheat cannot - scientists are working to give wheat this ability.

The reality of self-fertilizing crops is greater in 1985 than in 1970 with the science base much increased and several of the required technologies demonstrated. However, my temporal projections at the First International Symposium on Nitrogen Fixation in 1974 about practical impacts of this research have not been met. The science productivity has been good and in some cases superb; the conversion of this science to useful products and processes for agriculture and forestry has been slower than expected. The latter represents the underinvestment of the public and private sectors in development work.

Private sector commercial activity is expanding in recent years. There is still a need for at least one public sector effort of a critical mass to pursue applications where the low economic incentive or the high risk preclude private sector activities. Recent discussions at the Tennessee Valley Authority explored ways to meet this critical need in the public sector.

Continued long term support for the science of nitrogen fixation will need to be balanced with significant applications development. My comments will seek to provide a balanced optimistic view of applications from the molecular biological studies of nitrogen fixation.

KNOWLEDGE BASE/TECHNOLOGY STATUS

During the last twenty-five years, the biochemistry of

nitrogenase has become well understood. Additional detailed knowledge will be useful but is not necessary for many initial applications. Characterization of the nif genes and their regulation in Klebsiella pneumoniae is very advanced and is in progress in several other major diazotrophs as the presentations and posters of this meeting document. Nodulation genes in symbiotic plant and microbial diazotrophs are being identified. Overall the science base is strong.

The technology base has improved dramatically in the past few years. Transformation systems for plant nuclear and chloroplast genes are in hand. Plant cell and/or tissue culture is well advanced. Regeneration to whole plants is limited and varies with the plant at the cellular level but tissue regeneration which is more widespread will probably be adequate. New bioprocesses are available for improved microbe production. Sophisticated technologies such as site specific mutagenesis (SSM) will allow the design of genes to produce synthetic nitrogenase with improved functional characteristics such as possibly decreased ATP requirement or elimination of wasteful H_2 evolution. SSM has already been applied to the genes that code for enzymes that inactivate penicillin and photosynthetically fix CO_2 and should be applied aggressively to nitrogenase.

REGULATION

The current major regulatory concern is the environmental impact of the deliberate release of microbes or seeds that have been genetically engineered by molecular methods. In the U.S. the lead agency for regulation of diazotrophic microbes probably will be The Environmental Protection Agency using the Toxic Substance Control Act and for diazotrophic seeds will be the U.S. Department of Agriculture (USDA). The challenge is to regulate realistically. In the case of Rhizobium, there is a major favorable, long-term record for the deliberate release of Rhizobium genetically engineered by organismal methods. The USDA has strong internal professional strength to evaluate risks of both genetically engineered microbes and seeds. Rhizobium bacteria genetically engineered by molecular techniques are expected to be in R&D field tests in early 1986. Seeds genetically engineered by molecular techniques to provide agrichemical resistance will be in R&D field tests about the same time but seeds with N_2-fixing capabilities will not be available for several years.

COMMERCIAL ACTIVITY

Developing and established companies are recognizing increasingly the economic opportunities that exist for gentically engineered N_2-fixing organisms. One recent estimate of only the U.S. market gave a $1 billion (U.S.) opportunity for diazotrophic microbes and over $4 billion (U.S.) for diazotrophic seeds. About twenty companies located in Canada, France, Israel, U.K. and U.S are stated to have biotechnology programs on N_2 fixation. The majority - about 80% - are in developing biotechnology companies. Commercial activity is expected to expand.

BIOTECHNOLOGY PRODUCTS/PROCESSES

Several potential products or processes from application of N_2 fixation molecular biology can be proposed. The most probable - catalysts, diazotrophic inoculum - seed or seedling, agrichemicals, seed and delivery systems - are outlined in Table 1 with respect to nature, product, favorable and questionable characteristics, probability and timing, and extent of impact.

TABLE 1

Projected Biotechnological Products and Processes

A. New Catalysts

Nature - Designed Improved Nitrogenase Biomimics

Fixed N Product - Ammonia

Favorable Characteristics - Mild Conditions, Low Cost Reactor, Small Scale

Questionable Characteristics - Energy Cost, Dilute Aqueous Product

Probability and Timing - High in Long Term, Very Low in Short Term

Extent of Impact - Dependent on Economics,· In the Maximum Could Become Major New Source of Synthetic Fertilizer N

B. Diazotrophic Inoculum - Seed/Seedling

Nature - Rhizobium - Legumes, Azospirillum - Cereal Grains, Frankia - Forest Trees, Mycorrhiza - Crops/Forest Trees, Diazotrophs Genetically Engineered for e.g., Increased N_2 Fixation, Stress Tolerance, Competitiveness, Energy Efficiency, Insensitivity to Fixed N etc.

Fixed N Form - Ammonia, Amides, Ureides

Favorable Characteristics - Rhizobium - Legumes and Frankia - Trees are Established Major Agronomic and Forestry N-Input Systems, Uses Solar Energy, Couples N_2 Input Directly to Crop/Forest

Questionable Characteristics - Photosynthate Cost,
Competitiveness, Energy Efficiency, Yield Response to
Inoculation

Probability and Timing - Rhizobium Inoculation Established
Practice, High for Major Improvement of Rhizobium in Short
Term, Frankia and Mycorrhiza Inoculation Could Grow in
Importance, Unclear as to Useful System of Azospirillum
for Cereals

Extent of Impact - Increased N_2 Input and Yield of Major
Grass and Forage Legumes and Some Forest Species,
Fertilizer Use Could Decrease, Not Change or Increase

C. Agrichemicals

Nature - Designed Agrichemicals to Regulate nif , Improve
Photosynthate - Nitrogen Couple in Legumes, etc.

Fixed N Form - Ammonia, Amides, Ureides

Favorable Characteristics - Production and Marketing System
Exists, Strong Proprietariness

Questionable Characteristics - Toxicity and Residue, Regulation

Probability and Timing - High in Long Term, Low in Short Term

Extent of Impact - Dependent on Economics

D. Seed/Seedling

Nature - Cereals Genetically Engineered to Enable Rhizobium
Symbioses and Trees Genetically Engineered for Frankia
Symbioses, Agronomic, Horticultural and Forest Crops
Genetically Engineered with nif or Designed nif and other
Necessary Genes

Fixed N Product - Ammonia, Amides, Ureides

Favorable Characteristics - Single Seed/Seedling Product
Eliminates Problems of Microbe - Seed Products,
Established Distribution and Planting Equipment System for
Seed, Easily Applicable to Less Developed as Well As
Developed Economies, Direct Coupling N_2 Input Within
Plant

Questionable Characteristics - Yield Penalty Compared to Cost
of Synthetic Fertilizer

Probability and Timing - Very High in Long Term, Very Low in
    Short to Mid Term
Extent of Impact - Could Replace Need for Synthetic Nitrogen
    Fertilizer for Agronomic and Horticultural Crops and Trees

E. Delivery Systems

Nature - Genetically Engineered <u>Rhizobium</u>, <u>Frankia</u>, Mycorrhiza
    as Systems to Deliver Desired Molecules to Agronomic,
    Horticultural or Forestry Crop

Product - Genes, crop Protectants, Growth regulators

Favorable Characteristics - Exploits Existing Symabiotic
    Relationships, Timed Internal Delivery, May Be Cost
    Effective

Questionable Characteristics - Many Unknowns

Probability and Timing - May Be High in Long Term, Highly
    Speculative

Extent of Impact - Dependent on Science and Economics

SUMMARY

The molecular biology N_2 fixation science base exists
for model systems and is developing rapidly for agronomic and
less rapidly for tree systems. Most first generation required
technologies are in hand. Regulatory requirements are expected
to produce at most minor delays based on the extensive
favorable long term experience with deliberately released
<u>Rhizobium</u> engineered by organismal processes. There is an
increasingly recognized large economic opportunity.
Substanital commerical activity is occuring in both development
and established companies. A public sector center is needed
with a mission of developing applications for areas not
appropriate for commercial activity. In the short term,
improved genetically engineered diazotrophic inoculum for seeds
and seedlings will be in field tests and is expected to
increase legume yields. In the longer term, designed
agrichemicals, new catalysts, nitrogen self-fertilizing crops
and possibly microbial delivery systems are suggested to be the
major applications with the impact ranging from modest to
potential total replacement of synthetic nitrogen fertilizer
with expected benefits to include decreased costs of production
and/or increased production and decreased ground water
pollution. The prognosis for the applications of N_2 fixation
molecular biology in agriculture and forestry has never been
better.

INFLUENCE OF THREE MANURIAL SOURCES AND LEVELS ON NITROGEN FIXATION
IN VIGNA UNGUICULATA CV IFE BROWN

ADEWALE ADEBAYO
DEPT. OF SOIL SCIENCE
UNIVERSITY OF IFE
ILE-IFE, NIGERIA

The fertility and crop production of tropical soils are dependent upon
their organic matter content since little inorganic fertilizers are
applied in most agricultural practices. Efforts must therefore always be
geared towards the conservation and if possible, increasing the organic
matter contents of these soils as an alternative source to inorganic
fertilizers application. The organic matter could be in form of various
animal and plant residues.

Materials and Methods

Three manurial sources viZ: cowdung, poultry manure and household wastes
at 0, 2, 5 and 10% levels (dry weight basis) were each added separately
in triplicates to 5kg air-dried, unfertilized, 0-15cm surface Alfisol
soils and thoroughly mixed in the greenhouse. After a month of pre-incu-
bation (Olayinka and Adebayo, 1985), seeds of Ife brown cowpea were
planted. At 40days after planting, when 80 percent of the plants had
flowered, the plants were harvested, dried and analysed.

Results and Discussion

Results indicated that while poultry manure at all levels gave the
highest total dry matter accumulation (15.6g, 13.8g and 10.5g for
poultry, cowdung and household waste respectively at 10 percent level),
presumably as a result of the comparatively high nutrient content.
Cowdung at 10% gave the highest nodule dry mass and the highest nitrogen
fixed (2.20, 1.90 and 0.60 percent for cowdung, household waste and
poultry manure respectively). It seems therefore, that nitrogen fixation
in cowpea could benefit from organic matter additions especially tropical
soils which are generally low in organic matter and acidic in reaction.
Such a beneficial effect would depend on the amount of combined nitrogen
in the organic matter and the level of addition.

Olayinka, A and A. Adebayo 1985: Plant Soil 86: 47 - 56.

N_2 FIXATION AND CONSERVATION OF SOIL MINERAL NITROGEN BY LUPINUS ANGUSTIFOLIUS

JEFFREY EVANS[a], G.L. TURNER[b], G. O'CONNOR[a], F.J. BERGERSEN[b].
a, AGRICULTURAL RESEARCH INSTITUTE, PMB, WAGGA WAGGA, NSW 2650 AUSTRALIA
b, C.S.I.R.O., CANBERRA, ACT 2601, AUSTRALIA

1. INTRODUCTION

Lupin (L. angustifolius) is grown in rotation with cereal wheat (Triticum aestivum) to sustain yields and fertility in southern Australia. The N budget associated with cultivation of lupin is essential to effective management of the rotation for stable N fertility.

2. PROCEDURE

2.1 Materials and Methods

Accretion of N (N_A) by lupin is assessed as the difference in N_2 fixed (N_F) and the N in harvested grain (N_G). The proportion of N_2 fixed (N_{FP}) is determined by isotope dilution (Fried, Middelboe, 1977) using enrichment of microplots ($0.9m^2$) with $K^{15}NO_3$ (95 atms% ^{15}N) at 0.6-0.7 kg $N.ha^{-1}$ and the reference species T. aestivum cv. Osprey sown juxtaposed to lupin (cv. Illyarrie). Plant N (N_T) and grain N (N_G) are determined from Kjeldahl digest. Exploitation of soil N is assessed as N_T (wheat) and $N_T - N_F$ (lupin). Soil mineral N (Ns) is estimated to 1.0 metre as nitrate plus ammonium, by autoanalysis.

3. RESULTS

3.1 In 1983, on a red clay soil which received ca. 600 mm H_2O as rain, N budget (kgN/ha) estimates are shown in the table.

N_T (lupin) = 306 ± 14	N_G = 120 ± 11
N_T (wheat) = 112 ± 8	N_A = 139 ± 20
N_{FP} = 0.84 ± 0.03	N_T (lupin) - N_F = 49

The difference N_T (wheat) - (N_T lupin - N_F), vis. the difference in exploitation of soil N by wheat and lupin was 60 (±14) kg $N.ha^{-1}$, which suggested conservation of mineral N under lupin. This was confirmed by soil analysis which showed an additional 41 (±4) kg $N.ha^{-1}$ under lupin compared with wheat at harvest.

4. CONCLUSION

As well as a potential to return a substantial net addition of organic N, lupin also conserves N mineralised from soil sources. Accretion of N at the site was sufficient to balance the export of N in 7000 kg of wheat seed. This represents about 2 years of cereal production in the region.

5. REFERENCES

Fried M and Middelboe V (1977) Plant Soil 27, 713-715.

EFFECT OF NITROGEN FERTILIZATION ON SYMBIOTIC FIXATION AND GROWTH OF BROAD BEAN

ESSAM M.GEWAILY and ADEL K. KHEDER, Agric. Botany Dept., College of Agric., Zagazig University, Egypt

INTRODUCTION

Leguminous plants are very important crops for improving the physio-chemical properties of soils. The importance of legumes in such soils is further enhanced by their ability to fix atmospheric nitrogen in root nodules formed by soil bacteria of genus Rhizobium, and their capability to use these nitrogen compounds for growth. In addition, a crop that produces food rich in protein without addition of any nitrogenous fertilizers is especially advantageous in developing countries, where both protein and fertilizers are scarce and expensive.

MATERIALS AND METHODS

In a pot culture experiment, surface disinfected broad bean seeds (Vicia faba L.) cultivars (4-52 and 20-72) were inoculated with Rhizobium leguminosarum strains, 202, 1049 or R.I. and planted in different sterilized soils from Bakrajow, Syngar, Yayje and Askikalac in Northern Iraq. Plants were harvested after 7 weeks of growth. Subsequently, the best variety of broad bean was used to find the best level of nitrogen fertilizer as urea on efficiency of nitrogen fixation.

SUMMARY AND CONCLUSIONS

Inoculated broad bean gave a higher value of nitrogen content and dry matter yield especially in cultivar (20-72). Efficiency of the three tested strains varied with different soils and no significant differences were observed between the soils in total nitrogen content of plants. Addition of urea showed a marked increase in dry weight and total nitrogen in inoculated plants. High response was noted when 10 kg N/donum was added to Bakrajow soil inoculated with strain No. 202 while the other two strains were rather less efficient. Rhizobium inoculation without any addition of urea increased the efficiency of nitrogen fixation. Strain No. 202 was best in Yayje, while strain No. 1049 was promising in Syngar soil. In the case of Askikalak soil, 20 kg N/don. increased symbiotic nitrogen fixation. Among the strains R.I. was more efficient than the other two strains.

When different varieties of broad beans were studied for nitrogen fixation, it was found that variety 20-72 gave higher response to inoculation. It is therefore recommended for some areas of Iraq. Also, any application of urea to soil under broad bean should take into consideration such characteristics of the soil as organic matter, sulphate, calcium carbonate and soluble contents in addition to its texture. There is no correlation between soils and nitrogen fixation at 0.01 and 0.05 levels while significant correlations were found within plant varieties and rhizobia strains.

REFERENCES

Magu SP, Sen AN (1969) Arch. Microbiol. 68, 355-361.
Rice WA et al (1977) Can. J. Soil Sci. 57, 197-203.
Richards LA (1954) USDA Handbook No. 60, p 160.
Walker RH et al (1934) Soil Sci. 37, 387-401.

THE INFLUENCE OF PLANTING DENSITY ON DINITROGEN FIXATION AND YIELD
OF SOYBEANS

LAWRENCE A. KAPUSTKA and KENNETH G. WILSON, Dept. of Botany, Miami
University, Oxford, OH   45056, U.S.A.

We hypothesized that decreasing the interplant distance would result
in increasing rates of dinitrogen fixation in soybean. Wayne variety
soybeans were planted at 10-, 15-, 20-, and 30-cm equidistant intervals.
Morphological and physiological measures at 3-, 6-, and 9-weeks post-
emergence revealed that as planting density increased there were (1)
altered morphology and growth rates, (2) increased apparent specific
nitrogenase activity, (3) decreased nodule number and mass, and (4)
nearly constant fruit and seed production per plant. Expressed on
a unit area basis, N influx and yield increased geometrically as planting density increased,with maximum values obtained at the 10-cm plantings. Four different varieties of soybean (Pellas, Williams, Golden
Harvest 1285, and Stein 30-30s) were planted in a field at 10-, 20-,
and 30-cm equidistant intervals. Measurements of stem and leaf mass,
and plant height indicate that these three parameters increased geometrically as density increased.

Based on the literature survey and our experiments we conclude the
following:
1. Soybeans have higher net canopy photosynthesis at high-density
   plantings than at low density plantings.
2. Soybeans have altered morphology as a function of planting density
   expressed as differences in total plant height, pattern of inter-
   node elongation, leaf orientation and branching tendencies. At
   high-density plantings (10-cm spacing), the stem strength is
   decreased which could lead to lodging in some field conditions.
3. Dinitrogen fixation per g nodule is higher at high-density planting
   than at low-density planting, but this is compensated by greater
   nodule numbers and mass in the low-density planting.
4. Fruit and seed production is nearly constant in non-nodulated
   plants, whereas fruit and seed production by nodulated plants
   increases at 25- and 30-cm spacing. The consequence is that
   yield per area increases geometrically as spacing intervals
   decrease.
5. Discounting the possible problem of lodging, the data do not
   indicate that the minimum spacing interval has been achieved
   at 10-cm.
6. In the greenhouse, by separation of root-root interactions from
   shoot-shoot interactions (planting in individual pots), we have
   demonstrated that most of the observations of others may be
   explained on the basis of canopy interactions.
7. Vegetative growth under field conditions was consistent with
   the patterns shown in the greenhouse experiments. Reproductive
   effort of the field grown plants remains to be assessed.

STUDIES ON MICRONUTRIENTS IN CERTAIN GRAIN LEGUMES

K. SHIVASHANKAR, UNIVERSITY OF AGRICULTURAL SCIENCES, BANGALORE, 560024, INDIA

Molybdenum seed treatment at 4g Na2OMO4 increased the grain yield of soybeans by over a ton per ha. This was further studied in other grain legumes of groundnut, green-gram, black gram and cowpea in field and laboratory studies.

Mo as Na2OMO4 was highly toxic in burning away the root-tip and in causing production of 3-4 fresh lateral roots. Whereas P2O5 at 37.5kg/ha raised groundnut nodule weight from 228 to 278mg, Mo+P2O5 raised it to 311mg. Groundnut yields were the same with either 75kg P2O5 or 6gMo+37.5kgP2O5 with 8g Na2OMO4 seed treatment, yields of green gram raised from 7.4 to 12.1q and that of black gram from 8.3 to 12.3q/ha. Double coating of lime and Mo improved grain yields of cowpea

Generally, Mo had a toxic effect on the initial root growth suppressing the tap root system and causing proliferation of lateral roots resulting in increased root density, root lets and hairs. This enhanced nodulation and improved growth and yield of these crops. Zn and Ca based Mo compound had the same beneficial effect as that of Na2OMO4 thus bringing in economy. The same compound with vitamins improved root growth of soybeans remarkably. These studies, thus indicate a bright potentiality of making a breakthrough in yield barriers in grain legumes by suitably monitoring the use of Mo and Mo-compounds for seed treatment.

DROUGHT STRESS LIMITATIONS TO NITROGEN FIXATION IN GRAIN LEGUMES.

THOMAS R. SINCLAIR
USDA-ARS, UNIVERSITY OF FLORIDA, GAINESVILLE, FL 32611, USA

Nitrogen fixation (Acetylene Reduction Rates, ARR) of grain legumes decreases in response to drought stress. This study was undertaken to express ARR as a function of the fraction of total transpirable water (FTW) remaining in soil so that the results may be more readily extrapolated to field conditions. ARR was measured on glasshouse-grown plants of soybean (<u>Glycine max</u>), cowpea (<u>Vigna unguicalata</u>), and black gram (<u>Vigna mungo</u>) subjected to drought stress.

MATERIALS AND METHODS

Plants were grown in plastic pots (13 cm dia and either 14.5 or 23 cm tall) filled with a sandy loam soil. Once plants had attained at least 0.4 m^2 leaf area, the drought treatment was imposed. The plants were sealed in perspex chambers (15 cm dia x 25 cm tall) with the stem through the chamber lid. Plant-chamber units were weighed each day to determine transpirational water loss. Water equal to the loss was added to the well-watered plants and the droughted plants were left unwatered.

ARR were determined by flowing (8.3 ml/s) a 0.1 atm C_2H_2-air mixture through the chambers, beginning 30 min before gas sampling was initiated. The chambers were purged following analysis and no effects attributable to the C_2H_2 exposure were observed. ARR of the droughted plants were normalized to the rates of the well-watered plants.

RESULTS AND DISCUSSION

Transpiration was unaffected by soil dehydration until 0.3 FTW was reached. Substantial declines in transpiration rates occurred below 0.2 FTW. All grain legumes showed similar responses. Even though ARR data were more scattered than the transpiration data, it was apparent that the two processes responded quite differently to drought. For soybean, ARR began declining when there was still about 0.6 FTW and substantial reductions occurred after 0.4 FTW was reached. While there was a large variability in the data for black gram and cowpea, depressed ARR were also observed at high FTW. Substantial reductions in ARR were found above 0.3 FTW. These data suggest that one very important limitation to high nitrogen fixation rates in the field may be sensitivity to drought. In fact, reductions in nitrogen fixation rates at 0.5 FTW would occur well in advance of conditions usually considered droughty as gauged by stomata closure or reduction in leaf growth.

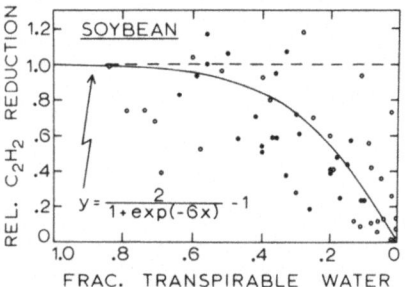

EFFECT OF PESTICIDE APPLICATION ON NITROGEN FIXATION IN TEMPERATE PASTURES

K.W. STEELE[a], R.N. WATSON[a], P.M. BONISH[a], G.W. YEATES[b]
[a]RUAKURA SOIL & PLANT RESEARCH STATION, HAMILTON, NEW ZEALAND
[b]N.Z. SOIL BUREAU, D.S.I.R., LOWER HUTT, NEW ZEALAND.

Inputs of symbiotically fixed N to pastoral agriculture in New Zealand, 1.1 m t annually, are well below their potential. Factors which limit fixation need to be identified because symbiotic N_2-fixation supplies more than 98 percent of total N inputs. In 1982 a research programme was commenced to determine the impact of invertebrate pests on N_2-fixation.

A reduction of invertebrates in sixteen pastures in the central North Island was obtained by application to soil of either 20 kg/ha oxamyl in spring + 10 kg/ha in autumn, or 10 kg/ha fenamiphos in spring and autumn, and foliar application of 175 g ai/ha omethoate when necessary as determined by sweep net sampling. All pastures were permanent pastures in which individual *Trifolium repens* L. feeding pests (*Costelytra zealandica; Graphognathus leucoloma; Sminthurus viridis; Heterodera trifolii; Meloidogyne hapla* and *Pratylenchus*) were usually present at population densities below recognised damage thresholds for herbage production.

Pesticide application resulted in an average increase in acetylene reduction activity per unit soil volume of 57 percent due to an increased growth of *Trifolium repens* and an increase in the number of nodules >0.5 mm/g root. ^{15}N isotopic dilution studies showed a high proportion of N in *Trifolium repens* herbage ($\bar{x}$ = 0.91) was derived from symbiotic N_2-fixation under the cutting regime imposed and this was not affected by pesticide application. Pesticide application, however, increased total N uptake in pasture herbage by up to 118 kg N/ha/yr (Steele et al. 1985).

Invertebrate pests markedly reduced the competitiveness of *Trifolium repens* with associated grass species preventing movement towards legume dominance in pastures exhibiting severe N deficiency. It appears that a reduction of soil nematode populations was a major factor in the results reported (Watson et al. 1985). It is evident that there is a large potential for increasing symbiotic N_2-fixation in pastures through selection and breeding *Trifolium repens* cultivars which have increased tolerance or resistance to invertebrate pests.

References:

Steele KW et al. (1985) XVth Int. Grassld. Cong. Kyoto. In press.
Watson RN et al. (1985) 4th Australasian Conference on Pasture
        Invertebrate Ecology, Christchurch, N.Z. In press.

SURVIVAL OF COWPEA RHIZOBIA IN JAMAICAN PEAT AND JAMAICAN SOILS

S.R. Aarons and M.H. Ahmad, Department of Biochemistry, University of the West Indies, Mona, Kingston 7, Jamaica.

Legumes are grown extensively in Jamaica. The yield is low and inoculants have never been used, possibly due to their unavailability. Also, indigenous rhizobial populations of red peas and cowpeas are low in Jamaican soils[2]. Peat is one of the most suitable carriers for inoculant production and large deposits of peat are found in Jamaica. To determine the suitability of Jamaican peat as a carrier for inoculant production, prior to starting inoculation programmes in Jamaica, we examined the survival of four cowpea rhizobia strains in Jamaican peat and in two sterile and non-sterile Jamaican soils (clay loam and bauxitic sandy loam). MATERIALS AND METHODS: Spontaneous antibiotic resistant mutants of cowpea rhizobia were isolated according to Kuykendall and Weber (1978) and Pugashetti and Wagner (1980). Survival of rhizobia in peat and soil was examined according to Roughley (1976) and Pena-Cabriales and Alexander (1979) respectively. Details about the materials and methods are given elsewhere[1].
RESULTS: All four cowpea rhizobia strains (JRC29, MI-50A, JRW3, IRC291) survived well in Jamaican peat: greater than 10^7 cells/g peat were recovered after long-term storage for six months. (The initial numbers of rhizobia were about 10^8 cells/g peat). The native strains JRC29 and JRW3 survived better than the introduced strains MI-50A and IRC291. We studied rhizobial survival in peat stored at two temperatures (4°C and 30°C) and found that long-term survival of rhizobia in peat was better at 4°C.
Survival of cowpea of rhizobia was determined in two soils (clay loam and bauxitic) subjected to slow drying at 37°C and/or 30°C. In sterile soils at both temperatures, all four strains increased in numbers (from approximately log 8.0 to log 11.0) for up to fifteen to twenty one days before a decline was observed. Rhizobia were recovered (log 3.0) from sterile soils even after 168 days of drying at both temperatures. Survival of the rhizobia strains (except IRC291) was similar in both soils at 30°C. At 37°C, the same was true except for strains JRW3 and IRC291. In non-sterile soils, none of the strains increased in numbers. However, all strains declined in numbers in clay loam soils.
CONCLUSION: Jamaican peat is suitable as a carrier for inoculant production. Cowpea rhizobia survived better in bauxitic soils (which form a large part of the arable soil in Jamaica) than in the clay loam soils.
REFERENCES: [1]Aarons, S.R. (1985). MPhil thesis: Survival of cowpea rhizobia in soils, in peat and on seeds. University of the West Indies, Jamaica. [2]Ahmad, M.H. and W. McLaughlin. (1985). Biotech. Adv. (in press). Kuykendall, L.D. and D.F. Weber. (1978). Appl. Environ. Microbiol. 36:915-919. Pena-Cabriales, J.J. and M. Alexander. (1979). Soil Sci. Soc. Am. J. 43:962-966. Pugashetti, B.K. and G.H. Wagner. (1980). Plant Soil 56:217-227. Roughley, R.J. (1976). In P.S. Nutman (ed.), Symbiotic Nitrogen Fixation in Plants. p. 125.

BROAD HOST RANGE SPECIFICITY OF Rhizobium STRAINS ISOLATED FROM LEGUME TREE ROOT NODULES

EULOGIO J. BEDMAR, MIGUEL A. HERRERA and JOSE OLIVARES
DEPARTAMENTO DE MICROBIOLOGIA, ESTACION EXPERIMENTAL DEL ZAIDIN, CSIC,
18008-GRANADA, SPAIN.

INTRODUCTION

Rhizobium strains involved in association with nitrogen fixing trees show a broad host range specificity (Dreyfus, Dommergues 1981) and frecuently the out-of-group nodulation is ineffective (Trinick 1980). Cross inoculation experiments involving woody or herbaceous legumes and Rhizobium strains isolated from legume tree root nodules were carried out to asses the specificity of such strains and the susceptibility of these plants to be infected by other rhizobia.

MATERIAL AND METHODS

Rhizobium strains GRH2, GRH3, GRH5 and GRH9 were isolated from root nodules of Acacia cyanophylla, A. melanoxylon, Prosopis chilensis and Sophora microphylla plants, respectively. Well characterized R. meliloti, R. trifolii, R. phaseoli, R. leguminosarum, R. lupini, R. japonicum, R. cowpea and the corresponding host plants were used. Plants were grown under controlled conditions either in test tubes, Leonard-jars or pots. Nitrogenase activity was measured as acetylene reduction, hydrogen evolution by gas chromatography and hydrogen uptake by amperometry.

RESULTS AND DISCUSSION

Rhizobium strains isolated from legume tree root nodules infected woody and herbaceous legumes and were also effective on at least one host in addition to the host from which they were isolated. There were significant differences ($P \leq 0.05$) in nitrogenase activity between herbaceous and tree legumes inoculated with their own Rhizobium strain, being greater for symbiosis estabilshed with herbaceous legumes. However, acetylene reduction values for P. chilensis-GRH2 symbiosis were similar to those found for the association involving herbaceous legumes. Relative efficiency of this symbiosis varied from 0.3 to 1 during the light time of the photoperiod but no hydrogenase activity could be detected either using hydrogen or methylene blue as final electron acceptor.

Uninoculated or GRH2-inoculated P. chilensis plants grown on non-sterile soil presented when they were 8 month old a similar number of nodules. However, nitrogenase activity and plant dry weight were much greater for those plants inoculated with GRH2. Therefore the need for inoculation of this legume with the corresponding bacterial strain to attaine maxima yields.

REFERENCES

Dreyfus BL and Dommergues YR (1981) Appl. Environ. Microbiol. 41, 97-99.
Trinick MJ (1980) J. Appl. Bacteriol. 49, 39-53.

AKNOWLEDGEMENT

This work has been supported by Comisión Asesora de Investigación Científica y Técnica, Grant No. 1764/82.

A RAPID ONE-STEP ELISA FOR RHIZOBIUM STRAIN DETECTION

P.E. Olsen and W.A. Rice, Agriculture Canada, Beaverlodge, Alberta

The enzyme-linked immunosorbent assay is ordinarily applied as an extremely sensitive technique for the detection of minute quantities of antigen among large amounts of contaminating antigen. We describe here a very rapid, single antibody-conjugate variation of the ELISA suitable for identification of Rhizobium strains in either plate count colonies or nodules squashes. Both of these cell sources provide large amounts of relatively pure and concentrated antigen. The procedure is complete within one-half work day as compared to the double-antibody sandwich ELISA which is more selective of specific antigen, but requires two work days.

Antibodies specific to heat-stable cell surface antigens were prepared and conjugated with alkaline phosphatase as previously described (Olsen, et al. 1983) and an antigen solid phase provided by evaporation of cell suspensions in microtiter plate wells (Olsen, Rice 1984). Specific antibody conjugate was applied directly to the bound cells, incubated, removed, and binding detected by color development in enzyme substrate. Even very small nodules (to 0.4 mg wet weight) provide sufficient cells for clear cut strain (and even dual-strain inhabitant) detection by the single antibody-conjugate ELISA.

Cell suspension (100uL) was pipetted into individual microtiter plate wells and the plates dried at 85 C in a force-draft oven for 1 h. Plates were washed repeatedly in a plate washer and 100uL of dilute specific IgG-alkaline phosphatase conjugate added to each well. Following incubation for 90 min and plate washing, 150 uL substrate was added. Substrate incubation for 30 min yielded strongly developed positive reactions (A410nm typically > 1.0) and plates were read and recorded with a computer controlled dual wavelength automatic plate reader. PBS negative controls gave values uniformly less tha A410m = 0.005, and heterologous R. meliloti strains gave values less than A410nm = 0.050. The single antibody-conjugate used per strain has been found to be useful at concentrations of approximately 0.1 times those required for the double antibody sandwich ELISA, presumably because greater amounts of antigen are bound to the plate wells by the evaporation prodedure. Since no sensitizing antibody is required, the procedure is extremely economical in terms of antibody as well as time.

REFERENCES

Olsen PE et al. (1983) Can. J. Microbiol. 29, 225-230.
Olsen PE and Rice WA (1984) Can. J. Microbiol. 30, 1093-1099.

INTERACTIONS OF Bradyrhizobium japonicum ISOLATED FROM FLORIDA SOILS WITH SOUTHERN SOYBEAN CULTIVARS.

A. R. ZIMET, S. L. ALBRECHT AND T. R. SINCLAIR
USDA-ARS AND UNIVERSITY OF FLORIDA, GAINESVILLE, FL   32611 USA.

Little information is available on the interactions of subtropical Bradyrhizobium japonicum strains with soybean [Glycine max (L.) Merr.] cultivars developed for the southeastern United States. Reports of soybean cultivars and B. japonicum strain interactions indicate that differences in strain effectivity do exist (Caldwell & Vest, 1970).   The objectives of this research were to examine 15 B. japonicum strains isolated in Florida and estimate which perform most effectively on two southern soybean cultivars. B. japonicum strains were isolated from soybean nodules taken from fields near Gainesville, Florida.   The strains were serotyped by fluorescent antibody and agglutination techniques. Plant-Bradyrhizobium combinations were grown in modified Leonard jar assemblies, in an unshaded greenhouse. Midday photosynthetically active radiation was 1400 $\mu E\ m^{-2}\ sec^{-1}$. Nitrogenase activity was determined by acetylene reduction. Relative efficiencies were calculated by the method of Schubert and Evans (1976). Bragg consistently showed greater nodule number and nodule dry weight, averaging (over 3 experiments) 280 nodules and 0.6948 grams of nodules per plant, compared to 193 nodules and 0.5453 grams of dry nodule weight for Centennial. Nitrogenase activity on a whole plant basis was also greater for Bragg, 10.97 $\mu moles$ ethylene formed per hour per plant as compared with 7.37 $\mu moles$ for Centennial. Four serogroups were identified among the local B. japonicum strains. The strains did not differ in their effect on shoot dry weight, nodule number, nodule dry weight or whole plant nitrogenase activity. Small differences in specific nitrogenase and hydrogenase activity and their relative efficiencies were found. In contrast with the Florida isolates, strains USDA 110 and USDA 122 significantly increased shoot and root dry weights. Although they yielded no measurable increase in nodule number or nodule dry weight, soybeans inoculated with USDA 110 and USDA 122 showed greater nitrogenase activity than plants inoculated with the Florida isolates. In terms of enhanced nodulation, Bragg showed greater nodule number and dry weight than Centennial. Among the local isolates, only minor differences were observed in those physiological characteristics examined. In monoxenic culture, strains USDA 110 and USDA 122 appear to stimulate plant growth and nodule activity under the conditions of this study. Although USDA 110 and USDA 122 appear to be superior to locally isolated B. japonicum strains, their prevalence, persistance and competitiveness with locally adapted strains is yet to be determined.

REFERENCES

Caldwell BE, G Vest (1970) Crop Sci. 10, 19-21.
Schubert KR, HJ Evans (1976) Proc. Natl. Acad. Sci. USA. 73, 1207-1211.

ISOLATION AND CHARACTERIZATION OF FRANKIA FROM NODULES OF CASUARINA

NANTAKORN BOONKERD, VITHAYA THANANUSONT, JITRAPORN POODPONG and YENCHAI VASUVAT, Soil Science Division, Department of Agriculture, Bangkok, 10900 Thailand

Attempting to isolate Frankia from nodules of actinorhizal plants has been performed in different methods: enzymatic digestion (Callaham D et al., 1978), microdissection (Burry A, Torrey JG, 1979), sucrose density fractionation (Baker D, et al 1979a), sephadex fractionation (Baker D et al 1979b), serial dilution (Diem HG et al, 1982), and superficially sterilized nodules directly incubated onto the nutrient medium (Diem HG, Dommergues Y, 1982). The method of isolation developed by Diem and Dommergues (Diem HG and Dommergues Y, 1982) was claimed successfully obtaining isolates which are capable of nodulating Casuarina seedlings.

The objective of these studies was to demonstrate another technique of isolation and characterization of the endophytes. Nodules collected from Casuarina equisetifolia were used as inoculum for C. equisetifolia seedlings grown sterily in Leonard jars containing sand and supplemented with N-depleted plant nutrient solution. Nodules from 4 month old seedlings were collected for isolation. Isolation technique was as follows:

1. Dissect nodule lobes and immerse in clorox (5.25% NaOCl) for 20 min.
2. Rinse three times in sterile distilled water.
3. Incubate surface-sterilized nodule lobes in a YEM (without manni-tol for 7 or 10 days.
4. Collect the nodule pieces from the tubes which do not show contamination and break up, then transfer to BAP medium (Murry M, et al (1984) containing 0.3% agar.
5. Observe for growth after 3 weeks, then transfer from the growing zone 0.1 ml by pipette to the BAP medium broth. The white-fluffy colonies adhering to the tube wall were observed after 4 weeks.

Frankia isolates maintained on BAP medium broth were shaken by vortex mixer to break up the colonies before inoculation. Two milliliters of the Frankia suspension were inoculated on C. equisetifolia seedlings grown in Leonard jars. The plants were placed in a glasshouse. Only 20% of the nodule pieces which were incubated in YEM broth were not contaminated. The uncontaminated crushed nodule piece was put into BAP-semisolid agar medium in 1 cm depth. The Frankia from the semisolid agar was transferred into BAP broth. The white puffy colonies were observed within three weeks. The colonies of Frankia were observed under SEM, it was found that they were typical of actinomycete. Two isolates of Frankia were used to inoculate C. equisetifolia and found that they could form nodules on C. equisetifolia seedlings.

Callaham D et al (1978) Science 199:899-902.
Burry A and Torrey JB (1979) In Gordon JC, Wheeler CT and Perry DA, eds, Symbiotic Nitrogen Fixation in Management of Temperate Forests, pp 69-83, Oregon State University, Corvallis.
Baker D et al (1979a) Nature 281, 76-78.
Baker D et al (1979b) Bot. Gaz. 140, 549-551.
Diem HG et al (1982) Can. J. Microbiol. 28, 526-530.
Diem HG and Dommergues Y (1982) Can J. Bot. 61, 2822-2825.
Murry MA et al (1984) Plant and Soil 79, 71-78.

COWANIA SUBINTEGRA (ROSACEAE): A NEW ACTINORHIZAL SPECIES

CATHERYN B. PERRY, JEAN C. STUTZ, Division of Agriculture, Arizona
State University, Tempe, AZ 85282 and TIMOTHY L. RIGHETTI, Department
of Horticulture, Oregon State University, Corvallis, OR 97331 U.S.A.

Nitrogen-fixing actinorhizal nodules are known to occur on rosaceous
plants in several genera. In the genera Cowania, nodulation and
nitrogen fixation has only been reported for Cowania stansburiana
(Righetti, Munns 1980). This report presents evidence of a new
actinorhizal species, Cowania subintegra Kearney (Arizona cliffrose), a
perennial shrub which only occurs in two isolated populations in
Arizona.

MATERIAL AND METHODS

Seeds of Arizona cliffrose were collected from wild populations in
Graham Co. AZ and planted in a mix of soils collected from five Cowania
communities in Arizona. Plants were maintained in the greenhouse and
harvested after 4 months by gently washing the plant roots. Nodules
were excised and prepared for transmission electron microscopy or
stained with 0.1% aniline blue lactophenol for light microscopy.

RESULTS AND DISCUSSION

All Arizona cliffrose seedlings formed corolloid nodules which were
indeterminate, dichotomously branched and up to several centimeters in
diameter. Light microscopy revealed that nodules were root-like in
structure with the endophyte confined to the nodule cortex. Endophyte
vesicles formed in enlarged cortical cells. The histology of the
nodules was similar to that reported for C. stansburiana (Nelson,
Schuttler 1984). Ultrastructural features of the endophyte including
septate, branched prokaroytic hyphae and terminal, non-septate vesicles
were similar to those reported for Frankia in nodules of other
rosaceous actinorhizal plants (Bond 1976, Newcomb 1981, Newcomb, Heisey
1984). Acetylene-dependent ethylene production is currently being
measured. Isolation and characterization of the endophyte from nodules
have not yet been attempted.

REFERENCES
Bond G (1976) Proc. R. Soc. London B 193, 127-135
Nelson DL and Schuttler PL (1984) Northwest Sci. 58, 49-56
Newcomb W (1981) Can. J. Bot 59, 2500-2514
Newcomb W and Heisey RM (1984) Can. J. Bot. 62, 1697-1707
Righetti TL and Munns DN (1980) Plant Physiol. 65, 411-412

ACKNOWLEDGEMENTS: This research was supported by NSF grant PCM-8204885
and PCM-840078.

FURTHER STUDIES ON THE NODULATING POTENTIAL OF RUBUS ELLIPTICUS BY THE ACTINOMYCETE, FRANKIA

Mark D. Stowers
NPI, University Research Park, Salt Lake City, Utah 84108 USA

1. INTRODUCTION

Root nodules on Rubus ellipticus were first reported by Bond (1976) during the International Biological Program survey of symbiotic nitrogen-
fixing plants. Further work by Becking (1979; 1984) established that
these nodules fixed dinitrogen and that the microsymbiont was the actinomycete, Frankia. Given the economic opportunities of increased N_2 fixation work was initiated to: 1) locate a nitrogen-fixing R. ellipticus;
2) confirm its ability to form nodules and fix dinitrogen; and 3) develop
this trait for Rubus spp. (raspberries).

2. PROCEDURE

Field surveys were conducted in Nepal, Indonesia and Hawaii using guidelines established by Stowers (1985). Seed and stem cuttings were made
from plants collected in Indonesia and Hawaii. Germinated seedlings or
rooted-stem-cuttings were transplanted to pots containing soil mix and
inoculated as previously described (Stowers, Smith, 1985). Frankia
strains included isolates from plants of the Betulaceae, Myricaceae,
Rhamnaceae and Rosaceae. Soil inoculum was collected from the site of
the only reported N_2-fixing R. ellipticus.

3. RESULTS AND DISCUSSION

Five sites of R. ellipticus were examined for root nodules in Cibodas
Mountain Gardens, Java, Indonesia (7/84; 2/85). None possessed root
nodules. At one site (the site of the reported N_2-fixing R. ellipticus)
many nodules were observed, but could be traced back to several Myrica
rubra trees. Additional field surveys in Nepal (1/85) and Hawaii (7/84;
8/84) failed to reveal root nodules on R. ellipticus. In greenhouse experiments no inoculum treatment (pure cultures of Frankia or soil) produced nodules on R. ellipticus regardless of plant origin, soil mix or
fertilization regime.

From the data presented here it is established that R. ellipticus does
not readily form N_2-fixing nodules under greenhouse or field conditions.
This information coupled to the disappearance or destruction of the only
N_2-fixing R. ellipticus (Becking, 1984), the inconsistencies of Becking's
reports (1979; 1984), the failure of either report to show an intact
whole plant with nodules and the presence of M. rubra root nodules at
the site of the reported N_2-fixing R. ellipticus provides considerable
evidence against the presence of symbiotic nitrogen fixation capability
between R. ellipticus and Frankia.

4. REFERENCE

Becking JH (1979) Plant Soil 53, 541-545.
Becking JH (1984) Plant Soil 78, 105-128.
Bond G (1976) In Symbiotic Nitrogen Fixation in Plants. Int. Biol Progr.
    Vol. 7, Ed. P.S. Nutman. Cambridge Univ. Press, Cambridge, UK p. 443.
Stowers MD (1985) In Practical Symbiotic Nitrogen Fixation Methodology.
    Ed. G.H. Elkan. Dekker Publishers, New York. (in press).
Stowers MD, Smith JE (1985) Plant Soil (in press).

NITROGEN:PHOSPHORUS RATIOS AS A REGULATOR OF N_2 FIXATION RATES IN TERRESTRIAL CYANOBACTERIA

K.A. EISELE/D.S. SCHIMEL/J.D. DUBOIS/L.A. KAPUSTKA
NREL, COLORADO STATE UNIV., FT. COLLINS, CO, DEPT. OF BIOCHEMISTRY, UNIV. OF WISCONSIN, MADISON, WI, DEPT. OF BOTANY, MIAMI UNIV., OXFORD, OH.

Cyanobacteria, principle contributors of biologically fixed N in natural grassland ecosystems, are generally restricted to aggregates which are correlated to soil phosphate levels. In our efforts to develop predictive models of dinitrogen fixation rates in grasslands, we have assessed the role of phosphate in relation to acetylene reduction (AR), chlorophyll (Chl) content, and total protein content. Experiments conducted with suspension cultures of Anabaena and Nostoc revealed a correlation between N:P ratios (varied from 10:1 to 1:10) and AR, Chl content, and total protein. Absolute amounts of N and P (ranging from 10-3 to 10-6 M) did not correlate with these parameters. Subsequent experiments with terrestrial Nostoc colonies were conducted on soil environments with varying additions of N and/or P to achieve ratios ranging from 6:1 to 1:6. Linear regression analyses of data from day 19 and 33 revealed correlations between ln (N:P) and mg Chl ($r=0.96$, 0.97); mg protein ($r=0.84$, 0.95); and AR ($r=0.91$, 0.97). That elemental ratios are better predictors of physiological response than are absolute amounts of N or P suggests that neither C nor N are limiting to these photosynthetic diazotrophs. The physiological basis for the apparent regulatory control exerted by phosphate remains obscure.

REFERENCES

Cole CV and Heil RD (1981) In Clark FE and Rosswall T, eds, Terrestrial Nitrogen Cycles Ecol. Bull. (Stockholm) 33:363-374.
Cole JJ and Howarth RW (1985) Amer. J. Bot. 70:8-16.
DuBois JD and Kapustka LA (1983) Amer. J. Bot. 70:8-16.
Griffith WF (1978) In Phosphate for Agriculture: A Situation Analysis, pp. 80-84, Potash/Phosphate Institute.

EFFECTS OF INOCULATION WITH ENDOMYCORRHIZAS AND ASYMBIOTIC N_2-FIXERS ON GROWTH OF WHEAT PLANTS

Y.Z. ISHAC, M. EL-HADDAD, E. SALEH, M. EL-BOROLLOSY, M. EL-DEMERDASH and A.A. REFAAT, Unit of Biofertilizers, Fac. Agric., Ain-shams Univ., Shoubra El-Kheima, Cairo, Egypt

It is well known that inoculation with asymbiotic N_2-fixers improves the growth and yield of many crop plants (Brown et al, 1969; Dobereiner, Day, 1976; Ishac et al, 1984) due to supplementing the growing plants with fixed nitrogen and growth promoting substances. Some synergistic effects of dual inoculation of wheat with endomycorrhizae and asymbiotic N_2-fixers. The present work shows the combined effect of inoculation with endomycorrhizae and asymbiotic N_2-fixers (Azotobacter and/or Azospirillum) on the growth of wheat plants and N_2-ase activity in a pot experiment. This was carried out in the presence of pulverized maize stalks at the rate of 0.2% of either super or rock phosphate. Results obtained indicate that inoculation with a mixture of Azotobacter, Azospirillum and endomycorrhizae in the presence of rock phosphate gave the highest growth of wheat plants, grain and straw yields. N_2-ase activity in the rhizosphere of wheat plants was also higher in this case at nearly all growth stages under investigation. Therefore, it can be concluded that application of organic manure and inoculation with asymbiotic N_2-fixers and endomycorrhizae in the presence of rock phosphate and only half the normal field rate of inorganic N-fertilizer gave a wheat crop of high straw and grain yields. Such application can save a high proportion of agricultural costs by reducing the amounts of inorganic nitrogen and phosphorus fertlizers.

References

Brown EM et al. (1964) Plant and Soil 20, 1984-214.
Dobereiner J and Day JM (1976) In Newton WE and Nymon CJ (eds) Proceedings of the First International Symposium on Nitrogen Fixation, Washington State University Press, Pullman.
Ishac YZ et al (1985) In Skinner FA and Umala P (eds) Nitrogen Fixation with Non-Legumes, Martinus Nijhoff/Junk, The Hague, (in press).

RICE PLANT-ASSOCIATED N$_2$ FIXATION AND SELECTION OF RICE VARIETIES WITH HIGHER N$_2$ FIXATION

J. K. LADHA, A. C. TIROL, W. VENTURA, G. CALDO, M. L. DAROY, and I. WATANABE, THE INTERNATIONAL RICE RESEARCH INSTITUTE, P.O. BOX 933, MANILA, PHILIPPINES.

The flooded rice ecosystem favors the maintenance and multiplication of N$_2$-fixing autotrophs and heterotrophs. Nitrogen fixed by free-living microorganisms normally enters the soil N pool after the microorganisms die. On the other hand, a part of nitrogen fixed by bacteria associated with rice plant is immediately available to the plant. Therefore, two components should be considered in studying varietal differences in their capacity to stimulate nitrogen fixation: (1) efficient uptake of soil N, and (2) a higher rate of plant-associated N$_2$ fixation. Significant varietal differences in their capacity to stimulate N$_2$ fixation have been shown by measuring nitrogen balance by Kjeldahl method in 90 rice varieties and wild Oryza species grown in pots (IRRI, 1982). N balance is tedious and time-consuming, and is not suitable for large-scale screening. N$_2$ fixation by several rice varieties grown in the field and in outdoor concrete boxes was measured by modified acetylene reduction assay and ^{15}N dilution, respectively. The relations between rice plant-associated acetylene-reducing activity (ARA) and plant growth components at several growth stages were measured. At rice heading stage, differences in ARA were most evident, coefficient of variation of ARA among plot was lowest, and correlations between ARA and total, root, and shoot dry weight and N uptake were significant. Nitrogen harvest index (NHI) and nitrogen remobilization efficiency (NRE) were significantly different -- traditional tall varieties showed lower NHI and NRE. ARA, shoot dry weight, and tiller number in long- and short-duration varieties showed similar trends, however root dry weight and leaf area showed different trends. Long-duration varieties had higher ARA per plant at heading than short-duration varieties, mainly due to differences in plant biomass. Based on data from limited survey of 16 rice varieties, IR42 showed the highest ARA among long-duration varieties and IR50 among the short-duration varieties. For large-scale varietal screening, plant ARA at heading, along with characterization of phenotypic traits, may be the most useful parameters to measure.

The possibility of adopting ^{15}N-isotope dilution technique, which provides an integrated estimate of the amount of fixed N accumulated by a crop over the growing season, was also studied. Rice varieties were grown in soil in outdoor concrete boxes. Soil was labeled with ^{15}N ammonium sulfate and one crop of rice was grown to stabilize ^{15}N abundance in the soil. The plants of the second rice crop grown without further addition of ^{15}N labeled substrate to the soil were analyzed for ^{15}N atom % excess.

ACTIVITIES OF NITROGEN-FIXING MICROORGANISMS IN EGYPTIAN DESERT SOILS

S.H. SALEM[*], T. EL-HADIDY[**] and F.A. HASHEM[**], * - Botany Dept., Faculty of Agriculture, Zagazig University; ** - Soil Dept., Desert Institute Egypt.

INTRODUCTION

One of the outstanding characteristics of the desert sand and calcareous soils under reclamation in Egypt is their lacking in some factors important to fertility and physical properties.

In this work, the effect of application of different kinds of organic substances as well as soil conditioners (CMC) on the activities of the N_2 fixers of the two desert soil types in Egypt was studied.

MATERIALS AND METHODS

The two soil types were representative of the calcareous virgin soil from Maryout and the sandy soil from Wadi El-Natrun regions in Egypt. The materials used in the experiments were either organic (sheep dung, barley straw, Egyptian clover) or soil conditioners (carboxy methyl cellulose) and applied in different levels.

RESULTS AND DISCUSSION

The study showed that the N_2 fixers (<u>Azotobacter</u> and clostridia) were greatly influenced with the organic matter additions. Their counts and activities in N_2 fixation have been increased markedly due to manuring of the soils. Supplementing the soil with organic matter or soil conditioner (CMC) increased the nitrogen fixation by non-symbiotic nitrogen fixers in both sandy and calcareous soils. However, gains in nitrogen varied according to the quantity as well as the type of the organic material added and soil type. The nitrogen gain was more pronounced in all manured soils than in the control soil, and of high parameters in sandy soils than their respectives in calcareous ones. Addition of organic materials of wide C/N ratios stimulated the N_2 fixation. Narrow C/N ratios resulted in the lowest rates of nitrogen fixed. The same conclusion has been reached by El-Malek et al (1961), Ishac (1961) El-Hadidy et al (1967) and Monib et al (1974).

REFERENCES

Abdel Malek Y, Abdel Salam MA, Monib M and El-Hadidy T (1961) Bull. Inst. Desert, Egypt 11, 41.
Ishac YZ (1961) Ph.D. Thesis, Cairo University.
El-Hadidy T, Abdel-Salam MA, Monib M, and Abdel Malek Y (1967) Bull. Inst. Desert, Egypt, 17 (1), 81.
Monib M, Abdel Malek Y, Hosny T, El-Hadidy TT and Ragab M (1974) Zbl. Bakt. Abt. II Bd. 129, 448.

IMPROVEMENT OF GROWTH AND YIELD OF NON-IRRIGATED SORGHUM BICOLOR BY AZOSPIRILLUM INOCULATION

SHLOMO SARIG* YAACOV OKON* and ABRAHAM BLUM**
* DEPARTMENT OF PLANT PATHOLOGY AND MICROBIOLOGY, THE HEBREW UNIVERSITY
OF JERUSALEM, REHOVOT 76100, ISRAEL
** DEPARTMENT OF FIELD CROPS, VOLCANI CENTER, ARO, POB 6, BET DAGAN,
ISRAEL

The objectives of this field study were to determine the benefits of
Azospirillum inoculation in non-irrigated sorghum growing under semi-
arid conditions, on stored soil moisture (ca. 200-280 mm), after the
rainy season. Un-irrigated sorghum is known in Israel as a crop that
does not respond to N fertilization.

In an experiment carried out in 1980 near Kibbutz Sede Yoav with S.
bicolor cv. 6078 for forage, inoculation significantly increased dry
matter yield by 19% and the total N yield by 90% as compared to non-
inoculated controls (Sarig et al., 1984). With S. bicolor cv 610 (in
1981) increases in grain yield (17%), significant increases in plant dry
weight, mineral content (N, P and K) and panicle number were obtained by
inoculation with Azospirillum (Sarig et al., 1984).

An inoculation experiment with Sorghum bicolor cv 610 was carried out
near Sede Yoav in 1983. Azospirillum increased total dry weight accumulation and N and K accumulation in plant parts. The plant water
status was improved as indicated by relatively higher leaf water potentials, at two growth stages. There were differences in some morphological parameters of the inoculated plants as compared to controls, such
as increased stem diameter and panicle length. There were indications
that root systems were more developed in inoculated plants as measured
by the average force needed for pulling out the plant from the ground.

In a field experiment with Sorghum bicolor cv 610 carried out in 1984
at Bet Dagan, inoculation with Azospirillum significantly enhanced top
dry matter accumulation. There was a clear improvement in plant-water
status as measured by the leaf water potential, osmotic potential at
full turgor, stomatal conductance and canopy temperature. Grain yield
was significantly increased by 14%. It was concluded that Azospirillum
inoculation benefits plant growth and yield of sorghum and other forage
grain grasses (Okon, 1984) by mechanisms involved with root development,
mineral uptake and plant water relationships. Sorghum grown under semi-
arid conditions was an excellent plant model to study Azospirillum
effects.

REFERENCES

Okon Y (1984) In Veeger C and Newton WE, eds.
Advances in Nitrogen Fixation Research, pp 303-309,
Martinus Nijhoff/Junk, The Hague
Sarig S et al (1984) Exp. Agric. 20, 59-66.

EFFECT OF INOCULATION WITH AN INEFFECTIVE *RHIZOBIUM* ON N_2-FIXATION IN CHICKPEA

K.E. GILLER, M.R. SUDARSHANA, J.A. THOMPSON and O.P. RUPELA
ROTHAMSTED EXPERIMENTAL STATION, HARPENDEN, HERTS, U.K. and
ICRISAT, PATANCHERU P.O., ANDHRA PRADESH, INDIA.

Two genotypes of chickpea (*Cicer arietinum* L.) were grown in the field with and without inoculation of an ineffective chickpea *Rhizobium*. The vertisol soil contained a large population of chickpea *Rhizobia* at the beginning of the experiment. Strain IC2094 formed more than twice as many of the nodules recovered from K850 than from G130, demonstrating a marked host-strain preference for nodulation. Nodule mass and ARA were consistently greater in uninoculated K850 than G130. Inoculation with ineffective strain IC2094 resulted in a large and consistent reduction in nodule mass and ARA in genotype K850. A small but consistent decrease in both of these parameters was also found with G130.

Isotope dilution estimates of N_2-fixation indicated that K850 fixed more nitrogen than G130. These absolute estimates may be inaccurate as the N-accumulation pattern of the safflower reference was different from that of chickpea (cf. Witty 1983). However, the N-uptake patterns of the chickpea genotypes were similar and comparative estimates between chickpea treatments are probably reliable. N accumulation was consistently greater in uninoculated than inoculated treatments. Isotope dilution indicated no significant differences in % N-fixed with inoculation despite the large consistent differences in nodule mass and ARA.

Thus there was a marked discrepancy between nodule mass and ARA compared with N-uptake and isotope dilution measurements when the effect of the ineffective strain on N_2-fixation is examined. If inoculated plants had formed nodules on secondary roots to compensate for the ineffective nodules, and these nodules had not been fully recovered, then this may account for the differences between ARA and isotope dilution estimates. In the light of these results it seems unlikely that ineffective *Rhizobium* can be used to produce non-fixing legume plants for control purposes.

REFERENCES

Witty JF (1983) Soil Biol. Biochem. 15, 631-639.

ESTIMATING N_2-FIXATION FROM VARIATIONS IN THE NATURAL ABUNDANCE OF ^{15}N.

DANIEL H. KOHL, GEORGIA SHEARER, Washington University, St. Louis, MO
63130 U.S.A., and ROSS A. VIRGINIA, University of California, Riverside,
CA 92521 U.S.A.

The natural abundance of ^{15}N of soil N is usually greater than
that of atmospheric N_2. This difference is reflected in a difference
between non-N_2-fixing and N_2-fixing plants, with the latter often
exhibiting the lower ^{15}N abundance. Such differences serve as the
basis for estimating the fractional contribution of N_2-fixation to
N_2-fixing plants. The method, henceforth referred to as the $\delta^{15}N$
method ($\delta^{15}N$ being the per mill ^{15}N excess over atmospheric N_2), is an
isotope dilution method at the level of natural ^{15}N abundance. The
$\delta^{15}N$ method has been tested under experimental field conditions and
gives estimates comparable to those provided by more conventional
methods. Because of the small differences in ^{15}N abundance, estimates
based on the $\delta^{15}N$ method are not highly precise. Nevertheless, this
method has certain advantages over other methods, especially in natural
ecosystems where it is desirable to avoid experimental manipulation and
where root systems are often too disperse to obtain representative
nodule samples for acetylene reduction assays. The $\delta^{15}N$ method gives a
time-integrated estimate of N_2-fixation and does not require surviving
tissue. This poster presents details of the $\delta^{15}N$ method of estimating
N_2-fixation, a discussion of limitations of the method, and estimates,
based on this method, of N_2-fixation by tree and herbaceous Dalea
species in warm desert ecosystems in the Philip L. Boyd Deep Canyon
Research Center and the Living Desert Reserve near Palm Springs,
California. D. spinosa, a tree legume which grows in sandy wash
woodlands, fixed 70 to 90 percent of its nitrogen. The contribution of
fixed nitrogen to low growing D. mollisima and D. schottii, which occur
on rocky slopes, was 80 to 90 percent. Irrigation had no detectable
effect on the fractional contribution of fixed N to D. spinosa.
However, visual inspection suggested that irrigation increased
productivity as would be expected. Hence irrigation probably enhanced
the total amount of nitrogen introduced into the ecosystem by
N_2-fixation. These data on the fractional contribution of
symbiotically fixed nitrogen to Dalea species, in combination with data
on nitrogen productivity being collected by our colleagues at the
University of California, will permit placing upper and lower bounds on
the quantity of nitrogen introduced to the ecosystem by N_2-fixation.

NITROGEN EFFICIENCY AND FIXATION IN A LONGTERM FERTILIZER POT
EXPERIMENT WITH RICE

GÜNTER TROLLDENIER
LANDWIRTSCHAFTLICHE FORSCHUNGSANSTALT BÜNTEHOF, D-3000 HANNOVER 71, FRG

The high fertility of paddy fields has been ascribed in part to microbial
nitrogen fixation (cit. in Trolldenier 1975) and was verified later by
the acetylene reduction assay. Its inadequacy for measuring N_2 fixation
in paddy soil has renewed the interest in nitrogen balance sheets from
flooded rice (Greenland and Watanabe 1977). Avoiding problems associated
with field experiments a longterm pot experiment was conducted under controlled conditions in the greenhouse.

METHODS
Growth of phototrophic diazotrophs in the flood water was prevented. The
experiment comprised treatments with dry (DF) and wet fallow (WF) between
cropping seasons. Each group received varying amounts of N and K, while P
and Mg were applied at equal rates. During rice growth the soil surface
was kept flooded by addition of deionized water. The pots were planted
to 2 hills of rice cv. Cigalon. In situ acetylene reduction assays were
conducted throughout the rice growing season (Trolldenier 1979).

RESULTS AND DISCUSSION
Straw and grain yields were generally higher with higher N and/or K supply. DF was superior to WF. Though yields were increased by higher N
application, nitrogen efficiency was lower, especially at low K supply.
N removal by straw and grain
was equivalent to N applied
at high N and low K (N_2K_1).
In all other treatments N removal was higher than N supply by fertilization, more
with DF than with WF (Table 1).
N removal exceeded fertilization most at N_1K_2DF. A
slight increase of N_t in soil
was found during the experimental period. Depending on
fertilization and water regime, N_2 fixation is soil may
thus have contributed considerably to N nutrition of
the rice plants.

Table 1
Nitrogen removal by 4 crops of rice as
related to NK nutrition and water regime

| Treatment | | N removal | N removal- | |
| Ferti-
lizer | Fallow | removal | N application | |
		g/pot		%
N_1K_1	wet	5.2	1.3	33.3
	dry	6.9	3.0	76.9
K_2	wet	5.5	1.6	41.0
	dry	7.1	3.2	82.0˙
N_2K_1	wet	7.8	0.0	0.0
	dry	8.4	0.6	7.7
K_2	wet	8.6	0.8	10.3
	dry	10.6	2.8	35.9

Total N application N_1 3.9 g N; N_2 7.8 g N

Although in situ acetylene reduction assays by far underestimated N_2 fixation, N_2ase activity paralleled N removal in the different fertilizer
treatments. Conversion of data to the field leads to N_2 fixation of 50 kg
N/ha per year by rhizosphere and soil bacteria in the N_1K_2DF treatment
which received 61 kg mineral N and yielded 3.6 t/ha grain.

REFERENCES
Greenland D J and Watanabe I (1982) Transact. 12th Int. Congr. Soil Sci.
Lee K K and Watanabe I (1977) Soil Sci. Plant Nutr. 24, 1-13.
Trolldenier G (1977) Proc. 11th Coll. Int. Potash Inst., pp 287-292.
Trolldenier G (1979) Mitteilgn.Dtsch.Bodenkundl. Gesellsch. 29, 339-344.

AGRICULTURE AND FORESTRY POSTER DISCUSSION

TIMOTHY L. RIGHETTI/DAVID B. HANNAWAY
OREGON STATE UNIVERSITY, CORVALLIS, OR   97331

Subjects presented and discussed emphasized applications rather than basic processes. Discussion covered the following categories: Methods of Quantifying Nitrogen Fixation, Importance and Applications of Free Living Nitrogen Fixing Organisms, Importance and Applications of Non-Legume Symbiotic Systems, and Inoculation Technology and Management Factors Related to Legume Performance.

Although there was general agreement on the need to better quantify nitrogen fixation, there was no consensus on a single approach. In general, approaches involving ^{15}N were preferred, but advantages and disadvantages of natural abundance vs. enriched or depleted amendments were not resolved. Defining an unbiased non-fixing control remains a substantial barrier. Some felt that trials conducted on fields where ^{15}N amendments have equilibrated with soil organic N pools are less sensitive to differences between non-fixing and fixing components. A similar argument was presented supporting the advantages of natural abundance approaches. There was general agreement that acetylene reduction approaches were not appropriate estimations of seasonal nitrogen input and some felt acetylene reduction data should be interpreted with caution even when making comparative evaluations between treatments. Although acetylene reduction may have limitations, alternative approaches that are inexpensive, simple and reproducible do not exist, therefore, its widespread use will likely continue. Yield and biomass differences between treatments are the most important measurements in applied experiments and there was some concern that overemphasizing quantification of nitrogen inputs in experiments where yield or biomass gains are negligible distorts the importance of the nitrogen fixing process.

Limited nitrogen input from heterotrophic free living nitrogen fixing systems in agriculture was cited as a reason for placing less emphasis on free living systems in the future. This view was countered with the argument that in marginal natural systems even small nitrogen inputs are ecologically important. Furthermore, nitrogen fixation is obviously important to the survival and growth of the organisms themselves, regardless of the impact their fixed nitrogen has on natural or agricultural systems. This is especially important in view of several reports where substantial yield advantages were associated with the presence of various free living nitrogen fixing organisms, even though nitrogen input from these organisms was apparently not the most important factor. Autotrophic free living nitrogen fixing organisms may not have the same constraints. Potential applications are possible especially in rice production. In any case, free living organisms will continue to be an important tool in basic studies on nitrogen fixation.

Discussion of commercial applications of non-legume symbiotic nitrogen fixing systems centered on <u>Frankia</u> associations. There was general agreement that evolving technology will make commercial inoculation products feasible in the near future. Problems in production of inoculum and development of meaningful quality control were not considered severe. There was some disagreement on the most important applications. Trade-offs between the advantages of using interplanted nitrogen fixing trees versus the obvious competition with more valuable timber species have not been resolved. There was more optimism with regard to fuel wood production in developing countries. There was general agreement that the small additional cost of producing trees that are nodulated with the appropriate Frankia is worth the effort if the entire expense of planting actinorhizal trees is considered. Some felt that the need for inoculation in forest ecosystems may be small because of the general presence of infective native Frankia. Since Frankia applications would likely involve nodulated trees rather than seed inoculation, problems due to competition with native strains may be less severe than with Rhizobia-legume systems. However, the longevity of nodules introduced at planting and the contribution of initial Frankia strains to long term stand fixation is unknown. Frankia inoculations or the establishment of nodulated trees on disturbed sites lacking native Frankia is an obvious application, but reclamation provides only a very small market for commercial firms to service.

The importance and potential of inoculation with Rhizobia to increase crop production dominated legume discussions. Some felt that indigenous Rhizobia dominate current production locations and responses to inoculation are generally rare. This group felt that most inoculation programs are relatively inexpensive and can best be viewed as an insurance program to guarantee adequate nodulation rather than as an effective means of enhancing production. Others felt responses to inoculation are more widespread and there is cause for more optimism. In any case, a better understanding of the ecology of Rhizobia in complex microbial soil populations and the basis for competitiveness among Rhizobium strains was viewed as one of the most important research goals to complement any Rhizobium strain improvement that occurs in the future.

Long term possibilities of introducing nitrogen fixation into additional hosts was briefly addressed. Possiblities involving both Rhizobium and Frankia systems were discussed and participants differed in their enthusiams for such an effort. For example, the wide taxonomic host range of Frankia systems is cause for optimism, but it is not clear what the eventual target should be. Although there was some speculation that Rosaceous fruit tree species and nut trees species in the Betulaceae would be logical choices, nitrogen fertilizer is only a small portion of the total costs in these systems. Unless growers are assured that nitrogen fixing trees will perform as well as nitrogen fertilized ones, there would probably be little interest. The most logical current and future application may be in low input agriculture, reclamation, or forestry, where nitrogen fertilizer options are not economic or are unavailable.

CURRENT THOUGHTS ON
NITROGEN FIXATION

CURRENT THOUGHTS ON NITROGEN FIXATION

John Postgate FRS
AFRC Unit of Nitrogen Fixation, University of Sussex, Brighton, BN1 9RQ, U.K.

This is the sixth of the series of International Symposia which started at Pullman, Washington, in 1974 (Newton, Nyman 1976) and the organisers have asked me to comment on the present status of nitrogen fixation research, on the directions it is taking and on the function of these meetings. It is certainly salutary to pause and look at the state of one's subject and I shall try to do this adequately, despite the handicap imposed by the fact that I am required to prepare this contribution in substantial unawareness of what advances will be presented at this meeting.

The strategic background

The study of nitrogen fixation is an immensely important area of strategic research, for reasons which are familiar. Its practical importance stems from the World's population explosion. We know that, short of global catastrophe, the World's population will increase to nearly double its 1985 size by the early decades of the next millenium. This prognosis is true even if all current measures for population control are successful, because the children who will mate and produce the new offspring are already among us. It is self-evident that the 7.5 to 8 billion people must be fed, which means, as Ralph Hardy (Hardy 1976) pointed out at Pullman, that the N-contents of the World's agricultural produce must increase proportionately. To obtain such a net increase in N-output, the N-input into the World's agricultural soils must increase disproportionately: Hardy (1976) assumed a 50% loss of added N-fertilizer in his calculations of future N requirements. Hardy (1976) and Pimental (1976) surveyed present and future scenarios for N-inputs at that meeting, with such cogency that their calculations and data remain valid over a decade later. Energy and fertilizer demands have fluctuated, so that fertilizer capacity has outstripped demand and demand has outstripped capacity at various times, but the foreseeable needs remain as they were in 1974. Today one might give greater weight to the energy cost of packaging and transport of N-fertilizer as a component of the economics of fertilizer use, and the environmental impacts of chemical N-inputs are more subtle than they then seemed. Nevertheless, the conclusion still remains that both biological and chemical N-inputs are essential over the next few decades. I shall not pursue this topic, because it has been perceived by politicians and administrators in most parts of the World and is accepted by scientists and technologists, a fact reflected in the ever-increasing size of these meetings.

Essentially, one can reach the satisfying, but nonetheless challenging, conclusion that the strategic basis of this area of scientific research has not changed in the past decade. We still need basic knowledge of diazotrophy to be able to tackle present and future problems of nitrogen inputs effectively.

The scientific background

Hardy (1976) surveyed research advances up to about 1973 and I (Postgate 1980a,b) surveyed research to about 1980. Numerous briefer surveys exist in the introductions to reviews and articles. Developments over the past quarter century have indeed been spectacular and do not need rehearsal here; the important message is that fundamental, seemingly curiosity-orientated science has underpinned the majority of these developments by providing the research tools and scientific insights enabling them to take place. I shall therefore look at some of the outstanding fundamental questions which remain.

Some questions in biological chemistry

A vexing problem is still the nature of the N_2-binding site in the MoFe-protein. We do not know for sure that it is a metal, but most research and thinking is based on the assumption that it is an Mo atom in the FeMoco fragment. Despite substantial research in the U.S.A., U.K. and China, the stoichiometry and character of this unstable material is elusive. Even the matter of its organic content, when isolated in N-methylformamide, is disputed. One of the newer grips we have on this moiety is the nifV mutation in K. pneumoniae, which leads to a nitrogenase of altered substrate specificity assignable to a change in FeMoco (see Smith et al., this meeting). The chemical differences between normal and nifV FeMoco should be very interesting; spectrometry and analyses of crude preparations so far show no substantial differences.

The chemistry of N_2 complexes now provides precedents for several probable steps in nitrogenase function: N_2-binding from water; displacement of H_2 by N_2; protonation via a hydrazide(2-) (M-NNH$_2$) intermediate to NH_3. However, the majority of protonation reactions of metal-bound N_2 to NH_3 (or to N_2H_4) are associated with a large valency change in the metal: sequential addition of electrons is not known except in the recently described system of Pickett, Talarmin (this meeting) in which cyclic electronation and protonation of a tungsten-dinitrogen complex at an electrode system was developed. The ready displacement of two hydride groups by N_2 to form a dinitrogen complex is the basis of plausible theories of nitrogenase action, accounting for the minimum H_2:NH_3 ratio of 0.5 characteristic of the enzyme (Simpson, Burris 1984) and accounting also, with some philosophical twisting, for N_2-dependent HD formation (Chatt 1980; Thorneley, Lowe 1985). Yet N_2 can be bound from water to a transition metal perfectly well without H_2 displacement (eg: to [Ru(NH$_3$)$_5$H$_2$O]) so it seems strangely inefficient for Nature to have 'chosen' a pathway which loses an energetically useful molecule. Most mechanisms assume that N_2 is bound end-on to the enzyme binding site (cheerfully ignoring (a) that C_2H_2 is not readily bound end-on, and (b) that its cis-deuteration (Dilworth 1966) implies side-on binding). However, as Dr. Shilov reminded us at the 5th meeting, there is still no convincing evidence that two metal atoms are not involved in binding N_2 to nitrogenase. One chemical system for binding N_2 from water leads to the binuclear complex [(NH$_3$)$_5$RuN$_2$Ru(NH$_3$)$_5$]; if the FeMoco fragment is indeed a cluster, a 2-metal site might readily occur in the enzyme.

The chemistry of N_2 reduction to the hydrazide(2-) level is becoming clearer but subsequent steps are more obscure (Henderson et al. 1983).

However, a totally new bull has entered the ring with the alternative, Mo-repressible nitrogenase of <u>Azotobacter vinelandii</u> studied by Bishop and his colleagues (see Premakumar et al. 1984 and references therein). Unequivocal evidence that the dinitrogenase component of the alternative system is different became available with the construction of a <u>nifKDH</u> deletion of <u>A. vinelandii</u> which shows Mo-repressed nitrogenase activity (P. Bishop, D. Dean, R. Eady, personal communications). If the alternative dinitrogenase is truly Mo-free, which Bishop et al. have not actually claimed, then the nature of the presumptively non-FeMoco N_2-binding site becomes a fascinating problem in dinitrogen chemistry.

The distribution of the alternative system and the nature of its dinitrogenase provides so obvious a new set of biochemical questions that little more need be said: its substrate specificity, oxygen sensitivity and H_2 relations show differences from the conventional system (Bishop, Eady, this meeting). Its ATP relations, metal and S content all need study, as does the character of its reductase and any ancillary proteins. Does the second copy of <u>nifH</u> (Kennedy et al., this meeting) specify its electron acceptor? Are any products from the regular <u>nif</u> cluster needed? This is an exciting area in which fascinating developments may be expected.

The conventional system still presents the perennial problems of the specific activities of its components and their molar contents of metal and sulfur atoms. Setting these aside, it also offers a number of paradoxes.

<u>O_2 sensitivity</u>. Although O_2 might well interfere with nitrogenase function, there is no logical reason why the proteins <u>per se</u> should be destroyed by O_2. What are the O_2-sensitive sites which are screened by the protective protein of Azotobacter? Could site-directed mutagenesis yield an O_2-tolerant enzyme?

<u>ATP requirement</u>. The myth that the ATP requirement of nitrogenase is a metabolic 'burden' on a diazotrophic aerobe or symbiosis has, I hope, died a well-deserved death. It has a modest cost, comparable to that of reducing nitrate. The puzzle is that there should be any ATP requirement at all and here progress has been slow: is it simply a matter of conferring specificity on the electron donor? Does it prevent reversibility or limit escape of H_2? Could it be a leaving group on FeMoco for proton rather than N_2-binding?

<u>Inefficiency</u>. If the many kinetic studies are to be believed, nitrogenase is a remarkably inefficient enzyme. Its turnover number is so slow that its synthesis and maintenance must be a greater burden on the diazotroph than its ATP consumption. Any perturbation of its component ratio, physical environment, ATP:ADP ratio or temperature causes preferential allocation of electrons to H_2 evolution rather than N_2 fixation. Thorneley, Lowe (1985) gave compelling reasons why the kinetic parameters of conventional nitrogenase impose these properties; they stem largely from the demonstration that the rate-limiting step is dissociation of the two proteins. Their view, that slow dissociation is necessary to ensure transfer of electrons to the bound substrate and not to ambient protons, implies that the environment of the prosthetic site cycles from protic to aprotic and back. Would not a single molecule, a co-integrate of the Fe-protein and the MoFe-protein, be kinetically more efficient?

Some molecular physiological questions

Advances in enzymology and genetics have, in many areas of biology, enabled physiological questions to be approached at an analytical or molecular level. This development as it applies to research on diazotrophy was illustrated elsewhere (Postgate et al. 1981). Exciting questions now arise at the interface of genetics and biochemistry. For example, nif regulation, basically a physiological problem, would have grated to a halt a dozen years ago without the input of genetic regulation studies and, particularly, the use of nif-lac fusions to dissect the regulatory apparatus. Among burning questions of the day in this area are: the universality of the ntr system and the precise role of ntrA in the recognition of its special promoter sequences; the nature of the signal substances responding to O_2 or NH_4^+; the functional analogies between ntrC and nifA, possibly between ntrB and nifL; the regulatory role, if any, of Mo compounds; the mesophily of the system. The latter two of these topics have acquired new dimensions. Firstly, although Shah et al. (1984) concluded that Mo derivatives are not regulators of nif expression in K. pneumoniae or E. coli constructs (assigning Dixon et al.'s (1980) suggestion of autoregulation by a molybdoprotein to secondary effects of Mo deprivation), Mo derivative(s) re-surface as negative regulators of the alternative system (Premakumar et al. 1984) in A. vinelandii. Secondly, a truly thermophilic diazotroph has now been reported among the Archebacteria (Belay et al. 1984) - does it possess a thermostable equivalent of the nifA product? Regulation in response to environmental factors such as O_2, NH_4^+, NO_3^- or Mo is widely recognized; a neglected area, in my view, is the long known phenomenon called 'hyperinduction': the apparent over-expression of a metabolic system in response to its limiting substrate (eg: hyperexpression of lac in response to lactose limitation in E. coli - see Horiachi et al. 1962). Nitrogenase activity in several diazotrophs has long been known to be augmented at low N_2 tensions (see Postgate 1982) and quantitative examination of Rhodopseudomonas showed that this is true hyperexpression of nif (Arp, Zumft 1983). The elucidation of this phenomenon at the molecular level opens prospects of constructing hyperactive diazotrophs as well as providing deeper insight into nif regulation.

Conventional physiological problems still remain, again I list a few.

Oxygen exclusion. (Not, please, oxygen protection!). Oelze and his colleagues (see Dingler, Oelze 1985) rightly point out that the original view of respiratory protection as a wholly kinetic process in Azotobacter is not entirely adequate; again, genetics is giving new leads: the phenotypes of O_2-sensitive (Fos⁻) mutants seem to indicate that enhanced respiration is supplemented by more subtle changes of K_m for O_2 (Ramos, Robson 1985).

Ammonia switch-off and covalently inactivated Fe protein. The rapid switch-off of nitrogenase activity by ammonia occurs in organisms such as Rhodospirillum which show the Mn^{2+}-dependent activation of the Fe-protein, yet that process is not necessarily involved since NH_3 switch-off occurs in Azotobacter, which shows no covalent inactivation. Why does either process occur at all? They could have value in economizing in nitrogenase function, but if this were a serious physiological problem, would they not be universal?

Hydrogenase.   The role, if any, of hydrogenase in diazotrophic efficiency has proved a difficult problem, probably because most research has been on the plant-nodule system, which is physiologically very complex to analyse experimentally.   If the intrinsically simpler Azotobacter cell is a useful guide, hydrogenase has but a trivial influence on diazotrophy in the steady state but is profoundly important for its initiation, and also in stress (Aguilar et al. 1984).   It is difficult to prescribe experiments with plant systems which would have revealed such information, yet it is easy now to envisage how possession of hydrogenase could, at certain stages of the legume symbiosis, influence yields considerably.

The alternative system.   Can the ability to fix N_2 in local conditions of Mo deficiency be the sole raison d'être of this system?

Physiological role of nif,nod products.   That genetical advances constantly raise new physiological problems is impressively illustrated in the study of the legume symbiosis.   Here a clutch of nod genes is associated with the fix (which include nif) genes and the functions, later the modes of action, of their products will require determined teamwork from plant and microbial scientists.   Even with the relatively well worked-over K. pneumoniae system only five genes (nifKDH, nifF, nifJ) specify products whose physiological role is reasonably well understood; we have hints of the roles of most (not all) of the remaining 17 though little evidence of how they perform their processing and/or regulatory roles.   Indeed, there is no need to assume that these products have single functions: bifunctionality has been proposed for nifA, H, KD products, for example.

On the more strictly genetical side, analysis of nif has revealed a variety of structures and locations (see Veeger, Newton, 1984, for documentation).   That the nitrogenase genes are chromosomal in some microbes and plasmid-borne in others is now well established and we are now familiar with the once surprising discovery of repetitions: second copies of nifH in Azotobacter and Anabaena, for example, re-iterations in rhizobia and extra but silent nifKDH clusters in R. phaseoli and Rhodopseudomonas.   A tightly-linked nif cluster such as the 17-gene regulon of K. pneumoniae is far from being the rule: K, D and H are not contiguous in Anabaena or Bradyrhizobium; in A. chroococcum and fast-growing rhizobia the nif cluster is larger, separated by nod genes in the latter.   Haselkorn et al. (1985) demonstrated a fascinating change in the Anabaena genome in which nifK, separated from nifDH in the vegetative cell, rearranges to become part of a Klebsiella-type nifKDH operon during derepression and heterocyst formation.   The physiological significance of so complex a process is elusive: could it be a means of amplifying low level basal activity?   For the nif genes in the vegetative cells are indeed expressed, as Smith, Evans (1970) claimed long ago.   Or might the truncated nifD product of vegetative cells specify a sort of alternative nitrogenase?

Some ecological and evolutionary questions

The widespread homology of nitrogenase genes from diverse microbes, particularly H, underpins the value of pSA30 and its fragments as probes for nif in a variety of organisms.   Clearly nif has undergone rather little evolution in its history, consistent with the view that it is a linkage group of fairly recent origin, still dispersing laterally among

the prokaryotes (Postgate 1974; Ruvkun, Ausubel 1980), a view
contradicted by limited 16S-RNA comparisons (H. Hennecke, personal
communication). Certainly K. pneumoniae nif has some of the properties
of a transposable genetic element. I pointed out at the fourth meeting
that the lateral dispersion view led to the conclusion that new
diazotrophs, and new symbioses, might be emerging today where appropriate
selection pressure exists. The unequivocal demonstration of diazotrophy
in a methanogen (Murray, Zinder 1984) together with evidence for homology
of methanogen DNA to nif DNA from Klebsiella (Sibold et al. 1985) suggests
that nif has spread to (or from) a wholly different Kingdom of living
things. It adds nothing to discussions of the age of diazotrophy but it
again raises the question why nif is not in the third Kingdom: why there
are no diazotrophic eukaryotes. Of course, examples of diazotrophic
eukaryotes are reported almost annually in the literature but none has yet
survived rigid criticism on grounds of chemical or microbial contamination:
so far they are all 'ghosts'. Yet there is still no clear physiological
reason why diazotrophic plants should not exist: the concept of the
chloroplast as a vestigial prokaryotic endosymbiont provides a precedent
for contemplating a diazotrophic organelle and the apparent ease with
which endosymbiotic L-forms of bacteria can persist in plant cells
(Aloysius, Paton 1984) suggests a mechanism whereby, in response to
appropriate selection pressure, such an organelle ought already to have
emerged. Perhaps it is on its way? Perhaps diazotrophy is so young a
biochemical process that there has been insufficient time for the
diazotrophic organelle to develop? Such evolutionary youth would be
consistent with the chemical crudity, instability and inefficiency of the
enzyme and with the fact that the only alternative nitrogenase we know so
far has emerged in azotobacter, which is physiologically one of the most
highly evolved of diazotrophs (see Postgate 1982). Yet it may be that
the diazotrophic organelle exists already but has not been found. Will
the remarkable claims of Turcin et al. (1963) for 'silent' diazotrophic
systems in plants prove to have substance? Perhaps such forms will
emerge spontaneously before the genetic engineers have constructed their
equivalents.

The study of nitrogen fixation has, over the past quarter century,
revealed so many surprises that one is tempted by the lapidary
pronouncement of my colleague Dr. Susan Hill: "In this area you have got
to believe everything until it is proved wrong".

Envoi

I have presented a restricted and biased view of the current state
of the subject, compelled by timing, space and ignorance to leave out
whole areas which are fascinating to me and to others. The major
message, however, is independent of my insights and prejudices, and it
reflects on the nature of these meetings. Although we are conditioned
to consider diazotrophy in terms of relevant chemistry, biochemistry,
physiology, genetics and ecology, this is not the way the subject develops.
We have seen in the past few years a mingling of chemistry and
biochemistry, of genetics and physiology and the beginnings of a serious
input from genetics to ecology. For a strategic research topic such as
the fixation of nitrogen, an open interdisciplinary approach is mandatory,
and this principle, which motivated the founders of this series of
meetings, has wholly justified itself. The greatest value of these
meetings is that they supplement more specialized gatherings. They

provide the opportunity for chemists, biochemists, geneticists, physiologists, microbiologists, ecologists, plant molecular biologists to stand back from their specialisms and to relate their researches to the subject as a whole.

ACKNOWLEDGEMENTS

I thank my colleagues in the UNF for provoking these 'thoughts' and particularly P.E. Bishop, R.A. Dixon, S. Hill and B.E. Smith for constructive criticism of this manuscript.

REFERENCES

Aguilar OM et al (1984) In Veeger C and Newton WE (1984) below, p 213.
Aloysius SKD and Paton AM (1984) J. appl. Bact. 56, 465-478.
Arp DJ and Zumft WG (1983) J. Bact. 153, 1322-1330.
Belay N et al (1984) Nature 312, 286-288.
Chatt J (1980) In Stewart WDP and Gallon JR, eds, Nitrogen Fixation, pp 1-18, Acad. Press, USA.
Dilworth MJ (1966) Biochim. biophys. Acta 127, 285-294.
Dingler C and Oelze J (1985) Arch. Microbiol. 141, 80-84.
Dixon R et al (1980) Nature 286, 128-132.
Hardy RWF (1976) In Newton WE and Nyman CJ (1974) below, pp 693-717.
Haselkorn R et al (1985) In Ludden PW and Burris JE, eds, Nitrogen fixation and CO_2 metabolism, pp 83-90, Elsevier, USA.
Henderson R et al (1983) Adv. Inorg. Chem. Radiochem. 27, 198-292.
Horiachi T et al (1962) Biochim. biophys. Acta 55, 152-163.
Murray PA and Zinder SH (1984) Nature 312, 284-286.
Newton WE and Nyman CJ (1976), eds, Nitrogen Fixation, Vols. 1 and 2, Washington State U. Press, USA.
Pimental D (1976) In Newton WE and Nyman CJ (1976), pp 656-673.
Postgate JR (1974) In Carlile MJ and Skehel JJ, eds, Evolution in the Microbial World, pp 263-292, Cambridge U. Press, UK.
Postgate JR (1980a) In Stewart WDP and Gallon JR, eds, Nitrogen Fixation, pp 423-434, Acad. Press, USA.
Postgate JR (1980b) In Subba Rao NS, ed, Recent Advances in Biological Nitrogen Fixation, pp v-xiv, Arnold, UK.
Postgate JR (1982) The Fundamentals of Nitrogen Fixation, Camb. U. Press.
Postgate JR et al (1981) In Bothe H and Trebst A, eds, Biology of Inorganic Nitrogen and Sulfur, pp 103-115, Springer-Verlag, FRG.
Premakumar R et al (1984) Biochim. biophys. Acta 797, 64-70.
Ramos JL and Robson RL (1985) J. gen. Microbiol. 131, 1449-1458.
Ruvkun GB and Ausubel F (1980) Proc. Nat. Acad. Sci. USA 77, 191-195.
Shah VK et al (1984) Annu. Rev. Biochem. 53, 231-257.
Sibold L et al (1985) Mol. gen. Genet. 200, 40-46.
Simpson FB and Burris RH (1984) Science 224, 1095-1097.
Smith RV and Evans MCW (1970) Nature 225, 1253-1254.
Thorneley RNF and Lowe DJ (1985) In Spiro T, ed, Molybdenum Enzymes, Vol. 7, Chap. 5, Acad. Press, USA.
Turcin FV et al (1963) C.R. Acad. Sci. USSR 149, 731-734.
Veeger C and Newton WE, eds (1984) Advances in Nitrogen Fixation Research, Nijhoff/Junk, The Hague.

AUTHOR INDEX